Combustion

THIRD EDITION

Combustion

THIRD EDITION

Irvin Glassman

Department of Mechanical and Aerospace Engineering
Princeton University
Princeton, New Jersey

ACADEMIC PRESS

San Diego   London   Boston   New York   Sydney   Tokyo   Toronto

Academic Press, Inc.
525 B Street, Suite 1900, San Diego, California 92101-4495, USA
http://www.apnet.com

Academic Press Limited
24-28 Oval Road, London NW1 7DX, UK
http://www.hbuk.co.uk/ap/

Library of Congress Cataloging-in-Publication Data

D
541.361
GLA

Glassman, Irvin.
    Combustion / by Irvin Glassman. -- 3rd ed.
        p.   cm.
    Includes bibliographical references and index.
    ISBN 0-12-285852-2 (alk. paper)
    1. Combustion.   I. Title.
QD516.G55   1996
541.3'61-dc20   96-3069

PRINTED IN THE UNITED STATES OF AMERICA
96 97 98 99 00 01 QW 9 8 7 6 5 4 3 2 1

To Those Who Helped

Bev
who was very much part of all of this,
on
our 45th wedding anniversary

Tony Bozowski
(Rest in Peace)

and

Joe Sivo
who gave so much of themselves in caring
about students and creating an enjoyable atmosphere
for learning in our laboratory

Princeton
a remarkable institution that
gave me opportunities I never dreamed
I would ever have

No man can reveal to you aught but that which already lies half asleep in the dawning of your knowledge.

If he (the teacher) is wise he does not bid you to enter the house of his wisdom, but leads you to the threshold of your own mind.

The astronomer may speak to you of his understanding of space, but he cannot give you his understanding.

And he who is versed in the science of numbers can tell of the regions of weight and measures, but he cannot conduct you hither.

For the vision of one man lends not its wings to another man.

Gibran, *The Prophet*

The reward to the educator lies in his pride in his students' accomplishments. The richness of that reward is the satisfaction in knowing the frontiers of knowledge have been extended.

D. F. Othmer

Contents

3 — Explosive and General Oxidative Characteristics of Fuels

4 Flame Phenomena in Premixed Combustible Gases

5 Detonation

6 —— Diffusion Flames

7 —— Ignition

8 ══ Environmental Combustion Considerations

9 — Combustion of Nonvolatile Fuels

Preface

Although the size of this new edition of *Combustion* has increased, the basic approach of providing students and practicing professionals with the underlying physical and chemical principles of the combustion field has not changed from the previous editions. The emphasis on clarity of concepts remains. This edition, however, has been enhanced as a reference work by the addition of numerous figures developed from computational combustion codes now available and by the expansion of the appendixes to include the most useful combustion data. Indeed, perhaps the most significant contributions to the field of combustion in the last decade have been the development of these codes and the recent supporting thermochemical, thermophysical, and kinetic rate data — all of which will be found in the new appendixes that serve as an essentially separate reference volume.

The new figures are designed to offer further insight into the basic developments presented in the text. Through these figures one may better understand the composition changes as a gas mixture proceeds through a laminar flame, how the flame thickness changes with pressure, why the temperature of a metal burning in oxygen is limited to the boiling point of the metal oxide, the effect of fuel molecular weight on detonation parameters, and the like. In addition, many of the previous figures from other sources remain even though more recent presentations are available. The reason for this choice is that these figures reveal points pertinent to the text.

The text in every chapter has been modified from the second edition to varying degrees to assist the reader in obtaining better understanding of the subject. Although no attempt has been made to discuss specific practical applications, the relationships of combustion fundamentals to applications are now frequently discussed where appropriate. Evidence of this concern will be found in the application of the basic material on thermodynamics, kinetics, and the oxidation of fuels in Chapters 1–3. Chapter 3 has been extended to include the latest information on the oxidation mechanisms of aromatic compounds. Chapter 4 has been largely rewritten to account for the current understanding of flame phenomena in premixed combustible gases with regard to flame structure, stability, and effects of turbulence. Computer calculations of detonation parameters as a function of initial conditions formed the main basis for modifications made in Chapter 5. The chapter on diffusion flames has been reorganized and consolidated with material previously in Chapter 9. In particular, this change permits a clearer picture of the compositional changes around a fuel droplet burning in air under quiescent conditions. Only modest changes will be found in Chapter 7 on ignition. The major additions to Chapter 8 on environmental combustion phenomena reflect the recent advances made in nitrogen oxide chemistry, in the understanding of soot processes, and in techniques for postcombustion emission reduction. The stratospheric ozone part of this chapter remains unchanged and, considering the extensive developments in atmospheric chemistry, is only retained as a simple introduction to chemical effects on the ozone layer. Chapter 9 on the combustion of low-volatile fuels has been enlarged to include metal combustion and combustion synthesis of refractory materials. This new material concentrates on the thermodynamics of the synthesis processes to provide a rationale for their use. Additional material on the oxidation of carbonaceous particles also has been included.

Because the author was unable to review the proofs to the second edition, numerous printing errors appeared in that edition. Apologies are due for the inconveniences caused. Much care has been exercised in the printing of this edition. SI units are used where appropriate throughout, particularly for thermodynamic data so that these data are now consistent with the JANAF tables in the appendix. In some instances where certain cgs databases have not been updated and where cgs units are so ingrained that SI may have proved an inconvenience, cgs units have prevailed. The table of conversion factors in the appendix should reduce any inconveniences.

This edition will be the last. It is the author's hope that many will find it a useful and lasting contribution.

Irvin Glassman

Acknowledgments

After teaching the combustion course at Princeton almost continuously for 30 years, I gave up the responsibility to Paul Ronney, now at the University of Southern California. Throughout his teaching span of the same course, he continually recommended improvements to the second edition of this text and found printing errors that had been overlooked. Later, he supplied me with extensive comments for modifications to sections in the text on turbulent combustion and flammability limits. His interest in this book and his recommendations are most gratefully acknowledged. Further, I acknowledge with thanks Paul Papas and Jackie Sung, who performed the computer calculations that constitute the many figures in the various chapters. Special thanks are due to Richard Yetter, who assumed the responsibility of updating and expanding the appendixes. Many conversations were held with Yetter on many different aspects of the material covered herein.

Kenneth Brezinsky deserves similar thanks for taking up more than his share of the burdens of operating our laboratory. He has always been a testing ground for my ideas and writing and contributed significantly to particular aspects of the text. The substance of many stimulating conversations with my colleagues C. K. Law and S. -H. Lam also has been incorporated into the text.

As I reflect on this, the last, edition of this book, I must thank the late Joseph Masi, Bernard Wolfson and Julian Tishkoff of AFOSR, and William Kirchoff and Alan Laufer of DOE for their continual support of my research in combustion and for their confidence in the contributions they thought my laboratory could make. Again, I hope this edition is another such contribution.

The onerous task of typing the manuscript and the numerous equations thought necessary fell to my long time editorial assistant, Elizabeth Adam. She did a marvelous and dedicated job and I am most grateful. Thanks are also due to my editor at Academic Press, David Packer, who encouraged me to undertake this third edition; and to Ellen Caprio, production editor, and Emily Thompson, copy editor, both of whom greatly improved my writing in many instances.

Finally, the dedication acknowledges very special people who contributed in many ways to a very special place to work.

1

Chemical Thermodynamics and Flame Temperatures

A. INTRODUCTION

The parameters essential for the evaluation of combustion systems are the equilibrium product temperature and composition. If all the heat evolved in the reaction is employed solely to raise the product temperature, this temperature is called the adiabatic flame temperature. Because of the importance of the temperature and gas composition in combustion considerations, it is appropriate to review those aspects of the field of chemical thermodynamics that deal with these subjects.

B. HEATS OF REACTION AND FORMATION

All chemical reactions are accompanied either by an absorption or evolution of energy, which usually manifests itself as heat. It is possible to determine this amount of heat—and hence the temperature and product composition—from very basic principles. Spectroscopic data and statistical calculations permit one to determine the internal energy of a substance. The internal energy of a given substance is found to be dependent upon its temperature, pressure, and state and is independent of the means by which the state is attained. Likewise the change in internal energy, ΔE, of a system that results from any physical change or chemical reaction depends only on the initial and final state of the system. The total change in internal energy will be the same, regardless of whether the energy is evolved as heat, energy, or work.

If a flow reaction proceeds with negligible changes in kinetic energy and potential energy and involves no form of work beyond that required for flow, the heat added is equal to the increase of enthalpy of the system

$$Q = \Delta H$$

where Q is the heat added and H is the enthalpy. For a nonflow reaction proceeding at constant pressure, the heat added is also equal to the gain in enthalpy

$$Q = \Delta H$$

and if heat is evolved,

$$Q = -\Delta H$$

Most thermochemical calculations are made for closed thermodynamic systems, and the stoichiometry is most conveniently represented in terms of the molar quantities as determined from statistical calculations. In dealing with compressible flow problems in which it is essential to work with open thermodynamic systems, it is best to employ mass quantities. Throughout this text uppercase symbols will be used for molar quantities and lowercase symbols for mass quantities.

One of the most important thermodynamic facts to know about a given chemical reaction is the change in energy or heat content associated with the reaction at some specified temperature, where each of the reactants and products is in an appropriate standard state. This change is known either as the energy or as the heat of reaction at the specified temperature.

The standard state means that for each state a reference state of the aggregate exists. For gases, the thermodynamic standard reference state is the ideal gaseous state at atmospheric pressure at each temperature. The ideal gaseous state is the case of isolated molecules which give no interactions and which obey the equation of state of a perfect gas. The standard reference state for pure liquids and solids at a given temperature is the real state of the substance at a pressure of one atmosphere. As discussed in Chapter 9, understanding this definition of the standard reference state is very important when considering the case of high-temperature combustion in which the product composition contains a substantial mole fraction of a condensed phase, such as a metal oxide.

The thermodynamic symbol that represents the property of the substance in the standard state at a given temperature is written, for example, as H_T°, E_T°, etc., where the "degree sign" superscript $^\circ$ specifies the standard state and the subscript T, the specific temperature. Statistical calculations actually permit the determination of $E_T - E_0$, which is the energy content at a given temperature referred to the energy content at 0 K. For one mole in the ideal gaseous state,

$$PV = RT \qquad (1)$$

$$H^\circ = E^\circ + (PV)^\circ = E^\circ + RT \qquad (2)$$

which at 0 K reduces to

$$H_0^\circ = E_0^\circ \qquad (3)$$

Thus the heat content at any temperature referred to the heat or energy content at 0 K is known and

$$\left(H^\circ - H_0^\circ\right) = \left(E^\circ - E_0^\circ\right) + RT = \left(E^\circ - E_0^\circ\right) + PV \qquad (4)$$

The value $(E^\circ - E_0^\circ)$ is determined from spectroscopic information and is actually the energy in the internal (rotational, vibrational, and electronic) and external (translational) degrees of freedom of the molecule. Enthalpy $(H^\circ - H_0^\circ)$ has meaning only when there is a group of molecules, a mole for instance; it is thus the ability of a group of molecules with internal energy to do PV work. In this sense, then, a single molecule can have internal energy, but not enthalpy. The use of the lowercase symbol will signify values on a mass basis. Since flame temperatures are calculated for a closed thermodynamic system and molar conservation is not required, working on a molar basis is most convenient. In flame propagation or reacting flows through nozzles, conservation of mass is a requirement for convenient solution; thus when these systems are considered, the per-unit mass basis of the thermochemical properties is used.

From the definition of the heat of reaction, Q_p will depend on the temperature T at which the reaction and product enthalpies are evaluated. The heat of reaction at one temperature T_0 can be related to that at another temperature T_1. Consider the reaction configuration shown in Fig. 1. According to the First Law of thermodynamics, the heat changes that proceed from reactants at temperatures T_0 to products at temperature T_1 by either path A or path B must be the same. Path A raises the reactants from temperature T_0 to T_1 and reacts at T_1. Path B reacts at T_0 and raises the products from T_0 to T_1. This energy equality, which relates the heats of reaction at the two different temperatures, is written as

$$\left\{\sum_{j,\text{react}} n_j \left[\left(H_{T_1}^\circ - H_0^\circ\right) - \left(H_{T_0}^\circ - H_0^\circ\right)\right]_j\right\} + \Delta H_{T_1}$$

$$= \Delta H_{T_0} + \left\{\sum_{i,\text{prod}} n_i \left[\left(H_{T_1}^\circ - H_0^\circ\right) - \left(H_{T_0}^\circ - H_0^\circ\right)\right]_i\right\} \qquad (5)$$

where n specifies the number of moles of the ith product or jth reactant. Any phase changes can be included in the heat content terms. Thus, by knowing the

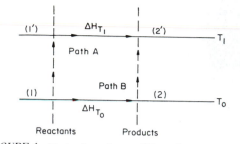

FIGURE 1   Heats of reactions at different base temperatures.

difference in energy content at the different temperatures for the products and reactants, it is possible to determine the heat of reaction at one temperature from the heat of reaction at another.

If the heats of reaction at a given temperature are known for two separate reactions, the heat of reaction of a third reaction at the same temperature may be determined by simple algebraic addition. This statement is the Law of Heat Summation. For example, reactions (6) and (7) can be carried out conveniently in a calorimeter at constant pressure:

$$C_{graphite} + O_2(g) \xrightarrow[298\ K]{} CO_2(g), \qquad Q_p = +393.52 \text{ kJ} \qquad (6)$$

$$CO(g) + \tfrac{1}{2}O_2(g) \xrightarrow[298\ K]{} CO_2(g), \qquad Q_p = +283.0 \text{ kJ} \qquad (7)$$

Subtracting these two reactions, one obtains

$$C_{graphite} + \tfrac{1}{2}O_2(g) \xrightarrow[298\ K]{} CO(g), \qquad Q_p = +110.52 \text{ kJ} \qquad (8)$$

Since some of the carbon would burn to CO_2 and not solely to CO, it is difficult to determine calorimetrically the heat released by reaction (8).

It is, of course, not necessary to have an extensive list of heats of reaction to determine the heat absorbed or evolved in every possible chemical reaction. A more convenient and logical procedure is to list the standard heats of formation of chemical substances. The standard heat of formation is the enthalpy of a substance in its standard state referred to its elements in their standard states at the same temperature. From this definition it is obvious that heats of formation of the elements in their standard states are zero.

The value of the heat of formation of a given substance from its elements may be the result of the determination of the heat of one reaction. Thus, from the calorimetric reaction for burning carbon to CO_2 [Eq. (6)], it is possible to write the heat of formation of carbon dioxide at 298 K as

$$\left(\Delta H_f^{\circ}\right)_{298,CO_2} = -393.52 \text{ kJ/mol}$$

The superscript to the heat of formation symbol ΔH_f° represents the standard state, and the subscript number represents the base or reference temperature. From the example for the Law of Heat Summation, it is apparent that the heat of formation of carbon monoxide from Eq. (8) is

$$\left(\Delta H_f^\circ\right)_{298,CO} = -110.52 \text{ kJ/mol}$$

It is evident that, by judicious choice, the number of reactions that must be measured calorimetrically will be about the same as the number of substances whose heats of formation are to be determined.

The logical consequence of the preceding discussion is that, given the heats of formation of the substances comprising any particular reaction, one can directly determine the heat of reaction or heat evolved at the reference temperature T_0, most generally T_{298}, as follows:

$$\Delta H_{T_0} = \sum_{i \text{ prod}} n_i \left(\Delta H_f^\circ\right)_{T_{0,i}} - \sum_{j \text{ react}} n_j \left(\Delta H_f^\circ\right)_{T_{0,j}} = -Q_p \tag{9}$$

Extensive tables of standard heats of formation are available, but they are not all at the same reference temperature. The most convenient are the compilations known as the JANAF [1] and NBS Tables [2], both of which use 298 K as the reference temperature. Table 1 lists some values of the heat of formation taken from the JANAF Thermochemical Tables. Actual JANAF tables are reproduced in Appendix A. These tables, which represent only a small selection from the JANAF volume, were chosen as those commonly used in combustion and to aid in solving the problem sets throughout this book. Note that, although the developments throughout this book take the reference state as 298 K, the JANAF Tables also list ΔH_f°'s for all temperatures.

When the products are measured at a temperature T_2 different from the reference temperature T_0 and the reactants enter the reaction system at a temperature T_0' different from the reference temperature, the heat of reaction becomes

$$\Delta H = \sum_{i \text{ prod}} n_i \left[\left\{\left(H_{T_2}^\circ - H_0^\circ\right) - \left(H_{T_0}^\circ - H_0^\circ\right)\right\} + \left(\Delta H_f^\circ\right)_{T_0}\right]_i$$

$$- \sum_{j \text{ react}} n_j \left[\left\{\left(H_{T_0'}^\circ - H_0^\circ\right) - \left(H_{T_0}^\circ - H_0^\circ\right)\right\} + \left(\Delta H_f^\circ\right)_{T_0}\right]_j$$

$$= -Q_p(\text{evolved}) \tag{10}$$

The reactants in most systems are considered to enter at the standard reference temperature 298 K. Consequently, the enthalpy terms in the braces for the reactants disappear. The JANAF Tables tabulate, as a putative convenience, $(H_T^\circ - H_{298}^\circ)$ instead of $(H_T^\circ - H_0^\circ)$. This type of tabulation is unfortunate since the reactants for systems using cryogenic fuels and oxidizers, such as those used in rockets, can enter the system at temperatures lower than the reference temperature. Indeed, the fuel and oxidizer individually could enter at different temperatures. Thus the summation in Eq. (10) is handled most conveniently by realizing that T_0' may vary with the substance j.

TABLE 1   Heats of Formation at 298 K

Chemical	Name	State	ΔH_f° (kJ/mol)	Δh_f° (kJ/g)
C	Carbon	Vapor	716.67	59.72
N	Nitrogen atom	Gas	472.68	33.76
O	Oxygen atom	Gas	249.17	15.57
C_2H_2	Acetylene	Gas	227.06	8.79
H	Hydrogen atom	Gas	218.00	218.00
O_3	Ozone	Gas	142.67	2.97
NO	Nitric oxide	Gas	90.29	3.01
C_6H_6	Benzene	Gas	82.96	1.06
C_6H_6	Benzene	Liquid	49.06	0.63
C_2H_4	Ethene	Gas	52.38	1.87
N_2H_4	Hydrazine	Liquid	50.63	1.58
OH	Hydroxyl radical	Gas	38.99	2.29
O_2	Oxygen	Gas	0	0
N_2	Nitrogen	Gas	0	0
H_2	Hydrogen	Gas	0	0
C	Carbon	Solid	0	0
NH_3	Ammonia	Gas	−45.90	−2.70
C_2H_4O	Ethylene oxide	Gas	−51.08	−0.86
CH_4	Methane	Gas	−74.87	−4.68
C_2H_6	Ethane	Gas	−84.81	−2.83
CO	Carbon monoxide	Gas	−110.53	−3.95
C_4H_{10}	Butane	Gas	−124.90	−2.15
CH_3OH	Methanol	Gas	−201.54	−6.30
CH_3OH	Methanol	Liquid	−239.00	−7.47
H_2O	Water	Gas	−241.83	−13.44
C_8H_{18}	Octane	Liquid	−250.31	−0.46
H_2O	Water	Liquid	−285.10	−15.84
SO_2	Sulfur dioxide	Gas	−296.84	−4.64
$C_{12}H_{16}$	Dodecane	Liquid	−347.77	−2.17
CO_2	Carbon dioxide	Gas	−393.52	−8.94
SO_3	Sulfur trioxide	Gas	−395.77	−4.95

The values of heats of formation reported in Table 1 are ordered so that the largest positive value of the heats of formation per mole are the highest and those with negative heats of formation are the lowest. In fact, this table is similar to a potential energy chart. As species at the top react to form species at the bottom, heat is released, and one has an exothermic system. Even a species that has a negative heat of formation can react to form products of still lower negative heats of formation species, thereby releasing heat. Since some fuels that have negative heats of formation form many moles of product species having negative heats of formation, the heat release in such cases can be large. Equation (9) shows this result clearly. Indeed, the first summation in Eq. (9) is generally much greater than the second. Thus the characteristic of the reacting species or the fuel that significantly determines the heat release is its chemical composition and not its heat of formation. Nonradical species that have positive heats of formation, such as acetylene and benzene can decompose to their elements and release heat. Thus they can be considered unstable—and may even be considered monopropellants. However, kinetic factors prevent some of these fuels from following a self-sustained decomposition, so that these fuels are in actuality stable. These kinetic aspects are discussed in Chapter 2.

When all the heat evolved is used to heat up the product gases, ΔH and Q_p become zero. The product temperature T_2 in this case is called the adiabatic flame temperature and Eq. (10) becomes

$$\sum_{i \text{ prod}} n_i \left[\left\{ \left(H^{\circ}_{T_2} - H^{\circ}_0 \right) - \left(H^{\circ}_{T_0} - H^{\circ}_0 \right) \right\} + \left(\Delta H^{\circ}_f \right)_{T_0} \right]_i$$

$$= \sum_{j \text{ react}} n_j \left[\left\{ \left(H^{\circ}_{T'_0} - H^{\circ}_0 \right) - \left(H^{\circ}_{T_0} - H^{\circ}_0 \right) \right\} + \left(\Delta H^{\circ}_f \right)_{T_0} \right]_j \qquad (11)$$

Again, note that T'_0 can be different for each reactant. Since the heats of formation throughout this text will always be considered as those evaluated at the reference temperature $T_0 = 298$ K, the expression in braces becomes $[(H^{\circ}_T - H^{\circ}_0) - (H^{\circ}_T - H^{\circ}_0)] = (H^{\circ}_T - H^{\circ}_{T_0})$, which is the value listed in the JANAF tables (see Appendix A).

If the products n_i of this reaction are known, Eq. (11) can be solved for the flame temperature. For a reacting lean system whose product temperature is less than 1250 K, the products are the normal stable species CO_2, H_2O, N_2, and O_2, whose molar quantities can be determined from simple mass balances. However, most combustion systems reach temperatures appreciably greater than 1250 K, and dissociation of the stable species occurs. Since the dissociation reactions are quite endothermic, a small percentage of dissociation can lower the flame temperature substantially. The stable products from a C–H–O reaction system can dissociate by any of the following reactions:

$$CO_2 \rightleftarrows CO + \tfrac{1}{2}O_2$$

$$CO_2 + H_2 \rightleftarrows CO + H_2O$$

$$H_2O \rightleftarrows H_2 + \tfrac{1}{2}O_2$$

$$H_2O \rightleftarrows H + OH$$

$$H_2O \rightleftarrows \tfrac{1}{2}H_2 + OH$$

$$H_2 \rightleftarrows 2H$$

$$O_2 \rightleftarrows 2O, \text{ etc.}$$

Each of these dissociation reactions also specifies a definite equilibrium concentration of each product at a given temperature; consequently, the reactions are written as equilibrium reactions. In the calculation of the heat of reaction of low-temperature combustion experiments the products could be specified from the chemical stoichiometry; but with dissociation, the specification of the product concentrations becomes much more complex and the n_i's in the flame temperature equation [Eq. (11)] are as unknown as the flame temperature itself. In order to solve the equation for the n_i's and T_2, it is apparent that one needs more than mass balance equations. The necessary equations are found in the equilibrium relationships that exist among the product composition in the equilibrium system.

C. FREE ENERGY AND THE EQUILIBRIUM CONSTANTS

The condition for equilibrium is determined from the combined form of the first and second laws of thermodynamics; i.e.,

$$dE = T\,dS - P\,dV \tag{12}$$

where S is the entropy. This condition applies to any change affecting a system of constant mass in the absence of gravitational, electrical, and surface forces. However, the energy content of the system can be changed by introducing more mass. Consider the contribution to the energy of the system on adding one molecule i to be μ_i. The introduction of a small number dn_i of the same type contributes a gain in energy of the system of $\mu_i dn_i$. All the possible reversible increases in the energy of the system due to each type of molecule i can be summed to give

$$dE = T\,dS - P\,dV + \sum_i \mu_i n_i \tag{13}$$

It is apparent from the definition of enthalpy H and the introduction of the concept of the Gibbs free energy G

$$G \equiv H - TS \tag{14}$$

that

$$dH = T\,dS + V\,dP + \sum_i \mu_i\,dn_i \tag{15}$$

and

$$dG = -S\,dT + V\,dP + \sum_i \mu_i\,dn_i \tag{16}$$

Recall that P and T are intensive properties that are independent of the size of mass of the system, whereas E, H, G, and S (as well as V and n) are extensive properties that increase in proportion to mass or size. By writing the general relation for the total derivative of G with respect to the variables in Eq. (16), one obtains

$$dG = \left(\frac{\partial G}{\partial T}\right)_{P,n_i} dT + \left(\frac{\partial G}{\partial P}\right)_{T,n_i} dP + \sum_i \left(\frac{\partial G}{\partial n_i}\right)_{P,T,n_{j(j\neq i)}} dn_i \qquad (17)$$

Thus,

$$\mu_i = \left(\frac{\partial G}{\partial n_i}\right)_{T,P,n_j} \qquad (18)$$

or, more generally, from dealing with the equations for E and H

$$\mu_i = \left(\frac{\partial G}{\partial n_i}\right)_{T,P,n_j} = \left(\frac{\partial E}{\partial n_i}\right)_{S,V,n_j} = \left(\frac{\partial H}{\partial n_i}\right)_{S,P,n_j} \qquad (19)$$

where μ_i is called the chemical potential or the partial molal free energy. The condition of equilibrium is that the entropy of the system have a maximum value for all possible configurations that are consistent with constant energy and volume. If the entropy of any system at constant volume and temperature is at its maximum value, the system is at equilibrium; therefore, in any change from its equilibrium state dS is zero. It follows then from Eq. (13) that the condition for equilibrium is

$$\sum \mu_i \, dn_i = 0 \qquad (20)$$

The concept of the chemical potential is introduced here because this property plays an important role in reacting systems. In this context one may consider that a reaction moves in the direction of decreasing chemical potential, reaching equilibrium only when the potential of the reactants equals that of the products [3].

Thus, from Eq. (16) the criterion for equilibrium for combustion products of a chemical system at constant T and P is

$$(dG)_{T,P} = 0 \qquad (21)$$

and it becomes possible to determine the relationship between the Gibbs free energy and the equilibrium partial pressures of a combustion product mixture.

One deals with perfect gases so that there are no forces of interactions between the molecules except at the instant of reaction; thus, each gas acts as if it were in a container alone. Let G, the total free energy of a product mixture, be represented by

$$G = \sum n_i G_i, \qquad i = A, B, \ldots, R, S \ldots \qquad (22)$$

for an equilibrium reaction among arbitrary products:

$$aA + bB + \cdots \rightleftharpoons rR + sS + \cdots \qquad (23)$$

Note that A, B, ..., R, S, ... represent substances in the products only and $a, b, \ldots, r, s, \ldots$ are the stoichiometric coefficients that govern the proportions by which different substances appear in the arbitrary equilibrium system chosen. The n_i's represent the instantaneous number of each compound. Under the ideal gas assumption the free energies are additive, as shown above. This assumption permits one to neglect the free energy of mixing. Thus, as stated earlier,

$$G(p, T) = H(T) - TS(p, T) \tag{24}$$

Since the standard state pressure for a gas is $p_0 = 1$ atm, one may write

$$G^\circ(p_0, T) = H^\circ(T) - TS^\circ(p_0, T) \tag{25}$$

Subtracting the last two equations, one obtains

$$G - G^\circ = (H - H^\circ) - T(S - S^\circ) \tag{26}$$

Since H is not a function of pressure, $H - H^\circ$ must be zero, and then

$$G - G^\circ = -T(S - S^\circ) \tag{27}$$

Equation (27) relates the difference in free energy for a gas at any pressure and temperature to the standard state condition at constant temperature. Here $dH = 0$, and from Eq. (15) the relationship of the entropy to the pressure is found to be

$$S - S^\circ = -R \ln(p/p_0) \tag{28}$$

Hence, one finds that

$$G(T, p) = G^\circ + RT \ln(p/p_0) \tag{29}$$

An expression can now be written for the total free energy of a gas mixture. In this case p is the partial pressure p_i of a particular gaseous component and obviously has the following relationship to the total pressure P:

$$p_i = \left(n_i / \sum_i n_i \right) P \tag{30}$$

where $(n_i / \sum_i n_i)$ is the mole fraction of gaseous species i in the mixture. Equation (29) thus becomes

$$G(T, P) = \sum_i n_i \{ G_i^\circ + RT \ln(p_i/p_0) \} \tag{31}$$

As determined earlier [Eq. (21)], the criterion for equilibrium is $(dG)_{T,P} = 0$. Taking the derivative of G in Eq. (31), one obtains

$$\sum_i G_i^\circ \, dn_i + RT \sum_i (dn_i) \ln(p_i/p_0) + RT \sum_i n_i (dp_i/p_i) = 0 \tag{32}$$

Evaluating the last term of the left-hand side of Eq. (32), one has

$$\sum_i n_i \frac{dp_i}{p_i} = \sum_i \left(\frac{\sum_i n_i}{P}\right) dp_i = \frac{\sum_i n_i}{P} \sum_i dp_i = 0 \tag{33}$$

since the total pressure is constant, and thus $\sum_i dp_i = 0$. Now consider the first term in Eq. (32):

$$\sum_i G_i^\circ \, dn_i = (dn_A)G_A^\circ + (dn_B)G_B^\circ + \cdots - (dn_R^\circ)G_R - (dn_s^\circ)G_S + \cdots \tag{34}$$

By the definition of the stoichiometric coefficients,

$$dn_i \sim a_i, \qquad dn_i = k a_i \tag{35}$$

where k is a proportionality constant. Hence

$$\sum_i G_i^\circ \, dn_i = k\{a G_A^\circ + b G_B^\circ + \cdots - r G_R^\circ - s G_S^\circ \cdots\} \tag{36}$$

Similarly, the proportionality constant k will appear as a multiplier in the second term of Eq. (32). Since Eq. (32) must equal zero, the third term already has been shown equal to zero, and k cannot be zero, one obtains

$$(a G_A^\circ + b G_B^\circ + \cdots - r G_R^\circ - s G_S^\circ - \cdots) = RT \ \ln\left\{\frac{(p_R/p_0)^r (p_S/p_0)^s}{(p_A/p_0)^a (p_B/p_0)^b}\right\} \tag{37}$$

One then defines

$$-\Delta G^\circ = a G_A^\circ + b G_B^\circ + \cdots - r G_R^\circ - s G_S^\circ - \cdots \tag{38}$$

where ΔG° is called the standard state free energy change and $p_0 = 1$ atm. This name is reasonable since ΔG° is the change of free energy for reaction (23) if it takes place at standard conditions and goes to completion to the right. Since the standard state pressure p_0 is 1 atm, the condition for equilibrium becomes

$$-\Delta G^\circ = RT \ \ln(p_R^r p_S^s / p_A^a p_B^b) \tag{39}$$

where the partial pressures are measured in atmospheres. One then defines the equilibrium constant at constant pressure from Eq. (39) as

$$K_p \equiv p_R^r p_S^s / p_A^a p_B^b$$

Then

$$-\Delta G^\circ = RT \ \ln K_p, \qquad K_p = \exp(-\Delta G^\circ / RT) \tag{40}$$

where K_p is not a function of the total pressure, but rather a function of temperature alone. It is a little surprising that the free energy change at the standard state pressure (1 atm) determines the equilibrium condition at all other pressures.

Equations (39) and (40) can be modified to account for nonideality in the product state; however, because of the high temperatures reached in combustion systems, ideality can be assumed even under rocket chamber pressures.

The energy and mass conservation equations used in the determination of the flame temperature are more conveniently written in terms of moles; thus it is best to write the partial pressure in K_p in terms of moles and the total pressure P. This conversion is accomplished through the relationship between partial pressure p and total pressure P, as given by Eq. (30). Substituting this expression for p_i [Eq. (30)] in the definition of the equilibrium constant [Eq. (40)], one obtains

$$K_p = (n_R^r n_S^s / n_A^a p_B^b)(P/\textstyle\sum n_i)^{r+s-a-b} \tag{41}$$

which is sometimes written as

$$K_p = K_N (P/\textstyle\sum n_i)^{r+s-a-b} \tag{42}$$

where

$$K_N \equiv n_R^r n_S^s / n_A^a p_B^b \tag{43}$$

When

$$r + s - a - b = 0 \tag{44}$$

the equilibrium reaction is said to be pressure-insensitive. Again, however, it is worth repeating that K_p is not a function of pressure; however, Eq. (42) shows that K_N can be a function of pressure.

The equilibrium constant based on concentration (in moles per cubic centimeter) is sometimes used, particularly in chemical kinetic analyses (to be discussed in the next chapter). This constant is found by recalling the perfect gas law, which states that

$$PV = \textstyle\sum n_i RT \tag{45}$$

or

$$(P/\textstyle\sum n_i) = (RT/V) \tag{46}$$

where V is the volume. Substituting for $(P/\sum n_i)$ in Eq. (42) gives

$$K_p = [(n_R^r n_S^s)/(n_A^a n_B^b)] \left(\frac{RT}{V}\right)^{r+s-a-b} \tag{47}$$

or

$$K_p = \frac{(n_R/V)^r (n_S/V)^s}{(n_A/V)^a (n_B/V)^b}(RT)^{r+s-a-b} \tag{48}$$

Equation (48) can be written as

$$K_p = (C_R^r C_S^s / C_A^a C_B^b)(RT)^{r+s-a-b} \tag{49}$$

where $C = n/V$, a molar concentration. From Eq. (49) it is seen that the definition of the equilibrium constant for concentration is

$$K_C = C_R^r C_S^s / C_A^a C_B^b \tag{50}$$

K_C is a function of pressure, unless $r + s - a - b = 0$. Given a temperature and pressure, all the equilibrium constants (K_p, K_N, and K_C) can be determined thermodynamically from $\Delta G°$ for the equilibrium reaction chosen.

How the equilibrium constant varies with temperature can be of importance. Consider first the simple derivative

$$\frac{d(G/T)}{dT} = \frac{T(dG/dT) - G}{T^2} \tag{51}$$

Recall that the Gibbs free energy may be written as

$$G = E + PV - TS \tag{52}$$

or, at constant pressure,

$$\frac{dG}{dT} = \frac{dE}{dT} + P\frac{dV}{dT} - S - T\frac{dS}{dT} \tag{53}$$

At equilibrium from Eq. (12) for the constant pressure condition

$$T\frac{dS}{dT} = \frac{dE}{dT} + P\frac{dV}{dT} \tag{54}$$

Combining Eqs. (53) and (54) gives

$$\frac{dG}{dT} = -S \tag{55}$$

Hence Eq. (51) becomes

$$\frac{d(G/T)}{dT} = \frac{-TS - G}{T^2} = -\frac{H}{T^2} \tag{56}$$

This expression is valid for any substance under constant pressure conditions. Applying it to a reaction system with each substance in its standard state, one obtains

$$d(\Delta G°/T) = -(\Delta H°/T^2)\, dT \tag{57}$$

where $\Delta H°$ is the standard state heat of reaction for any arbitrary reaction

$$aA + bB + \cdots \rightarrow rR + sS + \cdots$$

at temperature T (and, of course, a pressure of 1 atm). Substituting the expression for $\Delta G°$ given by Eq. (40) into Eq. (57), one obtains

$$d \ln K_p / dT = \Delta H° / RT^2 \qquad (58)$$

If it is assumed that $\Delta H°$ is a slowly varying function of T, one obtains

$$\ln \left(\frac{K_{p_2}}{K_{p_1}} \right) = -\frac{\Delta H°}{R} \left(\frac{1}{T_2} - \frac{1}{T_1} \right) \qquad (59)$$

Thus for small changes in T

$$\left(K_{p_2} \right) > \left(K_{p_1} \right) \qquad \text{when} \quad T_2 > T_1$$

In the same context as the heat of formation, the JANAF Tables have tabulated most conveniently the equilibrium constants of formation for practically every substance of concern in combustion systems. The equilibrium constant of formation $(K_{p,f})$ is based on the equilibrium equation of formation of a species from its elements in their normal states. Thus by algebraic manipulation it is possible to determine the equilibrium constant of any reaction. In flame temperature calculations, by dealing only with equilibrium constants of formation, there is no chance of choosing a redundant set of equilibrium reactions. Of course, the equilibrium constant of formation for elements in their normal state is zero.

Consider the following three equilibrium reactions of formation:

$$H_2 + \tfrac{1}{2}O_2 \rightleftharpoons H_2O \qquad K_{p,f(H_2O)} = \frac{p_{H_2O}}{(p_{H_2})(p_{O_2})^{1/2}}$$

$$\tfrac{1}{2}H_2 \rightleftharpoons H \qquad K_{p,f(H)} = \frac{p_H}{(p_{H_2})^{1/2}}$$

$$\tfrac{1}{2}O_2 + \tfrac{1}{2}H_2 \rightleftharpoons OH \qquad K_{p,f(OH)} = \frac{p_{OH}}{(p_{O_2})^{1/2}(p_{H_2})^{1/2}}$$

The equilibrium reaction is always written for the formation of one mole of the substances other than the elements. Now if one desires to calculate the equilibrium constant for reactions such as

$$H_2O \rightleftharpoons H + OH \quad \text{and} \quad H_2O \rightleftharpoons \tfrac{1}{2}H_2 + OH$$

one finds the respective K_p's from

$$K_p = \frac{p_H p_{OH}}{p_{H_2O}} = \frac{K_{p,f(H)} K_{p,f(OH)}}{K_{p,f(H_2O)}}, \qquad K_p = \frac{p_{H_2} p_{OH}}{p_{H_2O}} = \frac{K_{p,f(OH)}}{K_{p,f(H_2O)}}$$

Because of this type of result and the thermodynamic expression

$$\Delta G^\circ = -RT \ln K_p$$

the JANAF Tables list log $K_{p,f}$. Note the base 10 logarithm.

For those compounds that contain carbon and a combustion system in which solid carbon is found, the thermodynamic handling of the K_p is somewhat more difficult. The equilibrium reaction of formation for CO_2 would be

$$C_{graphite} + O_2 \rightleftarrows CO_2, \qquad K_p = \frac{p_{CO_2}}{p_{O_2} p_C}$$

However, since the standard state of carbon is the condensed state, carbon graphite, the only partial pressure it exerts is its vapor pressure (p_{vp}), a known thermodynamic property that is also a function of temperature. Thus, the preceding formation expression is written as

$$K_p(T) p_{vp,C}(T) = \frac{p_{CO_2}}{p_{O_2}} = K_p'$$

The $K_{p,f}$'s for substances containing carbon tabulated by JANAF are in reality K_p', and the condensed phase is simply ignored in evaluating the equilibrium expression. The number of moles of carbon (or any other condensed phase) is not included in the $\sum n_j$ since this summation is for the gas phase components contributing to the total pressure.

D. FLAME TEMPERATURE CALCULATIONS

1. Analysis

If one examines the equation for the flame temperature [Eq. (11)], one can make an interesting observation. Given the values in Table 1 and the realization that many moles of product form for each mole of the reactant fuel, one can see that the sum of the molar heats of the products will be substantially greater than the sum of the molar heats of the reactants; i.e.,

$$\sum_{i \; prod} n_i \left(\Delta H_f^\circ \right)_i \gg \sum_{j \; react} n_j \left(\Delta H_f^\circ \right)_j$$

Consequently, it would appear that the flame temperature is determined not by the specific reactants, but only by the atomic ratios and the specific atoms that are introduced. It is the atoms that determine what products will form. Only ozone and acetylene have positive molar heats of formation high enough to cause a noticeable variation (rise) in flame temperature. Ammonia has a negative heat of formation low enough to lower the final flame temperature. One can normalize for the effects of total moles of products formed by considering the heats of formation per gram (Δh_f°); these values are given for some fuels and oxidizers in Table 1.

The variation of Δh_f° among most hydrocarbon fuels is very small. This fact will be used later in correlating the flame temperatures of hydrocarbons in air.

One can draw the further conclusion that the product concentrations are also functions only of temperature, pressure, and the C/H/O ratio and not the original source of atoms. Thus, for any C–H–O system, the products will be the same; i.e., they will be CO_2, H_2O, and their dissociated products. The dissociation reactions listed earlier give some of the possible "new" products. A more complete list would be

$$CO_2, \quad H_2O, \quad CO, \quad H_2, \quad O_2, \quad OH, \quad H, \quad O, \quad O_3, \quad C, \quad CH_4$$

For a C, H, O, N system, the following could be added:

$$N_2, \quad N, \quad NO, \quad NH_3, \quad NO^+, \quad e^-$$

Nitric oxide has a very low ionization potential and could ionize at flame temperatures. For a normal composite solid propellant containing C–H–O–N–Cl–Al, many more products would have to be considered. In fact if one lists all the possible number of products for this system, the solution to the problem becomes more difficult, requiring the use of advanced computers and codes for exact results. However, knowledge of thermodynamic equilibrium constants and kinetics allows one to eliminate many possible product species. Although the computer codes listed in Appendix H essentially make it unnecessary to eliminate any product species, the following discussion gives one the opportunity to estimate which products can be important without running any computer code.

Consider a C–H–O–N system. For an overoxidized case, an excess of oxygen converts all the carbon and hydrogen present to CO_2 and H_2O by the following reactions:

$$CO_2 \rightleftharpoons CO + \tfrac{1}{2}O_2, \qquad Q_p = -283.2 \text{ kJ}$$

$$H_2O \rightleftharpoons H_2 + \tfrac{1}{2}O_2, \qquad Q_p = -242.2 \text{ kJ}$$

$$H_2O \rightleftharpoons H + OH, \qquad Q_p = -284.5 \text{ kJ}$$

where the Q_p's are calculated at 298 K. This heuristic postulate is based upon the fact that at these temperatures and pressures at least 1% dissociation takes place. The pressure enters into the calculations through Le Chatelier's principle that the equilibrium concentrations will shift with the pressure. The equilibrium constant, although independent of pressure, can be expressed in a form that contains the pressure. A variation in pressure shows that the molar quantities change. Since the reactions noted above are quite endothermic, even small concentration changes must be considered. If one initially assumes that certain products of dissociation are absent and calculates a temperature that would indicate 1% dissociation of the species, then one must reevaluate the flame temperature by including in the product mixture the products of dissociation; i.e., one must indicate the presence of CO, H_2, and OH as products.

Concern about emissions from power plant sources has raised the level of interest in certain products whose concentrations are much less than 1%, even though such concentrations do not affect the temperature even in a minute way. The major pollutant of concern in this regard is nitric oxide NO. To make an estimate of the amount of NO found in a system at equilibrium, one would use the equilibrium reaction of formation of NO

$$\tfrac{1}{2}N_2 + \tfrac{1}{2}O_2 \rightleftharpoons NO$$

As a rule of thumb, any temperature above 1700 K gives sufficient NO to be of concern. The NO formation reaction is pressure-insensitive, so there is no need to specify the pressure.

If in the overoxidized case $T_2 > 2400$ K at $P = 1$ atm and $T_2 > 2800$ K at $P = 20$ atm, the dissociation of O_2 and H_2 becomes important; namely,

$$H_2 \rightleftharpoons 2H, \qquad Q_p = -436.6 \text{ kJ}$$
$$O_2 \rightleftharpoons 2O, \qquad Q_p = -499.0 \text{ kJ}$$

Although these dissociation reactions are written to show the dissociation of one mole of the molecule, recall that the $K_{p,f}$'s are written to show the formation of one mole of the radical. These dissociation reactions are highly endothermic, and even very small percentages can affect the final temperature. The new products are H and O atoms. Actually, the presence of O atoms could be attributed to the dissociation of water at this higher temperature according to the equilibrium step

$$H_2O \rightleftharpoons H_2 + O, \qquad Q_p = -498.3 \text{ kJ}$$

Since the heat absorption is about the same in each case, Le Chatelier's principle indicates a lack of preference in the reactions leading to O. Thus in an overoxidized flame, water dissociation introduces the species H_2, O_2, OH, H, and O.

At even higher temperatures, the nitrogen begins to take part in the reactions and to affect the system thermodynamically. At $T > 3000$ K, NO forms mostly from the reaction

$$\tfrac{1}{2}N_2 + \tfrac{1}{2}O_2 \rightleftharpoons NO, \qquad Q_p = -90.5 \text{ kJ}$$

rather than

$$\tfrac{1}{2}N_2 + H_2O \rightleftharpoons NO + H_2, \qquad Q_p = -332.7 \text{ kJ}$$

If $T_2 > 3500$ K at $P = 1$ atm or $T > 3600$ K at 20 atm, N_2 starts to dissociate by another highly endothermic reaction:

$$N_2 \rightleftarrows 2N, \qquad Q_p = -946.9 \text{ kJ}$$

Thus the complexity in solving for the flame temperature depends on the number of product species chosen. For a system whose approximate temperature range is known, the complexity of the system can be reduced by the approach discussed earlier. Computer programs and machines are now available that can handle the most complex systems, but sometimes a little thought allows one to reduce the complexity of the problem and hence the machine time.

Equation (11) is now examined closely. If the n_i's (products) total a number μ, one needs $(\mu + 1)$ equations to solve for the μ n_i's and T_2. The energy equation is available as one equation. Furthermore, one has a mass balance equation for each atom in the system. If there are α atoms, then $(\mu - \alpha)$ additional equations are required to solve the problem. These $(\mu - \alpha)$ equations come from the equilibrium equations, which are basically nonlinear. For the C–H–O–N system one must simultaneously solve five linear equations and $(\mu - 4)$ nonlinear equations in which one of the unknowns, T_2, is not even present explicitly. Rather, it is present in terms of the enthalpies of the products. This set of equations is a difficult one to solve and can be done only with modern computational codes.

Consider the reaction between octane and nitric acid taking place at a pressure P as an example. The stoichiometric equation is written as

$$n_{C_8H_{18}}C_8H_{18} + n_{HNO_3}HNO_3 \rightarrow n_{CO_2}CO_2 + n_{H_2O}H_2O + n_{H_2}H_2$$

$$+ n_{CO}CO + n_{O_2}O_2 + n_{N_2}N_2 + n_{OH}OH + n_{NO}NO$$

$$+ n_OO + n_CC_{solid} + n_HH$$

Since the mixture ratio is not specified explicitly for this general expression, no effort is made to eliminate products and $\mu = 11$. Thus the new mass balance equations ($\alpha = 4$) are

$$N_H = 2n_{H_2} + 2n_{H_2O} + n_{OH} + n_H$$

$$N_O = 2n_{O_2} + n_{H_2O} + 2n_{CO_2} + n_{CO} + n_{OH} + n_O + n_{NO}$$

$$N_N = 2n_{N_2} + n_{NO}$$

$$N_C = n_{CO_2} + n_{CO} + n_C$$

where

$$N_H = 18n_{C_8H_{18}} + n_{HNO_3}$$

$$N_O = 3n_{HNO_3}$$

$$N_C = 8n_{C_8H_{18}}$$

$$N_N = n_{HNO_3}$$

The seven ($\mu - \alpha = 11 - 4 = 7$) equilibrium equations needed would be

$$C + O_2 \rightleftharpoons CO_2, \qquad K_{p,f} = n_{CO_2}/n_{O_2} \tag{i}$$

$$H_2 + \tfrac{1}{2}O_2 \rightleftharpoons H_2O, \qquad K_{p,f} = \left(n_{H_2O}/n_{H_2}n_{O_2}^{1/2}\right)\left(P/\sum n_i\right)^{-1/2} \tag{ii}$$

$$C + \tfrac{1}{2}O_2 \rightleftharpoons CO, \qquad K_{p,f} = \left(n_{CO}/n_{O_2}^{1/2}\right)\left(P/\sum n_i\right)^{+1/2} \tag{iii}$$

$$\tfrac{1}{2}H_2 + \tfrac{1}{2}O_2 \rightleftharpoons OH, \qquad K_{p,f} = n_{OH}/n_{H_2}^{1/2}n_{O_2}^{1/2} \tag{iv}$$

$$\tfrac{1}{2}O_2 + \tfrac{1}{2}N_2 \rightleftharpoons NO, \qquad K_{p,f} = \frac{n_{NO}}{n_{O_2}^{1/2}n_{N_2}^{1/2}} \tag{v}$$

$$\tfrac{1}{2}O_2 \rightleftharpoons O, \qquad K_{p,f} = \frac{n_O}{n_{O_2}^{1/2}}\left(P/\sum n_i\right)^{1/2} \tag{vi}$$

$$\tfrac{1}{2}H_2 \rightleftharpoons H, \qquad K_{p,f} = \left(n_H/n_{H_2}^{1/2}\right)\left(P/\sum n_i\right)^{1/2} \tag{vii}$$

In these equations $\sum n_i$ includes only the gaseous products; that is, it does not include n_C. One determines n_C from the equation for N_C.

The reaction between the reactants and products is considered irreversible, so that only the products exist in the system being analyzed. Thus, if the reactants were H_2 and O_2, H_2 and O_2 would appear on the product side as well. In dealing with the equilibrium reactions, one ignores the molar quantities of the reactants H_2 and O_2. They are given or known quantities. The amounts of H_2 and O_2 in the product mixture would be unknowns. This point should be considered carefully, even though it is obvious. It is one of the major sources of error in first attempts to solve flame temperature problems by hand.

There are various mathematical approaches for solving these equations by numerical methods [3–5]. The most commonly used program is that of Gordon and McBride [4] described in Appendix H.

As mentioned earlier, to solve explicitly for the temperature T_2 and the product composition, one must consider α mass balance equations, ($\mu - \alpha$) nonlinear equilibrium equations, and an energy equation in which one of the unknowns T_2 is not even explicitly present. Since numerical procedures are used to solve the problem on computers, the thermodynamic functions are represented in terms of power series with respect to temperature.

In the general iterative approach, one first determines the equilibrium state for the product composition at an initially assumed value of the temperature and pressure, and then one checks to see whether the energy equation is satisfied. Chemical equilibrium is usually described by either of two equivalent formulations— equilibrium constants or minimization of free energy. For such simple problems as determining the decomposition temperature of a monopropellant having few exhaust products or examining the variation of a specific species with temperature or pressure, it is most convenient to deal with equilibrium constants. For complex problems the problem reduces to the same number of interactive equations whether one uses equilibrium constants or minimization of free energy. However, when one uses equilibrium constants, one encounters more computational bookkeeping, more numerical difficulties with the use of components, more difficulty in testing for the presence of some condensed species, and more difficulty in extending the generalized methods to conditions that require nonideal equations of state [3, 4].

The condition for equilibrium may be described by any of several thermodynamic functions, such as the minimization of the Gibbs or Helmholtz free energy or the maximization of entropy. If one wishes to use temperature and pressure to characterize a thermodynamic state, one finds that the Gibbs free energy is most easily minimized, inasmuch as temperature and pressure are its natural variables. Similarly, the Helmholtz free energy is most easily minimized if the thermodynamic state is characterized by temperature and volume (density) [4].

As stated, the most commonly used procedure for temperature and composition calculations is the versatile computer program of Gordon and McBride [4], who use the minimization of the Gibbs free energy technique and a descent Newton–Raphson method to solve the equations iteratively. A similar method for solving the equations when equilibrium constants are used is shown in Ref. [5].

2. Practical Considerations

The flame temperature calculation is essentially the solution to a chemical equilibrium problem. Reynolds [6] has developed a more versatile approach to the solution. This method uses theory to relate mole fractions of each species to quantities called element potentials:

> There is one element potential for each independent atom in the system, and these element potentials, plus the number of moles in each phase, are the only variables that must be adjusted for the solution. In large problems there is a much smaller number than the number of species, and hence far fewer variables need to be adjusted.   [6]

The program, called Stanjan [6] (see Appendix H), is readily handled even on the most modest computers. Like the Gordon–McBride program, it is available free from the author (see Appendix H). Both approaches use the JANAF thermochemical data base [1].

In combustion calculations, one primarily wants to know the variation of the temperature with the ratio of oxidizer to fuel. Therefore, in solving flame temperature problems, it is normal to take the number of moles of fuel as 1 and the number of moles of oxidizer as that given by the oxidizer/fuel ratio. In this manner the reactant coefficients are 1 and a number normally larger than 1. Plots of flame temperature versus oxidizer/fuel ratio peak about the stoichiometric mixture ratio, generally (as will be discussed later) somewhat on the fuel-rich side of stoichiometric. If the system is overoxidized, the excess oxygen must be heated to the product temperature; thus, the product temperature drops from the stoichiometric value. If too little oxidizer is present—that is, the system is underoxidized—there is not enough oxygen to burn all the carbon and hydrogen to their most oxidized state, so the energy released is less and the temperature drops as well. More generally, the flame temperature is plotted as a function of the equivalence ratio (Fig. 2), where the equivalence ratio is defined as the fuel/oxidizer ratio divided by the stoichiometric fuel/oxidizer ratio. The equivalence ratio is given the symbol ϕ. For fuel-rich systems, there is more than the stoichiometric amount of fuel, and $\phi > 1$. For overoxidized, or fuel-lean systems, $\phi < 1$. Obviously, at the stoichiometric amount, $\phi = 1$. Since most combustion systems use air as the oxidizer, it is desirable to be able to conveniently determine the flame temperature of any fuel with air at any equivalence ratio. This objective is possible given the background developed in this chapter. As discussed earlier, Table 1 is similar to a potential energy diagram in that movement from the top of the table to products at the bottom indicates energy release. Moreover, as the size of most hydrocarbon fuel molecules increases, so does its negative heat of formation. Thus, it is possible to have fuels whose negative heats of formation approach that of carbon dioxide. It would appear, then, that heat release would be minimal. Heats of formation of hydrocarbons range from 227.1 kJ/mol for acetylene to -456.3 kJ/mol for *n*-ercosane ($C_{20}H_{42}$). However, the greater the number of carbon atoms in a hydrocarbon fuel, the greater the number of moles of CO_2, H_2O, and, of course, their formed dissociation products. Thus, even though a fuel may have a large negative heat of formation, it may form many moles of combustion products without necessarily having a low flame temperature. Then, in order to estimate the contribution of the heat of formation of the fuel to the flame temperature, it is more appropriate to examine the heat of formation on a unit mass basis rather than a molar basis. With this consideration, one finds that practically every hydrocarbon fuel has a heat of formation between -1.5 and 1.0 kJ/g. In fact, most fall in the range -2.1 to $+2.1$ kJ/g. Acetylene and methyl acetylene are the only exceptions, with values of 2.90 and 4.65 kJ/g, respectively.

    In considering the flame temperatures of fuels in air, it is readily apparent that the major effect on flame temperature is the equivalence ratio. Of almost equal importance is the H/C ratio, which determines the ratio of water vapor, CO_2, and their formed dissociation products. Since the heats of formation per unit mass of olefins do not vary much and the H/C ratio is the same for all, it is not surprising that flame temperature varies little among the monoolefins. When discussing fuel–air mixture temperatures, one must always recall the presence of the large number of moles of nitrogen.

    With these conceptual ideas it is possible to develop simple graphs that give the adiabatic flame temperature of any hydrocarbon fuel in air at any equivalence

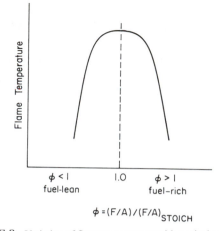

$$\phi = (F/A)/(F/A)_{\text{STOICH}}$$

FIGURE 2   Variation of flame temperature with equivalence ratio ϕ.

ratio [7]. Such graphs are shown in Figs. 3, 4, and 5. These graphs depict the flame temperatures for a range of hypothetical hydrocarbons that have heats of formation ranging from -1.5 to 1.0 kcal/g (i.e., from -6.3 to 4.2 kJ/g). The hydrocarbons chosen have the formulas CH_4, CH_3, $CH_{2.5}$, CH_2, $CH_{1.5}$, and CH_1; that is, they have H/C ratios of 4, 3, 2.5, 2.0, 1.5, and 1.0. These values include every conceivable hydrocarbon, except the acetylenes. The values listed, which were calculated from the standard Gordon–McBride computer program, were determined for all species entering at 298 K for a pressure of 1 atm. As a matter of interest, also plotted in the figures are the values of CH_0, or a H/C ratio of 0. Since the only possible species with this H/C ratio is carbon, the only meaningful points from a physical point of view are those for a heat of formation of 0. The results in the figures plot the flame temperature as a function of the chemical enthalpy content of the reacting system in kilocalories per gram of reactant fuel. Conversion to kilojoules per gram can be made readily. In the figures there are lines of constant H/C ratio grouped according to the equivalence ratio ϕ. For most systems the enthalpy used as the abscissa will be the heat of formation of the fuel in kilocalories per gram, but there is actually greater versatility in using this enthalpy. For example, in a cooled flat flame burner, the measured heat extracted by the water can be converted on a unit fuel flow basis to a reduction in the heat of formation of the fuel. This lower enthalpy value is then used in the graphs to determine the adiabatic flame temperature. The same kind of adjustment can be made to determine the flame temperature when either the fuel or the air or both enter the system at a temperature different from 298 K.

    If a temperature is desired at an equivalence ratio other than that listed, it is best obtained from a plot of T versus ϕ for the given values. The errors in extrapolating in this manner or from the graph are trivial, less than 1%. The reason for separate

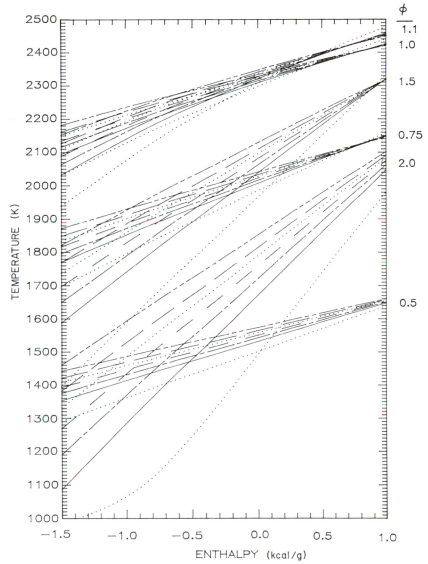

FIGURE 3 Flame temperatures (in kelvins) of hydrocarbons and air as a function of the total enthalpy content of reactions (in kilocalories per gram) for various equivalence and H/C ratios at 1 atm pressure. Reference sensible enthalpy related to 298 K. The H/C ratios are in the following order:

— · · —   H/C = 4
———   H/C = 3
· · · · · ·   H/C = 2.5
— —   H/C = 2.0
— · —   H/C = 1.5
———   H/C = 1.0
· · · · · ·   H/C = 0

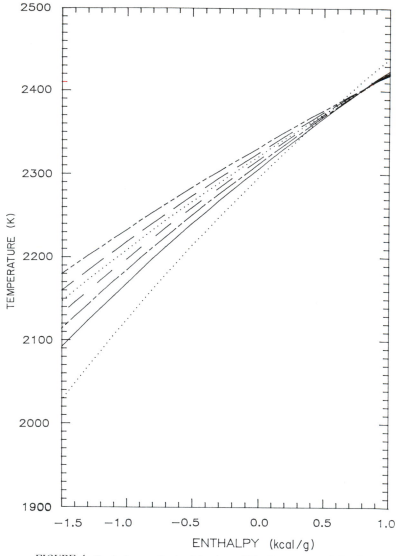

FIGURE 4   Equivalence ratio $\phi = 1.0$ values of Fig. 3 on an expanded scale.

Figs. 4 and 5 is that the values for $\phi = 1.0$ and $\phi = 1.1$ overlap to a great extent. For Fig. 5, $\phi = 1.1$ was chosen because the flame temperature for many fuels peaks not at the stoichiometric value, but between $\phi = 1.0$ and 1.1 owing to lower mean specific heats of the richer products. The maximum temperature for acetylene–air peaks, for example, at a value of $\phi = 1.3$ (see Table 2).

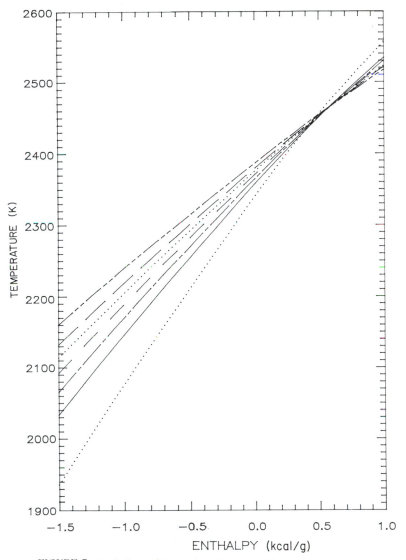

FIGURE 5   Equivalence ratio $\phi = 1.1$ values of Fig. 3 on an expanded scale.

The flame temperature values reported in Fig. 3 show some interesting trends. The H/C ratio has a greater effect in rich systems. One can attribute this trend to the fact that there is less nitrogen in the rich cases as well as to a greater effect of the mean specific heat of the combustion products. For richer systems the mean specific heat of the product composition is lower owing to the preponderance of the

TABLE 2   Approximate Flame Temperatures of Various Stoichiometric Mixtures, Critical Temperature 298 K

Fuel	Oxidizer	Pressure (atm)	Temperature (K)
Acetylene	Air	1	2600[a]
Acetylene	Oxygen	1	3410[b]
Carbon monoxide	Air	1	2400
Carbon monoxide	Oxygen	1	3220
Heptane	Air	1	2290
Heptane	Oxygen	1	3100
Hydrogen	Air	1	2400
Hydrogen	Oxygen	1	3080
Methane	Air	1	2210
Methane	Air	20	2270
Methane	Oxygen	1	3030
Methane	Oxygen	20	3460

[a]This maximum exists at $\phi = 1.3$.
[b]This maximum exists at $\phi = 1.7$.

diatomic molecules CO and H_2 in comparison to the triatomic molecules CO_2 and H_2O. The diatomic molecules have lower molar specific heats than the triatomic molecules. For a given enthalpy content of reactants, the lower the mean specific heat of the product mixture, the greater the final flame temperature. At a given chemical enthalpy content of reactants, the larger the H/C ratio, the higher the temperature. This effect also comes about from the lower specific heat of water and its dissociation products compared to that of CO_2 together with the higher endothermicity of CO_2 dissociation. As one proceeds to more energetic reactants, the dissociation of CO_2 increases and the differences diminish. At the highest reaction enthalpies, the temperature for many fuels peaks not at the stoichiometric value, but, as stated, between $\phi = 1.0$ and 1.1 owing to lower mean specific heats of the richer products.

At the highest temperatures and reaction enthalpies, the dissociation of the water is so complete that the system does not benefit from the heat of formation of the combustion product water. There is still a benefit from the heat or formation of CO, the major dissociation product of CO_2, so that the lower the H/C ratio, the higher the temperature. Thus for equivalence ratios around unity and very high energy content, the lower the H/C ratio, the greater the temperature; that is, the H/C curves intersect.

As the pressure is increased in a combustion system, the amount of dissociation decreases and the temperature rises, as shown in Fig. 6. This observation follows directly from Le Chatelier's principle. The effect is greatest, of course, at the stoichiometric air–fuel mixture ratio where the amount of dissociation is greatest. In a system that has little dissociation, the pressure effect on temperature

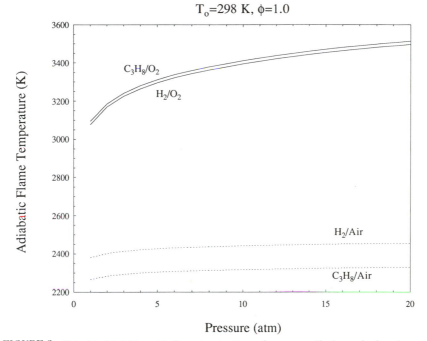

FIGURE 6   Calculated stoichiometric flame temperatures of propane and hydrogen in air and oxygen as a function of pressure.

is small. As one proceeds to a very lean operation, the temperatures and degree of dissociation are very low compared to the stoichiometric values; thus the temperature rise due to an increase in pressure is also very small. Figure 6 reports the calculated stoichiometric flame temperatures for propane and hydrogen in air and

TABLE 3   Equilibrium Product Composition of Propane–Air Combustion

ϕ	0.6	0.6	1.0	1.0	1.5	1.5
P (atm)	1	10	1	10	1	10
Species						
CO	0	0	0.0125	0.0077	0.1041	0.1042
CO_2	0.072	0.072	0.1027	0.1080	0.0494	0.0494
H	0	0	0.0004	0.0001	0.0003	0.0001
H_2	0	0	0.0034	0.0019	0.0663	0.0664
H_2O	0.096	0.096	0.1483	0.1512	0.1382	0.1383
NO	0.002	0.002	0.0023	0.0019	0	0
N_2	0.751	0.751	0.7208	0.7237	0.6415	0.6416
O	0	0	0.0003	0.0001	0	0
OH	0.0003	0.0003	0.0033	0.0020	0.0001	0
O_2	0.079	0.079	0.0059	0.0033	0	0
T (K)	1701	1702	2267	2318	1974	1976
Dissociation (%)			3	2		

TABLE 4   Equilibrium Product Composition of Propane–Oxygen Combustion

ϕ	0.6	0.6	1.0	1.0	1.5	1.5
P (atm)	1	10	1	10	1	10
Species						
CO	.090	.078	.200	.194	.307	.313
CO_2	.165	.184	.135	.151	.084	.088
H	.020	.012	.052	.035	.071	.048
H_2	.023	.016	.063	.056	.154	.155
H_2O	.265	.283	.311	.338	.307	.334
O	.054	.041	.047	.037	.014	.008
OH	.089	.089	.095	.098	.051	.046
O_2	.294	.299	.097	.091	.012	.008
T (K)	2970	3236	3094	3411	3049	3331
Dissociation (%)	27	23	55	51		

in pure oxygen as a function of pressure. Tables 3–6 list the product compositions of these fuels for three stoichiometries and pressures of 1 and 10 atm. As will be noted in Tables 3 and 5, the dissociation is minimal, amounting to about 3% at 1 atm and 2% at 10 atm. Thus one would not expect a large rise in temperature for this 10-fold increase in pressure, as indeed Tables 3 and 5 and Fig. 7 reveal. This small variation is due mainly to the presence of large quantities of inert nitrogen. The results for pure oxygen (Tables 4 and 5) show a substantial degree of dissociation and about a 15% rise in temperature as the pressure increases from 1 to 10 atm. The effect of nitrogen as a diluent can be noted from Table 2, where the maximum flame temperatures of various fuels in air and pure oxygen are compared. Comparisons for methane in particular show very interesting effects. First,

TABLE 5   Equilibrium Product Composition of Hydrogen–Air Combustion

ϕ	0.6	0.6	1.0	1.0	1.5	1.5
P (atm)	1	10	1	10	1	10
Species						
H	0	0	.002	0	.003	.001
H_2	0	0	.015	.009	.147	.148
H_2O	.223	.224	.323	.333	.294	.295
NO	.003	.003	.003	.002	0	0
N_2	.700	.700	.644	.648	.555	.556
O	0	0	.001	0	0	0
OH	.001	0	.007	.004	.001	0
O_2	.073	.073	.005	.003	0	0
T (K)	1838	1840	2382	2442	2247	2252
Dissociation (%)			3	2		

TABLE 6   Equilibrium Product Composition of Hydrogen–Oxygen Combustion

ϕ	0.6	0.6	1.0	1.0	1.5	1.5
P (atm)	1	10	1	10	1	10
Species						
H	.033	.020	.076	.054	.087	.060
H_2	.052	.040	.152	.141	.309	.318
H_2O	.554	.580	.580	.627	.535	.568
O	.047	.035	.033	.025	.009	.005
OH	.119	.118	.107	.109	.052	.045
O_2	.205	.207	.052	.044	.007	.004
T (K)	2980	3244	3077	3394	3003	3275
Dissociation (%)			42	37		

at 1 atm for pure oxygen the temperature rises about 37%; at 20 atm, over 50%. The rise in temperature for the methane–air system as the pressure is increased from 1 to 20 atm is only 2.7%, whereas for the oxygen system over the same pressure range the increase is about 14.2%. Again, these variations are due to the differences in the degree of dissociation. The dissociation for the equilibrium

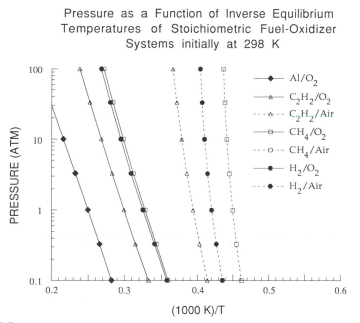

Pressure as a Function of Inverse Equilibrium
Temperatures of Stoichiometric Fuel-Oxidizer
Systems initially at 298 K

FIGURE 7   The variation of the stoichiometric flame temperature of various fuels in oxygen as a function of pressure in the form log P versus $(1/T_f)$, where the initial system temperature is 298 K.

calculations is determined from the equilibrium constants of formation; moreover, from Le Chatelier's principle, the higher the pressure the lower the amount of dissociation. Thus it is not surprising that a plot of ln P_{total} versus $(1/T_f)$ gives mostly straight lines, as shown in Fig. 7. Recall that the equilibrium constant is equal to $\exp(-\Delta G°/RT)$.

Many experimental systems in which nitrogen may undergo some reactions employ artificial air systems, replacing nitrogen with argon on a mole-for-mole basis. In this case the argon system creates much higher system temperatures because it absorbs much less of the heat of reaction owing to its lower specific heat as a monotomic gas. The reverse is true, of course, when the nitrogen is replaced with a triatomic molecule such as carbon dioxide.

PROBLEMS

1. Suppose that methane and air in stoichiometric proportions are brought into a calorimeter at 500 K. The product composition is brought to the ambient temperature (298 K) by the cooling water. The pressure in the calorimeter is assumed to remain at 1 atm, but the water formed has condensed. Calculate the heat of reaction.

2. Calculate the flame temperature of normal octane (liquid) burning in air at an equivalence ratio of 0.5. For this problem assume there is no dissociation of the stable products formed. All reactants are at 298 K and the system operates at a pressure of 1 atm. Compare the results with those given by the graphs in the text. Explain any differences.

3. Carbon monoxide is oxidized to carbon dioxide in an excess of air (1 atm) in an afterburner so that the final temperature is 1300 K. Under the assumption of no dissociation, determine the air–fuel ratio required. Report the results on both a molar and mass basis. For the purposes of this problem assume that air has the composition of 1 mol of oxygen to 4 mol of nitrogen. The carbon monoxide and air enter the system at 298 K.

4. The exhaust of a carbureted automobile engine, which is operated slightly fuel-rich, has an efflux of unburned hydrocarbons entering the exhaust manifold. Assume that all the hydrocarbons are equivalent to ethylene (C_2H_4) and all the remaining gases are equivalent to inert nitrogen (N_2). On a molar basis there are 40 mol of nitrogen for every mole of ethylene. The hydrocarbons are to be burned over an oxidative catalyst and converted to carbon dioxide and water vapor only. In order to accomplish this objective, ambient (298 K) air must be injected into the manifold before the catalyst. If the catalyst is to be maintained at 1000 K, how many moles of air per mole of ethylene must be added if the temperature of the manifold gases before air injection is 400 K and the composition of air is 1 mol of oxygen to 4 mol of nitrogen?

5. A combustion test was performed at 20 atm in a hydrogen–oxygen system. Analysis of the combustion products, which were considered to be in equilibrium, revealed the following:

Compound	Mole fraction
H_2O	0.493
H_2	0.498
O_2	0
O	0
H	0.020
OH	0.005

What was the combustion temperature in the test?

6. Whenever carbon monoxide is present in a reacting system, it is possible for it to disproportionate into carbon dioxide according to the equilibrium

$$2CO \rightleftharpoons C_s + CO_2$$

Assume that such an equilibrium can exist in some crevice in an automotive cylinder or manifold. Determine whether raising the temperature decreases or increases the amount of carbon present. Determine the K_p for this equilibrium system and the effect of raising the pressure on the amount of carbon formed.

7. Determine the equilibrium constant K_p at 1000 K for the following reaction:

$$2CH_4 \rightleftharpoons 2H_2 + C_2H_4$$

8. The atmosphere of Venus is said to contain 5% carbon dioxide and 95% nitrogen by volume. It is possible to simulate this atmosphere for Venus reentry studies by burning gaseous cyanogen (C_2N_2) and oxygen and diluting with nitrogen in the stagnation chamber of a continuously operating wind tunnel. If the stagnation pressure is 20 atm, what is the maximum stagnation temperature that could be reached while maintaining Venus atmosphere conditions? If the stagnation pressure was 1 atm, what would the maximum temperature be? Assume all gases enter the chamber at 298 K. Take the heat of formation of cyanogen as $(\Delta H_f^\circ)_{298} = 374$ kJ/mol.

9. A mixture of 1 mol of N_2 and 0.5 mol O_2 is heated to 4000 K at 0.5 atm, affording an equilibrium mixture of N_2, O_2, and NO only. If the O_2 and N_2 were initially at 298 K and the process is one of steady heating, how much heat is required to bring the final mixture to 4000 K on the basis of one initial mole at N_2?

10. Calculate the adiabatic decomposition temperature of benzene under the constant pressure condition of 20 atm. Assume that benzene enters the decomposition chamber in the liquid state at 298 K and decomposes into the following products: carbon (graphite), hydrogen, and methane.

11. Calculate the flame temperature and product composition of liquid ethylene oxide decomposing at 20 atm by the irreversible reaction

$$C_2H_4O(liq) \rightarrow aCO + bCH_4 + cH_2 + dC_2H_4$$

The four products are as specified. The equilibrium known to exist is

$$2CH_4 \rightleftharpoons 2H_2 + C_2H_4$$

The heat of formation of liquid ethylene oxide is

$$\Delta H^\circ_{f,298} = -76.7 \text{ kJ/mol}$$

It enters the decomposition chamber at 298 K.

12. Liquid hydrazine (N_2H_4) decomposes exothermically in a monopropellant rocket operating at 100 atm chamber pressure. The products formed in the chamber are N_2, H_2, and ammonia (NH_3) according to the irreversible reaction

$$N_2H_4(liq) \rightarrow aN_2 + bH_2 + cNH_3$$

Determine the adiabatic decomposition temperature and the product composition a, b, and c. Take the standard heat of formation of liquid hydrazine as 50.07 kJ/mol. The hydrazine enters the system at 298 K.

13. Gaseous hydrogen and oxygen are burned at 1 atm under the rich conditions designated by the following combustion reaction:

$$O_2 + 5H_2 \rightarrow aH_2O + bH_2 + cH$$

The gases enter at 298 K. Calculate the adiabatic flame temperature and the product composition a, b, and c.

14. The liquid propellant rocket combination nitrogen tetroxide (N_2O_4) and UDMH (unsymmetrical dimethyl hydrazine) has optimum performance at an oxidizer-to-fuel weight ratio of 2 at a chamber pressure of 67 atm. Assume that the products of combustion of this mixture are N_2, CO_2, H_2O, CO, H_2, O, H, OH, and NO. Set down the equations necessary to calculate the adiabatic combustion temperature and the actual product composition under these conditions. These equations should contain all the numerical data in the description of the problem and in the tables in the appendices. The heats of formation of the reactants are

$$N_2O_4(liq), \qquad \Delta H_{f,298} = -2.1 \text{ kJ/mol}$$

$$UDMH(liq), \qquad \Delta H_{f,298} = +53.2 \text{ kJ/mol}$$

The propellants enter the combustion chamber at 298 K.

REFERENCES

1. "JANAF Thermochemical Tables," 3rd Ed., *J. Phys. Chem. Ref. Data* **14** (1985).
2. *Nat. Bur. Stand. (U.S.), Circ.* No. C461 (1947).
3. Huff, V. N., and Morell, V. E., *Natl. Advis. Comm. Aeronaut., Tech. Note* **NACA TN-1113** (1950).
4. Gordon, S., and McBride, B. J., *NASA (Spec. Publ.), SP* **NASA SP-272**, Int. Rev. (Mar. 1976).
5. Glassman, I., and Sawyer, R. F., "The Performance of Chemical Propellants," Chap. 11. Technivision, London, 1970.
6. Reynolds, W. C., "Stanjan," *Dep. Mech. Eng.*, Stanford Univ., Stanford, California, 1986.
7. Glassman, I., and Clark, G., *East. States Combust. Inst. Meet., Providence, RI* Pap. No. 12 (Nov. 1983).

2

Chemical Kinetics

A. INTRODUCTION

Flames will propagate through only those chemical mixtures that are capable of reacting quickly enough to be considered explosive. Indeed, the expression "explosive" essentially specifies very rapid reaction. From the standpoint of combustion, the interest in chemical kinetic phenomena has generally focused on the conditions under which chemical systems undergo explosive reaction. Recently, however, great interest has developed in the rates and mechanisms of steady (nonexplosive) chemical reactions, since most of the known complex pollutants form in zones of steady, usually lower-temperature, reactions during, and even after, the combustion process.

These essential features of chemical kinetics, which occur frequently in combustion phenomena, are reviewed in this chapter. For a more detailed understanding of any of these aspects and a thorough coverage of the subject, refer to any of the books on chemical kinetics, such as those listed in Refs. [1, 1a].

B. RATES OF REACTIONS AND THEIR TEMPERATURE DEPENDENCE

All chemical reactions, whether of the hydrolysis, acid–base, or combustion type, take place at a definite rate and depend on the conditions of the system.

The most important of these conditions are the concentration of the reactants, the temperature, radiation effects, and the presence of a catalyst or inhibitor. The rate of the reaction may be expressed in terms of the concentration of any of the reacting substances or of any reaction product; i.e., the rate may be expressed as the rate of decrease of the concentration of a reactant or the rate of increase of a reaction product.

A stoichiometric relation describing a one-step chemical reaction of arbitrary complexity can be represented by the equation [2, 3]

$$\sum_{j=1}^{n} v'_j \left(M_j \right) = \sum_{j=1}^{n} v''_j \left(M_j \right) \tag{1}$$

where v'_j is the stoichiometric coefficient of the reactants, v''_j is the stoichiometric coefficient of the products, M is an arbitrary specification of all chemical species, and n is the total number of species involved. If a species represented by M_j does not occur as a reactant or product, its v_j equals zero. Consider, as an example, the recombination of H atoms in the presence of H atoms, i.e., the reaction

$$H + H + H \rightarrow H_2 + H$$

$$n = 2, \qquad M_1 = H, \qquad M_2 = H_2;$$

$$v'_1 = 3, \qquad v''_1 = 1, \qquad v'_2 = 0, \qquad v''_2 = 1$$

The reason for following this complex notation will become apparent shortly. The law of mass action, which is confirmed experimentally, states that the rate of disappearance of a chemical species i, defined as RR_i, is proportional to the product of the concentrations of the reacting chemical species, where each concentration is raised to a power equal to the corresponding stoichiometric coefficient; i.e.,

$$RR_i \sim \prod_{j=1}^{n} \left(M_j \right)^{v'_j}, \qquad RR_i = k \prod_{j=1}^{n} \left(M_j \right)^{v'_j} \tag{2}$$

where k is the proportionality constant called the specific reaction rate coefficient. In Eq. (2) $\sum v'_j$ is also given the symbol n, which is called the overall order of the reaction; v'_j itself would be the order of the reaction with respect to species j. In an actual reacting system, the rate of change of the concentration of a given species i is given by

$$\frac{d \left(M_i \right)}{dt} = \left[v''_i - v'_i \right] RR = \left[v''_i - v'_i \right] k \prod_{j=1}^{n} \left(M_j \right)^{v'_j} \tag{3}$$

since v''_i moles of M_i are formed for every v'_i moles of M_i consumed. For the previous example, then, $d(H)/dt = -2k(H)^3$. The use of this complex representation prevents error in sign and eliminates confusion when stoichiometric coefficients are different from 1.

In many systems M_j can be formed not only from a single-step reaction such as that represented by Eq. (3), but also from many different such steps, leading to a rather complex formulation of the overall rate. However, for a single-step reaction such as Eq. (3), $\Sigma v'_j$ not only represents the overall order of the reaction, but also the molecularity, which is defined as the number of molecules that interact in the reaction step. Generally the molecularity of most reactions of interest will be 2 or 3. For a complex reaction scheme the concept of molecularity is not appropriate and the overall order can take various values including fractional ones.

1. The Arrhenius Rate Expression

In most chemical reactions the rates are dominated by collisions of two species that may have the capability to react. Thus, most simple reactions are second-order. Other reactions are dominated by a loose bond-breaking step and thus are first-order. Most of these latter type reactions fall in the class of decomposition processes. Isomerization reactions are also found to be first-order. According to Lindemann's theory [1, 4] of first-order processes, first-order reactions occur as a result of a two-step process. This point will be discussed in a subsequent section.

An arbitrary second-order reaction may be written as

$$A + B \rightarrow C + D \qquad (4)$$

where a real example would be the reaction of oxygen atoms with nitrogen molecules

$$O + N_2 \rightarrow NO + N$$

For the arbitrary reaction (4), the rate expression takes the form

$$-RR = \frac{d(A)}{dt} = -k(A)(B) = -\frac{d(C)}{dt} = -\frac{d(0)}{dt} \qquad (5)$$

The convention used throughout this book is that parentheses around a chemical symbol signify the concentration of that species in moles or mass per cubic centimeter. Specifying the reaction in this manner does not infer that every collision of the reactants A and B would lead to products or cause the disappearance of either reactant. Arrhenius [5] put forth a simple theory that accounts for this fact and gives a temperature dependence of k. According to Arrhenius, only molecules that possess energy greater than a certain amount, E, will react. Molecules acquire the additional energy necessary from collisions induced by the thermal condition that exists. These high-energy activated molecules lead to products. Arrhenius'

postulate may be written as

$$RR = Z_{AB} \exp(-E/RT) \qquad (6)$$

where Z_{AB} is the gas kinetic collision frequency and $\exp(-E/RT)$ is the Boltzmann factor. Kinetic theory shows that the Boltzmann factor gives the fraction of all collisions that have an energy greater than E.

The energy term in the Boltzmann factor may be considered as the size of the barrier along a potential energy surface for a system of reactants going to products, as shown schematically in Fig. 1. The state of the reacting species at this activated energy can be regarded as some intermediate complex that leads to the products. This energy is referred to as the activation energy of the reaction and is generally given the symbol E_A. In Fig. 1, this energy is given the symbol E_f, to distinguish it from the condition in which the product species can revert to reactants by a backward reaction. The activation energy of this backward reaction is represented by E_b and is obviously much larger than E_f for the forward step.

Figure 1 shows an exothermic condition for reactants going to products. The relationship between the activation energy and the heat of reaction has been developed [1a]. Generally, the more exothermic a reaction is, the smaller the activation energy. In complex systems, the energy release from one such reaction can sustain other, endothermic reactions, such as that represented in Fig. 1 for products reverting back to reactants. For example, once the reaction is initiated, acetylene will decompose to the elements in a monopropellant rocket in a sustained fashion because the energy release of the decomposition process is greater than the activation energy of the process. In contrast a calculation of the decomposition of benzene shows the process to be exothermic, but the activation energy of the benzene decomposition process is so large that it will not sustain monopropellant decomposition. For this reason, acetylene is considered an unstable species and benzene a stable one.

Considering again Eq. (6) and referring to E as an activation energy, attention is focused on the collision rate Z_{AB}, which from simple kinetic theory can be

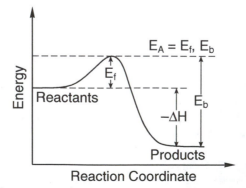

FIGURE 1   Energy as a function of a reaction coordinate for a reacting system.

represented by

$$Z_{AB} = (A)(B)\sigma_{AB}^2 [8\pi k_B T/\mu]^{1/2} \tag{7}$$

where σ_{AB} is the hard sphere collision diameter, k_B the Boltzmann constant, μ is the reduced mass $[m_A m_B/(m_A + m_B)]$, and m is the mass of the species. Z_{AB} may be written in the form

$$Z_{AB} = Z'_{AB}(A)(B) \tag{7'}$$

where $Z'_{AB} = \sigma_{AB}^2 [8\pi k_B T/\mu]^{1/2}$. Thus, the Arrhenius form for the rate is

$$RR = Z'_{AB}(A)(B) \exp(-E/RT)$$

When one compares this to the reaction rate written from the law of mass action [Eq. (2)], one finds that

$$k = Z'_{AB} \exp(-E/RT) = Z''_{AB} T^{1/2} \exp(-E/RT) \tag{8}$$

Thus, the important conclusion is that the specific reaction rate constant k is dependent on temperature alone and is independent of concentration. Actually, when complex molecules are reacting, not every collision has the proper steric orientation for the specific reaction to take place. To include the steric probability, one writes k as

$$k = Z''_{AB} T^{1/2} \left[\exp(-E/RT)\right] \wp \tag{9}$$

where $\wp$ is a steric factor, which can be a very small number at times. Most generally, however, the Arrhenius form of the reaction rate constant is written as

$$k = \text{const } T^{1/2} \exp(-E/RT) = A \exp(-E/RT) \tag{10}$$

where the constant A takes into account the steric factor and the terms in the collision frequency other than the concentrations and is referred to as the kinetic pre-exponential A factor. The factor A as represented in Eq. (10) has a very mild $T^{1/2}$ temperature dependence that is generally ignored when plotting data. The form of Eq. (10) holds well for many reactions, showing an increase in k with T that permits convenient straight-line correlations of data on $\ln k$ versus $(1/T)$ plots. Data that correlate as a straight line on a $\ln k$ versus $(1/T)$ plot are said to follow Arrhenius kinetics, and plots of the logarithm of rates or rate constants as a function of $(1/T)$ are referred to as Arrhenius plots. The slopes of lines on these plots are equal to $(-E/R)$; thus the activation energy may be determined readily (see Fig. 2). Low activation energy processes proceed faster than high activation energy processes at low temperatures and are much less temperature-sensitive. At high temperatures, high activation energy reactions can prevail because of this temperature sensitivity.

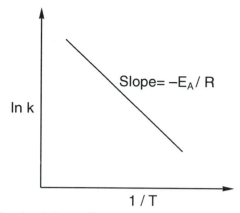

FIGURE 2   Arrhenius plot of the specific reaction rate constant as a function of the reciprocal temperature.

2. Transition State and Recombination Rate Theories

There are two classes of reactions for which Eq. (10) is not suitable. Recombination reactions and low activation energy free-radical reactions in which the temperature dependence in the pre-exponential term assumes more importance. In this low-activation, free-radical case the approach known as absolute or transition state theory of reaction rates gives a more appropriate correlation of reaction rate data with temperature. In this theory the reactants are assumed to be in equilibrium with an activated complex. One of the vibrational modes in the complex is considered loose and permits the complex to dissociate to products. Figure 1 is again an appropriate representation, where the reactants are in equilibrium with an activated complex, which is shown by the curve peak along the extent of the reaction coordinate. When the equilibrium constant for this situation is written in terms of partition functions and if the frequency of the loose vibration is allowed to approach zero, a rate constant can be derived in the following fashion.

The concentration of the activated complex may be calculated by statistical thermodynamics in terms of the reactant concentrations and an equilibrium constant [1, 6]. If the reaction scheme is written as

$$A + BC \rightleftharpoons (ABC)^{\#} \rightarrow AB + C \tag{11}$$

the equilibrium constant with respect to the reactants may be written as

$$K_{\#} = \frac{(ABC)^{\#}}{(A)(BC)} \tag{12}$$

where the symbol # refers to the activated complex. As discussed in Chapter 1, since $K_{\#}$ is expressed in terms of concentration, it is pressure-dependent. Statistical thermodynamics relates equilibrium constants to partition functions; thus for the case in question, one finds [6]

$$K_\# = \frac{(Q_T)^\#}{(Q_T)_A (Q_T)_{BC}} \exp\left(-\frac{E}{RT}\right) \tag{13}$$

where Q_T is the total partition function of each species in the reaction. Q_T can be considered separable into vibrational, rotational, and translation partition functions.

However, one of the terms in the vibrational partition function part of $Q^\#$ is different in character from the rest because it corresponds to a very loose vibration that allows the complex to dissociate into products. The complete vibrational partition function is written as

$$Q_{vib} = \prod_i \left[1 - \exp(-h\nu_i / k_B T)\right]^{-1} \tag{14}$$

where h is Planck's constant and ν_i is the vibrational frequency of the ith mode. For the loose vibration, one term of the complete vibrational partition function can be separated and its value employed when ν tends to zero,

$$\lim_{\nu \to 0} \left[1 - \exp(-h\nu / k_B T)\right]^{-1} = (k_B T / h\nu) \tag{15}$$

Thus

$$\{(ABC)^\# / [(A)(BC)]\} = \{\left[(Q_{T-1})^\#(k_B T / h\nu)\right] / [(Q_T)_A (Q_T)_{BC}]\}$$
$$\times \exp(-E/RT) \tag{16}$$

which rearranges to

$$\nu(ABC)^\# = \{\left[(A)(BC)(k_B T / h)(Q_{T-1})^\#\right] / [(Q_T)_A (Q_T)_{BC}]\}$$
$$\times \exp(-E/RT) \tag{17}$$

where $(Q_{T-1})^\#$ is the partition function of the activated complex evaluated for all vibrational frequencies except the loose one. The term $\nu(ABC)^\#$ on the left-hand side of Eq. (17) is the frequency of the activated complex in the degree of freedom corresponding to its decomposition mode and is therefore the frequency of decomposition. Thus,

$$k = (k_B T / h) \left[(Q_{T-1})^\# / (Q_T)_A (Q_T)_{BC}\right] \exp(-E_A / RT) \tag{18}$$

is the expression for the specific reaction rate as derived from transition state theory.

If species A is only a diatomic molecule, the reaction scheme can be represented by

$$A \rightleftarrows A^\# \to products \tag{19}$$

Thus $(Q_{T-1})^{\#}$ goes to 1. There is only one bond in A, so

$$Q_{vib,A} = \left[1 - \exp(-h\nu_A/k_B T)\right]^{-1} \tag{20}$$

Then

$$k = (k_B T/h)\left[1 - \exp(-h\nu_A/k_B T)\right]\exp(-E/RT) \tag{21}$$

If ν_A of the stable molecule is large, which it normally is in decomposition systems, then the term in square brackets goes to 1 and

$$k = (k_B T/h)\exp(-E/RT) \tag{22}$$

Note that the term $(k_B T/h)$ gives a general order of the pre-exponential term for these dissociation processes.

Although the rate constant will increase monotonically with T for Arrhenius' collision theory, examination of Eqs. (18) and (22) reveals that a nonmonotonic trend can be found [7] for the low activation energy processes represented by transition state theory. Thus, data that show curvature on an Arrhenius plot probably represent a reacting system in which an intermediate complex forms and in which the activation energy is low. As the results from Problem 1 of this chapter reveal, the term $(k_B T/h)$ and the Arrhenius pre-exponential term given by Eq. (7′) are approximately the same and/or about 10^{14} cm^3 mol s^{-1} at 1000 K. This agreement is true when there is little entropy change between the reactants and the transition state and is nearly true for most cases. Thus one should generally expect pre-exponential values to fall in a range near 10^{13} to 10^{14} cm^3 mol s^{-1}. When quantities far different from this range are reported, one should conclude that the representative expression is an empirical fit to some experimental data over a limited temperature range. The earliest representation of an important combustion reaction that showed curvature on an Arrhenius plot was for the CO + OH reaction by the author's group [7], which, by application of transition state theory, correlated a wide temperature range of experimental data. Since then, consideration of transition state theory has been given to many other reactions important to combustion [8].

The use of transition state theory as a convenient expression of rate data is obviously complex owing to the presence of the temperature-dependent partition functions. Most researchers working in the area of chemical kinetic modeling have found it necessary to adopt a uniform means of expressing the temperature variation of rate data and consequently have adopted a modified Arrhenius form

$$k = AT^n \exp(-E/RT) \tag{23}$$

where the power of T accounts for all the pre-exponential temperature-dependent terms in Eqs. (10), (18), and (22). Since most elementary binary reactions exhibit Arrhenius behavior over modest ranges of temperature, the temperature dependence can usually be incorporated with sufficient accuracy into the exponential alone; thus, for most data $n = 0$ is adequate, as will be noted for the extensive listing in the appendixes. However, for the large temperature ranges found in combustion, "non-Arrhenius" behavior of rate constants tends to be the rule rather than the exception, particularly for processes that have a small energy barrier. It should be noted that for these processes the pre-exponential factor that contains the ratio of partition functions (which are weak functions of temperature compared to an exponential) corresponds roughly to a T^n dependence with n in the $\pm$ 1–2 range [9]. Indeed the values of n for the rate data expressions reported in the appendixes fall within this range. Generally the values of n listed apply only over a limited range of temperatures and they may be evaluated by the techniques of thermochemical kinetics [10].

The units for the reaction rate constant k when the reaction is of order n (different from the n power of T) will be $[(conc)^{n-1} (time)]^{-1}$ Thus, for a first-order reaction the units of k are in reciprocal seconds (s^{-1}), and for a second-order reaction process the units are in moles per cubic centimeter per second $(mol\ cm^{-3}\ s^{-1})$.

Radical recombination is another class of reactions in which the Arrhenius expression will not hold. When simple radicals recombine to form a product, the energy liberated in the process is sufficiently great to cause the product to decompose into the original radicals. Ordinarily, a third body is necessary to remove this energy to complete the recombination. If the molecule formed in a recombination process has a large number of internal (generally vibrational) degrees of freedom, it can redistribute the energy of formation among these degrees, so a third body is not necessary. In some cases the recombination process can be stabilized if the formed molecule dissipates some energy radiatively (chemiluminescence) or collides with a surface and dissipates energy in this manner.

If one follows the approach of Landau and Teller [11], who in dealing with vibrational relaxation developed an expression by averaging a transition probability based on the relative molecular velocity over the Maxwellian distribution, one can obtain the following expression for the recombination rate constant [6]:

$$k \sim \exp(C/T)^{1/3} \tag{24}$$

where C is a positive constant that depends on the physical properties of the species of concern [6]. Thus, for radical recombination, the reaction rate constant decreases mildly with the temperature, as one would intuitively expect. In dealing with the recombination of radicals in nozzle flow, one should keep this mild temperature dependence in mind. Recall the example of H atom recombination given earlier. If one writes M as any (or all) third body in the system, the equation takes the form

$$H + H + M \rightarrow H_2 + M \tag{25}$$

The rate of formation of H_2 is third-order and given by

$$d(H_2)/dt = k(H)^2(M) \tag{25a}$$

Thus, in expanding dissociated gases through a nozzle, the velocity increases and the temperature and pressure decrease. The rate constant for this process thus increases, but only slightly. The pressure affects the concentrations and since the reaction is third-order, it enters the rate as a cubed term. In all, then, the rate of recombination in the high-velocity expanding region decreases owing to the pressure term. The point to be made is that third-body recombination reactions are mostly pressure-sensitive, generally favored at higher pressure, and rarely occur at very low pressures.

C. SIMULTANEOUS INTERDEPENDENT REACTIONS

In complex reacting systems, such as those in combustion processes, a simple one-step rate expression will not suffice. Generally, one finds simultaneous, interdependent reactions or chain reactions.

The most frequently occurring simultaneous, interdependent reaction mechanism is the case in which the product, as its concentration is increased, begins to dissociate into the reactants. The classical example is the hydrogen–iodine reaction:

$$H_2 + I_2 \underset{k_b}{\overset{k_f}{\rightleftharpoons}} 2HI \tag{26}$$

The rate of formation of HI is then affected by two rate constants, k_f and k_b, and is written as

$$d(HI)/dt = 2k_f(H_2)(I_2) - 2k_b(HI)^2 \tag{27}$$

in which the numerical value 2 should be noted. At equilibrium, the rate of formation of HI is zero, and one finds that

$$2k_f(H_2)_{eq}(I_2)_{eq} - 2k_b(HI)^2_{eq} = 0 \tag{28}$$

where the subscript eq designates the equilibrium concentrations. Thus,

$$\frac{k_f}{k_b} = \frac{(HI)^2_{eq}}{(H_2)_{eq}(I_2)_{eq}} \equiv K_c \tag{29}$$

i.e., the forward and backward rate constants are related to the equilibrium constant K based on concentrations (K_c). The equilibrium constants are calculated from basic thermodynamic principles as discussed in Section 1.B, and the relationship $(k_f/k_b) \equiv K_c$ holds for any reacting system. The calculation of the equilibrium constant is much more accurate than experimental measurements of specific reaction rate constants. Thus, given a measurement of a specific reaction rate constant, the reverse rate constant is determined from the relationship $K_c \equiv (k_f/k_b)$. For the particular reaction in Eq. (29), K_c is not pressure-dependent as there is a concentration squared in both the numerator and denominator. Indeed, K_c equals $(k_f/k_b) = K_p$ only when the concentration powers cancel.

With this equilibrium consideration the rate expression for the formation of HI becomes

$$\frac{d(\text{HI})}{dt} = 2k_f(\text{H}_2)(\text{I}_2) - 2\frac{k_f}{K_c}(\text{HI})^2 \tag{30}$$

which shows there is only one independent rate constant in the problem.

D. CHAIN REACTIONS

In most instances, two reacting molecules do not react directly as H_2 and I_2 do; rather one molecule dissociates first to form radicals. These radicals then initiate a chain of steps. Interestingly, this procedure occurs in the reaction of H_2 with another halogen, Br_2. Experimentally, Bodenstein [12] found that the rate of formation of HBr obeys the expression

$$\frac{d(\text{HBr})}{dt} = \frac{k'_{exp}(\text{H}_2)(\text{Br}_2)^{1/2}}{1 + k''_{exp}[(\text{HBr})/(\text{Br}_2)]} \tag{31}$$

This expression shows that HBr is inhibiting to its own formation.

Bodenstein explained this result by suggesting that the H_2–Br_2 reaction was chain in character and initiated by a radical (Br·) formed by the thermal dissociation of Br_2. He proposed the following steps:

(1) $\text{M} + \text{Br}_2 \xrightarrow{k_1} 2\text{Br}\cdot + \text{M}$ } chain initiating step

(2) $\text{Br}\cdot + \text{H}_2 \xrightarrow{k_2} \text{HBr} + \text{H}\cdot$
(3) $\text{H}\cdot + \text{Br}_2 \xrightarrow{k_3} \text{HBr} + \text{Br}\cdot$ } chain carrying or propagating steps
(4) $\text{H}\cdot + \text{HBr}\cdot \xrightarrow{k_4} \text{H}_2 + \text{Br}\cdot$

(5) $\text{M} + 2\text{Br}\cdot \xrightarrow{k_5} \text{Br}_2 + \text{M}$ } chain terminating step

The Br_2 bond energy is approximately 189 kJ/mol and the H_2 is approximately 427 kJ/mol. Consequently, over all the temperatures but the very highest, Br_2 dissociation will be the initiating step. These dissociation steps follow Arrhenius kinetics and form a plot similar to that shown in Fig. 2. In Fig. 3 two Arrhenius plots are shown, one for a high activation step and another for a low activation energy step. One can readily observe that for low temperature, the smaller E_A step prevails.

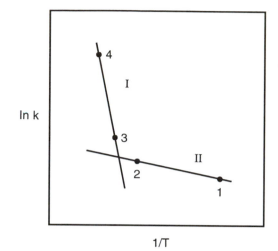

1/T

FIGURE 3   Plot of ln k vs $1/T$. Region I denotes a high activitation energy process and Region II a low activation energy process. Numerals designate conditions to be discussed in Chapter 3.

Perhaps the most important of the various chain types is the chain step that is necessary to achieve nonthermal explosions. This chain step, called chain branching, is one in which two radicals are created for each radical consumed. Two typical chain branching steps that occur in the H_2–O_2 reaction system are

$$H\cdot + O_2 \rightarrow \cdot OH + \cdot O\cdot$$

$$\cdot O\cdot + H_2 \rightarrow \cdot OH + H\cdot$$

where the dot next to or over a species is the convention for designating a radical. Such branching will usually occur when the monoradical (such as H·) formed by breaking a single bond reacts with a species containing a double bond type structure (such as that in O_2) or when a biradical (such as ·O·) formed by breaking a double bond reacts with a saturated molecule (such as H_2 or RH where R is any organic radical). For an extensive discussion of chain reactions, refer to the monograph by Dainton [13].

As shown in the H_2–Br_2 example, radicals are produced by dissociation of a reactant in the initiation process. These types of dissociation reactions are highly endothermic and therefore quite slow. The activation energy of these processes would be in the range of 160–460 kJ/mol. Propagation reactions similar to reactions (2)–(4) in the H_2–Br_2 example are important because they determine the rate at which the chain continues. For most propagation reactions of importance in combustion, activation energies normally lie between 0 and 40 kJ/mol. Obviously, branching chain steps are a special case of propagating steps and, as mentioned, these are the steps that lead to explosion. Branching steps need not necessarily occur rapidly because of the multiplication effect; thus, their activation energies may be higher than those of the linear propagation reactions with which they compete [14].

Termination occurs when two radicals recombine; they need not be similar to those shown in the H_2–Br_2 case. Termination can also occur when a radical reacts with a molecule to give either a molecular species or a radical of lower activity that cannot propagate a chain. Since recombination processes are exothermic, the energy developed must be removed by another source, as discussed previously. The source can be another gaseous molecule M, as shown in the example, or a wall. For the gaseous case, a termolecular or third-order reaction is required; consequently, these reactions are slower than other types except at high pressures.

In writing chain mechanisms note that backward reactions are often written as an individual step; that is, reaction (4) of the H_2–Br_2 scheme is the backward step of reaction (2). The inverse of reaction (3) proceeds very slowly; it is therefore not important in the system and is usually omitted for the H_2–Br_2 example.

From the five chain steps written for the H_2–Br_2 reaction, one can write an expression for the HBr formation rate:

$$\frac{d(\text{HBr})}{dt} = k_2(\text{Br})(\text{H}_2) + k_3(\text{H})(\text{Br}_2) - k_4(\text{H})(\text{HBr}) \tag{32}$$

In experimental systems, it is usually very difficult to measure the concentration of the radicals that are important intermediates. However, one would like to be able to relate the radical concentrations to other known or measurable quantities. It is possible to achieve this objective by the so-called steady-state approximation for the reaction's radical intermediates. The assumption is that the radicals form and react very rapidly so that the radical concentration changes only very slightly with time, thereby approximating a steady-state concentration. Thus, one writes the equations for the rate of change of the radical concentration, then sets them equal to zero. For the H_2–Br_2 system, then, one has for (H) and (Br)

$$\frac{d(\text{H})}{dt} = k_2(\text{Br})(\text{H}_2) - k_3(\text{H})(\text{Br}_2) - k_4(\text{H})(\text{HBr}) \cong 0 \tag{33}$$

$$\frac{d(\text{Br})}{dt} = 2k_1(\text{Br}_2) - k_2(\text{Br})(\text{H}_2) + k_3(\text{H})(\text{Br}_2)$$

$$+ k_4(\text{H})(\text{HBr}) - 2k_5(\text{Br})^2 \cong 0 \tag{34}$$

Writing these two equations equal to zero does not imply that equilibrium conditions exist, as was the case for Eq. (28). It is also important to realize that the steady-state approximation does not imply that the rate of change of the radical concentration is necessarily zero, but rather that the rate terms for the expressions of radical formation and disappearance are much greater than the radical concentration rate term. That is, the sum of the positive terms and the sum of the negative terms on the right-hand side of the equality in Eqs. (33) and (34) are, in absolute magnitude, very much greater than the term on the left of these equalities [3].

Thus in the H_2–Br_2 experiment it is assumed that steady-state concentrations of radicals are approached and the concentrations for H and Br are found to be

$$(Br) = (k_1/k_5)^{1/2}(Br_2)^{1/2} \tag{35}$$

$$(H) = \frac{k_2(k_1/k_5)^{1/2}(H_2)(Br_2)^{1/2}}{k_3(Br_2) + k_4(HBr)} \tag{36}$$

By substituting these values in the rate expression for the formation of HBr [Eq. (32)], one obtains

$$\frac{d(HBr)}{dt} = \frac{2k_2(k_1/k_5)^{1/2}(H_2)(Br_2)^{1/2}}{1 + [k_4(HBr)/k_3(Br_2)]} \tag{37}$$

which is the exact form found experimentally [Eq. (31)]. Thus,

$$k'_{exp} = 2k_2(k_1/k_5)^{1/2}, \qquad k''_{exp} = k_4/k_3$$

Consequently, it is seen, from the measurement of the overall reaction rate and the steady-state approximation, that values of the rate constants of the intermediate radical reactions can be determined without any measurement of radical concentrations. Values k'_{exp} and k''_{exp} evolve from the experimental measurements and the form of Eq. (31). Since (k_1/k_5) is the inverse of the equilibrium constant for Br_2 dissociation and this value is known from thermodynamics, k_2 can be found from k'_{exp}. The value of k_4 is found from k_2 and the equilibrium constant that represents reactions (2) and (4), as written in the H_2–Br_2 reaction scheme. From the experimental value of k''_{exp} and the calculated value of k_4, the value k_3 can be determined.

The steady-state approximation, found to be successful in application to this straight-chain process, can be applied to many other straight-chain processes, chain reactions with low degrees of branching, and other types of nonchain systems. Because the rates of the propagating steps greatly exceed those of the initiation and termination steps in most, if not practically all, of the straight-chain systems, the approximation always works well. However, the use of the approximation in the initiation or termination phase of a chain system, during which the radical concentrations are rapidly increasing or decreasing, can lead to substantial errors.

E. PSEUDO–FIRST-ORDER REACTIONS AND THE "FALL-OFF" RANGE

As mentioned earlier, practically all reactions are initiated by bimolecular collisions; however, certain bimolecular reactions exhibit first-order kinetics. Whether a reaction is first- or second-order is particularly important in combustion because of the presence of large radicals that decompose into a stable species and a smaller radical (primarily the hydrogen atom). A prominent combustion example is the decay of a paraffinic radical to an olefin and a H atom. The order of such reactions, and hence the appropriate rate constant expression, can change with the pressure. Thus, the rate expression developed from one pressure and temperature range may not be applicable to another range. This question of order was first addressed by Lindemann [4], who proposed that first-order processes occur as a result of a two-step reaction sequence in which the reacting molecule is activated by collisional processes, after which the activated species decomposes to products. Similarly, the activated molecule could be deactivated by another collision before it decomposes. If A is considered the reactant molecule and M its nonreacting collision partner, the Lindemann scheme can be represented as follows:

$$A + M \underset{k_b}{\overset{k_f}{\rightleftharpoons}} A^* + M \tag{38}$$

$$A^* \overset{k_p}{\longrightarrow} \text{products} \tag{39}$$

The rate of decay of species A is given by

$$\frac{d(A)}{dt} = -k_f(A)(M) + k_b(A^*)(M) \tag{40}$$

and the rate of change of the activated species A^* is given by

$$\frac{d(A^*)}{dt} = k_f(A)(M) - k_b(A^*)(M) - k_p(A^*) \cong 0 \tag{41}$$

Applying the steady-state assumption to the activated species equation gives

$$(A^*) = \frac{k_f(A)(M)}{k_b(M) + k_p} \tag{42}$$

Substituting this value of (A^*) into Eq. (40), one obtains

$$-\frac{1}{(A)}\frac{d(A)}{dt} = \frac{k_f k_p(M)}{k_b(M) + k_p} = k_{diss} \tag{43}$$

where k_{diss} is a function of the rate constants and the collision partner concentration—that is, a direct function of the total pressure if the effectiveness of all collision partners is considered the same. Owing to size, complexity, and the possibility of resonance energy exchange, the effectiveness of a collision partner (third body) can vary. Normally, collision effectiveness is not a concern, but for some reactions specific molecules may play an important role [15].

At high pressures, $k_b(M) \gg k_p$ and

$$k_{diss,\infty} \equiv \frac{k_f k_p}{k_b} = K k_p \tag{44}$$

where $k_{\mathrm{diss},\infty}$ becomes the high-pressure-limit rate constant and K is the equilibrium constant $(k_{\mathrm{f}}/k_{\mathrm{b}})$. Thus at high pressures the decomposition process becomes overall first-order. At low pressure, $k_{\mathrm{b}}(M) \ll k_{\mathrm{p}}$ as the concentrations drop and

$$k_{\mathrm{diss},0} \equiv k_{\mathrm{f}}(M) \qquad (45)$$

where $k_{\mathrm{diss},0}$ is the low-pressure-limit rate constant. The process is then second-order by Eq. (43), simplifying to $-d(A)/dt = k_{\mathrm{f}}(M)(A)$. Note the presence of the concentration (A) in the manner in which Eq. (43) is written.

Many systems fall in a region of pressures (and temperatures) between the high- and low-pressure limits. This region is called the "fall-off range," and its importance to combustion problems has been very adequately discussed by Troe [16]. The question, then, is how to treat rate processes in the fall-off range. Troe proposed that the fall-off range between the two limiting rate constants be represented using a dimensionless pressure scale

$$\left(k_{\mathrm{diss},0}/k_{\mathrm{diss},\infty}\right) = k_{\mathrm{b}}(M)/k_{\mathrm{p}} \qquad (46)$$

in which one must realize that the units of k_{b} and k_{p} are different so that the right-hand side of Eq. (46) is dimensionless. Substituting Eq. (44) into Eq. (43), one obtains

$$\frac{k_{\mathrm{diss}}}{k_{\mathrm{diss},\infty}} = \frac{k_{\mathrm{b}}(M)}{k_{\mathrm{b}}(M) + k_{\mathrm{p}}} = \frac{k_{\mathrm{b}}(M)/k_{\mathrm{p}}}{\left[k_{\mathrm{b}}(M)/k_{\mathrm{p}}\right] + 1} \qquad (47)$$

or, from Eq. (46),

$$\frac{k_{\mathrm{diss}}}{k_{\mathrm{diss},\infty}} = \frac{k_{\mathrm{diss},0}/k_{\mathrm{diss},\infty}}{1 + \left(k_{\mathrm{diss},0}/k_{\mathrm{diss},\infty}\right)} \qquad (48)$$

For a pressure (or concentration) in the center of the fall-off range, $\left(k_{\mathrm{diss},0}/k_{\mathrm{diss},\infty}\right) = 1$ and

$$k_{\mathrm{diss}} = 0.5 k_{\mathrm{diss},\infty} \qquad (49)$$

Since it is possible to write the products designated in Eq. (39) as two species that could recombine, it is apparent that recombination reactions can exhibit pressure sensitivity; so an expression for the recombination rate constant similar to Eq. (48) can be developed [16].

The preceding discussion stresses the importance of properly handling rate expressions for thermal decomposition of polyatomic molecules, a condition that prevails in many hydrocarbon oxidation processes. For a detailed discussion on evaluation of low- and high-pressure rate constants, again refer to Ref. [16].

Another example in which a pseudo–first-order condition can arise in evaluating experimental data is the case in which one of the reactants (generally the oxidizer in a combustion system) is in large excess. Consider the arbitrary process

$$A + B \rightarrow D \qquad (50)$$

where $(B) \gg (A)$. The rate expression is

$$\frac{d(A)}{dt} = -\frac{d(D)}{dt} = -k(A)(B) \tag{51}$$

Since $(B) \gg (A)$, the concentration of B does not change appreciably and $k(B)$ would appear as a constant. Then Eq. (51) becomes

$$\frac{d(A)}{dt} = -\frac{d(D)}{dt} = -k'(A) \tag{52}$$

where $k' = k(B)$. Equation (52) could represent experimental data because there is little dependence on variations in the concentration of the excess component B. The reaction, of course, appears overall first-order. One should keep in mind, however, that k' contains a concentration and is pressure-dependent. This pseudo–first-order concept arises in many practical combustion systems that are very fuel-lean; i.e., O_2 is present in large excess.

F. THE PARTIAL EQUILIBRIUM ASSUMPTION

As will be discussed in the following chapter, most combustion systems entail oxidation mechanisms with numerous individual reaction steps. Under certain circumstances a group of reactions will proceed rapidly and reach a quasi-equilibrium state. Concurrently, one or more reactions may proceed slowly. If the rate or rate constant of this slow reaction is to be determined and if the reaction contains a species difficult to measure, it is possible through a partial equilibrium assumption to express the unknown concentrations in terms of other measurable quantities. Thus, the partial equilibrium assumption is very much like the steady-state approximation discussed earlier. The difference is that in the steady-state approximation one is concerned with a particular species and in the partial equilibrium assumption one is concerned with particular reactions. Essentially then, partial equilibrium comes about when forward and backward rates are very large and the contribution that a particular species makes to a given slow reaction of concern can be compensated for by very small differences in the forward and backward rates of those reactions in partial equilibrium.

A specific example can illustrate the use of the partial equilibrium assumption. Consider, for instance, a complex reacting hydrocarbon in an oxidizing medium. By the measurement of the CO and CO_2 concentrations, one wants to obtain an estimate of the rate constant of the reaction

$$CO + OH \rightarrow CO_2 + H \tag{53}$$

The rate expression is

$$\frac{d(CO_2)}{dt} = -\frac{d(CO)}{dt} = k(CO)(OH) \tag{54}$$

Then the question is how to estimate the rate constant k without a measurement of the OH concentration. If we assume that equilibrium exists between the H_2–O_2 chain species, one can develop the following equilibrium reactions of formation from the complete reaction scheme:

$$\tfrac{1}{2}H_2 + \tfrac{1}{2}O_2 \rightleftarrows OH, \qquad H_2 + \tfrac{1}{2}O_2 \rightleftarrows H_2O$$

$$K_{C,f,OH}^2 = \frac{(OH)_{eq}^2}{(H_2)_{eq}(O_2)_{eq}}, \qquad K_{C,f,H_2O} = \frac{(H_2O)_{eq}}{(H_2)_{eq}(O_2)_{eq}^{1/2}} \tag{55}$$

Solving the two latter expressions for $(OH)_{eq}$ and eliminating $(H_2)_{eq}$, one obtains

$$(OH)_{eq} = (H_2O)^{1/2}(O_2)^{1/4} \left[K_{C,f,OH}^2 / K_{C,f,H_2O} \right]^{1/2} \tag{56}$$

and the rate expression becomes

$$\frac{d(CO_2)}{dt} = -\frac{d(CO)}{dt} = k \left[K_{C,f,OH}^2 / K_{C,f,H_2O} \right]^{1/2} (CO)(H_2O)^{1/2}(O_2)^{1/4} \tag{57}$$

Thus, one observes that the rate expression can be written in terms of readily measurable stable species. One must, however, exercise care in applying this assumption. Equilibria do not always exist among the H_2–O_2 reactions in a hydrocarbon combustion system—indeed, there is a question if equilibrium exists during CO oxidation in a hydrocarbon system. Nevertheless, it is interesting to note the availability of experimental evidence that shows the rate of formation of CO_2 to be $(1/4)$-order with respect to O_2, $(1/2)$-order with respect to water, and first-order with respect to CO [17, 18]. The partial equilibrium assumption is more appropriately applied to NO formation in flames, as will be discussed in Chapter 8.

G. PRESSURE EFFECT IN FRACTIONAL CONVERSION

In combustion problems, one is interested in the rate of energy conversion or utilization. Thus it is more convenient to deal with the fractional change of a particular substance rather than the absolute concentration. If (M) is used to denote the concentrations in a chemical reacting system of arbitrary order n, the rate expression is

$$\frac{d(M)}{dt} = -k(M)^n \tag{58}$$

Since (M) is a concentration, it may be written in terms of the total density ρ and the mole or mass fraction ε; i.e.,

$$(M) = \rho\varepsilon \tag{59}$$

It follows that at constant temperature

$$\rho(d\varepsilon/dt) = -k(\rho\varepsilon)^n \tag{60}$$

$$(d\varepsilon/dt) = -k\varepsilon^n \rho^{n-1} \tag{61}$$

For a constant-temperature system, $\rho \sim P$ and

$$(d\varepsilon/dt) \sim P^{n-1} \tag{62}$$

That is, the fractional change is proportional to the pressure raised to the reaction order -1.

PROBLEMS

1. For a temperature of 1000 K, calculate the pre-exponential factor in the specific reaction rate constant for (a) any simple bimolecular reaction and (b) any simple unimolecular decomposition reaction following transition state theory.
2. The decomposition of acetaldehyde is found to be overall first-order with respect to the acetaldehyde and to have an overall activation energy of 60 kcal/mol. Assume the following hypothetical sequence to be the chain decomposition mechanism of acetaldehyde:

    (1) $CH_3CHO \xrightarrow{k_1} 0.5CH_3\dot{C}O + 0.5\dot{C}H_3 + 0.5CO + 0.5H_2$
    (2) $CH_3\dot{C}O \xrightarrow{k_2} \dot{C}H_3 + CO$
    (3) $\dot{C}H_3 + CH_3\dot{C}O \xrightarrow{k_3} CH_4 + CH_3\dot{C}O$
    (4) $\cdot CH_3 + CH_3\dot{C}O \xrightarrow{k_4}$ minor products

    For these conditions,

    (a) List the type of chain reaction and the molecularity of each of the four reactions.
    (b) Show that these reaction steps would predict an overall reaction order of 1 with respect to the acetaldehyde
    (c) Estimate the activation energy of reaction (2), if $E_1 = 80$, $E_3 = 10$, and $E_4 = 5$ kcal/mol.

    Hint: E_1 is much larger than E_2, E_3, and E_4.
3. Assume that the steady state of (Br) is formally equivalent to partial equilibrium for the bromine radical chain-initiating step and recalculate the form of Eq. (37) on this basis.
4. Many early investigators interested in determining the rate of decomposition of ozone performed their experiments in mixtures of ozone and oxygen. Their

observations led them to write the following rate expression:

$$d\left(O_3\right)/dt = k_{\text{exp}} \left[\left(O_3\right)^2/\left(O_2\right)\right]$$

The overall thermodynamic equation for ozone conversion to oxygen is

$$2O_3 \rightarrow 3O_2$$

The inhibiting effect of the oxygen led many to expect that the decomposition followed the chain mechanism

$$M + O_3 \xrightarrow{k_1} O_2 + O + M$$

$$M + O + O_2 \xrightarrow{k_2} O_3 + M$$

$$O + O_3 \xrightarrow{k_3} 2O_2$$

(a) If the chain mechanisms postulated were correct and if k_2 and k_3 were nearly equal, would the initial mixture concentration of oxygen have been much less than or much greater than that of ozone?

(b) What is the effective overall order of the experimental result under these conditions?

(c) Given that k_{exp} was determined as a function of temperature, which of the three elementary rate constants is determined? Why?

(d) What type of additional experiment should be performed in order to determine all the elementary rate constants?

REFERENCES

1. Benson, S. W., "The Foundations of Chemical Kinetics." McGraw-Hill, New York, 1960; Weston, R. E., and Schwartz, H. A., "Chemical Kinetics." Prentice-Hall, Englewood Cliffs, New Jersey, 1972; Smith, I. W. M., "Kinetics and Dynamics of Elementary Reactions." Butterworth, London, 1980.

1a. Laidler, K. J., "Theories of Chemical Reaction Rates." McGraw-Hill, New York, 1969.

2. Penner, S. S., "Introduction to the Study of the Chemistry of Flow Processes," Chap. 1. Butterworth, London, 1955.

3. Williams, F. A., "Combustion Theory," 2nd Ed., Appendix B. Benjamin-Cummings, Menlo Park, California, 1985.

4. Lindemann, F. A., *Trans. Faraday Soc.* **17**, 598 (1922).

5. Arrhenius, S., *Phys. Chem.* **4**, 226 (1889).

6. Vincenti, W. G., and Kruger, C. H., Jr., "Introduction to Physical Gas Dynamics," Chap. 7. Wiley, New York, 1965.

7. Dryer, F. L., Naegeli, D. W., and Glassman, I., *Combust. Flame* **17**, 270 (1971).

8. Fontijn, A., and Zellner, R., *in* "Reactions of Small Transient Species" (A. Fontijn and M.A.A. Clyne, eds.), Academic Press, Orlando, Florida, 1983.

9. Kaufman, F., *Int. Symp. Combust., 15th* p. 752. Combust. Inst., Pittsburgh, Pennsylvania, 1974.

10. Golden, D. M., *in* "Fossil Fuel Combustion" (W. Bartok and A. F. Sarofim, eds.), p. 49. Wiley (Interscience), New York, 1991.

11. Landau, L., and Teller, E., *Phys. Z. Sowjet.* **10**(1), 34 (1936).
12. Bodenstein, M., *Phys. Chem.* **85**, 329 (1913).
13. Dainton, F. S., "Chain Reactions: An Introduction." Methuen, London, 1956.
14. Bradley, J. N., "Flame and Combustion Phenomena," Chap. 2. Methuen, London, 1969.
15. Yetter, R. A., Dryer, F. L., and Rabitz, H., *Int. Symp. Gas Kinet., 7th,* p. 231. Gottingen, 1982.
16. Troe, J., *Int. Symp. Combust., 15th* p. 667. Combust. Inst., Pittsburgh, Pennsylvania, 1974.
17. Dryer, F. L., and Glassman, I., *Int. Symp. Combust., 14th,* p. 987. Combust., Inst. Pittsburgh, Pennsylvania, 1972.
18. Williams, G. C., Hottel, H. C., and Morgan, A. G., *Int. Symp. Combust., 12th,* p. 913. Combust. Inst., Pittsburgh, Pennsylvania, 1968.

3

Explosive and General Oxidative Characteristics of Fuels

A. INTRODUCTION

In the previous chapters, the fundamental areas of thermodynamics and chemical kinetics were reviewed. These areas provide the background for the study of very fast reacting systems, termed explosions. In order for flames (deflagrations) or detonations to propagate, the reaction kinetics must be fast—i.e., the mixture must be explosive.

B. CHAIN BRANCHING REACTIONS AND CRITERIA FOR EXPLOSION

Consider a mixture of hydrogen and oxygen stored in a vessel in stoichiometric proportions and at a total pressure of 1 atm. The vessel is immersed in a thermal bath kept at 500°C (773 K), as shown in Fig. 1.

If the vessel shown in Fig. 1 is evacuated to a few millimeters of mercury (torr) pressure, an explosion will occur. Similarly, if the system is pressurized to 2 atm, there is also an explosion. These facts suggest explosive limits.

If H_2 and O_2 react explosively, it is possible that such processes could occur in a flame, which indeed they do. A fundamental question then is: What governs the conditions that give explosive mixtures? In order to answer this question, it is

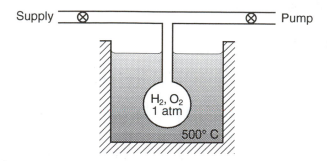

FIGURE 1   Experimental configuration for the determination of H_2–O_2 explosion limits.

useful to reconsider the chain reaction as it occurs in the H_2 and Br_2 reaction:

$$H_2 + Br_2 \rightarrow 2HBr \qquad \text{(the overall reaction)}$$

$$M + Br_2 \rightarrow 2Br + M \qquad \text{(chain initiating step)}$$

$$\left.\begin{array}{l} Br + H_2 \rightarrow HBr + H \\[1em] H + Br_2 \rightarrow HBr + Br \\[1em] H + HBr \rightarrow H_2 + Br \end{array}\right\} \quad \text{(chain propagating steps)}$$

$$M + 2Br \rightarrow Br_2 + M \qquad \text{(chain terminating step)}$$

There are two means by which the reaction can be initiated—thermally or photochemically. If the H_2–Br_2 mixture is at room temperature, a photochemical experiment can be performed by using light of short wavelength; i.e., high enough $h\nu$ to rupture the Br —Br bond through a transition to a higher electronic state. In an actual experiment, one makes the light source as weak as possible and measures the actual energy. Then one can estimate the number of bonds broken and measure the number of HBr molecules formed. The ratio of HBr molecules formed per Br atom created is called the photoyield. In the room-temperature experiment one finds that

$$(HBr)/(Br) \sim 0.01 \ll 1$$

and, of course, no explosive characteristic is observed. No explosive characteristic is found in the photolysis experiment at room temperature because the reaction

$$Br + H_2 \rightarrow HBr + H$$

is quite endothermic and therefore slow. Since the reaction is slow, the chain effect is overtaken by the recombination reaction

$$M + 2Br \rightarrow Br_2 + M$$

Thus, one sees that competitive reactions appear to determine the overall character of this reacting system and that a chain reaction can occur without an explosion.

For the H_2–Cl_2 system, the photoyield is of the order 10^4 to 10^7. In this case the chain step is much faster because the reaction

$$Cl + H_2 \rightarrow HCl + H$$

has an activation energy of only 25 kJ/mol compared to 75 kJ/mol for the corresponding bromine reaction. The fact that in the iodine reaction the corresponding step has an activation energy of 135 kJ/mol gives credence to the notion that the iodine reaction does not proceed through a chain mechanism, whether it is initiated thermally or photolytically.

It is obvious, then, that only the H_2–Cl_2 reaction can be exploded photochemically, that is, at low temperatures. The H_2–Br_2 and H_2–I_2 systems can support only thermal (high-temperature) explosions. A thermal explosion occurs when a chemical system undergoes an exothermic reaction during which insufficient heat is removed from the system so that the reaction process becomes self-heating. Since the rate of reaction, and hence the rate of heat release, increases exponentially with temperature, the reaction rapidly runs away; that is, the system explodes. This phenomenon is the same as that involved in ignition processes and is treated in detail in the chapter on thermal ignition (Chapter 7).

Recall that in the discussion of kinetic processes it was emphasized that the H_2–O_2 reaction contains an important, characteristic chain branching step, namely,

$$H + O_2 \rightarrow OH + O$$

which leads to a further chain branching system

$$O + H_2 \rightarrow OH + H$$

$$OH + H_2 \rightarrow H_2O + H$$

The first two of these three steps are branching, in that two radicals are formed for each one consumed. Since all three steps are necessary in the chain system, the multiplication factor, usually designated α, is seen to be greater than 1 but less than 2. The first of these three reactions is strongly endothermic; thus it will not proceed rapidly at low temperatures. So, at low temperatures an H atom can survive many collisions and can find its way to a surface to be destroyed. This result explains why there is steady reaction in some H_2–O_2 systems where H radicals are introduced. Explosions occur only at the higher temperatures, where the first step proceeds more rapidly.

It is interesting to consider the effect of the multiplication as it may apply in a practical problem such as that associated with automotive knock. However extensive the reacting mechanism in a system, most of the reactions will be bimolecular. The pre-exponential term in the rate constant for such reactions has been found to depend on the molecular radii and temperature, and will generally be between 4×10^{13} and 4×10^{14} cm^3 mol^{-1} s^{-1}. This appropriate assumption provides a ready means for calculating a collision frequency. If the state quantities in the knock regime lie in the vicinity of 1200 K and 20 atm and if nitrogen is assumed to be the major component in the gas mixture, the density of this mixture is of the order of 6 kg m^{-3} or approximately 200 mol m^{-3}. Taking the rate constant pre-exponential as 10^{14} cm^3 mol^{-1} s^{-1} or 10^8 m^3 mol^{-1} s^{-1}, an estimate of the collision frequency between molecules in the mixture is

$$(10^8 \mathrm{m^3 mol^{-1} s^{-1}})(200 \text{ mol m}^{-3}) = 2 \times 10^{10} \text{ collisions per second}$$

For arithmetic convenience, 10^{10} will be assumed to be the collision frequency in a chemical reacting system such as the knock mixture loosely defined.

Now consider that a particular straight-chain propagating reaction ensues, that the initial chain particle concentration is simply 1, and that 1 mol or 10^{19} molecules/cm^3 exist in the system. Thus all the molecules will be consumed in a straight-chain propagation mechanism in a time given by

$$\frac{10^{19} \text{molecules/cm}^3}{1 \text{ molecule/cm}^3} \times \frac{1}{10^{10} \text{collisions/s}} = 10^9 \mathrm{s}$$

or approximately 30 years, a preposterous result.

Specifying α as the chain branching factor, then, the previous example was for the condition $\alpha = 1$. If, however, pure chain branching occurs under exactly the same conditions, then $\alpha = 2$ and every radical initiating the chain system creates two, which create four, and so on. Then 10^{19} molecules/cm^3 are consumed in the following number of generations (N):

$$2^N = 10^{19}$$

or

$$N = 63$$

Thus the time to consume all the particles is

$$\frac{63}{1} \times \frac{1}{10^{10}} = 63 \times 10^{-10} \mathrm{s}$$

or roughly 6 nanoseconds.

If the system is one of both chain branching and propagating steps, α could equal 1.01, which would indicate that one out of a hundred reactions in the system is chain branching. Moreover, hidden in this assumption is the effect of the ordinary activation energy in that not all collisions cause reaction. Nevertheless, this point does not invalidate the effect of a small amount of chain branching. Then, if $\alpha = 1.01$, the number of generations N to consume the mole of reactants is

$$1.01^N = 10^{19}$$

$$N \cong 4400$$

Thus the time for consumption is 44×10^{-8} s or approximately half a microsecond. For $\alpha = 1.001$, or one chain branching step in a thousand, $N \cong 43{,}770$ and the time for consumption is approximately 4 milliseconds.

From this analysis one concludes that if one radical is formed at a temperature in a prevailing system that could undergo branching and if this branching system includes at least one chain branching step and if no chain terminating steps prevent run away, then the system is prone to run away; that is, the system is likely to be explosive.

To illustrate the conditions under which a system that includes chain propagating, chain branching, and chain terminating steps can generate an explosion, one chooses a simplified generalized kinetic model. The assumption is made that for the state condition just prior to explosion, the kinetic steady-state assumption with respect to the radical concentration is satisfactory. The generalized mechanism is written as follows:

$$M \xrightarrow{k_1} R \tag{1}$$

$$R + M \xrightarrow{k_2} \alpha R + M' \tag{2}$$

$$R + M \xrightarrow{k_3} P + R \tag{3}$$

$$R + M \xrightarrow{k_4} I \tag{4}$$

$$R + O_2 + M \xrightarrow{k_5} RO_2 + M \tag{5}$$

$$R \xrightarrow{k_6} I' \tag{6}$$

Reaction (1) is the initiation step, where M is a reactant molecule forming a radical R. Reaction (2) is a particular representation of a collection of propagation steps and chain branching to the extent that the overall chain branching ratio can be represented as α. M' is another reactant molecule and α has any value greater than 1. Reaction (3) is a particular chain propagating step forming a product P. It will be shown in later discussions of the hydrocarbon–air reacting system that this step is similar, for example, to the following important exothermic steps in hydrocarbon oxidation:

$$\left. \begin{array}{l} H_2 + OH \rightarrow H_2O + H \\ CO + OH \rightarrow CO_2 + H \end{array} \right\} \tag{3a}$$

Since a radical is consumed and formed in reaction (3) and since R represents any radical chain carrier, it is written on both sides of this reaction step. Reaction (4) is a gas phase termination step forming an intermediate stable molecule I, which can react further, much as M does. Reaction (5), which is not considered particularly important, is essentially a chain terminating step at high pressures. In step (5), R is generally an H radical and RO_2 is HO_2, a radical much less effective in reacting with stable (reactant) molecules. Thus reaction (5) is considered to be a third-order chain termination step. Reaction (6) is a surface termination step which forms minor intermediates (I') not crucial to the system. For example, tetraethyllead forms lead oxide particles during automotive combustion; if these particles act as a surface sink for radicals, reaction (6) would represent the effect of tetraethyllead. The automotive cylinder wall would produce an effect similar to that of tetraethyllead.

The question to be considered is what value of α is necessary for the system to be explosive. This explosive condition is determined by the rate of formation of a major product, and P from reaction (3) is the obvious selection for purposes here. Thus

$$\frac{d(P)}{dt} = k_3(R)(M) \tag{3b}$$

The steady-state assumption discussed in the consideration of the H_2–Br_2 chain system is applied for determination of the chain carrier concentration (R):

$$\frac{d(R)}{dt} = k_1(M) + k_2(\alpha - 1)(R)(M) - k_4(R)(M)$$
$$- k_5(O_2)(R)(M) - k_6(R) = 0 \tag{7}$$

Thus the steady-state concentration of (R) is found to be

$$(R) = \frac{k_1(M)}{k_4(M) + k_5(O_2)(M) + k_6 - k_2(\alpha - 1)(M)} \tag{8}$$

Substituting Eq. (8) into Eq. (3b), one obtains

$$\frac{d(P)}{dt} = \frac{k_1 k_3 (M)^2}{k_4(M) + k_5(O_2)(M) + k_6 - k_2(\alpha - 1)(M)} \tag{9}$$

The rate of formation of the product P can be considered to be infinite—i.e., the system explodes—when the denominator of Eq. (9) equals zero, It is as if the radical concentration is at a point where it can race to infinity. Note that k_1, the reaction rate constant for the initiation step, determines the rate of formation of P, but does not affect the condition of explosion. The condition under which the denominator would become negative implies that the steady-state approximation is not valid. The rate constant k_3, although regulating the major product-forming and energy-producing step, affects neither the explosion-producing step nor the explosion criterion. Solving for α when the denominator of Eq. (9) is zero gives the critical value for explosion; namely,

$$\alpha_{\text{crit}} = 1 + \frac{k_4(M) + k_5(O_2)(M) + k_6}{k_2(M)} \qquad (10)$$

Assuming there are no particles or surfaces to cause heterogeneous termination steps, then

$$\alpha_{\text{crit}} = 1 + \frac{k_4(M) + k_5(O_2)(M)}{k_2(M)} = 1 + \frac{k_4 + k_5(O_2)}{k_2} \qquad (11)$$

Thus for a temperature and pressure condition where $\alpha_{\text{react}} > \alpha_{\text{crit}}$, the system becomes explosive; for the reverse situation, the termination steps dominate and the products form by slow reaction.

Whether or not either Eq. (10) or Eq. (11) is applicable to the automotive knock problem may be open to question, but the results appear to predict qualitatively some trends observed with respect to automotive knock. α_{react} can be regarded as the actual chain branching factor for a system under consideration, and it may also be the appropriate branching factor for the temperature and pressure in the end gas in an automotive system operating near the knock condition. Under the concept just developed, the radical pool in the reacting combustion gases increases rapidly when $\alpha_{\text{react}} > \alpha_{\text{crit}}$, so the steady-state assumption no longer holds and Eq. (9) has no physical significance. Nevertheless, the steady-state results of Eq. (10) or Eq. (11) essentially define the critical temperature and pressure condition for which the presence of radicals will drive a chain reacting system with one or more chain branching steps to explosion, provided there are not sufficient chain termination steps. Note, however, that the steps in the denominator of Eq. (9) have various temperature and pressure dependences. It is worth pointing out that the generalized reaction scheme put forth cannot achieve an explosive condition, even if there is chain branching, if the reacting radical for the chain branching step is not regenerated in the propagating steps and this radical's only source is the initiation step.

Even though k_2 is a hypothetical rate constant for many reaction chain systems within the overall network of reactions in the reacting media and hence cannot be evaluated to obtain a result from Eq. (10), it is still possible to extract some qualitative trends, perhaps even with respect to automotive knock. Most importantly, Eq. (9) establishes that a chemical explosion is possible only when there is chain branching. Earlier developments show that with small amounts of chain branching, reaction times are extremely small. What determines whether the system will explode or not is whether chain termination is faster or slower than chain branching.

The value of α_{crit} in Eq. (11) is somewhat pressure-dependent through the oxygen concentration. Thus it seems that as the pressure rises, α_{crit} would increase and the system would be less prone to explode (knock). However, as the pressure increases, the temperature also rises. Moreover, k_4, the rate constant for a bond forming step, and k_5, a rate constant for a three-body recombination step, can be expected to decrease slightly with increasing temperature. The overall rate constant k_2, which includes branching and propagating steps, to a first approximation, increases exponentially with temperature. Thus, as the cylinder pressure in an automotive engine rises, the temperature rises, resulting in an α_{crit} that makes the system more prone to explode (knock).

The α_{crit} from Eq. (10) could apply to a system that has a large surface area. Tetraethyllead forms small lead oxide particles with a very large surface area, so the rate constant k_6 would be very large. A large k_6 leads to a large value of α_{crit} and hence a system unlikely to explode. This analysis supports the argument that tetraethylleads suppress knock by providing a heterogeneous chain terminating vehicle.

It is also interesting to note that, if the general mechanism [Eqs. (1)–(6)] were a propagating system with $\alpha = 1$, the rate of change in product concentration (P) would be

$$[d(P)/dt] = [k_1 k_3 (M)^2]/[k_4(M) + k_5(O_2)(M) + k_6]$$

Thus, the condition for fast reaction is

$$\left\{ k_1 k_3 (M)^2 / \left[k_4(M) + k_5(O_2)(M) + k_6 \right] \right\} \gg 1$$

and an explosion is obtained at high pressure and/or high temperature (where the rates of propagation reactions exceed the rates of termination reactions). In the photochemical experiments described earlier, the explosive condition would not depend on k_1, but on the initial perturbed concentration of radicals.

Most systems of interest in combustion include numerous chain steps. Thus it is important to introduce the concept of a chain length, which is defined as the average number of product molecules formed in a chain cycle or the product reaction rate divided by the system initiation rate [1]. For the previous scheme, then, the chain length (cl) is equal to Eq. (9) divided by the rate expression k_1 for reaction (1); i.e.,

$$\mathrm{cl} = \frac{k_3(M)}{k_4(M) + k_5(O_2)(M) + k_6 - k_2(\alpha - 1)(M)} \tag{12}$$

and if there is no heterogeneous termination step,

$$\mathrm{cl} = \frac{k_3}{k_4 + k_5(O_2) - k_2(\alpha - 1)} \tag{12a}$$

If the system contains only propagating steps, $\alpha = 1$, so the chain length is

$$\mathrm{cl} = \frac{k_3(M)}{k_4(M) + k_5(O_2)(M) + k_6} \tag{13}$$

and, again, if there is no heterogeneous termination,

$$cl = \frac{k_3}{k_4 + k_5(O_2)} \tag{13a}$$

Considering that for a steady system, the termination and initiation steps must be in balance, the definition of chain length could also be defined as the rate of product formation divided by the rate of termination. Such a chain length expression would not necessarily hold for the arbitrary system of reactions (1)–(6), but would hold for such systems as that written for the H_2–Br_2 reaction. When chains are long, the types of products formed are determined by the propagating reactions alone, and one can ignore the initiation and termination steps.

C. EXPLOSION LIMITS AND OXIDATION CHARACTERISTICS OF HYDROGEN

Many of the early contributions to the understanding of hydrogen–oxygen oxidation mechanisms developed from the study of explosion limits. Many extensive treatises were written on the subject of the hydrogen–oxygen reaction and, in particular, much attention was given to the effect of walls on radical destruction (a chain termination step) [2]. Such effects are not important in the combustion processes of most interest here; however, Appendix B details a complex modern mechanism based on earlier thorough reviews [3, 4].

Flames of hydrogen in air or oxygen exhibit little or no visible radiation, what radiation one normally observes being due to trace impurities. Considerable amounts of OH can be detected, however, in the ultraviolet region of the spectrum. In stoichiometric flames, the maximum temperature reached in air is about 2400 K and in oxygen about 3100 K. The burned gas composition in air shows about 5–7% conversion to water, the radicals H, O, and OH comprising about one-quarter of the remainder [5]. In static systems practically no reactions occur below 675 K, and above 850 K explosion occurs spontaneously in the moderate pressure ranges. At very high pressures the explosion condition is moderated owing to a third-order chain terminating reaction, reaction (5), as will be explained in the following paragraphs.

It is now important to stress the following points in order to eliminate possible confusion with previously held concepts and certain subjects to be discussed later. The explosive limits are not flammability limits. *Explosion limits* are the pressure–temperature boundaries for a specific fuel–oxidizer mixture ratio that separate the regions of slow and fast reaction. For a given temperature and pressure, *flammability limits* specify the lean and rich fuel–oxidizer mixture ratio beyond which no flame will propagate. Next, recall that one must have fast reactions for a flame to propagate. A stoichiometric mixture of H_2 and O_2 at standard conditions will support a flame because an ignition source initially brings a local mixture into the explosive regime, whereupon the established flame by diffusion heats fresh mixture to temperatures high enough to be explosive. Thus, in the early stages of any flame, the fuel–air mixture may follow a low-temperature steady reaction system and in the later stages, an explosive reaction system. This point is significant, especially in hydrocarbon combustion, because it is in the low-temperature regime that particular pollutant-causing compounds are formed.

Figure 2 depicts the explosion limits of a stoichiometric mixture of hydrogen and oxygen. Explosion limits can be found for many different mixture ratios. The point X on Fig. 2 marks the conditions (773 K; 1 atm) described at the very beginning of this chapter in Fig. 1. It now becomes obvious that either increasing or decreasing the pressure at constant temperature can cause an explosion.

Certain general characteristics of this curve can be stated. First, the third limit portion of the curve is as one would expect from simple density considerations. Next, the first, or lower, limit reflects the wall effect and its role in chain destruction. For example, HO_2 radicals combine on surfaces to form H_2O and O_2. Note the expression developed for α_{crit} [Eq. (9)] applies to the lower limit only when the wall effect is considered as a first-order reaction of chain destruction, since $R \xrightarrow{\frac{k_6}{wall}}$ destruction was written.

Although the features of the movement of the boundaries are not explained fully, the general shape of the three limits can be explained by reasonable hypotheses of mechanisms. The manner in which the reaction is initiated to give the boundary designated by the curve in Fig. 2 suggests, as was implied earlier, that the explosion is in itself a branched chain phenomenon. Thus, one must consider possible branched chain mechanisms to explain the limits.

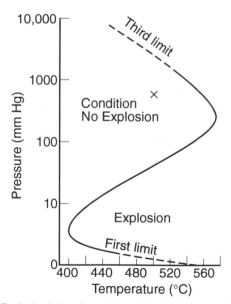

FIGURE 2   Explosion limits of a stoichiometric H_2–O_2 mixture (after Ref. [2]).

Basically, only thermal, not photolytic, mechanisms are considered. The dissociation energy of hydrogen is less than that of oxygen, so the initiation can be related to hydrogen dissociation. Only a few radicals are required to initiate the explosion in the region of temperature of interest, i.e., about 675 K. If hydrogen dissociation is the chain's initiating step, it proceeds by the reaction

$$H_2 + M \rightarrow 2H + M \tag{14}$$

which requires about 435 kJ/mol.

The early modeling literature suggested the initiation step

$$M + H_2 + O_2 \rightarrow H_2O_2 + M$$
$$\downarrow \tag{15}$$
$$2OH$$

because this reaction requires only 210 kJ/mol, but this trimolecular reaction has been evaluated to have only a very small rate [6]. Because in modeling it accurately reproduces experimental ignition delay measurements under shock tube and detonation conditions [7], the most probable initiation step, except at the very highest temperature at which reaction (14) would prevail, could be

$$H_2 + O_2 \rightarrow HO_2 + H \tag{16}$$

where HO_2 is the relatively stable hydroperoxy radical which has been identified by mass spectroscopic analysis.

The essential feature of the initiation step is to provide a radical for the chain system and, as discussed in the previous section, the actual initiation step is not important in determining the explosive condition, nor is it important in determining the products formed. Either reaction (14) or (16) provides an H radical that develops a radical pool of OH, O, and H by the chain reactions

$$H + O_2 \rightarrow O + HO \tag{17}$$

$$O + H_2 \rightarrow H + OH \tag{18}$$

$$H_2 + OH \rightarrow H_2O + H \tag{19}$$

$$O + H_2O \rightarrow OH + OH \tag{20}$$

Reaction (17) is chain branching and 66 kJ/mol endothermic. Reaction (18) is also chain branching and 8 kJ/mol exothermic. Note that the H radical is regenerated in the chain system and there is no chemical mechanism barrier to prevent the system from becoming explosive. Since radicals react rapidly, their concentration levels in many systems are very small; consequently, the reverse of reactions (17), (18), and (20) can be neglected. Normally, reactions between radicals are not considered, except in termination steps late in the reaction when the concentrations are high and only stable product species exist. Thus, the reverse reactions (17), (18), and (20) are not important for the determination of the second limit [i.e., $(M) = 2k_{17}/k_{21}$]; nor are they important for the steady-slow H_2/O_2 and $CO/H_2O/O_2$ reactions. However, they are generally important in all explosive H_2/O_2 and $CO/H_2O/O_2$ reactions. The importance of these radical–radical reactions in these cases is verified by the existence of superequilibrium radical concentrations and the validity of the partial equilibrium assumption.

The sequence [Eqs. (17)–(20)] is of great importance in the oxidation reaction mechanisms of any hydrocarbon in that it provides the essential chain branching and propagating steps as well as the radical pool for fast reaction.

The important chain termination steps in the static explosion experiments (Fig. 1) are

$$H \rightarrow \text{wall destruction}$$

$$OH \rightarrow \text{wall destruction}$$

Either or both of these steps explain the lower limit of explosion, since it is apparent that wall collisions become much more predominant at lower pressure than molecular collisions. The fact that the limit is found experimentally to be a function of the containing vessel diameter is further evidence of this type of wall destruction step.

The second explosion limit must be explained by gas phase production and destruction of radicals. This limit is found to be independent of vessel diameter. For it to exist, the most effective chain branching reaction [reaction (17)] must be overridden by another reaction step. When a system at a fixed temperature moves from a lower to higher pressure, the system goes from an explosive to a steady reaction condition, so the reaction step that overrides the chain branching step must be more pressure-sensitive. This reasoning leads one to propose a third-order reaction in which the species involved are in large concentration [2]. The accepted reaction that satisfies these prerequisites is

$$H + O_2 + M \rightarrow HO_2 + H \tag{21}$$

where M is the usual third body that takes away the energy necessary to stabilize the combination of H and O_2. At higher pressures it is certainly possible to obtain proportionally more of this trimolecular reaction than the binary system represented by reaction (17). The hydroperoxy radical HO_2 is considered to be relatively unreactive so that it is able to diffuse to the wall and thus become a means for effectively destroying H radicals.

The upper (third) explosion limit is due to a reaction that overtakes the stability of the HO_2 and is possibly the sequence

$$HO_2 + H_2 \rightarrow H_2O_2 + H$$
$$\downarrow$$
$$2OH$$
$$\tag{22}$$

The reactivity of HO_2 is much lower than that of OH, H, or O; therefore, somewhat higher temperatures are necessary for sequence (22) to become effective [6a]. Water vapor tends to inhibit explosion due to the effect of reaction (21) in that H_2O has a high third-body efficiency, which is most probably due to some resonance energy exchange with the HO_2 formed.

Since reaction (21) is a recombination step requiring a third body, its rate decreases with increasing temperature, whereas the rate of reaction (17) increases with temperature. One then can generally conclude that reaction (17) will dominate at higher temperatures and lower pressures, while reaction (21) will be more effective at higher pressures and lower temperatures. Thus, in order to explain the limits in Fig. 2 it becomes apparent that at temperatures above 875 K, reaction (17) always prevails and the mixture is explosive for the complete pressure range covered.

In this higher temperature regime and in atmospheric-pressure flames, the eventual fate of the radicals formed is dictated by recombination. The principal gas phase termination steps are

$$H + H + M \rightarrow H_2 + M \qquad (23)$$

$$O + O + M \rightarrow O_2 + M \qquad (24)$$

$$H + O + M \rightarrow OH + M \qquad (25)$$

$$H + OH + M \rightarrow H_2O + M \qquad (26)$$

In combustion systems other than those whose lower-temperature explosion characteristics are represented in Fig. 2, there are usually ranges of temperature and pressure in which the rates of reactions (17) and (21) are comparable. This condition can be specified by the simple ratio

$$\frac{k_{17}}{k_{21}(M)} = 1$$

Indeed, in developing complete mechanisms for the oxidation of CO and hydrocarbons applicable to practical systems over a wide range of temperatures and high pressures, it is important to examine the effect of the HO_2 reactions when the ratio is as high as 10 or as low as 0.1. Considering that for air combustion the total concentration (M) can be that of nitrogen, the boundaries of this ratio are depicted in Fig. 3, as derived from the data in Appendix B. These modern rate data indicate that the second explosion limit, as determined by glass vessel experiments and many other experimental configurations, as shown in Fig. 2, has been extended (Fig. 4) and verified experimentally [6a]. Thus, to be complete for the H_2–O_2 system and other oxidation systems containing hydrogen species, one must also consider reactions of HO_2. Sometimes HO_2 is called a metastable species because it is relatively unreactive as a radical. Its concentrations can build up in a reacting system. Thus, HO_2 may be consumed in the H_2–O_2 system by various radicals

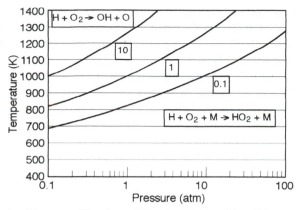

FIGURE 3   Ratio of the rates of $H + O_2 \rightarrow OH + O$ to $H + O_2 + M \rightarrow HO_2 + M$ at various total pressures.

according to the following reactions [4]:

$$HO_2 + H \rightarrow H_2 + O_2 \tag{27}$$

$$HO_2 + H \rightarrow OH + OH \tag{28}$$

$$HO_2 + H \rightarrow H_2O + O \tag{29}$$

$$HO_2 + O \rightarrow O_2 + OH \tag{30}$$

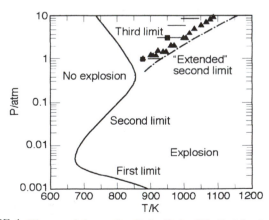

FIGURE 4   The extended second explosion limit of H_2–O_2 (after Ref. [6a]).

The recombination of HO_2 radicals by

$$HO_2 + HO_2 \rightarrow H_2O_2 + O_2 \qquad (31)$$

yields hydrogen peroxide (H_2O_2), which is consumed by reactions with radicals and by thermal decomposition according to the following sequence:

$$H_2O_2 + OH \rightarrow H_2O + HO_2 \qquad (32)$$

$$H_2O_2 + H \rightarrow H_2O + OH \qquad (33)$$

$$H_2O_2 + H \rightarrow HO_2 + H_2 \qquad (34)$$

$$H_2O_2 + M \rightarrow 2OH + M \qquad (35)$$

From the sequence of reactions (32)–(35) one finds that although reaction (21) terminates the chain under some conditions, under other conditions it is part of a chain propagating path consisting essentially of reactions (21) and (28) or reactions (21), (31), and (35). It is also interesting to note that, as are most HO_2 reactions, these two sequences of reactions are very exothermic; i.e.,

$$H + O_2 + M \rightarrow HO_2 + M$$

$$HO_2 + H \rightarrow 2OH$$

$$\overline{2H + O_2 \rightarrow 2OH - 350 \text{ kJ/mol}}$$

and

$$H + O_2 + M \rightarrow HO_2 + M$$

$$HO_2 + HO_2 \rightarrow H_2O_2 + O_2$$

$$H_2O_2 + M \rightarrow 2OH + M$$

$$\overline{H + HO_2 \rightarrow 2OH - 156 \text{ kJ/mol}}$$

Hence they can significantly affect the temperature of an (adiabatic) system and thereby move the system into an explosive regime. The point to be emphasized is that slow competing reactions can become important if they are very exothermic.

It is apparent that the fate of the H atom (radical) is crucial in determining the rate of the H_2–O_2 reaction or, for that matter, the rate of any hydrocarbon oxidation mechanism. From the data in Appendix B one observes that at temperatures encountered in flames the rates of reaction between H atoms and many hydrocarbon species are considerably larger than the rate of the chain branching reaction (17). Note the comparisons in Table 1. Thus, these reactions compete very effectively with reaction (17) for H atoms and reduce the chain branching rate. For this reason,

TABLE 1   Rate Constants of Specific Radical Reactions

Rate Constant	1000 K	2000 K
$k(C_3H_8 + OH)$	5.0×10^{12}	1.6×10^{13}
$k(H_2 + OH)$	1.6×10^{12}	6.0×10^{12}
$k(CO + OH)$	1.7×10^{11}	3.5×10^{11}
$k(H + C_3H_8) \rightarrow iC_3H_7$	7.1×10^{11}	9.9×10^{12}
$k(H + O_2)$	4.7×10^{10}	3.2×10^{12}

hydrocarbons act as inhibitors for the H_2–O_2 system [4]. As implied, at highly elevated pressures ($P \geq 20$ atm) and relatively low temperatures ($T \cong 1000$ K), reaction (21) will dominate over reaction (17); and as shown, the sequence of reactions (21), (31), and (35) provides the chain propagation. Also, at higher temperatures, when $H + O_2 \rightarrow OH + O$ is microscopically balanced, reaction (21) ($H + O_2 + M \rightarrow HO_2 + M$) can compete favorably with reaction (17) for H atoms since the net removal of H atoms from the system by reaction (17) may be small due to its equilibration. In contrast, when reaction (21) is followed by the reaction of the fuel with HO_2 to form a radical and hydrogen peroxide and then by reaction (35), the result is chain branching. Therefore, under these conditions increased fuel will accelerate the overall rate of reaction and will act as an inhibitor at lower pressures due to competition with reaction (17) [4].

The detailed rate constants for all the reactions discussed in this section are given in Appendix B. The complete mechanism for CO or any hydrocarbon or hydrogen-containing species should contain the appropriate reactions of the H_2–O_2 steps listed in Appendix B; one can ignore the reactions containing Ar as a collision partner in real systems. It is important to understand that, depending on the temperature and pressure of concern, one need not necessarily include all the H_2–O_2 reactions. It should be realized as well that each of these reactions is a set comprising a forward and a backward reaction; but, as the reactions are written, many of the backward reactions can be ignored. Recall that the backward rate constant can be determined from the forward rate constant and the equilibrium constant for the reaction system.

D. EXPLOSION LIMITS AND OXIDATION CHARACTERISTICS OF CARBON MONOXIDE

Early experimental work on the oxidation of carbon monoxide was confused by the presence of any hydrogen-containing impurity. The rate of CO oxidation in the presence of species such as water is substantially faster than the "bone dry" condition. It is very important to realize that very small quantities of hydrogen, even of the order of 20 ppm, will increase the rate of CO oxidation substantially [8]. Generally, the mechanism with hydrogen-containing compounds present is referred to as the "wet" carbon monoxide condition. Obviously, CO oxidation will proceed through this so-called wet route in most practical systems.

It is informative, however, to consider the possible mechanisms for dry CO oxidation. Again the approach is to consider the explosion limits of a stoichiometric, dry CO–O_2 mixture. However, neither the explosion limits nor the reproducibility of these limits is well defined, principally because the extent of dryness in the various experiments determining the limits may not be the same. Thus, typical results for explosion limits for dry CO would be as depicted in Fig. 5.

Figure 5 reveals that the low-pressure ignition of CO–O_2 is characterized by an explosion peninsula very much like that in the case of H_2–O_2. Outside this peninsula one often observes a pale-blue glow, whose limits can be determined as well. A third limit has not been defined; and, if it exists, it lies well above 1 atm.

As in the case of H_2–O_2 limits, certain general characteristics of the defining curve in Fig. 5 may be stated. The lower limit meets all the requirements of wall destruction of a chain propagating species. The effects of vessel diameter, surface character, and condition have been well established by experiment [2].

Under dry conditions the chain initiating step is

$$CO + O_2 \rightarrow CO_2 + O \tag{36}$$

which is mildly exothermic, but slow at combustion temperatures. The succeeding steps in this oxidation process involve O atoms, but the exact nature of these steps is not fully established. Lewis and von Elbe [2] suggested that chain branching would come about from the step

$$O + O_2 + M \rightarrow O_3 + M \tag{37}$$

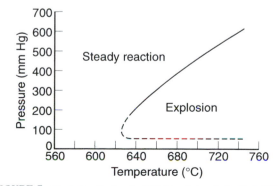

FIGURE 5   Explosion limits of a CO–O_2 mixture (after Ref. [2]).

This reaction is slow, but could build up in supply. Ozone O_3 is the metastable species in the process (like HO_2 in H_2–O_2 explosions) and could initiate chain branching, thus explaining the explosion limits. The branching arises from the reaction

$$O_3 + CO \rightarrow CO_2 + 2O \tag{38}$$

Ozone destruction at the wall to form oxygen molecules would explain the lower limit. Lewis and von Elbe explain the upper limit by the third-order reaction

$$O_3 + CO + M \rightarrow CO_2 + O_2 + M \tag{39}$$

However, O_3 does not appear to react with CO below 523 K. Since CO is apparently oxidized by the oxygen atoms formed by the decomposition of ozone [the reverse of reaction (37)], the reaction must have a high activation energy (>120 kJ/mol). This oxidation of CO by O atoms was thought to be rapid in the high-temperature range, but one must recall that it is a three-body recombination reaction.

Analysis of the glow and emission spectra of the CO–O_2 reaction suggests that excited carbon dioxide molecules could be present. If it is argued that O atoms cannot react with oxygen (to form ozone), then they must react with the CO. A suggestion of Semenov was developed further by Gordon and Knipe [9], who gave the following alternative scheme for chain branching:

$$CO + O \rightarrow CO_2^* \tag{40}$$

$$CO_2^* + O_2 \rightarrow CO_2 + 2O \tag{41}$$

where CO_2^* is the excited molecule from which the glow appears. This process is exothermic and might be expected to occur. Gordon and Knipe counter the objection that CO_2^* is short-lived by arguing that through system crossing in excited states its lifetime may be sufficient to sustain the process. In this scheme the competitive three-body reaction to explain the upper limit is the aforementioned one:

$$CO + O + M \rightarrow CO_2 + M \tag{42}$$

Because these mechanisms did not explain shock tube rate data, Brokaw [8] proposed that the mechanism consists of reaction (36) as the initiation step with subsequent large energy release through the three-body reaction (42) and

$$O + O + M \rightarrow O_2 + M \tag{43}$$

The rates of reactions (36), (42), and (43) are very small at combustion temperatures, so that the oxidation of CO in the absence of any hydrogen-containing material is very slow. Indeed it is extremely difficult to ignite and have a flame propagate through a bone-dry, impurity-free CO–O_2 mixture.

Very early, from the analysis of ignition, flame speed, and detonation velocity data, investigators realized that small concentrations of hydrogen-containing materials would appreciably catalyze the kinetics of $CO–O_2$. The H_2O-catalyzed reaction essentially proceeds in the following manner:

$$CO_2 + O_2 \rightarrow CO_2 + 2O \qquad (36)$$

$$O + H_2O \rightarrow 2OH \qquad (18)$$

$$CO + OH \rightarrow CO_2 + H \qquad (44)$$

$$H + O_2 \rightarrow OH + O \qquad (15)$$

If H_2 is the catalyst, the steps

$$O + H_2 \rightarrow OH + H \qquad (18)$$

$$OH + H_2 \rightarrow H_2O + H \qquad (19)$$

should be included. It is evident then that all of the steps of $H_2–O_2$ reaction scheme should be included in the so-called wet mechanism of CO oxidation. As discussed in the previous section, the reaction.

$$H + O_2 + M \rightarrow HO_2 + M \qquad (21)$$

enters and provides another route for the conversion of CO to CO_2 by

$$CO + HO_2 \rightarrow CO_2 + OH \qquad (45)$$

At high pressures or in the initial stages of hydrocarbon oxidation, high concentrations of HO_2 can make reaction (45) competitive to reaction (44), so reaction (45) is rarely as important as reaction (44) in most combustion situations [4]. Nevertheless, any complete mechanism for wet CO oxidation must contain all the $H_2–O_2$ reaction steps. Again, a complete mechanism means both the forward and backward reactions of the appropriate reactions in Appendix B. In developing an understanding of hydrocarbon oxidation, it is important to realize that any high-temperature hydrocarbon mechanism involves H_2 and CO oxidation kinetics, and that most, if not all, of the CO_2 that is formed results from reaction (44).

The very important reaction (44) actually proceeds through a four-atom activated complex [10, 11] and is not a simple reaction step like reaction (17). As shown in Fig. 6, the Arrhenius plot exhibits the curvature [10]. And because the reaction proceeds through an activated complex, the reaction rate exhibits some pressure dependence [12].

Just as the fate of H radicals is crucial in determining the rate of the $H_2–O_2$ reaction sequence in any hydrogen-containing combustion system, the concentration

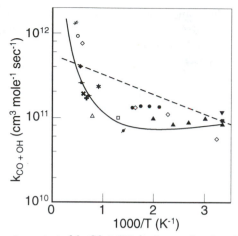

FIGURE 6   Reaction rate constant of the CO + OH reaction as a function of reciprocal temperature based on transition state (———) and Arrhenius (– –) theories compared with experimental data (after Ref. [10]).

of hydroxyl radicals is also important in the rate of CO oxidation. Again, as in the $H_2–O_2$ reaction, the rate data reveal that reaction (44) is slower than the reaction between hydroxyl radicals and typical hydrocarbon species; thus one can conclude—correctly—that hydrocarbons inhibit the oxidation of CO (see Table 1).

It is apparent that in any hydrocarbon oxidation process CO is the primary product and forms in substantial amounts. However, substantial experimental evidence indicates that the oxidation of CO to CO_2 comes late in the reaction scheme [13]. The conversion to CO_2 is retarded until all the original fuel and intermediate hydrocarbon fragments have been consumed [4, 13]. When these species have disappeared, the hydroxyl concentration rises to high levels and converts CO to CO_2. Further examination of Fig. 6 reveals that the rate of reaction (44) does not begin to rise appreciably until the reaction reaches temperatures above 1100 K. Thus, in practical hydrocarbon combustion systems whose temperatures are of the order of 1100 K and below, the complete conversion of CO to CO_2 may not take place.

E. EXPLOSION LIMITS AND OXIDATION CHARACTERISTICS OF HYDROCARBONS

To establish the importance of the high-temperature chain mechanism through the $H_2–O_2$ sequence, the oxidation of H_2 was discussed in detail. Also, because CO conversion to CO_2 is the highly exothermic portion of any hydrocarbon oxidation system, CO oxidation was then detailed. Since it will be shown that all carbon atoms in alkyl hydrocarbons and most in aromatics are converted to CO through the radical of formaldehyde (H_2CO) called formyl (HCO), the oxidation of aldehydes will be the next species to be considered. Then the sequence of oxidation reactions of the C_1 to C_5 alkyl hydrocarbons is considered. These systems provide the backdrop for consideration of the oxidation of the hydrocarbon oxygenates—alcohols, ethers, ketones, etc. Finally, the oxidation of the highly stabilized aromatics will be analyzed. This hierarchical approach should facilitate the understanding of the oxidation of most hydrocarbon fuels.

The approach is to start with analysis of the smallest of the hydrocarbon molecules, methane. It is interesting that the combustion mechanism of methane was for a long period of time the least understood. In recent years, however, there have been many studies of methane, so that to a large degree its specific oxidation mechanisms are known over various ranges of temperatures. Now among the best understood, these mechanisms will be detailed later in this chapter.

The higher-order hydrocarbons, particularly propane and above, oxidize much more slowly than hydrogen and are known to form metastable molecules that are important in explaining the explosion limits of hydrogen and carbon monoxide. The existence of these metastable molecules makes it possible to explain qualitatively the unique explosion limits of the complex hydrocarbons and to gain some insights into what the oxidation mechanisms are likely to be.

Mixtures of hydrocarbons and oxygen react very slowly at temperatures below 200°C; as the temperature increases, a variety of oxygen-containing compounds can begin to form. As the temperature is increased further, CO and H_2O begin to predominate in the products and H_2O_2 (hydrogen peroxide), CH_2O (formaldehyde), CO_2, and other compounds begin to appear. At 300–400°C, a faint light often appears, and this light may be followed by one or more blue flames that successively traverse the reaction vessel. These light emissions are called cool flames and can be followed by an explosion. Generally, the presence of aldehydes is revealed.

In discussing the mechanisms of hydrocarbon oxidation and, later, in reviewing the chemical reactions in photochemical smog, it becomes necessary to identify compounds whose structure and nomenclature may seem complicated to those not familiar with organic chemistry. One need not have a background in organic chemistry to follow the combustion mechanisms; one should, however, study the following section to obtain an elementary knowledge of organic nomenclature and structure.

1. Organic Nomenclature

No attempt is made to cover all the complex organic compounds that exist. The classes of organic compounds reviewed are those that occur most frequently in combustion process and photochemical smog.

Alkyl Compounds

Paraffins
  (alk*anes*:
  single bonds)

Olefins
  (alk*enes*:
  contain double
  bonds)

Cycloparaffins
  (cycloalkanes:
  single bonds)

Acetylenes
  (alk*ynes*:
  contain triple
  bonds)

—C≡C—

CH_4, C_2H_6, C_3H_8, C_4H_{10}, ..., C_nH_{2n+2}
meth*ane*, eth*ane*, prop*ane*, but*ane*, ..., straight-chain
isobutane, branched chain

All are saturated (i.e., no more hydrogen can be added
   to any of the compounds)
Radicals deficient in one H atom take the names
   methyl, ethyl, propyl, etc.

C_2H_4, C_3H_6, C_4H_8, ..., C_nH_{2n}
eth*ene*, prop*ene*, but*ene*
(ethylene, propylene, butylene)

Diolefins contain two double bonds
The compounds are unsaturated since C_nH_{2n} can be
   saturated to C_nH_{2n+2}

C_nH_{2n}—no double bonds
cycloprop*ane*, cyclobut*ane*, cyclopent*ane*

Compounds are unsaturated since ring can be broken
$C_nH_{2n} + H_2 \rightarrow C_nH_{2n+2}$

C_2H_2, C_3H_4, C_4H_6, ..., C_nH_{2n-2}
eth*yne*, prop*yne*, but*yne*
(acetylene, methyl acetylene, ethyl acetylene)

Unsaturated compounds

Aromatic Compounds

The building block for the aromatics is the ring-structured benzene C_6H_6,
which has many resonance structures and is therefore very stable:

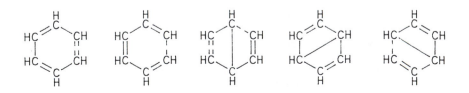

The ring structure of benzene is written in shorthand as either

 or ϕ

Thus

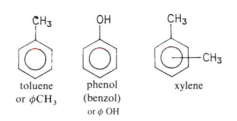

toluene   phenol   xylene
or ϕCH_3 (benzol)
or ϕ OH

xylene being *ortho, meta,* or *para* according to whether methyl groups are separated by one, two, or three carbon atoms, respectively.

Alcohols

Those organic compounds that contain a hydroxyl group (—OH) are called alcohols and follow the simple naming procedure.

$$CH_3OH \qquad C_2H_5OH$$
methanol   ethanol
(methyl alcohol) (ethyl alcohol)

The bonding arrangement is always

Aldehydes

The aldehydes contain the characteristic formyl radical group

and can be written as

where R can be a hydrogen atom or an organic radical. Thus

  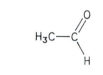

formaldehyde   acetaldehyde   proprionaldehyde

Ketones

The ketones contain the characteristic group

and can be written more generally as

where R and R' are always organic radicals. Thus

is methyl ethyl ketone.

Organic Acids

Organic acids contain the group

carboxyl radical

and are generally written as

where R can be a hydrogen atom or an organic radical. Thus

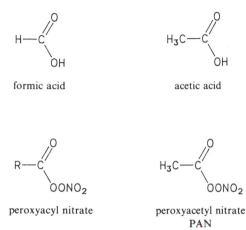

formic acid   acetic acid

Organic Salts

peroxyacyl nitrate   peroxyacetyl nitrate
                                              PAN

Other

Ethers take the form R—O—R', where R and R' are organic radicals. The peroxides take the form R—O—O—R' or R—O—O—H, in which case the term hydroperoxide is used.

2. Explosion Limits

At temperatures around 300–400°C and slightly higher, explosive reactions in hydrocarbon–air mixtures can take place. Thus, explosion limits exist in hydrocarbon oxidation. A general representation of the explosion limits of hydrocarbons is shown in Fig. 7.

The shift of curves, as shown in Fig. 7, is unsurprising since the larger fuel molecules and their intermediates tend to break down more readily to form radicals that initiate fast reactions. The shape of the propane curve suggests that branched chain mechanisms are possible for hydrocarbons. One can conclude that the character of the propane mechanisms is different from that of the H_2–O_2 reaction when one compares this explosion curve with the H_2–O_2 pressure peninsula. The island in the propane–air curve drops and goes slightly to the left for higher-order paraffins; e.g., for hexane it occurs at 1 atm. For the reaction of propane with pure oxygen, the curve drops to about 0.5 atm.

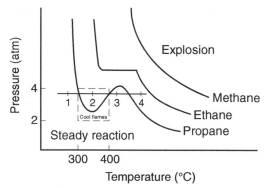

FIGURE 7   General explosion limit characteristics of stoichiometric hydrocarbon–air mixture. The dashed box denotes cool flame region.

Hydrocarbons exhibit certain experimental combustion characteristics that are consistent both with the explosion limit curves and with practical considerations; these characteristics are worth reviewing:

(a) Hydrocarbons exhibit induction intervals that are followed by a very rapid reaction rate. Below 400°C, these rates are of the order of 1 s or a fraction thereof, and below 300°C they are of the order of 60 s.

(b) Their rate of reaction is inhibited strongly by adding surface (therefore, an important part of the reaction mechanism must be of the free-radical type).

(c) They form aldehyde groups, which appear to have an influence (formaldehyde is the strongest). These groups accelerate and shorten the ignition lags .

(d) They exhibit cool flames, except in the cases of methane and ethane.

(e) They exhibit negative temperature coefficients of reaction rate.

(f) They exhibit two-stage ignition, which may be related to the cool flame phenomenon.

(g) Their reactions are explosive without appreciable self-heating (branched chain explosion without steady temperature rise). Explosion usually occurs when passing from region 1 to region 2 in Fig. 5. Explosions may occur in other regions as well, but the reactions are so fast that we cannot tell whether they are self-heating or not.

a. The Negative Coefficient of Reaction Rate

Semenov [14] explained the long induction period by hypothesizing unstable, but long-lived species that form as intermediates and then undergo different reactions according to the temperature. This concept can be represented in the form of the following competing fuel (A) reaction routes after the formation of the

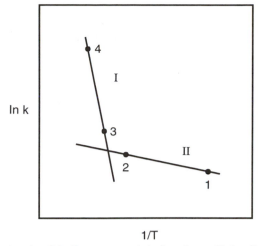

1/T

FIGURE 8   Arrhenius plot of the Semenov steps in hydrocarbon oxidation. Points 1–4 correspond to the same points as in Fig. 7.

unstable intermediate M^*:

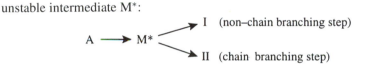

Route I is controlled by an activation energy process larger than that of II.

Figure 8 shows the variation of the reaction rate of each step as a function of temperature. The numbers in Fig. 8 correspond to the temperature position designation in Fig. 7. At point 1 in Fig. 8 one has a chain branching system since the temperature is low and α_{crit} is large; thus, $\alpha < \alpha_{crit}$ and the system is nonexplosive. As the temperature is increased (point 2), the rate constants of the chain steps in the system increase and α_{crit} drops; so $\alpha > \alpha_{crit}$ and the system explodes. At a still higher temperature (point 3), the non-chain branching route I becomes faster. Although this step is faster, α is always less than α_{crit}; thus the system cannot explode. Raising temperatures along route I still further leads to a reaction so fast that it becomes self-heating and hence explosive again (point 4).

The temperature domination explains the peninsula in the *P–T* diagram (Fig. 7), and the negative coefficient of reaction rate is due to the shift from point 2 to 3.

b. Cool Flames

The cool-flame phenomenon [15] is generally a result of the type of experiment performed to determine the explosion limits and the negative temperature coefficient feature of the explosion limits. The chemical mechanisms used to explain these phenomena are now usually referred to as cool-flame chemistry.

Most explosion limit experiments are performed in vessels immersed in iso-thermal liquid baths (see Fig. 1). Such systems are considered to be isothermal within the vessel itself. However, the cool gases that must enter will become hotter at the walls than in the center. The reaction starts at the walls and then propagates to the center of the vessel. The initial reaction volume, which is the hypothetical outermost shell of gases in the vessel, reaches an explosive condition (point 2). However, owing to the exothermicity of the reaction, the shell's temperature rises and moves the reacting system to the steady condition point 3; and because the reaction is slow at this condition, not all the reactants are consumed. Each succes-sive inner (shell) zone is initiated by the previous zone and progresses through the steady reaction phase in the same manner. Since some chemiluminescence occurs during the initial reaction stages, it appears as if a flame propagates through the mixture. Indeed, the events that occur meet all the requirements of an ordinary flame, except that the reacting mixture loses its explosive characteristic. Thus there is no chance for the mixture to react completely and reach its adiabatic flame temperature. The reactions in the system are exothermic and the temperatures are known to rise about 200°C—hence the name "cool flames."

After the complete vessel moves into the slightly higher temperature zone, it begins to be cooled by the liquid bath. The mixture temperature drops, the system at the wall can move into the explosive regime again, and the phenomenon can repeat itself since all the reactants have not been consumed. Depending on the specific experimental conditions and mixtures under study, as many as five cool flames have been known to propagate through a given single mixture. Cool flames have been observed in flow systems, as well [16].

3. "Low-Temperature" Hydrocarbon Oxidation Mechanisms

It is essential to establish the specific mechanisms that explain the cool flame phenomenon, as well as the hydrocarbon combustion characteristics mentioned earlier. Semenov [14] was the first to propose the general mechanism that formed the basis of later research, which clarified the processes taking place. This mech-anism is written as follows:

$$RH + O_2 \rightarrow \dot{R} + HO_2 \qquad \text{(initiation)} \qquad (46)$$

$$\dot{R} + O_2 \rightarrow \text{olefin} + H\dot{O}_2 \qquad (47)$$

$$\dot{R} + O_2 \rightarrow \dot{R}O_2 \qquad (48)$$

$$R\dot{O}_2 + RH \rightarrow ROOH + \dot{R} \qquad \text{(chain propagating)} \qquad (49)$$

$$R\dot{O}_2 \rightarrow R'CHO + R''\dot{O} \qquad (50)$$

$$H\dot{O}_2 + RH \rightarrow H_2O_2 + \dot{R} \qquad (51)$$

$$ROOH \rightarrow R\dot{O} + OH \qquad \text{(degenerate branching)} \qquad (52)$$

$$R'CHO + O_2 \rightarrow R'\dot{C}O + H\dot{O}_2 \tag{53}$$

$$\dot{R}O_2 \rightarrow \text{destruction} \qquad \text{(chain terminating)} \tag{54}$$

where the dot above a particular atom designates the radical position. This scheme is sufficient for all hydrocarbons with a few carbon atoms, but for multicarbon (> 5) species, other intermediate steps must be added, as will be shown later.

Since the system requires the buildup of ROOH and $\dot{R}'CHO$ before chain branching occurs to a sufficient degree to dominate the system, Semenov termed these steps degenerate branching. This buildup time, indeed, appears to account for the experimental induction times noted in hydrocarbon combustion systems. It is important to emphasize that this mechanism is a low-temperature scheme and consequently does not include the high temperature $H_2–O_2$ chain branching steps.

All first, the question of the relative importance of ROOH versus aldehydes as intermediates was much debated; however, recent work indicates that the hydroperoxide step dominates. Aldehydes are quite important as fuels in the cool-flame region, but they do not lead to the important degenerate chain branching step as readily. The $\dot{R}O$ compounds form ROH species, which play no role with respect to the branching of concern.

Owing to its high endothermicity, the chain initiating reaction is not an important route to formation of the radical R once the reaction system has created other radicals. Obviously, the important generation step is a radical attack on the fuel, and the fastest rate of attack is by the hydroxyl radicals since this reaction step is highly exothermic owing to the creation of water as a product. So the system for obtaining R comes from the reactions

$$RH + X \rightarrow \dot{R} + XH \tag{55}$$

$$RH + \dot{O}H \rightarrow \dot{R} + HOH \tag{56}$$

where X represents any radical. It is the fate of the hydrocarbon radical that determines the existence of the negative temperature coefficient and cool flames. The alkyl peroxy radical $\dot{R}O_2$ forms via reaction (48). The structure of this radical can be quite important. The H abstracted from RH to form the radical $\dot{R}$ comes from a preferential position. The weakest C—H bonding is on a tertiary carbon; and, if such C atoms exist, the O_2 will preferentially attack this position. If no tertiary carbon atoms exist, the preferential attack occurs on the next weakest C—H bonds, which are those on the second carbon atoms from the ends of the chain. (Refer to Appendix C for all bond strengths.) Then, as the hydroxyl radical pool builds, $\dot{O}H$ becomes the predominant attacker of the fuel. Because of the energetics of the hydroxyl step (56), for all intents and purposes, it is relatively nonselective in hydrogen abstraction.

It is known that when O_2 attaches to the radical, it forms a near $90°$ angle with the carbon atoms. (The realization of this steric condition will facilitate understanding of certain reactions to be depicted later). The peroxy radical abstracts a H from any fuel molecule or other hydrogen donor to form the hydroperoxide (ROOH) [reaction (49)]. Tracing the steps, one realizes that the amount of hydroperoxy radical that will form depends on the competition of reaction (48) with reaction (47), which forms the stable olefin together with $H\dot{O}_2$. The $H\dot{O}_2$ that forms from reaction (47) then forms the peroxide H_2O_2 through reaction (51). At high temperatures H_2O_2 dissociates into two hydroxyl radicals; however, at the temperatures of concern here, this dissociation does not occur and the fate of the H_2O_2 (usually heterogeneous) is to form water and oxygen. Thus, reaction (47) essentially leads only to steady reaction. In brief, then, under low-temperature conditions it is the competition between reactions (47) and (48) that determines whether the fuel–air mixture will become explosive or not. Its capacity to explode depends on whether the chain system formed is sufficiently branching to have an α greater than α_{crit}.

a. Competition between Chain Branching and Steady Reaction Steps

Whether the sequence given as reactions (46)–(54) becomes chain branching or not depends on the competition between the reactions

$$\dot{R} + O_2 \rightarrow \text{olefin} + HO_2 \tag{47}$$

and

$$\dot{R} + O_2 \rightarrow \dot{R}O_2 \tag{48}$$

Some evidence [17, 17a] suggests that both sets of products develop from a complex via a process that can be written as

$$\dot{R} + O_2 \rightleftharpoons \dot{R}O_2^* \rightarrow R_{-H}O_2H^* \rightarrow \text{olefin} + HO_2$$

$$\downarrow \text{[M]}$$

$$\dot{R}O_2 \tag{57}$$

At low temperatures and modest pressures, a significant fraction of the complex dissociates back to reactants. A small fraction of the complex at low pressures then undergoes the isomerization

$$\dot{R}O_2 \rightarrow R_{-H}O_2H \tag{58}$$

and subsequent dissociation to the olefin and HO_2. Another small fraction is stabilized to form $\dot{R}O_2$:

$$\dot{R}O_2 \xrightarrow{\text{[M]}} \dot{R}O_2 \tag{59}$$

With increasing pressure, the fraction of the activated complex that is stabilized will approach unity [17]. As the temperature increases, the route to the olefin becomes favored. The direct abstraction leading to the olefin reaction (47) must therefore become important at some temperature higher than 1000 K [17a].

b. Importance of Isomerization in Large Hydrocarbon Radicals

With large hydrocarbon molecules an important isomerization reaction will occur. Benson [17b] has noted that with six or more carbon atoms, this reaction becomes a dominant feature in the chain mechanism. Since most practical fuels contain large paraffinic molecules, one can generalize the new competitive mechanisms as

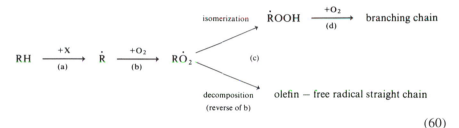

$$(60)$$

Note that the isomerization step is

$$R\dot{O}_2 \rightarrow \dot{R}OOH \tag{61}$$

while the general sequence of step (d) is

$$\dot{R}OOH \xrightarrow{O_2} \dot{O}OR^I - R^{II}OOH \xrightarrow{+RH} HOOR^I - R^{II}OOH + \dot{R}$$

$$\rightarrow R^{III}_{ketone}CO + R^{IV}_{aldehyde}CHO + 2\dot{O}H \tag{62}$$

where the roman numeral superscripts represent different hydrocarbon radicals R of smaller chain length than RH. It is this isomerization concept that requires one to add reactions to the Semenov mechanism to make this mechanism most general.

The oxidation reactions of 2-methylpentane provide a good example of how the hydroperoxy states are formed and why molecular structure is important in establishing a mechanism. The C—C bond angles in hydrocarbons are about

108°. The reaction scheme is then

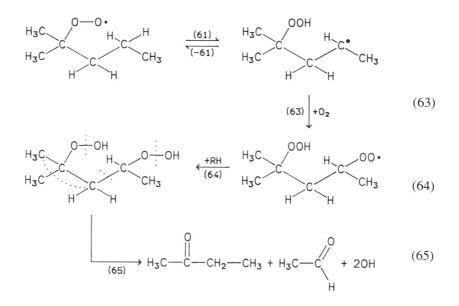

Here one notices that the structure of the 90° (COO) bonding determines the intermediate ketone, aldehyde, and hydroxyl radicals that form.

Although reaction (61) is endothermic and its reverse step reaction (−61) is faster, the competing step reaction (63) can be faster still; thus the isomerization [reaction (61)] step controls the overall rate of formation of RO̊O and subsequent chain branching. This sequence essentially negates the extent of reaction (−48). Thus the competition between RO̊O and olefin production becomes more severe and it is more likely that RO̊O would form at the higher temperatures.

It has been suggested [18] that the greater tendency for long-chain hydrocarbons to knock as compared to smaller and branched chain molecules may be a result of this internal, isomerization branching mechanism.

F. THE OXIDATION OF ALDEHYDES

The low-temperature hydrocarbon oxidation mechanism discussed in the previous section is incomplete because the reactions leading to CO were not included. Water formation is primarily by reaction (56). The CO forms by the conversion of aldehydes and their acetyl (and formyl) radicals, RĊO. The same type of conversion takes place at high temperatures; thus, it is appropriate, prior to considering high-temperature hydrocarbon oxidation schemes, to develop an understanding of the aldehyde conversion process.

As shown in Section 3.1, aldehydes have the structure

where R is either an organic radical or a hydrogen atom and HĊO is the formyl radical. The initiation step for the high-temperature oxidation of aldehydes is the thermolysis reaction

$$RCHO + M \rightarrow RĊO + H + M \qquad (66)$$

The CH bond in the formyl group is the weakest of all CH bonds in the molecule (see Appendix C) and is the one predominantly broken. The R—C bond is substantially stronger than this CH bond, so cleavage of this bond as an initiation step need not be considered. As before, at lower temperatures, high pressures, and under lean conditions, the abstraction initiation step

$$RCHO + O_2 \rightarrow RĊO + HO_2 \qquad (53)$$

must be considered. Hydrogen-labeling studies have shown conclusively that the formyl H is the one abstracted—a finding consistent with the bond energies.

The H atom introduced by reaction (66) and the OH, which arises from the HO_2, initiate the H radical pool that comes about from reactions (17)–(20). The subsequent decay of the aldehyde is then given by

$$RCHO + X \rightarrow RĊO + XH \qquad (67)$$

where X represents the dominant radicals OH, O, H, and CH_3. The methyl radical CH_3 is included not only because of its slow reactions with O_2, but also because many methyl radicals are formed during the oxidation of practically all aliphatic hydrocarbons. The general effectiveness of each radical is in the order $OH > O > H > CH_3$, where the hydroxyl radical reacts the fastest with the aldehyde. In a general hydrocarbon oxidation system these radicals arise from steps other than reaction (66) for combustion processes, so the aldehyde oxidation process begins with reaction (67).

An organic group R is physically much larger than an H atom, so the radical RCO is much more unstable than HCO, which would arise if R were a hydrogen atom. Thus one needs to consider only the decomposition of RCO in combustion systems; that is,

$$RĊO + M \rightarrow Ṙ + CO + M \qquad (68)$$

Similarly, HCO decomposes via

$$HCO + M \rightarrow H + CO + M \qquad (69)$$

but under the usual conditions, the following abstraction reaction must play some small part in the process:

$$HCO + O_2 \rightarrow CO + HO_2 \tag{70}$$

At high pressures the presence of the HO_2 radical also contributes via $HCO + HO_2 \rightarrow H_2O_2 + CO$, but HO_2 is the least effective of OH, O, and H, as the rate constants in Appendix B will confirm. The formyl radical reacts very rapidly with the OH, O, and H radicals. However, radical concentrations are much lower than those of stable reactants and intermediates, and thus formyl reactions with these radicals are considered insignificant relative to the other formyl reactions. As will be seen when the oxidation of large hydrocarbon molecules is discussed (Section 3.H), R is most likely a methyl radical, and the highest-order aldehydes to arise in high-temperature combustion are acetaldehyde and propionaldehyde. The acetaldehyde is the dominant form. Essentially, then, the sequence above was developed with the consideration that R was a methyl group.

G. THE OXIDATION OF METHANE

1. Low-Temperature Mechanism

Methane exhibits certain oxidation characteristics that are different from those of all other hydrocarbons. Tables of bond energy show that the first broken C—H bond in methane takes about 40 kilojoules more than the others, and certainly more than the C—H bonds in longer-chain hydrocarbons. Thus, it is not surprising to find various kinds of experimental evidence indicating that ignition is more difficult with methane/air (oxygen) mixtures than it is with other hydrocarbons. At low temperatures, even oxygen atom attack is slow. Indeed, in discussing exhaust emissions with respect to pollutants, the terms *total hydrocarbons* and *reactive hydrocarbons* are used. The difference between the two terms is simply methane, which reacts so slowly with oxygen atoms at atmospheric temperatures that it is considered unreactive.

The simplest scheme that will explain the lower-temperature results of methane oxidation is the following:

$$CH_4 + O_2 \rightarrow \dot{C}H_3 + H\dot{O}_2 \Big\} \qquad \text{(chain initiating)} \tag{71}$$

$$\dot{C}H_3 + O_2 \rightarrow CH_2O + \dot{O}H \tag{72}$$

$$\dot{O}H + CH_4 \rightarrow H_2O + \dot{C}H_3 \Big\} \qquad \text{(chain propagating)} \tag{73}$$

$$\dot{O}H + CH_2O \rightarrow H_2O + H\dot{C}O \tag{74}$$

$$CH_2O + O_2 \rightarrow H\dot{O}_2 + H\dot{C}O \Big\} \qquad \text{(chain branching)} \tag{75}$$

$$\left. \begin{array}{l} \dot{H}CO + O_2 \rightarrow CO + H\dot{O}_2 \\[6pt] H\dot{O}_2 + CH_4 \rightarrow H_2O_2 + CH_3 \\[6pt] H\dot{O}_2 + CH_2O \rightarrow H_2O_2 + H\dot{C}O \end{array} \right\} \quad \text{(chain propagating)}$$

$$\begin{array}{l} (76) \\[6pt] (77) \\[6pt] (78) \end{array}$$

$$\left. \begin{array}{l} \dot{O}H \rightarrow \text{wall} \\[6pt] CH_2 \rightarrow \text{wall} \\[6pt] H\dot{O}_2 \rightarrow \text{wall} \end{array} \right\} \quad \text{(chain terminating)}$$

$$\begin{array}{l} (79) \\[6pt] (80) \\[6pt] (81) \end{array}$$

There is no H_2O_2 dissociation to OH radicals at low temperatures. H_2O_2 dissociation does not become effective until temperature reaches about 900 K.

As before, reaction (71) is slow. Reactions (72) and (73) are faster since they involve a radical and one of the initial reactants. The same is true for reactions (75), (76), and (77). Reaction (75) represents the necessary chain branching step. Reactions (74) and (78) introduce the formyl radical known to exist in the low-temperature combustion scheme. Carbon monoxide is formed by reaction (76), and water by reaction (73) and the subsequent decay of the peroxides formed. A conversion step of CO to CO_2 is not considered because the rate of conversion by reaction (44) is too slow at the temperatures of concern here.

It is important to examine more closely reaction (72), which proceeds [18, 19] through a metastable intermediate complex—the methyl peroxy radical—in the following manner:

$$CH_3 + O_2 \rightleftarrows \underset{\substack{| \\ H}}{H - C} \overset{\substack{H \quad O \\ | \quad\quad |}}{} \underset{}{- O} \rightarrow H - \underset{\substack{| \\ H}}{C} - O + HO \qquad (82)$$

At lower temperatures the equilibrium step is shifted strongly toward the complex, allowing the formaldehyde and hydroxyl radical formation. The structure of the complex represented in reaction (82) is well established. Recall that when O_2 adds to the carbon atom in a hydrocarbon radical, it forms about a 90° bond angle. Perhaps more important, however, is the suggestion [18] that at temperatures of the order of 1000 K and above the equilibrium step in reaction (82) shifts strongly toward the reactants so that the overall reaction to form formaldehyde and hydroxyl cannot proceed. This condition would therefore pose a restriction on the rapid oxidation of methane at high temperatures. This possibility should come as no surprise as one knows that a particular reaction mechanism can change substantially as the temperature and pressure changes. There now appears to be evidence that another route to the aldehydes and OH formation by reaction (72) may be possible at high temperatures [6a, 19]; this route is discussed in the next section.

2. High-Temperature Mechanism

Many extensive models of the high-temperature oxidation process of methane have been published [20, 20a, 20b, 21]. Such models are quite complex and include hundreds of reactions. The availability of sophisticated computers and computer programs such as those described in Appendix H permits the development of these models, which can be used to predict flow reactor results, flame speeds, emissions, etc., and to compare these predictions with appropriate experimental data. Differences between model and experiment are used to modify the mechanisms and rate constants that are not firmly established. The purpose here is to point out the dominant reaction steps in these complex models of methane oxidation from a chemical point of view, just as modern sensitivity analysis [20, 20a, 20b] can be used to designate similar steps according to the particular application of the mechanism. The next section will deal with other, higher-order hydrocarbons.

In contrast to reaction (71), at high temperatures the thermal decomposition of the methane provides the chain initiation step, namely

$$CH_4 + M \rightarrow CH_3 + H + M \tag{83}$$

With the presence of H atoms at high temperature, the endothermic initiated H_2–O_2 branching and propagating scheme proceeds, and a pool of OH, O, and H radicals develops. These radicals, together with HO_2 [which would form if the temperature range were to permit reaction (71) as an initiating step], abstract hydrogen from CH_4 according to

$$CH_4 + X \rightarrow CH_3 + XH \tag{84}$$

where again X represents any of the radicals. The abstraction rates by the radicals OH, O, and H are all fast, with OH abstraction generally being the fastest. However, these reactions are known to exhibit substantial non-Arrhenius temperature behavior over the temperature range of interest in combustion. The rate of abstraction by O compared to H is usually somewhat faster, but the order could change according to the prevailing stoichiometry; that is, under fuel-rich conditions the H rate will be faster than the O rate owing to the much larger hydrogen atom concentrations under these conditions.

The fact that reaction (82) may not proceed as written at high temperatures may explain why methane oxidation is slow relative to that of other hydrocarbon fuels and why substantial concentrations of ethane are found [4] during the methane oxidation process. The processes consuming methyl radicals are apparently slow, so the methyl concentration builds up and ethane forms through simple recombination:

$$CH_3 + CH_3 \rightarrow C_2H_6 \tag{85}$$

Thus methyl radicals are consumed by other methyl radicals to form ethane, which must then be oxidized. The characteristics of the oxidation of ethane and the higher-order aliphatics are substantially different from those of methane (see Section H.1). For this reason, methane should not be used to typify hydrocarbon oxidation processes in combustion experiments. Generally, a third body is not written for reaction (85) since the ethane molecule's numerous internal degrees of freedom can redistribute the energy created by the formation of the new bond.

Brabbs and Brokaw [22] were among the first who suggested the main oxidation destruction path of methyl radicals to be

$$CH_3 + O_2 \rightarrow CH_3\dot{O} + \dot{O} \tag{86}$$

where $CH_3\dot{O}$ is the methoxy radical. Reaction (86) is very endothermic and has a relatively large activation energy ($\sim$ 120 kJ/mol [4]); thus it is quite slow for a chain step. There has been some question [23] as to whether reaction (72) could prevail even at high temperature, but reaction (86) is generally accepted as the major path of destruction of methyl radicals. Reaction (72) can be only a minor contribution at high temperatures. Other methyl radical reactions are [4]

$$CH_3 + O \rightarrow H_2CO + H \tag{87}$$

$$CH_3 + OH \rightarrow H_2CO + H_2 \tag{88}$$

$$CH_3 + OH \rightarrow CH_3O + H \tag{89}$$

$$CH_3 + H_2CO \rightarrow CH_3 + HCO \tag{90}$$

$$CH_3 + HCO \rightarrow CH_4 + CO \tag{91}$$

$$CH_3 + HO_2 \rightarrow CH_3O + OH \tag{92}$$

These are radical–radical reactions or reactions of methyl radicals with a product of a radical–radical reaction (owing to concentration effects) and are considered less important than reactions (72) and (86). However, reaction (72) and (86) are slow, and reaction (92) can become competitive to form the important methoxy radical, particularly at high pressures and in the lower-temperature region of flames (see Chapter 4).

The methoxy radical formed by reaction (86) decomposes primarily and rapidly via

$$CH_3O + M \rightarrow H_2CO + H + M \tag{93}$$

Although reactions with radicals to give formaldehyde and another product could be included, they would have only a very minor role. They have large rate constants, but concentration factors in reacting systems keep these rates slow.

Reaction (86) is relatively slow for a chain branching step; nevertheless, it is followed by the very rapid decay reaction for the methoxy [reaction (93)], and the products of this two-step process are formaldehyde and two very reactive radicals, O and H. Similarly, reaction (92) may be equally important and can contribute a reactive OH radical. These radicals provide more chain branching than the low-temperature step represented by reaction (72), which produces formaldehyde and a single hydroxyl radical. The added chain branching from the reaction path [reactions (86) and (93)] may be what produces a reasonable overall oxidation rate for methane at high temperatures. In summary, the major reaction paths for the high-temperature oxidation of methane are

$$CH_4 + M \rightarrow CH_3 + H + M \tag{83}$$

$$CH_4 + X \rightarrow CH_3 + XH \tag{84}$$

$$CH_3 + O_2 \rightarrow CH_3O + O \tag{86}$$

$$CH_3 + O_2 \rightarrow H_2CO + OH \tag{72}$$

$$CH_3O + M \rightarrow H_2CO + H + M \tag{93}$$

$$H_2CO + X \rightarrow HCO + XH \tag{67}$$

$$HCO + M \rightarrow H + CO + M \tag{69}$$

$$CH_3 + CH_3 \rightarrow C_2H_6 \tag{85}$$

$$CO + OH \rightarrow CO_2 + H \tag{44}$$

Of course, all the appropriate higher-temperature reaction paths for H_2 and CO discussed in the previous sections must be included. Again, note that when X is an H atom or OH radical, molecular hydrogen (H_2) or water forms from reaction (84). As previously stated, the system is not complete because sufficient ethane forms so that its oxidation path must be a consideration. For example, in atmospheric-pressure methane–air flames, Warnatz [24, 25] has estimated that for lean stoichiometric systems about 30% of methyl radicals recombine to form ethane, and for fuel-rich systems the percentage can rise as high as 80%. Essentially, then, there are two parallel oxidation paths in the methane system: one via the oxidation of methyl radicals and the other via the oxidation of ethane. Again, it is worthy of note that reaction (84) with hydroxyl is faster than reaction (44), so that early in the methane system CO accumulates; later, when the CO concentration rises, it effectively competes with methane for hydroxyl radicals and the fuel consumption rate is slowed.

The mechanisms of CH_4 oxidation covered in this section appear to be most appropriate, but are not necessarily definitive. Rate constants for various individual reactions could vary as the individual steps in the mechanism are studied further.

H. THE OXIDATION OF HIGHER-ORDER HYDROCARBONS

1. Aliphatic Hydrocarbons

The high-temperature oxidation of paraffins larger than methane is a fairly complicated subject owing to the greater instability of the higher-order alkyl

TABLE 2   Relative Importance of Intermediates in Hydrocarbon Combustion

Fuel	Relative hydrocarbon intermediate concentrations
Ethane	ethene ≥ methane
Propane	ethene > propene ≥ methane > ethane
Butane	ethene > propene ≥ methane > ethane
Hexane	ethene > propene > butene > methane ≥ pentene > ethane
2-Methyl pentane	propene > ethene > butene > methane ≥ pentene > ethane

radicals and the great variety of minor species that can form (see Table 2). But, as is the case with methane [20, 20a, 20b, 21], there now exist detailed models of ethane [26], propane [27], and many other higher-order aliphatic hydrocarbons (see Cathonnet [28]). Despite these complications, it is possible to develop a general framework of important steps that elucidate this complex subject.

a. Overall View

It is interesting to review a general pattern for oxidation of hydrocarbons in flames, as suggested very early by Fristrom and Westenberg [29]. They suggested two essential thermal zones: the primary zone, in which the initial hydrocarbons are attacked and reduced to products (CO, H_2, H_2O) and radicals (H, O, OH), and the secondary zone, in which CO and H_2 are completely oxidized. The intermediates are said to form in the primary zone. Initially, then, hydrocarbons of lower order than the initial fuel appear to form in oxygen-rich, saturated hydrocarbon flames according to

$$\{OH + C_nH_{2n+2} \rightarrow H_2O + [C_nH_{2n+1}] \rightarrow C_{n-1}H_{2n-2} + CH_3]$$

Because hydrocarbon radicals of higher order than ethyl are unstable, the initial radical C_nH_{2n+1} usually splits off CH_3 and forms the next lower-order olefinic compound, as shown. With hydrocarbons of higher order than C_3H_8, there is fission into an olefinic compound and a lower-order radical. Alternatively, the radical splits off CH_3. The formaldehyde that forms in the oxidation of the fuel and of the radicals is rapidly attacked in flames by O, H, and OH, so that formaldehyde is usually found only as a trace in flames.

Fristrom and Westenberg claimed that the situation is more complex in fuel-rich saturated hydrocarbon flames, although the initial reaction is simply the H abstraction analogous to the preceding OH reaction; e.g.,

$$H + C_nH_{2n+2} \rightarrow H_2 + C_2H_{2n+1}$$

Under these conditions the concentrations of H and other radicals are large enough that their recombination becomes important, and hydrocarbons of order higher than the original fuel are formed as intermediates.

The general features suggested by Fristrom and Westenberg were confirmed at Princeton [12, 30] by high-temperature flow-reactor studies. However, this work permits more detailed understanding of the high-temperature oxidation mechanism and shows that under oxygen-rich conditions the initial attack by O atoms must be considered as well as the primary OH attack. More importantly, however, it has been established that the paraffin reactants produce intermediate products that are primarily olefinic, and the fuel is consumed, to a major extent, before significant energy release occurs. The higher the initial temperature, the greater the energy release, as the fuel is being converted. This observation leads one to conclude that the olefin oxidation rate simply increases more appreciably with temperature; i.e., the olefins are being oxidized while they are being formed from the fuel. Typical flow-reactor data for the oxidation of ethane and propane are shown in Figs. 9 and 10.

The evidence in Figs. 9 and 10 [12, 30] indicates three distinct, but coupled zones in hydrocarbon combustion:

(1) Following ignition, the primary fuel disappears with little or no energy release and produces unsaturated hydrocarbons and hydrogen. A little of the hydrogen is concurrently oxidized to water.

(2) Subsequently, the unsaturated compounds are further oxidized to carbon monoxide and hydrogen. Simultaneously, the hydrogen present and formed is oxidized to water.

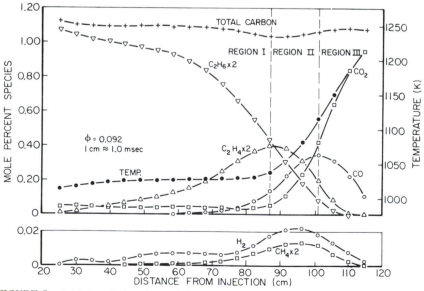

FIGURE 9   Oxidation of ethane in a turbulent flow reactor showing intermediate and final product formation (after Ref. [12]).

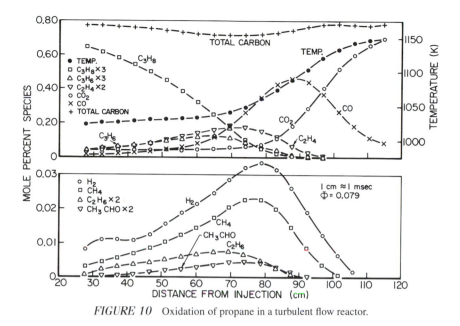

FIGURE 10 Oxidation of propane in a turbulent flow reactor.

(3) Finally, the large amounts of carbon monoxide formed are oxidized to carbon dioxide and most of the heat released from the overall reaction is obtained. Recall that the CO is not oxidized to CO_2 until most of the fuel is consumed owing to the rapidity with which OH reacts with the fuel compared to its reaction to CO (see Table 1).

b. Paraffin Oxidation

In the high-temperature oxidation of large paraffin molecules, the chain initiation step is one in which a CC bond is broken to form hydrocarbon radicals; namely,

$$RH(+M) \rightarrow R' + R''(+M) \qquad (94)$$

This step will undoubtedly dominate, since the CC bond is substantially weaker than any of the CH bonds in the molecule. As mentioned in the previous section, the radicals R' and R'' (fragments of the original hydrocarbon molecule RH) decay into olefins and H atoms. At any reasonable combustion temperature, some CH bonds are broken and H atoms appear owing to the initiation step

$$RH(+M) \rightarrow R + H(+M) \qquad (95)$$

For completeness, one could include a lower-temperature abstraction initiation step

$$RH + O_2 \rightarrow R + HO_2 \tag{96}$$

The essential point is that the initiation steps provide H atoms that react with the oxygen in the system to begin the chain branching propagating sequence that nourishes the radical reservoir of OH, O, and H; that is, the reaction sequences for the complete H_2–O_2 system must be included in any high-temperature hydrocarbon mechanism. Similarly, when CO forms, its reaction mechanism must be included as well.

Once the radical pool forms, the disappearance of the fuel is controlled by the reactions

$$RH + OH \rightarrow R + H_2O \tag{97}$$

$$RH + X \rightarrow R + XH \tag{98}$$

where, again, X is any radical. For the high-temperature condition, X is primarily OH, O, H, and CH_3. Since the RH under consideration is a multicarbon compound, the character of the radical R formed depends on which hydrogen in the molecule is abstracted. Furthermore, it is important to consider how the rate of reaction (98) varies as X varies, since the formation rates of the alkyl isomeric radicals may vary.

Data for the specific rate coefficients for abstraction from CH bonds have been derived from experiments with hydrocarbons with different distributions of primary, secondary, and tertiary CH bonds. A primary CH bond is one on a carbon that is only connected to one other carbon, that is, the end carbon in a chain or a branch of a chain of carbon atoms. A secondary CH bond is one on a carbon atom connected to two others, and a tertiary CH bond is on a carbon atom that is connected to three others. In a chain the CH bond strength on the carbons second from the ends is a few kilojoules less than other secondary atoms. The tertiary CH bond strength is still less, and the primary is the greatest. Assuming additivity of these rates, one can derive specific reaction rate constants for abstraction from the higher-order hydrocarbons by H, O, OH, and HO_2 [31].

From the rates given in Ref.[31], the relative magnitudes of rate constants for abstraction of H by H, O, OH, and HO_2 species from single tertiary, secondary, and primary CH bonds at 1080 K have been determined [32]. These relative magnitudes, which should not vary substantially over modest ranges of temperatures, were found to be as listed here:

	Tertiary	:	Secondary	:	Primary
H	13	:	4	:	1
O	10	:	5	:	1
OH	4	:	3	:	1
HO_2	10	:	3	:	1

Note that the OH abstraction reaction, which is more exothermic than the others, is the least selective of H atom position in its attack on large hydrocarbon molecules. There is also great selectivity by H, O, and HO_2 between tertiary and primary CH bonds. Furthermore, estimates of rate constants at 1080 K [31] and radical concentrations for a reacting hydrocarbon system [33] reveal that the k values for H, O, and OH are practically the same and that during early reaction stages, when concentrations of fuel are large, the radical species concentrations are of the same order of magnitude. Only the HO_2 rate constant departs from this pattern, being lower than the other three. Consequently, if one knows the structure of a paraffin hydrocarbon, one can make estimates of the proportions of various radicals that would form from a given fuel molecule [from the abstraction reaction (98)]. The radicals then decay further according to

$$R(+M) \rightarrow \text{olefin} + R'(+M) \tag{99}$$

where R' is a H atom or another hydrocarbon radical. The ethyl radical will thus become ethene and a H atom. Propane leads to an *n*-propyl and isopropyl radical:

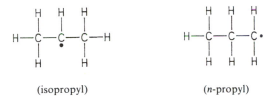

(isopropyl)   (*n*-propyl)

These radicals decompose according to the β-scission rule, which implies that the bond that will break is one position removed from the radical site, so that an olefin can form without a hydrogen shift. Thus the isopropyl radical gives propene and a H atom, while the *n*-propyl radical gives ethene and a methyl radical. The β-scission rule states that when there is a choice between a CC single bond and a CH bond, the CC bond is normally the one that breaks because it is weaker than the CH bond. Even though there are six primary CH bonds in propane and these are somewhat more tightly bound than the two secondary ones, one finds substantially more ethene than propene as an intermediate in the oxidation process. The experimental results [12] shown in Fig. 10 verify this conclusion. The same experimental effort found the olefin trends shown in Table 2. Note that it is possible to estimate the order reported from the principles just described.

If the initial intermediate or the original fuel is a large monoolefin, the radicals will abstract H from those carbon atoms that are singly bonded because the CH bond strengths of doubly bonded carbons are large (see Appendix C). Thus, the evidence [12, 32] is building that, during oxidation, all nonaromatic hydrocarbons primarily form ethene and propene (and some butene and isobutene) and that the oxidative attack that eventually leads to CO is almost solely from these small intermediates. Thus the study of ethene oxidation is crucially important for all alkyl hydrocarbons.

It is also necessary to explain why there are parentheses around the collision partner M in reactions (94), (95), and (99). When RH in reactions (94) and (95) is ethane and R in reaction (99) is the ethyl radical, the reaction order depends on the temperature and pressure range. Reactions (94), (95), and (99) for the ethane system are in the fall-off regime for most typical combustion conditions. Reactions (94) and (95) for propane may lie in the fall-off regime for some combustion conditions; however, around 1 atm, butane and larger molecules pyrolyze near their high-pressure limits [34] and essentially follow first-order kinetics. Furthermore, for the formation of the olefin, an ethyl radical in reaction (99) must compete with the abstraction reaction

$$C_2H_5 + O_2 \rightarrow C_2H_4 + HO_2 \qquad (100)$$

Owing to the great instability of the radicals formed from propane and larger molecules, reaction (99) is fast and effectively first-order; thus, competitive reactions similar to (100) need not be considered. Thus, in reactions (94) and (95) the M has to be included only for ethane and, to a small degree, propane; and in reaction (99) M is required only for ethane. Consequently, ethane is unique among all paraffin hydrocarbons in its combustion characteristics. For experimental purposes, then, ethane (like methane) should not be chosen as a typical hydrocarbon fuel.

c. Olefin and Acetylene Oxidation

Following the discussion from the preceding section, consideration will be given to the oxidation of ethene and propene (when a radical pool already exists) and, since acetylene is a product of this oxidation process, to acetylene as well. These small olefins and acetylene form in the oxidation of a paraffin or any large olefin. Thus, the detailed oxidation mechanisms for ethane, propane, and other paraffins necessarily include the oxidation steps for the olefins [28].

The primary attack on ethene is by addition of the biradical O, although abstraction by H and OH can play some small role. In adding to ethene, O forms an adduct [35] that fragments according to the scheme

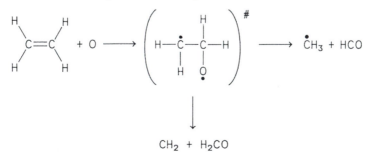

The primary products are methyl and formyl radicals [36, 37] because potential energy surface crossing leads to a H shift at combustion temperatures [35]. It is

rather interesting that the decomposition of cyclic ethylene oxide proceeds through a route in which it isomerizes to acetaldehyde and readily dissociates into CH_3 and HCO. Thus two primary addition reactions that can be written are

$$C_2H_4 + O \rightarrow CH_3 + HCO \qquad (101)$$

$$C_2H_4 + O \rightarrow CH_2 + H_2CO \qquad (102)$$

Another reaction—the formation of an adduct with OH—has also been suggested [38]:

$$C_2H_4 + OH \rightarrow CH_3 + H_2CO \qquad (103)$$

However, this reaction has been questioned [4] because it is highly endothermic.

OH abstraction via

$$C_2H_4 + OH \rightarrow C_2H_3 + H_2O \qquad (104)$$

could have a rate comparable to the preceding three addition reactions, and H abstraction

$$C_2H_4 + H \rightarrow C_2H_3 + H_2 \qquad (105)$$

could also play a minor role. Addition reactions generally have smaller activation energies than abstraction reactions; so at low temperatures the abstraction reaction is negligibly slow, but at high temperatures the abstraction reaction can dominate. Hence the temperature dependence of the net rate of disappearance of reactants can be quite complex.

The vinyl radical (C_2H_3) decays to acetylene primarily by

$$C_2H_3 + M \rightarrow C_2H_2 + H + M \qquad (106)$$

but, again, under particular conditions the abstraction reaction

$$C_2H_3 + O_2 \rightarrow C_2H_2 + HO_2 \qquad (107)$$

must be included. Other minor steps are given in Appendix B.

Since the oxidation mechanisms of CH_3, H_2CO (formaldehyde), and CO have been discussed, only the fate of C_2H_2 and CH_2 (methylene) remains to be determined.

The most important means of consuming acetylene for lean, stoichiometric, and even slightly rich conditions is again by reaction with the biradical O [37, 39, 39a] to form a methylene radical and CO,

$$C_2H_2 + O \rightarrow CH_2 + CO \qquad (108)$$

through an adduct arrangement as described for ethene oxidation. The rate constant for reaction (108) would not be considered large in comparison with that for reaction of O with either an olefin or a paraffin. Mechanistically, reaction (108) is of significance. Since the C_2H_2 reaction with H atoms is slower than $H + O_2$, the oxidation of acetylene does not significantly inhibit the radical pool formation. Also, since its rate with OH is comparable to that of CO with OH, C_2H_2—unlike the other fuels discussed—will not inhibit CO oxidation. Therefore substantial amounts of C_2H_2 can be found in the high-temperature regimes of flames. Reaction (108) states that acetylene consumption depends on events that control the O atom concentration. As discussed in Chapter 8, this fact has implications for acetylene as the soot-growth species in premixed flames. Acetylene–air flame speeds and detonation velocities are fast primarily because high temperatures evolve, not necessarily because acetylene reaction mechanisms contain steps with favorable rate constants. The primary candidate to oxidize the methylene formed is O_2 via

$$CH_2 + O_2 \rightarrow H_2CO + O \tag{109}$$

however, some uncertainty attaches to the products as specified.

Numerous other possible reactions can be included in a very complete mechanism of any of the oxidation schemes of any of the hydrocarbons discussed. Indeed, the very fact that hydrocarbon radicals form is evidence that higher-order hydrocarbon species can develop during an oxidation process. All these reactions play a very minor, albeit occasionally interesting, role; however, their inclusion here would detract from the major steps and important insights necessary for understanding the process.

With respect to propene, it has been suggested [35] that O atom addition is the dominant decay route through an intermediate complex in the following manner:

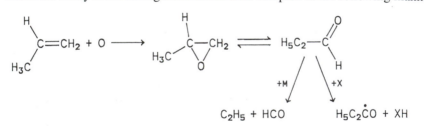

For the large activated propionaldehyde molecule, the pyrolysis step appears to be favored and the equilibrium with the propylene oxide shifts in its direction. The products given for this scheme appear to be consistent with experimental results [38]. The further reaction history of the products has already been discussed.

Essentially, the oxidation chemistry of the aliphatics higher than C_2 has already been discussed since the initiation step is mainly CC bond cleavage with some CH bond cleavage. But the initiation steps for pure ethene or acetylene oxidation are somewhat different. For ethene the major initiation steps are [4, 39a]

$$C_2H_4 + M \rightarrow C_2H_2 + H_2 + M \tag{110}$$

$$C_2H_4 + M \rightarrow C_2H_3 + H + M \tag{111}$$

Reaction (110) is the fastest, but reaction (111) would start the chain. Similarly, the acetylene initiation steps [4] are

$$C_2H_2 + M \rightarrow C_2H + H + M \tag{112}$$

$$C_2H_2 + C_2H_2 \rightarrow C_4H_3 + H + M \tag{113}$$

Reaction (112) dominates under dilute conditions and reaction (113) is more important at high fuel concentrations [4].

The subsequent history of C_2H and C_4H_3 is not important for the oxidation scheme once the chain system develops. Nevertheless, the oxidation of C_2H could lead to chemiluminescent reactions that form CH and C_2, the species responsible for the blue-green appearance of hydrocarbon flames. These species may be formed by the following steps [40]:

$$C_2H + O \rightarrow CH^* + CO$$

$$CH + H \rightarrow C + H_2$$

$$CH + CH \rightarrow C_2^* + H_2$$

where the asterisk (*) represents electronically excited species.

Taking all these considerations into account, it is possible to postulate a general mechanism for the oxidation of aliphatic hydrocarbons; namely,

$$RH + \begin{Bmatrix} M \\ M \\ O_2 \end{Bmatrix} \rightarrow \begin{Bmatrix} R^I + R^{II}(+M) \\ R^I + H(+M) \\ R^I + HO_2 \end{Bmatrix}$$

where the H creates the radical pool (X=H, O, and OH)

$$RH + X \rightarrow \dot{R} + XH$$

$$\dot{R} + \begin{Bmatrix} O_2 \\ M \end{Bmatrix} \rightarrow \text{olefin} + \begin{cases} HO_2 & (\dot{R} = \text{ethyl only}) \\ H + M & (\dot{R} = \text{ethyl and propyl only}) \\ \dot{R}^{III}(M) \end{cases}$$

$$\downarrow$$

$$\text{olefin} \qquad (\text{except for } \dot{R} = CH_3)$$

$$\text{olefin (ethene)} + \begin{Bmatrix} O \\ H \\ OH \end{Bmatrix} \rightarrow \dot{R}^{IV} + \begin{pmatrix} \text{formyl or acetyl radical,} \\ \text{formaldehyde, acetylene, or CO} \end{pmatrix}$$

$$C_2H_2 + O \rightarrow CH_2 + CO$$

$$CO + OH \rightarrow CO_2 + H$$

As a matter of interest, the oxidation of the diolefin butadiene appears to occur through O atom addition to a double bond as well as through abstraction reactions involving OH and H. Oxygen addition leads to 3-butenal and finally alkyl radicals and CO. The alkyl radical is oxidized by O atoms through acrolein to form CO, acetylene, and ethene. The abstraction reactions lead to a butadienyl radical and then vinyl acetylene. The butadienyl radical is now thought to be important in aromatic ring formation processes in soot generation [41–43]. Details of butadiene oxidation are presented in Ref. [44].

2. Alcohols

Consideration of the oxidation of alcohol fuels follows almost directly from Refs. [45, 46].

The presence of the OH group in alcohols makes alcohol combustion chemistry an interesting variation of the analogous paraffin hydrocarbon. Two fundamental pathways can exist in the initial attack on alcohols. In one, the OH group can be displaced while an alkyl radical also remains as a product. In the other, the alcohol is attacked at a different site and forms an intermediate oxygenated species, typically an aldehyde. The dominant pathway depends on the bond strengths in the particular alcohol molecule and on the overall stoichiometry that determines the relative abundance of the reactive radicals.

For methanol, the alternative initiating mechanisms are well established [47–50]. The dominant initiation step is the high-activation process

$$CH_3OH + M \rightarrow CH_3 + OH + M \qquad (114)$$

which contributes little to the products in the intermediate (~ 1000 K) temperature range [49]. By means of deuterium labeling, Aders [51] has demonstrated the occurrence of OH displacement by H atoms:

$$CH_3OH + H \rightarrow CH_3 + H_2O \qquad (115)$$

This reaction may account for as much as 20% of the methanol disappearance under fuel-rich conditions [49]. The chain branching system originates from the reactions

$$CH_3OH + M \rightarrow CH_2OH + H \qquad and \qquad CH_3OH + H \rightarrow CH_2OH + H_2$$

which together are sufficient, with reaction (117) below, to provide the chain. As in many hydrocarbon processes, the major oxidation route is by radical abstraction. In the case of methanol, this yields the hydroxymethyl radical and, ultimately, formaldehyde via

$$CH_3OH + X \rightarrow CH_2OH + XH \qquad (116)$$

$$CH_2OH + \left\{ \begin{matrix} M \\ O_2 \end{matrix} \right\} \rightarrow H_2CO + \left\{ \begin{matrix} H + M \\ HO_2 \end{matrix} \right\} \qquad (117)$$

where as before X represents the radicals in the system. Radical attack on CH_2OH is slow because the concentrations of both radicals are small owing to the rapid rate of reaction (116). These radical steps are given in Appendix B.

The mechanism of ethanol oxidation is less well established, but it apparently involves two mechanistic pathways of approximately equal importance that lead to acetaldehyde and ethene as major intermediate species. Although in flow reactor studies [45] acetaldehyde appears earlier in the reaction than does ethene, both species are assumed to form directly from ethanol. Studies of acetaldehyde oxidation [52] do not indicate any direct mechanism for the formation of ethene from acetaldehyde.

Because C—C bonds are weaker than the C—OH bond, ethanol, unlike methanol, does not lose the OH group in an initiation step. The dominant initial step is

$$C_2H_5OH + M \rightarrow CH_3 + CH_2OH + M \tag{118}$$

As in all long-chain fuel processes, this initiation step does not appear to contribute significantly to the product distribution and, indeed, no formaldehyde is observed experimentally as a reaction intermediate.

It appears that the reaction sequence leading to acetaldehyde would be

$$C_2H_3OH + X \rightarrow CH_3CHOH + XH \tag{119}$$

$$CH_3CHOH + \left\{ \begin{matrix} M \\ O_2 \end{matrix} \right\} \rightarrow CH_3CHO + \left\{ \begin{matrix} H+M \\ HO_2 \end{matrix} \right\} \tag{120}$$

By analogy with methanol, the major source of ethene may be the displacement reaction

$$C_2H_5OH + H \rightarrow C_2H_5 + H_2O \tag{121}$$

with the ethyl radical decaying into ethene.

Because the initial oxygen concentration determines the relative abundance of specific abstracting radicals, ethanol oxidation, like methanol oxidation, shows a variation in the relative concentration of intermediate species according to the overall stoichiometry. The ratio of acetaldehyde to ethene increases for lean mixtures.

As the chain length of the primary alcohols increases, thermal decomposition through fracture of C—C bonds becomes more prevalent. In the pyrolysis of n-butanol, following the rupture of the C_3H_7—CH_2OH bond, the species found are primarily formaldehyde and small hydrocarbons. However, because of the relative weakness of the C—OH bond at a tertiary site, t-butyl alcohol loses its OH group quite readily. In fact, the reaction

$$t\text{-}C_4H_9OH \rightarrow i\text{-}C_4H_8 + H_2O \tag{122}$$

serves as a classic example of unimolecular thermal decomposition.

In the oxidation of t-butanol, acetone and isobutene appear [46] as intermediate species. Acetone can arise from two possible sequences. In one,

$$(CH_3)_3COH \rightarrow (CH_3)_2COH + CH_3 \qquad (123)$$

$$(CH_3)_2COH + X \rightarrow CH_3COCH_3 + XH \qquad (124)$$

and in the other, H abstraction leads to β-scission and a H shift as

$$C_4H_9OH + X \rightarrow C_4H_8OH + XH \qquad (125)$$

$$C_4H_8OH \rightarrow CH_3COCH_3 + CH_3 \qquad (126)$$

Reaction (123) may be fast enough at temperatures above 1000 K to be competitive with reaction (122) [53].

3. Aromatic Hydrocarbons

As discussed by Brezinsky [54], the oxidation of benzene and alkylated aromatics poses a problem different from the oxidation of aliphatic fuels. The aromatic ring provides a site for electrophilic addition reactions which effectively compete with the abstraction of H from the ring itself or from the side chain. When the abstraction reactions involve the side chain, the aromatic ring can strongly influence the degree of selectivity of attack on the side chain hydrogens. At high enough temperatures the aromatic ring thermally decomposes and thereby changes the whole nature of the set of hydrocarbon species to be oxidized. As will be discussed in Chapter 4, in flames the attack on the fuel begins at temperatures below those where pyrolysis of the ring would be significant. As the following sections will show, the oxidation of benzene can follow a significantly different path than that of toluene and other higher alkylated aromatics. In the case of toluene, its oxidation bears a resemblance to that of methane; thus it, too, is different from benzene and other alkylated aromatics.

In order to establish certain terms used in defining aromatic reactions, consider

the following, where the structure of benzene is represented by the symbol ⬡.

Abstraction reaction:

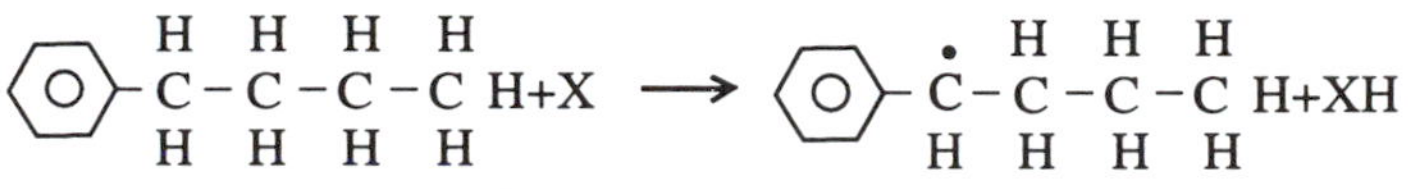

Displacement reaction:

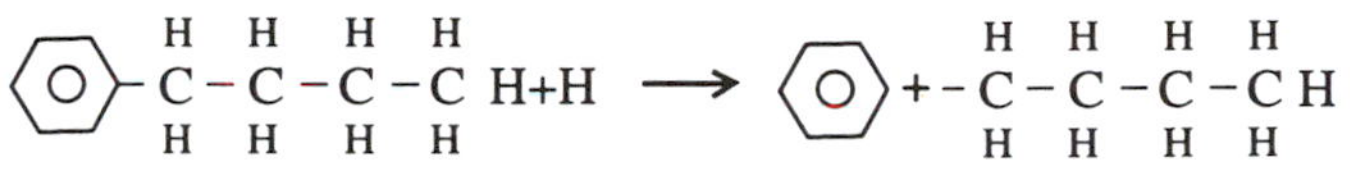

Homolysis reaction:

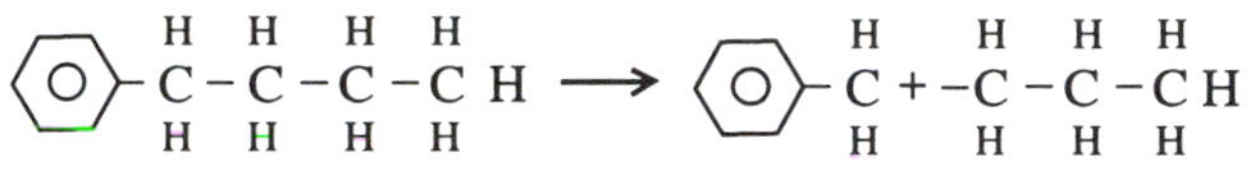

Addition reaction:

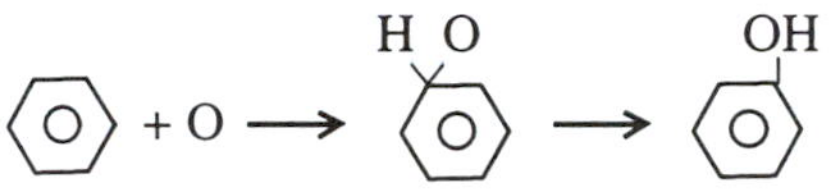

a. Benzene Oxidation

Based on the early work of Norris and Taylor [55] and Bernard and Ibberson [56], who confirmed the theory of multiple hydroxylation, a general low-temperature oxidation scheme was proposed [57, 58]; namely,

$$C_6H_6 + O_2 \quad \rightarrow \quad C_6H_5 + HO_2 \tag{127}$$

$$C_6H_5 + O_2 \quad \rightleftharpoons \quad C_6H_5OO \tag{128}$$

$$C_6H_5OO + C_6H_6 \quad \rightarrow \quad C_6H_5OOH + C_6H_5 \tag{129}$$

$$\downarrow$$

$$C_6H_5O + OH$$

$$C_6H_5O + C_6H_6 \quad \rightarrow \quad C_6H_5OH + C_6H_5 \tag{130}$$

$$C_6H_5OH + O_2 \xrightarrow[\text{as above}]{\text{sequence}} C_6H_4(OH)_2 \tag{131}$$

There are two dihydroxy benzenes that can result from reaction (131)—hydroquinone and pyrocatechol. It has been suggested that they react with oxygen in the following manner [55]:

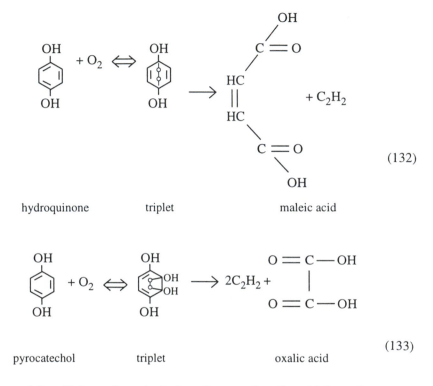

hydroquinone   triplet   maleic acid   (132)

pyrocatechol   triplet   oxalic acid   (133)

Thus maleic acid forms from the hydroquinone and oxalic acid forms from pyrocatechol. However, the intermediate compounds are triplets, so the intermediate steps are "spin-resistant" and may not proceed in the manner indicated. The intermediate maleic acid and oxalic acid are experimentally detected in this low-temperature oxidation process. Although many of the intermediates were detected in low-temperature oxidation studies, Benson [59] determined that the ceiling temperature for bridging peroxide molecules formed from aromatics was of the order of 300°C; that is, the reverse of reaction (128) was favored at higher temperatures.

It is interesting to note that maleic acid dissociates to two carboxyl radicals and acetylene

$$\text{maleic acid} \rightarrow 2(\text{HO}-\dot{\text{C}}{=}\text{O}) + \text{C}_2\text{H}_2 \qquad (134)$$

while oxalic acid dissociates into two carboxyl radicals

$$\text{oxalic acid} \rightarrow 2(\text{HO}-\dot{\text{C}}{=}\text{O}) \qquad (135)$$

Under this low-temperature condition the carboxylic radical undergoes attack

$$HO\text{---}\dot{C}\text{=}O + \begin{Bmatrix} M \\ O_2 \\ X \end{Bmatrix} \rightarrow CO_2 + \begin{Bmatrix} H + M \\ HO_2 \\ XH \end{Bmatrix} \qquad (136)$$

to produce CO_2 directly rather than through the route of CO oxidation by OH characteristic of the high-temperature oxidation of hydrocarbons.

High-temperature flow-reactor studies [60, 61] on benzene oxidation revealed a sequence of intermediates which followed the order: phenol, cyclopentadiene, vinyl acetylene, butadiene, ethene, and acetylene. Since the sampling techniques used in these experiments could not distinguish unstable species, the intermediates could have been radicals that reacted to form a stable compound, most likely by hydrogen addition in the sampling probe. The relative time order of the maximum concentrations, while not the only criterion for establishing a mechanism, has been helpful in the modeling of many oxidation systems [4, 13].

As stated earlier, the benzene molecule is stabilized by strong resonance; consequently, removal of a H from the ring by pyrolysis or O_2 abstraction is difficult and hence slow. It is not surprising, then, that the induction period for benzene oxidation is longer than that for alkylated aromatics. The high-temperature initiation step is similar to that of all the cases described before, i.e.,

$$\text{⬡} + \begin{Bmatrix} M \\ O_2 \end{Bmatrix} \longrightarrow \overset{\bullet}{\text{⬡}} + \begin{Bmatrix} H + M \\ HO_2 \end{Bmatrix} \qquad (137)$$

phenyl

but it probably plays a small role once the radical pool builds from the H obtained. Subsequent formation of the phenyl radical arises from the propagating step

$$\text{⬡} + \begin{Bmatrix} O \\ OH \\ H \end{Bmatrix} \longrightarrow \overset{\bullet}{\text{⬡}} + \begin{Bmatrix} OH \\ H_2O \\ H_2 \end{Bmatrix} \qquad (138)$$

The O atom could venture through a displacement and possibly an addition [60] reaction to form a phenoxyl radical and phenol according to the steps

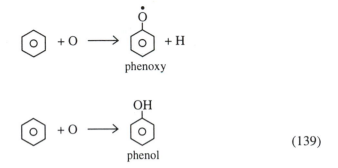

phenoxy

(139)

phenol

Phenyl radical reactions with O_2, O, or HO_2 seem to be the most likely candidates for the first steps in the aromatic ring-breaking sequence [54, 61]. A surprising metathesis reaction that is driven by the resonance stability of the phenoxy product has been suggested from flow reactor studies [54] as a key step in the oxidation of the phenyl radical:

$$\bigodot + O_2 \rightarrow \bigodot + O$$

(140)

In comparison, the analogous reaction written for the methyl radical is highly endothermic.

This chain branching step was found [61, 62] to be exothermic to the extent of approximately 46 kJ/mol, to have a low activation energy, and to be relatively fast. Correspondingly, the main chain branching step [reaction (17)] in the H_2–O_2 system is endothermic to about 65 kJ/mol. This rapid reaction (138) would appear to explain the large amount of phenol found in flow reactor studies. In studies [63] of near-sooting benzene flames, the low mole fraction of phenyl found could have required an unreasonably high rate of reaction (138). The difference could be due to the higher temperatures, and hence the large O atom concentrations, in the flame studies.

The cyclopentadienyl radical could form from the phenoxy radical by

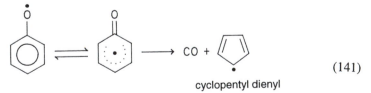

cyclopentyl dienyl

(141)

The expulsion of CO from ketocyclohexadienyl radical is also reasonable, not only in view of the data of flow reactor results, but also in view of other pyrolysis studies [64]. The expulsion indicates the early formation of CO in aromatic oxidation, whereas in aliphatic oxidation CO does not form until later in the reaction after the small olefins form (see Figs. 9 and 10). Since resonance makes the cyclopentadienyl radical very stable, its reaction with an O_2 molecule has a large endothermicity. One feasible step is reaction with O atoms; namely,

$$(142)$$

The butadienyl radical found in reaction (142) then decays along various paths [44], but most likely follows path (c) of reaction (143):

$$(143)$$

Although no reported work is available on vinylacetylene oxidation, oxidation by O would probably lead primarily to the formation of CO, H_2, and acetylene (via an intermediate methyl acetylene) [37]. The oxidation of vinylacetylene, or the cyclopentadienyl radical shown earlier, requires the formation of an adduct [as shown in reaction (142)]. When OH forms the adduct, the reaction is so exothermic that it drives the system back to the initial reacting species. Thus, O atoms become the primary oxidizing species in the reaction steps. This factor may explain why the fuel decay and intermediate species formed in rich and lean oxidation experiments follow the same trend, although rich experiments show much slower rates [65] because the concentrations of oxygen atoms are lower. Figure 11 is a summary of the reaction steps that form the general mechanism of benzene and the phenyl radical oxidation based on a modified version of a model proposed by Emdee

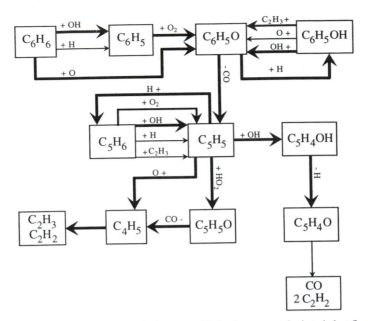

FIGURE 11   Molar rates of progress for benzene oxidation in an atmospheric turbulent flow reactor. The thicknesses of the lines represent the relative magnitudes of certain species as they pass through each reaction pathway.

et al. [61, 66]. Other models of benzene oxidation [67, 68], which are based on Ref. [61], place emphasis on different reactions.

b. Oxidation of Alkylated Aromatics

The initiation step in the high-temperature oxidation of toluene is the pyrolytic cleavage of a hydrogen atom from the methyl side chain, and at lower temperatures it is O_2 abstraction of an H from the side chain, namely

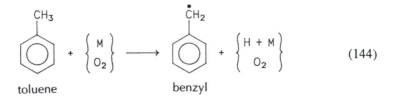

$$(144)$$

The H_2–O_2 radical pool that then develops begins the reactions that cause the fuel concentration to decay. The most effective attackers of the methyl side chain of toluene are OH and H. OH does not add to the ring, but rather abstracts a H from the methyl side chain. This side-chain H is called a benzylic H. The attacking H has been found not only to abstract the benzylic H, but also to displace the methyl group to form benzene and a methyl radical [69]. The reactions are then

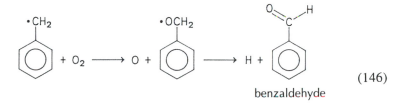

$$\text{(145)}$$

The early appearance of noticeable dibenzyl quantities in flow reactor studies certainly indicates that significant paths to benzyl exist and that benzyl is a stable radical intermediate.

The primary product of benzyl radical decay appears to be benzaldehyde [33, 61]:

benzaldehyde

$$\text{(146)}$$

The reaction of benzyl radicals with O_2 through an intermediate adduct may not be possible, as was found for reaction of methyl radical and O_2. (Indeed, one may think of benzyl as a methyl radical with one H replaced by a phenyl group.) However, it is to be noted that the reaction

$$\text{(147)}$$

has been shown [33] to be orders of magnitude faster than reaction (146). The fate of benzaldehyde is the same as that of any aldehyde in an oxidizing system, as shown by the following reactions which lead to phenyl radicals and CO:

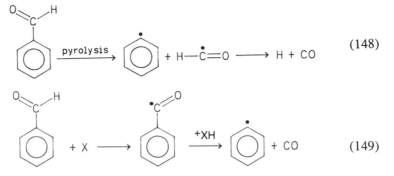

Reaction (149) is considered the major channel.

Reaction (147) is the dominant means of oxidizing benzyl radicals. It is a slow step, so the oxidation of toluene is overall slower than that of benzene, even though the induction period for toluene is shorter. The oxidation of the phenyl radical has been discussed, so one can complete the mechanism of the oxidation of toluene by referring to that section. Figure 12 from Ref. [66] is an appropriate summary of the reactions.

The first step of other high-order alkylated aromatics proceeds through pyrolytic cleavage of a CC bond. The radicals formed soon decay to give H atoms that initiate the H_2–O_2 radical pool. The decay of the initial fuel is dominated by radical attack by OH and H, or possibly O and HO_2, which abstract a H from the side chain. The benzylic H atoms (those attached to the carbon next to the ring) are somewhat easier to remove because of their lower bond strength. To some degree, the benzylic H atoms resemble tertiary or even aldehydic H atoms. As in the case of abstraction from these two latter sites, the case of abstraction of a single benzylic H can be quickly overwhelmed by the cumulative effect of a greater number of primary and secondary H atoms. The abstraction of a benzylic H creates a radical such as

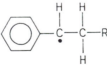

which by the β-scission rule decays to styrene and a radical [65]

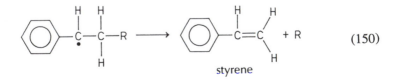

where R can, of course, be H if the initial aromatic is ethyl benzene. It is interesting

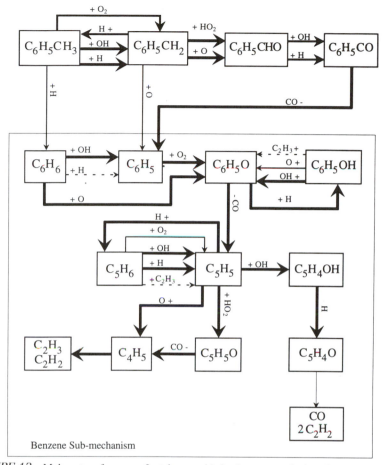

FIGURE 12   Molar rates of progress for toluene oxidation in an atmospheric turbulent flow reactor (cf. to Fig. 11). The benzene submechanism is outlined for toluene oxidation. Dashed arrows represent paths that are important for benzene oxidation, but not significant for toluene. From Ref. [66].

that in the case of ethyl benzene, abstraction of a primary H could also lead to styrene. Apparently, two approximately equally important processes occur during the oxidation of styrene. One is oxidative attack on the double-bonded side chain, most probably through O atom attacks in much the same manner that ethylene is oxidized. This direct oxidation of the vinyl side chain of styrene leads to a benzyl radical and probably a formyl radical. The other is the side chain displacement by H to form benzene and a vinyl radical. Indeed the displacement of the ethyl side chain by the same process has been found to be a major decomposition route for the parent fuel molecule.

If the side chain is in an "iso" form, a more complex aromatic olefin forms. Isopropyl benzene leads to a methyl styrene and styrene [70]. The long-chain alkylate aromatics decay to styrene, phenyl, benzyl, benzene, and alkyl fragments. The oxidation processes of the xylenes follow somewhat similar mechanisms [71, 72].

PROBLEMS

1. In hydrocarbon oxidation a negative reaction rate coefficient is possible. (a) What does this statement mean and when does the negative rate occur? (b) What is the dominant chain branching step in the high-temperature oxidation of hydrocarbons? (c) What are the four dominant overall steps in the oxidative conversion of aliphatic hydrocarbons to fuel products?

2. Explain in a concise manner what the essential differences in the oxidative mechanisms of hydrocarbons are under the following conditions:

   (a) The temperature is such that the reaction is taking place at a slow (measurable) rate–i.e., a steady reaction.
   (b) The temperature is such that the mixture has just entered the explosive regime.
   (c) The temperature is very high, like that obtaining in the latter part of a flame or in a shock tube.

   Assume that the pressure is the same in all three cases.

3. Draw the chemical structure of heptane, 3-octene, and isopropyl benzene.

4. What are the first two species to form during the thermal dissociation of each of the following radicals?

$$H_3C - \underset{\underset{\bullet}{\overset{\displaystyle CH_3}{|}}}{C} - CH_3 \qquad H_3C - \underset{\overset{\bullet}{|}}{\overset{\overset{\displaystyle H}{|}}{C}} - \underset{\overset{\displaystyle |}{H}}{\overset{\overset{\displaystyle H}{|}}{C}} - \underset{\overset{\displaystyle |}{H}}{\overset{\overset{\displaystyle H}{|}}{C}} - CH_3 \qquad H_3C - \underset{\overset{\displaystyle |}{H}}{\overset{\overset{\displaystyle H}{|}}{C}} - \underset{\overset{\displaystyle |}{H}}{\overset{\overset{\displaystyle CH_3}{|}}{C}} - \underset{\overset{\displaystyle |}{H}}{\overset{\overset{\displaystyle |}{\bullet}}{C}} - CH_3$$

5. Toluene is easier to ignite than benzene, yet its overall burning rate is slower. Explain why.

REFERENCES

1. Dainton, F. S.,"Chain Reactions: An Introduction," 2nd Ed. Methuen, London, 1966.
2. Lewis, B., and von Elbe, G., "Combustion, Flames and Explosions of Gases," 2nd Ed., Part 1. Academic Press, New York, 1961.
3. Gardiner, W. C., Jr., and Olson, D. B., *Annu. Rev. Phys. Chem.* **31**, 377 (1980).
4. Westbrook, C. K., and Dryer, F. L., *Prog. Energy Combust. Sci.* **10**, 1 (1984).
5. Bradley, J. N., "Flame and Combustion Phenomena," Chap. 2. Methuen, London, 1969.

6. Baulch, D. L., *et al.*, "Evaluated Kinetic Data for High Temperature Reactions," Vols. 1–3. Butterworth, London, 1973 and 1976.

6a. Yetter, R. A., Dryer, F. L., and Golden, D. M., *in* "Major Research Topics in Combustion" (M. Y. Hussaini, A. Kumar, and R. G. Voigt, eds.), p. 309. Springer-Verlag, New York, 1992.

7. Westbrook, C. K., *Combust. Sci. Technol.* **29**, 67 (1982).

8. Brokaw, R. S., *Int. Symp. Combust.*, *11th* p.1063. Combust. Inst., Pittsburgh, Pennsylvania, 1967.

9. Gordon, A. S., and Knipe, R., *J. Phys. Chem.* **S9**, 1160 (1955).

10. Dryer, F. L., Naegeli, D. W., and Glassman, I., *Combust. Flame* **17**, 270 (1971).

11. Smith, I. W. M., and Zellner, R., *J.C.S. Faraday Trans.* **269**, 1617 (1973).

12. Larson, C. W., Stewart, P. H., and Goldin, D. M., *Int. J. Chem. Kinet.* **20**, 27 (1988).

13. Dryer, F. L., and Glassman, I., *Prog. Astronaut. Aeronaut.* **62**, 55 (1979).

14. Semenov, N. N., "Some Problems in Chemical Kinetics and Reactivity," Chap. 7. Princeton Univ. Press, Princeton, New Jersey, 1958.

15. Minkoff, G. J., and Tipper, C. F. H., "Chemistry of Combustion Reactions." Butterworth, London, 1962.

16. Williams, F. W., and Sheinson, R. S., *Combust. Sci. Technol.* **7**, 85 (1973).

17. Bozzelli, J. W., and Dean, A. M., *J. Phys. Chem.* **94**, 3313 (1990).

17a. Wagner, A. F., Sleagle, I. R., Sarzynski, D., and Gutman, D., *J. Phys. Chem.* **94**, 1853 (1990).

17b. Benson, S. W., *Prog. Energy Combust. Sci.* **7**, 125 (1981).

18. Benson, S. W., *NBS Spec. Publ. (U.S.)* No. 359, p. 101 (1972).

19. Baldwin, A. C., and Golden, D. M., *Chem. Phys. Lett.* **55**, 350 (1978).

20. Dagaut, P., and Cathonnet, M., *J. Chem. Phys.* **87**, 221 (1990).

20a. Frenklach, M., Wang, H., and Rabinovitz, M. J., *Prog. Energy Combust. Sci.* **18**, 47 (1992).

20b. Hunter, J. B., Wang, H., Litzinger, T. A., and Frenklach, M., *Combust. Flame* **97**, 201 (1994).

21. GRI-Mech 2.11, available through the World Wide Web, http://www.gri.org.

22. Brabbs, T. A., and Brokaw, R. S., *Int. Symp. Combust.*, *15th* p. 893. Combust. Inst., Pittsburgh, Pennsylvania, 1975.

23. Yu, C.-L., Wang, C., and Frenklach, M., *J. Phys. Chem.* **99**, 1437 (1995).

24. Warnatz, J., *Int. Symp. Combust.*, *18th* p. 369, Combust. Inst., Pittsburgh, Pennsylvania, 1981.

25. Warnatz, J., *Prog. Astronaut. Aeronaut.* **76**, 501 (1981).

26. Dagaut, P., Cathonnet, M., and Boettner, J. C., *Int. J. Chem. Kinet.* **23**, 437 (1991).

27. Dagaut, P., Cathonnet, M., Boettner, J. C., and Guillard, F., *Combust. Sci. Technol.* **56**, 232 (1986).

28. Cathonnet, M., *Combust. Sci. Technol.* **98**, 265 (1994).

29. Fristrom, R. M., and Westenberg, A. A., "Flame Structure." Chap. 14. McGraw-Hill, New York, 1965.

30. Dryer, F. L., and Brezinsky, K., *Combust. Sci. Technol.* **45**, 199 (1986).

31. Warnatz, J. *in* "Combustion Chemistry" (W. C. Gardiner, Jr., ed.), Chap. 5. Springer-Verlag, New York, 1984.

32. Dryer, F. L., and Brezinsky, K., *West. States Sect. Combust. Inst.* Pap. No. 84–88 (1984).

33. Brezinsky, K., Litzinger, T. A., and Glassman, I., *Int. J. Chem. Kinet.* **16**, 1053 (1984).

34. Golden, D. M., and Larson, C. W., *Int. Symp. Combust.*, *20th* p. 595. Combust. Inst., Pittsburgh, Pennsylvania, 1985.

35. Hunziker, H. E., Kneppe, H., and Wendt, H. R., *J. Photochem.* **17**, 377 (1981).

36. Peters, J., and Mahnen, G., *in* "Combustion Institute European Symposium" (F. J. Weinberg, ed.), p. 53. Academic Press, New York, 1973.

37. Blumenberg, B., Hoyermann, K., and Sievert, R., *Int. Symp. Combust.*, *16th* p. 841. Combust. Inst., Pittsburgh, Pennsylvania, 1977.

38. Tully, F. P., *Phys. Chem. Lett.* **96**,148 (1983).

39. Miller, J. A., Mitchell, R. E., Smooke, M. D., and Kee, R. J., *Int. Symp. Combust.*, *19th* p. 181. Combust. Inst., Pittsburgh, Pennsylvania, 1982.

39a. Kiefer, J. H., Kopselis, S. A., Al-Alami, M. Z., and Budach, K. A., *Combust. Flame* **51**, 79, 1983.

40. Grebe, J., and Homann, R. H., *Ber. Busenges. Phys. Chem.* **86**, 587 (1982).

41. Glassman, I., Mech. Aerosp. Eng. Rep., No. 1450. Princeton Univ., Princeton, New Jersey (1979).

42. Cole, J. A., M.S. Thesis, Dep. Chem. Eng., MIT, Cambridge, Massachusetts, 1982.

43. Frenklach, M., Clary, D. W., Gardiner, W. C., Jr., and Stein, S. E., *Int. Symp. Combust., 20th* p. 887. Combust. Inst., Pittsburgh, Pennsylvania, 1985.

44. Brezinsky, K., Burke, E. J., and Glassman, I., *Int. Symp. Combust., 20th* p. 613. Combust. Inst., Pittsburgh, Pennsylvania, 1985.

45. Norton, T. S., and Dryer, F. L., *Combust. Sci. Technol.* **63**, 107 (1989).

46. Held, T. J., and Dryer, F. L., *Int. Symp. Combust., 25th* p. 901. Combust. Inst., Pittsburgh, Pennsylvania, 1994.

47. Aronowitz, D., Santoro, R. J., Dryer, F. L., and Glassman, I., *Int. Symp. Combust., 16th* p. 633. Combust. Inst., Pittsburgh, Pennsylvania, 1978.

48. Bowman, C. T., *Combust. Flame* **25**, 343 (1975).

49. Westbrook, C. K., and Dryer, F. L., *Combust. Sci. Technol.* **20**, 125 (1979).

50. Vandooren, J., and van Tiggelen, P. J., *Int. Symp. Combust., 18th* p. 473. Combust. Inst., Pittsburgh, Pennsylvania, 1981.

51. Aders, W. K., *in* "Combustion Institute European Symposium" (F. J. Weinberg, ed.), Academic Press, New York, p. 79. 1973.

52. Colket, M. B., III, Naegeli, D. W., and Glassman, I., *Int. Symp. Combust., 16th* p. 1023. Combust. Inst., Pittsburgh, Pennsylvania, 1977.

53. Tsang, W., *J. Chem. Phys.* **40**, 1498 (1964).

54. Brezinsky, K., *Prog. Energy Combust. Sci.* **12**, 1 (1986).

55. Norris, R. G. W., and Taylor, G. W., *Proc. R. Soc. London, Ser. A* **153**, 448 (1936).

56. Barnard, J. A., and Ibberson, V. J., *Combust. Flame* **9**, 81, 149 (1965).

57. Glassman, I., Mech. Aerosp. Eng. Rep., No. 1446. Princeton Univ., Princeton, New Jersey (1979).

58. Santoro, R. J., and Glassman, I., *Combust. Sci. Technol.* **19**, 161 (1979).

59. Benson, S. W., *J. Am. Chem. Soc.* **87**, 972 (1965).

60. Lovell, A. B., Brezinsky, K., and Glassman, I., *Int. Symp. Combust., 22nd* p. 1065. Combust. Inst., Pittsburgh, Pennsylvania, 1988.

61. Emdee, J. L., Brezinsky, K., and Glassman, I., *J. Phys. Chem.* **96**, 2151 (1992).

62. Frank, P., Herzler, J., Just, T., and Wahl, C., *Int. Symp. Combust., 25th*, p. 833. Combust. Inst., Pittsburgh, Pennsylvania, 1994.

63. Bittner, J. D., and Howard, J. B., *Int. Symp. Combust., 19th* p. 211. Combust. Inst., Pittsburgh, Pennsylvania, 1977.

64. Lin, C. Y., and Lin, M. C., *J. Phys. Chem.* **90**, 1125 (1986).

65. Litzinger, T. A., Brezinsky, K., and Glassman, I., *Combust. Flame* **63**, 251 (1986).

66. Davis, S. C., personal communication, Princeton Univ., Princeton, New Jersey, 1996.

67. Lindstedt, R. P., and Skevis, G., *Combust. Flame* **99**, 551 (1994).

68. Zhang, H.-Y., and McKinnon, J. T., *Combust. Sci. Technol.* **107**, 261 (1995).

69. Astholz, D. C., Durant, J., and Troe, J., *Int. Symp. Combust., 18th* p. 885. Combust. Inst., Pittsburgh, Pennsylvania, 1981.

70. Litzinger, T. A., Brezinsky, K., and Glassman, I., *J. Phys. Chem.* **90**, 508 (1986).

71. Emdee, J. L., Brezinsky, K., and Glassman, I., *Int. Symp. Combust., 23rd* p. 77. Combust. Inst., Pittsburgh, Pennsylvania, 1990.

72. Emdee, J. L., Brezinsky, K., and Glassman, I., *J. Phys. Chem.* **95**, 1616 (1991).

4

Flame Phenomena in Premixed Combustible Gases

A. INTRODUCTION

In the previous chapter, the conditions under which a fuel and oxidizer would undergo explosive reaction were discussed. Such conditions are strongly dependent on the pressure and temperature. Given a premixed fuel–oxidizer system at room temperature and ambient pressure, the mixture is essentially unreactive. However, if an ignition source applied locally raises the temperature substantially, or causes a high concentration of radicals to form, a region of explosive reaction can propagate through the gaseous mixture, provided that the composition of the mixture is within certain limits. These limits, called flammability limits, will be discussed in this chapter. Ignition phenomena will be covered in a later chapter.

When a premixed gaseous fuel–oxidizer mixture within the flammability limits is contained in a long tube, a combustion wave will propagate down the tube if an ignition source is applied at one end. When the tube is opened at both ends, the velocity of the combustion wave falls in the range of 20–200 cm/s. For example, the combustion wave for most hydrocarbon–air mixtures has a velocity of about 40 cm/s. The velocity of this wave is controlled by transport processes, mainly simultaneous heat conduction and diffusion of radicals; thus it is not surprising to find that the velocities observed are much less than the speed of sound in the unburned gaseous mixture. In this propagating combustion wave, subsequent reaction, after the ignition source is removed, is induced in the layer of gas ahead of the flame front by two mechanisms that are closely analogous to the thermal and chain branching mechanisms discussed in the preceding chapter for static systems [1]. This combustion wave is normally referred to as a flame; and since it can be treated as a flow entity, it may also be called a deflagration.

When the tube is closed at one end and ignited there, the propagating wave undergoes a transition from subsonic to supersonic speeds. The supersonic wave is called a detonation. In a detonation heat conduction and radical diffusion do not control the velocity; rather, the shock wave structure of the developed supersonic wave raises the temperature and pressure substantially to cause explosive reaction and the energy release that sustains the wave propagation.

The fact that subsonic and supersonic waves can be obtained under almost the same conditions suggests that more can be learned by regarding the phenomena as overall fluid-mechanical in nature. Consider that the wave propagating in the tube is opposed by the unburned gases flowing at a velocity exactly equal to the wave propagation velocity. The wave then becomes fixed with respect to the containing tube (Fig. 1). This description of wave phenomena is readily treated analytically by the following integrated conservation equations, where the subscript 1 specifies the unburned gas conditions and subscript 2 the burned gas conditions:

$$\rho_1 u_1 = \rho_2 u_2 \qquad\qquad \text{continuity} \qquad\qquad (1)$$

$$p_1 + \rho_1 u_1^2 = p_2 + \rho_2 u_2^2 \qquad\qquad \text{momentum} \qquad\qquad (2)$$

$$c_p T_1 + \tfrac{1}{2} u_1^2 + q = c_p T_2 + \tfrac{1}{2} u_2^2 \qquad\qquad \text{energy} \qquad\qquad (3)$$

$$p_1 = \rho_1 R T_1 \qquad\qquad \text{state} \qquad\qquad (4)$$

$$p_2 = \rho_2 R T_2 \qquad\qquad \text{state} \qquad\qquad (5)$$

Equation (4), which connects the known variables, unburned gas pressure, temperature, and density, is not an independent equation. In the coordinate system chosen, u_1 is the velocity fed into the wave and u_2 is the velocity coming out of the wave. In the laboratory coordinate system, the velocity ahead of the wave is zero, the wave velocity is u_1, and $(u_1 - u_2)$ is the velocity of the burned gases with respect to the tube. The unknowns in the system are u_1, u_2, p_2, T_2, and ρ_2. The chemical energy release is q, and the stagnation adiabatic combustion temperature is T_2 for $u_2 = 0$.

Notice that there are five unknowns and only four independent equations. Nevertheless, one can proceed by analyzing the equations at hand. Simple algebraic

FIGURE 1   Combustion wave fixed in the laboratory frame.

manipulations (detailed in Chapter 5) result in twó new equations:

$$\frac{\gamma}{\gamma - 1}\left(\frac{p_2}{\rho_2} - \frac{p_1}{\rho_1}\right) - \frac{1}{2}(p_2 - p_1)\left(\frac{1}{\rho_1} + \frac{1}{\rho_2}\right) = q \tag{6}$$

and

$$\gamma M_1^2 = \left(\frac{p_2}{p_1} - 1\right) / \left[1 - \frac{(1/\rho_2)}{(1/\rho_1)}\right] \tag{7}$$

where γ is the ratio of specific heats and M is the wave velocity divided by $(\gamma RT_1)^{1/2}$, the Mach number of the wave. For simplicity, the specific heats are assumed constant; i.e., $c_{p1} = c_{p2}$; however, γ is a much milder function of composition and temperature and the assumption that γ does not change between burned and unburned gases is an improvement.

Equation (6) is referred to as the Hugoniot relationship, which states that for given initial conditions (p_1, $1/\rho_1$, q) a whole family of solutions (p_2, $1/\rho_2$) is possible. The family of solutions lies on a curve on a plot of p_2 versus $1/\rho_2$, as shown in Fig. 2. Plotted on the graph in Fig. 2 is the initial point (p_1, $1/\rho_1$) and the two tangents through this point of the curve representing the family of solutions. One obtains a different curve for each fractional value of q. Indeed, a curve is obtained for $q = 0$, i.e., no energy release. This curve traverses the point, representing the initial condition and, since it gives the solution for simple shock waves, is referred to as the shock Hugoniot.

Horizontal and vertical lines are drawn through the initial condition point, as well. These lines, of course, represent the conditions of constant pressure and constant specific volume ($1/\rho$), respectively. They further break the curve into three sections. Sections I and II are further divided into sections by the tangent points (J and K) and the other letters defining particular points.

Examination of Eq. (7) reveals the character of M_1 for regions I and II. In region I, P_2 is much greater than P_1, so that the difference is a number much larger

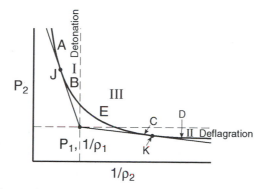

FIGURE 2   Reacting system ($q > o$) Hugoniot plot divided into five regimes A–E.

than 1. Furthermore, in this region $(1/\rho_2)$ is a little less than $(1/\rho_1)$, and thus the ratio is a number close to, but a little less than 1. Therefore, the denominator is very small, much less than 1. Consequently, the right-hand side of Eq. (7) is positive and very much larger than 1, certainly greater than 1.4. If one assumes conservatively that $\gamma = 1.4$, then M_1^2 and M_1 are both greater than 1. Thus, region I defines supersonic waves and is called the detonation region. Consequently, a detonation can be defined as a supersonic wave supported by energy release (combustion).

One can proceed similarly in region II. Since p_2 is a little less than p_1, the numerator of Eq. (7) is a small negative fraction. $(1/\rho_2)$ is much greater than $(1/\rho_1)$, and so the denominator is a negative number whose absolute value is greater than -1. The right-hand side of Eq. (7) for region II is less than 1; consequently, M_1 is less than 1. Thus region II defines subsonic waves and is called the deflagration region. Thus deflagration waves in this context are defined as subsonic waves supported by combustion.

In region III, $p_2 > p_1$ and $1/\rho_2 > 1/\rho_1$, the numerator of Eq. (7) is positive, and the denominator is negative. Thus M_1 is imaginary in region III and therefore does not represent a physically real solution.

It will be shown in Chapter 5 that for points on the Hugoniot curve higher than J, the velocity of sound in the burned gases is greater than the velocity of the detonation wave relative to the burned gases. Consequently, in any real physical situation in a tube, wall effects cause a rarefaction. This rarefaction wave will catch up to the detonation front, reduce the pressure, and cause the final values of P_2 and $1/\rho_2$ to drop to point J, the so-called Chapman–Jouguet point. Points between J and B are eliminated by considerations of the structure of the detonation wave or by entropy. Thus, the only steady-state solution in region II is given by point J. This unique solution has been found strictly by fluid-dynamic and thermodynamic considerations.

Furthermore, the velocity of the burned gases at J and K can be shown to equal the velocity of sound there; thus $M_2 = 1$ is a condition at both J and K. An expression similar to Eq. (7) for M_2 reveals that M_2 is greater than 1 as values past K are assumed. Such a condition cannot be real, for it would mean that the velocity of the burned gases would increase by heat addition, which is impossible. It is well known that heat addition cannot increase the flow of gases in a constant area duct past the sonic velocity. Thus region KD is ruled out. Unfortunately, there are no means by which to reduce the range of solutions that is given by region CK. In order to find a unique deflagration velocity for a given set of initial conditions, another equation must be obtained. This equation, which comes about from the examination of the structure of the deflagration wave, deals with the rate of chemical reaction or, more specifically, the rate of energy release.

The Hugoniot curve shows that in the deflagration region the pressure change is very small. Indeed, approaches seeking the unique deflagration velocity assume the pressure to be constant and eliminate the momentum equation.

The gases that flow in a Bunsen tube are laminar. Since the wave created in the horizontal tube experiment is so very similar to the Bunsen flame, it too is laminar. The deflagration velocity under these conditions is called the laminar flame velocity. The subject of laminar flame propagation is treated in the remainder of this section.

For those who have not studied fluid mechanics, the definition of a deflagration as a subsonic wave supported by combustion may sound oversophisticated; nevertheless, it is the only precise definition. Others describe flames in a more relative context. A flame can be considered a rapid, self-sustaining chemical reaction occurring in a discrete reaction zone. Reactants may be introduced into this reaction zone, or the reaction zone may move into the reactants, depending on whether the unburned gas velocity is greater than or less than the flame (deflagration) velocity.

B. LAMINAR FLAME STRUCTURE

Much can be learned by analyzing the structure of a flame in more detail. Consider, for example, a flame anchored on top of a single Bunsen burner as shown in Fig. 3. Recall that the fuel gas entering the burner induces air into the tube from its surroundings. As the fuel and air flow up the tube, they mix and, before the top of the tube is reached, the mixture is completely homogeneous. The flow velocity in the tube is considered to be laminar and the velocity across the tube is parabolic in nature. Thus the flow velocity near the tube wall is very low. This low flow velocity is a major factor, together with heat losses to the burner rim, in stabilizing the flame at the top.

The dark zone designated in Fig. 3 is simply the unburned premixed gases before they enter the area of the luminous zone where reaction and heat release take place. The luminous zone is less than 1 mm thick. More specifically, the luminous zone is that portion of the reacting zone in which the temperature is the highest; indeed, much of the reaction and heat release take place in this zone. The color of the luminous zone changes with fuel–air ratio. For hydrocarbon–air mixtures that are fuel-lean, a deep violet radiation due to excited CH radicals appears. When the mixture is fuel-rich, the green radiation found is due to excited C_2 molecules. The high-temperature burned gases usually show a reddish glow, which arises from CO_2 and water vapor radiation. When the mixture is adjusted to be very rich, an intense yellow radiation can appear. This radiation is continuous and attributable to the presence of the solid carbon particles. Although Planck's black-body curve peaks in the infrared for the temperatures that normally obtain in these flames, the response of the human eye favors the yellow part of the electromagnetic spectrum. However, non–carbon-containing hydrogen–air flames are nearly invisible.

Building on the foundation of the hydrocarbon oxidation mechanisms developed earlier, it is possible to characterize the flame as consisting of three zones [1]: a preheat zone, a reaction zone, and a recombination zone. The general structure of the reaction zone is made up of early pyrolysis reactions and a zone in which the intermediates, CO and H_2, are consumed. For a very stable molecule like methane, little or no pyrolysis can occur during the short residence time within

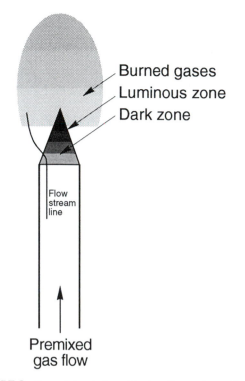

Burned gases
Luminous zone
Dark zone

Flow
stream
line

Premixed
gas flow

FIGURE 3   General description of laminar Bunsen burner flame.

the flame. But with the majority of the other saturated hydrocarbons, considerable degradation occurs, and the fuel fragments that leave this part of the reaction zone are mainly olefins, hydrogen, and the lower hydrocarbons. Since the flame temperature of the saturated hydrocarbons would also be very nearly the same (for reasons discussed in Chapter 1), it is not surprising that their burning velocities, which are very dependent on reaction rate, would all be of the same order (~45 cm/s for a stoichiometric mixture in air).

The actual characteristics of the reaction zone and the composition changes throughout the flame are determined by the convective flow of unburned gases toward the flame zone and the diffusion of radicals from the high-temperature reaction zone against the convective flow into the preheat region. This diffusion is dominated by H atoms; consequently, significant amounts of HO_2 form in the lower-temperature preheat region. Because of the lower temperatures in the preheat zone, reaction (21) in Chapter 3 proceeds readily to form the HO_2 radicals. At these temperatures the chain branching step ($H + O_2 \rightarrow OH + O$) does not occur. The HO_2 subsequently forms hydrogen peroxide. Since the peroxide does not dissociate at the temperatures in the preheat zone, it is convected into the reaction zone where it forms OH radicals at the higher temperatures that prevail there [2].

Owing to this large concentration of OH relative to O and H in the early part of the reaction zone, OH attack on the fuel is the primary reason for the fuel decay. Since the OH rate constant for abstraction from the fuel is of the same order as those for H and O, its abstraction reaction must dominate. The latter part of the reaction zone forms the region where the intermediate fuel molecules are consumed and where the CO is converted to CO_2. As discussed in Chapter 3, the CO conversion results in the major heat release in the system and is the reason the rate of heat release curve peaks near the maximum temperature. This curve falls off quickly because of the rapid disappearance of CO and the remaining fuel intermediates. The temperature follows a smoother, exponential-like rise because of the diffusion of heat back to the cooler gases.

The recombination zone falls into the burned gas or post-flame zone. Although recombination reactions are very exothermic, the radicals recombining have such low concentrations that the temperature profile does not reflect this phase of the overall flame system. Specific descriptions of hydrocarbon–air flames are shown later in this chapter.

C. THE LAMINAR FLAME SPEED

The flame velocity—also called the burning velocity, normal combustion velocity, or laminar flame speed—is more precisely defined as the velocity at which unburned gases move through the combustion wave in the direction normal to the wave surface.

The initial theoretical analyses for the determination of the laminar flame speed fell into three categories: thermal theories, diffusion theories, and comprehensive theories. The historical development followed approximately the same order.

The thermal theories date back to Mallard and Le Chatelier [3], who proposed that it is propagation of heat back through layers of gas that is the controlling mechanism in flame propagation. As one would expect, a form of the energy equation is the basis for the development of the thermal theory. Mallard and Le Chatelier postulated (as shown in Fig. 4) that a flame consists of two zones separated at the point where the next layer ignites. Unfortunately, this thermal theory requires the concept of an ignition temperature. But adequate means do not exist for the determination of ignition temperatures; moreover, an actual ignition temperature does not exist in a flame.

Later, there were improvements in the thermal theories. Probably the most significant of these is the theory proposed by Zeldovich and Frank-Kamenetskii. Because their derivation was presented in detail by Semenov [4], it is commonly called the Semenov theory. These authors included the diffusion of molecules as well as heat, but did not include the diffusion of free radicals or atoms. As a result, their approach emphasized a thermal mechanism and was widely used in correlations of experimental flame velocities. As in the Mallard–Le Chatelier theory, Semenov assumed an ignition temperature, but by approximations eliminated

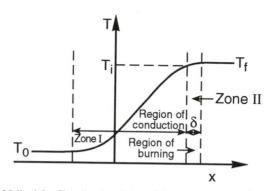

FIGURE 4   Mallard–Le Chatelier description of the temperature in a laminar flame wave.

it from the final equation to make the final result more useful. This approach is similar to what is now termed activation energy asymptotics.

The theory was advanced further when it was postulated that the reaction mechanism can be controlled not only by heat, but also by the diffusion of certain active species such as radicals. As described in the preceding section, low–atomic- and molecular-weight particles can readily diffuse back and initiate further reactions.

The theory of particle diffusion was first advanced in 1934 by Lewis and von Elbe [5] in dealing with the ozone reaction. Tanford and Pease [6] carried this concept further by postulating that it is the diffusion of radicals that is all important, not the temperature gradient as required by the thermal theories. They proposed a diffusion theory that was quite different in physical concept from the thermal theory. However, one should recall that the equations that govern mass diffusion are the same as those that govern thermal diffusion.

These theories fostered a great deal of experimental research to determine the effect of temperature and pressure on the flame velocity and thus to verify which of the theories were correct. In the thermal theory, the higher the ambient temperature, the higher is the final temperature and therefore the faster is the reaction rate and flame velocity. Similarly, in the diffusion theory, the higher the temperature, the greater is the dissociation, the greater is the concentration of radicals to diffuse back, and therefore the faster is the velocity. Consequently, data obtained from temperature and pressure effects did not give conclusive results.

Some evidence appeared to support the diffusion concept, since it seemed to best explain the effect of H_2O on the experimental flame velocities of $CO–O_2$. As described in the previous chapter, it is known that at high temperatures water provides the source of hydroxyl radicals to facilitate rapid reaction of CO and O_2.

Hirschfelder *et al.* [7] reasoned that no dissociation occurs in the cyanogen–oxygen flame. In this reaction the products are solely CO and N_2, no intermediate species form, and the C$=$O and N$\equiv$N bonds are difficult to break. It is apparent that the concentration of radicals is not important for flame propagation in this system, so one must conclude that thermal effects predominate. Hirschfelder *et al.* [7] essentially concluded that one should follow the thermal theory concept while including the diffusion of all particles, both into and out of the flame zone.

In developing the equations governing the thermal and diffusional processes, Hirschfelder obtained a set of complicated nonlinear equations that could be solved only by numerical methods. In order to solve the set of equations, Hirschfelder had to postulate some heat sink for a boundary condition on the cold side. The need for this sink was dictated by the use of the Arrhenius expressions for the reaction rate. The complexity is that the Arrhenius expression requires a finite reaction rate even at $x = -\infty$, where the temperature is that of the unburned gas.

In order to simplify the Hirschfelder solution, Friedman and Burke [8] modified the Arrhenius reaction rate equation so the rate was zero at $T = T_0$, but their simplification also required numerical calculations.

Then it became apparent that certain physical principles could be used to simplify the complete equations so they could be solved relatively easily. Such a simplification was first carried out by von Karman and Penner [9]. Their approach was considered one of the more significant advances in laminar flame propagation, but it could not have been developed and verified if it were not for the extensive work of Hirschfelder and his collaborators. The major simplification that von Karman and Penner introduced is the fact that the eigenvalue solution of the equations is the same for all ignition temperatures, whether it be near T_f or not. More recently, asymptotic analyses have been developed that provide formulas with greater accuracy and further clarification of the wave structure. These developments are described in detail in three books [10–12].

It is easily recognized that any exact solution of laminar flame propagation must make use of the basic equations of fluid dynamics modified to account for the liberation and conduction of heat and for changes of chemical species within the reaction zones. By use of certain physical assumptions and mathematical techniques, the equations have been simplified. Such assumptions have led to many formulations (see Refs. [10–12]), but the theories that will be considered here are an extended development of the simple Mallard–Le Chatelier approach and the Semenov approach. The Mallard–Le Chatelier development is given because of its historical significance and because this very simple thermal analysis readily permits the establishment of the important parameters in laminar flame propagation that are more difficult to interpret in the complex analyses. The Zeldovich–Frank-Kamenetskii–Semenov theory is reviewed because certain approximations related to the ignition temperature that are employed are useful in other problems in the combustion field and permit an introductory understanding to activation energy asymptotics.

1. The Theory of Mallard and Le Chatelier

Conceptually, Mallard and Le Chatelier stated that the heat conducted from zone II in Fig. 4 is equal to that necessary to raise the unburned gases to the ignition temperature (the boundary between zones I and II). If it is assumed that the slope of the temperature curve is linear, the slope can be approximated by the expression $[(T_f - T_i)/\delta]$, where T_f is the final or flame temperature, T_i is the ignition temperature, and δ is the thickness of the reaction zone. The enthalpy balance then becomes

$$\dot{m}c_p(T_i - T_0) = \lambda \frac{(T_f - T_i)}{\delta} A \tag{8}$$

where λ is the thermal conductivity, $\dot{m}$ is the mass rate of the unburned gas mixture into the combustion wave, T_0 is the temperature of the unburned gases, and A is the cross-sectional area taken as unity. Since the problem as described is fundamentally one-dimensional,

$$\dot{m} = \rho A u = \rho S_L A \tag{9}$$

where ρ is the density, u is the velocity of the unburned gases, and S_L is the symbol for the laminar flame velocity. Because the unburned gases enter normal to the wave, by definition

$$S_L = u \tag{10}$$

Equation (8) then becomes

$$\rho S_L c_p(T_i - T_0) = \lambda(T_f - T_i)/\delta \tag{11}$$

or

$$S_L = \left(\frac{\lambda(T_f - T_i)}{\rho c_p(T_i - T_0)} \frac{1}{\delta} \right) \tag{12}$$

Equation (12) is the expression for the flame speed obtained by Mallard and Le Chatelier. Unfortunately, in this expression δ is not known; therefore, a better representation is required.

Since δ is the reaction zone thickness, it is possible to relate δ to S_L. The total rate of mass per unit area entering the reaction zone must be the mass rate of consumption in that zone for the steady flow problem being considered. Thus

$$\rho u = \rho S_L = \dot{\omega}\delta \tag{13}$$

where $\dot{\omega}$ specifies the reaction rate in terms of concentration (in grams per cubic centimeter) per unit time. Equation (12) for the flame velocity then becomes

$$S_L = \left[\frac{\lambda}{\rho c_p} \frac{(T_f - T_i)}{(T_i - T_0)} \frac{\dot{\omega}}{\rho} \right]^{1/2} \sim \left(\alpha \frac{\dot{\omega}}{\rho} \right)^{1/2} \tag{14}$$

where it is important to understand that ρ is the unburned gas density and α is the thermal diffusivity. More fundamentally, the mass of reacting fuel mixture consumed by the laminar flame is represented by

$$\rho S_{\mathrm{L}} \sim \left(\frac{\lambda}{c_p}\dot{\omega}\right)^{1/2} \tag{15}$$

Combining Eqs. (13) and (15), one finds that the reaction thickness in the complete flame wave is

$$\delta \sim \alpha/S_{\mathrm{L}} \tag{16}$$

This adaptation of the simple Mallard–LeChatelier approach is most significant in that the result

$$S_{\mathrm{L}} \sim \left(\alpha \frac{\dot{\omega}}{\rho}\right)^{1/2}$$

is very useful in estimating the laminar flame phenomena as various physical and chemical parameters are changed.

Linan and Williams [13] review the description of the flame wave offered by Mikhelson [14], who equated the heat release in the reaction zone to the conduction of energy from the hot products to the cool reactants. Since the overall conservation of energy shows that the energy per unit mass (h) added to the mixture by conduction is

$$h = c_p(T_{\mathrm{f}} - T_0) \tag{17}$$

then

$$h\dot{\omega}\delta_{\mathrm{L}} = \lambda(T_{\mathrm{f}} - T_0)/\delta_{\mathrm{L}} \tag{18}$$

In this description δ_{L} represents not only the reaction zone thickness δ in the Mallard–Le Chatelier consideration, but also the total of zones I and II in Fig. 4. Substituting Eq. (17) into Eq. (18) gives

$$c_p(T_{\mathrm{f}} - T_0)\dot{\omega}\delta_{\mathrm{L}} = \lambda(T_{\mathrm{f}} - T_0)/\delta_{\mathrm{L}}$$

or

$$\delta_{\mathrm{L}} = \left(\frac{\lambda}{c_p}\frac{1}{\dot{\omega}}\right)^{1/2}$$

The conditions of Eq. (13) must hold, so that in this case

$$\rho S_{\mathrm{L}} = \dot{\omega}\delta_{\mathrm{L}}$$

and Eq. (18) becomes

$$S_{\mathrm{L}} = \left(\frac{\lambda}{\rho c_p}\frac{\dot{\omega}}{\rho}\right)^{1/2} = \left(\alpha\frac{\dot{\omega}}{\rho}\right)^{1/2} \tag{19}$$

Whereas the proportionality of Eq. (14) is the same as the equality in Eq. (19), the difference in the two equations is the temperature ratio

$$\left(\frac{T_f - T_i}{T_i - T_0} \right)^{1/2}$$

In the next section, the flame speed development of Zeldovich, Frank-Kamenetskii, and Semenov will be discussed. They essentially evaluate this term to eliminate the unknown ignition temperature T_i by following what is now the standard procedure of narrow reaction zone asymptotics, which assumes that the reaction rate decreases very rapidly with a decrease in temperature. Thus, in the course of the integration of the rate term $\dot{\omega}$ in the reaction zone, they extend the limits over the entire flame temperature range T_0 to T_f. This approach is, of course, especially valid for large activation energy chemical processes, which are usually the norm in flame studies. Anticipating this development, one sees that the temperature term essentially becomes

$$\frac{R T_f^2}{E(T_f - T_0)}$$

This term specifies the ratio δ_L/δ and has been determined explicitly by Linan and Williams [13] by the procedure they call activation energy asymptotics. Essentially, this is the technique used by Zeldovich, Frank-Kamenetskii, and Semenov [see Eq. (59)]. The analytical development of the asymptotic approach is not given here. For a discussion of the use of asymptotics, one should refer to the excellent books by Williams [12], Linan and Williams [13], and Zeldovich *et al.* [10]. Linan and Williams have called the term $R T_f^2 / E(T_f - T_0)$ the Zeldovich number and give this number the symbol β in their book. Thus

$$\beta = (\delta_L/\delta)$$

It follows, then, that Eq. (14) may be rewritten as

$$S_L = \left(\frac{\alpha}{\beta} \frac{\dot{\omega}}{\rho} \right)^{1/2} \tag{20}$$

and, from the form of Eq. (13), that

$$\delta_L = \beta\delta = \frac{\alpha\beta}{S_L} \tag{21}$$

The general range of hydrocarbon–air premixed flame speeds falls around 40 cm/s. Using a value of thermal diffusivity evaluated at a mean temperature of 1300 K, one can estimate δ_L to be close to 0.1 cm. Thus, hydrocarbon–air flames have a characteristic length of the order of 1 mm. The characteristic time is $(\alpha/S_L{}^2)$, and for these flames this value is estimated to be of the order of a few milliseconds. If one assumes that the overall activation energy of the hydrocarbon–air process is of the order 160 kJ/mol and that the flame temperature is 2100 K, then β is about 10, and probably somewhat less in actuality. Thus, it is estimated from this simple physical approach that the reaction zone thickness, δ, would be a small fraction of a millimeter.

The simple physical approaches proposed by Mallard and Le Chatelier [3] and Mikhelson [14] offer significant insight into the laminar flame speed and factors affecting it. Modern computational approaches now permit not only the calculation of the flame speed, but also a determination of the temperature profile and composition changes throughout the wave. These computational approaches are only as good as the thermochemical and kinetic rate values that form their data base. Since these approaches include simultaneous chemical rate processes and species diffusion, they are referred to as comprehensive theories which is the topic of Section 4.C.3.

Equation (20) permits one to establish various trends of the flame speed as various physical parameters change. Consider, for example, how the flame speed should change with a variation of pressure. If the rate term $\dot\omega$ follows second-order kinetics, as one might expect from a hydrocarbon–air system, then the concentration terms in $\dot\omega$ would be proportional to p^2. However, the density term in $\alpha (= \lambda/\rho c_p)$ and the other density term in Eq. (20) also give a p^2 dependence. Thus for a second-order reaction system the flame speed appears independent of pressure. A more general statement of the pressure dependence in the rate term is that $\dot\omega \sim p^n$, where n is the overall order of the reaction. Thus it is found that

$$S_{\mathrm{L}} \sim (p^{n-2})^{1/2} \tag{22}$$

For a first order dependence such as that observed for a hydrazine decomposition flame, $S_{\mathrm{L}} \sim p^{-1/2}$. As will be shown in Section 4.C.5, although hydrocarbon–air oxidation kinetics are approximately second order, many hydrocarbon–air flame speeds decrease as the pressure rises. This trend is due to the increasing role of the third-order reaction $H + O_2 + M \rightarrow HO_2 + M$ in effecting the chain branching and slowing the rate of energy release. Although it is now realized that S_{L} in these hydrocarbon systems may decrease with pressure, it is important to recognize that the mass burning rate ρS_{L} increases with pressure. Essentially, then, one should note that

$$\dot m_0 \equiv \rho S_{\mathrm{L}} \sim p^{n/2} \tag{23}$$

where $\dot m_0$ is the mass flow rate per unit area of the unburned gases. Considering β a constant, the flame thickness δ_{L} decreases as the pressure rises since

$$\delta_{\mathrm{L}} \sim \frac{\alpha}{S_{\mathrm{L}}} \sim \frac{\lambda}{c_{\mathrm{p}}\rho S_{\mathrm{L}}} \sim \frac{\lambda}{c_p \dot m_0} \tag{24}$$

Since (λ/c_p) does not vary with pressure and $\dot m_0$ increases with pressure as specified by Eq. (23), then Eq. (24) verifies that the flame thickness must decrease with pressure. It follows from Eq. (24) as well that

$$\dot{m}_0 \delta_L \sim \frac{\lambda}{c_p} \tag{25}$$

or that $\dot{m}_0 \delta_L$ is essentially equal to a constant, and that for most hydrocarbon–air flames in which nitrogen is the major species and the reaction product molar compositions do not vary greatly, $\dot{m}_0 \delta_L$ is the same. How these conclusions compare with the results of comprehensive theory calculations will be examined in Section 4.C.5.

The temperature dependence in the flame speed expression is dominated by the exponential in the rate expression $\dot{\omega}$; thus, it is possible to assume that

$$S_L \sim [\exp(-E/RT)]^{1/2} \tag{26}$$

The physical reasoning used indicates that most of the reaction and heat release must occur close to the highest temperature if high activation energy Arrhenius kinetics controls the process. Thus the temperature to be used in the above expression is T_f and one rewrites Eq. (26) as

$$S_L \sim [\exp(-E/RT_f)]^{1/2} \tag{27}$$

Thus, the effect of varying the initial temperature is found in the degree to which it alters the flame temperature. Recall that, due to chemical energy release, a $100°$ rise in initial temperature results in a rise of flame temperature that is much smaller. These trends due to temperature have been verified experimentally.

2. The Theory of Zeldovich, Frank-Kamenetskii, and Semenov

As implied in the previous section, the Russian investigators Zeldovich, Frank-Kamenetskii, and Semenov derived an expression for the laminar flame speed by an important extension of the very simplified Mallard–Le Chatelier approach. Their basic equation included diffusion of species as well as heat. Since their initial insight was that flame propagation was fundamentally a thermal mechanism, they were not concerned with the diffusion of radicals and its effect on the reaction rate. They were concerned with the energy transported by the diffusion of species.

As in the Mallard–Le Chatelier approach, an ignition temperature arises in this development, but it is used only as a mathematical convenience for computation. Because the chemical reaction rate is an exponential function of temperature according to the Arrhenius equation, Semenov assumed that the ignition temperature, above which nearly all reaction occurs, is very near the flame temperature. With this assumption, the ignition temperature can be eliminated in the mathematical development. Since the energy equation is the one to be solved in this approach, the assumption is physically correct. As described in the previous section for hydrocarbon flames, most of the energy release is due to CO oxidation, which takes place very late in the flame where many hydroxyl radicals are available.

For the initial development, although these restrictions are partially removed in further developments, two other important assumptions are made. The assumptions are that the c_p and λ are constant and that

$$(\lambda/c_p) = D\rho$$

where D is the mass diffusivity. This assumption is essentially that

$$\alpha = D$$

Simple kinetic theory of gases predicts

$$\alpha = D = \nu$$

where ν is kinematic viscosity (momentum diffusivity). The ratios of these three diffusivities give some of the familiar dimensionless similarity parameters,

$$\text{Pr} = \nu/\alpha, \qquad \text{Sc} = \nu/D, \qquad \text{Le} = \alpha/D$$

where Pr, Sc, and Le are the Prandtl, Schmidt, and Lewis numbers, respectively. The Prandtl number is the ratio of momentum to thermal diffusion, the Schmidt number is momentum to mass diffusion, and the Lewis number is thermal to mass diffusion. Elementary kinetic theory of gases then predicts as a first approximation

$$\text{Pr} = \text{Sc} = \text{Le} = 1$$

With this approximation, one finds

$$(\lambda/c_p) = D\rho \neq f(P)$$

that is, neither (λ/c_p) nor $D\rho$ is a function of pressure.

Consider the thermal wave given in Fig. 4. If a differential control volume is taken within this one-dimensional wave and the variations as given in the figure are in the x direction, then the thermal and mass balances are as shown in Fig. 5. In Fig. 5, a is the mass of reactant per cubic centimeter, $\dot{\omega}$ is the rate of reaction, Q is the heat of reaction per unit mass, and ρ is the total density. Since the problem is a steady one, there is no accumulation of species or heat with respect to time, and the balance of the energy terms and the species terms must each be equal to zero.

The amount of mass convected into the volume $A \Delta x$ (where A is the area usually taken as unity) is

$$\dot{m} \left[\left(\frac{a}{\rho} \right) + \frac{d(a/\rho)}{dx} \Delta x \right] A - \dot{m} \left(\frac{a}{\rho} \right) A = \dot{m} \frac{d(a/\rho)}{dx} A \, \Delta x \qquad (28)$$

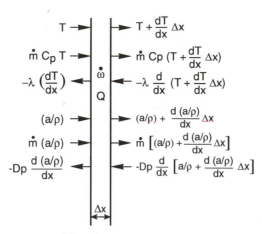

FIGURE 5   Balances across a differential element in a thermal wave describing a laminar flame.

For this one-dimensional configuration $\dot{m} = \rho_0 S_L$. The amount of mass diffusing into the volume is

$$-\frac{d}{dx}\left[D\rho\left(\frac{a}{\rho} + \frac{d(a/\rho)}{dx}\,\Delta x \right) \right] A - \left(-D\rho\frac{d(a/\rho)}{dx} \right) A = -(D\rho)\frac{d^2(a/\rho)}{dx}\,A\,\Delta x \tag{29}$$

The number of moles reacting (disappearing) in the volume is

$$\dot{\omega} A\,\Delta x$$

and it is to be noted that $\dot{\omega}$ is a negative quantity. Thus the continuity equation for the reactant is

$$\underset{\text{(diffusion term)}}{-(D\rho)\frac{d^2(a/\rho)}{dx^2}} + \underset{\text{(convective term)}}{\dot{m}\,\frac{d(a/\rho)}{dx}} + \underset{\text{(generation term)}}{\dot{\omega}} = 0 \tag{30}$$

The energy equation is determined similarly and is

$$-\lambda\frac{d^2 T}{dx^2} + \dot{m}c_p\,\frac{dT}{dx} - \dot{\omega} Q = 0 \tag{31}$$

Because $\dot{\omega}$ is negative and the overall term must be positive since there is heat release, the third term has a negative sign. The state equation is written as

$$(\rho/\rho_0) = (T_0/T)$$

New variables are defined as

$$\tilde{T} = \frac{c_p(T - T_0)}{Q}$$

$$\tilde{a} = (a_0/\rho_0) - (a/\rho)$$

where the subscript 0 designates initial conditions. Substituting the new variables in Eqs. (28) and (29), one obtains two new equations:

$$D\rho \frac{d^2\tilde{a}}{dx^2} - \dot{m}\frac{d\tilde{a}}{dx} + \dot{\omega} = 0 \tag{32}$$

$$\frac{\lambda}{c_p}\frac{d^2\tilde{T}}{dx^2} - \dot{m}\frac{d\tilde{T}}{dx} + \dot{\omega} = 0 \tag{33}$$

The boundary conditions for these equations are

$$x = -\infty, \qquad \tilde{a} = 0, \qquad \tilde{T} = 0$$

$$x = +\infty, \qquad \tilde{a} = a_0/\rho_0, \qquad \tilde{T} = [c_p(T_f - T_0)]/Q \tag{34}$$

where T_f is the final or flame temperature. For the condition $D\rho = (\lambda/c_p)$, Eqs. (32) and (33) are identical in form. If the equations and boundary conditions for $\tilde{a}$ and $\tilde{T}$ coincide; i.e., if $\tilde{a} = \tilde{T}$ over the entire interval, then

$$c_p T_0 + (a_0 Q/\rho_0) = c_p T_f = c_p T + (a Q/\rho) \tag{35}$$

The meaning of Eq. (35) is that the sum of the thermal and chemical energies per unit mass of the mixture is constant in the combustion zone; i.e., the relation between the temperature and the composition of the gas mixture is the same as that for the adiabatic behavior of the reaction at constant pressure.

Thus, the variable defined in Eq. (35) can be used to develop a new equation in the same manner as Eq. (30), and the problem reduces to the solution of only one differential equation. Indeed, either Eq. (30) or (31) can be solved; however, Semenov chose to work with the energy equation.

In the first approach it is assumed, as well, that the reaction proceeds by zero order. Since the rate term $\dot{\omega}$ is not a function of concentration, the continuity equation is not required so we can deal with the more convenient energy equation. Semenov, like Mallard and Le Chatelier, examined the thermal wave as if it were made up of two parts. The unburned gas part is a zone of no chemical reaction, and the reaction part is the zone in which the reaction and diffusion terms dominate and the convective term can be ignored. Thus, in the first zone (I), the energy equation reduces to

$$\frac{d^2T}{dx^2} - \frac{\dot{m}c_p}{\lambda}\frac{dT}{dx} = 0 \tag{36}$$

with the boundary conditions

$$x = -\infty, \qquad T = T_0; \qquad x = 0, \qquad T = T_i \qquad (37)$$

It is apparent from the latter boundary condition that the coordinate system is so chosen that T_i is at the origin. The reaction zone extends a small distance δ, so that in the reaction zone (II) the energy equation is written as

$$\frac{d^2 T}{dx^2} + \frac{\dot{\omega} Q}{\lambda} = 0 \qquad (38)$$

with the boundary conditions

$$x = 0, \qquad T = T_i; \qquad x = \delta, \qquad T = T_f$$

The added condition, which permits the determination of the solution (eigenvalue), is the requirement of the continuity of heat flow at the interface of the two zones:

$$\lambda \left(\frac{dT}{dx} \right)_{x=0,\mathrm{I}} = \lambda \left(\frac{dT}{dx} \right)_{x=0,\mathrm{II}} \qquad (39)$$

The solution to the problem is obtained by initially considering Eq. (38). First, recall that

$$\frac{d}{dx} \left(\frac{dT}{dx} \right)^2 = 2 \left(\frac{dT}{dx} \right) \frac{d^2 T}{dx^2} \qquad (40)$$

Now, Eq. (38) is multiplied by $2(dT/dx)$ to obtain

$$2 \left(\frac{dT}{dx} \right) \frac{d^2 T}{dx^2} = -2 \frac{\dot{\omega} Q}{\lambda} \left(\frac{dT}{dx} \right) \qquad (41)$$

$$\frac{d}{dx} \left(\frac{dT}{dx} \right)^2 = -2 \frac{\dot{\omega} Q}{\lambda} \left(\frac{dT}{dx} \right) \qquad (42)$$

Integrating Eq. (42), one obtains

$$- \left(\frac{dT}{dx} \right)^2_{x=0} = -2 \frac{Q}{\lambda} \int_{T_i}^{T_f} \dot{\omega} \, dT \qquad (43)$$

since $(dT/dx)^2$, evaluated at $x = \delta$ or $T = T_f$, is equal to zero. But from Eq. (36), one has

$$\frac{d}{dx} \left(\frac{dT}{dx} \right) = \frac{\dot{m} c_p}{\lambda} \frac{dT}{dx} \qquad (44)$$

Integrating Eq. (44), one gets

$$dT/dx = (\dot{m} c_p / \lambda) T + \text{const}$$

Since at $x = -\infty$, $T = T_0$ and $(dT/dx) = O$,

$$\text{const} = -(\dot{m}c_p/\lambda)T_0 \tag{45}$$

and

$$dT/dx = [\dot{m}c_p(T - T_0)]/\lambda \tag{46}$$

Evaluating the expression at $x = 0$ where $T = T_i$, one obtains

$$(dT/dx)_{x=0} = \dot{m}c_p(T_i - T_0)/\lambda \tag{47}$$

The continuity of heat flux permits this expression to be combined with Eq. (43) to obtain

$$\frac{\dot{m}c_p(T_i - T_0)}{\lambda} = \left(\frac{2Q}{\lambda}\int_{T_i}^{T_f}\dot{\omega}\,dT\right)^{1/2}$$

Since Arrhenius kinetics dominate, it is apparent that T_i is very close to T_f, so the last expression is rewritten as

$$\frac{\dot{m}c_p(T_f - T_0)}{\lambda} = \left(\frac{2Q}{\lambda}\int_{T_i}^{T_f}\dot{\omega}\,dT\right)^{1/2} \tag{48}$$

For $\dot{m} = S_L\rho_0$ and (a_0/ρ_0), Q can be taken equal to $c_p(T_f - T_0)$ [from Eq. (36)], and one obtains

$$S_L = \left[2\left(\frac{\lambda}{\rho c_p}\right)\frac{I}{(T_f - T_0)}\right]^{1/2} \tag{49}$$

where

$$I = \frac{1}{a_0}\int_{T_i}^{T_f}\dot{\omega}\,dT \tag{50}$$

and $\dot{\omega}$ is a function of T and not of concentration for a zero-order reaction. Thus it may be expressed as

$$\dot{\omega} = Z'e^{-E/RT} \tag{51}$$

where Z' is the pre-exponential term in the Arrhenius expression.

For sufficiently large energy of activation such as that for hydrocarbon–oxygen mixtures where E is of the order of 160 kJ/mol, $(E/RT) > 1$. Thus most of the energy release will be near the flame temperature, T_i will be very near the flame temperature, and zone II will be a very narrow region. Consequently, it is possible to define a new variable σ such that

$$\sigma = (T_f - T) \tag{52}$$

The values of σ will vary from

$$\sigma_i = (T_f - T_i) \tag{53}$$

to zero. Since

$$\sigma < T_f$$

then

$$(E/RT) = [E/R(T_f - \sigma)] = [E/RT_f(1 - \sigma/T_f)]$$

$$= (E/RT_f)[1 + (\sigma/T_f)] = (E/RT_f) + (E\sigma/RT_f^2)$$

Thus the integral I becomes

$$I = \frac{Z'e^{-E/RT_f}}{\alpha_0} \int_{T_i}^{T_f} e^{-E\sigma/RT_f^2} \, dT = -\frac{Z'e^{-E/RT_f}}{\alpha_0} \int_{\sigma_i}^{0} e^{-E\sigma/RT_f^2} \, d\sigma \tag{54}$$

Defining still another variable β as

$$\beta = E\sigma/RT_f^2 \tag{55}$$

the integral becomes

$$I = \frac{Z'e^{-E/RT_f}}{\alpha_0} \left[\int_0^{\beta_i} e^{-\beta} \, d\beta \right] \frac{RT_f^2}{E} \tag{56}$$

With sufficient accuracy one may write

$$j = \int_0^{\beta_i} e^{-\beta} \, d\beta = (1 - e^{-\beta_i}) \cong 1 \tag{57}$$

since $(E/RT_f) > 1$ and $(\sigma_i/T_f) \cong 0.25$. Thus,

$$I = \left(\frac{Z'}{a_0} \right) \left(\frac{RT_f^2}{E} \right) e^{-E/RT_f} \tag{58}$$

and

$$S_L = \left[\frac{2}{a_0} \left(\frac{\lambda}{\rho_0 c_p} \right) (Z'e^{-E/RT_f}) \left(\frac{RT_f^2}{E(T_f - T_0)} \right) \right]^{1/2} \tag{59}$$

In the preceding development, it was assumed that the number of moles did not vary during reaction. This restriction can be removed to allow the number to change in the ratio (n_r/n_p), which is the number of moles of reactant to product. Furthermore, the Lewis number equal to one restriction can be removed to allow

$$(\lambda/c_p)D\rho = A/B$$

where A and B are constants. With these restrictions removed, the result for a first-order reaction becomes

$$S_L = \left\{ \frac{2\lambda_f(c_p)_f Z'}{\rho_0 \bar{c}_p^2} \left(\frac{T_0}{T_f} \right) \left(\frac{n_r}{n_p} \right) \left(\frac{A}{B} \right) \left(\frac{RT_f^2}{E} \right)^2 \frac{e^{-E/RT_f}}{(T_f - T_0)^2} \right\}^{1/2} \qquad (60a)$$

and for a second-order reaction

$$S_L = \left\{ \frac{2\lambda c_{pf}^2 Z' \alpha_0}{\rho_0 (\bar{c}_p)^3} \left(\frac{T_0}{T_f} \right)^2 \left(\frac{n_r}{n_p} \right) \left(\frac{A}{B} \right)^2 \left(\frac{RT_f^2}{E} \right)^3 \frac{e^{-E/RT_f}}{(T_f - T_0)^3} \right\}^{1/2} \qquad (60b)$$

where c_{pf} is the specific heat evaluated at T_f and $\bar{c}_p$ is the average specific heat between T_0 and T_f.

Notice that α_0 and ρ_0 are both proportional to pressure and S_L is independent of pressure. Furthermore, this complex development shows that

$$S_L \sim \left(\frac{\lambda c_{pf}^2}{\rho_0 (\bar{c}_p)^3} \alpha_0 Z' e^{-E/RT_f} \right)^{1/2}, \qquad S_L \sim \left(\frac{\lambda}{\rho_0 c_p} RR \right)^{1/2} \sim (\alpha RR)^{1/2} \qquad (61)$$

as was obtained from the simple Mallard–Le Chatelier approach.

3. Comprehensive Theory and Laminar Flame Structure Analysis

To determine the laminar flame speed and flame structure, it is now possible to solve by computational techniques the steady-state comprehensive mass, species, and energy conservation equations with a complete reaction mechanism for the fuel–oxidizer system which specifies the heat release. The numerical code for this simulation of a freely propagating, one-dimensional, adiabatic premixed flame is based on the scheme of Kee *et al.* [15]. The code uses a hybrid time-integration/Newton-iteration technique to solve the equations. The mass burning rate of the flame is calculated as an eigenvalue of the problem and, since the unburned gas mixture density is known, the flame speed S_L is determined ($\dot{m} = \rho_0 S_L$). In addition, the code permits one to examine the complete flame structure and the sensitivities of all reaction rates on the temperature and species profiles as well as on the mass burning rate. Generally, two preprocessors are used in conjunction with the freely propagating flame code. The first, CHEMKIN, is used to evaluate the thermodynamic properties of the reacting mixture and to process an established chemical reaction mechanism of the particular fuel–oxidizer system [16]. The second is a molecular property package which provides the transport properties of the mixture [17]. See Appendix H.

In order to evaluate the flame structure of characteristic fuels, this procedure was applied to propane–, methane–, and hydrogen–air flames at the stoichiometric equivalence ratio and unburned gas conditions of 298.1 K and 1 atm. The fuels were chosen because of their different kinetic characteristics. Propane is characteristic of most of the higher-order hydrocarbons. As discussed in the previous

TABLE 1   Flame Properties at $\phi = 1^a$

Fuel/Air	S_L $(cm\,s^{-1})$	$\dot{m}_0 = \rho S_L$ $(g\,cm^{-2}\,s^{-1})$	δ_L [cm (est.)]	$\dot{m}_0 \delta_L$ $(g\,cm^{-1}\,s^{-1})$	$(\dot{m}_0 \delta_L)/(\lambda/c_p)_0$
H_2	219.7	0.187	0.050 (Fig. 11)	0.0093	0.73
CH_4	36.2	0.041	0.085 (Fig. 9)	0.0035	1.59
C_3H_8	46.3	0.055	0.057 (Fig. 6)	0.0031	1.41

$^a T_0 = 298$ K, $P = 1$ atm

chapter, methane is unique with respect to hydrocarbons, and hydrogen is a non-hydrocarbon which exhibits the largest mass diffusivity trait. Table 1 reports the calculated values of the mass burning rate and laminar flame speed, and Figs. 6–12 report the species, temperature, and heat release rate distributions. These figures and Table 1 reveal much about the flame structure and confirm much of what was described in this and preceding chapters. The δ_L reported in Table 1 was estimated by considering the spatial distance of the first perceptible rise of the temperature or reactant decay in the figures and the extrapolation of the $\dot{q}$ curve decay branch to the axis. This procedure eliminates the gradual curvature of the decay branch near the point where all fuel elements are consumed and which is due to radical recombination. Since for hydrocarbons one would expect (λ/c_p) to be approximately the same, the values of $\dot{m}_0 \delta_L$ for CH_4 and C_3H_8 in Table 1 should be quite close, as indeed they are. Since the thermal conductivity of H_2 is much larger than that of gaseous hydrocarbons, it is not surprising that its value of $\dot{m}_0 S_L$ is larger than those for CH_4 and C_3H_8. What the approximation $\dot{m}_0 S_L \sim (\lambda/c_p)$ truly states is that $\dot{m}_0 S_L/(\lambda/c_p)$ is of order 1. This order simply arises from the fact that if the thermal equation in the flame speed development is nondimensionalized with δ_L and S_L as the critical dimension and velocity, then $\dot{m}_0 S_L/(\lambda/c_p)$ is the Peclet number (Pe) before the convection term in this equation. This point can be readily seen from

$$\frac{\dot{m}_0 \delta_L}{(\lambda/c_p)} = \frac{\rho_0 \delta_L S_L}{(\lambda/c_p)} = \frac{S_L \delta_L}{\alpha_0} = \text{Pe}$$

Since $\dot{m}_0 = \rho_0 S_L$, the term (λ/c_p) above and in Table 1 is evaluated at the unburned gas condition. Considering that δ_L has been estimated from graphs, the value for all fuels in the last column of Table 1 can certainly be considered O(1).

Figures 6–8 are the results for the stoichiometric propane–air flame. Figure 6 reports the variance of the major species, temperature, and heat release; Figure 7 reports the major stable propane fragment distribution due to the proceeding reactions; and Figure 8 shows the radical and formaldehyde distributions—all as a function of a spatial distance through the flame wave. As stated, the total wave thickness is chosen from the point at which one of the reactant mole fractions begins to decay to the point at which the heat release rate begins to taper off sharply.

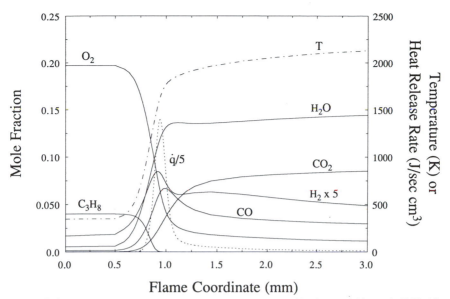

FIGURE 6   Composition, temperature, and heat-release rate profiles for a stoichiometric C_3H_8/air laminar flame at 1 atm and $T_0 = 298$ K.

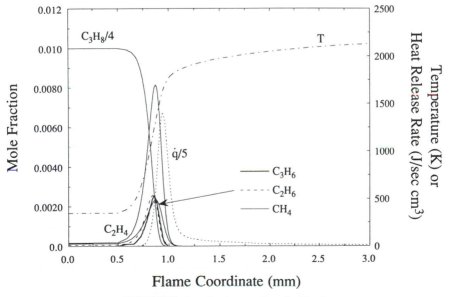

FIGURE 7   Reaction intermediates for Fig. 6.

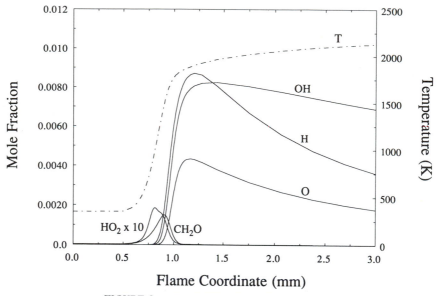

FIGURE 8   Radical distribution profiles for Fig. 6.

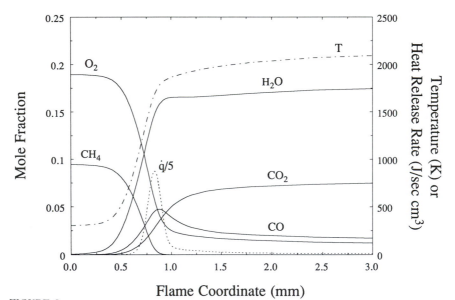

FIGURE 9   Composition, temperature, and heat-release rate profiles for a stoichiometric CH_4/air laminar flame at 1 atm and $T_0 = 298$ K.

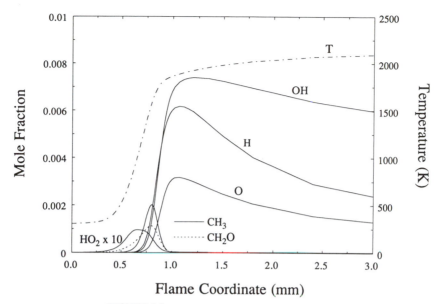

FIGURE 10   Radical distribution profiles for Fig. 9.

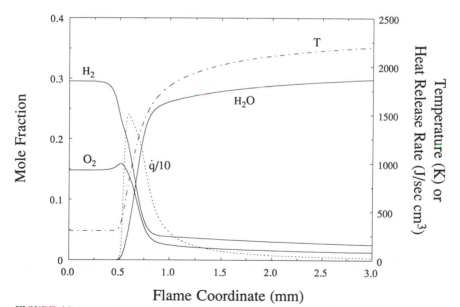

FIGURE 11   Composition, temperature, and heat-release rate profiles for a stoichiometric H_2/air laminar flame at 1 atm and $T_0 = 298$ K.

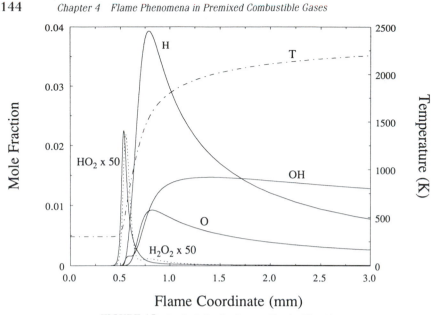

FIGURE 12   Radical distribution profiles for Fig. 11.

Since the point of initial reactant decay corresponds closely to the initial perceptive rise in temperature, the initial thermoneutral period is quite short. The heat release rate curve would ordinarily drop to zero sharply except that the recombination of the radicals in the burned gas zone contribute some energy. The choice of the position that separates the preheat zone and the reaction zone has been made to account for the slight exothermicity of the fuel attack reactions by radicals which have diffused into the preheat zone, and the reaction of the resulting species to form water. Note that water and hydrogen exist in the preheat zone. This choice of operation is then made at the point where the heat release rate curve begins to rise sharply. At this point, as well, there is noticeable CO. This certainly establishes the lack of a sharp separation between the preheat and reaction zones discussed earlier in this chapter and indicates that in the case of propane–air flames the zones overlap. On the basis just described, the thickness of the complete propane–air flame wave is about 0.6 mm and the preheat and reaction zones are about 0.3 mm each. Thus, although maximum heat release rate occurs near the maximum flame temperature (if it were not for the radicals recombining), the ignition temperature in the sense of Mallard–Le Chatelier and Zeldovich–Frank-Kemenetskii–Semenov is not very close to the flame temperature.

Consistent with the general conditions that occur in flames, the HO_2 formed by H atom diffusion upstream maximizes just before the reaction zone. H_2O_2 would begin to form and decompose to OH radicals. This point is in the 900–1000 K range known to be the thermal condition for H_2O_2 decomposition. As would be expected, this point corresponds to the rapid decline of the fuel mole fraction and the onset of radical chain branching. Thus the rapid rise of the radical mole fractions and the formation of the olefins and methane intermediates occur at this point as well (see Figs. 7 and 8). The peak of the intermediates is followed by those of formaldehyde, CO, and CO_2 in the order described from flow reactor results.

Propane disappears well before the end of the reaction zone to form as major intermediates ethene, propene and methane in magnitudes that the β-scission rule and the type and number of C—H bonds would have predicted. Likewise, owing to the greater availability of OH radicals after the fuel disappearance, the CO_2 concentration begins to rise sharply as the fuel concentration decays.

It is not surprising that the depth of the methane–air flame wave is thicker than that of propane–air (Fig. 9). Establishing the same criteria for estimating this thickness, the methane-air wave thickness appears to be about 0.9 mm. The thermal thickness is estimated to be 0.5 mm, and the reaction thickness is about 0.4 mm. Much of what was described for the propane–air flame holds for methane–air except as established from the knowledge of methane–air oxidation kinetics; the methane persists through the whole reaction zone and there is a greater overlap of the preheat and reaction zones. Figure 10 reveals that at the chosen boundary between the two zones, the methyl radical mole fraction begins to rise sharply. The formaldehyde curve reveals the relatively rapid early conversion of these forms of methyl radicals; that is, as the peroxy route produces ample OH, the methane is more rapidly converted to methyl radical while simultaneously the methyl is converted to formaldehyde. Again, initially, the large mole fraction increases of OH, H, and O is due to H_2–O_2 chain branching at the temperature corresponding to this boundary point. In essence, this point is where explosive reaction occurs and the radical pool is more than sufficient to convert the stable reactants and intermediates to products.

If the same criteria are applied to the analysis of the H_2–air results in Figs. 11 and 12, some initially surprising conclusions are reached. At best, it can be concluded that the flame thickness is approximately 0.5 mm. At most, if any preheat zone exists, it is only 0.1 mm. In essence, then, because of the formation of large H atom concentrations, there is extensive upstream H atom diffusion which causes the sharp rise in HO_2. This HO_2 reacts with the H_2 fuel to form H atoms and H_2O_2, which immediately dissociates into OH radicals. Furthermore, even at these low temperatures, the OH reacts with the H_2 to form water and an abundance of H atoms. This reaction is about 50 kJ exothermic. What appears as a rise in the O_2 is indeed only a rise in mole fraction and not in mass.

Figure 13 reports the results of varying the pressure from 0.5 to 8 atm on the structure of stoichiometric methane–air flames, and Table 2 gives the corresponding flame speeds and mass burning rates. Note from Table 2 that, as the pressure increases, the flame speed decreases and the mass burning rate increases for the reasons discussed in Section 4.C.1. The fact that the temperature profiles in Fig. 13 steepen as the pressure rises and that the flame speed results in Table 2 decline with pressure would at first appear counterintuitive in light of the simple thermal theories. However, the thermal diffusivity is also pressure-dependent and

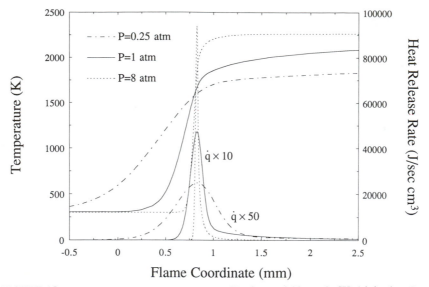

FIGURE 13   Heat-release rate and temperature profiles for a stoichiometric CH_4/air laminar flame at various pressures and $T_0 = 298$ K

is inversely proportional to the pressure. Thus the thermal diffusivity effect overrides the effect of pressure on the reaction rate and the energy release rate, which affects the temperature distribution. The mass burning rate does increase with pressure although for a very few particular reacting systems either the flame speed or the mass burning rate might not follow the trends shown. However, for most hydrocarbon–air systems the trends described would hold.

As discussed for Table 1 and considering that $(\lambda/c_p) \neq f(P)$, it is not surprising that $\dot{m}_0\delta_L$ and $(\dot{m}_0S_L)/(\lambda/c_p)_0$ in Table 2 essentially do not vary with pressure and remain of order 1.

TABLE 2   Flame Properties as a Function of Pressure[a]

P (atm)	S_L ($cm\ s^{-1}$)	$\dot{m}_0\rho S_L$ ($g\ cm^{-2}\ s^{-1}$)	δ_L [$cm\ (est.)$][b]	$\dot{m}_0\delta_L$ ($g\ cm^{-1}\ s^{-1}$)	$(\dot{m}_0\delta_L)/(\lambda/c_p)_0$
0.25	54.51	0.015	0.250	0.0038	1.73
1.00	36.21	0.041	0.085	0.0035	1.59
8.00	18.15	0.163	0.022	0.0036	1.64

[a]CH_4/Air, $\phi = 1$, $T_0 = 298$ K
[b]Fig. 13

4. The Laminar Flame and the Energy Equation

An important point about laminar flame propagation—one that has not previously been discussed—is worth stressing: It has become common to accept that reaction rate phenomena dominate in premixed homogeneous combustible gaseous mixtures and diffusion phenomena dominate in initially unmixed fuel–oxidizer systems. (The subject of diffusion flames will be discussed in Chapter 6.) In the case of laminar flames, and indeed in most aspects of turbulent flame propagation, it should be emphasized that it is the diffusion of heat (and mass) that causes the flame to propagate; i.e., flame propagation is a diffusional mechanism. The reaction rate determines the thickness of the reaction zone and, thus, the temperature gradient. The temperature effect is indeed a strong one, but flame propagation is still attributable to the diffusion of heat and mass. The expression $S_L \sim (\alpha R R)^{1/2}$ says it well—the propagation rate is proportional to the square root of the diffusivity and the reaction rate.

5. Flame Speed Measurements

For a long time there was no interest in flame speed measurements. Sufficient data and understanding were thought to be at hand. But as lean burn conditions became popular in spark ignition engines, the flame speed of lean limits became important. Thus, interest has been rekindled in measurement techniques.

Flame velocity has been defined as the velocity at which the unburned gases move through the combustion wave in a direction normal to the wave surface. If, in an infinite plane flame, the flame is regarded as stationary and a particular flow tube of gas is considered, the area of the flame enclosed by the tube does not depend on how the term "flame surface or wave surface" in which the area is measured is defined. The areas of all parallel surfaces are the same, whatever property (particularly temperature) is chosen to define the surface; and these areas are all equal to each other and to that of the inner surface of the luminous part of the flame. The definition is more difficult in any other geometric system. Consider, for example, an experiment in which gas is supplied at the center of a sphere and flows radially outward in a laminar manner to a stationary spherical flame. The inward movement of the flame is balanced by the outward flow of gas. The experiment takes place in an infinite volume at constant pressure. The area of the surface of the wave will depend on where the surface is located. The area of the sphere for which $T = 500°C$ will be less than that of one for which $T = 1500°C$. So if the burning velocity is defined as the volume of unburned gas consumed per second divided by the surface area of the flame, the result obtained will depend on the particular surface selected. The only quantity that does remain constant in this system is the product $u_r \rho_r A_r$, where u_r is the velocity of flow at the radius r, where the surface area is A_r, and the gas density is ρ_r. This product equals $\dot{m}_r$, the mass flowing through the layer at r per unit time, and must be constant for all values of r. Thus, u_r varies with r the distance from the center in the manner shown in Fig. 14.

It is apparent from Fig. 14 that it is difficult to select a particular linear flow rate of unburned gas up to the flame and regard this velocity as the burning velocity.

If an attempt is made to define burning velocity strictly for such a system, it is found that no definition free from all possible objections can be formulated. Moreover, it is impossible to construct a definition that will, of necessity, determine the same value as that found in an experiment using a plane flame. The essential difficulties are as follow: (1) Over no range of r values does the linear velocity of the gas have even an approximately constant value; and (2) in this ideal system, the temperature varies continuously from the center of the sphere outward and approaches the flame surface asymptotically as r approaches infinity. So no spherical surface can be considered to have a significance greater than any other.

In Fig. 14, u_x, the velocity of gas flow at x for a plane flame, is plotted on the same scale against x, the space coordinate measured normal to the flame front. It is assumed that over the main part of the rapid temperature rise, u_r and u_x coincide. This correspondence is likely to be true if the curvature of the flame is large compared with the flame thickness. The burning velocity is then, strictly speaking, the value to which u_x approaches asymptotically as x approaches $-\infty$. However, because the temperature of the unburned gas varies exponentially with x, the value of u_x becomes effectively constant only a very short distance from the flame. The value of u_r on the low-temperature side of the spherical flame will not at any point be as small as the limiting value of u_x. In fact, the difference, although not zero, will probably not be negligible for such flames. This value of u_r could be determined using the formula

$$u_r = \dot{m}/\rho_r A_r$$

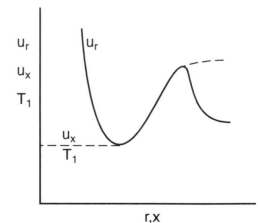

r,x

FIGURE 14   Velocity and temperature variations through non-one-dimensional flame systems.

Since the layer of interest is immediately on the unburned side of the flame, ρ_r will be close to ρ_u, the density of the unburned gas, and $\dot{m}/\rho$, will be close to the volume flow rate of unburned gas.

So, to obtain, in practice, a value for burning velocity close to that for the plane flame, it is necessary to locate and measure an area as far on the unburned side of the flame as possible.

Systems such as Bunsen flames are in many ways more complicated than either the plane case or the spherical case. Before proceeding, consider the methods of observation. The following methods have been most widely used to observe the flame:

(a) The luminous part of the flame is observed, and the side of this zone, which is toward the unburned gas, is used for measurement (direct photograph).
(b) A shadowgraph picture is taken.
(c) A Schlieren picture is taken.
(d) Interferometry (a less frequently used method).

Which surface in the flame does each method give? Again consider the temperature distribution through the flame as given in Fig. 15. The luminous zone comes late in the flame and thus is generally not satisfactory.

A shadowgraph picture measures the derivative of the density gradient $(\partial \rho / \partial x)$ or $(-1/T^2)(\partial T / \partial x)$; i.e., it evaluates $\{\partial [(-1/T^2)(\partial T / \partial x)]/\partial x\} = (2/T^3)(\partial T / \partial x)^2 - (1/T^2)(\partial^2 T / \partial x^2)$. Shadowgraphs, therefore measure the earliest variational front and do not precisely specify a surface. Actually, it is possible to define two shadowgraph surfaces—one at the unburned side and one on the burned side. The inner cone is much brighter than the outer cone, since the absolute value for the expression above is greater when evaluated at T_0 than at T_f.

Schlieren photography gives simply the density gradient $(\partial \rho / \partial x)$ or $(-1/T^2)(\partial T / \partial x)$, which has the greatest value about the inflection point of the temperature curve; it also corresponds more closely to the ignition temperature. This surface

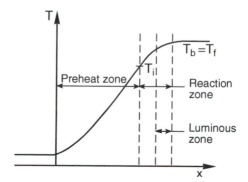

FIGURE 15 Temperature regimes in a laminar flame.

lies quite early in the flame, is more readily definable than most images, and is recommended and preferred by many workers. Interferometry, which measures density or temperature directly, is much too sensitive and can be used only on two-dimensional flames. In an exaggerated picture of a Bunsen tube flame, the surfaces would lie as shown in Fig. 16.

The various experimental configurations used for flame speeds may be classified under the following headings:

(a) Conical stationary flames on cylindrical tubes and nozzles
(b) Flames in tubes
(c) Soap bubble method
(d) Constant volume explosion in spherical vessel
(e) Flat flame methods

The methods are listed in order of decreasing complexity of flame surface and correspond to an increasing complexity of experimental arrangement. Each has certain advantages that attend its usage.

a. Burner Method

In this method premixed gases flow up a jacketed cylindrical tube long enough to ensure streamline flow at the mouth. The gas burns at the mouth of the tube, and the shape of the Bunsen cone is recorded and measured by various means and in various ways. When shaped nozzles are used instead of long tubes, the flow is uniform instead of parabolic and the cone has straight edges. Because of the complicated flame surface, the different procedures used for measuring the flame cone have led to different results.

The burning velocity is not constant over the cone. The velocity near the tube wall is lower because of cooling by the walls. Thus, there are lower temperatures, which lead to lower reaction rates and, consequently, lower flame speeds. The top

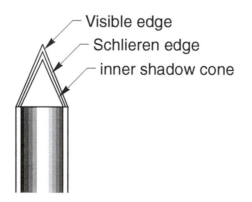

Visible edge
Schlieren edge
inner shadow cone

FIGURE 16   Optical fronts in a Bunsen burner flame.

of the cone is crowded owing to the large energy release; therefore, reaction rates are too high.

It has been found that 30% of the internal portion of the cone gives a constant flame speed when related to the proper velocity vector, thereby giving results comparable with other methods. Actually, if one measures S_L at each point, one will see that it varies along every point for each velocity vector, so it is not really constant. This variation is the major disadvantage of this method.

The earliest procedure of calculating flame speed was to divide the volume flow rate (cm^3/s^{-1}) by the area (cm^2) of flame cone:

$$S_L = \frac{Q}{A} \frac{cm^3 \, s^{-1}}{cm^2} = cm \, s^{-1}$$

It is apparent, then, that the choice of cone surface area will give widely different results. Experiments in which fine magnesium oxide particles are dispersed in the gas stream have shown that the flow streamlines remain relatively unaffected until the Schlieren cone, then diverge from the burner axis before reaching the visible cone. These experiments have led many investigators to use the Schlieren cone as the most suitable one for flame speed evaluation.

The shadow cone is used by many experimenters because it is much simpler than the Schlieren techniques. Moreover, because the shadow is on the cooler side, it certainly gives more correct results than the visible cone. However, the flame cone can act as a lens in shadow measurements, causing uncertainties to arise with respect to the proper cone size.

Some investigators have concentrated on the central portion of the cone only, focusing on the volume flow through tube radii corresponding to this portion. The proper choice of cone is of concern here also.

The angle the cone slant makes with the burner axis can also be used to determine S_L (see Fig. 17). This angle should be measured only at the central portion of the cone. Thus $S_L = u_u \sin \alpha$.

Two of the disadvantages of the burner methods are

1. Wall effects can never be completely eliminated.
2. A steady source of gas supply is necessary, which is hard to come by for rare or pure gases.

The next three methods to be discussed make use of small amounts of gas.

b. Cylindrical Tube Method

In this method, a gas mixture is placed in a horizontal tube opened at one end; then the mixture is ignited at the open end of the tube. The rate of progress of the flame into the unburned gas is the flame speed. The difficulty with this method is that, owing to buoyancy effects, the flame front is curved. Then the question arises

FIGURE 17   Velocity vectors in a Bunsen core flame.

as to which flame area to use. The flame area is no longer a geometric image of the tube; if it is hemispherical, $S_L A_f = u_m \pi R^2$. Closer observation also reveals quenching at the wall. Therefore, the unaffected center mixes with an affected peripheral area.

Because a pressure wave is established by the burning (recall that heating causes pressure change), the statement that the gas ahead of the flame is not affected by the flame is incorrect. This pressure wave causes a velocity in the unburned gases, so one must account for this movement. Therefore, since the flame is in a moving gas, this velocity must be subtracted from the measured value. Moreover, friction effects downstream generate a greater pressure wave; therefore, length can have an effect. One can deal with this by capping the end of the tube, drilling a small hole in the cap, and measuring the efflux with a soap solution [18]. The rate of growth of the resultant soap bubble is used to obtain the velocity exiting the tube, and hence the velocity of unburned gas. A restriction at the open end minimizes effects due to the back flow of the expanding burned gases.

These adjustments permit relatively good values to be obtained, but still there are errors due to wall effects and distortion due to buoyancy. This buoyancy effect can be remedied by turning the tube vertically.

c. Soap Bubble Method

In an effort to eliminate wall effects, two spherical methods were developed. In the one discussed here, the gas mixture is contained in a soap bubble and ignited at the center by a spark so that a spherical flame spreads radially through the mixture. Because the gas is enclosed in a soap film, the pressure remains constant. The growth of the flame front along a radius is followed by some photographic means. Because, at any stage of the explosion, the burned gas behind the flame occupies a larger volume than it did as unburned gas, the fresh gas into which the flame is burning moves outward. Then

$$S_L A \rho_0 = u_r A \rho_f$$

$$\left.\begin{pmatrix} \text{amount of material} \\ \text{that must go into} \\ \text{flame to increase} \\ \text{volume} \end{pmatrix}\right) = \text{velocity observed}$$

$$S_L = u_r(\rho_f/\rho_0)$$

The great disadvantage is the large uncertainty in the temperature ratio T_0/T_f necessary to obtain ρ_f/ρ_0. Other disadvantages are the facts that (1) the method can only by used for fast flames to avoid the convective effect of hot gases, and (2) the method cannot work with dry mixtures.

d. Closed Spherical Bomb Method

The bomb method is quite similar to the bubble method except that the constant volume condition causes a variation in pressure. One must, therefore, follow the pressure simultaneously with the flame front.

As in the soap bubble method, only fast flames can be used because the adiabatic compression of the unburned gases must be measured in order to calculate the flame speed. Also, the gas into which the flame is moving is always changing; consequently, both the burning velocity and flame speed vary throughout the explosion. These features make the treatment complicated and, to a considerable extent, uncertain.

The following expression has been derived for the flame speed [19]:

$$S_L = \left[1 - \frac{R^3 - r^3}{3 p \gamma_u r^2} \frac{dp}{dr} \right] \frac{dr}{dt}$$

where R is the sphere radius and r is the radius of spherical flames at any moment. The fact that the second term in the brackets is close to 1 makes it difficult to attain high accuracy.

e. Flat Flame Burner Method

The flame burner method is usually attributed to Powling [20]. Because it offers the simplest flame front—one in which the area of shadow, Schlieren, and visible fronts are all the same—it is probably the most accurate.

By placing either a porous metal disk or a series of small tubes (1 mm or less in diameter) at the exit of the larger flow tube, one can create suitable conditions for flat flames. The flame is usually ignited with a high flow rate, then the flow or composition is adjusted until the flame is flat. Next, the diameter of the flame is measured, and the area is divided into the volume flow rate of unburned gas. If the velocity emerging is greater than the flame speed, one obtains a cone due to the larger flame required. If velocity is too slow, the flame tends to flash back and is quenched. In order to accurately define the edges of the flame, an inert gas is usually flowed around the burners. By controlling the rate of efflux of burned gases with a grid, a more stable flame is obtained. This experimental apparatus is illustrated in Fig. 18.

As originally developed by Powling, this method was applicable only to mixtures having low burning velocities of the order of 15 cm/s and less. At higher burning velocities, the flame front positions itself too far from the burner and takes a multiconical form.

Later, however, Spalding and Botha [21] extended the flat flame burner method to higher flame speeds by cooling the plug. The cooling draws the flame front closer to the plug and stabilizes it. Operationally, the procedure is as follows. A flow rate giving a velocity greater than the flame speed is set, and the cooling is controlled until a flat flame is obtained. For a given mixture ratio many cooling rates are used. A plot of flame speed versus cooling rate is made and extrapolated to zero cooling rate (Fig. 19). At this point the adiabatic flame speed S_L is obtained. This procedure can be used for all mixture ratios within the flammability limits. This procedure is superior to the other methods because the heat that is generated leaks to the porous plug, not to the unburned gases as in the other model. Thus, quenching occurs all along the plug, not just at the walls.

The temperature at which the flame speed is measured is calculated as follows. For the approach gas temperature, one calculates what the initial temperature would have been if there were no heat extraction. Then the velocity of the mixture, which would give the measured mass flow rate at this temperature, is determined. This velocity is S_L at the calculated temperature. Detailed descriptions of various burned systems and techniques are to be found in Ref. [22].

A similar flat flame technique—one that does not require a heat loss correction—is the so-called opposed jet system. This approach to measuring flame speeds was introduced to determine the effect of flame stretch on the measured laminar flame velocity. The concept of *stretch* was introduced in attempts to understand the effects of turbulence on the mass burning rate of premixed systems. (This subject is considered in more detail in Section 4.E.) The technique uses two

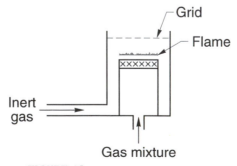

FIGURE 18   Flat flame burner apparatus.

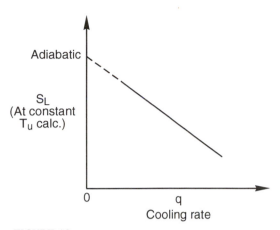

FIGURE 19   Cooling effect in flat flame burner apparatus.

opposing jets of the same air–fuel ratio to create an almost planar stagnation plane with two flat flames on both sides of the plane. For the same mixture ratio, stable flames are created for different jet velocities. In each case, the opposing jets have the same exit velocity. The velocity leaving a jet decreases from the jet exit toward the stagnation plane. This velocity gradient is related to the stretch affecting the flames: the larger the gradient, the greater the stretch. Measurements are made for different gradients for a fixed mixture. A plot is then made of the flame velocity as a function of the calculated stress function (velocity gradient), and the values are extrapolated to zero velocity gradient. The extrapolated value is considered to be the flame velocity free from any stretch effects—a value that can be compared to theoretical calculations which do not account for the stretch factor. The same technique is used to evaluate diffusion flames in which one jet contains the fuel and the other the oxidizer. Figures depicting opposed jet systems are shown in Chapter 6. The effect of stretch on laminar premixed flame speeds is generally slight for most fuels in air.

6. Experimental Results—Physical and Chemical Effects

The Mallard–Le Chatelier development for the laminar flame speed permits one to determine the general trends with pressure and temperature. When an overall rate expression is used to approximate real hydrocarbon oxidation kinetics experimental results, the activation energy of the overall process is found to be quite high—of the order of 160 kJ/mol. Thus, the exponential in the flame speed equation is quite sensitive to variations in the flame temperature. This sensitivity is the dominant temperature effect on flame speed. There is also, of course, an effect of temperature on the diffusivity; generally, the diffusivity is considered to vary with the temperature to the 1.75 power.

The pressure dependence of flame speed as developed from the thermal approaches was given by the expression

$$S_L \sim (p^{(n-2)})^{1/2} \tag{22}$$

where n was the overall order of reaction. Thus, for second-order reactions the flame speed appears independent of pressure. In observing experimental measurements of flame speed as a function of pressure, one must determine whether the temperature was kept constant with inert dilution. As the pressure is increased, dissociation decreases and the temperature rises. This effect must be considered in the experiment. For hydrocarbon–air systems, however, the temperature varies little from atmospheric pressure and above due to a minimal amount of dissociation. There is a more pronounced temperature effect at subatmospheric pressures.

To a first approximation one could perhaps assume that hydrocarbon–air reactions are second-order. Although it is impossible to develop a single overall rate expression for the complete experimental range of temperatures and pressures used by various investigators, values have been reported and hold for the limited experimental ranges of temperature and pressure from which the expression was derived. The overall reaction orders reported range from 1.5 to 2.0, and most results are around 1.75 [2, 23]. Thus, it is not surprising that experimental results show a decline in flame speed with increasing pressure [2].

As briefly mentioned earlier, with the background developed in the detailed studies of hydrocarbon oxidation, it is possible to explain this pressure trend more thoroughly. Recall that the key chain branching reaction in any hydrogen-containing system is the following reaction [Chapter 3, reaction (15)]:

$$H + O_2 \rightarrow O + OH \tag{62}$$

Any process that reduces the H atom concentration and any reaction that competes with reaction (62) for H atoms will tend to reduce the overall oxidation rate; that is, it will inhibit combustion. As discussed in Chapter 3 [reaction (21)], reaction (63)

$$H + O_2 + M \rightarrow HO_2 + M \tag{63}$$

competes directly with reaction (62). Reaction (63) is third-order and therefore much more pressure-dependent than reaction (62). Consequently, as pressure is increased, reaction (63) essentially inhibits the overall reaction and reduces the flame speed. Figure 20 reports the results of some analytical calculations of flame speeds in which detailed kinetics were included; the results obtained are quite consistent with recent measurements [2]. For pressures below atmospheric, there is only a very small decrease in flame speed as the pressure is increased; and at higher pressure (1–5 atm), the decline in S_L with increasing pressure becomes more pronounced. The reason for this change of behavior is twofold. Below atmospheric pressure, reaction (63) does not compete effectively with reaction (62) and any decrease due to reaction (63) is compensated by a rise in temperature. Above

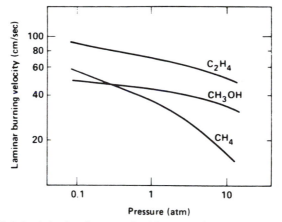

FIGURE 20   Variation in laminar flame speeds with pressure for some stoichiometric fuel–air mixtures (after Westbrook and Dryer [2]).

1 atm reaction (63) competes very effectively with reaction (62); temperature variation with pressure in this range is slight, and thus a steeper decline in S_L with pressure is found. Since the kinetic and temperature trends with pressure exist for all hydrocarbons, the same pressure effect on S_L will exist for all such fuels.

Even though S_L decreases with increasing pressure for the conditions described, $\dot{m}_0$ increases with increasing pressure because of the effect of pressure on ρ_0. And for higher O_2 concentrations, the temperature rises substantially, about 30% for pure O_2; thus the point where reaction (63) can affect the chain branching step reaction (62) goes to much higher pressure. Consequently, in oxygen-rich systems S_L almost always increases with pressure.

The variation of flame speed with equivalence ratio follows the variation with temperature. Since flame temperatures for hydrocarbon–air systems peak slightly on the fuel-rich side of stoichiometric (as discussed in Chapter 1), so do the flame speeds. In the case of hydrogen–air systems, the maximum S_L falls well on the fuel-rich side of stoichiometric, since excess hydrogen increases the thermal diffusivity substantially. Hydrogen gas with a maximum value of 325 cm/s has the highest flame speed in air of any other fuel.

Reported flame speed results for most fuels vary somewhat with the measurement technique used. Most results, however, are internally consistent. Plotted in Fig. 21 are some typical flame speed results as a function of the stoichiometric mixture ratio. Detailed data, which were given in recent combustion symposia, are available in the extensive tabulations of Refs. [24–26]. The flame speeds for many fuels in air have been summarized from these references and are listed in Appendix D. Since most paraffins, except methane, have approximately the same flame temperature in air, it is not surprising that their flame speeds are about the same (~45 cm/s). Methane has a somewhat lower speed (< 40 cm/s). Attempts [24]

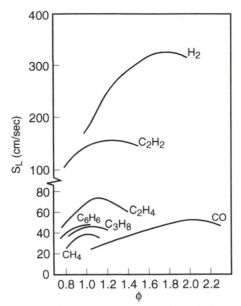

FIGURE 21   General variation in laminar flame speeds with equivalence ratio ϕ for various fuel–air systems at $P = 1$ atm and $T_0 = 298$ K.

have been made to correlate flame speed with hydrocarbon fuel structure and chain length, but these correlations appear to follow the general trends of temperature. Olefins, having the same C/H ratio, have the same flame temperature (except for ethene, which is slightly higher) and have flame speeds of approximately 50 cm/s. In this context ethene has a flame speed of approximately 75 cm/s. Owing to its high flame temperature, acetylene has a maximum flame speed of about 160 cm/s. Molecular hydrogen peaks far into the fuel-rich region because of the benefit of the fuel diffusivity. Carbon monoxide favors the rich side because the termination reaction $H + CO + M \rightarrow HCO + M$ is a much slower step than the termination step $H + O_2 + M \rightarrow HO_2 + M$ which would prevail in the lean region.

   The variation of flame speed with oxygen concentration poses further questions about the factors that govern the flame speed. Shown in Fig. 22 is the flame speed of a fuel in various oxygen–nitrogen mixtures relative to its value in air. Note the 10-fold increase for methane between pure oxygen and air, the 7.5-fold increase for propane, the 3.4-fold increase for hydrogen, and the 2.4-fold increase for carbon monoxide. From the effect of temperature on the overall rates and diffusivities, one would expect about a 5-fold increase for all these fuels. Since the CO results contain a fixed amount of hydrogen additives [24], the fact that the important OH radical concentration does not increase as much as expected must play a role in the lower rise. Perhaps for general considerations the hydrogen values are near enough to a general estimate. Indeed, there is probably a sufficient

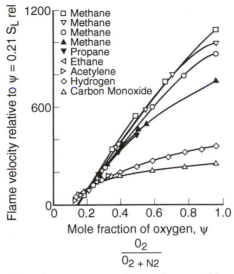

FIGURE 22   Relative effect of oxygen concentrations on flame speed for various fuel–air systems at $P = 1$ atm and $T_0 = 298$ K (after Zebatakis [25]).

radical pool at all oxygen concentrations. For the hydrocarbons, the radical pool concentration undoubtedly increases substantially as one goes to pure oxygen for two reasons—increased temperature and no nitrogen dilution. Thus, applying the same general rate expression for air and oxygen just does not suffice.

The effect of the initial temperature of a premixed fuel–air mixture on the flame propagation rate again appears to be reflected through the final flame temperature. Since the chemical energy release is always so much greater than the sensible energy of the reactants, small changes of initial temperature generally have little effect on the flame temperature. Nevertheless, the flame propagation expression contains the flame temperature in an exponential term; thus, as discussed many times previously, small changes in flame temperature can give noticeable changes in flame propagation rates. If the initial temperatures are substantially higher than normal ambient, the rate of reaction (63) can be reduced in the preheat zone. Since reaction (63) is one of recombination, its rate decreases with increasing temperature, and so the flame speed will be attenuated even further.

Perhaps the most interesting set of experiments to elucidate the dominant factors in flame propagation was performed by Clingman *et al.* [27]. Their results clearly show the effect of the thermal diffusivity and reaction rate terms. These investigators measured the flame propagation rate of methane in various oxygen–inert gas mixtures. The mixtures of oxygen to inert gas were 0.21/0.79 on a volumetric basis, the same as that which exists for air. The inerts chosen were nitrogen (N_2), helium (He), and argon (Ar). The results of these experiments are shown in Fig. 23.

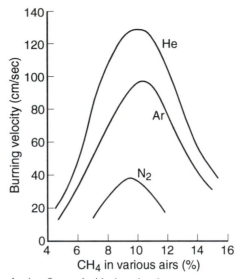

FIGURE 23   Methane laminar flame velocities in various inert gas–oxygen mixtures (after Clingman *et al.* [27]).

   The trends of the results in Fig. 23 can be readily explained. Argon and nitrogen have thermal diffusivities that are approximately equal. However, Ar is a monatomic gas whose specific heat is lower than that of N_2. Since the heat release in all systems is the same, the final (or flame) temperature will be higher in the Ar mixture than in the N_2 mixture. Thus, S_L will be higher for Ar than for N_2. Argon and helium are both monatomic, so their final temperatures are equal. However, the thermal diffusivity of He is much greater than that of Ar. Helium has a higher thermal conductivity and a much lower density than argon. Consequently, S_L for He is much greater than that for Ar.
   The effect of chemical additives on the flame speed has also been explored extensively. Leason [28] has reported the effects on flame velocity of small concentrations of additive ($< 3\%$) and other fuels. He studied the propane–air flame. Among the compounds considered were acetone, acetaldehyde, benzaldehyde, diethyl ether, benzene, and carbon disulfide. In addition, many others were chosen from those classes of compounds that were shown to be oxidation intermediates in low-temperature studies; these compounds were expected to decrease the induction period and, thus, increase the flame velocity. Despite differences in apparent oxidation properties, all the compounds studied changed the flame velocity in exactly the same way that dilution with excess fuel would on the basis of oxygen requirement. These results support the contention that the laminar flame speed is controlled by the high-temperature reaction region. The high temperatures generate more than ample radicals via chain branching, so it is unlikely that any additive could contribute any reaction rate accelerating feature.

There is, of course, a chemical effect in carbon monoxide flames. This point was mentioned in the discussion of carbon monoxide explosion limits. Studies have shown that CO flame velocities increase appreciably when small amounts of hydrogen, hydrogen-containing fuels, or water are added. For 45% CO in air, the flame velocity passes through a maximum after approximately 5% by volume of water has been added. At this point, the flame velocity is 2.1 times the value with 0.7% H_2O added. After the 5% maximum is attained a dilution effect begins to cause a decrease in flame speed. The effect and the maximum arise because a sufficient steady-state concentration of OH radicals must be established for the most effective explosive condition.

Although it may be expected that the common antiknock compounds would decrease the flame speed, no effects of antiknocks have been found in constant pressure combustion. The effect of the inhibition of the preignition reaction on flame speed is of negligible consequence. There is no universal agreement on the mechanism of antiknocks, but it has been suggested that they serve to decrease the radical concentrations by absorption on particle surfaces (see Chapter 2). The reduction of the radical concentration in the preignition reactions or near the flammability limits can severely affect the ability to initiate combustion. In these cases the radical concentrations are such that the chain branching factor is very close to the critical value for explosion. Any reduction could prevent the explosive condition from being reached. Around the stoichiometric mixture ratio, the radical concentrations are normally so great that it is most difficult to add any small amounts of additives that would capture enough radicals to alter the reaction rate and the flame speed.

Certain halogen compounds, such as the Freons, are known to alter the flammability limits of hydrocarbon–air mixtures. The accepted mechanism is that the halogen atoms trap hydrogen radicals necessary for the chain branching step. Near the flammability limits, conditions exist in which the radical concentrations are such that the chain branching factor α is just above α_{crit}. Any reduction in radicals and the chain branching effects these radicals engender could eliminate the explosive (fast reaction rate and larger energy release rate) regime. However, small amounts of halogen compounds do not seem to affect the flame speed in a large region around the stoichiometric mixture ratio. The reason is, again, that in this region the temperatures are so high and radicals so abundant that elimination of some radicals does not affect the reaction rate.

It has been found that some of the larger halons (the generic name for the halogenated compounds sold under commercial names such as Freon) are effective flame suppressants. Also, some investigators have found that inert powders are effective in fire fighting. Fundamental experiments to evaluate the effectiveness of the halons and powders have been performed with various types of apparatus that measure the laminar flame speed. Results have indicated that the halons and the powders reduce flame speeds even around the stoichiometric air–fuel ratio. The investigators performing these experiments have argued that those agents are effective because they reduce the radical concentrations. However, this explanation could be questioned. The quantities of these added agents are great enough that they could absorb sufficient amounts of heat to reduce the temperature and hence the flame speed. Both halons and powders have large total heat capacities.

D. STABILITY LIMITS OF LAMINAR FLAMES

There are two types of stability criteria associated with laminar flames. The first is concerned with the ability of the combustible fuel–oxidizer mixture to support flame propagation and is strongly related to the chemical rates in the system. In this case a point can be reached for a given limit mixture ratio in which the rate of reaction and its subsequent heat release are not sufficient to sustain reaction and, thus, propagation. This type of stability limit includes (1) flammability limits in which gas-phase losses of heat from limit mixtures reduce the temperature, rate of heat release, and the heat feedback, so that the flame is not permitted to propagate; and (2) quenching distances in which the loss of heat to a wall and radical quenching at the wall reduce the reaction rate so that it cannot sustain a flame in a confined situation such as propagation in a tube.

The other type of stability limit is associated with the mixture flow and its relationship to the laminar flame itself. This stability limit, which includes the phenomena of flashback, blowoff, and the onset of turbulence, describes the limitations of stabilizing a laminar flame in a real experimental situation.

1. Flammability Limits

The explosion limit curves presented earlier and most of those that appear in the open literature are for a definite fuel–oxidizer mixture ratio, usually stoichiometric. For the stoichiometric case, if an ignition source is introduced into the mixture even at a very low temperature and at reasonable pressures (e.g., $\sim$1 atm), the gases about the ignition source reach a sufficient temperature so that the local mixture moves into the explosive region and a flame propagates. This flame, of course, continues to propagate even after the ignition source is removed. There are mixture ratios, however, that will not self-support the flame after the ignition source is removed. These mixture ratios fall at the lean and rich end of the concentration spectrum. The leanest and richest concentrations that will just self-support a flame are called the lean and rich flammability limits, respectively. The primary factor that determines the flammability limit is the competition between the rate of heat generation, which is controlled by the rate of reaction and the heat of reaction for the limit mixture, and the external rate of heat loss by the flame. The literature reports flammability limits in both air and oxygen. The lean limit rarely differs for air or oxygen, as the excess oxygen in the lean condition has the same thermophysical properties as nitrogen.

Some attempts to standardize the determination of flammability limits have been made. Coward and Jones [29] recommended that a 2-inch glass tube about

4 feet long be employed; such a tube should be ignited by a spark a few millimeters long or by a small naked flame. The high-energy starting conditions are such that weak mixtures will be sure to ignite. The large tube diameter is selected because it gives the most consistent results. Quenching effects may interfere in tubes of small diameter. Large diameters create some disadvantages since the quantity of gas is a hazard and the possibility of cool flames exists. The 4-foot length is chosen in order to allow an observer to truly judge whether the flame will propagate indefinitely or not.

It is important to specify the direction of flame propagation. Since it may be assumed as an approximation that a flame cannot propagate downward in a mixture contained within a vertical tube if the convection current it produces is faster than the speed of the flame, the limits for upward propagation are usually slightly wider than those for downward propagation or those for which the containing tube is in a horizontal position.

Table 3 lists some upper and lower flammability limits (in air) taken from Refs. [24 and 25] for some typical combustible compounds. Data for other fuels are given in Appendix E.

In view of the accelerating effect of temperature on chemical reactions, it is reasonable to expect that limits of flammability should be broadened if the temperature is increased. This trend is confirmed experimentally. The increase is slight and it appears to give a linear variation for hydrocarbons.

As noted from the data in Appendix E, the upper limit for many fuels is about three times stoichiometric and the lower limit is about 50% of stoichiometric. Generally, the upper limit is higher than that for detonation. The lower (lean) limit of a gas is the same in oxygen as in air owing to the fact that the excess oxygen has the same heat capacity as nitrogen. The higher (rich) limit of all flammable gases is much greater in oxygen than in air, due to higher temperature, which comes about from the absence of any nitrogen. Hence, the range of flammability is always greater in oxygen. Table 4 shows this effect.

TABLE 3 Flammability Limits of Some Fuels in Air[a]

	Lower (lean)	Upper (rich)	Stoichiometric
Methane	5	15	9.47
Heptane	1	6.7	1.87
Hydrogen	4	75	29.2
Carbon Monoxide	12.5	74.2	29.5
Acetaldehyde	4.0	60	7.7
Acetylene	2.5	100	7.7
Carbon Disulfide	1.3	50	7.7
Ethylene Oxide	3.6	100	7.7

[a] Volume percent.

TABLE 4   Comparison of Oxygen and Air Flammability Limits[a]

	Lean			Rich	
	Air	O_2		Air	O_2
H_2	4	4		75	94
CO	12	16		74	94
NH_3	15	15		28	79
CH_4	5	5		15	61
C_3H_8	2	2		10	55

[a]Fuel volume percent.

As increasing amounts of an incombustible gas or vapor are added to the atmosphere, the flammability limits of a gaseous fuel in the atmosphere approach one another and finally meet. Inert diluents such as CO_2, N_2, or Ar merely replace part of the O_2 in the mixture, but these inert gases do not have the same extinction power. It is found that the order of efficiency is the same as that of the heat capacities of these three gases:

$$CO_2 > N2 > Ar \,(or\, He)$$

For example, the minimum percentage of oxygen that will permit flame propagation in mixtures of CH_4, O_2, and CO_2 is 14.6%; if N_2 is the diluent, the minimum percentage of oxygen is less and equals 12.1%. In the case of Ar, the value is 9.8%. As discussed, when a gas of higher specific heat is present in sufficient quantities, it will reduce the final temperature, thereby reducing the rate of energy release that must sustain the rate of propagation over other losses.

It is interesting to examine in more detail the effect of additives as shown in Fig. 24 [25]. As discussed, the general effect of the nonhalogenated additives follows the variation in the molar specific heat; that is, the greater the specific heat of an inert additive, the greater the effectiveness. Again, this effect is strictly one of lowering the temperature to a greater extent; this was verified as well by flammability measurements in air where the nitrogen was replaced by carbon dioxide and argon. Figure 24, however, represents the condition in which additional species were added to the air in the fuel–air mixture. As noted in Fig. 24, rich limits are more sensitive to inert diluents than lean limits; however, species such as halogenated compounds affect both limits and this effect is greater than that expected from heat capacity alone. Helium addition extends the lean limit somewhat because it increases the thermal diffusivity and, thus, the flame speed.

That additives affect the rich limit more than the lean limit can be explained by the important competing steps for possible chain branching. When the system is rich [Chapter 3, reaction (23)],

$$H + H + M \rightarrow H_2 + M \tag{64}$$

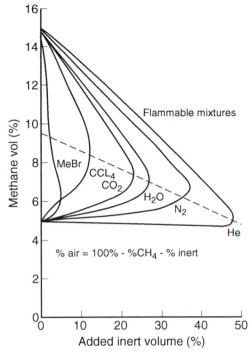

FIGURE 24 Limits of flammability of various methane–inert gas–air mixtures at $P = 1$ and $T_0 = 298$ K (after Zebatakis [25]).

competes with [Chapter 3, reaction (15)]

$$H + O_2 \rightarrow OH + O \tag{62}$$

The recombination [reaction (64)] increases with decreasing temperature and increasing concentration of the third body M. Thus, the more diluent added, the faster this reaction is compared to the chain branching step [reaction (62)]. This aspect is also reflected in the overall activation energy found for rich systems compared to lean systems. Rich systems have a much higher overall activation energy and therefore a greater temperature sensitivity.

The effect of all halogen compounds on flammability limits is substantial. The addition of only a few percent can make some systems nonflammable. These observed results support the premise that the effect of halogen additions is not one of dilution alone, but rather one in which the halogens act as catalysts in reducing the H atom concentration necessary for the chain branching reaction sequence. Any halogen added—whether in elemental form, as hydrogen halide, or bound in an organic compound—will act in the same manner. Halogenated hydrocarbons have weak carbon–halogen bonds that are readily broken in flames of any respectable temperature, thereby placing a halogen atom in the reacting system. This halogen atom rapidly abstracts a hydrogen from the hydrocarbon fuel to form the hydrogen halide; then the following reaction system, in which X represents any of the halogens F, Cl, Br, or I, occurs:

$$HX + H \rightarrow H_2 + X \tag{65}$$

$$X + X + M \rightarrow X_2 + M \tag{66}$$

$$X_2 + H \rightarrow HX + X \tag{67}$$

Reactions (65)–(67) total overall to

$$H + H \rightarrow H_2$$

and thus it is seen that X is a homogeneous catalyst for recombination of the important H atoms. What is significant in the present context is that the halide reactions above are fast compared to the other important H atom reactions such as

$$H + O_2 \rightarrow O + OH \tag{62}$$

or

$$H + RH \rightarrow R + H_2 \tag{68}$$

This competition for H atoms reduces the rate of chain branching in the important $H + O_2$ reaction. The real key to this type of inhibition is the regeneration of X_2, which permits the entire cycle to be catalytic.

Because sulfur dioxide (SO_2) essentially removes O atoms catalytically by the mechanism

$$SO_2 + O + M \rightarrow SO_3 + M \tag{69}$$

$$SO_3 + O \rightarrow SO_2 + O_2 \tag{70}$$

and also by H radical removal by the system

$$SO_2 + H + M \rightarrow HSO_2 + M \tag{71}$$

$$HSO_2 + OH \rightarrow SO_2 + H_2O \tag{72}$$

and by

$$SO_2 + O + M \rightarrow SO_3 + M \tag{73}$$

$$SO_3 + H + M \rightarrow HSO_3 + M \tag{74}$$

$$HSO_3 + H \rightarrow SO_2 + H_2O \tag{75}$$

SO_2 is similarly a known inhibitor that affects flammability limits. These catalytic cycles [reactions (69)–(70), reactions (71)–(72), and reactions (73)–(75)]

are equivalent to

$$O + O \rightarrow O_2$$

$$H + OH \rightarrow H_2O$$

$$H + H + O \rightarrow H_2O$$

The behavior of flammability limits at elevated pressures can be explained somewhat satisfactorily. For simple hydrocarbons (ethane, propane, . . . , pentane), it appears that the rich limits extend almost linearly with increasing pressure at a rate of about 0.13 vol%/atm; the lean limits, on the other hand, are at first extended slightly and thereafter narrowed as pressure is increased to 6 atm. In all, the lean limit appears not to be affected appreciably by the pressure. Figure 25 for natural gas in air shows the pressure effect for conditions above atmospheric.

Most early studies of flammability limits at reduced pressures indicated that the rich and lean limits converge as the pressure is reduced until a pressure is reached below which no flame will propagate. However, this behavior appears to be due to wall quenching by the tube in which the experiments were performed. As shown in Fig. 26, the limits are actually as wide at low pressure as at 1 atm, provided the tube is sufficiently wide and an ignition source can be found to ignite the mixtures. Consequently, the limits obtained at reduced pressures are not generally true limits of flammability, since they are influenced by the tube diameter. Therefore, these limits are not physicochemical constants of a given fuel. Rather, they are limits of flame propagation in a tube of specified diameter.

In examining the effect of high pressures on flammability limits, it has been assumed that the limit is determined by a critical value of the rate of heat loss to the rate of heat development. Consider, for example, a flame anchored on a Bunsen

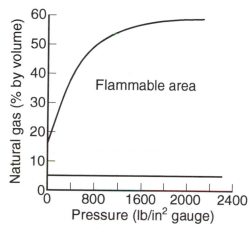

FIGURE 25 Effect of pressure increase above atmospheric pressure on flammability limits of natural gas–air mixtures (from Lewis and von Elbe [5]).

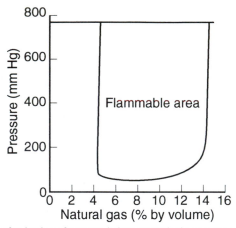

FIGURE 26   Effect of reduction of pressure below atmospheric pressure on flammability limits of natural gas–air mixtures (from Lewis and von Elbe [5]).

tube. The loss to the anchoring position is small, and thus the radiation loss must be assumed to be the major heat loss condition. This radiative loss is in the infrared, due primarily to the band radiation systems of CO_2, H_2O and CO. The amount of product composition changes owing to dissociation at the flammability limits is indeed small, so there is essentially no increase in temperature with pressure. Even so, with temperatures near the limits and wavelengths of the gaseous radiation, the radiation bands lie near or at the maximum of the energy density radiation distribution given by Planck's law. If λ is the wavelength, then $\lambda_{max}T$ equals a constant, by Wien's law. Thus the radiant loss varies as T^5. But for most hydrocarbon systems the activation energy of the reaction media and temperature are such that the variation of $\exp(-E/RT)$ as a function of temperature is very much like a T^5 variation [30]. Thus, any effect of pressure on temperature shows a balance of these loss and gain terms, except that the actual radiation contains an emissivity term. Due to band system broadening and emitting gas concentration, this emissivity is approximately proportional to the total pressure for gaseous systems. Then, as the pressure increases, the emissivity and heat loss increase monotonically. On the fuel-rich side the reaction rate is second-order and the energy release increases with P^2 as compared to the heat loss that increases with P. Thus the richness of the system can be increased as the pressure increases before extinction occurs [30]. For the methane flammability results reported in Fig. 25, the rich limit clearly broadens extensively and then begins to level off as the pressure is increased over a range of about 150 atm. The leveling-off happens when soot formation occurs. The soot increases the radiative loss. The lean limit appears not to change with pressure, but indeed broadens by about 25% over the same pressure range. Note that over a span of 28 atm, the rich limit broadens about 300% and the lean limit only about 1%. There is no definitive explanation of this difference; but, considering the size, it could possibly be related to the temperature because of its exponential effect on the energy release rate and the emissive power of the product gases. The rule of thumb quoted earlier that the rich limit is about three times the stoichiometric value and the lean limit half the stoichiometric value can be rationalized from the temperature effect. Burning near the rich limit generates mostly diatomic products—CO and H_2—and some H_2O. Burning near the lean limit produces CO_2 and H_2O exclusively. Thus for the same percentage composition change, regardless of the energy effect, the fuel-rich side temperature will be higher than the lean side temperature. As was emphasized in Chapter 1, for hydrocarbons the maximum flame temperature occurs on the fuel-rich side of stoichiometric owing to the presence of diatomics, particularly H_2. Considering percentage changes due to temperature, the fuel side flammability limit can broaden more extensively as one increases the pressure to account for the reaction rate compensation necessary to create the new limit. Furthermore the radiative power of the fuel-rich side products is substantially less than that of the lean side because the rich side contains only one diatomic radiator and a little water, whereas the lean side contains exclusively triatomic radiators.

The fact that flammability limits have been found [29] to be different for upward and downward propagation of a flame in a cylindrical tube if the tube is large enough could be an indication that heat losses [30, 31] are not the dominant extinction mechanism specifying the limit. Directly following a discourse by Ronney [32], it is well first to emphasize that buoyancy effects are an important factor in the flammability limits measured in large cylindrical tubes. Extinction of upward-propagating flames for a given fuel–oxidizer mixture ratio is thought to be due to flame stretch at the tip of the rising hemispherical flame [33, 34]. For downward propagation, extinction is thought to be caused by a cooling, sinking layer of burned gases near the tube wall that overtakes the weakly propagating flame front whose dilution leads to extinction [35, 36]. For small tubes, heat loss to walls can be the primary cause for extinction; indeed, such wall effects can quench the flames regardless of mixture ratio. Thus, as a generalization, flammability limits in tubes are probably caused by the combined influences of heat losses to the tube wall, buoyancy-induced curvature and strain, and even Lewis number effects. Because of the difference in these mechanisms, it has been found that the downward propagation limits can sometimes be wider than the upward limits, depending upon the degree of buoyancy and Lewis number.

It is interesting that experiments under microgravity conditions [37, 38] reveal that the flammability limits are different from those measured for either upward or downward propagation in tubes at normal gravity. Upon comparing theoretical predictions [30] to such experimental measurements as the propagation rate at the limit and the rate of thermal decay in the burned gases, Ronney [39] suggested that radiant heat loss is probably the dominant process leading to flame extinction at microgravity.

Ronney [39] concludes that, while surprising, the completely different processes dominating flammability limits at normal gravity and microgravity are readily understandable in light of the time scales of the processes involved. He showed that the characteristic loss rate time scale for upward-propagating flames in tubes (τ_u), downward-propagating flames (τ_d), radiative losses (τ_r), and conductive heat losses to the wall (τ_c) scale as $(d/g)^{1/2}$, α/g^2, $\rho c_p T_f/E$, and d^2/α, respectively. The symbols not previously defined are d, the tube diameter; g, the gravitational acceleration; α, the thermal diffusivity, and E, the radiative heat loss per unit volume. Comparison of these time scales indicates that for any practical gas mixture, pressure, and tube diameter, it is difficult to obtain $\tau_r < \tau_u$ or $\tau_r < \tau_d$ at normal gravity; thus, radiative losses are not as important as buoyancy-induced effects under this condition. At microgravity, τ_u and τ_d are very large, but still τ_r must be less than τ_c, so radiant effects are dominant. In this situation, large tube diameters are required.

2. Quenching Distance

Wall quenching affects not only flammability limits, but also ignition phenomena (see Chapter 7). The quenching diameter, d_T, which is the parameter given the greatest consideration, is generally determined experimentally in the following manner. A premixed flame is established on a burner port and the gas flow is suddenly stopped. If the flame propagates down the tube into the gas supply source, a smaller tube is substituted. The tube diameter is progressively decreased until the flame cannot propagate back to the source. Thus the quenching distance, or diameter d_T, is the diameter of the tube that just prevents flashback.

A flame is quenched in a tube when the two mechanisms that permit flame propagation—diffusion of species and of heat—are affected. Tube walls extract heat: the smaller the tube, the greater is the surface area to volume ratio within the tube and hence the greater is the volumetric heat loss. Similarly, the smaller the tube, the greater the number of collisions of the active radical species that are destroyed. Since the condition and the material composition of the tube wall affect the rate of destruction of the active species [5], a specific analytical determination of the quenching distance is not feasible.

Intuition would suggest that an inverse correlation would obtain between flame speed and quenching diameter. Since flame speed S_L varies with equivalence ratio ϕ, so should d_T vary with ϕ; however, the curve of d_T would be inverted compared to that of S_L, as shown in Fig. 27.

One would also expect, and it is found experimentally, that increasing the temperature would decrease the quenching distance. This trend arises because the heat losses are reduced with respect to heat release and species are not as readily deactivated. However, sufficient data are not available to develop any specific correlation.

It has been concretely established and derived theoretically [30] that quenching distance increases as pressure decreases; in fact, the correlation is almost exactly

$$d_T \sim 1/P$$

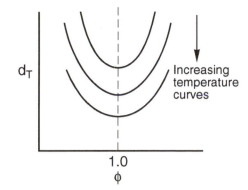

FIGURE 27   Variation of quenching diameter d_T as a function of equivalence ratio ϕ and trend with initial temperature.

for many compounds. For various fuels, P sometimes has an exponent somewhat less than 1. An exponent close to 1 in the $d_T \sim P$ relationship can be explained as follows. The mean free path of gases increases as pressure decreases; thus there are more collisions with the walls and more species are deactivated. Pressure results are generally represented in the form given in Fig. 28, which also shows that when measuring flammability limits as a function of subatmospheric pressures, one must choose a tube diameter that is greater than the d_T given for the pressure. The horizontal dot-dash line in Fig. 28 specifies the various flammability limits that would be obtained at a given subatmospheric pressure in tubes of different diameters.

3. Flame Stabilization (Low Velocity)

In the introduction to this chapter a combustion wave was considered to be propagating in a tube. When the cold premixed gases flow in a direction opposite

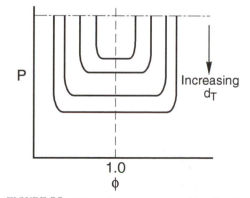

FIGURE 28   Effect of pressure on quenching diameter.

to the wave propagation and travel at a velocity equal to the propagation velocity (i.e., the laminar flame speed), the wave (flame) becomes stationary with respect to the containing tube. Such a flame would possess only neutral stability, and its actual position would drift [1]. If the velocity of the unburned mixture is increased, the flame will leave the tube and, in most cases, fix itself in some form at the tube exit. If the tube is in a vertical position, then a simple burner configuration, as shown in Fig. 29, is obtained. In essence, such burners stabilize the flame. As described earlier, these burners are so configured that the fuel and air become a homogeneous mixture before they exit the tube. The length of the tube and physical characteristics of the system are such that the gas flow is laminar in nature. In the context to be discussed here, a most important aspect of the burner is that it acts as a heat and radical sink, which stabilizes the flame at its exit under many conditions. In fact, it is the burner rim and the area close to the tube that provide the stabilization position.

When the flow velocity is increased to a value greater than the flame speed, the flame becomes conical in shape. The greater the flow velocity, the smaller is the cone angle of the flame. This angle decreases so that the velocity component of the flow normal to the flame is equal to the flame speed. However, near the burner rim the flow velocity is lower than that in the center of the tube; at some point in this area the flame speed and flow velocity equalize and the flame is anchored by this point. The flame is quite close to the burner rim and its actual speed is controlled by heat and radical loss to the wall. As the flow velocity is increased, the flame edge moves further from the burner, losses to the rim decrease and the flame speed increases so that another stabilization point is reached. When the flow is such that the flame edge moves far from the rim, outside air is entrained, a lean mixture is diluted, the flame speed drops, and the flame reaches its blowoff limit.

If, however, the flow velocity is gradually reduced, this configuration reaches a condition in which the flame speed is greater than the flow velocity at some point

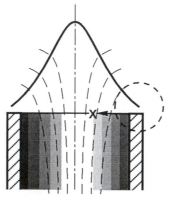

FIGURE 29   Gas mixture streamlines through a Bunsen cone flame.

across the burner. The flame will then propagate down into the burner, so that the flashback limit is reached. Slightly before the flashback limit is reached, tilted flames may occur. This situation occurs because the back pressure of the flame causes a disturbance in the flow so that the flame can enter the burner only in the region where the flow velocity is reduced. Because of the constraint provided by the burner tube, the flow there is less prone to distortion; so further propagation is prevented and a tilted flame such as that shown in Fig. 30 is established [1].

Thus it is seen that the laminar flame is stabilized on burners only within certain flow velocity limits. The following subsections treat the physical picture just given in more detail.

a. Flashback and Blowoff

Assume Poiseuille flow in the burner tube. The gas velocity is zero at the stream boundary (wall) and increases to a maximum in the center of the stream. The linear dimensions of the wall region of interest are usually very small; in slow-burning mixtures such as methane and air, they are of the order of 1 mm. Since the burner tube diameter is usually large in comparison, as shown in Fig. 31, the gas velocity near the wall can be represented by an approximately linear vector profile. Figure 31 represents the conditions in the area where the flame is anchored by the burner rim. Further assume that the flow lines of the fuel jet are parallel to

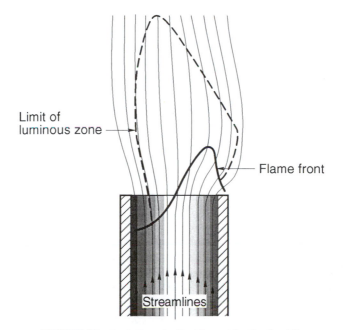

FIGURE 30   Formation of a tilted flame (after Bradley [1]).

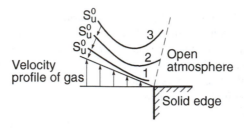

FIGURE 31   Stabilization positions of a Bunsen burner flame (after Lewis and von Elbe [5]).

the tube axis, that a combustion wave is formed in the stream, and that the fringe of the wave approaches the burner rim closely. Along the flame wave profile, the burning velocity attains its maximum value S_L°. Toward the fringe, the burning velocity decreases as heat and chain carriers are lost to the rim. If the wave fringe is very close to the rim (position 1 in Fig. 31), the burning velocity in any flow streamline is smaller than the gas velocity and the wave is driven farther away by gas flow. As the distance from the rim increases, the loss of heat and chain carriers decreases and the burning velocity becomes larger. Eventually, a position is reached (position 2 in Fig. 31) in which the burning velocity is equal to the gas velocity at some point of the wave profile. The wave is now in equilibrium with respect to the solid rim. If the wave is displaced to a larger distance (position 3 in Fig. 31), the burning velocity at the indicated point becomes larger than the gas velocity and the wave moves back to the equilibrium position.

Consider Fig. 32, a graph of flame velocity S_L as a function of distance, for a wave inside a tube. In this case, the flame has entered the tube. The distance from the burner wall is called the penetration distance d_p (half the quenching diameter d_T). If $\bar{u}_1$ is the mean velocity of the gas flow in the tube and the line labeled $\bar{u}_1$ is the graph of the velocity near the tube wall, the local flame velocity is not greater than the local gas velocity at any point; therefore, any flame that finds itself inside the tube will then blow out of the tube. At a lower velocity $\bar{u}_2$, which is just tangent to the S_L curve, a stable point is reached. Then $\bar{u}_2$ is the minimum mean velocity before flashback occurs. The line for the mean velocity $\bar{u}_3$ indicates a region where the flame speed is greater than the velocity in the tube represented by $\bar{u}_3$; in this case, the flame does flash back. The gradient for flashback, g_F, is S_L/d_p. Analytical developments [30] show that

$$d_p \approx (\lambda/c_p\rho)(1/S_L) \approx (\alpha/S_L)$$

Similar reasoning can apply to blowoff, but the arguments are somewhat different and less exact because nothing similar to a boundary layer exists. However, a free boundary does exist.

When the gas flow in the tube is increased, the equilibrium position shifts away from the rim. With increasing distance from the rim, a lean gas mixture

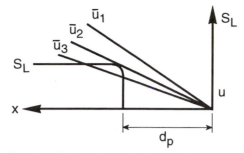

FIGURE 32   General burning velocity and gas velocity profiles inside a Bunsen burner tube (from Lewis and von Elbe [5]).

becomes progressively diluted by interdiffusion with the surrounding atmosphere, and the burning velocity in the outermost streamlines decreases correspondingly. This effect is indicated by the increasing retraction of the wave fringe for flame positions 1 to 3 in Fig. 33. But, as the wave moves farther from the rim, it loses less heat and fewer radicals to the rim, so it can extend closer to the hypothetical edge. However, an ultimate equilibrium position of the wave exists beyond which the effect of dilution overbalances the effect of increased distance from the burner rim everywhere on the burning velocity. If the boundary layer velocity gradient is so large that the combustion wave is driven beyond this position, the gas velocity exceeds the burning velocity along every streamline and the combustion wave blows off.

These trends are represented diagrammatically in Fig. 33. The diagram follows the postulated trends in which S_L° is the flame velocity after the gas has been diluted because the flame front has moved slightly past $\bar{u}_3$. Thus, there is blowoff and $\bar{u}_3$ is the blowoff velocity.

b. Analysis and Results

The topic of concern here is the stability of laminar flames fixed to burner tubes. The flow profile of the premixed gases flowing up the tube in such a system

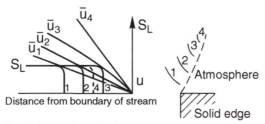

FIGURE 33   Burning velocity and gas velocity profiles above a Bunsen burner tube rim (from Lewis and von Elbe [5]).

must be parabolic; that is, Poiseuille flow exists. The gas velocity along any streamline is given by

$$u = n(R^2 - r^2)$$

where R is the tube radius. Since the volumetric flow rate, $Q(\text{cm}^3/\text{s})$ is given by

$$Q = \int_0^R 2\pi r u \, dr$$

then n must equal

$$n = 2Q/\pi R^4$$

The gradient for blowoff or flashback is defined as

$$g_{F,B} \equiv - \lim_{r \to R}(du/dr)$$

then

$$g_{F,B} = \frac{4Q}{\pi R^3} = 4\frac{\bar{u}_{av}}{R} = 8\frac{\bar{u}_{av}}{d}$$

where d is the diameter of the tube.

Most experimental data on flashback are plotted as a function of the average flashback velocity, $u_{av,F}$, as shown in Fig. 34. It is possible to estimate penetration distance (quenching thickness) from the burner wall in graphs such as Fig. 34 by observing the cut-off radius for each mixture.

The development for the gradients of flashback and blowoff suggests a more appropriate plot of $g_{B,F}$ versus ϕ, as shown in Figs. 35 and 36. Examination of these figures reveals that the blowoff curve is much steeper than that for flashback. For rich mixtures the blowoff curves continue to rise instead of decreasing after the stoichiometric value is reached. The reason for this trend is that experiments

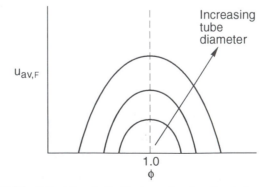

FIGURE 34   Critical flow for flashback as a function of equivalence ratio ϕ.

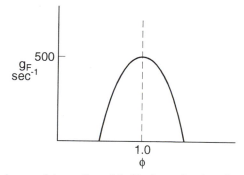

FIGURE 35 Typical curve of the gradient of flashback as a function of equivalence ratio ϕ. The value of $\phi = 1$ is for natural gas.

are performed in air, and the diffusion of air into the mixture as the flame lifts off the burner wall increases the local flame speed of the initially fuel-rich mixture. Experiments in which the surrounding atmosphere was not air, but nitrogen, verify this explanation and show that the g_B would peak at stoichiometric.

The characteristics of the lifted flame are worthy of note as well. Indeed, there are limits similar to those of the seated flame [1]. When a flame is in the lifted position, a dropback takes place when the gas velocity is reduced, and the flame takes up its normal position on the burner rim. When the gas velocity is increased instead, the flame will blow out. The instability requirements of both the seated and lifted flames are shown in Fig. 37.

4. Stability Limits and Design

The practicality of understanding stability limits is uniquely obvious when one considers the design of Bunsen tubes and cooking stoves using gaseous fuels.

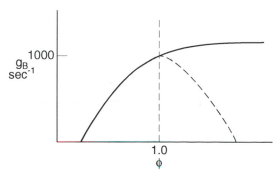

FIGURE 36 Typical curves of the gradient of blowoff as a function of equivalence ratio ϕ. The value at $\phi = 1$ is for natural gas.

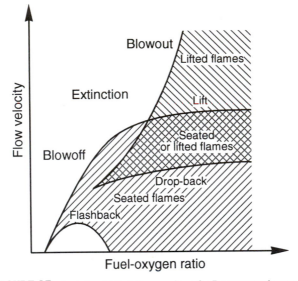

FIGURE 37   Seated and lifted flame regimes for Bunsen type burners.

In the design of a Bunsen burner, it is desirable to have the maximum range of volumetric flow without encountering stability problems. What, then, is the optimum size tube for maximum flexibility? First, the diameter of the tube must be at least twice the penetration distance, i.e., greater than the quenching distance. Second, the average velocity must be at least twice S_L; otherwise, a precise Bunsen cone would not form. Experimental evidence shows further that if the average velocity is five times S_L, the fuel penetrates the Bunsen cone tip. If the Reynolds number of the combustible gases in the tube exceeds 2000, the flow can become turbulent, destroying the laminar characteristics of the flame. Of course, there are the limitations of the gradients of flashback and blowoff. If one graphs u_{av} versus d for these various limitations, one obtains a plot similar to that shown in Fig. 38. In this figure the dotted area represents the region that has the greatest flow variability without some stabilization problem. Note that this region d maximizes at about 1 cm; consequently, the tube diameter of Bunsen burners is always about 1 cm.

The burners on cooking stoves are very much like Bunsen tubes. The fuel induces air and the two gases premix prior to reaching the burner ring with its flame holes. It is possible to idealize this situation as an ejector. For an ejector, the total gas mixture flow rate can be related to the rate of fuel admitted to the system through continuity of momentum

$$m_m u_m = m_f u_f$$

$$u_m(\rho_m A_m u_m) = (\rho_f A_f u_f)u_f$$

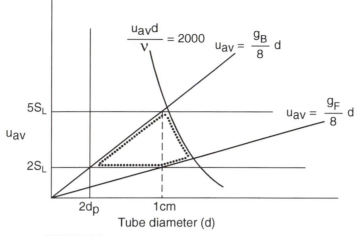

FIGURE 38 Stability and operation limits of a Bunsen burner.

where the subscript m represents the conditions for the mixture (A_m is the total area) and the subscript f represents conditions for the fuel. The ejector is depicted in Fig. 39. The momentum expression can written as

$$\rho_m u_m^2 = a \rho_f \rho_f^2$$

where a is the area ratio.

If one examines the g_F and g_B on the graph shown in Fig. 40, one can make some interesting observations. The burner port diameter is fixed such that a rich-mixture ratio is obtained at a value represented by the dashed line on Fig. 40. When the mixture ratio is set at this value, the flame can never flash back into the stove and burn without the operator's noticing the situation. If the fuel is changed, difficulties may arise. Such difficulties arose many decades ago when the gas industry switched from manufacturer's gas to natural gas, and could arise again

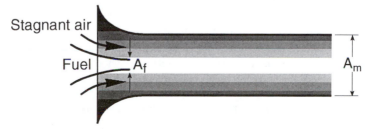

FIGURE 39 Fuel-jet ejector system for premixed fuel–air burners.

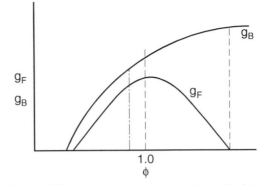

FIGURE 40   Flame stability diagram for an operating fuel gas–air mixture burner system.

if the industry is ever compelled to switch to synthetic gas or to use synthetic or petroleum gas as an additive to natural gas.

The volumetric fuel–air ratio in the ejector is given by

$$(F/A) = (u_f A_f)/(u_m A_m)$$

It is assumed here that the fuel–air (F/A) mixture is essentially air; that is, the density of the mixture does not change as the amount of fuel changes. From the momentum equation, this fuel–air mixture ratio becomes

$$(F/A) = (\rho_m/\rho_f)^{1/2} a^{1/2}$$

The stoichiometric molar (volumetric) fuel–air ratio is strictly proportional to the molecular weight of the fuel for two common hydrocarbon fuels; i.e.,

$$(F/A)_{stoich} \sim 1/MW_f \sim 1/\rho_f$$

The equivalence ratio then is

$$\phi = \frac{(F/A)}{(F/A)_{stoich}} \sim \frac{a^{1/2}(\rho_m/\rho_f)^{1/2}}{(1/\rho_f)}$$

Examining Fig. 40, one observes that in converting from a heavier fuel to a lighter fuel, the equivalence ratio drops, as indicated by the new dot-dash operating line. Someone adjusting the same burner with the new lighter fuel would have a very consistent flashback–blowoff problem. Thus, when the gas industry switched to natural gas, it was required that every fuel port in every burner on every stove be drilled open, thereby increasing a to compensate for the decreased ρ_f. Synthetic gases of the future will certainly be heavier than methane (natural gas). They will probably be mostly methane with some heavier components, particularly ethane. Consequently, today's burners will not present a stability problem; however, they will operate more fuel-rich and thus be more wasteful of energy. It would be logical to make the fuel ports smaller by inserting caps so that the operating line would be moved next to the rich flashback cut-off line.

E. TURBULENT REACTING FLOWS AND TURBULENT FLAMES

Most practical combustion devices create flow conditions so that the fluid state of the fuel and oxidizer, or fuel–oxidizer mixture, is turbulent. Nearly all mobile and stationary power plants operate in this manner because turbulence increases the mass consumption rate of the reactants, or reactant mixture, to values much greater than those that can be obtained with laminar flames. A greater mass consumption rate increases the chemical energy release rate and hence the power available from a given combustor or internal engine of a given size. Indeed, few combustion engines could function without the increase in mass consumption during combustion that is brought about by turbulence. Another example of the importance of turbulence arises with respect to spark timing in automotive engines. As the RPM of the engines increases, the level of turbulence increases, whereupon the mass consumption rate (or turbulent flame speed) of the fuel–air mixture increases. This explains why spark timing does not have to be altered as the RPM of the engine changes with a given driving cycle.

As has been shown, the mass consumption rate per unit area in premixed laminar flames is simply ρS_L, where ρ is the unburned gas mixture density. Correspondingly, for power plants operating under turbulent conditions, a similar consumption rate is specified as ρS_T, where S_T is the turbulent burning velocity. Whether a well defined turbulent burning velocity characteristic of a given combustible mixture exists as S_L does under laminar conditions will be discussed later in this section. What is known is that the mass consumption rate of a given mixture varies with the state of turbulence created in the combustor. Explicit expressions for a turbulent burning velocity S_T will be developed, and these expressions will show that various turbulent fields increase S_T to values much larger than S_L. However, increasing turbulence levels beyond a certain value increases S_T very little, if at all, and may lead to quenching of the flame [39, 40].

To examine the effect of turbulence on flames, and hence the mass consumption rate of the fuel mixture, it is best to first recall the tacit assumption that in laminar flames the flow conditions alter neither the chemical mechanism nor the associated chemical energy release rate. Now one must acknowledge that, in many flow configurations, there can be an interaction between the character of the flow and the reaction chemistry. When a flow becomes turbulent, there are fluctuating components of velocity, temperature, density, pressure, and concentration. The degree to which such components affect the chemical reactions, heat release rate, and flame structure in a combustion system depends upon the relative characteristic times associated with each of these individual parameters. In a general sense, if the characteristic time (τ_c) of the chemical reaction is much shorter than a characteristic time (τ_m) associated with the fluid-mechanical fluctuations, the chemistry is essentially unaffected by the flow field. But if the contra condition ($\tau_c > \tau_m$) is true, the fluid mechanics could influence the chemical reaction rate, energy release rates, and flame structure.

The interaction of turbulence and chemistry, which constitutes the field of turbulent reacting flows, is of importance whether flame structures exist or not. The concept of turbulent reacting flows encompasses many different meanings and depends on the interaction range, which is governed by the overall character of the flow environment. Associated with various flows are different characteristic times, or, as more commonly used, different characteristic lengths.

There are many different aspects to the field of turbulent reacting flows. Consider, for example, the effect of turbulence on the rate of an exothermic reaction typical of those occuring in a turbulent flow reactor. Here, the fluctuating temperatures and concentrations could affect the chemical reaction and heat release rates. Then, there is the situation in which combustion products are rapidly mixed with reactants in a time much shorter than the chemical reaction time. (This latter example is the so-called *stirred reactor*, which will be discussed in more detail in the next section.) In both of these examples, no flame structure is considered to exist.

Turbulence–chemistry interactions related to premixed flames comprise another major stability category. A turbulent flow field dominated by large-scale, low-intensity turbulence will affect a premixed laminar flame so that it appears as a wrinkled laminar flame. The flame would be contiguous throughout the front. As the intensity of turbulence increases, the contiguous flame front is partially destroyed and laminar flamelets exist within turbulent eddies. Finally, at very high-intensity turbulence, all laminar flame structure disappears and one has a distributed reaction zone. Time-averaged photographs of these three flames show a very bushy flame front that looks very thick in comparison to the smooth thin zone that characterizes a laminar flame. However, when a very fast response thermocouple is inserted into these three flames, the fluctuating temperatures in the first two cases show a bimodal probability density function with well-defined peaks at the temperatures of the unburned and completely burned gas mixtures. But a bimodal function is not found for the distributed reaction case.

Under premixed fuel–oxidizer conditions the turbulent flow field causes a mixing between the different fluid elements, so the characteristic time was given the symbol τ_m. In general with increasing turbulent intensity, this time approaches the chemical time, and the associated length approaches the flame or reaction zone thickness. Essentially the same is true with respect to non-premixed flames. The fuel and oxidizer (reactants) in non-premixed flames are not in the same flow stream; and, since different streams can have different velocities, a gross shear effect can take place and coherent structures (eddies) can develop throughout this mixing layer. These eddies enhance the mixing of fuel and oxidizer. The same type of shear can occur under turbulent premixed conditions when large velocity gradients exist.

The complexity of the turbulent reacting flow problem is such that it is best to deal first with the effect of a turbulent field on an exothermic reaction in a plug flow reactor. Then the different turbulent reacting flow regimes will be described more precisely in terms of appropriate characteristic lengths, which will be developed from a general discussion of turbulence. Finally, the turbulent premixed flame will be examined in detail.

1. The Rate of Reaction in a Turbulent Field

As an excellent, simple example of how fluctuating parameters can affect a reacting system, one can examine how the mean rate of a reaction would differ from the rate evaluated at the mean properties when there are no correlations among these properties. In flow reactors, time-averaged concentrations and temperatures are usually measured, and then rates are determined from these quantities. Only by optical techniques or very fast response thermocouples could the proper instantaneous rate values be measured, and these would fluctuate with time.

The fractional rate of change of a reactant can be written as

$$\dot{\omega} = -k\rho^{n-1}Y_i^n = -Ae^{-E/RT}(P/R)^{n-1}T^{1-n}Y_i^n$$

where the Y_i's are the mass fractions of the reactants. The instantaneous change in rate is given by

$$d\dot{\omega} = -A(P/R)^{n-1}[(E/RT^2)e^{-E/RT}T^{1-n}Y_i^n\,dT$$

$$+ (1-n)T^{-n}e^{-E/RT}Y_i^n\,dT$$

$$+ ne^{-E/RT}T^{1-n}Y_i^{n-1}\,dY]$$

$$d\dot{\omega} = (E/RT)\dot{\omega}(dT/T) + (1-n)\dot{\omega}(dT/T) + \dot{\omega}n(dY_i/Y_i)$$

or

$$d\dot{\omega}/\dot{\omega} = \{E/RT + (1-n)\}(dT/T) + n(dY_i/Y_i)$$

For most hydrocarbon flame or reacting systems the overall order of reaction is about 2, E/R is approximately 20,000 K, and the flame temperature is about 2000 K. Thus,

$$(E/RT) + (1-n) \cong 9$$

and it would appear that the temperature variation is the dominant factor. Since the temperature effect comes into this problem through the specific reaction rate constant, the problem simplifies to whether the mean rate constant can be represented by the rate constant evaluated at the mean temperature.

In this hypothetical simplified problem one assumes further that the temperature T fluctuates with time around some mean represented by the form

$$T(t)/\bar{T} = 1 + a_n f(t)$$

where a_n is the amplitude of the fluctuation and $f(t)$ is some time-varying function in which

$$-1 \le f(t) \le +1$$

and

$$\bar{T} = \frac{1}{\tau} \int_0^\tau T(t)\, dt$$

over some time interval τ. $T(t)$ can be considered to be composed of $\bar{T} + T'(t)$, where T' is the fluctuating component around the mean. Ignoring the temperature dependence in the pre-exponential, one writes the instantaneous-rate constant as

$$k(T) = A \exp(-E/RT)$$

and the rate constant evaluated at the mean temperature as

$$k(\bar{T}) = A \exp(-E/R\bar{T})$$

Dividing the two expressions, one obtains

$$\{k(T)/k(\bar{T})\} = \exp\{(E/R\bar{T})[1 - (\bar{T}/T)]\}$$

Obviously, then, for small fluctuations

$$1 - (\bar{T}/T) = [a_n f(t)]/[1 + a_n f(t)] \approx a_n f(t)$$

The expression for the mean rate is written as

$$\frac{\overline{k(T)}}{k(\bar{T})} = \frac{1}{\tau} \int_0^\tau \frac{k(T)}{k\bar{T}}\, dt = \frac{1}{\tau} \int_0^\tau \exp\left(\frac{E}{R\bar{T}} a_n f(t)\right) dt$$

$$= \frac{1}{\tau} \int_0^\tau \left[1 + \frac{E}{R\bar{T}} a_n f(t) + \frac{1}{2}\left(\frac{E}{R\bar{T}} a_n f(t)\right)^2 + \cdots\right] dt$$

But recall

$$\int_0^\tau f(t)\, dt = 0 \qquad \text{and } 0 \le f^2(t) \le 1$$

Examining the third term, it is apparent that

$$\frac{1}{\tau} \int_0^\tau a_n^2 f^2(t)\, dt \le a_n^2$$

since the integral of the function can never be greater than 1. Thus,

$$\frac{\overline{k(T)}}{k(\bar{T})} \le 1 + \frac{1}{2}\left(\frac{E}{R\bar{T}} a_n\right)^2 \qquad \text{or} \qquad \Delta = \frac{\overline{k(T)} - k(\bar{T})}{k(\bar{T})} \le \frac{1}{2}\left(\frac{E}{R\bar{T}} a_n\right)^2$$

If the amplitude of the temperature fluctuations is of the order of 10% of the mean temperature, one can take $a_n \approx 0.1$; and if the fluctuations are considered sinusoidal, then

$$\frac{1}{\tau} \int_0^\tau \sin^{-2} t \, dt = \frac{1}{2}$$

Thus for the example being discussed,

$$\Delta = \frac{1}{4} \left(\frac{E}{R\bar{T}} a_n \right)^2 = \frac{1}{4} \left(\frac{40,000 \times 0.1}{2 \times 2000} \right)^2, \qquad \Delta \cong \frac{1}{4}$$

or a 25% difference in the two rate constants.

This result could be improved by assuming a more appropriate distribution function of T' instead of a simple sinusoidal fluctuation; however, this example—even with its assumptions—usefully illustrates the problem. Normally, probability distribution functions are chosen. If the concentrations and temperatures are correlated, the rate expression becomes very complicated. Bilger [41] has presented a form of a two-component mean-reaction rate when it is expanded about the mean states, as follows:

$$-\bar{\omega} = \bar{\rho}^2 \bar{Y}_i \bar{Y}_j \exp(-E/R\bar{T})\{1 + (\overline{\rho'^2}/\bar{\rho}^2) + (\overline{Y_i'Y_j'})/(\overline{Y_i'Y_j'})$$

$$+ 2(\overline{\rho'Y_i'}/\bar{\rho}\bar{Y}_i) + 2(\overline{\rho'Y_j'}/\bar{\rho}\bar{Y}_j)$$

$$+ (E/R\bar{T})(\overline{Y_i'T'}/\bar{Y}_i\bar{T})(\overline{Y_j'T'}/\bar{Y}_j\bar{T})$$

$$+ [(E/2R\bar{T}) - 1](\overline{T'^2}/\bar{T}^2) + \cdots\}$$

2. Regimes of Turbulent Reacting Flows

The previous example epitomizes how the reacting media can be affected by a turbulent field. To understand the detailed effect, one must understand the elements of the field of turbulence. When considering turbulent combustion systems in this regard, a suitable starting point is the consideration of the quantities that determine the fluid characteristics of the system. The material presented subsequently has been mostly synthesized from Refs. [42 and 43].

Most flows have at least one characteristic velocity, U, and one characteristic length scale, L, of the device in which the flow takes place. In addition there is at least one representative density ρ_0 and one characteristic temperature T_0, usually the unburned condition when considering combustion phenomena. Thus, a characteristic kinematic viscosity $\nu_0 \equiv \mu_0/\rho_0$ can be defined, where μ_0 is the coefficient of viscosity at the characteristic temperature T_0. The Reynolds number for the system is then Re $= UL/\nu_0$. It is interesting that ν is approximately proportional to T^2. Thus, a change in temperature by a factor of 3 or more, quite modest by combustion standards, means a drop in Re by an order of magnitude.

Thus, energy release can damp turbulent fluctuations. The kinematic viscosity ν is inversely proportional to the pressure p, and changes in p are usually small; the effects of such changes in ν typically are much less than those of changes in T.

Even though the Reynolds number gives some measure of turbulent phenomena, flow quantities characteristic of turbulence itself are of more direct relevance to modeling turbulent reacting systems. The turbulent kinetic energy $\bar{q}$ may be assigned a representative value $\bar{q}_0$ at a suitable reference point. The relative intensity of the turbulence is then characterized by either $\bar{q}_0/(\frac{1}{2}U^2)$ or U'/U, where $U' = (2\bar{q}_0)^{1/2}$ is a representative root-mean-square velocity fluctuation. Weak turbulence corresponds to $U'/U < 1$ and intense turbulence has U'/U of the order unity.

Although a continuous distribution of length scales is associated with the turbulent fluctuations of velocity components and of state variables (p, ρ, T), it is useful to focus on two widely disparate lengths that determine separate effects in turbulent flows. First, there is a length l_0 which characterizes the large eddies, those of low frequencies and long wavelengths; this length is sometimes referred to as the integral scale. Experimentally, l_0 can be defined as a length beyond which various fluid-mechanical quantities become essentially uncorrelated; typically, l_0 is less than L but of the same order of magnitude. This length can be used in conjunction with U' to define a turbulent Reynolds number

$$R_l = U'l_0/\nu_0$$

which has more direct bearing on the structure of turbulence in flows than does Re. Large values of R_l can be achieved by intense turbulence, large-scale turbulence, and small values of ν produced, for example, by low temperatures or high pressures. The cascade view of turbulence dynamics is restricted to large values of R_l. From the characterization of U' and l_0, it is apparent that $R_l <$ Re.

The second length scale characterizing turbulence is that over which molecular effects are significant; it can be introduced in terms of a representative rate of dissipation of velocity fluctuations, essentially the rate of dissipation of the turbulent kinetic energy. This rate of dissipation, which is given by the symbol ε_0, is

$$\varepsilon_0 = \frac{\bar{q}_0}{t} \approx \frac{(U')^2}{(l_0/U')} \approx \frac{(U')^3}{l_0}$$

This rate estimate corresponds to the idea that the time scale over which velocity fluctuations (turbulent kinetic energy) decay by a factor of $(1/e)$ is the order of the turning time of a large eddy. The rate ε_0 increases with turbulent kinetic energy (which is due principally to the large-scale turbulence) and decreases with increasing size of the large-scale eddies. For the small scales at which molecular dissipation occurs, the relevant parameters are the kinematic viscosity, which causes the dissipation, and the rate of dissipation. The only length scale that can be constructed from these two parameters is the so-called Kolmogorov length (scale):

$$l_k = \left(\frac{\nu^3}{\varepsilon_0}\right)^{1/4} = \left[\frac{cm^6\,s^{-3}}{(cm^3\,s^{-3})(1\,cm^{-1})}\right]^{1/4} = (cm^4)^{1/4} = 1\,cm$$

However, note that

$$l_k = [\nu^3 l_0/(U')^3]^{1/4} = [(\nu^3 l_0^4)/(U')^3 l_0^3]^{1/4} = (l_0^4/R_l^3)^{1/4}$$

Therefore

$$l_k = l_0/R_l^{3/4}$$

This length is representative of the dimension at which dissipation occurs and defines a cut-off of the turbulence spectrum. For large R_l there is a large spread of the two extreme lengths characterizing turbulence. This spread is reduced with the increasing temperature found in combustion of the consequent increase in ν_0.

Considerations analogous to those for velocity apply to scalar fields as well, and lengths analogous to l_k have been introduced for these fields. They differ from l_k by factors involving the Prandtl and Schmidt numbers, which differ relatively little from unity for representative gas mixtures. Therefore, to a first approximation for gases, l_k may be used for all fields and there is no need to introduce any new corresponding lengths.

An additional length, intermediate in size between l_0 and l_k, which often arises in formulations of equations for average quantities in turbulent flows is the Taylor length (λ), which is representative of the dimension over which strain occurs in a particular viscous medium. The strain can be written as (U'/l_0). As before, the length that can be constructed between the strain and the viscous forces is

$$\lambda = [\nu/(U'/l_0)]^{1/2}$$

$$\lambda = (\nu l_0/U') = (\nu l_0^2/U'l_0) = (l_0^2/R_l)$$

In a sense, the Taylor microscale is similar to an average of the other scales, l_0 and l_k, but heavily weighted toward l_k.

Recall that there are length scales associated with laminar flame structures in reacting flows. One is the characteristic thickness of a premixed flame, δ_L, given by

$$\delta_L \approx \left(\frac{\alpha}{\dot{\omega}/\rho}\right)^{1/2} = \left(\frac{cm^2\,s^{-1}}{1\,s^{-1}}\right)^{1/2} = 1\,cm$$

The derivation is, of course, consistent with the characteristic velocity in the flame speed problem. This velocity is obviously the laminar flame speed itself, so that

$$S_L \approx \frac{\nu}{\delta_L} = \frac{cm^2\,s^{-1}}{1\,cm} = 1\,cm\,s^{-1}$$

As discussed in an earlier section, δ_L is the characteristic length of the flame and includes the thermal preheat region and that associated with the zone of rapid chemical reaction. This reaction zone is the rapid heat release flame segment at the high-temperature end of the flame. The earlier discussion of flame structure from detailed chemical kinetic mechanisms revealed that the heat release zone need not be narrow compared to the preheat zone. Nevertheless, the magnitude of δ_L does not change, no matter what the analysis of the flame structure is. It is then possible to specify the characteristic time of the chemical reaction in this context to be

$$\tau_c \approx \left(\frac{1}{\dot{\omega}/\rho}\right) \approx \frac{\delta_L}{S_L}$$

It may be expected, then, that the nature of the various turbulent flows, and indeed the structures of turbulent flames, may differ considerably and their characterization would depend on the comparison of these chemical and flow scales in a manner specified by the following inequalities and designated flame type:

$\delta_L < l_k$;	$l_k < \delta_L < \lambda$;	$\lambda < \delta_L < l_0$;	$l_0 < \delta_L$
wrinkled flame	severely wrinkled flame	flamelets in eddies	distributed reaction front

The nature, or more precisely the structure, of a particular turbulent flame implied by these inequalities cannot be exactly established at this time. The reason is that values of δ_L, l_k, λ, or l_0 cannot be explicitly measured under a given flow condition or analytically estimated. Many of the early experiments with turbulent flames appear to have operated under the condition $\delta_L < l_k$, so the early theories that developed specified this condition in expressions for S_T. The flow conditions under which δ_L would indeed be less than l_k has been explored analytically in detail and will be discussed subsequently.

To expand on the understanding of the physical nature of turbulent flames, it is also beneficial to look closely at the problem from a chemical point of view, exploring how heat release and its rate affect turbulent flame structure.

One begins with the characteristic time for chemical reaction designated τ_c, which was defined earlier. (Note that this time would be appropriate whether a flame existed or not.) Generally, in considering turbulent reacting flows, chemical lengths are constructed to be $U\tau_c$ or $U'\tau_c$. Then comparison of an appropriate chemical length with a fluid dynamic length provides a nondimensional parameter that has a bearing on the relative rate of reaction. Nondimensional numbers of this type are called Damkohler numbers and are conventionally given the symbol Da. An example appropriate to the considerations here is

$$\text{Da} = (l_0/U'\tau_c) = (\tau_m/\tau_c) = (l_0 S_L/U'\delta_L)$$

where τ_m is a mixing (turbulent) time defined as (l_0/U'), and the last equality in the expression applies when there is a flame structure. Following the earlier development, it is also appropriate to define another turbulent time based on the Kolmogorov scale $\tau_k = (\nu/\varepsilon)^{1/2}$.

For large Damkohler numbers, the chemistry is fast (i.e., reaction time is short) and reaction sheets of various wrinkled types may occur. For small Da numbers, the chemistry is slow and well-stirred flames may occur.

Two other nondimensional numbers relevant to the chemical reaction aspect of this problem [42] have been introduced by Frank-Kamenetskii and others. These Frank-Kamenetskii numbers (FK) are the nondimensional heat release $FK_1 \equiv (Q_p/c_p T_f)$, where Q_p is the chemical heat release of the mixture and T_f is the flame (or reaction) temperature; and the nondimensional activation energy $FK_2 \equiv (T_a/T_f)$, where the activation temperature $T_a = (E_A/R)$. Combustion, in general, and turbulent combustion, in particular, are typically characterized by large values of these numbers. When FK_1 is large, chemistry is likely to have a large influence on turbulence. When FK_2 is large, the rate of reaction depends strongly on the temperature. It is usually true that the larger the FK_2, the thinner will be the region in which the principal chemistry occurs. Thus, irrespective of the value of the Damkohler number, reaction zones tend to be found in thin, convoluted sheets in turbulent flows, for both premixed and non-premixed systems having large FK_2. For premixed flames, the thickness of the reaction region has been shown to be of the order δ_L/FK_2. Different relative sizes of δ_L/FK_2 and fluid-mechanical lengths, therefore, may introduce additional classes of turbulent reacting flows.

The flames themselves can alter the turbulence. In simple open Bunsen flames whose tube Reynolds number indicates that the flow is in the turbulent regime, some results have shown that the temperature effects on the viscosity are such that the resulting flame structure is completely laminar. Similarly, for a completely laminar flow in which a simple wire is oscillated near the flame surface, a wrinkled flame can be obtained (Fig. 41). Certainly, this example is relevant to $\delta_L < l_k$; that is, a wrinkled flame. Nevertheless, most open flames created by a turbulent fuel jet exhibit a wrinkled flame type of structure. Indeed, short-duration Schlieren photographs suggest that these flames have continuous surfaces. Measurements of flames such as that shown in Figs. 42a and 42b have been taken at different time intervals and the instantaneous flame shapes verify the continuous wrinkled flame structure. A plot of these instantaneous surface measurements results in a thick flame region (Fig. 43), just as the eye would visualize that a larger number of these measurements would result in a thick flame. Indeed, turbulent premixed flames are described as bushy flames. The thickness of this turbulent flame zone appears to be related to the scale of turbulence. Essentially, this case becomes that of severe wrinkling and is categorized by $l_k < \delta_L < \lambda$. Increased turbulence changes the character of the flame wrinkling, and flamelets begin to form. These flame elements take on the character of a fluid-mechanical vortex rather than a simple distorted wrinkled front, and this case is specified by $\lambda < \delta_L < l_0$. For $\delta_L \ll l_0$, some of the flamelets fragment from the front and the flame zone becomes highly wrinkled with pockets of combustion. To this point, the flame is considered practically contiguous. When $l_0 > \delta_L$, contiguous flames no longer exist and a distributed

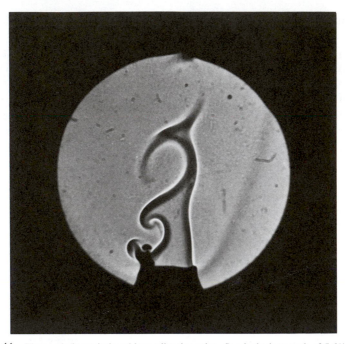

FIGURE 41   Flow turbulence induced by a vibrating wire. Spark shadowgraph of 5.6% propane-air flame (after G.H. Markstein, *Int. Symp. Combust., 7th* p. 289. Combust. Inst., Pittsburgh, Pennsylvania, 1959).

reaction front forms. Under these conditions, the fluid mixing processes are very rapid with respect to the chemical reaction time and the reaction zone essentially approaches the condition of a stirred reactor. In such a reaction zone, products and reactants are continuously intermixed.

For a better understanding of this type of flame occurrence and for more explicit conditions that define each of these turbulent flame types, it is necessary to introduce the flame stretch concept. This will be done shortly, at which time the regions will be more clearly defined with respect to chemical and flow rates with a graph that relates the nondimensional turbulent intensity, Reynold numbers, Damkohler number, and characteristic lengths *l*.

First, however, consider that in turbulent Bunsen flames the axial component of the mean velocity along the centerline remains almost constant with height above the burner; but away from the centerline, the axial mean velocity increases with height. The radial outflow component increases with distance from the centerline and reaches a peak outside the flame. Both axial and radial components of turbulent velocity fluctuations show a complex variation with position and include peaks and troughs in the flame zone. Thus, there are indications of both generation and removal of turbulence within the flame. With increasing height above the burner,

FIGURE 42   Short durations in Schlieren photographs of open turbulent flames [after Fox and Weinberg, *Proc. Roy. Soc., London, A* **268**, 222 (1962)].

the Reynolds shear stress decays from that corresponding to an initial pipe flow profile.

In all flames there is a large increase in velocity as the gases enter the burned gas state. Thus, it should not be surprising that the heat release itself can play a role in inducing turbulence. Such velocity changes in a fixed combustion configuration can cause shear effects that contribute to the turbulence phenomenon. There is no better example of some of these aspects than the case in which turbulent flames are stabilized in ducted systems. The mean axial velocity field of ducted flames involves considerable acceleration resulting from gas expansion engendered by heat release. Typically, the axial velocity of the unburned gas doubles before it is entrained into the flame, and the velocity at the centerline at least doubles again. Large mean velocity gradients are therefore produced. The streamlines in the unburned gas are deflected away from the flame.

The growth of axial turbulence in the flame zone of these ducted systems is attributed to the mean velocity gradient resulting from the combustion. The production of turbulence energy by shear depends on the product of the mean velocity gradient and the Reynolds stress. Such stresses provide the most plausible mechanism for the modest growth in turbulence observed.

Now it is important to stress that, whereas the laminar flame speed is a unique thermochemical property of a fuel–oxidizer mixture ratio, a turbulent flame speed is a function not only of the fuel–oxidizer mixture ratio, but also of the flow characteristics and experimental configuration. Thus, one encounters great difficulty

FIGURE 43 Superimposed contours of instantaneous flame boundaries in a turbulent flame [after Fox and Weinberg, *Proc. Roy. Soc., London, A* **268**, 222 (1962)].

in correlating the experimental data of various investigators. In a sense, there is no flame speed in a turbulent stream. Essentially, as a flow field is made turbulent for a given experimental configuration, the mass consumption rate (and hence the rate of energy release) of the fuel–oxidizer mixture increases. Therefore, some researchers have found it convenient to define a turbulent flame speed S_T as the mean mass flux per unit area (in a coordinate system fixed to the time-averaged motion of the flame) divided by the unburned gas density ρ_0. The area chosen is the smoothed surface of the time-averaged flame zone. However, this zone is thick and curved; thus the choice of an area near the unburned gas edge can give quite a different result than one in which a flame position is taken in the center or the burned-gas side of the bushy flame. Therefore, a great deal of uncertainty is associated with the various experimental values of S_T reported. Nevertheless, definite trends have been reported. These trends can be summarized as follows:

1. S_T is always greater than S_L. This trend would be expected once the increased area of the turbulent flame allows greater total mass consumption.
2. S_T increases with increasing intensity of turbulence ahead of the flame. Many have found the relationship to be approximately linear. (This point will be discussed later.)
3. Some experiments show S_T to be insensitive to the scale of the approach flow turbulence.

4. In open flames, the variation of S_T with composition is generally much the same as for S_L, and S_T has a well-defined maximum close to stoichiometric. Thus, many report turbulent flame speed data as the ratio of S_T/S_L.
5. Very large values of S_T may be observed in ducted burners at high approach flow velocities. Under these conditions, S_T increases in proportion to the approach flow velocities, but is insensitive to approach flow turbulence and composition. It is believed that these effects result from the dominant influence of turbulence generated within the stabilized flame by the large velocity gradients.

The definition of the flame speed as the mass flux through the flame per unit area of the flame divided by the unburned gas density ρ_0 is useful for turbulent nonstationary and oblique flames as well.

Now with regard to stretch, consider first a plane oblique flame. Because of the increase in velocity demanded by continuity, a streamline through such an oblique flame is deflected toward the direction of the normal to the flame surface. The velocity vector may be broken up into a component normal to the flame wave and a component tangential to the wave (Fig. 44). Because of the energy release, the continuity of mass requires that the normal component increase on the burned gas side while, of course, the tangential component remains the same. A consequence of the tangential velocity is that fluid elements in the oblique flame surface move along this surface. If the surface is curved, adjacent points traveling along the flame surface may move either farther apart (flame stretch) or closer together (flame compression).

An oblique flame is curved if the velocity U of the approach flow varies in a direction y perpendicular to the direction of the approach flow. Strehlow showed that the quantity

$$K_1 = (\delta_L/U)(\partial U/\partial y)$$

which is known as the Karlovitz flame stretch factor, is approximately equal to the ratio of the flame thickness δ_L to the flame curvature. The Karlovitz school has argued that excessive stretching can lead to local quenching of the reaction. Klimov [44], and later Williams [45], analyzed the propagation of a laminar flame

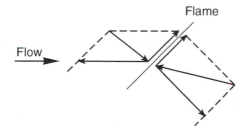

FIGURE 44   Deflection of the velocity vector through an oblique flame.

in a shear flow with velocity gradient in terms of a more general stretch factor

$$K_2 = (\delta_L/S_L)(1/\Lambda)d\Lambda/dt$$

where Λ is the area of an element of flame surface, $d\Lambda/dt$ is its rate of increase, and δ_L/S_L is a measure of the transit time of the gases passing through the flame. Stretch ($K_2 > 0$) is found to reduce the flame thickness and to increase reactant consumption per unit area of the flame and large stretch ($K_2 \gg 0$) may lead to extinction. On the other hand, compression ($K_2 < 0$) increases flame thickness and reduces reactant consumption per unit incoming reactant area. These findings are relevant to laminar flamelets in a turbulent flame structure.

Since the concern here is with the destruction of a contiguous laminar flame in a turbulent field, consideration must also be given to certain inherent instabilities in laminar flames themselves. There is a fundamental hydrodynamic instability as well as an instability arising from the fact that mass and heat can diffuse at different rates; i.e., the Lewis number (Le) is nonunity. In the latter mechanism, a flame instability can occur when the Le number (D/α) is less than 1.

Consider initially the hydrodynamic instability—i.e., the one due to the flow—first described by Darrieus [46], Landau [47], and Markstein [48]. If no wrinkle occurs in a laminar flame, the flame speed S_L is equal to the upstream unburned gas velocity U_0. But if a minor wrinkle occurs in a laminar flame, the approach flow streamlines will either diverge or converge as shown in Fig. 45. Considering the two middle streamlines, one notes that, because of the curvature due to the wrinkle, the normal component of the velocity, with respect to the flame, is less than U_0. Thus, the streamlines diverge as they enter the wrinkled flame front. Since there must be continuity of mass between the streamlines, the unburned gas velocity at the front must decrease owing to the increase of area. Since S_L is now greater than the velocity of unburned approaching gas, the flame moves farther downstream and the wrinkle is accentuated. For similar reasons, between another pair of streamlines if the unburned gas velocity increases near the flame front, the flame bows in the upstream direction. It is not clear why these instabilities do not keep growing. Some have attributed the growth limit to nonlinear effects that arise in hydrodynamics.

When the Lewis number is nonunity, the mass diffusivity can be greater than the thermal diffusivity. This discrepancy in diffusivities is important with respect to the reactant that limits the reaction. Ignoring the hydrodynamic instability, consider again the condition between a pair of streamlines entering a wrinkle in a laminar flame. This time, however, look more closely at the flame structure that these streamlines encompass, noting that the limiting reactant will diffuse into the flame zone faster than heat can diffuse from the flame zone into the unburned mixture. Thus, the flame temperature rises, the flame speed increases, and the flame wrinkles bow further in the downstream direction. The result is a flame that looks very much like the flame depicted for the hydrodynamic instability in Fig. 45. The flame surface breaks up continuously into new cells in a chaotic

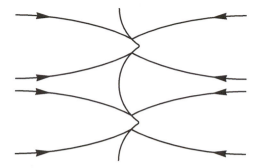

FIGURE 45   Convergence–divergence of the flow streamlines due to a wrinkle in a laminar flame.

manner, as photographed by Markstein [48]. There appears to be, however, a higher-order stabilizing effect. The fact that the phenomenon is controlled by a limiting reactant means that this cellular condition can occur when the unburned premixed gas mixture is either fuel-rich or fuel-lean. It should not be surprising, then, that the most susceptible mixture would be a lean hydrogen–air system.

Earlier it was stated that the structure of a turbulent velocity field may be presented in terms of two parameters—the scale and the intensity of turbulence. The intensity was defined as the square root of the turbulent kinetic energy, which essentially gives a root-mean-square velocity fluctuation U'. Three length scales were defined: the integral scale l_0, which characterizes the large eddies; the Taylor microscale λ, which is obtained from the rate of strain; and the Kolmogorov microscale l_k, which typifies the smallest dissipative eddies. These length scales and the intensity can be combined to form not one, but three turbulent Reynolds numbers: $R_l = U'l_0/\nu$, $R_\lambda = U'\lambda/\nu$, and $R_k = U'l_k/\nu$. From the relationship between l_0, λ, and l_k previously derived it is found that $R_l \approx R_\lambda^2 \approx R_k^4$.

There is now sufficient information to relate the Damkohler number Da and the length ratios l_0/δ_L, l_k/δ_L and l_0/l_k to a nondimensional velocity ratio U'/S_L and the three turbulence Reynolds numbers. The complex relationships are given in Fig. 46 and are very informative. The right-hand side of the figure has $R_\lambda > 100$ and ensures the length-scale separation that is characteristic of high Reynolds number behavior. The largest Damkohler numbers are found in the bottom right corner of the figure.

Using this graph and the relationship it contains, one can now address the question of whether and under what conditions a laminar flame can exist in a turbulent flow. As before, if allowance is made for flame front curvature effects, a laminar flame can be considered stable to a disturbance of sufficiently short wavelength; however, intense shear can lead to extinction. From solutions of the laminar flame equations in an imposed shear flow, Klimov [44] and Williams [45] showed that a conventional propagating flame may exist only if the stretch factor K_2 is less than a critical value of unity. Modeling the area change term in the

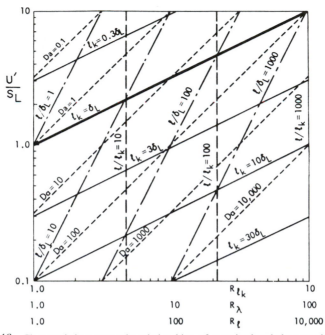

FIGURE 46 Characteristic parametric relationships of premixed turbulent combustion.   The Klimov–Williams criterion is satisfied below the heavy line $l_k = \delta_L$.

stretch expression as

$$(1/\Lambda)d\Lambda/dt \approx U'/\lambda$$

and recalling that

$$\delta_L \approx \nu/S_L$$

one can define the Karlovitz number for stretch in turbulent flames as

$$K_2 \approx \frac{\delta_L U'}{S_L \Lambda}$$

with no possibility of negative stretch. Thus

$$K_2 \approx \frac{\delta_L}{S_L}\frac{U'}{\lambda} = \frac{\delta_L^2}{\nu}\frac{U'}{\lambda} = \frac{\delta_L^2}{\nu}\frac{U'}{l_0}R_l^{1/2}\frac{l_0}{l_0} = \frac{\delta_L^2}{l_0^2}R_l^{3/2}$$

But as shown earlier

$$l_k = l_0/R_l^{3/4} \qquad \text{or} \qquad l_0^2 = l_k^2/R_l^{3/2}$$

so that

$$K_2 = \delta_L^2/l_k^2 = (\delta_L/l_k)^2$$

Thus, the criterion to be satisfied if a laminar flame is to exist in a turbulent flow is that the laminar flame thickness δ_L be less than the Kolmogorov microscale l_k of the turbulence.

The heavy line in Fig. 46 indicates the conditions $\delta_L = l_k$. This line is drawn in this fashion since

$$K_2 \approx \frac{\delta_L U'}{S_L \lambda} \approx \frac{\nu}{S_L^2 \lambda} \approx \frac{\nu}{S_L^2} \frac{U'}{\lambda} \cdot \frac{U'}{U'} \approx \frac{(U')^2}{S_L^2} \frac{1}{R_\lambda} \approx 1 \quad \text{or} \quad \left(\frac{U'}{\delta_L}\right)^2 \approx R_\lambda$$

Thus for $(U'/\delta_L) = 1$, $R_\lambda = 1$; and for $(U'/\delta_L) = 10$, $R_\lambda = 100$. The other Reynolds numbers follow from $R_k^4 = R_\lambda^2 = R_l$.

Below and to the right of this line, the Klimov–Williams criterion is satisfied and wrinkled laminar flames may occur. The figure shows that this region includes both large and small values of turbulence Reynolds numbers and velocity ratios (U'/S_L) both greater and less than 1, but predominantly large Da.

Above and to the left of the criterion line is the region in which $l_k < \delta_L$. According to the Klimov–Williams criterion, the turbulent velocity gradients in this region, or perhaps in a region defined with respect to any of the characteristic lengths, are sufficiently intense that they may destroy a laminar flame. The figure shows $U' \geq S_L$ in this region and Da is predominantly small. At the highest Reynolds numbers the region is entered only for very intense turbulence, $U' \geq S_L$. The region has been considered a distributed reaction zone in which reactants and products are somewhat uniformly dispersed throughout the flame front. Reactions are still fast everywhere, so that unburned mixture near the burned-gas side of the flame is completely burned before it leaves what would be considered the flame front. An instantaneous temperature measurement in this flame would yield a normal probability density function—more importantly, one that is not bimodal.

3. The Turbulent Flame Speed

Although a laminar flame speed S_L is a physicochemical and chemical kinetic property of the unburned gas mixture that can be assigned, a turbulent flame speed S_T is, in reality, a mass consumption rate per unit area divided by the unburned gas mixture density. Thus, S_T must depend on the properties of the turbulent field in which it exists and the method by which the flame is stabilized. Of course, difficulty arises with this definition of S_T because the time-averaged turbulent flame is bushy (thick) and there is a large difference between the area on the unburned-gas side of the flame and that on the burned-gas side. Nevertheless, many experimental data points are reported as S_T.

In his attempts to analyze the early experimental data, Damkohler [49] considered that large-scale, low-intensity turbulence simply distorts the laminar flame while the transport properties remain the same; thus, the laminar flame structure would not be affected. Essentially, his concept covered the range of the wrinkled and severely wrinkled flame cases defined earlier. Whereas a planar laminar flame would appear as a simple Bunsen cone, that cone is distorted by turbulence as shown in Fig. 43. It is apparent, then, that the area of the laminar flame will increase due to a turbulent field. Thus, Damkohler [49] proposed for large-scale, small-intensity turbulence that

$$(S_T/S_L) = (A_L/A_T)$$

where A_L is the total area of laminar surface contained within an area of turbulent flame whose time-averaged area is A_T. Damkohler further proposed that the area ratio could be approximated by

$$(A_L/A_T) = 1 + (U'_0/S_L)$$

which leads to the results

$$S_T = S_L + U'_0 \quad \text{or} \quad (S_T/S_L) = (U'_0/S_L) + 1$$

where U'_0 is the turbulent intensity of the unburned gases ahead of the turbulent flame front.

Many groups of experimental data have been evaluated by semiempirical correlations of the type

$$(S_T/S_L) = A(U'_0/S_L) + B$$

and

$$S_T = A\,\text{Re} + B$$

The first expression here is very similar to the Damkohler result for A and B equal to 1. Since the turbulent exchange coefficient (eddy diffusivity) correlates well with $l_0 U'$ for tube flow and, indeed, l_0 is essentially constant for the tube flow characteristically used for turbulent premixed flame studies, it follows that

$$U' \sim \varepsilon \sim \text{Re}$$

where Re is the tube Reynolds number. Thus, the latter expression has the same form as the Damkohler result except that the constants would have to equal 1 and S_L, respectively, for similarity.

Schelkin [50] also considered large-scale, small-intensity turbulence. He assumed that flame surfaces distort into cones with bases proportional to the square of the average eddy diameter (i.e., proportional to l_0). The height of the cone was assumed proportional to U' and to the time t during which an element of the wave is associated with an eddy. Thus, time can be taken as equal to (l_0/S_L). Schelkin then proposed that the ratio of S_T/S_L (average) equals the ratio of the average cone area to the cone base. From the geometry thus defined

$$A_C = A_B[1 + (4h^2/l_0^2)]^{1/2}$$

where A_C is the surface area of the cone and A_B is the area of the base. Therefore,

$$S_T = S_L[1 + (2U'/S_L)^2]^{1/2}$$

For large values of (U'/S_L), i.e., high-intensity turbulence, the preceding expression reduces to that developed by Damkohler: $S_T \sim U'$.

A more rigorous development of wrinkled turbulent flames led Clavin and Williams [51] to the following result where isotropic turbulence is assumed:

$$(S_T/S_L) \sim \{1 + [\overline{(U')^2}/S_L^2]\}^{1/2}$$

This result differs from Shelkin's heuristic approach only by the factor of 2 in the second term. The Clavin–Williams expression is essentially restricted to the case of $(U'/S_L) \ll 1$. For small (U'/S_L), the Clavin–Williams expression simplifies to

$$(S_T/S_L) \sim 1 + \tfrac{1}{2}[\overline{(U')^2}/S_L]$$

which is quite similar to the Damkohler result. Kerstein and Ashurst [52], in a reinterpretation of the physical picture of Clavin and Williams, proposed the expression

$$(S_T/S_L) \sim \{1 + (U'/S_L)^2\}^{1/2}$$

Using a direct numerical simulation, Yakhot [53] proposed the relation

$$\left(\frac{S_T}{S_L}\right) = \exp\left[(U'/S_L)^2 \Big/ \left(\frac{S_T}{S_L}\right)^2\right]$$

For small-scale, high-intensity turbulence, Damkohler reasoned that the transport properties of the flame are altered from laminar kinetic theory viscosity ν_0 to the turbulent exchange coefficient ε so that

$$(S_T/S_L) = (\varepsilon/\nu)^{1/2}$$

This expression derives from $S_L \sim \alpha^{1/2} \sim \nu^{1/2}$. Then, realizing that $\varepsilon \sim U'l_0$,

$$S_T \sim S_L[1 + (\lambda_T/\lambda_L)]^{1/2}$$

Schelkin [50] also extended Damkohler's model by starting from the fact that the transport in a turbulent flame could be made up of molecular movements (laminar λ_L) and turbulent movements, so that

$$S_T \sim [(\lambda_L + \lambda_T)/\tau_c]^{1/2} \sim \{(\lambda_L/\tau_c)[1 + (\lambda_T/\lambda_L)]\}^{1/2}$$

where the expression is again analogous to that for S_L. (Note that λ is the thermal conductivity in this equation, not the Taylor scale.) Then it would follow that

$$(\lambda_L/\tau_c) \sim S_L$$

and

$$S_T \sim S_L[1 + (\lambda_T/\lambda_L)]^{1/2}$$

or essentially

$$(S_T/S_L) \sim [1 + (\nu_T/\nu_L)]^{1/2}$$

The Damkohler turbulent exchange coefficient ε is the same as ν_T, so that both expressions are similar, particularly in that for high-intensity turbulence $\varepsilon \gg \nu$. The Damkohler result for small-scale, high-intensity turbulence that

$$(S_T/S_L) \sim Re^{1/2}$$

is significant, for it reveals that (S_T/S_L) is independent of (U'/S_L) at fixed Re. Thus, as stated earlier, increasing turbulence levels beyond a certain value increases S_T very little, if at all. In this regard, it is well to note that Ronney [39] reports "smoothed" experimental data from Bradley [40] in the form (S_T/S_L) versus (U'/S_L) for Re = 1000. Ronney's correlation of these data are reinterpreted in Fig. 47. Recall that all the expressions for small-intensity, large-scale turbulence were developed for small values of (U'/S_L) and reported a linear relationship between (U'/S_L) and (S_T/S_L). It is not surprising, then, that a plot of these expressions—and even some more advanced efforts which also show linear relations—do not correlate well with the curve in Fig. 47. Furthermore, most developments do not take into account the effect of stretch on the turbulent flame. Indeed, the expressions reported here hold and show reasonable agreement with experiment only for $(U'/S_L) \ll 1$. Bradley [40] suggests that burning velocity is reduced with stretch, i.e., as the Karlovitz stretch factor K_2 increases. There are also some Lewis number effects. Refer to Ref. [39] for more details and further insights.

The general data representation in Fig. 47 does show a rapid rise of (S_T/S_L) for values of $(U'/S_L) \ll 1$. It is apparent from the discussion to this point that as (U'/S_L) becomes greater than 1, the character of the turbulent flame varies; and under the appropriate turbulent variables, it can change as depicted in Fig. 48, which essentially comes from Borghi [54] as presented by Abdel–Gayed *et al.* [55].

F. STIRRED REACTOR THEORY

In the discussion of premixed turbulent flames, the case of infinitely fast mixing of reactants and products was introduced. Generally this concept is referred to as a stirred reactor. Many investigators have employed stirred reactor theory

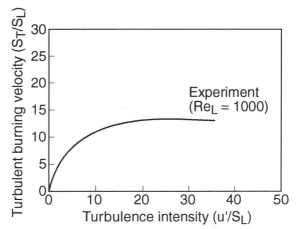

FIGURE 47 General trend of experimental turbulent burning velocity (S_T/S_L) data as a function of turbulent intensity (U'/S_L) for $R_l = 1000$ (from Ronney [39]).

not only to describe turbulent flame phenomena, but also to determine overall reaction kinetic rates [23] and to understand stabilization in high-velocity streams [56]. Stirred reactor theory is also important from a practical point of view because it predicts the maximum energy release rate possible in a fixed volume at a particular pressure.

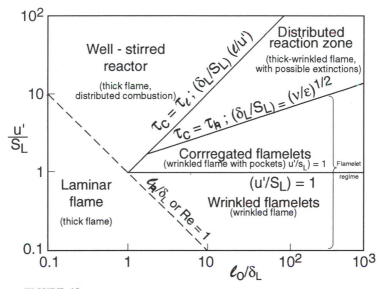

FIGURE 48 Turbulent combustion regimes (from Abdul-Gayed et al. [55]).

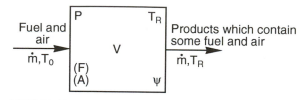

FIGURE 49   Variables of a stirred reactor system of fixed volume.

Consider a fixed volume V into which fuel and air are injected at a fixed total mass flow rate $\dot{m}$ and temperature T_0. The fuel and air react in the volume and the injection of reactants and outflow of products (also equal to $\dot{m}$) are so oriented that within the volume there is instantaneous mixing of the unburned gases and the reaction products (burned gases). The reactor volume attains some steady temperature T_R and pressure P. The temperature of the gases leaving the reactor is, thus, T_R as well. The pressure differential between the reactor and the exit is generally considered to be small. The mass leaving the reactor contains the same concentrations as those within the reactor and thus contains products as well as fuel and air. Within the reactor there exists a certain concentration of fuel (F) and air (A), and also a fixed unburned mass fraction, ψ. Throughout the reactor volume, T_R, P, (F), (A), and ψ are constant and fixed; i.e., the reactor is so completely stirred that all elements are uniform everywhere. Figure 49 depicts the stirred reactor concept in a generalized manner.

The stirred reactor may be compared to a plug flow reactor in which premixed fuel–air mixtures flow through the reaction tube. In this case, the unburned gases enter at temperature T_0 and leave the reactor at the flame temperature T_f. The system is assumed to be adiabatic. Only completely burned products leave the reactor. This reactor is depicted in Fig. 50.

The volume required to convert all the reactants to products for the plug flow reactor is greater than that for the stirred reactor. The final temperature is, of course, higher than the stirred reactor temperature.

It is relatively straightforward to develop the controlling parameters of a stirred reactor process. If ψ is defined as the unburned mass fraction, it must follow that

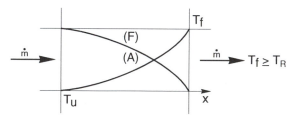

FIGURE 50   Variables in plug-flow reactor.

the fuel–air mass rate of burning $\dot{R}_{\mathrm{B}}$ is

$$\dot{R}_{\mathrm{B}} = \dot{m}(1 - \psi)$$

and the rate of heat evolution $\dot{q}$ is

$$\dot{q} = q\dot{m}(1 - \psi)$$

where q is the heat reaction per unit mass of reactants for the given fuel–air ratio. Assuming that the specific heat of the gases in the stirred reactor can be represented by some average quantity $\bar{c}_p$, one can write an energy balance as

$$\dot{m}q(1 - \psi) = \dot{m}\bar{c}_p(T_{\mathrm{R}} - T_0)$$

For the plug flow reactor or any similar adiabatic system, it is also possible to define an average specific heat that takes its explicit definition from

$$\bar{c}_p \equiv q/(T_{\mathrm{f}} - T_0)$$

To a very good approximation these two average specific heats can be assumed equal. Thus, it follows that

$$(1 - \psi) = (T_{\mathrm{R}} - T_0)/(T_{\mathrm{f}} - T_0), \qquad \psi = (T_{\mathrm{f}} - T_{\mathrm{R}})/(T_{\mathrm{f}} - T_0)$$

The mass burning rate is determined from the ordinary expression for chemical kinetic rates; i.e., the fuel consumption rate is given by

$$d(\mathrm{F})/dt = -(\mathrm{F})(\mathrm{A})Z'e^{-E/RT_{\mathrm{R}}} = -(\mathrm{F})^2(\mathrm{A/F})Z'e^{-E/RT_{\mathrm{R}}}$$

where $(\mathrm{A/F})$ represents the air–fuel ratio. The concentration of the fuel can be written in terms of the total density and unburned mass fraction

$$(\mathrm{F}) = \frac{(\mathrm{F})}{(\mathrm{A}) + (\mathrm{F})}\rho\psi = \frac{1}{(\mathrm{A/F}) + 1}\rho\psi$$

which permits the rate expression to be written as

$$\frac{d(\mathrm{F})}{dt} = -\frac{1}{[(\mathrm{A/F}) + 1]^2}\rho^2\psi^2\left(\frac{\mathrm{A}}{\mathrm{F}}\right)Z'e^{-E/RT_{\mathrm{R}}}$$

Now the great simplicity in stirred reactor theory is realizable. Since (F), (A), and T_{R} are constant in the reactor, the rate of conversion is constant. It is now possible to represent the mass rate of burning in terms of the preceding chemical kinetic expression:

$$\dot{m}(1 - \psi) = -V\frac{(\mathrm{A}) + (\mathrm{F})}{(\mathrm{F})}\frac{d(\mathrm{F})}{dt}$$

or

$$\dot{m}(1 - \psi) = +V \left[\left(\frac{A}{F}\right) + 1 \right] \frac{1}{[(A/F) + 1]^2} \left(\frac{A}{F}\right) \rho^2 \psi^2 Z' e^{-E/RT_R}$$

From the equation of state, by defining

$$B = \frac{Z'}{[(A/F) + 1]}$$

and substituting for $(1 - \psi)$ in the last rate expression, one obtains

$$\left(\frac{\dot{m}}{V}\right) = \left(\frac{A}{F}\right) \left(\frac{P}{RT_R}\right)^2 \left[\frac{(T_f - T_R)^2}{T_f - T_0} \right] \frac{B e^{-E/RT_R}}{T_R - T_0}$$

By dividing through by P^2, one observes that

$$(\dot{m}/V P^2) = f(T_R) = f(A/F)$$

or

$$(\dot{m}/V P^2)(T_R - T_0) = f(T_R) = f(A/F)$$

which states that the heat release is also a function of T_R.

This derivation was made as if the overall order of the air–fuel reaction were 2. In reality, this order is found to be closer to 1.8. The development could have been carried out for arbitrary overall order n, which would give the result

$$(\dot{m}/V P^n)(T_R - T_0) = f(T_R) = f(A/F)$$

A plot of $(\dot{m}/V P^2)(T_R - T_0)$ versus T_R reveals a multivalued graph that exhibits a maximum as shown in Fig. 51. The part of the curve in Fig. 51 that approaches the value T_x asymptotically cannot exist physically since the mixture could not be ignited at temperatures this low. In fact, the major part of the curve, which is to the left of T_{opt}, has no physical meaning. At fixed volume and pressure it is not possible for both the mass flow rate and temperature of the reactor to rise. The only stable region exists between T_{opt} to T_f. Since it is not possible to mix some unburned gases with the product mixture and still obtain the adiabatic flame temperature, the reactor parameter must go to zero when $T_R = T_f$.

The value of T_R, which gives the maximum value of the heat release, is obtained by maximizing the last equation. The result is

$$T_{R,opt} = \frac{T_f}{1 + (2RT_f/E)}$$

For hydrocarbons, the activation energy falls within a range of 120–160 kJ/mol and the flame temperature in a range of 2000–3000 K. Thus,

$$(T_{R,opt}/T_f) \sim 0.75$$

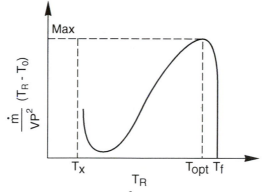

FIGURE 51   Stirred reactor parameter $(\dot{m}/VP^2)(T_R - T_0)$ as a function of reactor temperature T_R.

Stirred reactor theory reveals a fixed maximum mass loading rate for a fixed reactor volume and pressure. Any attempts to overload the system will quench the reaction. Attempts have been made to determine chemical kinetic parameters from stirred reactor measurements; however, the usefulness of such measurements is limited. First, the analysis is based on the assumption that a hydrocarbon–air system can be represented by a simple one-step overall order kinetic expression. Recent evidence indicates that such an assumption is not realistic. Second, the analysis is based on the assumption of complete instantaneous mixing, which is impossible to achieve experimentally.

In a positive sense, however, it is worth noting that the analysis does give the maximum overall energy release rate that is possible for a fuel–oxidizer mixture in a fixed volume at a given pressure.

G. FLAME STABILIZATION IN HIGH-VELOCITY STREAMS

The values of laminar flame speeds for hydrocarbon fuels in air are rarely greater than 45 cm/s. Hydrogen is unique in its flame velocity, which approaches 240 cm/s. If one could attribute a turbulent flame speed to hydrocarbon mixtures, it would be at most a few hundred centimeters per second. However, in many practical devices, such as ramjet and turbojet combustors in which high volumetric heat release rates are necessary, the flow velocities of the fuel–air mixture are of the order of 50 m/s. Furthermore, for such velocities, the boundary layers are too thin in comparison to the quenching distance for stabilization to occur by the same means as that obtaining in Bunsen burners. Thus, some other means for stabilization is necessary. In practice, stabilization is accomplished by causing some of the combustion products to recirculate, thereby continuously igniting the fuel mixture. Of course, continuous ignition could be obtained by inserting small pilot flames. Because pilot flames are an added inconvenience—and because they can blow themselves out—they are generally not used in fast flowing turbulent streams.

Recirculation of combustion products can be obtained by several means: (1) by inserting solid obstacles in the stream, as in ramjet technology (bluff-body stabilization); (2) by directing part of the flow or one of the flow constituents, usually air, opposed or normal to the main stream, as in gas turbine combustion chambers (aerodynamic stabilization), or (3) by using a step in the wall enclosure (step stabilization), as in the so-called "dump" combustors. These modes of stabilization are depicted in Fig. 52. Complete reviews of flame stabilization of premixed turbulent gases appear in Refs. [57 and 58].

Photographs of ramjet-type burners, which use rods as bluff obstacles, show that the regions behind the rods recirculate part of the flow that reacts to form hot combustion products; in fact, the wake region of the rod acts as a pilot flame. Nicholson and Fields [59] very graphically showed this effect by placing small aluminum particles in the flow (Fig. 53). The wake pilot condition initiates flame spread. The flame spread process for a fully developed turbulent wake has been depicted [57], as shown in Fig. 54. The theory of flame spread in a uniform laminar flow downstream from a laminar mixing zone has been fully developed [12, 57] and reveals that the angle of flame spread is $\sin^{-1}(S_L/U)$, where U is the main stream flow velocity. For a turbulent flame one approximates the spread angle by replacing S_L by an appropriate turbulent flame speed S_T. The limitations in defining S_T in this regard were described in Section 4.E.

The types of obstacles used in stabilization of flames in high-speed flows could be rods, vee gutters, toroids, disks, strips, etc. But in choosing the bluff-body stabilizer, the designer must consider not only the maximum blowoff velocity

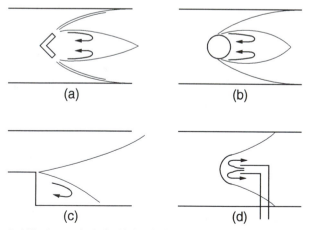

FIGURE 52   Stabilization methods for high-velocity streams: (a) vee gutter, (b) rodoz sphere, (c) step or sudden expansion, and (d) opposed jet (after Strehlow, "Combustion Fundamentals," McGraw-Hill, New York, 1985).

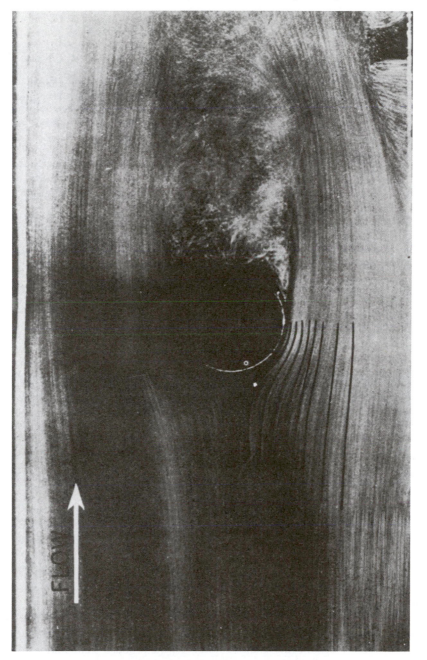

FIGURE 53   Flow past a 0.5-cm rod at a velocity of 50 ft/sec as depicted by an aluminum powder technique. Solid lines are experimental flow streamlines [59].

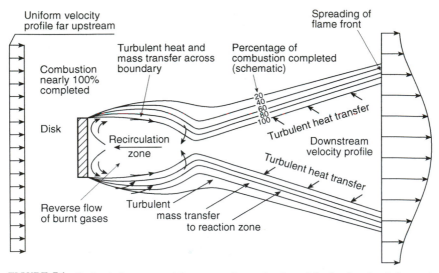

FIGURE 54   Recirculation zone and flame-spreading region for a fully developed turbulent wake behind a bluff body (after Williams [57]).

the obstacle will permit for a given flow, but also pressure drop, cost, ease of manufacture, etc.

Since the combustion chamber should be of minimum length, it is rare that a single rod, toroid, etc., is used. In Fig. 55, a schematic of flame spreading from multiple flame holders is given. One can readily see that multiple units can appreciably shorten the length of the combustion chamber. However, flame holders cause a stagnation pressure loss across the burner, and this pressure loss must be added to the large pressure drop due to burning. Thus, there is an optimum between the number of flame holders and pressure drop. It is sometimes difficult to use aerodynamic stabilization when large chambers are involved because the flow creating the recirculation would have to penetrate too far across the main stream. Bluff-body stabilization is not used in gas turbine systems because of the required combustor shape and the short lengths. In gas turbines a high weight penalty is paid for even the slightest increase in length. Because of reduced pressure losses, step stabilization has at times commanded attention. A wall heating problem associated with this technique would appear solvable by some transpiration cooling.

In either case, bluff-body or aerodynamic, blowout is the primary concern. In ramjets, the smallest frontal dimension for the highest flow velocity to be used is desirable; in turbojets, it is the smallest volume of the primary recirculation zone that is of concern; and in dump combustors, it is the least severe step.

There were many early experimental investigations of bluff-body stabilization. Most of this work [60] used premixed gaseous fuel–air systems and typically plotted the blowoff velocity as a function of the air–fuel ratio for various stabilized

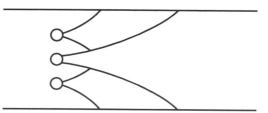

FIGURE 55 Flame spreading interaction behind multiple bluff-body flame stabilizers.

sizes, as shown in Fig. 56. Early attempts to correlate the data appeared to indicate that the dimensional dependence of blowoff velocity was different for different bluff-body shapes. Later, it was shown that the Reynolds number range was different for different experiments and that a simple independent dimensional dependence did not exist. Furthermore, the state of turbulence, the temperature of the stabilizer, incoming mixture temperature, etc., also had secondary effects. All these facts suggest that fluid mechanics plays a significant role in the process.

Considering that fluid mechanics does play an important role, it is worth examining the cold flow field behind a bluff body (rod) in the region called the wake. Figure 57 depicts the various stages in the development of the wake as the Reynolds number of the flow is increased. In region (1), there is only a slight slowing behind the rod and a very slight region of separation. In region (2), eddies start to form and the stagnation points are as indicated. As the Reynolds number increases, the eddy (vortex) size increases and the downstream stagnation point moves farther away from the rod. In region (3), the eddies become unstable, shed alternately, as shown in the figure, and $(h/a) \sim 0.3$. As the velocity u increases, the frequency N of shedding increases; $N \sim 0.3(u/d)$. In region (4), there is a complete turbulent wake behind the body. The stagnation point must pass 90° to

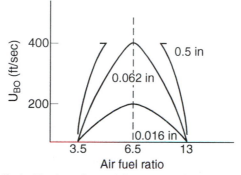

FIGURE 56 Blowoff velocities for various rod diameters as a function of air–fuel ratio. Short duct using premixed fuel–air mixtures 0.5 in data limited by choking of duct (after Scurlock [60]).

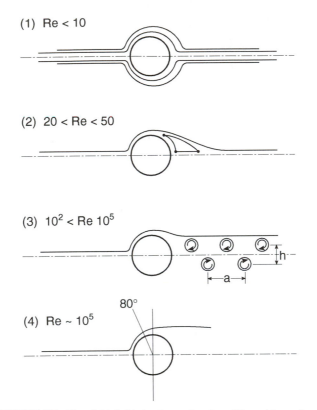

(1) Re < 10

(2) 20 < Re < 50

(3) 10^2 < Re 10^5

(4) Re ~ 10^5

$80°$

FIGURE 57   Flow fields behind rods as a function of Reynolds number.

about $80°$ and the boundary layer is also turbulent. The turbulent wake behind the body is eventually destroyed downstream by jet mixing.

The flow fields described in Fig. 57 are very specific in that they apply to cold flow over a cylindrical body. When spheres are immersed in a flow, region (3) does not exist. More striking, however, is the fact that when combustion exists over this Reynolds number range of practical interest, the shedding eddies disappear and a well-defined, steady vortex is established. The reason for this change in flow pattern between cold flow and a combustion situation is believed to be due to the increase in kinematic viscosity caused by the rise in temperature. Thus, the Reynolds number affecting the wake is drastically reduced, as discussed in the section of premixed turbulent flow. Then, it would be expected that in region (2) the Reynolds number range would be $10 <$ Re $< 10^5$. Flame holding studies by Zukoski and Marble [61, 62] showed that the ratio of the length of the wake (recirculation zone) to the diameter of the cylindrical flame holder was independent of the approach flow Reynolds number above a critical value of about 10^4. These Reynolds numbers are based on the critical dimension of the bluff body; that is, the diameter of the cylinder. Thus, one may assume that for an approach flow Reynolds number greater than 10^4, a fully developed turbulent wake would exist during combustion.

Experiments [57] have shown that any original ignition source located upstream, near or at the flame holder, appears to establish a steady ignition position from which a flame spreads from the wake region immediately behind the stabilizer. This ignition position is created by the recirculation zone that contains hot combustion products near the adiabatic flame temperature [62]. The hot combustion products cause ignition by transferring heat across the mixing layer between the free-stream gases and the recirculation wake. Based on these physical concepts, two early theories were developed that correlated the existing data well. One was proposed by Spalding [63] and the other, by Zukoski and Marble [61, 62]. Another early theory of flame stabilization was proposed by Longwell *et al.* [64], who considered the wake behind the bluff body to be a stirred reactor zone.

Considering the wake of a flame holder as a stirred reactor may be inconsistent with experimental data. It has been shown [57] that as blowoff is approached, the temperature of the recirculating gases remains essentially constant; furthermore, their composition is practically all products. Both of these observations are contrary to what is expected from stirred reactor theory. Conceivably, the primary zone of a gas turbine combustor might approach a state that could be considered completely stirred. Nevertheless, as will be shown, all three theories give essentially the same correlation.

Zukoski and Marble [61, 62] held that the wake of a flame holder establishes a critical ignition time. Their experiments, as indicated earlier, established that the length of the recirculating zone was determined by the characteristic dimension of the stabilizer. At the blowoff condition, they assumed that the free-stream combustible mixture flowing past the stabilizer had a contact time equal to the ignition time associated with the mixture; i.e., $\tau_w = \tau_i$, where τ_w is the flow contact time with the wake and τ_i is the ignition time. Since the flow contact time is given by

$$\tau_w = L/U_{BO}$$

where L is the length of the recirculating wake and U_{BO} is the velocity at blowoff, they essentially postulated that blowoff occurs when the Damkohler number has the critical value of 1; i.e.,

$$Da = (L/U_{BO})(1/\tau_i) = 1$$

The length of the wake is proportional to the characteristic dimension of the stabilizer, the diameter d in the case of a rod, so that

$$\tau_w \sim (d/U_{BO})$$

Thus it must follow that

$$(U_{BO}/d) \sim (1/\tau_i)$$

For second-order reactions, the ignition time is inversely proportional to the pressure. Writing the relation between pressure and time by referencing them to a standard pressure P_0 and time τ_0, one has

$$(\tau_0/\tau_i) = (P/P_0)$$

where P is the actual pressure in the system of concern.

The ignition time is a function of the combustion (recirculating) zone temperature, which, in turn, is a function of the air–fuel ratio (A/F). Thus,

$$(U_{BO}/dP) \sim (1/\tau_0 P_0) = f(T) = f(A/F)$$

Spalding [63] considered the wake region as one of steady-state heat transfer with chemical reaction. The energy equation with chemical reaction was developed and nondimensionalized. The solution for the temperature profile along the outer edge of the wake zone, which essentially heats the free stream through a mixing layer, was found to be a function of two nondimensional parameters that are functions of each other. Extinction or blowout was considered to exist when these dimensionless groups were not of the same order. Thus, the functional extinction condition could be written as

$$(U_{BO}d/\alpha) = f(Z'P^{n-1}d^2/\alpha)$$

where d is, again, the critical dimension, α is the thermal diffusivity, Z' is the pre-exponential in the Arrhenius rate constant, and n is the overall reaction order.

From laminar flame theory, the relationship

$$S_L \sim (\alpha R R)^{1/2}$$

was obtained, so that the preceding expression could be modified by the relation

$$S_L \sim (\alpha Z'P^{n-1})^{1/2}$$

Since the final correlations have been written in terms of the air–fuel ratio, which also specifies the temperature, the temperature dependences were omitted. Thus, a new proportionality could be written as

$$(S_L^2/\alpha) \sim Z'P^{n-1}$$

$$(Z'P^{n-1}d^2/\alpha) = (S_L^2 d^2/\alpha^2)$$

and the original functional relation would then be

$$(U_{BO}d/\alpha) \sim f(S_L d/\alpha) \sim (U_{BO}d/v)$$

Both correlating parameters are in the form of Peclet numbers, and the air–fuel ratio dependence is in S_L. Figure 58 shows the excellent correlation of data by the above expression developed from the Spalding analysis. Indeed, the power dependence of d with respect to blowoff velocity can be developed from the slopes of the lines in Fig. 58. Notice that the slope is 2 for values $(U_{BO}d/\alpha) > 10^4$, which was found experimentally to be the range in which a fully developed turbulent wake exists. The correlation in this region should be compared to the correlation developed from the work of Zukoski and Marble.

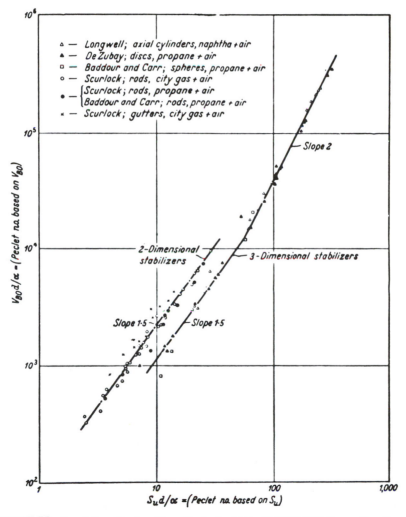

FIGURE 58   Correlation of various blowoff velocity data by Spalding [63]; $V_{BO} = U_{BO}$, $S_u = S_L$.

Stirred reactor theory was initially applied to stabilization in gas turbine combustor cans in which the primary zone was treated as a completely stirred region. This theory has sometimes been extended to bluff-body stabilization, even though aspects of the theory appear inconsistent with experimental measurements made in the wake of a flame holder. Nevertheless, it would appear that stirred reactor theory gives the same functional dependence as the other correlations developed. In the previous section, it was found from stirred reactor considerations that

$$(\dot{m}/VP^2) \cong f(\text{A/F})$$

for second-order reactions. If $\dot{m}$ is considered to be the mass entering the wake and V its volume, then the following proportionalities can be written:

$$\dot{m} = \rho A U \sim P d^2 U_{\text{BO}}, \qquad V \sim d^3$$

where A is an area. Substituting these proportionalities in the stirred reactor result, one obtains

$$[(P d^2 U_{\text{BO}})/(d^3 P^2)] = f(\text{A/F}) = (U_{\text{BO}}/dP)$$

which is the same result as that obtained by Zukoski and Marble. Indeed, in the turbulent regime, Spalding's development also gives the same form since in this regime the correlation can be written as the equality

$$(U_{\text{BO}}d/\alpha) = \text{const}(S_{\text{L}}d/\alpha)^2$$

Then it follows that

$$(U_{\text{BO}}/d) \sim (S_{\text{L}}^2/\alpha) \sim P^{n-1}f(T) \qquad \text{or} \qquad (U_{\text{BO}}/dP^{n-1}) \sim f(T)$$

Thus, for a second-order reaction

$$(U_{\text{BO}}dP) \sim f(T) \sim f(\text{A/F})$$

From these correlations it would be natural to expect that the maximum blowoff velocity as a function of air–fuel ratio would occur at the stoichiometric mixture ratio. For premixed gaseous fuel–air systems, the maxima do occur at this mixture ratio, as shown in Fig. 56. However, in real systems liquid fuels are injected upstream of the bluff-body flame holder in order to allow for mixing. Results [60] for such liquid injection systems show that the maximum blowoff velocity is obtained on the fuel-lean side of stoichiometric. This trend is readily explained by the fact that liquid droplets impinge on the stabilizer and enrich the wake. Thus, a stoichiometric wake undoubtedly occurs for a lean upstream liquid-fuel injection system. That the wake can be modified to alter blowoff characteristics was proved experimentally by Fetting *et al.* [65]. The trends of these experiments can be explained by the correlations developed in this section.

When designed to have sharp leading edges, recesses in combustor walls cause flow separation, as shown in Fig. 52c. During combustion, the separated regions establish recirculation zones of hot combustion products much like the wake of bluff-body stabilizers. Studies [66] of turbulent propane–air mixtures stabilized by wall recesses in a rectangular duct showed stability limits significantly wider than that of a gutter bluff-body flame holder and lower pressure drops. The observed blowoff limits for a variety of symmetrically located wall recesses showed [57] substantially the same results, provided: (1) the recess was of sufficient depth to support an adequate amount of recirculating gas, (2) the slope of the recess at the upstream end was sharp enough to produce separation, and (3) the geometric construction of the recess lip was such that flow oscillations were not induced.

The criterion for blowoff from recesses is essentially the same as that developed for bluff bodies, and L is generally taken to be proportional to the height of the recess [66]. The length of the recess essentially serves the same function as the length of the bluff-body recirculation zone unless the length is large enough for flow attachment to occur within the recess, in which case the recirculation length depends on the depth of the recess [12]. This latter condition applies to the so-called dump combustor, in which a duct with a small cross-sectional area exhausts coaxially with a right-angle step into a duct with a larger cross section. The recirculation zone forms at the step.

Recess stabilization appears to have two major disadvantages. The first is due to the large increase in heat transfer in the step area, and the second to flame spread angles smaller than those obtained with bluff bodies. Smaller flame spread angles demand longer combustion chambers.

Establishing a criterion for blowoff during opposed jet stabilization is difficult owing to the sensitivity of the recirculation region formed to its stoichiometry. This stoichiometry is well defined only if the main stream and opposed jet compositions are the same. Since the combustor pressure drop is of the same order as that found with bluff bodies [67], the utility of this means of stabilization is questionable.

PROBLEMS

1. A stoichiometric fuel–air mixing flowing in a Bunsen burner forms a well-defined conical flame. The mixture is then made leaner. For the same flow velocity in the tube, how does the cone angle change for the leaner mixture? That is, does the cone angle become larger or smaller than the angle for the stoichiometric mixture? Explain.
2. Sketch a temperature profile that would exist in a one-dimensional laminar flame. Superimpose on this profile a relative plot of what the rate of energy release would be through the flame as well. Below the inflection point in the temperature profile, large amounts of HO_2 are found. Explain why. If flame was due to a first-order, one-step decomposition reaction, could rate data be obtained directly from the existing temperature profile?
3. In which of the two cases would the laminar flame speed be greater: (1) oxygen in a large excess of a wet equimolar CO–CO_2 mixture or (2) oxygen

in a large excess of a wet equimolar CO-N_2 mixture? Both cases are ignitable, contain the same amount of water, and have the same volumetric oxygen–fuel ratio. Discuss the reasons for the selection made.

4. A gas mixture is contained in a soap bubble and ignited by a spark in the center so that a spherical flame spreads radially through the mixture. It is assumed that the soap bubble can expand. The growth of the flame front along a radius is followed by some photographic means. Relate the velocity of the flame front as determined from the photographs to the laminar flame speed as defined in the text. If this method were used to measure flame speeds, what would be its advantages and disadvantages?

5. On what side of stoichiometric would you expect the maximum flame speed of hydrogen–air mixtures? Why?

6. A laminar flame propagates through a combustible mixture in a horizontal tube 3 cm in diameter. The tube is open at both ends. Due to buoyancy effects, the flame tilts at a 45° angle to the normal and is planar. Assume the tilt is a straight flame front. The normal laminar flame speed for the combustible mixture is 40 cm/s. If the unburned gas mixture has a density of 0.0015 gm/cm³, what is the mass burning rate of the mixture in grams per second under this laminar flow condition?

7. The flame speed for a combustible hydrocarbon–air mixture is known to be 30 cm/s. The activation energy of such hydrocarbon reactions is generally assumed to be 160 kJ/mol. The true adiabatic flame temperature for this mixture is known to be 1600 K. An inert diluent is added to the mixture to lower the flame temperature to 1450 K. Since the reaction is of second order, the addition of the inert can be considered to have no other effect on any property of the system. Estimate the flame speed after the diluent is added.

8. A horizontal long tube 3 cm in diameter is filled with a mixture of methane and air in stoichiometric proportions at 1 atm and 27°C. The column is ignited at the left end and a flame propagates at uniform speed from left to right. At the left end of the tube there is a convergent nozzle which has a 2-cm diameter opening. At the right end there is a similar nozzle 0.3 cm in diameter at the opening. Calculate the velocity of the flame with respect to the tube in centimeters per second. Assume the following:

    (a) The effect of pressure increase on the burning velocity can be neglected; similarly, the small temperature increase due to adiabatic compression has a negligible effect.

    (b) The entire flame surface consumes combustible gases at the same rate as an ideal one-dimensional flame.

    (c) The molecular weight of the burned gases equals that of the unburned gases.

    (d) The flame temperature is 2100 K.

    (e) The normal burning velocity at stoichiometric is 40 cm/s.

Hint: Assume that the pressure in the burned gases is essentially 1 atm. In calculating the pressure in the cold gases make sure the value is correct to many decimal places.

9. A continuous-flow stirred reactor operates off the decomposition of gaseous ethylene oxide fuel. If the fuel injection temperature is 300 K, the volume of the reactor 1500 cm^3, and the operating pressure is 20 atm, calculate the maximum rate of heat evolution possible in the reactor. Assume that the ethylene oxide follows homogeneous first-order reaction kinetics and that values of the reaction rate constant k are

$$k = 3.5 \, \text{s}^{-1} \text{ at } 980 \, \text{K}$$

$$k = 50 \, \text{s}^{-1} \text{ at } 1000 \, \text{K}$$

$$k = 600 \, \text{s}^{-1} \text{ at } 1150 \, \text{K}$$

Develop any necessary rate data from these values. You are given that the adiabatic decomposition temperature of gaseous ethylene oxide is 1300 K. The heat of formation of gaseous ethylene oxide at 300 K is 50 kJ/mol. The overall reaction is

$$C_2H_4O \rightarrow CH_4 + CO.$$

10. What are the essential physical processes that determine the flammability limit?

11. You want to measure the laminar flame speed at 273 K of a homogeneous gas mixture by the Bunsen burner tube method. If the mixture to be measured is 9% natural gas in air, what size would you make the tube diameter? Natural gas is mostly methane. The laminar flame speed of the mixture can be taken as 34 cm/s at 298 K. Other required data can be found in standard reference books.

12. A ramjet has a flame stabilized in its combustion chamber by a single rod whose diameter is 1.25 cm. The mass flow of the unburned fuel air mixture entering the combustion chamber is 22.5 kg /s, which is the limit amount that can be stabilized at a combustor pressure of 3 atm for the cylindrical configuration. The ramjet is redesigned to fly with the same fuel–air mixture and a mass flow rate twice the original mass flow in the same size (cross section) combustor. The inlet diffusion is such that the temperature entering the combustor is the same as in the original case, but the pressure has dropped to 2 atm. What is the minimum size rod that will stabilize the flame under these new conditions?

13. A laminar flame propagates through a combustible mixture at 1 atm pressure, has a velocity of 50 cm s^{-1} and a mass burning rate of 0.1 g s^{-1}cm^2. The overall chemical reaction rate is second-order in that it depends only on the fuel and oxygen concentrations. Now consider a turbulent flame propagating through the same combustible mixture at a pressure of 10 atm. In this situation the turbulent intensity is such that the turbulent diffusivity is 10 times the laminar diffusivity. Estimate the turbulent flame propagation and mass burning rates.

14. Discuss the difference between explosion limits and flammability limits. Why is the lean flammability limit the same for both air and oxygen?
15. Explain briefly why halogen compounds are effective in altering flammability limits.

REFERENCES

1. Bradley, J. N., "Flame and Combustion Phenomena," Chap. 3. Methuen, London, 1969.
2. Westbrook, C. K., and Dryer, F. L., *Prog. Energy Combust. Sci.* **10**, 1 (1984).
3. Mallard, E., and Le Chatelier, H. L., *Ann. Mines* **4**, 379 (1883).
4. Semenov, N. N., *NACA Tech. Memo.* No. 1282 (1951).
5. Lewis, B., and von Elbe, G., "Combustion, Flames and Explosion of Gases," 2nd Ed., Chap. 5. Academic Press, New York, 1961.
6. Tanford, C., and Pease, R. N., *J. Chem. Phys.* **15**, 861 (1947).
7. Hirschfelder, J. O., Curtiss, C. F., and Bird, R. B., "The Molecular Theory of Gases and Liquids," Chap. 11. Wiley, New York, 1954.
8. Friedman, R., and Burke, E., *J. Chem. Phys.* **21**, 710 (1953).
9. von Karman, T., and Penner, S. S., "Selected Combustion Problems" (AGARD Combust. Colloq.), p. 5. Butterworth, London, 1954.
10. Zeldovich, Y. H., Barenblatt, G. I., Librovich, V. B., and Makviladze, G. M., "The Mathematical Theory of Combustion and Explosions." Nauka, Moscow, 1980 [Engl. transl., Plenum, New York, 1985].
11. Buckmaster, J. D., and Ludford, G. S. S., "The Theory or Laminar Flames." Cambridge Univ. Press, Cambridge, England, 1982.
12. Williams, F. A., "Combustion Theory," 2nd Ed. Benjamin-Cummins, Menlo Park, California, 1985.
13. Linan, A., and Williams , F. A., "Fundamental Aspects of Combustion." Oxford Univ. Press, Oxford, England, 1994.
14. Mikhelson, V. A., Ph.D. Thesis, Moscow Univ., Moscow, 1989.
15. Kee, R. J., Grcar, J. F., Smooke, M. D., and Miller, J. A., "A Fortran Program for Modeling Steady Laminar One-Dimensional Premixed Flames," Sandia Rep., SAND85-8240, (1985).
16. Kee, R. J., Rupley, F. M., and Miller, J. A., "CHEMKIN II: A Fortran Chemical Kinetics Package for the Analysis of Gas Phase Chemical Kinetics," Sandia Rep., SAND89-8009B (1989).
17. Kee, R. J., Dixon-Lewis, G., Warnatz, J., Coltrin, M. E., and Miller, J. A., "A Fortran Computer Code Package for the Evaluation of Gas-Phase Multicomponent Transport," Sandia Rep., SAND86-8246 (1986).
18. Gerstein, M., Levine, O., and Wong, E. L., *J. Am. Chem. Soc.* **73**, 418 (1951).
19. Flock, E. S., Marvin, C. S., Jr., Caldwell, F. R., and Roeder, C. H., *NACA Rep.* No. 682 (1940).
20. Powling, J., *Fuel* **28**, 25 (1949).
21. Spalding, D. B., and Botha, J. P., *Proc. R. Soc. London, Ser. A* **225**, 71 (1954).
22. Fristrom, R. M., "Flame Structure and Processes." Oxford Univ. Press, New York, 1995.
23. Hottel, H. C., Williams, G. C., Nerheim, N. M., and Schneider, G. R., *Int. Symp. Combust., 10th* p. 111. Combust. Inst., Pittsburgh, Pennsylvania, 1965.
24. *Natl. Advis. Comm. Aeronaut., Rep.* No. 1300, Chap. 4 (1959).
25. Zebatakis, K. S., *Bull.—U.S. Bur. Mines* No. 627 (1965).
26. Gibbs, G. J., and Calcote, H. F., *J. Chem. Eng. Data* **5**, 226 (1959).
27. Clingman, W. H., Jr., Brokaw, R. S., and Pease, R. D., *Int. Symp. Combust., 4th* p. 310. Williams & Wilkins, Baltimore, Maryland, 1953.
28. Leason, D. B., *Int. Symp. Combust., 4th* p. 369. Williams & Wilkins, Baltimore, Maryland, 1953.
29. Coward, H. F., and Jones, O. W., *Bull.—U.S. Bur. Mines* No. 503 (1951).
30. Spalding, D. B., *Proc. R. Soc. London, Ser. A* **240**, 83 (1957).

31. Jarosinsky, J., *Combust. Flame* **50**, 167 (1983).
32. Ronney, P. D., personal communication, 1995.
33. Buckmaster, J. D., and Mikolaitis, D., *Combust. Flame* **45**, 109 (1982).
34. Strehlow, R. A., and Savage, L. D., *Combust. Flame* **31**, 209 (1978).
35. Jarosinsky, J., Strehlow, R. A., and Azarbarzin, A., *Int. Symp. Combust., 19th* p. 1549. Combust. Inst., Pittsburgh, Pennsylvania, 1982.
36. Patnick, G. and Kailasanath, K., *Int. Symp. Combust., 24th* p. 189. Combust. Inst., Pittsburgh, Pennsylvania, 1994.
37. Noe, K., and Strehlow, R. A., *Int. Symp. Combust., 21st* p. 1899. Combust. Inst., Pittsburgh, Pennsylvania, 1986.
38. Ronney, P. D., *Int. Symp. Combust., 22nd* p. 1615. Combust. Inst., Pittsburgh, Pennsylvania, 1988.
39. Ronney, P. D., *in* "Lecture Notes in Physics" (T. Takeno and J. Buckmaster, eds.), p. 3. Springer-Verlag, New York, 1995.
40. Bradley, D., *Int. Symp. Combust., 24th* p. 247. Combust. Inst., Pittsburgh, Pennsylvania, 1992.
41. Bilger, R. W., *in* "Turbulent Reacting Flows" (P. A. Libby and F. A. Williams, eds.), p. 65. Springer-Verlag, Berlin, 1980.
42. Libby, P. A., and Williams, F. A., *in* "Turbulent Reacting Flows" (P. A. Libby and F. A. Williams, eds.), p. 1. Springer-Verlag, Berlin, 1980.
43. Bray, K. N. C., *in* "Turbulent Reacting Flows" (P. A. Libby and F. A. Williams, eds.), p. 115. Springer-Verlag, Berlin, 1980.
44. Klimov, A. M., *Zh. Prikl. Mekh. Tekh. Fiz.* **3**, 49 (1963).
45. Williams, F. A., *Combust. Flame* **26**, 269 (1976).
46. Darrieus, G., "Propagation d'un Front de Flamme." Congr. Mec. Appl., Paris, 1945.
47. Landau, L. D., *Acta Physicochem. URSS* **19**, 77(1944).
48. Markstein, G. H., *J. Aeronaut. Sci.* **18**, 199 (1951).
49. Damkohler, G., *Z. Elektrochem.* **46**, 601 (1940).
50. Schelkin, K. L., *NACA Tech. Memo* No. 1110 (1947).
51. CIavin, P., and Williams, F. A., *J. Fluid Mech.* **90**, 589 (1979).
52. Kerstein, A. R., and Ashurst, W. T., *Phys. Rev. Lett.* **68**, 934 (1992).
53. Yakhot, V., *Combust. Sci. Technol.* **60**, 191 (1988).
54. Borghi, R., *in* "Recent Advances in the Aerospace Sciences" (C. Bruno and C. Casci, eds.), Plenum, New York, 1985.
55. Abdel-Gayed, R. G., Bradley, D., and Lung, F. K.-K., *Combust. Flame* **76**, 213 (1989).
56. Longwell, J. P., and Weiss, M. A., *Ind. Eng. Chem.* **47**, 1634 (1955).
57. Williams, F. A., *in* "Applied Mechanics Surveys" (N. N. Abramson, H. Licbowita, J. M. Crowley, and S. Juhasz, eds.), p. 1158. Spartan Books, Washington, D.C., 1966.
58. Edelman, R. B., and Harsha, P. T., *Prog. Energ Combust. Sci.* **4**, 1 (1978).
59. Nicholson, H. M., and Fields, J. P., *Int Symp. Combust., 3rd* p. 44. Combust. Inst., Pittsburgh, Pennsylvania, 1949.
60. Scurlock, A. C., MIT Fuel Res. Lab. Meterol. Rep., No. 19 (1948).
61. Zukoski, E. E., and Marble, F. E., "Combustion Research and Reviews," p. 167. Butterworth, London, 1955.
62. Zukoski, E. E., and Marble, F. E., *Proc. Gas Dyn. Symp. Aerothermochem.* Evanston, IL., p. 205 (1956).
63. Spalding, D. B., "Some Fundamentals of Combustion," Chap. 5. Butterworth, London, 1955.
64. Longwell, J. P., Frost, E. E., and Weiss, M. A., *Ind. Eng. Chem.* **45**, 1629 (1953).
65. Fetting, F., Choudbury, P. R., and Wilhelm, R. H., *Int. Symp. Combust., 7th* p. 621. Combust. Inst., Pittsburgh, Pennsylvania, 1959.
66. Choudbury, P. R., and Cambel, A. B., *Int. Symp. Combust., 4th* p. 743. Williams & Wilkins, Baltimore, Maryland, 1953.
67. Putnam, A. A., *Jet Propul.* **27**, 177 (1957).

5
Detonation

A. INTRODUCTION

Established usage of certain terms related to combustion phenomena can be misleading, for what appear to be synonyms are not really so. Consequently, this chapter begins with a slight digression into the semantics of combustion, with some brief mention of subjects to be covered later.

1. Premixed and Diffusion Flames

The previous chapter covered primarily laminar flame propagation. By inspecting the details of the flow, particularly high-speed or higher Reynolds number flow, it was possible to consider the subject of turbulent flame propagation. These subjects (laminar and turbulent flames) are concerned with gases in the premixed state only. The material presented is not generally adaptable to the consideration of the combustion of liquids and solids, or systems in which the gaseous reactants diffuse toward a common reacting front.

Diffusion flames can best be described as the combustion state controlled by mixing phenomena—i.e., the diffusion of fuel into oxidizer, or vice versa— until some flammable mixture ratio is reached. According to the flow state of the individual diffusing species, the situation may be either laminar or turbulent. It will be shown later that gaseous diffusion flames exist, that liquid burning proceeds by a diffusion mechanism, and that the combustion of solids and some solid propellants falls in this category as well.

2. Explosion, Deflagration, and Detonation

Explosion is a term that corresponds to rapid heat release (or pressure rise). An explosive gas or gas mixture is one that will permit rapid energy release, as compared to most steady, low-temperature reactions. Certain gas mixtures (fuel and oxidizer) will not propagate a burning zone or combustion wave. These gas mixtures are said to be outside the flammability limits of the explosive gas.

Depending upon whether the combustion wave is a deflagration or detonation, there are limits of flammability or detonation.

In general, the combustion wave is considered as a deflagration only, although the detonation wave is another class of the combustion wave. The detonation wave is, in essence, a shock wave that is sustained by the energy of the chemical reaction in the highly compressed explosive medium existing in the wave. Thus, a *deflagration* is a subsonic wave sustained by a chemical reaction and a *detonation* is a supersonic wave sustained by chemical reaction. In the normal sense, it is common practice to call a combustion wave a "flame," so *combustion wave*, *flame*, and *deflagration* have been used interchangeably.

It is a very common error to confuse a pure explosion and a detonation. An explosion does not necessarily require the passage of a combustion wave through the exploding medium, whereas an explosive gas mixture must exist in order to have either a deflagration or a detonation. That is, both deflagrations and detonations require rapid energy release; but explosions, though they too require rapid energy release, do not require the presence of a waveform.

The difference between deflagration and detonation is described qualitatively, but extensively, by Table 1 (from Friedman [1]).

3. The Onset of Detonation

Depending upon various conditions, an explosive medium may support either a deflagration or a detonation wave. The most obvious conditions are confinement, mixture ratio, and ignition source.

Original studies of gaseous detonations have shown no single sequence of events due primarily to what is now known as the complex cellular structure of a detonation wave. The primary result of an ordinary thermal initiation always appears to be a flame that propagates with subsonic speed. When conditions are such that the flame causes adiabatic compression of the still unreacted mixture ahead of it, the flame velocity increases. According to some early observations, the speed of the flame seems to rise gradually until it equals that of a detonation wave. Normally, a discontinuous change of velocity is observed from the low flame velocity to the high speed of detonation. In still other observations, the detonation wave has been seen to originate apparently spontaneously some distance ahead

TABLE 1 Qualitative Differences between Detonations and Deflagration in Gases

	Usual magnitude of ratio	
Ratio	Detonation	Deflagration
u_u/c_u^a	5–10	0.0001–0.03
u_b/u_u	0.4–0.7	4–16
P_b/P_u	13–55	0.98–0.976
T_b/T_u	8–21	4–16
ρ_b/ρ_u	1.4–2.6	0.06–0.25

$^a c_u$ is the acoustic velocity in the unburned gases. u_u/c_u is the Mach number of the wave

of the flame front. The place of origin appears to coincide with the location of a shock wave sent out by the expanding gases of the flame. Modern experiments and analysis have shown that these seemingly divergent observations were in part attributable to the mode of initiation. In detonation phenomena one can consider that two modes of initiation exist: a slower mode, sometimes called thermal initiation, in which there is transition from deflagration; and a fast mode brought about by an ignition blast or strong shock wave. Some [2] refer to these modes as self-ignition and direct ignition, respectively.

When an explosive gas mixture is placed in a tube having one or both ends open, a combustion wave can propagate when the tube is ignited at an open end. This wave attains a steady velocity and does not accelerate to a detonation wave.

If the mixture is ignited at one end that is closed, a combustion wave is formed; and, if the tube is long enough, this wave can accelerate to a detonation. This thermal initiation mechanism is described as follows. The burned gas products from the initial deflagration have a specific volume of the order of 5–15 times that of the unburned gases ahead of the flame. Since each preceding compression wave that results from this expansion tends to heat the unburned gas mixture somewhat, the sound velocity increases and the succeeding waves catch up with the initial one. Furthermore, the preheating tends to increase the flame speed, which then accelerates the unburned gas mixture even further to a point where turbulence is developed in the unburned gases. Then, a still greater velocity and acceleration of the unburned gases and compression waves are obtained. This sequence of events forms a shock that is strong enough to ignite the gas mixture ahead of the front. The reaction zone behind the shock sends forth a continuous compression wave that keeps the shock front from decaying, and so a detonation is obtained. At the point of shock formation a detonation forms and propagates back into the unburned gases [2, 3]. Transverse vibrations associated with the onset of detonation have been noticed, and this observation has contributed to the understanding of the cellular structure of the detonation wave. Photographs of the onset of detonation have been taken by Oppenheim and co-workers [3] using a strobscopic-laser Schlieren technique.

The reaction zone in a detonation wave is no different from that in other flames, in that it supplies the sustaining energy. A difference does exist in that the detonation front initiates chemical reaction by compression, by diffusion of both heat and species, and thus inherently maintains itself. A further, but not essential, difference worth noting is that the reaction takes place with extreme rapidity in highly compressed and preheated gases.

The transition length for deflagration to detonation is of the order of a meter for highly reactive fuels such as acetylene, hydrogen, and ethylene, but larger for most other hydrocarbon–air mixtures. Consequently, most laboratory results for detonation are reported for acetylene and hydrogen. Obviously, this transition length is governed by many physical and chemical aspects of the experiments. Such elements as overall chemical composition, physical aspects of the detonation tube, and initiation ignition characteristics can all play a role. Interestingly, some question exists as to whether methane will detonate at all.

According to Lee [2], direct initiation of a detonation can occur only when a strong shock wave is generated by a source and this shock retains a certain minimum strength for some required duration. Under these conditions "the blast and reaction front are always coupled in the form of a multiheaded detonation wave that starts at the (ignition) source and expands at about the detonation velocity"[2]. Because of the coupling phenomena necessary, it is apparent that reaction rates play a role in whether a detonation is established or not. Thus ignition energy is one of the dynamic detonation parameters discussed in the next section. However, no clear quantitative or qualitative analysis exists for determining this energy, so this aspect of the detonation problem will not be discussed further.

B. DETONATION PHENOMENA

Scientific studies of detonation phenomena date back to the end of the 19th century and persist as an active field of investigation. A wealth of literature has developed over this period; consequently, no detailed reference list will be presented. For details and extensive references the reader should refer to books on detonation phenomena [4], Williams' book on combustion [5], and the review by Lee [6].

Since the discussion of the detonation phenomena to be considered here will deal extensively with premixed combustible gases, it is appropriate to introduce much of the material by comparison with deflagration phenomena. As the data in Table 1 indicate, deflagration speeds are orders of magnitude less than those of detonation. The simple solution for laminar flame speeds given in Chapter 4 was essentially obtained by starting with the integrated conservation and state equations. However, by establishing the Hugoniot relations and developing a Hugoniot plot, it was shown that deflagration waves are in the very low Mach number regime; then it was assumed that the momentum equation degenerates and the situation through the wave is one of uniform pressure. The degeneration of the momentum equation ensures that the wave velocity to be obtained from the integrated equations used will be small. In order to obtain a deflagration solution, it was necessary to have some knowledge of the wave structure and the chemical reaction rates that affected this structure.

As will be shown, the steady solution for the detonation velocity does not involve any knowledge of the structure of the wave. The Hugoniot plot discussed in Chapter 4 established that detonation is a large Mach number phenomenon. It is apparent, then, that the integrated momentum equation is included in obtaining a solution for the detonation velocity. However, it was also noted that there are four integrated conservation and state equations and five unknowns. Thus, other considerations were necessary to solve for the velocity. Concepts proposed by Chapman [7] and Jouguet [8] provided the additional insights that permitted the mathematical solution to the detonation velocity problem. The solution from the integrated conservation equations is obtained by assuming the detonation wave to be steady, planar, and one-dimensional; this approach is called Chapman–Jouguet theory. Chapman and Jouguet established for these conditions that the flow behind the supersonic detonation is sonic. The point on the Hugoniot curve that represents this condition is called the Chapman–Jouguet point and the other physical conditions of this state are called the Chapman–Jouguet conditions. What is unusual about the Chapman–Jouguet solution is that, unlike the deflagration problem, it requires no knowledge of the structure of the detonation wave and equilibrium thermodynamic calculations for the Chapman–Jouguet state suffice. As will be shown, the detonation velocities which result from this theory agree very well with experimental observations, even in near-limit conditions when the flow structure near the flame front is highly three-dimensional [6].

Reasonable models for the detonation wave structure have been presented by Zeldovich [9], von Neumann [10], and Döring [11]. Essentially, they constructed the detonation wave to be a planar shock followed by a reaction zone initiated after an induction delay. This structure, which is generally referred to as the ZND model, will be discussed further in a later section.

As in consideration of deflagration phenomena, other parameters are of import in detonation research. These parameters—detonation limits, initiation energy, critical tube diameter, quenching diameter, and thickness of the supporting reaction zone—require a knowledge of the wave structure and hence of chemical reaction rates. Lee [6] refers to these parameters as "dynamic" to distinguish them from the equilibrium "static" detonation states, which permit the calculation of the detonation velocity by Chapman–Jouguet theory.

Calculation of the dynamic parameters using a ZND wave structure model do not agree with experimental measurements, mainly because the ZND structure is unstable and is never observed experimentally except under transient conditions. This disagreement is not surprising, as numerous experimental observations show that all self-sustained detonations have a three-dimensional cell structure that comes about because reacting blast "wavelets" collide with each other to form a series of waves transverse to the direction of propagation. Currently, there are no suitable theories that define this three-dimensional cell structure.

The next section deals with the calculation of the detonation velocity based on Chapman–Jouguet theory. The subsequent section discusses the ZND model in detail, and the last deals with the dynamic detonation parameters.

C. HUGONIOT RELATIONS AND THE HYDRODYNAMIC THEORY OF DETONATIONS

If one is to examine the approach to calculating the steady, planar, one-dimensional gaseous detonation velocity, one should consider a system configuration similar to that given in Chapter 4. For the configuration given there, it is important to understand the various velocity symbols used. Here, the appropriate velocities are defined in Fig. 1. With these velocities, the integrated conservation and static equations are written as

$$\rho_1 u_1 = \rho_2 u_2 \tag{1}$$

$$P_1 + \rho_1 u_1^2 = P_2 + \rho_2 u_2^2 \tag{2}$$

$$c_p T_1 + \tfrac{1}{2} u_1^2 + q = c_p T_2 + \tfrac{1}{2} u_2^2 \tag{3}$$

$$P_1 = \rho_1 R T_1 \qquad \text{(connects known variables)} \tag{4}$$

$$P_2 = \rho_2 R T_2$$

In this type of representation, all combustion events are collapsed into a discontinuity (the wave). Thus, the unknowns are u_1, u_2, ρ_2, T_2, and P_2. Since there are four equations and five unknowns, an eigenvalue cannot be obtained. Experimentally it is found that the detonation velocity is uniquely constant for a given mixture. In order to determine all unknowns, one must know something about the internal structure (rate of reaction), or one must obtain another necessary condition, which is the case for the detonation velocity determination.

Velocities with wave fixed in lab space

$-u_2$	0	$-u_1$
Burned gas	Wave direction in lab flame	Unburned gas
$u_1 - u_2$	u_1	0

Actual laboratory velocities
or velocities with respect to the tube

FIGURE 1   Velocities used in analysis of detonation problem.

1. Characterization of the Hugoniot Curve and the Uniqueness of the Chapman–Jouguet Point

The method of obtaining a unique solution, or the elimination of many of the possible solutions, will be deferred at present. In order to establish the argument for the nonexistence of various solutions, it is best to pinpoint or define the various velocities that arise in the problem and then to develop certain relationships that will prove convenient.

First, one calculates expressions for the velocities, u_1 and u_2. From Eq. (1),

$$u_2 = (\rho_1/\rho_2)u_1$$

Substituting in Eq. (2), one has

$$\rho_1 u_1^2 - (\rho_1^2/\rho_2)u_1^2 = (P_2 - P_1)$$

Dividing by ρ_1^2, one obtains

$$u_1^2 \left(\frac{1}{\rho_1} - \frac{1}{\rho_2} \right) = \frac{P_2 - P_1}{\rho_1^2},$$

$$u_1^2 = \frac{1}{\rho_1^2} \left[(P_2 - P_1) / \left(\frac{1}{\rho_1} - \frac{1}{\rho_2} \right) \right] \tag{5}$$

Note that Eq. (5) is the equation of the Rayleigh line which can also be derived without involving any equation of state. Since $(\rho_1 u_1)^2$ is always a positive value, it follows that if $\rho_2 > \rho_1$, $P_2 > P_1$ and vice versa. Since the sound speed c can be written as

$$c_1^2 = \gamma R T_1 = \gamma P_1(1/\rho_1)$$

where γ is the ratio of specific heats,

$$\gamma M_1^2 = \left(\frac{P_2}{P_1} - 1 \right) / \left[1 - \frac{(1/\rho_2)}{(1/\rho_1)} \right] \tag{5'}$$

Substituting Eq. (5) into Eq. (1), one obtains

$$u_2^2 = \frac{1}{\rho_2^2} \left[(P_2 - P_1) / \left(\frac{1}{\rho_1} - \frac{1}{\rho_2} \right) \right] \tag{6}$$

and

$$\gamma M_2^2 = \left(1 - \frac{P_1}{P_2} \right) / \left[\frac{(1/\rho_1)}{(1/\rho_2)} - 1 \right] \tag{6'}$$

A relationship called the Hugoniot equation, which is used throughout these developments, is developed as follows. Recall that

$$(c_p/R) = \gamma/(\gamma - 1), \qquad c_p = R[\gamma/(\gamma - 1)]$$

Substituting in Eq. (3), one obtains

$$R[\gamma/(\gamma - 1)]T_1 + \tfrac{1}{2}u_1^2 + q = R[\gamma/(\gamma - 1)]T_2 + \tfrac{1}{2}u_2^2$$

Implicit in writing the equation in this form is the assumption that c_p and γ are constant throughout. Since $RT = P/\rho$, then

$$\frac{\gamma}{\gamma - 1}\left(\frac{P_2}{\rho_2} - \frac{P_1}{\rho_1}\right) - \tfrac{1}{2}(u_1^2 - u_2^2) = q \tag{7}$$

One then obtains from Eqs. (5) and (6)

$$u_1^2 - u_2^2 = \left(\frac{1}{\rho_1^2} - \frac{1}{\rho_2^2}\right)\left[\frac{P_2 - P_1}{(1/\rho_1) - (1/\rho_2)}\right] = \frac{\rho_2^2 - \rho_1^2}{\rho_1^2 \rho_2^2}\left[\frac{P_2 - P_1}{(1/\rho_1) - (1/\rho_2)}\right]$$

$$= \left(\frac{1}{\rho_1^2} - \frac{1}{\rho_2^2}\right)\left[\frac{P_2 - P_1}{(1/\rho_1) - (1/\rho_2)}\right] = \left(\frac{1}{\rho_1} + \frac{1}{\rho_2}\right)(P_2 - P_1) \tag{8}$$

Substituting Eq. (8) into Eq. (7), one obtains the Hugoniot equation

$$\frac{\gamma}{\gamma - 1}\left(\frac{P_2}{\rho_2} - \frac{P_1}{\rho_1}\right) - \tfrac{1}{2}(P_2 - P_1)\left(\frac{1}{\rho_1} + \frac{1}{\rho_2}\right) = q \tag{9}$$

This relationship, of course, will hold for a shock wave when q is set equal to zero. The Hugoniot equation is also written in terms of the enthalpy and internal energy changes. The expression with internal energies is particularly useful in the actual solution for the detonation velocity u_1. If a total enthalpy (sensible plus chemical) in unit mass terms is defined such that

$$h \equiv c_p T + h°$$

where $h°$ is the heat of formation in the standard state and per unit mass, then a simplification of the Hugoniot equation evolves. Since by this definition

$$q = h_1° - h_2°$$

Eq. (3) becomes

$$\tfrac{1}{2}u_1^2 + c_p T_1 + h_1° = c_p T_2 + h_2° + \tfrac{1}{2}u_2^2 \quad \text{or} \quad \tfrac{1}{2}(u_1^2 - u_2^2) = h_2 - h_1$$

Or further from Eq. (8), the Hugoniot equation can also be written as

$$\tfrac{1}{2}(P_2 - P_1)[(1/\rho_1) + (1/\rho_2)] = h_2 - h_1 \tag{10}$$

To develop the Hugoniot equation in terms of the internal energy, one proceeds by first writing

$$h = e + RT = e + (P/\rho)$$

where e is the total internal energy (sensible plus chemical) per unit mass. Substituting for h in Eq. (10), one obtains

$$\frac{1}{2}\left[\left(\frac{P_2}{\rho_1} + \frac{P_2}{\rho_2}\right) - \left(\frac{P_1}{\rho_1} + \frac{P_1}{\rho_2}\right)\right] = e_2 + \frac{P_2}{\rho_2} - e_1 - \frac{P_1}{\rho_1}$$

$$\frac{1}{2}\left(\frac{P_2}{\rho_1} - \frac{P_2}{\rho_2} + \frac{P_1}{\rho_1} - \frac{P_1}{\rho_2}\right) = e_2 - e_1$$

Another form of the Hugoniot equation is obtained by factoring:

$$\tfrac{1}{2}(P_2 + P_1)[(1/\rho_1) - (1/\rho_2)] = e_2 - e_1 \tag{11}$$

If the energy equation [Eq. (3)] is written in the form

$$h_1 + \tfrac{1}{2}u_1^2 = h_2 + \tfrac{1}{2}u_2^2$$

the Hugoniot relations [Eqs. (10) and (11)] are derivable without the perfect gas and constant c_p and γ assumptions and thus are valid for shocks and detonations in gases, metals, etc.

There is also interest in the velocity of the burned gases with respect to the tube, since as the wave proceeds into the medium at rest, it is not known whether the burned gases proceed in the direction of the wave (i.e., follow the wave) or proceed away from the wave. From Fig. 1 it is apparent that this velocity, which is also called the particle velocity (Δu), is

$$\Delta u = u_1 - u_2$$

and from Eqs. (5) and (6)

$$\Delta u = \{[(1/\rho_1) - (1/\rho_2)](P_2 - P_1)\}^{1/2} \tag{12}$$

Before proceeding further, it must be established which values of the velocity of the burned gases are valid. Thus, it is now best to make a plot of the Hugoniot equation for an arbitrary q. The Hugoniot equation is essentially a plot of all the possible values of $(1/\rho_2, P_2)$ for a given value of $(1/\rho_1, P_1)$ and a given q. This point $(1/\rho_1, P_1)$, called A, is also plotted on the graph.

The regions of possible solutions are constructed by drawing the tangents to the curve that go through $A[(1/\rho_1, P_1)]$. Since the form of the Hugoniot equation obtained is a hyperbola, there are two tangents to the curve through A, as shown in Fig. 2. The tangents and horizontal and vertical lines through the initial condition A divide the Hugoniot curve into five regions, as specified by Roman numerals (I–V). The horizontal and vertical through A are, of course, the lines of constant P and $1/\rho$. A pressure difference for a final condition can be determined very readily from the Hugoniot relation [Eq. (9)] by considering the conditions along the vertical through A, i.e., the condition of constant $(1/\rho_1)$ or constant volume

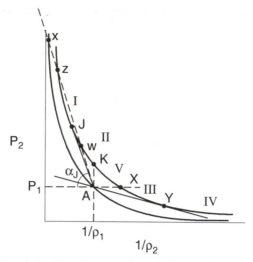

FIGURE 2   Hugoniot relationship with energy release divided into five regions (I–V) and the shock Hugoniot.

heating:

$$\frac{\gamma}{\gamma - 1}\left(\frac{P_2 - P_1}{\rho}\right) - \left(\frac{P_2 - P_1}{\rho}\right) = q$$

$$\left[\left(\frac{\gamma}{\gamma - 1}\right) - 1\right]\left(\frac{P_2 - P_1}{\rho}\right) = q, \quad (P_2 - P_1) = \rho(\gamma - 1)q \quad (13)$$

From Eq. (13), it can be considered that the pressure differential generated is proportional to the heat release q. If there is no heat release ($q = 0$), $P_1 = P_2$ and the Hugoniot curve would pass through the initial point A. As implied before, the shock Hugoniot curve must pass through A. For different values of q, one obtains a whole family of Hugoniot curves.

The Hugoniot diagram also defines an angle α_J such that

$$\tan \alpha_J = \frac{(P_2 - P_1)}{(1/\rho_1) - (1/\rho_2)}$$

From Eq. (5) then

$$u_1 = (1/\rho_1)(\tan \alpha_J)^{1/2}$$

Any other value of a obtained, say by taking points along the curve from J to K and drawing a line through A, is positive and the velocity u_1 is real and possible.

However, from K to X, one does not obtain a real velocity due to negative α_J. Thus, region V does not represent real solutions and can be eliminated. A result in this region would require a compression wave to move in the negative direction — an impossible condition.

Regions III and IV give expansion waves, which are the low-velocity waves already classified as deflagrations. That these waves are subsonic can be established from the relative order of magnitude of the numerator and denominator of Eq. (6′), as has already been done in Chapter 4.

Regions I and II give compression waves, high velocities, and are the regions of detonation (also as established in Chapter 4).

One can verify that regions I and II give compression waves and regions III and IV give expansion waves by examining the ratio of Δu to u_1 obtained by dividing Eq. (12) by the square root of Eq. (5):

$$\frac{\Delta u}{u_1} = \frac{(1/\rho_1) - (1/\rho_2)}{(1/\rho_1)} = 1 - \frac{(1/\rho_2)}{(1/\rho_1)} \tag{14}$$

In regions I and II, the detonation branch of the Hugoniot curve, $1/\rho_2 < 1/\rho_1$ and the right-hand side of Eq. (14) is positive. Thus, in detonations, the hot gases follow the wave. In regions III and IV, the deflagration branch of the Hugoniot curve, $1/\rho_2 > 1/\rho_1$ and the right-hand side of Eq. (14) is negative, Thus, in deflagrations the hot gases move away from the wave.

Thus far in the development, the deflagration and detonation branches of the Hugoniot curve have been characterized and region V has been eliminated. There are some specific characteristics of the tangency point J that were initially postulated by Chapman [7] in 1889. Chapman established that the slope of the adiabat is exactly the slope through J; i.e.,

$$\left[\frac{(P_2 - P_1)}{(1/\rho_1) - (1/\rho_2)}\right]_J = -\left\{\left[\frac{\partial P_2}{\partial(1/\rho_2)}\right]_s\right\}_J \tag{15}$$

The proof of Eq. (15) is a very interesting one and is verified in the following development. From thermodynamics one can write for every point along the Hugoniot curve

$$T_2\,ds_2 = de_2 + P_2\,d(1/\rho_2) \tag{16}$$

where s is the entropy per unit mass. Differentiating Eq. (11), the Hugoniot equation in terms of e is

$$de_2 = -\tfrac{1}{2}(P_1 + P_2)\,d(1/\rho_2) + [(1/\rho_1) - (1/\rho_2)]\,dP_2$$

since the initial conditions e_1, P_1, and $(1/\rho_1)$ are constant. Substituting this result in Eq. (16), one obtains

$$T_2\,ds_2 = -\tfrac{1}{2}(P_1 + P_2)\,d(1/\rho_2) + \tfrac{1}{2}[(1/\rho_1) - (1/\rho_2)]\,d P_2 + P_2 d(1/\rho_2)$$

$$= -\tfrac{1}{2}(P_1 - P_2)\,d(1/\rho_2) + \tfrac{1}{2}[(1/\rho_1) - (1/\rho_2)]\,d P_2 \qquad (17)$$

It follows from Eq. (17) that along the Hugoniot curve,

$$T_2\left[\frac{ds_2}{d(1/\rho_2)}\right]_H = \frac{1}{2}\left(\frac{1}{\rho_1} - \frac{1}{\rho_2}\right)\left\{-\frac{P_1 - P_2}{(1/\rho_1) - (1/\rho_2)} + \left[\frac{d P_2}{d(1/\rho_2)}\right]_H\right\} \qquad (18)$$

The subscript H is used to emphasize that derivatives are along the Hugoniot curve. Now, somewhere along the Hugoniot curve, the adiabatic curve passing through the same point has the same slope as the Hugoniot curve. There, ds_2 must be zero and Eq. (18) becomes

$$\left\{\left[\frac{d P_2}{d(1/\rho_2)}\right]_H\right\}_S = \frac{(P_1 - P_2)}{(1/\rho_1) - (1/\rho_2)} \qquad (19)$$

But notice that the right-hand side of Eq. (19) is the value of the tangent that also goes through point A; therefore, the tangency point along the Hugoniot curve is J. Since the order of differentiation on the left-hand side of Eq. (19) can be reversed, it is obvious that Eq. (15) has been developed.

Equation (15) is useful in developing another important condition at point J. The velocity of sound in the burned gas can be written as

$$c_2^2 = \left(\frac{\partial P_2}{\partial \rho_2}\right)_S = -\frac{1}{\rho_2^2}\left[\frac{\partial P_2}{\partial(1/\rho_2)}\right]_S \qquad (20)$$

Cross-multiplying and comparing with Eq. (15), one obtains

$$\rho_2^2 c_2^2 = -\left[\frac{\partial P_2}{\partial(1/\rho_2)}\right]_S = \left[\frac{(P_2 - P_1)}{(1/\rho_1) - (1/\rho_2)}\right]_J$$

or

$$[c_2^2]_J = \frac{1}{\rho_2^2}\left[\frac{(P_2 - P_1)}{(1/\rho_1) - (1/\rho_2)}\right]_J = [u_2^2]_J$$

Therefore

$$[u_2]_J = [c_2]_J \quad \text{or} \quad [M_2]_J = 1$$

Thus, the important result is obtained that at J the velocity of the burned gases (u_2) is equal to the speed of sound in the burned gases. Furthermore, an identical analysis would show, as well, that

$$[M_2]_Y = 1$$

Recall that the velocity of the burned gas with respect to the tube (Δu) is written as

$$\Delta u = u_1 - u_2$$

or at J

$$u_1 = \Delta u + u_2, \qquad u_1 = \Delta u + c_2 \tag{21}$$

Thus, at J the velocity of the unburned gases moving into the wave, i.e., the detonation velocity, equals the velocity of sound in the gases behind the detonation wave plus the mass velocity of these gases (the velocity of the burned gases with respect to the tube). It will be shown presently that this solution at J is the only solution that can exist along the detonation branch of the Hugoniot curve for actual experimental conditions.

Although the complete solution of u_1 at J will not be attempted at this point, it can be shown readily that the detonation velocity has a simple expression now that u_2 and c_2 have been shown to be equal. The conservation of mass equation is rewritten to show that

$$\rho_1 u_1 = \rho_2 u_2 = \rho_2 c_2 \qquad \text{or} \qquad u_1 = \frac{\rho_2}{\rho_1} c_2 = \frac{(1/\rho_1)}{(1/\rho_2)} c_2 \tag{21a}$$

Then from Eq. (20) for c_2, it follows that

$$u_1 = \frac{(1/\rho_1)}{(1/\rho_2)} (1/\rho_2) \left\{ -\left[\frac{\partial P_2}{\partial(1/\rho_2)} \right]_S \right\}^{1/2} = \left(\frac{1}{\rho_1} \right) \left\{ -\left[\frac{\partial P_2}{\partial(1/\rho_2)} \right] \right\}^{1/2} \tag{22}$$

With the condition that $u_2 = c_2$ at J, it is possible to characterize the different branches of the Hugoniot curve in the following manner:

Region I: Strong detonation since $P_2 > P_J$ (supersonic flow to subsonic)
Region II: Weak detonation since $P_2 < P_J$ (supersonic flow to supersonic)
Region III: Weak deflagration since $P_2 > P_Y$ (subsonic flow to subsonic)
Region IV: Strong deflagration since $P_2 < P_Y$ ($1/\rho_2 > 1/\rho_Y$) (subsonic flow to supersonic)

At points above J, $P_2 > P_J$; thus, $u_2 < u_{2,J}$. Since the temperature increases somewhat at higher pressures, $c_2 > c_{2,J}$ [$c = (\gamma R T)^{1/2}$]. More exactly, it is shown in the next section that above J, $c_2 > u_2$. Thus, M_2 above J must be less than 1. Similar arguments for points between J and K reveal $M_2 > M_{2,J}$ and hence supersonic flow behind the wave. At points past Y, $1/\rho_2 > 1/\rho_1$, or the velocities are greater than $u_{2,Y}$. Also past Y, the sound speed is about equal to the value at Y. Thus, past Y, $M_2 > 1$. A similar argument shows that $M_2 < 1$ between X and Y. Thus, past Y, the density decreases; therefore, the heat addition prescribes that there be supersonic outflow. But, in a constant area duct, it is not possible to have heat addition and proceed past the sonic condition. Thus, region IV is not a physically possible region of solutions and is ruled out.

Region III (weak deflagration) encompasses the laminar flame solutions that were treated in Chapter 4.

There is no condition by which one can rule out strong detonation; however, Chapman stated that in this region only velocities corresponding to J are valid. Jouguet [8] gave the following analysis.

If the final values of P and $1/\rho$ correspond to a point on the Hugoniot curve higher than the point J, it can be shown (next section) that the velocity of sound in the burned gases is greater than the velocity of the detonation wave relative to the burned gases because, above J, c_2 is shown to be greater than u_2. (It can also be shown that the entropy is a minimum at J and that M_J is greater than values above J.) Consequently, if a rarefaction wave due to any reason whatsoever starts behind the wave, it will catch up with the detonation front; $u_1 - \Delta u = u_2$. The rarefaction will then reduce the pressure and cause the final value of P_2 and $1/\rho_2$ to drop and move down the curve toward J. Thus, points above J are not stable. Heat losses, turbulence, friction, etc., can start the rarefaction. At the point J, the velocity of the detonation wave is equal to the velocity of sound in the burned gases plus the mass velocity of these gases, so that the rarefaction will not overtake it; thus, J corresponds to a "self-sustained" detonation. The point and conditions at J are referred to as the Chapman–Jouguet results.

Thus, it appears that solutions in region I are possible, but only in the transient state, since external effects quickly break down this state. Some investigators have claimed to have measured strong detonations in the transient state. There also exist standing detonations that are strong. Overdriven detonations have been generated by pistons, and some investigators have observed oblique detonations which are overdriven.

The argument used to exclude points on the Hugoniot curve below J is based on the structure of the wave. If a solution in region II were possible, there would be an equation that would give results in both region I and region II. The broken line in Fig. 2 representing this equation would go through A and some point, say Z, in region I and another point, say W, in region II. Both Z and W must correspond to the same detonation velocity. The same line would cross the shock Hugoniot curve at point X. As will be discussed in Section E, the structure of the detonation is a shock wave followed by chemical reaction. Thus, to detail the structure of the detonation wave on Fig. 2, the pressure could rise from A to X, and then be reduced along the broken line to Z as there is chemical energy release. To proceed to the weak detonation solution at W, there would have to be further energy release. However, all the energy is expended for the initial mixture at point Z. Hence, it is physically impossible to reach the solution given by W as long as the structure requires a shock wave followed by chemical energy release. Therefore, the condition of tangency at J provides the additional condition necessary to specify the detonation velocity uniquely. The physically possible solutions represented by the Hugoniot curve, thus, are only those shown in Fig. 3.

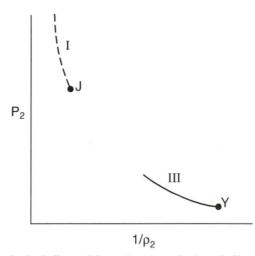

FIGURE 3   The only physically possible steady-state results along the Hugoniot—the point J and region III. The broken line represents transient conditions.

2. Determination of the Speed of Sound in the Burned Gases for Conditions above the Chapman–Jouguet Point

a. Behavior of the Entropy along the Hugoniot Curve

Equation (18) was written as

$$T_2 \left[\frac{d^2 s}{d(1/\rho_2)^2} \right]_H = \frac{1}{2} \left(\frac{1}{\rho_1} - \frac{1}{\rho_2} \right) \left\{ \left[\frac{d P_2}{d(1/\rho_2)} \right]_H - \frac{P_1 - P_2}{(1/\rho_1) - (1/\rho_2)} \right\}$$

with the further consequence that $[ds_2/d(1/\rho_2)] = 0$ at points Y and J (the latter is the Chapman–Jouguet point for the detonation condition).

Differentiating again and taking into account the fact that

$$[ds_2/d(1/\rho_2)] = 0$$

at point J, one obtains

$$\left[\frac{d^2 s}{d(1/\rho_2)^2} \right]_{H \text{ at } J \text{ or } Y} = \frac{(1/\rho_1) - (1/\rho_2)}{2 T_2} \left[\frac{d^2 P_2}{d(1/\rho_2)^2} \right] \tag{23}$$

Now $[d^2 P_2/d(1/\rho_2)^2] > 0$ everywhere, since the Hugoniot curve has its concavity directed toward the positive ordinates (see formal proof later).

Therefore, at point J, $[(1/\rho_1) - (1/\rho_2)] > 0$, and hence the entropy is minimum at J. At point Y, $[(1/\rho_1) - (1/\rho_2)] < 0$, and hence s_2 goes through a maximum.

When $q = 0$, the Hugoniot curve represents an adiabatic shock. Point $1(P_1, \rho_1)$ is then on the curve and Y and J are 1. Then $[(1/\rho_1) - (1/\rho_2)] = 0$,

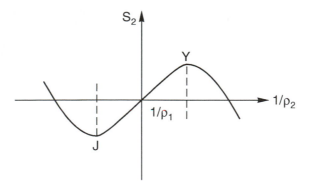

FIGURE 4   Variation of entropy along the Hugoniot.

and the classical result of the shock theory is found; i.e., the shock Hugoniot curve osculates the adiabat at the point representing the conditions before the shock.

Along the detonation branch of the Hugoniot curve, the variation of the entropy is as given in Fig. 4. For the adiabatic shock, the entropy variation is as shown in Fig. 5.

b. The Concavity of the Hugoniot Curve

Solving for P_2 in the Hugoniot relation, one obtains

$$P_2 = \frac{a + b(1/\rho_2)}{c + d(1/\rho_2)}$$

where

$$a = q + \frac{\gamma + 1}{2(\gamma - 1)} \frac{P_1}{\rho_1}, \quad b = -\frac{1}{2} P_1, \quad c = -\frac{1}{2} \rho_1^{-1}, \quad d = \frac{\gamma + 1}{2(\gamma - 1)} \tag{24}$$

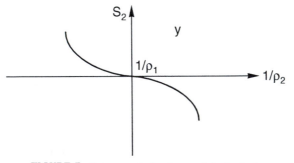

FIGURE 5   Entropy variation for an adiabatic shock.

From this equation for the pressure, it is obvious that the Hugoniot curve is a hyperbola. Its asymptotes are the lines

$$\frac{1}{\rho_2} = \left(\frac{\gamma-1}{\gamma+1}\right)\left(\frac{1}{\rho_1}\right) > 0, \qquad P_2 = -\frac{\gamma-1}{\gamma+1}P_1 < 0$$

The slope is

$$\left[\frac{dP_2}{d(1/\rho_2)}\right]_{\mathrm{H}} = \frac{bc-ad}{[c+d(1/\rho_2)]^2}$$

where

$$bc - ad = -\left[\frac{\gamma+1}{2(\gamma-1)}q + \frac{P_1}{\rho_1}\frac{\gamma}{(\gamma-1)^2}\right] < 0 \tag{25}$$

since $q > 0$, $P_1 > 0$, and $\rho_1 > 0$. A complete plot of the Hugoniot curves with its asymptotes would be as shown in Fig. 6. From Fig. 6 it is seen, as could be seen from earlier figures, that the part of the hyperbola representing the strong detonation branch has its concavity directed upward. It is also possible to determine directly the sign of

$$\left[\frac{d^2 P_2}{d(1/\rho_2)^2}\right]_{\mathrm{H}}$$

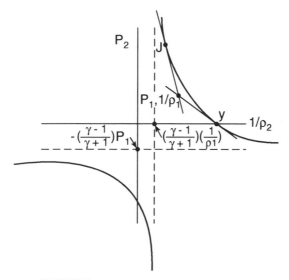

FIGURE 6 Asymptotes to the Hugoniot curves.

By differentiating Eq. (24), one obtains

$$\frac{d^2 P_2}{d(1/\rho_2)^2} = \frac{2d(ad - bc)}{[c + d(1/\rho_2)^3]}$$

Now, $d > 0$, $ad - bc > 0$ [Eq. (25)], and

$$c + d\left(\frac{1}{\rho_2}\right) = \frac{1}{2}\left[\frac{\gamma + 1}{\gamma - 1}\left(\frac{1}{\rho_2}\right) - \left(\frac{1}{\rho_1}\right)\right] > 0$$

The solutions lie on the part of the hyperbola situated on the right-hand side of the asymptote

$$(1/\rho_2) = [(\gamma - 1)/(\gamma + 1)](1/\rho_1)$$

Hence

$$[d^2 P_2/d(1/\rho_2)^2] > 0$$

c. The Burned Gas Speed

Here

$$ds = \left(\frac{\partial s}{\partial(1/\rho)}\right)_P d\left(\frac{1}{\rho}\right) + \left(\frac{\partial s}{\partial P}\right)_{1/\rho} dP \tag{26}$$

Since $ds = 0$ for the adiabat, Eq. (26) becomes

$$0 = \left[\frac{\partial s}{\partial(1/\rho)}\right]_P + \left(\frac{\partial s}{\partial P}\right)_{1/\rho}\left[\frac{\partial P}{\partial(1/\rho)}\right]_s \tag{27}$$

Differentiating Eq. (26) along the Hugoniot curve, one obtains

$$\left[\frac{ds}{d(1/\rho)}\right]_H = \left[\frac{\partial s}{\partial(1/\rho)}\right]_P + \left(\frac{\partial s}{\partial P}\right)_{1/\rho}\left[\frac{dP}{d(1/\rho)}\right]_H \tag{28}$$

Subtracting and transposing Eqs. (27) and (28), one has

$$\left[\frac{dP}{d(1/\rho)}\right]_H - \left[\frac{\partial P}{\partial(1/\rho)}\right]_s = \frac{[ds/d(1/\rho)]_H}{(\partial s/\partial P)_{1/\rho}} \tag{29}$$

A thermodynamic expression for the enthalpy is

$$dh = T\,ds + dP/\rho \tag{30}$$

With the conditions of constant c_p and an ideal gas, the expressions

$$dh = c_p\,dT, \qquad T = P/R\rho, \qquad c_p = [\gamma/(\gamma - 1)]R$$

are developed and substituted in

$$dh = \left(\frac{\partial h}{\partial P}\right) dP + \frac{\partial h}{\partial(1/\rho)} d\left(\frac{1}{\rho}\right)$$

to obtain

$$dh = \left(\frac{\gamma}{\gamma - 1}\right) R \left[\left(\frac{1}{R\rho}\right) dP + \frac{P}{R} d\left(\frac{1}{\rho}\right)\right] \tag{31}$$

Combining Eqs. (30) and (31) gives

$$ds = \frac{1}{T} \left[\left(\frac{\gamma}{\gamma - 1}\right) \frac{1}{\rho} dP - \frac{dP}{\rho} + \left(\frac{\gamma}{\gamma - 1}\right) P d\left(\frac{1}{\rho}\right)\right]$$

$$= \frac{1}{T} \left[\left(\frac{1}{\gamma - 1}\right) \frac{dP}{\rho} + \left(\frac{\gamma}{\gamma - 1}\right) \rho R T d\left(\frac{1}{\rho}\right)\right]$$

$$= \frac{dP}{(\gamma - 1)\rho T} + \left(\frac{\gamma}{\gamma - 1}\right) R\rho d\left(\frac{1}{\rho}\right)$$

Therefore,

$$(\partial s/\partial P)_{1/\rho} = 1/(\gamma - 1)\rho T \tag{32}$$

Then substituting in the values of Eq. (32) into Eq. (29), one obtains

$$\left[\frac{\rho P}{\partial(1/\rho)}\right]_{\text{H}} - \left[\frac{\partial P}{\partial(1/\rho)}\right]_{\text{S}} = (\gamma - 1)\rho T \left[\frac{\partial s}{\partial(1/\rho)}\right]_{\text{H}} \tag{33}$$

Equation (18) may be written as

$$\left[\frac{\partial P}{\partial(1/\rho)}\right]_{\text{H}} - \frac{P_1 - P_2}{(1/\rho_1) - (1/\rho_2)} = \frac{2T_2}{(1/\rho_1) - (1/\rho_2)} \left[\frac{\partial s_2}{\partial(1/\rho)}\right]_{\text{H}} \tag{34}$$

Combining Eqs. (33) and (34) gives

$$\left[-\frac{dP_2}{\rho(1/\rho_2)}\right]_{\text{S}} - \frac{P_2 - P_1}{(1/\rho_1) - (1/\rho_2)}$$

$$= \left[\frac{\partial s_2}{\partial(1/\rho_2)}\right]_{\text{H}} \left[-\frac{2T_2}{(1/\rho_1) - (1/\rho_2)} + (\gamma - 1)\rho_2 T_2\right]$$

or

$$\rho_2^2 c_2^2 - \rho_2^2 u_2^2 = \frac{P_2}{R} \left[\gamma - \frac{1 + (\rho_1/\rho_2)}{1 - (\rho_1/\rho_2)}\right] \left[\frac{\partial s_2}{\partial(1/\rho_2)}\right]_{\text{H}} \tag{35}$$

Since the asymptote is given by

$$1/\rho_2 = [(\gamma - 1)/(\gamma + 1)](1/\rho_1)$$

values of $(1/\rho_2)$ on the right-hand side of the asymptote must be

$$1/\rho_2 > [(\gamma - 1)/(\gamma + 1)](1/\rho_1)$$

which leads to

$$\left[\gamma - \frac{1 + (\rho_1/\rho_2)}{1 - (\rho_1/\rho_2)}\right] < 0$$

Since also $[\partial s/\partial(1/\rho_2)] < 0$, the right-hand side of Eq. (35) is the product of two negative numbers, or a positive number. If the right-hand side of Eq. (35) is positive, c_2 must be greater than u_2; i.e.,

$$c_2 > u_2$$

3. Calculation of the Detonation Velocity

With the background provided, it is now possible to calculate the detonation velocity for an explosive mixture at given initial conditions. Equation (22)

$$u_1 = \left(\frac{1}{\rho_1}\right)\left[-\frac{dP_2}{d(1/\rho_2)}\right]_S^{1/2} \tag{22}$$

shows the strong importance of density of the initial gas mixture, which is reflected more properly in the molecular weight of the products, as will be derived later.

For ideal gases, the adiabatic expansion law is

$$PV^\gamma = \text{const} = P_2(1/\rho_2)^{\gamma_2}$$

Differentiating this expression, one obtains

$$\left(\frac{1}{\rho_2}\right)^{\gamma_2} dP_2 + P_2(1/\rho_2)^{\gamma_2 - 1}\gamma_2\, d(1/\rho_2) = 0$$

which gives

$$-\left[\frac{dP_2}{d(1/\rho_2)}\right]_S = \frac{P_2}{(1/\rho_2)}\gamma_2 \tag{36}$$

Substituting Eq. (36) into Eq. (22), one obtains

$$u_1 = \frac{(1/\rho_1)}{(1/\rho_2)}[\gamma_2 P_2(1/\rho_2)]^{1/2} = \frac{(1/\rho_1)}{(1/\rho_2)}(\gamma_2 R T_2)^{1/2}$$

If one defines

$$\mu = (1/\rho_1)/(1/\rho_2)$$

then

$$u_1 = \mu(\gamma_2 R T_2)^{1/2} \tag{37}$$

Rearranging Eq. (5), it is possible to write

$$P_2 - P_1 = u_1^2 \frac{(1/\rho_1) - (1/\rho_2)}{(1/\rho_1)^2}$$

Substituting for u_1^2 from above, one obtains

$$(P_2 - P_1)\frac{(1/\rho_2)}{\gamma_2 P_2} = \left(\frac{1}{\rho_1} - \frac{1}{\rho_2}\right) \tag{38}$$

Now Eq. (11) was

$$e_2 - e_1 = \tfrac{1}{2}(P_2 + P_1)\left[(1/\rho_1) - (1/\rho_2)\right]$$

Substituting Eq. (38) into Eq. (11), one has

$$e_2 - e_1 = \frac{1}{2}(P_2 + P_1)\frac{(P_2 - P_1)(1/\rho_2)}{\gamma_2 P_2}$$

or

$$e_2 - e_1 = \frac{1}{2}\frac{(P_2^2 - P_1^2)(1/\rho_2)}{\gamma_2 P_2}$$

Since $P_2^2 > P_1^2$,

$$e_2 - e_1 = \frac{1}{2}\frac{P_2^2(1/\rho_2)}{\gamma_2 P_2} = \frac{1}{2}\frac{P_2(1/\rho_2)}{\gamma_2}$$

Recall that all expressions are in mass units; therefore, the gas constant R is not the universal gas constant. Indeed, it should now be written R_2 to indicate this condition. Thus

$$e_2 - e_1 = \frac{1}{2}\frac{P_2(1/\rho_2)}{\gamma_2} = \frac{1}{2}\frac{R_2 T_2}{\gamma_2} \tag{39}$$

Recall, as well, that e is the sum of the sensible internal energy plus the internal energy of formation. Equation (39) is the one to be solved in order to obtain T_2, and hence u_1. However, it is more convenient to solve this expression on a molar basis, because the available thermodynamic data and stoichiometric equations are in molar terms.

Equation (39) may be written in terms of the universal gas constant R' as

$$e_2 - e_1 = \tfrac{1}{2}(R'/MW_2)(T_2/\gamma_2) \tag{40}$$

where MW_2 is the average molecular weight of the products. The gas constant R used throughout this chapter must be the engineering gas constant since all the equations developed are in terms of unit mass, not moles. R' specifies the universal gas constant. If one multiplies through Eq. (40) with MW_1, the average molecular weight of the reactants, one obtains

$$(MW_1/MW_2)e_2(MW_2) - (MW_1)e_1 = \tfrac{1}{2}(R'T_2/\gamma_2)(MW_1/MW_2)$$

or

$$n_2 E_2 - E_1 = \tfrac{1}{2}(n_2 R'T_2/\gamma_2) \tag{41}$$

where the E's are the total internal energies per mole of all reactants or products and n_2 is (MW_1/MW_2), which is the number of moles of the product per mole of reactant. Usually, one has more than one product and one reactant; thus, the E's are the molar sums.

Now to solve for T_2, first assume a T_2 and estimate ρ_2 and MW_2, which do not vary substantially for burned gas mixtures. For these approximations, it is possible to determine $1/\rho_2$ and P_2.

If Eq. (38) is multiplied by $(P_1 + P_2)$,

$$(P_1 + P_2)\{(1/\rho_1) - (1/\rho_2)\} = (P_2^2 - P_1^2)(1/\rho_2)/\gamma_2 P_2$$

Again $P_2^2 \gg P_1^2$, so that

$$\frac{P_1}{\rho_1} - \frac{P_1}{\rho_2} + \frac{P_2}{\rho_1} - \frac{P_2}{\rho_2} = \frac{P_2(1/\rho_2)}{\gamma_2}$$

$$\frac{P_2}{\rho_1} - \frac{P_1}{\rho_2} = \frac{P_2(1/\rho_2)}{\gamma_2} + \frac{P_2}{\rho_2} - \frac{P_1}{\rho_1}$$

or

$$\frac{P_2\rho_2}{\rho_2\rho_1} - \frac{P_1\rho_1}{\rho_1\rho_2} = \frac{R_2 T_2}{\gamma_2} + R_2 T_2 - R_1 T_1$$

$$R_2 T_2 \left[\frac{(1/\rho_1)}{(1/\rho_2)} \right] - R_1 T_1 \left[\frac{(1/\rho_2)}{(1/\rho_1)} \right] = R_2 T_2 - R_1 T_1$$

In terms of μ,

$$R_2 T_2 \mu - R_1 T_1 (1/\mu) = [(1/\gamma_2) + 1]R_2 T_2 - R_1 T_1$$

which gives

$$\mu^2 - [(1/\gamma_2) + 1 - (R_1 T_1/R_2 T_2)]\mu - (R_1 T_1/R_2 T_2) = 0 \tag{42}$$

This quadratic equation can be solved for μ; thus, for the initial condition $(1/\rho_1)$, $(1/\rho_2)$ is known. P_2 is then determined from the ratio of the state equations at 2 and 1:

$$P_2 = \mu(R_2 T_2/R_1 T_1)P_1 = \mu\left(\frac{MW_1 T_2}{MW_2 T_1}\right)P_1 \tag{42a}$$

Thus, for the assumed T_2, P_2 is known. Then it is possible to determine the equilibrium composition of the burned gas mixture in the same fashion as described in Chapter 1. For this mixture and temperature, both sides of Eq. (39) or (41) are deduced. If the correct T_2 was assumed, both sides of the equation will be equal. If not, reiterate the procedure until T_2 is found. The correct γ_2 and MW_2 will be determined readily. For the correct values, u_1 is determined from Eq. (37) written as

$$u_1 = \mu \left(\frac{\gamma_2 R' T_2}{MW_2} \right)^{1/2} \tag{42b}$$

The physical significance of Eq. (42b) is that the detonation velocity is proportional to $(T_2/MW_2)^{1/2}$; thus it will not maximize at the stoichiometric mixture, but at one that is more likely to be fuel-rich.

The solution is simpler if the assumption $P_2 > P_1$ is made. Then from Eq. (38)

$$\left(\frac{1}{\rho_1} - \frac{1}{\rho_2} \right) = \frac{1}{\gamma_2} \left(\frac{1}{\rho_2} \right), \qquad \left(\frac{1}{\rho_2} \right) = \frac{\gamma_2}{1 + \gamma_2} \left(\frac{1}{\rho_1} \right), \qquad \mu = \frac{\gamma_2 + 1}{\gamma_2}$$

Since one can usually make an excellent guess of γ_2, one obtains μ immediately and, thus, P_2. Furthermore, μ does not vary significantly for most detonation mixtures, particularly when the oxidizer is air. It is a number close to 1.8, which means, as Eq. (21a) specifies, that the detonation velocity is 1.8 times the sound speed in the burned gases.

Gordon and McBride [12] present a more detailed computational scheme and the associated computational program.

D. COMPARISON OF DETONATION VELOCITY CALCULATIONS WITH EXPERIMENTAL RESULTS

In the previous discussion of laminar and turbulent flames, the effects of the physical and chemical parameters on flame speeds were considered and the trends were compared with the experimental measurements. It is of interest here to recall that it was not possible to calculate these flame speeds explicitly; but, as stressed throughout this chapter, it is possible to calculate the detonation velocity accurately. Indeed, the accuracy of the theoretical calculations, as well as the ability to measure the detonation velocity precisely, has permitted some investigators to calculate thermodynamic properties (such as the bond strength of nitrogen and heat of sublimation of carbon) from experimental measurements of the detonation velocity.

In their book, Lewis and von Elbe [13] made numerous comparisons between calculated detonation velocities and experimental values. This book is a source of such data. Unfortunately, most of the data available for comparison purposes were calculated long before the advent of digital computers. Consequently, the theoretical values do not account for all the dissociation that would normally take place. The data presented in Table 2 were abstracted from Lewis and von Elbe [13] and were so chosen to emphasize some important points about the factors that affect the detonation velocity. Although the agreement between the calculated and experimental values in Table 2 can be considered quite good, there is no doubt that the agreement would be much better if dissociation of all possible species had been allowed for in the final condition. These early data are quoted here because there have been no recent similar comparisons in which the calculated values were determined for equilibrium dissociation concentrations using modern computational techniques. Some data from Strehlow [14] are shown in Table 3, which provides a comparison of measured and calculated detonation velocities. The experimental data in both tables have not been corrected for the infinite tube diameter condition for which the calculations hold. This small correction would make the general agreement shown even better. Note that all experimental results are somewhat less than the calculated values. The calculated results in Table 3 are the more accurate ones because they were obtained by using the Gordon–McBride [12] computational program, which properly accounts for dissociation in the product composition. Shown in Table 4 are further calculations of detonation parameters for propane–air and H_2–air at various mixture ratios. Included in these tables are the adiabatic flame temperatures (T_{ad}) calculated at the pressure of the burned detonation gases (P_2). There are substantial differences between these values and the corresponding T_2's for the detonation condition. This difference is due to the nonisentropicity of the detonation process. The entropy change across the shock condition contributes to the additional energy term.

Variations in the initial temperature and pressure should not affect the detonation velocity for a given initial density. A rise in the initial temperature could only cause a much smaller rise in the final temperature. In laminar flame theory, a small rise in final temperature was important since the temperature was in an

TABLE 2 Detonation Velocities of Stoichiometric Hydrogen–Oxygen Mixtures[a]

Mixture	P_2 (atm)	T_2 (K)	u_1 (m/s) Calculated	u_1 (m/s) Experimental
$(2H_2 + O_2)$	18.05	3583	2806	2819
$(2H_2 + O_2) + 5O_2$	14.13	2620	1732	1700
$(2H_3 + O_2) + 5N_2$	14.39	2685	1850	1822
$(2H_2 + O_2) + 4H_2$	15.97	2975	3627	3527
$(2H_2 + O_2) + 5He$	16.32	3097	3617	3160
$(2H_2 + O_2) + 5Ar$	16.32	3097	1762	1700

[a] $P_0 = 1$ atm, $T_0 = 291$ K.

TABLE 3 Detonation Velocities of Various Mixtures[a]

	Measured velocity (m/s)	Calculated		
		Velocity (m/s)	P_2 (atm)	T_2 (K)
$4H_2 + O_2$	3390	3408	17.77	3439
$2H_2 + O_2$	2825	2841	18.56	3679
$H_2 + 3O_2$	1663	1737	14.02	2667
$CH_4 + O_2$	2528	2639	31.19	3332
$CH_2 + 1.5O_2$	2470	2535	31.19	3725
$0.7C_2N_2 + O_2$	2570	2525	45.60	5210

[a] $P_0 = 1$ atm, $T_0 = 298$ K.

exponential term. For detonation theory, recall that

$$u_1 = \mu(\gamma_2 R_2 T_2)^{1/2}$$

γ_2 does not vary significantly and μ is a number close to 1.8 for many fuels and stoichiometric conditions.

Examination of Table 2 leads one to expect that the major factor affecting u_1 is the initial density. Indeed, many investigators have stated that the initial density is one of the most important parameters in determining the detonation velocity. This point is best seen by comparing the results for the mixtures in which the helium and argon inerts are added. The lower-density helium mixture gives a much higher detonation velocity than the higher-density argon mixture, but identical values of P_2 and T_2 are obtained.

TABLE 4 Detonation Velocities of Fuel–Air Mixtures

Fuel–Air Mixture	Hydrogen–Air $\phi = 0.6$		Hydrogen–Air $\phi = 1.0$		Propane–Air $\phi = 0.60$	
	1	2	1	2	1	2
M	4.44	1.00	4.84	1.00	4.64	1.00
u(m/s)	1710	973	1971	1092	1588	906
P(atm)	1.0	12.9	1.0	15.6	1.0	13.8
T(K)	298	2430	2998	2947	298	2284
ρ/ρ_1	1.00	1.76	1.00	1.80	1.00	1.75
T_{ad} at P_1(K)		1838		2382		1701
T_{ad} at P_2(K)		1841		2452		1702

Notice as well that the addition of excess H_2 gives a larger detonation velocity than the stoichiometric mixture. The temperature of the stoichiometric mixture is higher throughout. One could conclude that this variation is a result of the initial density of the mixture. The addition of excess oxygen lowers both detonation velocity and temperature. Again, it is possible to argue that excess oxygen increases the initial density.

Whether the initial density is the important parameter should be questioned. The initial density appears in the parameter μ. A change in the initial density by species addition also causes a change in the final density, so that, overall, μ does not change appreciably. However, recall that

$$R_2 = R'/MW_2 \qquad \text{or} \qquad u_1 = \mu(\gamma_2 R' T_2/MW_2)^{1/2}$$

where R' is the universal gas constant and MW_2 is the average molecular weight of the burned gases. It is really MW_2 that is affected by initial diluents, whether the diluent is an inert or a light-weight fuel such as hydrogen. Indeed, the ratio of the detonation velocities for the excess helium and argon cases can be predicted almost exactly if one takes the square root of the inverse of the ratio of the molecular weights. If it is assumed that little dissociation takes place in these two burned gas mixtures, the reaction products in each case are two moles of water and five moles of inert. In the helium case, the average molecular weight is 9; in the argon case, the average molecular weight is 33.7. The square root of the ratio of the molecular weights is 2.05. The detonation velocity calculated for the argon mixtures is 1762. The estimated velocity for helium would be $2.05 \times 1762 = 3560$, which is very close to the calculated result of 3617. Since the c_p's of He and Ar are the same, T_2 remains the same.

The variation of the detonation velocity u_1, Mach number of the detonation M_1, and the physical parameters at the Chapman–Jouguet (burned gas) condition with equivalence ratio ϕ is most interesting. Figures 7–12 show this variation for hydrogen, methane, acetylene, ethene, propane, and octane detonating in oxygen and air. The data in Fig. 7 are interesting in that hydrogen in air or oxygen has a greater detonation velocity than any of the hydrocarbons. Indeed, there is very little difference in u_1 between the hydrocarbons considered. The slight differences coincide with heats of formation of the fuels and hence the final temperature T_2. No maximum for hydrogen was reached in Fig. 7 because at large ϕ, the molecular weight at the burned gas condition becomes very low and $(T_2/MW_2)^{1/2}$ becomes very large. As discussed later in Section F, the rich detonation limit for H_2 in oxygen occurs at $\phi \cong 4.5$. The rich limit for propane in oxygen occurs at $\phi \cong 2.5$. Since the calculations in Fig. 7 do not take into account the structure of the detonation wave, it is possible to obtain results beyond the limit. The same effect occurs for deflagrations in that, in a pure adiabatic condition, calculations will give values of flame speeds outside known experimental flammability limits.

The order of the Mach numbers for the fuels given in Fig. 8 is exactly the inverse of the u_1 values in Fig. 7. This trend is due to the sound speed in the initial

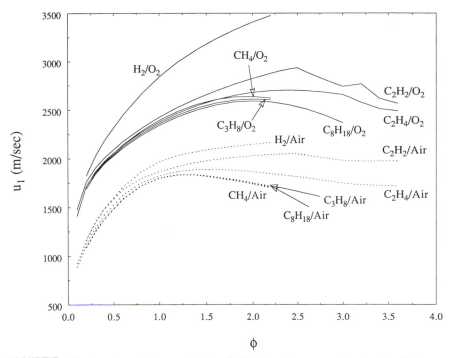

FIGURE 7   The detonation velocity u_1 of various fuels in air and oxygen as a function of equivalence ratio ϕ at initial conditions of $P_1 = 1$ atm and $T_1 = 298$ K.

mixture. Since for the calculations, T_1 was always 298 K and P_1 was 1 atm, the sound speed essentially varies with the inverse of the square root of the average molecular weight (MW_1) of the initial mixture. For H_2–O_2 mixtures, MW_1 is very low compared to that of the heavier hydrocarbons. Thus the sound speed in the initial mixture is very large. What is most intriguing is to compare Figs. 9 and 10. The ratio T_2/T_1 in the equivalence ratio range of interest does not vary significantly among the various fuels. However, there is significant change in the ratio P_2/P_1, particularly for the oxygen detonation condition. Indeed, there is even appreciable change in the air results; this trend is particularly important for ram accelerators operating on the detonation principle [15]. It is the P_2 that applies the force in these accelerators; thus one concludes it is best to operate at near stoichiometric mixture ratios with a high-molecular-weight fuel. Equation (42a) explicitly specifies this P_2 trend. The results of the calculated density ratio (ρ_2/ρ_1) again reveal there is very little difference in this value for the various fuels (see Fig. 11). The maximum values are close to 1.8 for air and 1.85 for oxygen at the stoichiometric condition. Since μ is approximately the same for all fuels and maximizes close to $\phi = 1$, and since the temperature ratio is nearly the same,

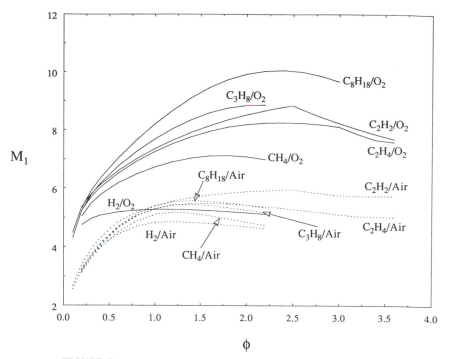

FIGURE 8   The detonation Mach number M_1 for the conditions in Fig. 7.

Eq. (42a) indicates that it is the ratio of $(MW_1/MW_2)^{1/2}$ that determines which fuel would likely have the greatest effect in determining P_2 (see Fig. 12). MW_2 decreases slightly with increasing ϕ for all fuels except H_2 and is approximately the same for all hydrocarbons. MW_1 decreases with ϕ mildly for hydrocarbons with molecular weights less than that of air or oxygen. Propane and octane, of course, increase with ϕ.

Equation (42a) clearly depicts what determines P_2, and indeed it appears that the average molecular weight of the unburned gas mixtures is a major factor [16]. A physical interpretation as to the molecular-weight effect can be related to M_1. As stated, the larger the molecular weight of the unburned gases, the larger the M_1. Considering the structure of the detonation to be discussed in the next section, the larger the M_1, the larger the pressure behind the driving shock of the detonation, which is given the symbol P_1'. Thus the reacting mixture that starts at a higher pressure is most likely to achieve the highest Chapman–Jouguet pressure P_2.

It is rather interesting that the maximum specific impulse of a rocket propellant system occurs when $(T_2/MW_2)^{1/2}$ is maximized, even though the rocket combustion process is not one of detonation [17].

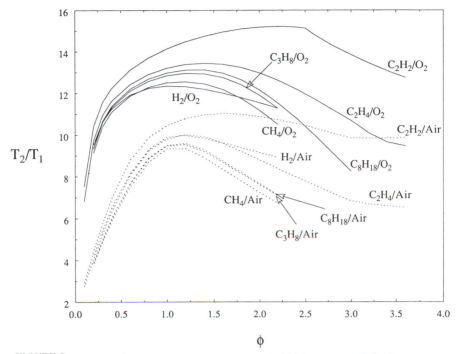

<figure>
T_2/T_1
</figure>

FIGURE 9   The ratio of the burned gas temperature T_2 to the initial temperature T_1 for the conditions in Fig. 7.

E. The ZND Structure of Detonation Waves

Zeldovich [9], von Neumann [10], and Döring [11] independently arrived at a theory for the structure of the detonation wave. The ZND theory states that the detonation wave consists of a planar shock that moves at the detonation velocity and leaves heated and compressed gas behind it. After an induction period, the chemical reaction starts; and as the reaction progresses, the temperature rises and the density and pressure fall until they reach the Chapman–Jouguet values and the reaction attains equilibrium. A rarefaction wave whose steepness depends on the distance traveled by the wave then sets in. Thus, behind the *C–J* shock, energy is generated by thermal reaction.

When the detonation velocity was calculated in the previous section, the conservation equations were used and no knowledge of the chemical reaction rate or structure was necessary. The wave was assumed to be a discontinuity. This assumption is satisfactory because these equations placed no restriction on the distance between a shock and the seat of the generating force.

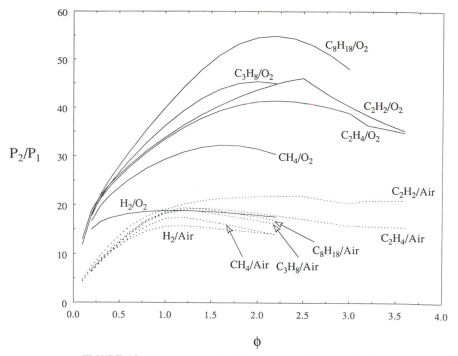

FIGURE 10   The pressure ratio P_2/P_1 for the conditions in Fig. 7.

But to look somewhat more closely at the structure of the wave, one must deal with the kinetics of the chemical reaction. The kinetics and mechanism of reaction give the time and spatial separation of the front and the $C-J$ plane.

The distribution of pressure, temperature, and density behind the shock depends upon the fraction of material reacted. If the reaction rate is exponentially accelerating (i.e., follows an Arrhenius law and has a relatively large overall activation energy like that normally associated with hydrocarbon oxidation), the fraction reacted changes very little initially; the pressure, density, and temperature profiles are very flat for a distance behind the shock front and then change sharply as the reaction goes to completion at a high rate.

Figure 13, which is a graphical representation of the ZND theory, shows the variation of the important physical parameters as a function of spatial distribution. Plane 1 is the shock front, plane $1'$ is the plane immediately after the shock, and plane 2 is the Chapman–Jouguet plane. In the previous section, the conditions for plane 2 were calculated and u_1 was obtained. From u_1 and the shock relationships or tables, it is possible to determine the conditions at plane $1'$. Typical results are shown in Table 5 for various hydrogen and propane detonation conditions. Note from this table that $(\rho_2/\rho_1) \cong 1.8$. Therefore, for many situations the approximation that μ_1 is 1.8 times the sound speed, c_2, can be used.

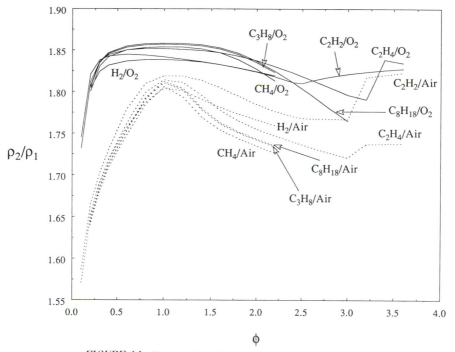

FIGURE 11   The density ratio ρ_2/ρ_1 for the conditions in Fig. 7.

Thus, as the gas passes from the shock front to the $C-J$ state, its pressure drops about a factor of 2, the temperature rises about a factor of 2, and the density drops by a factor of 3. It is interesting to follow the model on a Hugoniot plot, as shown in Fig. 14.

There are two alternative paths by which a mass element passing through the wave from $\varepsilon = 0$ to $\varepsilon = 1$ may satisfy the conservation equations and at the same time change its pressure and density continuously, not discontinuously, with a distance of travel.

The element may enter the wave in the state corresponding to the initial point and move directly to the $C-J$ point. However, this path demands that this reaction occur everywhere along the path. Since there is little compression along this path, there cannot be sufficient temperature to initiate any reaction. Thus, there is no energy release to sustain the wave. If on another path a jump is made to the upper point $(1')$, the pressure and temperature conditions for initiation of reaction are met. In proceeding from 1 to $1'$, the pressure does not follow the points along the shock Hugoniot curve.

The general features of the model in which a shock, or at least a steep pressure and temperature rise, creates conditions for reaction and in which the subsequent

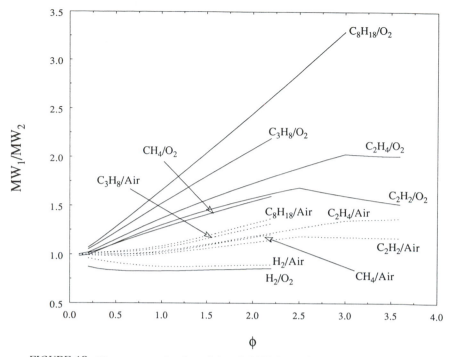

FIGURE 12   The average molecular weight ratio MW_1/MW_2 for the conditions in Fig. 7.

energy release causes a drop in pressure and density have been verified by measurements in a detonation tube [18]. Most of these experiments were measurements of density variation by x-ray absorption. The possible effect of reaction rates on this structure is depicted in Fig. 14 as well [19].

The ZND concepts consider the structure of the wave to be one-dimensional and are adequate for determining the "static" parameters μ, ρ_2, T_2, and P_2. However, there is now evidence that all self-sustaining detonations have a three-dimensional cellular structure.

F. THE STRUCTURE OF THE CELLULAR DETONATION FRONT AND OTHER DETONATION PHENOMENA PARAMETERS

1. The Cellular Detonation Front

An excellent description of the cellular detonation front, its relation to chemical rates and their effect on the dynamic parameters, has been given by Lee [6]. With permission, from the *Annual Review of Fluid Mechanics* Volume 16, © 1984 by Annual Reviews Inc., this description is reproduced almost verbatim here.

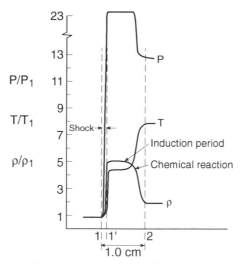

FIGURE 13   Variation of physical parameters through a typical detonation wave (see Table 4).

Figure 15 shows the pattern made by the normal reflection of a detonation on a glass plate coated lightly with carbon soot, which may be from either a wooden match or a kerosene lamp. The cellular structure of the detonation front is quite evident. If a similarly soot-coated polished metal (or mylar) foil is inserted into a detonation tube, the passage of the detonation wave will leave a characteristic "fish-scale" pattern on the smoked foil. Figure 16 is a sequence of laser-Schlieren

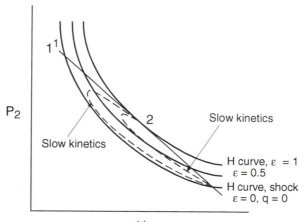

FIGURE 14   Effect of chemical reaction rates on detonation structures as viewed on Hugoniot curves; ε is fractional amount of chemical energy converted.

TABLE 5 Calculated Values of the Physical Parameters for Various Hydrogen– and Propane–Air/Oxygen Detonations

	1	1′	2
H_2/Air ($\phi = 1.2$)			
M	4.86	0.41	1.00
u (m/s)	2033	377	1129
P (atm)	1	28	16
T (K)	298	1548	2976
ρ/ρ_1	1.00	5.39	1.80
H_2/O_2 ($\phi = 1.1$)			
M	5.29	0.40	1.00
u (m/s)	2920	524	1589
P (atm)	1	33	19
T (K)	298	1773	3680
ρ/ρ_1	1.00	5.57	1.84
C_3H_8/Air ($\phi = 1.3$)			
M	5.45	0.37	1.00
u (m/s)	1838	271	1028
P (atm)	1	35	19
T (K)	298	1556	2805
ρ/ρ_1	1.00	6.80	1.79
C_3H_8/O_2 ($\phi = 2.0$)			
M	8.87	0.26	1.00
u (m/s)	2612	185	1423
P (atm)	1	92	45
T (K)	298	1932	3548
ρ/ρ_1	1.00	14.15	1.84
C_3H_8/O_2 ($\phi = 2.2$)			
M	8.87	0.26	1.00
u (m/s)	2603	179	1428
P (atm)	1	92	45
T (K)	298	1884	3363
ρ/ρ_1	1.00	14.53	1.82

records of a detonation wave propagating in a rectangular tube. One of the side windows has been coated with smoke, and the fish-scale pattern formed by the propagating detonation front itself is illustrated by the interferogram shown in Fig. 17. The direction of propagation of the detonation is toward the right. As can be seen in the sketch at the top left corner, there are two triple points. At the first triple point A, AI and AM represent the incident shock and Mach stem of the leading front, while AB is the reflected shock. Point B is the second triple point of another three-shock Mach configuration on the reflected shock AB: the entire shock pattern represents what is generally referred to as a double Mach reflection. The hatched lines denote the reaction front, while the dash–dot lines represent the shear discontinuities or slip lines associated with the triple-shock Mach configurations. The entire front $ABCDE$ is generally referred to as the transverse wave, and it propagates normal to the direction of the detonation motion (down in the present case) at about the sound speed of the hot product gases. It has been shown conclusively that it is the triple-point regions at A and B that "write" on the smoke foil. The exact mechanics of how the triple-point region does the writing is not clear. It has been postulated that the high shear at the slip discontinuity causes the soot particles to be erased. Figure 17 shows a schematic of the motion of the detonation front. The fish-scale pattern is a record of the trajectories of the triple points. It is important to note the cyclic motion of the detonation front. Starting at the apex of the cell at A, the detonation shock front is highly overdriven, propagating at about 1.6 times the equilibrium Chapman–Jouguet detonation velocity. Toward the end of the cell at D, the shock has decayed to about 0.6 times the Chapman–Jouguet velocity before it is impulsively accelerated back to its highly overdriven state when the transverse waves collide to start the next cycle again. For the first half of the propagation from A to BC, the wave serves as the Mach stem to the incident shocks of the adjacent cells. During the second half from BC to D, the wave then becomes the incident shock to the Mach stems of the neighboring cells. Details of the variation of the shock strength and chemical reactions inside a cell can be found in a paper by Libouton *et al.* [20]. AD is usually defined as the length L_c of the cell, and BC denotes the cell diameter (also referred to as the cell width or the transverse-wave spacing). The average velocity of the wave is close to the equilibrium Chapman–Jouguet velocity.

We thus see that the motion of a real detonation front is far from the steady and one-dimensional motion given by the ZND model. Instead, it proceeds in a cyclic manner in which the shock velocity fluctuates within a cell about the equilibrium Chapman–Jouguet value. Chemical reactions are essentially complete within a cycle or a cell length. However, the gas dynamic flow structure is highly three-dimensional; and full equilibration of the transverse shocks, so that the flow becomes essentially one-dimensional, will probably take an additional distance of the order of a few more cell lengths.

From both the cellular end-on or the axial fish-scale smoke foil, the average cell size λ can be measured. The end-on record gives the cellular pattern at one precise instant. The axial record, however, permits the detonation to be observed as it travels along the length of the foil. It is much easier by far to pick out the characteristic cell size λ from the axial record; thus, the end-on pattern is not used, in general, for cell-size measurements.

Early measurements of the cell size have been carried out mostly in low-pressure fuel–oxygen mixtures diluted with inert gases such as He, Ar, and N_2 [21]. The purpose of these investigations is to explore the details of the detonation structure and to find out the factors that control it. It was not until very recently that Bull *et al.* [22] made some cell-size measurements in stoichiometric fuel–air mixtures at atmospheric pressure. Due to the fundamental importance of the cell size in the correlation with the other dynamic parameters, a systematic program has been carried out by Kynstantas to measure the cell size of atmospheric fuel–air detonations in all the common fuels (e.g., H_2, C_2H_2, C_2H_4, C_3H_6, C_2H_6, C_3H_8, C_4H_{10}, and the welding fuel MAPP) over the entire range of fuel composition between the limits [23]. Stoichiometric mixtures of these fuels with pure oxygen,

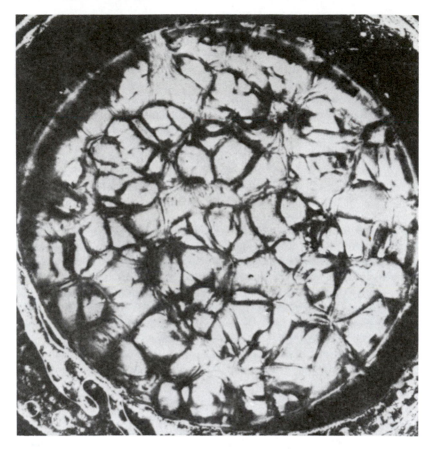

FIGURE 15   End-on pattern from the normal reflection of a cellular detonation on a smoked glass plate (after Lee [2]).

and with varying degrees of N_2 dilution at atmospheric pressures, were also studied [24]. To investigate the pressure dependence, Kynstantas *et al.* [24] have also measured the cell size in a variety of stoichiometric fuel–oxygen mixtures at initial pressures $10 \le p_0 \le 200$ torr. The minimum cell size usually occurs at about the most detonable composition ($\phi = 1$). The cell size λ is representative of the sensitivity of the mixture. Thus, in descending order of sensitivity, we have C_2H_2, H_2, C_2H_4, and the alkanes C_3H_8, C_2H_6, and C_4H_{10}. Methane (CH_4), although belonging to the same alkane family, is exceptionally insensitive to detonation, with an estimated cell size $\lambda \approx 33$ cm for stoichiometric composition as compared with the corresponding value of $\lambda \approx 5.35$ cm for the other alkanes. That the cell size λ is proportional to the induction time of the mixture had been suggested by Shchelkin and Troshin [25] long ago. However, to compute an induction time requires that the model for the detonation structure be known, and no theory exists as yet for the real three-dimensional structure. Nevertheless, one can use the classical ZND model for the structure and compute an induction

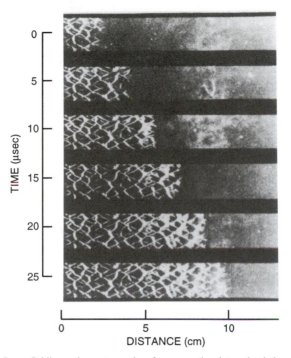

0

5

10

15

20

25

TIME (μsec)

0   5   10
DISTANCE (cm)

FIGURE 16   Laser-Schlieren chromatography of a propagating detonation in low-pressure mixtures with fish-scale pattern on a soot-covered window (courtesy of A. K. Oppenheim).

time or, equivalently, an induction-zone length l. While this is not expected to correspond to the cell size λ (or cell length L_c), it may elucidate the dependence of λ on l itself (e.g., a linear dependence $\lambda = Al$, as suggested by Shchelkin and Troshin). Westbrook [26, 27] has made computations of the induction-zone length l using the ZND model, but his calculations are based on a constant-volume process after the shock, rather than integration along the Rayleigh line. Very detailed kinetics of the oxidation processes are employed. By matching with one experimental point, the proportionality constant A can be obtained. The constant A differs for different gas mixtures (e.g., $A = 10.14$ for C_2H_4, $A = 52.23$ for H_2); thus, the three-dimensional gas dynamic processes cannot be represented by a single constant alone over a range of fuel composition for all the mixtures. The chemical reactions in a detonation wave are strongly coupled to the details of the transient gas dynamic processes, with the end product of the coupling being manifested by a characteristic chemical length scale λ (or equivalently L_c) or time scale $t_c = l/C_1$ (where C_1 denotes the sound speed in the product gases, which is approximately the velocity of the transverse waves) that describes the global rate of the chemical reactions. Since $\lambda \simeq 0.6L_c$ and $C_1 \simeq D$ is the Chapman–Jouguet detonation velocity, we have $0.5D$, where $\tau_c \simeq L_c/D$, which corresponds to the fact that the chemical reactions are essentially completed within one-cell length (or one cycle).

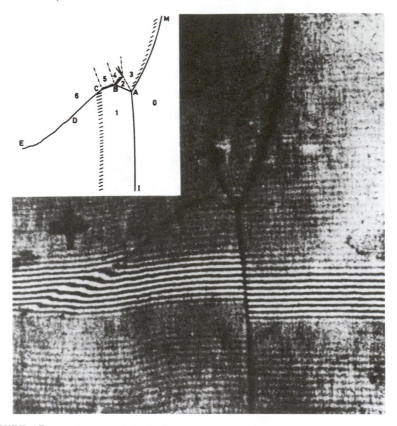

FIGURE 17   Interferogram of the detailed double Mach-reflection configurations of the structure of a cellular front (courtesy of D. H. Edwards).

2. The Dynamic Detonation Parameters

The extent to which a detonation will propagate from one experimental configuration into another determines the dynamic parameter called critical tube diameter. "It has been found that if a planar detonation wave propagating in a circular tube emerges suddenly into an unconfined volume containing the same mixture, the planar wave will transform into a spherical wave if the tube diameter d exceeds a certain critical value d_c (i.e., $d \geq d_c$). If $d < d_c$, the expansion waves will decouple the reaction zone from the shock, and a spherical deflagration wave results" [6].

Rarefaction waves are generated circumferentially at the tube as the detonation leaves; then they propagate toward the tube axis, cool the shock-heated gases, and, consequently, increase the reaction induction time. This induced delay decouples the reaction zone from the shock and a deflagration persists. The tube diameter must be large enough so that a core near the tube axis is not quenched and this core can support the development of a spherical detonation wave.

Some analytical and experimental estimates show that the critical tube diameter is 13 times the detonation cell size ($d_c \geq 13\lambda$) [6]. This result is extremely useful in that only laboratory tube measurements are necessary to obtain an estimate of d_c. It is a value, however, that could change somewhat as more measurements are made.

As in the case of deflagrations, a quenching distance exists for detonations; i.e., a detonation will not propagate in a tube whose diameter is below a certain size or between infinitely large parallel plates whose separation distance is again below a certain size. This quenching diameter or distance appears to be associated with the boundary layer growth in the retainer configuration [5]. According to Williams [5], the boundary layer growth has the effect of an area expansion on the reaction zone that tends to reduce the Mach number in the burned gases, so the quenching distance arises from the competition of this effect with the heat release that increases this Mach number. For the detonation to exist, the heat release effect must exceed the expansion effect at the C–J plane; otherwise, the subsonic Mach number and the associated temperature and reaction rate will decrease further behind the shock front and the system will not be able to recover to reach the C–J state. The quenching distance is that at which the two effects are equal. This concept leads to the relation [5]

$$\delta^* \approx (\gamma - 1)H/8$$

where δ^* is the boundary layer thickness at the C–J plane and H is the hydraulic diameter (four times the ratio of the area to the perimeter of a duct which is the diameter of a circular tube or twice the height of a channel). Order-of-magnitude estimates of quenching distance may be obtained from the above expression if boundary layer theory is employed to estimate δ^*; namely, $\delta^* \approx l/\sqrt{\mathrm{Re}}$ where Re is $\rho l'(u_1 - u_2)/\mu$ and l' is the length of the reaction zone; μ is evaluated at the C–J plane. Typically, $\mathrm{Re} \geq 10^5$ and l can be found experimentally and approximated as 6.5 times the cell size λ [28].

3. Detonation Limits

As is the case with deflagrations, there exist mixture ratio limits outside of which it is not possible to propagate a detonation. Because of the quenching distance problem, one could argue that two sets of possible detonation limits can be determined. One is based on chemical-rate–thermodynamic considerations and would give the widest limits since infinite confinement distance is inherently assumed; the other follows extension of the arguments with respect to quenching distance given in the preceding paragraph.

The quenching distance detonation limit comes about if the induction period or reaction zone length increases greatly as one proceeds away from the stoichiometric mixture ratio. Then the variation of δ^* or l will be so great that, no matter how large the containing distance, the quenching condition will be achieved for the given mixture ratio. This mixture is the detonation limit.

Belles [29] essentially established a pure chemical-kinetic–thermodynamic approach to estimating detonation limits. Questions have been raised about the approach, but the line of reasoning developed is worth considering. It is a fine example of coordinating various fundamental elements discussed to this point in order to obtain an estimate of a complex phenomenon.

Belles' prediction of the limits of detonability takes the following course. He deals with the hydrogen–oxygen case. Initially, the chemical kinetic conditions for branched-chain explosion in this system are defined in terms of the temperature, pressure, and mixture composition. The standard shock wave equations are used to express, for a given mixture, the temperature and pressure of the shocked gas before reaction is established (condition 1'). The shock Mach number (M) is determined from the detonation velocity. These results are then combined with the explosion condition in terms of M and the mixture composition in order to specify the critical shock strengths for explosion. The mixtures are then examined to determine whether they can support the shock strength necessary for explosion. Some cannot, and these define the limit.

The set of reactions that determine the explosion condition of the hydrogen–oxygen system is essentially

$$\mathrm{OH} + \mathrm{H}_2 \overset{k_1}{\rightarrow} \mathrm{H}_2\mathrm{O} + \mathrm{H}$$

$$\mathrm{H} + \mathrm{O}_2 \overset{k_2}{\rightarrow} \mathrm{OH} + \mathrm{O}$$

$$\mathrm{O} + \mathrm{H}_2 \overset{k_3}{\rightarrow} \mathrm{OH} + \mathrm{H}$$

$$\mathrm{H} + \mathrm{O}_2 + \mathrm{M'} \overset{k_4}{\rightarrow} \mathrm{HO}_2 + \mathrm{M'}$$

where M' specifies the third body. (The M' is used to distinguish this symbol from the symbol M used to specify the Mach number.) The steady-state solution shows that

$$d(\mathrm{H}_2\mathrm{O})/dt = \text{various terms}/[k_4(\mathrm{M'}) - 2k_2]$$

Consequently the criterion for explosion is

$$k_4(\mathrm{M'}) = 2k_2 \tag{43}$$

Using rate constants for k_2 and k_4 and expressing the third-body concentration (M') in terms of the temperature and pressure by means of the gas law, Belles rewrites Eq. (43) in the form

$$3.11 T e^{-8550/T}/f_x P = 1 \tag{44}$$

where f_x is the effective mole fraction of the third bodies in the formation reaction for HO_2. Lewis and von Elbe [13] give the following empirical relationship for

f_x:

$$f_x = f_{H_2} + 0.35 f_{O_2} + 0.43 f_{N_2} + 0.20 f_{Ar} + 1.47 f_{CO_2} \tag{45}$$

This expression gives a weighting for the effectiveness of other species as third bodies, as compared to H_2 as a third body. Equation (44) is then written as a logarithmic expression

$$(3.710/T) - \log_{10}(T/P) = \log_{10}(3.11/f_x) \tag{46}$$

This equation suggests that if a given hydrogen–oxygen mixture, which could have a characteristic value of f dependent on the mixture composition, is raised to a temperature and pressure that satisfy the equation, then the mixture will be explosive.

For the detonation waves, the following relationships for the temperature and pressure can be written for the condition (l') behind the shock front. It is these conditions that initiate the deflagration state in the detonation wave:

$$P_1/P_0 = (1/\alpha)\left[(M^2/\beta) - 1\right] \tag{47}$$

$$T_1/T_0 = \left[(M^2/\beta) - 1\right]\left[\beta M^2 + (1/\gamma)\right]/\alpha^2 \beta M^2 \tag{48}$$

where M is the Mach number, $\alpha = (\gamma + 1)/(\gamma - 1)$, and $\beta = (\gamma - 1)/2\gamma$. Shock strengths in hydrogen mixtures are sufficiently low so that one does not have to be concerned with the real gas effects on the ratio of specific heats γ, and γ can be evaluated at the initial conditions.

From Eq. (46) it is apparent that many combinations of pressure and temperature will satisfy the explosive condition. However, if the condition is specified that the ignition of the deflagration state must come from the shock wave, Belles argues that only one Mach number will satisfy the explosive condition. This Mach number, called the critical Mach number, is found by substituting Eqs. (47) and (48) into Eq. (46) to give

$$\frac{3.710\alpha^2 \beta M^2}{T_0\left[(M^2/\beta) - 1\right]\left[\beta M^2 + (1/\gamma)\right]} - \log_{10}\left[\frac{T_0\left[\beta M^2 + (1/\gamma)\right]}{P_0 \alpha \beta M^2}\right]$$

$$= f(T_0, P_0, \alpha, M) = \log_{10}(3.11 f_x) \tag{49}$$

This equation is most readily solved by plotting the left-hand side as a function of M for the initial conditions. The logarithm term on the right-hand side is calculated for the initial mixture and M is found from the plot.

The final criterion that establishes the detonation limits is imposed by energy considerations. The shock provides the mechanism whereby the combustion process is continuously sustained; however, the energy to drive the shock, i.e., to heat up the unburned gas mixture, comes from the ultimate energy release in the combustion process. But if the enthalpy increase across the shock that corresponds

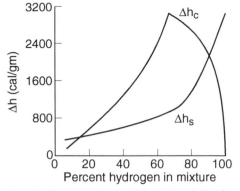

FIGURE 18   Heat of combustion per unit mass (Δh_c) and enthalpy rise across detonation shock (Δh_s) as a function of hydrogen in oxygen (after Belles [29]).

to the critical Mach number is greater than the heat of combustion, an impossible situation arises. No explosive condition can be reached, and the detonation cannot propagate. Thus the criterion for the detonation of a mixture is

$$\Delta h_s \leq \Delta h_c$$

where Δh_c is the heat of combustion per unit mass for the mixture and Δh_s is the enthalpy rise across the shock for the critical Mach number (M_c). Thus

$$h_{T_1'} - h_{T_0} = \Delta h_s \qquad \text{where} \qquad T_1' = T_0 \left[1 + \tfrac{1}{2}(\gamma - 1)M_c^2\right]$$

The plot of Δh_c and Δh_s for the hydrogen–oxygen case as given by Belles is shown in Fig. 18. Where the curves cross in Fig. 18, $\Delta h_c = \Delta h_s$, and the limits are specified. The comparisons with experimental data are very good, as is shown in Table 6.

Questions have been raised about this approach to calculating detonation limits, and some believe that the general agreement between experiments and the theory as shown in Table 6 is fortuitous. One of the criticisms is that a given Mach number specifies a particular temperature and a pressure behind the shock. The expression representing the explosive condition also specifies a particular pressure

TABLE 6   Hydrogen Detonation Limits in Oxygen and Air

	Lean limit (vol %)		Rich limit (vol %)	
System	*Experimental*	*Calculated*	*Experimental*	*Calculated*
H_2–O_2	15	16.3	90	92.3
H_2–Air	18.3	15.8	59.9	59.7

TABLE 7   Comparison of Deflagration and Detonation Limits

	Lean		Rich	
	Deflagration	Detonation	Deflagration	Detonation
H_2–O_2	4	15	94	90
H_2–Air	4	18	74	59
CO–O_2	16	38	94	90
NH_3–O_2	15	25	79	75
C_3H_8–O_2	2	3	55	37

and temperature. It is unlikely that there would be a direct correspondence of the two conditions from the different shock and explosion relationships. Equation (49) must give a unique result for the initial conditions because of the manner in which it was developed.

Detonation limits have been measured for various fuel–oxidizer mixtures. These values and comparison with the deflagration (flammability) limits are given in Table 7. It is interesting that the detonation limits are always narrower than the deflagration limits. But for H_2 and the hydrocarbons, one should recall that, because of the product molecular weight, the detonation velocity has its maximum near the rich limit. The deflagration velocity maximum is always very near the stoichiometric value and indeed has its minimum values at the limits. Indeed, the experimental definition of the deflagration limits would require this result.

G. DETONATIONS IN NONGASEOUS MEDIA

Detonations can be generated in solid propellants and solid and liquid explosives. Such propagation through these condensed phase media make up another important aspect of the overall subject of detonation theory. The general Hugoniot relations developed are applicable, but a major difficulty exists in obtaining appropriate solutions due to the lack of good equations of state necessary due to the very high (10^5 atm) pressures generated. For details on this subject the reader is referred to any [30] of a number of books.

Detonations will also propagate through liquid fuel droplet dispersions (sprays) in air and through solid–gas mixtures such as dust dispersions. Volatility of the liquid fuel plays an important role in characterizing the detonation developed. For low-volatility fuels, fracture and vaporization of the fuel droplets become important in the propagation mechanism, and it is not surprising that the velocities obtained are less than the theoretical maximum. Recent reviews of this subject can be found in Refs. [31 and 32]. Dust explosions and subsequent detonation generally occur when the dust particle size becomes sufficiently small that the heterogeneous surface reactions occur rapidly enough that the energy release rates will nearly support Chapman–Jouguet conditions. The mechanism of propagation of this type of detonation is not well understood. Some reported results of detonations in dust dispersions can be found in Refs. [33 and 34].

PROBLEMS

1. A mixture of hydrogen, oxygen, and nitrogen, having partial pressures in the ratio 2:1:5 in the order listed, is observed to detonate and produce a detonation wave that travels at 1890 m/s when the initial temperature is 292 K and the initial pressure is 1 atm. Assuming fully relaxed conditions, calculate the peak pressure in the detonation wave and the pressure and temperature just after the passage of the wave. Prove that u_2 corresponds to the Chapman–Jouguet condition. Reasonable assumptions should be made for this problem. That is, assume that no dissociation occurs, that the pressure after the wave passes is much greater than the initial pressure, that existing gas dynamic tables designed for air can be used to analyze processes inside the wave, and that the specific heats are independent of pressure.

2. Calculate the detonation velocity in a gaseous mixture of 75% ozone (O_3) and 25% oxygen (O_2) initially at 298 K and 1 atm pressure. The only products after detonation are oxygen molecules and atoms. Take the $\Delta H_f^{\circ}(O_3) = 140$ kJ/mol and all other thermochemical data from the JANAF tables in the appendixes.

    Report the temperature and pressure of the Chapman–Jouguet point as well.

    For the mixture described in the previous problem, calculate the adiabatic (deflagration) temperature when the initial cold temperature is 298 K and the pressure is the same as that calculated for the Chapman–Jouguet point.

    Compare and discuss the results for these deflagration and detonation temperatures.

3. Two mixtures (A and B) will propagate both a laminar flame and a detonation wave under the appropriate conditions:

$$A: \quad CH_4 + i(0.21\,O_2 + 0.79\,N_2)$$

$$B: \quad CH_4 + i(0.21\,O_2 + 0.79\,Ar)$$

    Which mixture will have the higher flame speed? Which will have the higher detonation velocity? Very brief explanations should support your answers. The stoichiometric coefficient i is the same for both mixtures.

4. What would be the most effective diluent to a detonable mixture to lower, or prevent, detonation possibility: carbon dioxide, helium, nitrogen, or argon? Order the expected effectiveness.

REFERENCES

1. Friedman, R., *J. Am. Rocket Soc.* **23**, 349 (1953).
2. Lee, J. H., *Annu. Rev. Phys. Chem.* **38**, 75 (1977).
3. Urtiew, P. A., and Oppenheim, A. K., *Proc. R. Soc. London, Ser. A* **295**, 13 (1966).
4. Fickett, W., and Davis, W. C., "Detonation." Univ. of California Press, Berkeley. 1979; Zeldovich, Y. B., and Kompaneets, A. S., "Theory of Detonation." Academic Press, New York, 1960.
5. Williams, F. A., "Combustion Theory," Chap. 6. Benjamin-Cummins, Menlo Park, California, 1985. See also Linan, A., and Williams, F. A., "Fundamental Aspects of Combustion," Oxford Univ. Press, Oxford, England, 1994.
6. Lee, J. H., *Annu. Rev. Fluid Mech.* **16**, 311 (1984).
7. Chapman, D. L., *Philos. Mag.* **47**, 90 (1899).
8. Jouguet, E., "Méchaniques des Explosifs." Dorn, Paris, 1917.
9. Zeldovich, Y. N., *NACA Tech. Memo* No. 1261 (1950).
10. von Neumann, J., *OSRD Rep.* No. 549 (1942).
11. Döring, W., *Ann. Phys.* **43**, 421 (1943).
12. Gordon, S., and McBride, B. V., *NASA [Spec. Publ.] SP* **NASA SP-273** (1971).
13. Lewis, B., and von Elbe, G., "Combustion, Flames and Explosions of Gases," 2nd Ed., Chap. 8. Academic Press, New York, 1961.
14. Strehlow, R. A., "Fundamentals of Combustion." International Textbook Co., Scranton, Pennsylvania, 1984.
15. Li, C., Kailasanath, K., and Oran, E. S., *Prog. Astronaut. Aeronaut.* **153**, 231 (1993).
16. Glassman, I., and Sung, C.-J., *East. States/Combust. Inst. Meet.*, Worcester, MA Pap. No. 28 (1995).
17. Glassman, I., and Sawyer, R., "The Performance of Chemical Propellants." Technivision, Slough, England, 1971.
18. Kistiakowsky, G. B., and Kydd, J. P., *J. Chem. Phys.* **25**, 824 (1956).
19. Hirschfelder, J. O., Curtiss, C. F., and Bird, R. B., "The Molecular Theory of Gases and Liquids." Wiley, New York, 1954.
20. Libouton, J. C., Dormal, M., and van Tiggelen, P. J., *Prog. Astronaut. Aeronaut.* **75**, 358 (1981).
21. Strehlow, R. A., and Engel, C. D., *AIAA J.* **7**, 492 (1969).
22. Bull, D. C., Elsworth, J. E., Shuff, P. J., and Metcalfe, E., *Combust. Flame* **45**, 7 (1982).
23. Kynstantas, R., Guirao, C., Lee, J. H. S., and Sulmistras, A., *Int. Colloq. Dyn. Explos. React. Syst., 9th, Poitiers, Fr.* (1983).
24. Kynstantas, R., Lee, J. H. S., and Guirao, C., *Combust. Flame* **48**, 63 (1982).
25. Shchelkin, K. I., and Troshin, Y. K., "Gasdynamics of Combustion." Mono Book Corp., Baltimore, Maryland, 1965.
26. Westbrook, C., *Combust. Flame* **46**, 191 (1982).
27. Westbrook, C., and Urtiew, P., *Int. Symp. Combust., 19th* p. 615. Combust. Inst., Pittsburgh, Pennsylvania, 1982.
28. Edwards, D. H., Jones, A. J., and Phillips, P. F., *J. Phys.* **D9**, 1331 (1976).
29. Belles, F. E., *Int. Symp. Combust., 7th* p. 745. Butterworth, London, (1959).
30. Bowden, F. P., and Yoffee, A. D., "Limitation and Growth of Explosions in Liquids and Solids." Cambridge Univ. Press, Cambridge, England, 1952.
31. Dabora, E. K., *in* "Fuel–Air Explosions" (J. H. S. Lee and C.M. Guirao, eds.), p. 245. Univ. of Waterloo Press, Waterloo, Canada, 1982.
32. Sichel, M., *in* "Fuel–Air Explosions" (J. H. S. Lee and C. M. Guirao, eds.), p. 265. Univ. of Waterloo Press, Waterloo, Canada, 1982.
33. Strauss, W. A., *AIAA J.* **6**, 1753 (1968).
34. Palmer, K. N., and Tonkin, P. S., *Combust. Flame* **17**, 159 (1971).

6

Diffusion Flames

A. INTRODUCTION

Earlier chapters were concerned with flames in which the fuel and oxidizer are homogeneously mixed. Even if the fuel and oxidizer are separate entities in the initial stages of a combustion event and mixing occurs rapidly compared to the rate of combustion reactions, or if mixing occurs well ahead of the flame zone (as in a Bunsen burner), the burning process may be considered in terms of homogeneous premixed conditions. There are also systems in which the mixing rate is slow relative to the reaction rate of the fuel and oxidizer, in which case the mixing controls the burning rate. Most practical systems are mixing-rate–controlled and lead to diffusion flames in which fuel and oxidizer come together in a reaction zone through molecular and turbulent diffusion. The fuel may be in the form of a gaseous fuel jet or a condensed medium (either liquid or solid), and the oxidizer may be a flowing gas stream or the quiescent atmosphere. The distinctive characteristic of a diffusion flame is that the burning (or fuel consumption) rate is determined by the rate at which the fuel and oxidizer are brought together in proper proportions for reaction.

Since diffusion rates vary with pressure and the rate of overall combustion reactions varies approximately with the pressure squared, at very low pressures the flame formed will exhibit premixed combustion characteristics even though

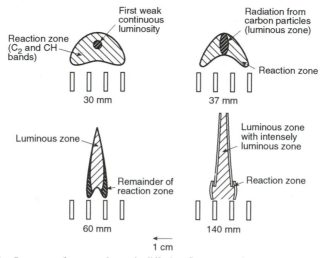

FIGURE 1   Structure of an acetylene–air diffusion flame at various pressures in mm Hg (after Gaydon and Wolfhard [1]).

the fuel and oxidizer may be separate concentric gaseous streams. Figure 1 details how the flame structure varies with pressure for such a configuration where the fuel is a simple higher-order hydrocarbon [1]. Normally, the concentric fuel–oxidizer configuration is typical of diffusion flame processes.

B. GASEOUS FUEL JETS

Diffusion flames have far greater practical application than premixed flames. Gaseous diffusion flames, unlike premixed flames, have no fundamental characteristic property, such as flame velocity, which can be measured readily; even initial mixture strength (the overall oxidizer-to-fuel ratio) has no practical meaning. Indeed, a mixture strength does not exist for a gaseous fuel jet issuing into a quiescent atmosphere. Certainly no mixture strength exists for a single small fuel droplet burning in the infinite reservoir of a quiescent atmosphere.

1. Appearance

Only the shape of the burning laminar fuel jet depends on the mixture strength. If in a concentric configuration the volumetric flow rate of air flowing in the outer annulus is in excess of the stoichiometric amount required for the volumetric flow rate of the inner fuel jet, the flame that develops takes a closed, elongated form. A similar flame forms when a fuel jet issues into the quiescent atmosphere. Such

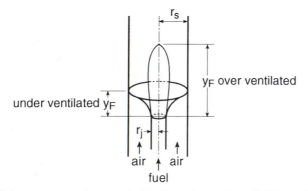

FIGURE 2   Appearance of gaseous fuel jet flames in a co-flow cylindrical configuration.

flames are referred to as being overventilated. If in the concentric configuration the air supply in the outer annulus is reduced below an initial mixture strength corresponding to the stoichiometric required amount, a fan-shaped, underventilated flame is produced. The general shapes of the underventilated and overventilated flame are shown in Fig. 2. As will be shown later in this chapter, the actual heights vary with the flow conditions.

The axial symmetry of the concentric configuration shown in Fig. 2 is not conducive to experimental analyses, particularly when some optical diagnostic tools or thermocouples are used. There are parametric variations in the r and y coordinates shown in Fig. 2. To facilitate experimental measurements on diffusion flames, the so-called Wolfhard–Parker two-dimensional gaseous fuel jet burner is used. Such a configuration is shown in Fig. 3, taken from Smyth *et al.* [2]; the screens shown in the figure are used to stabilize the flame. As can be seen in this figure, ideally there are no parametric variations along the length of the slot.

Another type of gaseous diffusion flame is created by opposed jets of fuel and oxidizer. The types most frequently used are shown in Fig. 4. Although these configurations are somewhat more complex to establish experimentally, they have definite advantages. The opposed jet configuration in which the fuel stream is injected through a porous cylinder has two major advantages compared to the concentric fuel jet or Wolfhard–Parker burners. First, there is no possibility of oxidizer diffusion to the fuel side through the quench zone at the jet-tube lip; second, the flow configuration is very amenable to analysis. Although the aerodynamic analysis of the configuration that produces the flat diffusion flame is somewhat more complex and the stability of the process is a little more sensitive to flow conditions, the creation of a flat flame is a definite experimental advantage when immersion or optical instrumentation is used.

The color of a hydrocarbon–air diffusion flame is distinctively different from that of its premixed counterpart. Whereas a premixed hydrocarbon–air flame is

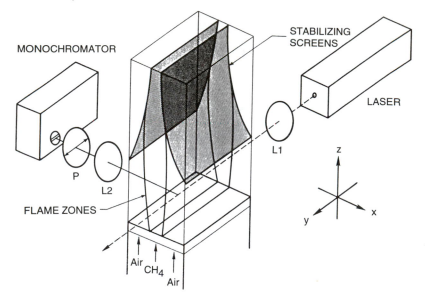

FIGURE 3   Two-dimensional Wolfhard–Parker fuel jet burner flame configuration (after Smyth *et al.* [2]).

violet or blue-green, the corresponding diffusion flame varies from bright yellow to orange. The color of the diffusion flame arises from the formation of soot on the fuel side of the flame. Soot particles, which flow through the reaction zone, reach the flame temperature and are usually burned in the reaction zone. Due to the flame temperatures that exist and the sensitivity of the eye to various wavelengths in the visible region of the electromagnetic spectrum, the hydrocarbon–air diffusion flame appears to be yellow or orange. In Wolfhard–Parker and opposed jet burners, some soot escapes without entering the flame zone and appears as black smoke. In the concentric jet configuration, if the volumetric flow rate of the fuel is increased substantially, the top of the flame will appear to open and a soot smoke trail appears. The presence of soot also confuses the selection of the actual height of a concentric diffusion flame. Many early investigators assumed that the diffusion flame height coincided with the height of the visible tip. However, soot particles penetrate the actual diffusion flame around its apex, burn, and radiate. Thus, the luminous height of most hydrocarbon diffusion flames is greater than the actual diffusion flame height [3–6]. The actual flame height thus must be determined from measurement of the flame gas composition. Non–soot-forming diffusion flames, such as those found with H_2, CO, and CH_3OH, are mildly visible and look very much like their premixed counterparts. Their heights can be estimated visually.

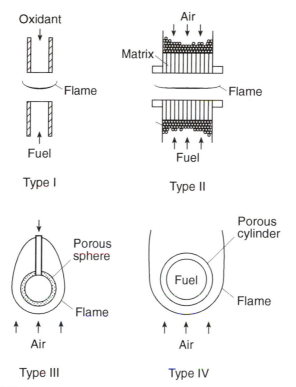

FIGURE 4   Various counterflow diffusion flame experimental configurations.

2. Structure

Unlike premixed flames, which have a very narrow reaction zone, diffusion flames have a wider region over which the composition changes and chemical reaction can take place. Obviously, these changes are due principally to some interdiffusion of reactants and products. Hottel and Hawthorne [7] were the first to make detailed measurements of species distributions in a concentric laminar H_2–air diffusion flame. Figure 5 shows the type of results they obtained for a radial distribution at a height corresponding to a cross-section of the overventilated flame depicted in Fig. 2. Smyth *et al*. [2] made very detailed and accurate measurements of temperature and species variation across a Wolfhard–Parker burner in which methane was the fuel. Their results are shown in Figs. 6 and 7.

The flame front can be assumed to exist at the point of maximum temperature, and indeed this point corresponds to that at which the maximum concentrations of major products (CO_2 and H_2O) exist. The same type of profiles would exist for a simple fuel jet issuing into quiescent air. The maxima arise due to diffusion of reactants in a direction normal to the flowing streams. It is most important

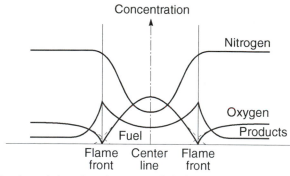

FIGURE 5   Species variations through a gaseous fuel–air diffusion flame at a fixed height above a fuel jet tube.

to realize that, for the concentric configuration, molecular diffusion establishes a bulk velocity component in the normal direction. In the steady state, the flame front produces a flow outward, molecular diffusion establishes a bulk velocity component in the normal direction, and oxygen plus a little nitrogen flows inward toward the centerline. In the steady state, the total volumetric rate of products is usually greater than the sum of the other two. Thus, the bulk velocity that one

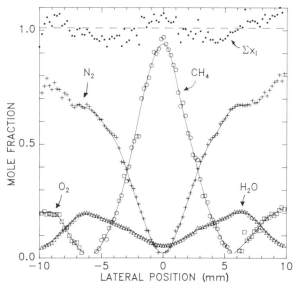

FIGURE 6   Species variations throughout a Wolfhard–Parker methane–air diffusion flame (after Smyth *et al.* [2]).

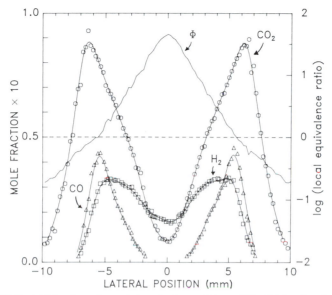

FIGURE 7   Additional species variations for the conditions of Fig. 6 (after Smyth *et al.* [2]).

would observe moves from the flame front outward. The oxygen between the outside stream and the flame front then flows in the direction opposite to the bulk flow. Between the centerline and the flame front, the bulk velocity must, of course, flow from the centerline outward. There is no sink at the centerline. In the steady state, the concentration of the products reaches a definite value at the centerline. This value is established by the diffusion rate of products inward and the amount transported outward by the bulk velocity.

Since total disappearance of reactants at the flame front would indicate infinitely fast reaction rates, a more likely graphical representation of the radial distribution of reactants should be that given by the dashed lines in Fig. 5. To stress this point, the dashed lines are drawn to grossly exaggerate the thickness of the flame front. Even with finite reaction rates, the flame front is quite thin. The experimental results shown in Figs. 6 and 7 indicate that in diffusion flames the fuel and oxidizer diffuse toward each other at rates that are in stoichiometric proportions. Since the flame front is a sink for both the fuel and oxidizer, intuitively one would expect this most important observation. Independent of the overall mixture strength, since fuel and oxidizer diffuse together in stoichiometric proportions, the flame temperature closely approaches the adiabatic stoichiometric flame temperature. It is probably somewhat lower due to finite reaction rates, i.e., approximately 90% of the adiabatic stoichiometric value [8] whether it is a hydrocarbon fuel or not. This observation establishes an interesting aspect of practical diffusion flames in that for an adiabatic situation two fundamental temperatures exist for a fuel: one that corresponds to its stoichiometric value and occurs at the flame front, and one that occurs after the products mix with the excess air to give an adiabatic temperature that corresponds to the initial mixture strength.

3. Theoretical Considerations

The theory of premixed flames essentially consists of an analysis of factors such as mass diffusion, heat diffusion, and the reaction mechanisms as they affect the rate of homogeneous reactions taking place. Inasmuch as the primary mixing processes of fuel and oxidizer appear to dominate the burning processes in diffusion flames, the theories emphasize the rates of mixing (diffusion) in deriving the characteristics of such flames.

It can be verified easily by experiments that in an ethylene–oxygen premixed flame, the average rate of consumption of reactants is abut 4 mol/cm^3 s, whereas for the diffusion flame (by measurement of flow, flame height, and thickness of reaction zone—a crude, but approximately correct approach), the average rate of consumption is only 6×10^{-5} mol/cm^3 s. Thus, the consumption and heat release rates of premixed flames are much larger than those of pure mixing-controlled diffusion flames.

The theoretical solution to the diffusion flame problem is best approached in the overall sense of a steady flowing gaseous system in which both the diffusion and chemical processes play a role. Even in the burning of liquid droplets, a fuel flow due to evaporation exists. This approach is much the same as that presented in Section 4.C.2, except that the fuel and oxidizer are diffusing in opposite directions and in stoichiometric proportions relative to each other. If one selects a differential element along the x direction of diffusion, the conservation balances for heat and mass may be obtained for the fluxes, as shown in Fig. 8.

In Fig. 8, j is the mass flux as given by a representation of Fick's law when there is bulk movement. From Fick's law

$$j = -D(\partial \rho_A / \partial x)$$

As will be shown in Section 6.B.2, the following form of j is exact; however, the same form can be derived if it is assumed that the total density does not vary with the distance x, as, of course, it actually does:

$$j = -D\rho \frac{\partial(\rho_A/\rho)}{\partial x} = -D\rho \frac{\partial m_A}{\partial x}$$

where m_A is the mass fraction of species A. In Fig. 8, q is the heat flux given by Fourier's law of heat conduction; $\dot{m}_A$ is the rate of decrease of mass of species A in the volumetric element $(\Delta x \cdot 1)(\text{g/cm}^3 \text{ s})$, and $\dot{H}$ is the rate of chemical energy release in the volumetric element $(\Delta x \cdot 1)(\text{cal/cm}^3 \text{ s})$.

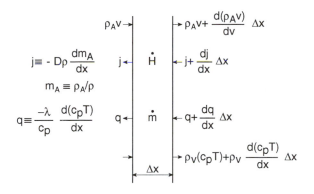

FIGURE 8   Balances across a differential element within a diffusion flame.

With the preceding definitions, for the one-dimensional problem defined in Fig. 8, the expression for conservation of a species A (say the oxidizer) is

$$\frac{\partial \rho_A}{\partial t} = \frac{\partial}{\partial x}\left[(D\rho)\frac{\partial m_A}{\partial x}\right] - \frac{\partial(\rho_A v)}{\partial x} - \dot{m}_A \tag{1}$$

where ρ is the total mass density, ρ_A is the partial density of species A, and v is the bulk velocity in direction x. Solving this time-dependent diffusion flame problem is outside the scope of this text. Indeed, most practical combustion problems have a steady fuel mass input. Thus, for the steady problem, which implies steady mass consumption and flow rates, one may not only take $\partial \rho_A/\partial t$ as zero, but also use the following substitution:

$$\frac{d(\rho_A v)}{dx} = \frac{d\left[(\rho v)(\rho_A/\rho)\right]}{dx} = (\rho v)\frac{dm_A}{dx} \tag{2}$$

The term (ρv) is a constant in the problem since there are no sources or sinks. With the further assumption from simple kinetic theory that $D\rho$ is independent of temperature, and hence of x, Eq. (1) becomes

$$D\rho\frac{d^2 m_A}{dx^2} - (\rho v)\frac{dm_A}{dx} = \dot{m}_A \tag{3}$$

Obviously, the same type of expression must hold for the other diffusing species B (say the fuel), even if its gradient is opposite to that of A so that

$$D\rho\frac{d^2 m_B}{dx^2} - (\rho v)\frac{d m_B}{dx} = \dot{m}_B = i\dot{m}_A \tag{4}$$

where $\dot{m}_B$ is the rate of decrease of species B in the volumetric element $(\Delta x \cdot 1)$ and i is the mass stoichiometric coefficient

$$i \equiv \dot{m}_B/\dot{m}_A$$

The energy equation evolves as it did in Section 4.C.2 to give

$$\frac{\lambda}{c_p}\frac{d^2(c_p T)}{dx^2} - (\rho v)\frac{d(c_p T)}{dx} = +\dot{H} = -i\dot{m}_A H \tag{5}$$

where $\dot{H}$ is the rate of chemical energy release per unit volume and H is the heat release per unit mass of fuel consumed (in joules per gram); i.e.,

$$-\dot{m}_B H = \dot{H}, \qquad -i\dot{m}_A H = \dot{H} \tag{6}$$

since $\dot{m}_B$ must be negative for heat release (exothermic reaction) to take place.

Although the form of Eqs. (3), (4), and (5) is the same as that obtained in dealing with premixed flames, an important difference lies in the boundary conditions that exist. Furthermore, in comparing Eqs. (3) and (4) here with Eqs. (28) and (29) in Chapter 4, one must realize that in Chapter 4, the mass change symbol $\dot{\omega}$ was always defined as a negative quantity.

Multiplying Eq. (3) by iH, then combining it with Eq. (5) for the condition Le = 1 or $D\rho = (\lambda/c_p)$, one obtains

$$D\rho\frac{d^2}{dx^2}(c_p T + \dot{m}_A H) - (\rho v)\frac{d}{dx}(c_p T + \dot{m}_A H) = 0 \tag{7}$$

This procedure is sometimes referred to as the Schvab–Zeldovich transformation. Mathematically, what has been accomplished is that the nonhomogeneous terms ($\dot{m}$ and $\dot{H}$) have been eliminated and a homogeneous differential equation [Eq. (7)] has been obtained.

The equations could have been developed for a generalized coordinate system. In a generalized coordinate system, they would have the form

$$\nabla \cdot \left[(\rho v)(c_p T) - (\lambda/c_p)\nabla(c_p T)\right] = -\dot{H} \tag{8}$$

$$\nabla \cdot \left[(\rho v)(m_j) - \rho D\nabla m_j\right] = +\dot{m}_j \tag{9}$$

These equations could be generalized even further (see Williams [9]) by simply writing $\sum h_j^\circ \dot{m}_j$ instead of $\dot{H}$, where h_j° is the heat of formation per unit mass at the base temperature of each species j. However, for notation simplicity—and because energy release is of most importance for most combustion and propulsion systems—an overall rate expression for a reaction of the type which follows will suffice:

$$\mathrm{F} + \phi \mathrm{O} = \mathrm{P} \tag{10}$$

where F is the fuel, O is the oxidizer, P is the product, and ϕ is the molar stoichiometric index. Then Eqs. (8) and (9) may be written as

$$\nabla \cdot \left[(\rho v) \frac{m_j}{\mathrm{MW}_j v_j} - (\rho D) \nabla \frac{m_j}{\mathrm{MW}_j v_j} \right] = \dot{M} \tag{11}$$

$$\nabla \cdot \left[(\rho v) \frac{c_p T}{H \mathrm{MW}_j v_j} - (\rho D) \nabla \frac{c_p T}{H \mathrm{MW}_j v_j} \right] = \dot{M} \tag{12}$$

where MW is the molecular weight, $\dot{M} = \dot{m}_j / \mathrm{MW}_j v_j$; $v_j = \phi$ for the oxidizer, and $v_j = 1$ for the fuel. Both equations have the form

$$\nabla \cdot [(\rho v) \alpha - (\rho D \nabla \alpha)] = \dot{M} \tag{13}$$

where $\alpha_\mathrm{T} = c_p T / H \mathrm{MW}_j v_j$ and $\alpha_j = m_j / \mathrm{MW}_j v_j$. They may be expressed as

$$L(\alpha) = \dot{M} \tag{14}$$

where the linear operation $L(\ )$ is defined as

$$L(\alpha) = \nabla \cdot [(\rho v) \alpha - (\rho v) \nabla \alpha] \tag{15}$$

The nonlinear term may be eliminated from all but one of the relationships $L(\alpha) = \dot{M}$. For example,

$$L(\alpha_1) = \dot{M} \tag{16}$$

can be solved for α_1, then the other flow variables can be determined from the linear equations for a simple coupling function Ω so that

$$L(\Omega) = 0 \tag{17}$$

where $\Omega = (\alpha_\mathrm{T} - \alpha_1) \equiv \Omega_j$ $(j \neq 1)$. Obviously if $1 = $ fuel and there is a fuel–oxidizer system only, $j = 1$ gives $\Omega = 0$ and shows the necessary redundancy.

4. The Burke–Schumann Development

With the development in the previous section, it is now possible to approach the classical problem of determining the shape and height of a burning gaseous fuel jet in a coaxial stream as first described by Burke and Schumann and presented in detail in Lewis and von Elbe [10].

This description is given by the following particular assumptions:

1. At the port position, the velocities of the air and fuel are considered constant, equal, and uniform across their respective tubes. Experimentally, this condition can be obtained by varying the radii of the tubes (see Fig. 2). The molar fuel rate is then given by the radii ratio:

$$r_j^2 / \left(r_s^2 - r_j^2 \right)$$

2. The velocity of the fuel and air up the tube in the region of the flame is the same as the velocity at the port.
3. The coefficient of interdiffusion of the two gas streams is constant.

Burke and Schumann [11] suggested that the effects of Assumptions 2 and 3 compensate for each other, thereby minimizing errors. Although D increases as $T^{1.67}$ and velocity increases as $T^{1.00}$, this disparity should not be the main objection. The main objection should be the variation of D with T in the horizontal direction due to heat conduction from the flame.

4. Interdiffusion is entirely radial.
5. Mixing is by diffusion only; i.e., there are no radial velocity components.
6. Of course, the general stoichiometric relation prevails.

With these assumptions one may readily solve the coaxial jet problem. The only differential equation that one is obliged to consider is

$$L(\Omega) = 0 \qquad \text{with} \qquad \Omega = \alpha_F - \alpha_O$$

where $\alpha_F = +m_F/\mathrm{MW}_F v_F$ and $\alpha_O = +m_O/\mathrm{MW}_O v_O$. In cylindrical coordinates the generation equation becomes

$$(v/D)(\partial\Omega/\partial y) - (1/r)(\partial/\partial r)(r\,\partial\Omega/\partial r) = 0 \qquad (18)$$

The cylindrical coordinate terms in $\partial/\partial\theta$ are set equal to zero because of the symmetry. The boundary conditions become

$$\Omega = \frac{m_{F,O}}{\mathrm{MW}_F v_F} \qquad \text{at } y = 0, \quad 0 \leq r \leq r_j$$

$$= -\frac{m_{O,O}}{\mathrm{MW}_O v_O} \qquad \text{at } y = 0, \quad r_j \leq r \leq r_s$$

and $\partial\Omega/\partial r = 0$ at $r = r_s$, $y > 0$. In order to achieve some physical insight to the coupling function Ω, one should consider it as a concentration or, more exactly, a mole fraction. At $y = 0$ the radial concentration difference is

$$\Omega_{r<r_j} - \Omega_{r>r_j} = \frac{m_{F,O}}{\mathrm{MW}_F v_F} - \left(\frac{-m_{O,O}}{\mathrm{MW}_O v_O} \right)$$

$$\Omega_{r<r_j} - \Omega_{r>r_j} = \frac{m_{F,O}}{\mathrm{MW}_F v_F} + \frac{m_{O,O}}{\mathrm{MW}_O v_O}$$

This difference in Ω reveals that the oxygen acts as it were a negative fuel concentration in a given stoichiometric proportion, or vice versa. This result is, of course, a consequence of the choice of the coupling function and the assumption that the fuel and oxidizer approach each other in stoichiometric proportion.

It is convenient to introduce dimensionless coordinates

$$\xi \equiv r/r_s, \qquad \eta \equiv yD/vr_s^2$$

and to define parameters $c \equiv r_j/r_s$ and an initial molar mixture strength v:

$$v \cong m_{O,0}\mathrm{MW_F}v_F/m_{F,0}\mathrm{MW_O}v_O$$

and the reduced variable

$$\gamma = \Omega(\mathrm{MW_F}v_F/\mathrm{M_{F,O}})$$

Equation (18) and the boundary condition then become

$$\partial\gamma/\partial\eta = (1/\xi)(\partial/\partial\xi)(\xi\,\partial\gamma/\partial\xi) \qquad \gamma = 1 \text{ at } \eta = 0, \quad 0 \le \xi < c;$$

$$\gamma = -v \text{ at } \eta = 0, c \le \xi < 1$$

and

$$\partial\gamma/\partial\xi = 0 \qquad \text{at} \qquad \xi = 1, \eta > 0 \tag{19}$$

Equation (19) with these new boundary conditions has the known solution

$$\gamma = (1+v)c^2 - v + 2(1+v)c\sum_{n=1}^{\infty}(1/\phi_n)\left\{J_1(c\phi_n)/\left[J_0(\phi_n)\right]^2\right\}$$

$$\times J_0(\phi_n\xi)\exp(-\phi_n^2\eta) \tag{20}$$

where J_0 and J_1 are Bessel functions of the first kind (of order 0 and 1, respectively) and the ϕ_n represent successive roots of the equation $J_1(\phi) = 0$ (with the ordering convention $\phi_n > \phi_{n-1}, \phi_0 = 0$). This equation gives the solution for Ω in the present problem.

The flame shape is defined at the point where the fuel and oxidizer disappear, i.e., the point that specifies the place where $\Omega = 0$. Hence, setting $\gamma = 0$ provides a relation between ξ and η that defines the locus of the flame surface.

The equation for the flame height is obtained by solving Eq. (20) for η after setting $\xi = 0$ for the overventilated flame and $\xi = 1$ for the underventilated flame (also $\gamma = 0$).

The resulting equation is still very complex. Since flame heights are large enough to cause the factor $\exp(-\phi_n^2\eta)$ to decrease rapidly as n increases at these values of η, it usually suffices to retain only the first few terms of the sum in the basic equation for this calculation. Neglecting all terms except $n = 1$, one obtains the rough approximation

$$\eta = (1/\phi_1^2) \ln\{2(1+v)c J_1(c\phi_1)/[v - (1+v)c]\phi_1 J_0(\phi_1)\} \qquad (21)$$

for the dimensionless height of the overventilated flame. The first zero of $J_1(\phi)$ is $\phi_1 = 3.83$.

The flame shapes and heights predicted by such expressions (see Fig. 9) are shown by Lewis and von Elbe [10] to be in good agreement with experimental results—surprisingly so, considering the basic drastic assumptions.

Indeed, it should be noted that Eq. (21) specifies that the dimensionless flame height can be written as

$$\eta = f(c, v) \qquad \frac{y_F D}{v r_s^2} = f(c, v). \qquad (22)$$

Thus since $c = (r_j/r_s)$, the flame height can also be represented by

$$y_F = \frac{v r_j^2}{Dc^2} f(c, v) = \frac{\pi r_j^2 v}{\pi D} f'(c, v) = \frac{Q}{\pi D} f'(c, v) \qquad (23)$$

where v is the average fuel velocity, Q is the volumetric flow rate of the fuel, and $f' = (f/c^2)$. Thus, one observes that the flame height of a circular fuel jet is directly proportional to the volumetric flow rate of the fuel.

In a pioneering paper [3], Roper vastly improved on the Burke–Schumann approach to determine flame heights not only for circular ports, but also for square ports and slot burners. Roper's work is significant because he used the fact that

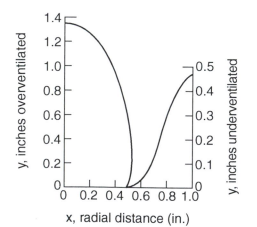

x, radial distance (in.)

FIGURE 9   Flame shapes as predicted by Burke–Schumann theory for cylindrical fuel jet systems (after Burke and Schumann [11]).

the Burke–Schumann approach neglects buoyancy and assumes the mass velocity should everywhere be constant and parallel to the flame axis to satisfy continuity [9], and then pointed out that resulting errors cancel for the flame height of a circular port burner but not for the other geometries [3, 4]. In the Burke–Schumann analysis the major assumption is that the velocities are everywhere constant and parallel to the flame axis. Considering the case in which buoyancy forces increase the mass velocity after the fuel leaves the burner port, Roper showed that continuity dictates decreasing streamline spacing as the mass velocity increases. Consequently, all volume elements move closer to the flame axis, the widths of the concentration profiles are reduced, and the diffusion rates are increased.

Roper [3] also showed that the velocity of the fuel gases is increased due to heating and that the gases leaving the burner port at temperature T_0 rapidly attain a constant value T_f in the flame regions controlling diffusion; thus the diffusivity in the same region is

$$D = D_0(T_f/T_0)^{1.67}$$

where D_0 is the ambient or initial value of the diffusivity. Then, considering the effect of temperature on the velocity, Roper developed the following relationship for the flame height:

$$y_F = \frac{Q}{4\pi D_0} \frac{1}{\ln[1 + (1/S)]} \left(\frac{T_0}{T_f} \right)^{0.67} \tag{24}$$

where S is the stoichiometric volume rate of air to volume rate of fuel. Although Roper's analysis does not permit calculation of the flame shape, it does produce for the flame height a much simpler expression than Eq. (21).

If due to buoyancy the fuel gases attain a velocity v_b in the flame zone after leaving the port exit, continuity requires that the effective radial diffusion distance be some value r_b. Obviously, continuity requires that ρ be the same for both cases, so that

$$r_j^2 v = r_b^2 v_b$$

Thus, one observes that regardless of whether the fuel jet is momentum- or buoyancy-controlled, the flame height y_F is directly proportional to the volumetric flow leaving the port exit.

Given the condition that buoyancy can play a significant role, the fuel gases start with an axial velocity and continue with a mean upward acceleration g due to buoyancy. The velocity of the fuel gases v is then given by

$$v = \left(v_0^2 + v_b^2 \right)^{1/2} \tag{25}$$

where v is the actual velocity, v_0 is the momentum-driven velocity at the port, and v_b is the velocity due to buoyancy. However, the buoyancy term can be closely approximated by

$$v_b^2 = 2g y_F \qquad (26)$$

where g is the acceleration due to buoyancy. If one substitutes Eq. (26) into Eq. (25) and expands the result in terms of a binomial expression, one obtains

$$v = v_0 \left[1 + \left(\frac{g y_F}{v_0^2} \right) - \frac{1}{2} \left(\frac{g y_F}{v_0^2} \right)^2 + \cdots \right] \qquad (27)$$

where the term in parentheses is the inverse of the modified Froude number

$$\mathrm{Fr} \equiv (v_0^2 / g y_F) \qquad (28)$$

Thus, Eq. (27) can be written as

$$v = v_0 [1 + (1/\mathrm{Fr}) - (1/2\mathrm{Fr}^2) + \cdots] \qquad (29)$$

For large Froude numbers, the diffusion flame height is momentum-controlled and $v = v_0$. However, most laminar burning fuel jets will have very small Froude numbers and $v = v_b$; that is, most laminar fuel jets are buoyancy-controlled.

Although the flame height is proportional to the fuel volumetric flow rate whether the flame is momentum- or buoyancy-controlled, the time to the flame tip does depend on what the controlling force is. The characteristic time for diffusion (t_D) must be equal to the time (t_s) for a fluid element to travel from the port to the flame tip; i.e., if the flame is momentum-controlled,

$$t_D \sim \left(r_j^2 / D \right) = (y_F / v) \sim t_s \qquad (30)$$

It follows from Eq. (30), of course, that

$$y_F \sim \frac{r_j^2 v}{D} \sim \frac{\pi r_j^2 v}{\pi D} \sim \frac{Q}{\pi D} \qquad (31)$$

Equation (31) shows the same dependence on Q as that developed from the Burke–Schumann approach [Eqs. (21)–(23)]. For a momentum-controlled fuel jet flame, the diffusion distance is r_j, the jet port radius; and from Eq. (30) it is obvious that the time to the flame tip is independent of the fuel volumetric flow rate. For a buoyancy-controlled flame t_s remains proportional to (y_F / v); however, since $v = (2g y_F)^{1/2}$,

$$t \sim \frac{y_F}{v} \sim \frac{y_F}{(y_F)^{1/2}} \sim y_F^{1/2} \sim Q^{1/2} \qquad (32)$$

Thus, the stay time of a fuel element in a buoyancy-controlled laminar diffusion flame is proportional to the square root of the fuel volumetric flow rate. This conclusion is significant with respect to the soot smoke height tests to be discussed in Chapter 8.

The preceding analyses hold only for circular fuel jets. Roper [3] has shown, and the experimental evidence verifies [4], that the flame height for a slot burner is not the same for momentum- and buoyancy-controlled jets. Consider a slot burner of the Wolfhard–Parker type in which the slot width is x and the length is L. As discussed earlier for a buoyancy-controlled situation, the diffusive distance would not be x, but some smaller width, say x_b. Following the terminology of Eq. (25), for a momentum-controlled slot burner,

$$t \sim \frac{x^2}{D} \sim \frac{y_F}{v_0} \tag{33}$$

and

$$y_F \sim \frac{v_0 x^2/D}{L/L} \sim \left(\frac{Q}{D}\right)\left(\frac{x_b}{L}\right) \tag{34}$$

For buoyancy-controlled slot burner,

$$t \sim \frac{(x_b)^2}{D} \sim \frac{y_F}{v_b} \tag{35}$$

$$y_F \sim \frac{v_b (x_b)^2/D}{L/L} \sim \left(\frac{Q}{D}\right)\left(\frac{x_b}{L}\right) \tag{36}$$

Recalling that x_b must be a function of the buoyancy, one has

$$x_b L v_b \sim Q$$

$$x \sim \frac{Q}{v_b L} \sim \frac{Q}{(2g y_F)^{1/2} L}$$

Thus Eq. (36) becomes

$$y_F \sim \left(\frac{Q^2}{D L^2 (2g)^{1/2}}\right)^{2/3} \sim \left(\frac{Q^4}{D^2 L^4 2g}\right)^{1/3} \tag{37}$$

Comparing Eqs. (34) and (37), one notes that under momentum-controlled conditions for a given Q, the flame height is directly proportional to the slot width while that under buoyancy-controlled conditions for a given Q, the flame height is independent of the slot width. Roper *et al.* [4] have verified these conclusions experimentally.

5. Turbulent Fuel Jets

The previous section considered the burning of a laminar fuel jet, and the essential result with respect to flame height was that

$$y_{F,L} \sim \left(r_j^2 v/D\right) \sim (Q/D) \tag{38}$$

where $y_{F,L}$ specifies the flame height. When the fuel jet velocity is increased to the extent that the flow becomes turbulent, the Froude number becomes large and the system becomes momentum-controlled—moreover, molecular diffusion considerations lose their validity under turbulent fuel jet conditions. One would intuitively expect a change in the character of the flame and its height. Turbulent flows are affected by the occurrence of flames, as discussed for premixed flame conditions, and they are affected by diffusion flames by many of the same mechanisms discussed for premixed flames. However, the diffusion flame in a turbulent mixing layer between the fuel and oxidizer streams steepens the maximum gradient of the mean velocity profile somewhat and generates vorticity of opposite signs on opposite sides of the high-temperature, low-density reaction region.

The decreased overall density of the mixing layer with combustion increases the dimensions of the large vortices and reduces the rate of entrainment of fluids into the mixing layer [12]. Thus it is appropriate to modify the simple phenomenological approach that led to Eq. (31) to account for turbulent diffusion by replacing the molecular diffusivity with a turbulent eddy diffusivity ε. Consequently, the turbulent form of Eq. (38) becomes

$$y_{F,T} \sim \left(r_j^2 v / \varepsilon \right)$$

where $y_{F,T}$ is the flame height of a turbulent fuel jet. But $\varepsilon \sim l U'$, where l is the integral scale of turbulence, which is proportional to the tube diameter (or radius r_j), and U' is the intensity of turbulence, which is proportional to the mean flow velocity v along the axis. Thus one may assume that

$$\varepsilon \sim r_j v \tag{39}$$

Combining Eq. (39) and the preceding equation, one obtains

$$y_{F,T} \sim r_j^2 v / r_j v \sim r_j \tag{40}$$

This expression reveals that the height of a turbulent diffusion flame is proportional to the port radius (or diameter) above, irrespective of the volumetric fuel flow rate or fuel velocity issuing from the burner! This important practical conclusion has been verified by many investigators.

The earliest verification of $y_{F,T} \sim r_j$ was by Hawthorne *et al.* [12a], who reported their results as $y_{F,T}$ as a function of jet exit velocity from a fixed tube exit radius. Thus varying the exit velocity is the same as varying the Reynolds number. The results in Ref. [12a] were represented by Linan and Williams [12] in the diagram duplicated here as Fig. 10. This figure clearly shows that as the velocity increases in the laminar range, the height increases linearly, in accordance with Eq. (38). After transition to turbulence, the height becomes independent of the velocity, in agreement with Eq. (40). Observing Fig. 10, one notes that the transition to turbulence begins near the top of the flame, then as the velocity increases, the turbulence rapidly recedes to the exit of the jet. At a high enough

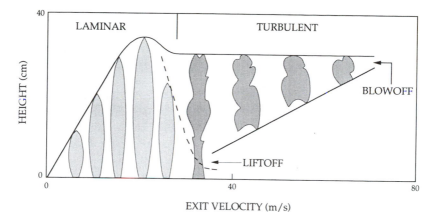

FIGURE 10 Variation of the character of a gaseous diffusion as a function of fuel jet velocity showing experimental flame lift off (after Linan and Williams [12] and Hawthorne *et al.* [12a]).

velocity, the flow in the fuel tube becomes turbulent and turbulence is observed everywhere in the flame. Depending on the fuel mixture, liftoff usually occurs after the flame becomes fully turbulent. As the velocity increases further after liftoff, the liftoff height (the axial distance between the fuel jet exit and the point where combustion begins) increases approximately linearly with the jet velocity [12]. After the liftoff height increases to such an extent that it reaches a value comparable to the flame diameter, a further increase in the velocity causes blowoff [12].

C. BURNING OF CONDENSED PHASES

When most liquids or solids are projected into an atmosphere so that a combustible mixture is formed, an ignition of this mixture produces a flame that surrounds the liquid or solid phase. Except at the very lowest of pressures, around 10^{-6} torr, this flame is a diffusion flame. If the condensed phase is a liquid fuel and the gaseous oxidizer is oxygen, the fuel evaporates from the liquid surface and diffuses to the flame front as the oxygen moves from the surroundings to the burning front. This picture of condensed phase burning is most readily applied to droplet burning, but can also be applied to any liquid surface.

The rate at which the droplet evaporates and burns is generally considered to be determined by the rate of heat transfer from the flame front to the fuel surface. Here, as in the case of gaseous diffusion flames, chemical processes are assumed to occur so rapidly that the burning rates are determined solely by mass and heat transfer rates.

Many of the early analytical models of this burning process considered a double-film model for the combustion of the liquid fuel. One film separated the droplet surface from the flame front and the other separated the flame front from the surrounding oxidizer atmosphere, as depicted in Fig. 11.

In some analytical developments the liquid surface was assumed to be at the normal boiling point of the fuel. Surveys of the temperature field in burning liquids by Khudyakov [13] indicated that the temperature is just a few degrees below the boiling point. In the approach to be employed here, the only requirement is that the droplet be at a uniform temperature at or below the normal boiling point. In the sf region of Fig. 11, fuel evaporates at the drop surface and diffuses toward the flame front where it is consumed. Heat is conducted from the flame front to the liquid and vaporizes the fuel. Many analyses assume that the fuel is heated to the flame temperature before it chemically reacts and that the fuel does not react until it reaches the flame front. This latter assumption implies that the flame front is a mathematically thin surface where the fuel and oxidizer meet in stoichiometric proportions. Some early investigators first determined T_f in order to calculate the fuel burning rate. However, in order to determine a T_f, the infinitely thin reaction zone at the stoichiometric position must be assumed.

In the film f∞, oxygen diffuses to the flame front, and combustion products and heat are transported to the surrounding atmosphere. The position of the boundary designated by ∞ is determined by convection. A stagnant atmosphere places the boundary at an infinite distance from the fuel surface.

Although most analyses assume no radiant energy transfer, as will be shown subsequently, the addition of radiation poses no mathematical difficulty in the solution to the mass burning rate problem.

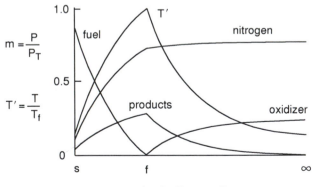

FIGURE 11   Characteristic parametric variations of dimensionless temperature T' and mass fraction m of fuel, oxygen, nitrogen, and products along a radius of a droplet diffusion flame in a quiescent atmosphere. T_f is the adiabatic, stoichiometric flame temperature, P is the partial pressure, and P_T is the total pressure. The estimated values derived for benzene are given in Section 6.C.2.b.

1. General Mass Burning Considerations and the Evaporation Coefficient

Three parameters are generally evaluated: the mass burning rate (evaporation), the flame position above the fuel surface, and the flame temperature. The most important parameter is the mass burning rate, for it permits the evaluation of the so-called evaporation coefficient, which is most readily measured experimentally.

The use of the term *evaporation coefficient* comes about from mass and heat transfer experiments without combustion—i.e., evaporation, as generally used in spray-drying and humidification. Basically, the evaporation coefficient β is defined by the following expression, which has been verified experimentally:

$$d^2 = d_0^2 - \beta t \tag{41}$$

where d_0 is the original drop diameter and d is the drop diameter after time t. It will be shown later that the same expression must hold for mass and heat transfer with chemical reaction (combustion).

The combustion of droplets is one aspect of a much broader problem, which involves the gasification of a condensed phase, i.e., a liquid or a solid. In this sense, the field of diffusion flames is rightly broken down into gases and condensed phases. Here the concern is with the burning of droplets, but the concepts to be used are just as applicable to other practical experimental properties such as the evaporation of liquids, sublimation of solids, hybrid burning rates, ablation heat transfer, solid propellant burning, transpiration cooling, and the like. In all categories, the interest is the mass consumption rate, or the rate of regression of a condensed phase material. In gaseous diffusion flames, there was no specific property to measure and the flame height was evaluated; but in condensed phase diffusion flames, a specific quantity is measurable. This quantity is some representation of the mass consumption rate of the condensed phase. The similarity to the case just mentioned arises from the fact that the condensed phase must be gasified; consequently, there must be an energy input into the condensed material. What determines the rate of regression or evolution of material is the heat flux at the surface. Thus, in all the processes mentioned,

$$q = \dot{r} \rho_f L_v' \tag{42}$$

where q is the heat flux to the surface in calories per square centimeter per second; $\dot{r}$ is the regression rate in centimeters per second; ρ_f is the density of the condensed phase; and L_v' is the overall energy per unit mass required to gasify the material. Usually, L_v' is the sum of two terms—the heat of vaporization, sublimation, or gasification plus the enthalpy required to bring the surface to the temperature of vaporization, sublimation, or gasification.

From the foregoing discussion, it is seen that the heat flux q, L_v', and the density determine the regression rate; but this statement does not mean that the heat flux is the controlling or rate-determining step in each case. In fact, it is generally not the controlling step. The controlling step and the heat flux are always interrelated, however. Regardless of the process of concern (assuming no radiation),

$$q = -\lambda (\partial T / \partial y)_s \qquad (43)$$

where λ is the thermal conductivity and the subscript s designates the fuel surface. This simple statement of the Fourier heat conduction law is of such great significance that its importance cannot be overstated.

This same equation holds whether or not there is mass evolution from the surface and whether or not convective effects prevail in the gaseous stream. Even for convective atmospheres in which one is interested in the heat transfer to a surface (without mass addition of any kind, i.e., the heat transfer situation generally encountered), one writes the heat transfer equation as

$$q = h(T_\infty - T_s) \qquad (44)$$

Obviously, this statement is shorthand for

$$q = -\lambda (\partial T / \partial y)_s = h(T_\infty - T_s) \qquad (45)$$

where T_∞ and T_s are the free-stream and surface temperatures, respectively; the heat transfer coefficient h is by definition

$$h \equiv \lambda / \delta \qquad (46)$$

where δ is the boundary layer thickness. Again by definition, the boundary layer is the distance between the surface and free-stream condition; thus, as an approximation,

$$q = \lambda (T_\infty - T_s) / \delta \qquad (47)$$

The $(T_\infty - T_s)/\delta$ term is the temperature gradient, which correlates $(\partial T / \partial y)_s$ through the boundary layer thickness. The fact that δ can be correlated with the Reynolds number and that the Colburn analogy can be applied leads to the correlation of the form

$$\mathrm{Nu} = f(\mathrm{Re}, \mathrm{Pr}) \qquad (48)$$

where Nu is the Nusselt number, hx/λ; Pr is the Prandtl number, $c_p \mu / \lambda$; and Re is the Reynolds number, $\rho v x / \mu$; here x is the critical dimension—the distance from the leading edge of a flat plate or the diameter of a tube.

Although the correlations given by Eq. (48) are useful for practical evaluation of heat transfer to a wall, one must not lose sight of the fact that the temperature gradient at the wall actually determines the heat flux there. In transpiration cooling problems, it is not so much that the injection of the transpiring fluid increases the boundary layer thickness, thereby decreasing the heat flux, but rather that the temperature gradient at the surface is decreased by the heat absorbing capability of the injected fluid. What Eq. (43) specifies is that regardless of the processes taking place, the temperature profile at the surface determines the regression rate—whether it be evaporation, solid propellant burning, etc. Thus, all the mathematical approaches used in this type of problem simply seek to evaluate the temperature gradient at the surface. The temperature gradient at the surface is different for the various processes discussed. Thus, the temperature profile from the surface to the source of energy for evaporation will differ from that for the burning of a liquid fuel, which releases energy when oxidized in a flame structure.

Nevertheless, a diffusion mechanism generally prevails; and because it is the slowest step, it determines the regression rate. In evaporation, the mechanism is the conduction of heat from the surrounding atmosphere to the surface; in ablation, it is the conduction of heat through the boundary layer; in droplet burning, it is the rates at which the fuel diffuses to approach the oxidizer; etc.

It is mathematically interesting that the gradient at the surface will always be a boundary condition to the mathematical statement of the problem. Thus, the mathematical solution is necessary simply to evaluate the boundary condition.

Furthermore, it should be emphasized that the absence of radiation has been assumed. Incorporating radiation transfer is not difficult if one assumes that the radiant intensity of the emitters is known and that no absorption occurs between the emitters and the vaporizing surfaces; i.e., it can be assumed that q_r, the radiant heat flux to the surface, is known. Then Eq. (43) becomes

$$q + q_r = -\lambda(\partial T/\partial y)_s + q_r = r\rho_f L_v' \tag{49}$$

Note that using these assumptions does not make the mathematical solution of the problem significantly more difficult, for again, q_r—and hence radiation transfer—is a known constant and enters only in the boundary condition. The differential equations describing the processes are not altered.

First, the evaporation rate of a single fuel droplet is calculated before considering the combustion of this fuel droplet; or, to say it more exactly, one calculates the evaporation of a fuel droplet in the absence of combustion. Since the concern is with diffusional processes, it is best to start by reconsidering the laws that hold for diffusional processes.

Fick's law states that if a gradient in concentration of species A exists, say (dn_A/dy), a flow or flux of A, say j_A, across a unit area in the y direction will be proportional to the gradient so that

$$j_A = -D \, dn_A/dy \tag{50}$$

where D is the proportionality constant called the molecular diffusion coefficient or, more simply, the diffusion coefficient; n_A is the number concentration of molecules per cubic centimeter; and j is the flux of molecules, in number of molecules per square centimeter per second. Thus, the units of D are square centimeters per second.

The Fourier law of heat conduction relates the flux of heat q per unit area, as a result of a temperature gradient, such that

$$q = -\lambda \, dT/dy$$

The units of q are calories per square centimeter per second and those of the thermal conductivity λ are calories per centimeter per second per degree Kelvin. It is not the temperature, an intensive thermodynamic property, that is exchanged, but energy content, an extensive property. In this case, the energy density and the exchange reaction, which show similarity, are written as

$$q = -\frac{\lambda}{\rho c_p} \rho c_p \frac{dT}{dy} = -\frac{\lambda}{\rho c_p} \frac{d(\rho c_p T)}{dy} = -\alpha \frac{dH}{dy} \tag{51}$$

where α is the thermal diffusivity whose units are square centimeters per second since $\lambda = \text{cal/cm s K}$, $c_p = \text{cal/g K}$, and $\rho = \text{g/cm}^3$; and H is the energy concentration in calories per cubic centimeter. Thus, the similarity of Fick's and Fourier's laws is apparent. The former is due to a number concentration gradient, and the latter to an energy concentration gradient.

A law similar to these two diffusional processes is Newton's law of viscosity, which relates the flux (or shear stress) τ_{yx} of the x component of momentum due to a gradient in u_x; this law is written as

$$\tau_{yx} = -\mu \, du_x/dy \tag{52}$$

where the units of the stress τ are dynes per square centimeter and those of the viscosity are grams per centimeter per second. Again, it is not velocity that is exchanged, but momentum; thus when the exchange of momentum density is written, similarity is again noted:

$$\tau_{yx} = -(\mu/\rho)[d(\rho u_x)/dy] = -\nu[d(\rho u_x)/dy] \tag{53}$$

where ν is the momentum diffusion coefficient or, more acceptably, the kinematic viscosity; ν is a diffusivity and its units are also square centimeters per second. Since Eq. (53) relates the momentum gradient to a flux, its similarity to Eqs. (50) and (51) is obvious. Recall, as stated in Chapter 4, that the simple kinetic theory for gases predicts $\alpha = D = \nu$. The ratios of these three diffusivities give some familiar dimensionless similarity parameters:

$$\text{Pr} = \nu/\alpha, \qquad \text{Sc} = \nu/D, \qquad \text{Le} = \alpha/D$$

where Pr, Sc, and Le are Prandtl, Schmidt, and Lewis numbers, respectively. Thus, for gases simple kinetic theory gives as a first approximation

$$\text{Pr} = \text{Sc} = \text{Le} = 1$$

2. Single Fuel Droplets in Quiescent Atmospheres

Since Fick's law will be used in many different forms in the ensuing development, it is best to develop those forms so that the later developments need not be interrupted.

Consider the diffusion of molecules A into an atmosphere of B molecules, i.e., a binary system. For a concentration gradient in A molecules alone, future developments can be simplified readily if Fick's law is now written

$$j_A = -D_{AB} \, d(n_A)/dy \tag{54}$$

where D_{AB} is the binary diffusion coefficient. Here, n_A is considered in number of moles per unit volume, since one could always multiply Eq. (50) through by Avogadro's number, and j_A is expressed in number of moles as well.

Multiplying through Eq. (54) by MW_A, the molecular weight of A, one obtains

$$(j_A MW_A) = -D_{AB} \, d(n_A MW_A)/dy = -D_{AB} \, d\rho_A/dy \tag{55}$$

where ρ_A is the mass density of A, ρ_B is the mass density of B, and n is the total number of moles

$$n = n_A + n_B \tag{56}$$

which is constant; and

$$dn/dy = 0, \qquad dn_A/dy = -dn_B/dy, \qquad j_A = -j_B \tag{57}$$

The result is a net flux of mass by diffusion equal to

$$\rho v = j_A MW_A + j_B MW_B \tag{58}$$

$$= j_A(MW_A - MW_B) \tag{59}$$

where v is the bulk direction velocity established by diffusion.

In problems dealing with the combustion of condensed matter, and hence regressing surfaces, there is always a bulk velocity movement in the gases. Thus, species are diffusing while the bulk gases are moving at a finite velocity. The diffusion of the species can be against or with the bulk flow (velocity). For mathematical convenience, it is best to decompose such flows into a flow with an average mass velocity v and a diffusive velocity relative to v.

When one gas diffuses into another, as A into B, even without the quasi–steady-flow component imposed by the burning, the mass transport of a species, say A, is made up of two components—the normal diffusion component and the component related to the bulk movement established by the diffusion process. This mass transport flow has a velocity Δ_A and the mass of A transported per unit area is $\rho_A \, \Delta_A$. The bulk velocity established by the diffusive flow is given by Eq. (58). The fraction of that flow is Eq. (58) multiplied by the mass fraction of A, ρ_A/ρ.

Thus,

$$\rho_A \, \Delta_A = -D_{AB}(d\rho_A \, dy)\{1 - [1 - (MW_B/MW_A)](\rho_A/\rho)\} \tag{60}$$

Since $j_A = -j_B$

$$MW_A j_A = -MW_A j_B = -(MW_B j_B)(MW_A/MW_B) \tag{61}$$

and

$$-d\rho_A/dy = (MW_A/MW_B)(d\rho_B/dy) \tag{62}$$

However,

$$(d\rho_A/dy) + (d\rho_B/dy) = d\rho/dy \tag{63}$$

which gives with Eq. (62)

$$(d\rho_A/dy) - (MW_B/MW_A)(d\rho_A/dy) = d\rho/dy \tag{64}$$

Multiplying through by ρ_A/r, one obtains

$$(\rho_A/\rho)[1 - (MW_B/MW_A)](d\rho_A \, dy) = (\rho_A/\rho)(d\rho/dy) \tag{65}$$

Substituting Eq. (65) into Eq. (60), one finds

$$\rho_A \, \Delta_A = -D_{AB} \, [(d\rho_A/dy) - (\rho_A/\rho)(d\rho/dy)] \tag{66}$$

or

$$\rho_A \, \Delta_A = -\rho D_{AB} \, d(\rho_A/\rho)/dy \tag{67}$$

Defining m_A as the mass fraction of A, one obtains the following proper form for the diffusion of species A in terms of mass fraction:

$$\rho_A \, \Delta_A = -\rho D_{AB} \, d(\rho_A/\rho)/dy \tag{68}$$

This form is that most commonly used in the conservation equation.

The total mass flux of A under the condition of the burning of a condensed phase, which imposes a bulk velocity developed from the mass burned, is then

$$\rho_A v_A = \rho_A v_A + \rho_A \, \Delta_A = \rho_A v - \rho D_{AB} \, dm_A/dy \tag{69}$$

where $\rho_A v$ is the bulk transport part and $\rho_A \Delta_A$ is the diffusive transport part. Indeed, in the developments of Chapter 4, the correct diffusion term was used without the proof just completed.

a. Heat and Mass Transfer without Chemical Reaction (Evaporation)— The Transfer Number B

Following Blackshear's [14] adaptation of Spalding's approach [15, 16], consideration will now be given to the calculation of the evaporation of a single fuel droplet in a nonconvective atmosphere at a given temperature and pressure. A major assumption is now made in that the problem is considered as a quasi-steady one. Thus, the droplet is of fixed size and retains that size by a steady flux of fuel. One can consider the regression as being constant; or, even better, one can think of the droplet as a porous sphere being fed from a very thin tube at a rate equal to the mass evaporation rate so that the surface of the sphere is always wet and any liquid evaporated is immediately replaced. The porous sphere approach shows that for the diffusion flame, a bulk gaseous velocity outward must exist; and although this velocity in the spherical geometry will vary radially, it must always be the value given by $\dot{m} = 4\pi r^2 \rho v$. This velocity is the one referred to in the last section. With this physical picture one may take the temperature throughout the droplet as constant and equal to the surface temperature as a consequence of the quasi-steady assumption.

In developing the problem, a differential volume in the vapor above the liquid droplet is chosen, as shown in Fig. 12. The surface area of a sphere is $4\pi r^2$. Since mass cannot accumulate in the element,

$$d(\rho A v) = 0, \qquad (d/dr)(4\pi r^2 \rho v) = 0 \tag{70}$$

which is essentially the continuity equation.

Consider now the energy equation of the evaporating droplet in spherical-symmetric coordinates in which c_p and λ are taken independent of temperature.

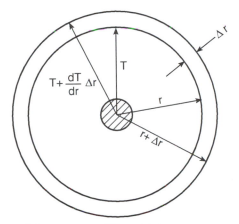

FIGURE 12   Temperature balance across a differential element of a diffusion flame in spherical symmetry.

The heat entering at the surface (i.e., the amount of heat convected in) is $\dot{m}c_p T$ or $(4\pi r^2 \rho v)c_p T$ (see Fig. 12). The heat leaving after $r + \Delta r$ is $\dot{m}c_p[T + (dT/dr)\,\Delta r]$ or $(4\pi r^2 \rho v)c_p[T + (dT/dr)\,\Delta r]$. The difference, then, is

$$-4\pi r^2 \rho v c_p (dT/dr)\,\Delta r$$

The heat diffusing from r toward the drop (out of the element) is

$$-\lambda 4\pi r^2 (dT/dr)$$

The heat diffusing into the element is

$$-\lambda 4\pi r^2 (r + \Delta r)^2 (d/dr)[T + (dT/dr)\,\Delta r]$$

or

$$-\left[\lambda 4\pi r^2 (dT/dr) + \lambda 4\pi r^2 (d^2 T/dr^2)\,\Delta r + \lambda 8\pi r^2\,\Delta r(dT/dr)\right]$$

plus two terms in Δr^2 and one in Δr^3 which are negligible. The difference in the two terms is

$$+\left[\lambda 4\pi r^2 (d^2 T/dr^2)\,\Delta r + 8\lambda\pi r(dT/dr)\right]$$

Heat could be generated in the volume element defined by Δr so one has

$$+(4\pi r^2\,\Delta r)\dot{H}$$

where $\dot{H}$ is the rate of enthalpy change per unit volume. Thus, for the energy balance

$$4\pi r^2 \rho v c_p(dT/dr)\,\Delta r - \lambda 4\pi r^2 (d^2 T/dr^2)\,\Delta r + \lambda 8\pi r\,\Delta r(dT/dr) =$$
$$-\,4\pi r^2\,\Delta r\dot{H} \tag{71}$$

$$4\pi r^2 \rho v c_p(dT/dr) = \lambda 4\pi r^2 (d^2 T/dr^2) + 2\lambda 4\pi r^2 (dT/dr)$$
$$+\,4\pi r^2\,\dot{H} \tag{72}$$

or

$$4\pi r^2 (\rho v)(dc_p T/dr) = d/dr\left[(\lambda 4\pi r^2 /c_p)(dc_p T/dr)\right] + 4\pi r^2\,\Delta r\dot{H} \tag{73}$$

Similarly, the conservation of the Ath species can be written as

$$4\pi r^2 \rho v(dm_A/dr) = (d/dr)\left[4\pi r^2 \rho D(dm_A/dr)\right] + 4\pi r^2\dot{m}_A \tag{74}$$

where $\dot{m}_A$ is the generation or disappearance rate of A due to reaction in the unit volume. According to the kinetic theory of gases, to a first approximation the product $D\rho$ (and hence λ/c_p) is independent of temperature and pressure; consequently, $D_s\rho_s = D\rho$, where the subscript s designates the condition at the droplet surface.

Consider a droplet of radius r. If the droplet is vaporizing, the fluid will leave the surface by convection and diffusion. Since at the liquid droplet surface only A exists, the boundary condition at the surface is

$$\underbrace{\rho_l \dot{r} = \rho_s v_s}_{\substack{\text{amount of material} \\ \text{leaving the surface}}} = \rho m_{As} - \rho D(dm_A/dr)_s \tag{75}$$

where ρ_l is the liquid droplet density and $\dot{r}$ is the rate of change of the liquid droplet radius. Equation (75) is, of course, explicitly Eq. (69) when $\rho_s v_s$ is the bulk mass movement, which at the surface is exactly the amount of A that is being convected (evaporated) written in terms of a gaseous density and velocity plus the amount of gaseous A that diffuses to or from the surface. Since products and inerts diffuse to the surface, m_{As} has a value less than 1. Equation (75), then, is written as

$$v_s = \frac{D(dm_A/dr)_s}{(m_{As} - 1)} \tag{76}$$

In the sense of Spalding, a new parameter b is defined:

$$b \equiv m_A/(m_{As} - 1)$$

Equation (76) thus becomes

$$v_s = D(db/dr)_s \tag{77}$$

and Eq. (74) written in terms of the new variable b and for the evaporation condition (i.e., $\dot{m}_A = 0$) is

$$r^2 \rho v(db/dr) = (d/dr)[r^2 \rho D(db/dr)] \tag{78}$$

The boundary condition at $r = \infty$ is $m_A = m_{A\infty}$ or

$$b = b_\infty \qquad \text{at } r \to \infty \tag{79}$$

From continuity

$$r^2 \rho v = r_s^2 \rho_s v_s \tag{80}$$

Since $r^2 \rho v = $ constant on the left-hand side of the equation, integration of Eq. (78) proceeds simply and yields

$$r^2 \rho v b = r^2 \rho D(db/dr) + \text{const} \tag{81}$$

Evaluating the constant at $r = r_s$, one obtains

$$r_s^2 \rho_s v_s b_s = r_s^2 \rho_s v_s + \text{const}$$

since from Eq. (77) $v_s = D(db/dr)_s$. Or, one has from Eq. (81)

$$r_s^2 \rho_s v_s(b - b_s + 1) = r^2 \rho D(db/dr) \tag{82}$$

By separating variables,

$$(r_s^2 \rho_s v_s / r^2 \rho D)\, dr = db/(b - b_s + 1) \tag{83}$$

assuming ρD constant, and integrating (recall that $\rho D = \rho_s D_s$), one has

$$-(r_s^2 \rho_s / r D_s) = \ln(b - b_s + 1) + \text{const} \tag{84}$$

Evaluating the constant at $r \rightarrow \infty$, one obtains

$$\text{const} = -\ln(b_\infty - b_s + 1)$$

or Eq. (84) becomes

$$r_s^2 v_s / r D_s = \ln\left[(b_\infty - b_s + 1)/(b - b_s + 1)\right] \tag{85}$$

The left-hand term of Eq. (84) goes to zero as $r \rightarrow \infty$. This point is significant because it shows that the quiescent spherical-symmetric case is the only mathematical case that does not blow up. No other quiescent case, such as that for cylindrical symmetry or any other symmetry, is tractable. Evaluating Eq. (85) at $r = r_s$ results in

$$(r_s v_s / D) = \ln(b_\infty - b_s + 1) = \ln(1 + B)$$
$$r_s v_s = D_s \ln(b_\infty - b_s + 1) = D_s \ln(1 + B)$$
$$= D_s \ln\left[(m_{A\infty} - m_{As})/(m_{As} - 1)\right] \tag{86}$$

Here $B \equiv b_\infty - b_s$ and is generally referred to as the Spalding transfer number. The mass consumption rate per unit area $G_A = \dot{m}_A/4\pi r_s^2$, where $\dot{m}_A = \rho_s v_s 4\pi r_s^2$, is then found by multiplying Eq. (86) by $4\pi r_s^2 \rho_s$ and cross-multiplying r_s to give

$$4\pi r_s^2 \rho_s v_s / 4\pi r_s^2 = (D_s \rho_s / r_s) \ln(1 + B)$$
$$G_A = \dot{m}_A/4\pi r_s^2 = (\rho_s D_s / r_s) \ln(1 + B) = (D\rho/r_s) \ln(1 + B) \tag{87}$$

Since the product $D\rho$ is independent of pressure, the evaporation rate is essentially independent of pressure. There is a mild effect of pressure on the transfer number, as will be discussed in more detail when the droplet burning case is considered. In order to find a solution for Eq. (87) or, more rightly, to evaluate the transfer number B, m_{As} must be determined. A reasonable assumption would be that the gas surrounding the droplet surface is saturated at the surface temperature T_s. Since vapor pressure data are available, the problem then is to determine T_s.

For the case of evaporation, Eq. (73) becomes

$$r_s^2 \rho_s v_s c_p\, dT/dr = (d/dr)\left[r^2 \lambda (dT/dr)\right] \tag{88}$$

which, upon integration, becomes

$$r_s^2 \rho_s v_s c_p T = r^2 \lambda (dT/dr) + \text{const} \tag{89}$$

The boundary condition at the surface is

$$[\lambda (dT/dr)]_s = \rho_s v_s L_v \tag{90}$$

where L_v is the latent heat of vaporization at the temperature T_s. Recall that the droplet is considered uniform throughout at the temperature T_s. Substituting Eq. (90) into Eq. (89), one obtains

$$\text{const} = r_s^2 \rho_s v_s [c_p T_s - L_v]$$

Thus, Eq. (89) becomes

$$r_s^2 \rho_s v_s c_p \left[T - T_s + (L_v/c_p) \right] = r^2 \lambda (dT/dr) \tag{91}$$

Integrating Eq. (91), one has

$$-r_s^2 \rho_s v_s c_p / r \lambda = \ln[T - T_s + (L_v/c_p)] + \text{const} \tag{92}$$

After evaluating the constant at $r \to \infty$, one obtains

$$\frac{r_s^2 \rho_s v_s c_p}{r \lambda} = \ln \left[\frac{T_\infty - T_s + (L_v/c_p)}{T - T_s + (L_v/c_p)} \right] \tag{93}$$

Evaluating Eq. (93) at the surface ($r = r_s; T = T_s$) gives

$$\frac{r_s^2 \rho_s v_s c_p}{\lambda} = \ln \left(\frac{c_p(T_\infty - T_s)}{L_v} + 1 \right) \tag{94}$$

And since $\alpha = \lambda / c_p \rho$,

$$r_s v_s = \alpha_s \ln \left(1 + \frac{c_p(T_\infty - T_s)}{L_v} \right) \tag{95}$$

Comparing Eqs. (86) and (95), one can write

$$r_s v_s = \alpha_s \ln \left(\frac{c_p(T_\infty - T_s)}{L_v} + 1 \right) = D_s \left(\frac{m_{A\infty} - m_{As}}{m_{As} - 1} \right)$$

or

$$r_s v_s = \alpha_s \ln(1 + B_T) = D_s \ln(1 + B_M) \tag{96}$$

where

$$B_T \equiv \frac{c_p(T_\infty - T_s)}{L_v}, \qquad B_M \equiv \frac{m_{A\infty} - m_{As}}{m_{As} - 1}$$

Again, since $\alpha = D$,

$$B_T = B_M$$

and

$$c_p(T_\infty - T_s)/L_v = (m_{A\infty} - m_{As})/(m_{As} - 1) \tag{97}$$

Although m_{As} is determined from the vapor pressure of A or the fuel, one must realize that m_{As} is a function of the total pressure since

$$m_{As} \equiv \rho_{As}/\rho = n_A MW_A/n MW = (P_A/P)(MW_A/MW) \tag{98}$$

where n_A and n are the number of moles of A and the total number of moles, respectively; MW_A and MW are the molecular weight of A and the average molecular weight of the mixture, respectively; and P_A and P are the vapor pressure of A and the total pressure, respectively.

In order to obtain the solution desired, a value of T_s is assumed, the vapor pressure of A is determined from tables, and m_{As} is calculated from Eq. (98). This value of m_{As} and the assumed value of T_s are inserted in Eq. (97). If this equation is satisfied, the correct T_s was chosen. If not, one must reiterate. When the correct value of T_s and m_{As} are found, B_T or B_M are determined for the given initial conditions T_∞, or $m_{A\infty}$. For fuel combustion problems, $m_{A\infty}$ is usually zero; however, for evaporation, say of water, there is humidity in the atmosphere and this humidity must be represented as $m_{A\infty}$. Once B_T and B_M are determined, the mass evaporation rate is determined from Eq. (87) for a fixed droplet size. It is, or course, much preferable to know the evaporation coefficient β from which the total evaporation time can be determined. Once B is known, the evaporation coefficient can be determined readily, as will be shown later.

b. Heat and Mass Transfer with Chemical Reaction (Droplet Burning Rates)

The previous developments also can be used to determine the burning rate, or evaporation coefficient, of a single droplet of fuel burning in a quiescent atmosphere. In this case, the mass fraction of the fuel, which is always considered to be the condensed phase, will be designated m_f, and the mass fraction of the oxidizer m_o. m_o is the oxidant mass fraction exclusive of inerts and i is used as the mass stoichiometric fuel–oxidant ratio, also exclusive of inerts. The same assumptions that hold for evaporation from a porous sphere hold here. Recall that the temperature throughout the droplet is considered to be uniform and equal to the surface temperature T_s. This assumption is sometimes referred to as one of infinite condensed phase thermal conductivity. For liquid fuels, this temperature is generally near, but slightly less than, the saturation temperature for the prevailing ambient pressure.

As in the case of burning gaseous fuel jets, it is assumed that the fuel and oxidant approach each other in stoichiometric proportions. The stoichiometric relations are written as

$$\dot{m}_f = \dot{m}_o i, \qquad \dot{m}_f H = \dot{m}_o H i = -\dot{H} \tag{99}$$

where $\dot{m}_f$ and $\dot{m}_o$ refer to the mass consumption rates per unit volume, H is the heat of reaction (combustion) of the fuel per unit mass, and $\dot{H}$ is the heat release rate per unit volume. The mass consumption rates $\dot{m}_f$ and $\dot{m}_o$ are decreasing.

There are now three fundamental diffusion equations: one for the fuel, one for the oxidizer, and one for the heat. Equation (74) is then written as two equations: one in terms of m_f and the other in terms of m_o. Equation (73) remains the same for the consideration here. As seen in the case of the evaporation, the solution to the equations becomes quite simple when written in terms of the b variable, which led to the Spalding transfer number B. As noted in this case the b variable was obtained from the boundary condition. Indeed, another b variable could have been obtained for the energy equation [Eq. (73)]—a variable having the form

$$b = (c_p T / L_v)$$

As in the case of burning gaseous fuel jets, the diffusion equations are combined readily by assuming $D\rho = (\lambda/c_p)$; i.e., Le = 1. The same procedure can be followed in combining the boundary conditions for the three droplet burning equations to determine the appropriate b variables to simplify the solution for the mass consumption rate.

The surface boundary condition for the diffusion of fuel is the same as that for pure evaporation [Eq. (75)] and takes the form

$$\rho_s v_s (m_{fs} - 1) = D\rho (dm_f/dr)_s \tag{100}$$

Since there is no oxidizer leaving the surface, the surface boundary condition for diffusion of oxidizer is

$$0 = \rho_s v_s m_{os} - D\rho (dm_o \, dr)_s \qquad \text{or} \qquad \rho_s v_s m_{os} = D\rho (dm_o \, dr)_s \tag{101}$$

The boundary condition for the energy equation is also the same as that would have been for the droplet evaporation case [Eq. (75)] and is written as

$$\rho_s v_s L_v = (\lambda/c_p) \left[d(c_p T)/dr \right]_s \tag{102}$$

Multiplying Eq. (100) by H and Eq. (101) by iH gives the new forms

$$H \left[\rho_s v_s (m_{fs} - 1) \right] = [D\rho (dm_f/dr)_s] H \tag{103}$$

and

$$H[i\{\rho_s v_s m_{os}\}] = [\{D\rho (dm_o/dr)_s\}i] H \tag{104}$$

By adding Eqs. (102) and (103) and recalling that $D\rho = (\lambda/c_p)$, one obtains

$$\rho_s v_s [H(m_{fs} - 1) + L_v] = D\rho [d(m_f H + c_p T)/dr]_s$$

and after transposing,

$$\rho_s v_s = D\rho \left\{ d \left[\frac{m_f H + c_p T}{H(m_{fs} - 1) + L_v} \right] / dr \right\}_s \tag{105}$$

Similarly, by adding Eqs. (102) and (104), one obtains

$$\rho_s v_s (L_v + H i m_{os}) = D\rho \left[(d/dr)(c_p T + m_{oi} H) \right]_s$$

or

$$\rho_s v_s = D\rho \left[d \left(\frac{m_{oi} H + c_p T}{H i m_{os} + L_v} \right) / dr \right]_s \tag{106}$$

And, finally, by subtracting Eq. (104) from (103), one obtains

$$\rho_s v_s \left[(m_{fs} - 1) - i m_{os} \right] = D\rho [d(m_f - i m_o)/dr]_s$$

or

$$\rho_s v_s = D\rho \left\{ d \left[\frac{m_f - i m_o}{(m_{fs} - 1) - i m_{os}} \right] / dr \right\}_s \tag{107}$$

Thus, the new b variables are defined as

$$b_{fq} \equiv \frac{m_f H + c_p T}{H(m_{fs} - 1) + L_v}, \qquad b_{oq} \equiv \frac{m_{oi} H + c_p T}{H i m_{os} + L_v},$$

$$b_{fo} \equiv \frac{m_f - i m_o}{(m_{fs} - 1) - i m_{os}} \tag{108}$$

The denominator of each of these three b variables is a constant. The three diffusion equations are transformed readily in terms of these variables by multiplying the fuel diffusion equation by H and the oxygen diffusion equation by iH. By using the stoichiometric relations [Eq. 99] and combining the equations in the same manner as the boundary conditions, one can eliminate the nonhomogeneous terms $\dot{m}_f$, $\dot{m}_o$, and $\dot{H}$. Again, it is assumed that $D\rho = (\lambda/c_p)$. The combined equations are then divided by the appropriate denominators from the b variables so that all equations become similar in form. Explicitly then, one has the following developments:

$$r^2 \rho v \frac{d(c_p T)}{dr} = \frac{d}{dr} \left[r^2 \frac{\lambda}{c_p} \frac{d(c_p T)}{dr} \right] - r^2 \dot{H} \tag{109}$$

$$H \left\{ r^2 \rho v \frac{dm_f}{dr} = \frac{d}{dr} \left[r^2 D\rho \frac{dm_f}{dr} \right] + r^2 \dot{m}_f \right\} \tag{110}$$

$$Hi \left\{ r^2 \rho v \frac{dm_o}{dr} = \frac{d}{dr} \left[r^2 D\rho \frac{dm_o}{dr} \right] + r^2 \dot{m}_o \right\} \tag{111}$$

$$\dot{m}_f H = \dot{m}_o Hi = -\dot{H} \tag{99}$$

Adding Eqs. (109) and (110), dividing the resultant equation through by $[H(m_{fs} - 1) + L_v]$, and recalling that $\dot{m}_f H = -\dot{H}$, one obtains

$$r^2 \rho v \frac{d}{dr} \left(\frac{m_f H + c_p T}{H(m_{fs} - 1) + L_v} \right) = \frac{d}{dr} \left[r^2 D\rho \frac{d}{dr} \left(\frac{m_f H + c_p T}{H(m_{fs} - 1) + L_v} \right) \right] \quad (112)$$

which is then written as

$$r^2 \rho v (db_{fq}/dr) = (d/dr)[r^2 D\rho (db_{fq}/dr)] \quad (113)$$

Similarly, by adding Eqs. (109) and (111) and dividing the resultant equation through by $[Him_{os} + L_v]$, one obtains

$$r^2 \rho v \frac{d}{dr} \left(\frac{m_o i H + c_p T}{Him_{os} + L_v} \right) = \frac{d}{dr} \left[r^2 D\rho \frac{d}{dr} \left(\frac{m_o i H + c_p T}{Him_{os} + L_v} \right) \right] \quad (114)$$

or

$$r^2 \rho v \frac{d}{dr} (b_{oq}) = \frac{d}{dr} \left[r^2 D\rho \frac{d}{dr} (b_{oq}) \right] \quad (115)$$

Following the same procedures by subtracting Eq. (111) from Eq. (110), one obtains

$$r^2 \rho v \frac{d}{dr} (b_{fo}) = \frac{d}{dr} \left[r^2 D\rho \frac{d}{dr} (b_{fo}) \right] \quad (116)$$

Obviously, all the combined equations have the same form and boundary conditions; i.e.,

$$r^2 \rho v \frac{d}{dr} (b)_{fo, fq, oq} = \frac{d}{dr} \left[r^2 D\rho \frac{d}{dr} (b)_{fo, fq, oq} \right]$$

$$\rho_s v_s = D\rho \left[\frac{d}{dr} (b)_{fo, fq, oq} \right]_s \qquad \text{at } r = r_s \quad (117)$$

$$b = b_\infty \qquad \text{at} \qquad r \to \infty$$

The equation and boundary conditions are the same as those obtained for the pure evaporation problem; consequently, the solution is the same. Thus, one writes

$$G_f = \dot{m}_f / 4\pi r_s^2 = (D\rho_s / r_s) \ln(1 + B) \qquad \text{where } B = b_\infty - b_s \quad (118)$$

It should be recognized that since $Le = Sc = Pr$, Eq. (118) can also be written as

$$G_f = (\lambda/c_p r_s) \ln(1 + B) = (\mu/r_s) \ln(1 + B) \quad (119)$$

In Eq. (118) the transfer number B can take any of the following forms:

Without combustion assumption   With combustion assumption

$$B_{\text{fo}} = \frac{(m_{\text{f}\infty} - m_{\text{fs}}) + (m_{\text{os}} - m_{\text{o}\infty})i}{(m_{\text{fs}} - 1) - im_{\text{os}}} = \frac{(im_{\text{o}\infty} + m_{\text{fs}})}{1 - m_{\text{fs}}}$$

$$B_{\text{fq}} = \frac{H(m_{\text{f}\infty} - m_{\text{fs}}) + c_p(T_\infty - T_s)}{L_v + H(m_{\text{fs}} - 1)} = \frac{c_p(T_\infty - T_s) - m_{\text{fs}}H}{L_v + H(m_{\text{fs}} - 1)} \qquad (120)$$

$$B_{\text{fq}} = \frac{Hi(m_{\text{o}\infty} - m_{\text{os}}) + c_p(T_\infty - T_s)}{L_v + im_{\text{os}}H} = \frac{c_p(T_\infty - T_s) + im_{\text{o}\infty}H}{L_v}$$

The combustion assumption in Eq. (120) is that $m_{\text{os}} = m_{\text{f}\infty} = 0$ since it is assumed that neither fuel nor oxidizer can penetrate the flame zone. This requirement is not that the flame zone be infinitely thin, but that all the oxidizer must be consumed before it reaches the fuel surface and that the quiescent atmosphere contain no fuel.

As in the evaporation case, in order to solve Eq. (118), it is necessary to proceed by first equating $B_{\text{fo}} = B_{\text{oq}}$. This expression

$$\frac{im_{\text{o}\infty} + m_{\text{fs}}}{1 - m_{\text{fs}}} = \frac{c_p(T_\infty - T_s) + im_{\text{o}\infty}H}{L_v} \qquad (121)$$

is solved with the use of the vapor pressure data for the fuel. The iteration process described in the solution of $B_{\text{M}} = B_{\text{T}}$ in the evaporation problem is used. The solution of Eq. (121) gives T_s and m_{fs} and, thus, individually B_{fo} and B_{oq}. With B known, the burning rate is obtained from Eq. (118).

For the combustion systems, B_{oq} is the most convenient form of B: $c_p(T_\infty - T_s)$ is usually much less than $im_{\text{o}\infty}H$ and to a close approximation $B_{\text{oq}} \cong (im_{\text{o}\infty}H/L_v)$. Thus the burning rate (and, as will be shown later, the evaporation coefficient β) is readily determined. It is not necessary to solve for m_{fs} and T_s. Furthermore, detailed calculations reveal that T_s is very close to the saturation temperature (boiling point) of most liquids at the given pressure. Thus, although $D\rho$ in the burning rate expression is independent of pressure, the transfer number increases as the pressure is raised because pressure rises concomitantly with the droplet saturation temperature. As the saturation temperature increases, the latent heat of evaporation L_v decreases. Of course, L_v approaches zero at the critical point. This increase in L_v is greater than the decrease of the $c_{p,\text{g}}(T_\infty - T_s)$ and $c_{pl}(T_s - T_l)$ terms; thus B increases with pressure and the burning follows some logarithmic trend through $\ln(1 + B)$.

The form of B_{oq} and B_{fq} presented in Eq. (120) is based on the assumption that the fuel droplet has infinite thermal conductivity; i.e., the temperature of the droplet is T_s throughout. But in an actual porous sphere experiment, the fuel enters the center of the sphere at some temperature T_l and reaches T_s at the sphere surface. For a large sphere, the enthalpy required to raise the cool entering liquid to the surface temperature is $c_{pl}(T_s - T_l)$ where c_{pl} is the specific heat of the liquid fuel.

To obtain an estimate of B that gives a conservative (lower) result of the burning rate for this type of condition, one could replace L_v by

$$L'_v = L_v + c_{pl}(T_s - T_l) \tag{122}$$

in the forms of B represented by Eq. (120).

Table 1, extracted from Kanury [17], lists various values of B for many condensed phase combustible substances burning in air. Examination of this table reveals that the variation of B for different combustible liquids is not great; rarely does one liquid fuel have a value of B a factor of 2 greater than another. Since the transfer number always enters the burning rate expression as a $\ln(1 + B)$ term, one may conclude that, as long as the oxidizing atmosphere is kept the same, neither the burning rate nor the evaporation coefficient of liquid fuels will vary greatly. The diffusivities and gas density dominate the burning rate. Whereas a tenfold variation in B results in an approximately twofold variation in burning rate, a tenfold variation of the diffusivity or gas density results in a tenfold variation in burning.

Furthermore, note that the burning rate has been determined without determining the flame temperature or the position of the flame. In the approach attributed to Godsave [18] and extended by others [19, 20], it was necessary to find the flame temperature, and the early burning rate developments largely followed this procedure. The early literature contains frequent comparisons not only of the calculated and experimental burning rates (or β), but also of the flame temperature and position. To their surprise, most experimenters found good agreement with respect to burning rate but poorer agreement in flame temperature and position. What they failed to realize is that, as shown by the discussion here, the burning rate is independent of the flame temperature. As long as an integrated approach is used and the gradients of temperature and product concentration are zero at the outer boundary, it does not matter what the reactions are or where they take place, provided they take place within the boundaries of the integration.

It is possible to determine the flame temperature T_f and position r_f corresponding to the Godsave-type calculations simply by assuming that the flame exists at the position where $im_o = m_f$. Equation (87) is written in terms of b_{fo} as

$$\frac{r_s^2 \rho_s v_s}{D_s \rho_s r} = \ln \left[\frac{m_{f\infty} - m_{fs} - i(m_{o\infty} - m_{os}) + (m_{fs} - 1) - im_{os}}{m_f - m_{fs} - i(m_o - m_{os}) + (m_{fs} - 1) - im_{os}} \right] \tag{123}$$

At the flame surface, $m_f = m_o = 0$ and $m_{f\infty} = m_{os} = 0$; thus Eq. (123) becomes

$$r_s^2 \rho_s v_s / D_s \rho_s r_f = \ln(1 + im_{o\infty}) \tag{124}$$

Since $\dot{m} = 4\pi r_s^2 \rho_s v_s$ is known, r_f can be estimated. Combining Eqs. (118) and (124) results in the ratio of the radii

$$\frac{r_f}{r_s} = \frac{\ln(1 + B)}{\ln(1 + im_{o\infty})} \tag{125}$$

TABLE 1   Transfer Numbers of Various Liquids in Air[a]

	B	B'	B''	B'''
n-Pentane	8.94	8.15	8.19	9.00
n-Hexane	8.76	6.70	6.82	8.83
n-Heptane	8.56	5.82	6.00	8.84
n-Octane	8.59	5.24	5.46	8.97
i-Octane	9.43	5.56	5.82	9.84
n-Decane	8.40	4.34	4.62	8.95
n-Dodecane	8.40	4.00	4.30	9.05
Octene	9.33	5.64	5.86	9.72
Benzene	7.47	6.05	6.18	7.65
Methanol	2.95	2.70	2.74	3.00
Ethanol	3.79	3.25	3.34	3.88
Gasoline	9.03	4.98	5.25	9.55
Kerosene	9.78	3.86	4.26	10.80
Light diesel	10.39	3.96	4.40	11.50
Medium diesel	11.18	3.94	4.38	12.45
Heavy diesel	11.60	3.91	4.40	13.00
Acetone	6.70	5.10	5.19	6.16
Toluene	8.59	6.06	6.30	8.92
Xylene	9.05	5.76	6.04	9.43

[a] $T_\infty = 20°C$.
Note. $B = [im_o H + c_p(T_\infty - T_s)]/L_v$; $B' = [im_o H + c_p(T_\infty - T_s)]/L_v'$; $B'' = (im_{o\infty} H/L_v')$; $B'''(im_{o\infty} H/L_v')$.

For the case of benzene (C_6H_6) burning in air, the mass stoichiometric index i is found to be

$$i = \frac{78}{7.5 \times 32} = 0.325$$

Since the mass fraction of oxygen in air is 0.23 and B' from Table 1 is given as 6, one has

$$\frac{r_f}{r_s} = \frac{\ln(1 + 6)}{[1 + (0.325 \times 0.23)]} \cong 27$$

This value is much larger than the value of about 2 to 4 observed experimentally. The large deviation between the estimated value and that observed is most likely due to the assumptions made with respect to the thermophysical properties and the Lewis number. This point is discussed in Section 6.C.2.c. Although this estimate does not appear suitable, it is necessary to emphasize that the results obtained for the burning rate and the combustion evaporation coefficient derived later give good comparisons with experimental data. The reason for this supposed anomaly is that the burning rate is obtained by integration between the infinite atmosphere and the droplet and is essentially independent of the thermophysical parameters necessary to estimate internal properties. Such parameters include the flame radius and the flame temperature, to be discussed in subsequent paragraphs.

For surface burning, as in an idealized case of char burning, it is appropriate to assume that the flame is at the particle surface and that $r_f = r_s$. Thus from Eq. (125), B must equal $im_{o\infty}$. Actually, the same result would arise from the first transfer number in Eq. (120):

$$B_{fo} = \frac{im_{o\infty} + m_{fs}}{1 - m_{fs}}$$

Under the condition of surface burning, $m_{fs} = 0$ and thus, again, $B_{fo} = im_{o\infty}$.

As in the preceding approach, an estimate of the flame temperature T_f at r_f can be obtained by writing Eq. (85) with $b = b_{oq}$. For the condition that Eq. (85) is determined at $r = r_f$, one makes use of Eq. (124) to obtain the expression

$$1 + m_{o\infty} = [(b_\infty - b_s + 1)/(b_f - b_s + 1)]$$

$$1 + m_{o\infty} = \left[Him_{o\infty} + c_p(T_\infty - T_s) + L_v \right] / [c_p(T_f - T_s) + L_v] \qquad (126)$$

or

$$c_p(T_\infty - T_s) = \left\{ \left[Him_{o\infty} + c_p(T_\infty - T_s) + L_v \right] /(1 + m_{o\infty}) \right\} - L_v$$

Again, comparisons of results obtained by Eq. (126) with experimental measurements show the same extent of deviation as that obtained for the determination of r_f/r_s from Eq. (125). The reasons for this deviation are also the same.

Irrespective of these deviations, it is also possible from this transfer number approach to obtain some idea of the species profiles in the droplet burning case. It is best to establish the conditions for the nitrogen profile for this quasi-steady fuel droplet burning case in air first. The conservation equation for the inert i (nitrogen) in droplet burning is

$$4\pi r^2 \rho v m_i = 4\pi r^2 D\rho \frac{dm_i}{dr} \qquad (127)$$

Integrating for the boundary condition that at $r = \infty$, $m_i = m_{i\infty}$, one obtains

$$-\frac{4\pi r^2 \rho_s v_s}{4\pi} \cdot \frac{1}{r} = D\rho \, \ln\left(\frac{m_i}{m_{i\infty}} \right) \qquad (128)$$

From either Eq. (86) or Eq. (118) it can be seen that

$$r_s \rho_s v_s = D_s \rho_s \, \ln(1 + B)$$

Then Eq. (128) can be written in the form

$$-r_s \rho_s v_s \frac{r_s}{r} = -[D\rho \ \ln(1 + B)]\frac{r_s}{r} = D\rho \ \ln\left(\frac{m_i}{m_{i\infty}}\right) \qquad (129)$$

For the condition at $r = r_s$, this last expression reveals that

$$\ln(1 + B) = \ln\left(\frac{m_i}{m_{i\infty}}\right)$$

or

$$1 + B = m_{i\infty}/m_{is}$$

Since the mass fraction of nitrogen in air is 0.77, for the condition $B \cong 6$

$$m_{is} = \frac{0.77}{7} = 0.11$$

Thus the mass fraction of the inert nitrogen at the droplet surface is 0.11.
    Evaluating Eq. (129) at $r = r_f$, one obtains

$$-\frac{r_s}{r_f} \ln(1 + B) = \ln\left(\frac{m_i}{m_{i\infty}}\right)$$

However, Eq. (125) gives

$$\frac{r_s}{r_f} \ln(1 + B) = \ln(1 + i m_{o\infty})$$

Thus

$$1 + i m_{o\infty} = \frac{m_{i\infty}}{m_{if}}$$

or for the case under consideration

$$m_{if} \cong \frac{0.77}{1.075} \cong 0.72$$

The same approach would hold for the surface burning of a carbon char except that i would equal 0.75 (see Chapter 9) and

$$m_{is} = \frac{0.77}{1.075 \times 0.22} \cong 0.66$$

which indicates that the mass fraction of gaseous products m_{ps} for this carbon case equals 0.34.
    For the case of a benzene droplet in which B is taken as 6, the expression that permits the determination of m_{fs} is

$$B = \frac{i m_{o\infty} + m_{fs}}{1 - m_{fs}} = 6$$

or

$$m_{fs} = 0.85$$

This transfer number approach for a benzene droplet burning in air reveals species profiles characteristic of most hydrocarbons and gives the results summarized below:

$$
\begin{array}{lll}
m_{fs} = 0.85 & m_{ff} = 0.00 & m_{f\infty} = 0.00 \\
m_{os} = 0.00 & m_{of} = 0.00 & m_{o\infty} = 0.23 \\
m_{is} = 0.11 & m_{if} = 0.72 & m_{i\infty} = 0.77 \\
m_{ps} = 0.04 & m_{pf} = 0.28 & m_{p\infty} = 0.00
\end{array}
$$

These results are graphically represented in Fig. 11. The product composition in each case is determined for the condition that the mass fractions at each position must total to 1.

It is now possible to calculate the burning rate of a droplet under the quasi-steady conditions outlined and to estimate, as well, the flame temperature and position; however, the only means to estimate the burning time of an actual droplet is to calculate the evaporation coefficient for burning, β. From the mass burning results obtained, β may be readily determined. For a liquid droplet, the relation

$$dm/dt = -\dot{m}_f = 4\pi\rho_l r_s^2 (dr_s/dt) \tag{130}$$

gives the mass rate in terms of the rate of change of radius with time. Here, ρ_l is the density of the liquid fuel. It should be evaluated at T_s.

Many experimenters have obtained results similar to those given in Fig. 13. These results confirm that the variation of droplet diameter during burning follows the same "law" as that for evaporation:

$$d^2 = d_o^2 - \beta t \tag{131}$$

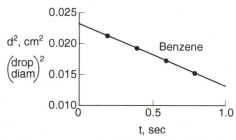

FIGURE 13   Diameter–time measurements of a benzene droplet burning in quiescent air showing diameter-squared dependence (after Godsave [18]).

Since $d = 2r_s$,

$$dr_s/dt = -\beta/8r_s \tag{132}$$

It is readily shown that Eqs. (118) and (131) verify that a "d^2" law should exist. Equation (131) is rewritten as

$$\dot{m} = -2\pi\rho_l r_s(dr_s^2/dt) = -(\pi\rho_l r_s/2)[d(d^2)/dt]$$

Making use of Eq. (118)

$$\dot{m}/4\pi r_s = (\lambda/c_p)\ln(1 + B) = -(\rho_l/8)[d(d^2)/dt] = +(\rho_l/8)\beta$$

shows that

$$d(d^2)/dt = \text{const}, \qquad \beta = (8/\rho_l)(\lambda/c_p)\ln(1 + B) \tag{133}$$

which is a convenient form since λ/c_p is temperature-insensitive. Sometimes β is written as

$$\beta = 8(\rho_s/\rho_l)\alpha \, \ln(1 + B)$$

to correspond more closely to expressions given by Eqs. (96) and (118). The proper response to Problem 9 of this chapter will reveal that the "d^2" law is not applicable in modest Reynolds number or convective flow.

c. Refinements of the Mass Burning Rate Expression

Some major assumptions were made in the derivation of Eqs. (118) and (133). First, it was assumed that the specific heat, thermal conductivity, and the product $D\rho$ were constants and that the Lewis number was equal to 1. Second, it was assumed that there was no transient heating of the droplet. Furthermore, the role of finite reaction kinetics was not addressed adequately. These points will be given consideration in this section.

Variation of Transport Properties   The transport properties used throughout the previous developments are normally strong functions of both temperature and species concentration. The temperatures in a droplet diffusion flame as depicted in Fig. 11 vary from a few hundred degrees Kelvin to a few thousand. In the regions of the droplet flame one essentially has fuel, nitrogen, and products. However, under steady-state conditions as just shown the major constituent in the sf region is the fuel. It is most appropriate to use the specific heat of the fuel and its binary diffusion coefficient. In the region f∞, the constituents are oxygen, nitrogen, and products. With similar reasoning, one essentially considers the properties of nitrogen to be dominant; moreover, as the properties of fuel are used in the sf region, the properties of nitrogen are appropriate to use in the f∞ region. To illustrate the importance of variable properties, Law [21] presented a simplified model in which the temperature dependence was suppressed, but the concentration dependence was represented by using λ, c_p and $D\rho$ with constant, but different, values in the sf and f∞ regions. The burning rate result of this model was

$$\frac{\dot{m}}{4\pi r_s} = \ln\left\{1 + \left[\frac{c_{pf}(T_f - T_s)}{L_v}\right]^{[\lambda/(c_p)_{sf}]}[1 + im_{o\infty}]^{(D\rho)_{f\infty}}\right\} \tag{134}$$

where T_f is obtained from expressions similar to Eq. (126) and found to be

$$(c_p)_{sf}(T_f - T_s) + \frac{(c_p)_{f\infty}(T_f - T_\infty)}{\left[(1 + im_{o\infty})^{1/(Le)_{f\infty}} - 1\right]} = H - L_v' \tag{135}$$

(r_f/r_s) was found to be

$$\frac{r_f}{r_s} = 1 + \left[\frac{(\lambda/c_p)_{sf}}{(D\rho)_{f\infty}} \frac{\ln[1 + (c_p)_{fs}(T_f - T_s)/L_v']}{\ln(1 + im_{o\infty})}\right]$$

Law [21] points out that since $im_{o\infty}$ is generally much less than 1, the denominator of the second term in Eq. (135) becomes $[im_{o\infty}/(Le)_{f\infty}]$, which indicates that the effect of $(Le)_{f\infty}$ is to change the oxygen concentration by a factor $(Le)_{f\infty}^{-1}$ as experienced by the flame. Obviously, then, for $(Le)_{f\infty} > 1$, the oxidizer concentration is effectively reduced and the flame temperature is also reduced from the adiabatic value obtained for $Le = 1$, the value given by Eq. (126). The effective Lewis number for the mass burning rate [Eq. (134)] is

$$Le = (\lambda/c_p)_{sf}(D\rho)_{f\infty}$$

which reveals that the individual Lewis numbers in the two regions play no role in determining $\dot{m}$. Law and Law [22] estimated that for heptane burning in air the effective Lewis number was between 1/3 and 1/2. Such values in Eq. (136) predict values of r_f/r_s in the right range of experimental values [21].

The question remains as to the temperature at which to evaluate the physical properties discussed. One could use an arithmetic mean for the two regions [23] or Sparrow's 1/3 rule [24], which gives

$$T_{sf} = \tfrac{1}{3}T_s + \tfrac{2}{3}T_f; \qquad T_{f\infty} = \tfrac{1}{3}T_f + \tfrac{2}{3}T_\infty$$

Transient Heating of Droplets   When a cold liquid fuel droplet is injected into a hot stream or ignited by some other source, it must be heated to its steady-state temperature T_s derived in the last section. Since the heat-up time can influence the "d^2" law, particularly for high–boiling-point fuels, it is of interest to examine the effect of the droplet heating mode on the main bulk combustion characteristic—the burning time.

For this case, the boundary condition corresponding to Eq. (102) becomes

$$\dot{m}L_v + \left(4\pi r^2 \lambda_l \frac{\partial T}{\partial r}\right)_{s^-} = \left(4\pi r^2 \lambda_g \frac{\partial T}{\partial r}\right)_{s^+} = \dot{m}L_v' \tag{136}$$

where the subscript s^- designates the liquid side of the liquid-surface–gas interface and s^+ designates the gas side. The subscripts l and g refer to liquid and gas, respectively.

At the initiation of combustion, the heat-up (second) term of Eq. (136) can be substantially larger than the vaporization (first) term. Throughout combustion the third term is fixed. Thus, some [25, 26] have postulated that droplet combustion can be considered to consist of two periods: namely, an initial droplet heating period of slow vaporization with

$$\dot{m} L_v' \approx \left(4\pi r^2 \lambda_l \frac{\partial T}{\partial r} \right)_{s^-}$$

followed by fast vaporization with almost no droplet heating so that

$$\dot{m} L_v' \approx \dot{m} L_v$$

The extent of the droplet heating time depends on the mode of ignition and the fuel volatility. If a spark is used, the droplet heating time can be a reasonable percentage (10–20%) of the total burning time [24]. Indeed, the burning time would be 10–20% longer if the value of B used to calculate β is calculated with $L_v' = L_v$, i.e., on the basis of the latent heat of vaporization. If the droplet is ignited by injection into a heated gas atmosphere, a long heating time will precede ignition; and after ignition, the droplet will be near its saturation temperature so the heat-up time after ignition contributes little to the total burn-up time.

To study the effects due to droplet heating, one must determine the temperature distribution $T(r, t)$ within the droplet. In the absence of any internal motion, the unsteady heat transfer process within the droplet is simply described by the heat conduction equation and its boundary conditions

$$\frac{\partial T}{\partial t} = \frac{\alpha_l}{r^2} \frac{\partial}{\partial r} \left(r^2 \frac{\partial T}{\partial r} \right), \qquad T(r, t = 0) = T_l(r), \qquad \left(\frac{\partial T}{\partial r} \right)_{r=0} = 0 \qquad (137)$$

The solution of Eq. (137) must be combined with the nonsteady equations for the diffusion of heat and mass. This system can only be solved numerically and the computing time is substantial. Therefore, a simpler alternative model of droplet heating is adopted [24, 25]. In this model, the droplet temperature is assumed to be spatially uniform at T_s and temporally varying. With this assumption Eq. (136) becomes

$$\dot{m} L_v' = \left(4\pi r_s^2 \lambda_g \frac{\partial T}{\partial r} \right)_{s^+} = \dot{m} L_v + \left(\frac{4}{3} \right) \pi r_s^2 \rho_l (c_p)_l \frac{\partial T_s}{\partial t} \qquad (138)$$

But since

$$\dot{m} = -(d/dt) \left[(4/3)\pi r_s^3 \rho_l \right]$$

Eq. (138) integrates to

$$\left(\frac{r_{\rm s}}{r_{\rm so}}\right)^3 = \exp\left\{-c_{pl}\int_{T_{\rm so}}^{T_{\rm s}}\left(\frac{dT_{\rm s}}{L_{\rm v}' - L_{\rm v}}\right)\right\} \tag{139}$$

where $L_{\rm v}' = L_{\rm v}'(T_{\rm s})$ is given by

$$L_{\rm v}' = \frac{(1 - m_{\rm fs})\left[im_{\rm o\infty}H + c_p(T_\infty - T_{\rm s})\right]}{(m_{\rm fs} + im_{\rm o\infty})} \tag{140}$$

and $r_{\rm so}$ is the original droplet radius.

Figure 14, taken from Law [26], is a plot of the nondimensional radius squared as a function of a nondimensional time for an octane droplet initially at 300 K burning under ambient conditions. Shown in this figure are the droplet histories calculated using Eqs. (137) and (138). Sirignano and Law [25] refer to the result of Eq. (137) as the diffusion limit and that of Eq. (138) as the distillation limit, respectively. Equation (137) allows for diffusion of heat into the droplet, whereas Eq. (138) essentially assumes infinite thermal conductivity of the liquid and has vaporization at $T_{\rm s}$ as a function of time. Thus, one should expect Eq. (137) to give a slower burning time.

Also plotted in Fig. 14 are the results from using Eq. (133) in which the transfer number was calculated with $L_{\rm v}' = L_{\rm v}$ and $L_{\rm v}' = L_{\rm v} + c_p(T_{\rm s} - T_l)$, [Eq. (122)]. As one would expect, $L_{\rm v}' = L_{\rm v}$ gives the shortest burning time, and Eq. (137) gives the longest burning time since it does not allow storage of heat in the droplet as it burns. All four results are remarkably close for octane. The spread could be

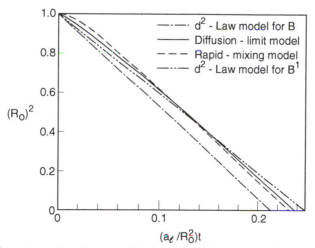

FIGURE 14 Variation of nondimensional droplet radius as a function of nondimensional time during burning with droplet heating and steady-state models (after Law [26]).

greater for a heavier fuel. For a conservative estimate of the burning time, use of B with L'_v evaluated by Eq. (122) is recommended.

The Effect of Finite Reaction Rates   When the fuel and oxidizer react at a finite rate, the flame front can no longer be considered infinitely thin. The reaction rate is then such that oxidizer and fuel can diffuse through each other and the reaction zone is spread over some distance. However, one must realize that although the reaction rates are considered finite, the characteristic time for the reaction is also considered to be much shorter than the characteristic time for the diffusional processes, particularly the diffusion of heat from the reaction zone to the droplet surface.

The development of the mass burning rate [Eq. (118)] in terms of the transfer number B [Eq. (120)] was made with the assumption that no oxygen reaches the fuel surface and no fuel reaches ∞, the ambient atmosphere. In essence, the only assumption made was that the chemical reactions in the gas-phase flame zone were fast enough so that the conditions $m_{os} = 0 = m_{f\infty}$ could be met. The beauty of the transfer number approach, given that the kinetics are finite but faster than diffusion and the Lewis number is equal to 1, is its great simplicity compared to other endeavors [18, 19].

For infinitely fast kinetics, then, the temperature profiles form a discontinuity at the infinitely thin reaction zone (see Fig. 11). Realizing that the mass burning rate must remain the same for either infinite or finite reaction rates, one must consider three aspects dictated by physical insight when the kinetics are finite: first, the temperature gradient at $r = r_s$ must be the same in both cases; second, the maximum temperature reached when the kinetics are finite must be less than that for the infinite kinetics case; third, if the temperature is lower in the finite case, the maximum must be closer to the droplet in order to satisfy the first aspect. Lorell *et al.* [20] have shown analytically that these physical insights as depicted in Fig. 15 are correct.

D. BURNING OF DROPLET CLOUDS

Current understanding of how particle clouds and sprays burn is still limited, despite numerous studies—both analytical and experimental—of burning droplet arrays. The main consideration in most studies has been the effect of droplet separation on the overall burning rate. It is questionable whether study of simple arrays will yield much insight into the burning of particle clouds or sprays.

An interesting approach to the spray problem has been suggested by Chiu and Liu [27], who consider a quasi-steady vaporization and diffusion process with infinite reaction kinetics. They show the importance of a group combustion number (G), which is derived from extensive mathematical analyses and takes the form

$$G = 3(1 + 0.276\text{Re}^{1/2}\text{Sc}^{1/2}\text{Le } N^{2/3})(R/S) \qquad (141)$$

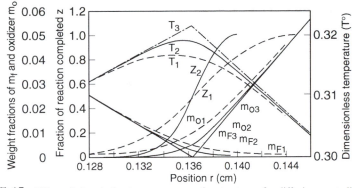

FIGURE 15   Effect of chemical rate processes on the structure of a diffusion-controlled droplet flame (after Lovell *et al* [20]).

where Re, Sc, and Le are the Reynolds, Schmidt, and Lewis numbers, respectively; N is the total number of droplets in the cloud, R is the instantaneous average radius, and S is the average spacing of the droplets.

The value of G was shown to have a profound effect upon the flame location and distribution of temperature, fuel vapor, and oxygen. Four types of behaviors were found for large G numbers. External sheath combustion occurs for the largest value; and as G is decreased, there is external group combustion, internal group combustion, and isolated droplet combustion.

Isolated droplet combustion obviously is the condition for a separate flame envelope for each droplet. Typically, a group number less that 10^{-2} is required. Internal group combustion involves a core with a cloud where vaporization exists such that the core is totally surrounded by flame. This condition occurs for G values above 10^{-2} and somewhere below 1. As G increases, the size of the core increases. When a single flame envelops all droplets, external group combustion exists. This phenomenon begins with G values close to unity. (Note that many industrial burners and most gas turbine combustors are in this range.) With external group combustion, the vaporization of individual droplets increases with distance from the center of the core. At very high G values (above 10^{-2}), only droplets in a thin layer at the edge of the cloud are vaporizing. This regime is called the external sheath condition.

E. BURNING IN CONVECTIVE ATMOSPHERES

1. The Stagnant Film Case

The spherical-symmetric fuel droplet burning problem is the only quiescent case that is mathematically tractable. However, the equations for mass burning may be readily solved in one-dimensional form for what may be considered the stagnant film case. If the stagnant film is of thickness δ, the free-stream conditions are thought to exist at some distance δ from the fuel surface (see Fig. 16).

Within the stagnant film, the energy equation can be written as

$$\rho v c_p (dT/dy) = \lambda (d^2 T/dy^2) + \dot{H} \tag{142}$$

With b defined as before, the solution of this equation and case proceeds as follows. Analogous to Eq. (117), for the one-dimensional case

$$\rho v (db/dy) = \rho D (d^2 b/dy^2) \tag{143}$$

Integrating Eq. (143), one has

$$\rho v b = \rho D \, db/dy + \text{const} \tag{144}$$

The boundary condition is

$$\rho D (db/dy)_o = \rho_s v_s = \rho v \tag{145}$$

Substituting this boundary condition into Eq. (144), one obtains

$$\rho v b_o = \rho v + \text{const}, \qquad \rho v (b_o - 1) = \text{const}$$

The integrated equation now becomes

$$\rho v (b - b_o + 1) = \rho D \, db/dy \tag{146}$$

which upon second integration becomes

$$\rho v y - \rho D (b - b_o + 1) + \text{const} \tag{147}$$

At $y = 0, b = b_o$; therefore, the constant equals zero so that

$$\rho v y = \rho D \, \ln(b - b_o + 1) \tag{148}$$

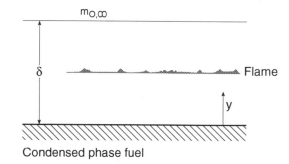

FIGURE 16   Stagnant film height representation for condensed phase burning.

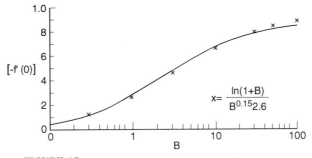

FIGURE 18 $[-f(0)]$ as a function of the transfer number B.

An interesting approximation to this result [Eq. (159)] can be made from the stagnant film result of the last section; i.e.,

$$G_f = (\rho D/\delta)\ln(1 + B) = (\lambda/c_p\delta)\ln(1 + B) = (\mu/\delta)\ln(1 + B) \quad (152)$$

If δ is assumed to be the Blasius boundary layer thickness δ_x, then

$$\delta_x = 5.2x\mathrm{Re}_x^{-1/2} \tag{160}$$

Substituting Eq. (160) into Eq. (152) gives

$$G_f x/\mu = (\mathrm{Re}_x^{1/2}/5.2)\ln(1 + B) \tag{161}$$

The values predicted by Eq. (161) are somewhat high compared to those predicted by Eq. (159). If Eq. (161) is divided by $B^{0.15}/2$ to give

$$G_f x/\mu = (\mathrm{Re}_x^{-1/2}/2.6)[\ln(1 + B)/B^{0.15}] \tag{162}$$

the agreement is very good over a wide range of B values. To show the extent of the agreement, the function

$$\ln(1 + B)/B^{0.15}2.6 \tag{163}$$

is plotted on Fig. 18 as well.

Obviously, these results do not hold at very low Reynolds numbers. As Re approaches zero, the boundary layer thickness approaches infinity. However, the burning rate is bounded by the quiescent results.

3. The Flowing Droplet Case

When droplets are not at rest relative to the oxidizing atmosphere, the quiescent results no longer hold so forced convection must again be considered. No one has solved this complex case. As discussed in Section 4.E, flow around a sphere can be complex, and at relatively high Re(> 20), there is a boundary layer flow around the front of the sphere and a wake or eddy region behind it.

For this burning droplet case, an overall heat transfer relationship could be written to define the boundary condition given by Eq. (90):

$$h(\Delta T) = G_f L_v \tag{164}$$

The thermal driving force is represented by a temperature gradient ΔT which is the ambient temperature T_∞ plus the rise in this temperature due to the energy release minus the temperature at the surface T_s, or

$$\Delta T = T_\infty + (im_{o\infty} H/c_p) - T_s = [im_{o\infty} H + c_p(T_\infty - T_s)]/c_p \tag{165}$$

Substituting Eq. (165) and Eq. (118) for G_f into Eq. (164), one obtains

$$h[im_{o\infty} H + c_p(T_\infty - T_s)]/c_p = [(D\rho/r)\ln(1+B)] L_v$$
$$= [(\lambda/c_p r)\ln(1+B)] L_v \tag{166}$$

where r is now the radius of the droplet. Upon cross-multiplication, Eq. (166) becomes

$$hr/\lambda = \ln(1+B)/B = \text{Nu} \tag{167}$$

since

$$B = [im_{o\infty} H + c_p(T_\infty - T_s)/L_v]$$

Since Eq. (118) was used for G_f, this Nusselt number [Eq. (167)] is for the quiescent case (Re $\rightarrow$ 0). For small B, $\ln(1+B) \approx B$ and the Nu $= 1$, the classical result for heat transfer without mass addition.

The term $[\ln(1+B)]/B$ has been used as an empirical correction for higher Reynolds number problems, and a classical expression for Nu with mass transfer is

$$\text{Nu}_r = [\ln(1+B)/B](1 + 0.39\text{Pr}^{1/3}\text{Re}_r^{1/2}) \tag{168}$$

As Re $\rightarrow$ 0, Eq. (168) approaches Eq. (167). For the case Pr $= 1$ and Re > 1, Eq. (163) becomes

$$\text{Nu}_r = (0.39)[\ln(1+B)/B]\text{Re}_r^{1/2} \tag{169}$$

The flat plate result of the preceding section could have been written in terms of a Nusselt number as well. In that case

$$\text{Nu}_x = [-f'(0)]\text{Re}_x^{1/2}/\sqrt{2}B \tag{170}$$

Thus, the burning rate expressions related to Eqs. (169) and (170) are, respectively,

$$G_f r/\mu = 0.39\text{Re}_r^{1/2}\ln(1+B) \tag{171}$$

$$= \text{Re}_r^{1/2}[-f'(0)]/\sqrt{2} \tag{172}$$

In convective flow a wake exists behind the droplet and droplet heat transfer in the wake may be minimal, so these equations are not likely to predict quantitative results. Note, however, that if the right-hand side of Eq. (171) is divided by $B^{0.15}$, the expressions given by Eqs. (171) and (172) follow identical trends; thus data can be correlated as

$$(G_f r/\mu)\{B^{0.15}/\ln(1+B)\} \quad \text{vs} \quad Re^{1/2} \tag{173}$$

If turbulent boundary layer conditions are achieved under certain conditions, the same type of expression should hold and Re should be raised to the 0.8 power.

If, indeed, Eqs. (171) and (172) adequately predict the burning rate of a droplet in laminar convective flow, the droplet will follow a "$d^{3/2}$" burning rate law for a given relative velocity between the gas and the droplet. In this case β will be a function of the relative velocity as well as B and other physical parameters of the system. This result should be compared to the "d^2" law [Eq. (172)] for droplet burning in quiescent atmospheres. In turbulent flow, droplets will appear to follow a burning rate law in which the power of the diameter is close to 1.

4. Burning Rates of Plastics: The Small B Assumption and Radiation Effects

Current concern with the fire safety of polymeric (plastic) materials has prompted great interest in determining the mass burning rate of plastics. For plastics whose burning rates are measured so that no melting occurs, or for non-melting plastics, the developments just obtained should hold. For the burning of some plastics in air or at low oxygen concentrations, the transfer number may be considered small compared to 1. For this condition, which of course would hold for any system in which $B \ll 1$,

$$\ln(1+B) \cong B \tag{174}$$

and the mass burning rate expression for the case of nontransverse air movement may be written as

$$G_f \cong (\lambda/c_p\delta)B \tag{175}$$

Recall that for these considerations the most convenient expression for B is

$$B = [im_{o\infty}H + c_p(T_\infty - T_s)]/L_v \tag{176}$$

In most cases

$$im_{o\infty}H > c_p(T_\infty - T_s) \tag{177}$$

so

$$B \cong im_{o\infty}H/L_v \tag{178}$$

and

$$G_f \cong (\lambda/c_p\delta)(im_{o\infty}H/L_v) \tag{179}$$

Equation (179) shows why good straight-line correlations are obtained when G_f is plotted as a function of $m_{o\infty}$ for burning rate experiments in which the dynamics of the air are constant or well controlled (i.e., δ is known or constant). One should realize that

$$G_f \sim m_{o\infty} \tag{180}$$

holds only for small B.

The consequence of this small B assumption may not be immediately apparent. One may obtain a physical interpretation by again writing the mass burning rate expression for the two assumptions made (i.e., $B \ll 1$ and $B = [im_{o\infty}H]/L_v$)

$$G_f \cong (\lambda/c_p\delta)(im_{o\infty}H/L_v) \tag{181}$$

and realizing that as an approximation

$$(im_{o\infty}H) \cong c_p(T_f - T_s) \tag{182}$$

where T_f is the diffusion flame temperature. Thus, the burning rate expression becomes

$$G_f \cong (\lambda/c_p\delta)[c_p(T_f - T_s)/L_v] \tag{183}$$

Cross-multiplying, one obtains

$$G_fL_v \cong \lambda(T_f - T_s)/\delta \tag{184}$$

which says that the energy required to gasify the surface at a given rate per unit area is supplied by the heat flux established by the flame. Equation (184) is simply another form of Eq. (164). Thus, the significance of the small B assumption is that the gasification from the surface is so small that it does not alter the gaseous temperature gradient determining the heat flux to the surface. This result is different from the result obtained earlier, which stated that the stagnant film thickness is not affected by the surface gasification rate. The small B assumption goes one step further, revealing that under this condition the temperature profile in the boundary layer is not affected by the small amount of gasification.

If flame radiation occurs in the mass burning process—or any other radiation is imposed, as is frequently the case in plastic flammability tests—one can obtain a convenient expression for the mass burning rate provided one assumes that only the gasifying surface, and none of the gases between the radiation source and the surface, absorbs radiation. In this case Fineman [30] showed that the stagnant film expression for the burning rate can be approximated by

$$G_f = (\lambda/c_p\delta)\ln[(1+B)/(1-E)] \qquad \text{where} \qquad E \equiv Q_R/G_f L_v$$

and Q_R is the radiative heat transfer flux.

This simple form for the burning rate expression is possible because the equations are developed for the conditions in the gas phase and the mass burning rate arises explicitly in the boundary condition to the problem. Since the assumption is made that no radiation is absorbed by the gases, the radiation term appears only in the boundary condition to the problem.

Notice that as the radiant flux Q_R increases, E and the term $B/(1-E)$ increase. When $E = 1$, the problem disintegrates because the equation was developed in the framework of a diffusion analysis. $E = 1$ means that the solid is gasified by the radiant flux alone.

PROBLEMS

1. A laminar fuel jet issues from a tube into air and is ignited to form a flame. The height of the flame is 8 cm. With the same fuel the diameter of the jet is increased by 50% and the laminar velocity leaving the jet is decreased by 50%. What is the height of the flame after the changes are made? Suppose the experiments are repeated but that grids are inserted in the fuel tube so that all flows are turbulent. Again for the initial turbulent condition it is assumed the flame height is 8 cm.

2. An ethylene oxide monopropellant rocket motor is considered part of a ram rocket power plant in which the turbulent exhaust of the rocket reacts with induced air in an afterburner. The exit area of the rocket motor is 8 cm². After testing it is found that the afterburner length must be reduced by 42.3%. What size must the exit port of the new rocket be to accomplish this reduction with the same afterburner combustion efficiency? The new rocket would operate at the same chamber pressure and area ratio. How many of the new rockets would be required to maintain the same level of thrust as the original power plant?

3. A spray of benzene fuel is burned in quiescent air at 1 atm and 298 K. The burning of the spray can be assumed to be characterized by single droplet burning. The (Sauter) mean diameter of the spray is 100 μm; that is, the burning time of the spray is the same as that of a single droplet of 100 μm. Calculate a mean burning time for the spray. For calculation purposes, assume whatever mean properties of the fuel, air, and product mixture are required. For some properties those of nitrogen would generally suffice. Also assume that the droplet is essentially of infinite conductivity. Report, as well, the steady-state temperature of the fuel droplet as it is being consumed.

4. Repeat the calculation of the previous problem for the initial condition that the air is at an initial temperature of 1000 K. Further calculate the burning time for the benzene in pure oxygen at 298 K. Repeat all calculations with ethanol as the fuel. Then discuss the dependence of the results obtained on ambient conditions and fuel type.

5. Two liquid sprays are evaluated in a single cylinder test engine. The first is liquid methanol which has a transfer number $B = 2.9$. The second is a regular diesel fuel which has a transfer number $B = 3.9$. The two fuels have approximately the same liquid density; however, the other physical characteristics of the diesel spray are such that its Sauter mean diameter is 1.5 times that of the methanol. Both are burning in air. Which spray requires the longer burning time and how much longer is it than the other?

6. Consider each of the condensed phase fuels listed to be a spherical particle burning with a perfect spherical flame front in air. From the properties of the fuels given, estimate the order of the fuels with respect to mass burning rate per unit area. List the fastest burning fuel first, etc.

	Latest heat of vaporization (cal gm^{-1})	Density (gm cm^{-3})	Thermal conductivity (cal s^{-1} cm^{-1} K^{-1})	Stoichiometric heat evolution in air per unit weight of fuel (cal g^{-1})	Heat capacity (cal g^{-1} K^{-1})
Aluminum	2570	2.7	0.48	1392	0.28
Methanol	263	0.8	0.51×10^{-3}	792	0.53
Octane	87	0.7	0.33×10^{-3}	775	0.51
Sulfur	420	2.1	0.60×10^{-3}	515	0.24

7. Suppose fuel droplets of various sizes are formed at one end of a combustor and move with the gas through the combustor at a velocity of 30m/s. It is known that the 50-μm droplets completely vaporize in 5 ms. It is desired to vaporize completely each droplet of 100 μm and less before they exit the combustion chamber. What is the minimum length of the combustion chamber allowable in design to achieve this goal?

8. A radiative flux Q_R is imposed on a solid fuel burning in air in a stagnation film mode. The expression for the burning rate is

$$G_f = (D\rho/\delta) \ln\left[(1 + B)/(1 - E)\right]$$

where $E = Q_R/G_f L_s$. Develop this expression. It is a one-dimensional problem as described.

9. Experimental evidence from a porous sphere burning rate measurement in a low Reynolds number laminar flow condition confirms that the mass burning rate per unit area can be represented by

$$(G_f r/\mu) = [f'(0, B)/\sqrt{2}]\mathrm{Re}_r^{1/2}$$

Would a real droplet of the same fuel follow a "d^2" law under the same conditions? If not, what type of power law should it follow?

10. Write what would appear to be the most important physical result in each of the following areas:

   (a) laminar flame propagation
   (b) laminar diffusion flames
   (c) turbulent diffusion flames
   (d) detonation
   (e) droplet diffusion flames

Explain the physical significance of the answers. Do not develop equations.

REFERENCES

1. Gaydon, A. G., and Wolfhard, H., "Flames," Chap. 6. Macmillan, New York, 1960.
2. Smyth, K. C., Miller, J. H., Dorfman, R. C., Mallard, W. G., and Santoro, R. J., *Combust. Flame* **62**, 157 (1985).
3. Roper, F. G., *Combust. Flame* **29**, 219 (1977).
4. Roper, F. G., Smith, C., and Cunningham, A. C., *Combust. Flame* **29**, 227 (1977).
5. Spengler, G., and Kern, J., *Brennst.-Chem.* **50**, 321 (1969).
6. Gomez, A., Sidebotham, G., and Glassman, I., *Combust. Flame* **58**, 45 (1984).
7. Hottel, H. C., and Hawthorne, W. R., *Int. Symp. Combust., 3rd* p. 255. Williams & Wilkins, Baltimore, Maryland, 1949.
8. Boedecker, L. R., and Dobbs, G. M., *Combust. Sci. Technol.* **46**, 301 (1986).
9. Williams, F. A., Combustion Theory," 2nd Ed., Chap. 3. Benjamin-Cummins, Menlo Park, California, 1985.
10. Lewis, B., and von Elbe, G., "Combustion, Flames and Explosion of Gases," 2nd Ed., Academic Press, New York, 1961.
11. Burke, S. P., and Schumann, T. E. W., *Ind. Eng. Chem.* **20**, 998 (1928).
12. Linan, A., and Williams, F. A., "Fundamental Aspects of Combustion." Oxford Univ. Press, Oxford, England, 1995.
12a. Hawthorne, W. R., Weddel, D. B., and Hottel, H. C., *Int. Symp. Combust., 3rd* p. 266. Williams & Wilkins, Baltimore, Maryland, 1946.
13. Khudyakov, L., *Chem. Abstr.* **46**, 10844e (1955).
14. Blackshear, P. L., Jr., "An Introduction to Combustion," Chap. 4. Dep. Mech. Eng., Univ. of Minnesota, Minneapolis, 1960.
15. Spalding, D. B., *Int. Symp. Combust., 4th* p. 846. Williams & Wilkins, Baltimore, Maryland, 1953.
16. Spalding, D. B., "Some Fundamentals of Combustion," Chap. 4. Butterworth, London, 1955.
17. Kanury, A. M., "Introduction to Combustion Phenomena," Chap. 5. Gordon & Breach, New York, 1975.
18. Godsave, G. A. E., *Int. Symp. Combust., 4th* p. 818. Williams & Wilkins, Baltimore, Maryland, 1953.
19. Goldsmith, M., and Penner, S. S., *Jet Propul.* **24**, 245 (1954).
20. Lorell, J., Wise, H., and Carr, R. E., *J. Chem. Phys.* **25**, 325 (1956).
21. Law, C. K., *Prog. Energy Combust. Sci.* **8**, 169 (1982).
22. Law, C. K., and Law, A. V., *Combust. Sci. Technol.* **12**, 207 (1976).
23. Law, C. K., and Williams, F. A., *Combust. Flame* **19**, 393 (1972).

24. Hubbard, G. L., Denny, V. E., and Mills, A. F., *Int. J. Heat Mass Transfer* **18**, 1003 (1975).
25. Sirignano, W. A., and Law, C. K., *Adv. Chem. Ser.* No. 166, p. 1 (1978).
26. Law, C. K., *Combust. Flame* **26**, 17 (1976).
27. Chiu, H H., and Liu, T. M., *Combust. Sci. Technol*, **17**, 1127 (1977).
28. Blasius, H., *Z. Math. Phys.* **56**, 1 (1956).
29. Emmons, H. W., *Z. Angew. Math. Mech.* **36**, 60 (1956).
30. Fineman, S., M. S. E. Thesis, Dep. Aeronaut. Eng., Princeton Univ., Princeton, New Jersey, 1962.

7

Ignition

A. CONCEPTS

If the concept of ignition were purely a chemical phenomenon, it would be treated more appropriately prior to the discussion of gaseous explosions (Chapter 3). However, thermal considerations are crucial to the concept of ignition. Indeed, thermal considerations play the key role in consideration of the ignition of condensed phases. The problem of storage of wet coal or large concentrations of solid materials (grain, pulverized coal, etc.) that can undergo slow exothermic decomposition is also one of ignition; i.e., the concept of spontaneous combustion is an element of the theory of thermal ignition.

It is appropriate to reexamine the elements discussed in analysis of the explosion limits of hydrocarbons. The explosion limits shown in Fig. 7 of Chapter 3 exist for particular conditions of pressure and temperature. When the thermal conditions for point 1 in this figure exist, some reaction begins; thus, some heat must be evolved. The experimental configuration is assumed to be such that the heat of reaction is dissipated infinitely fast at the walls of the containing vessel to retain the temperature at the initial value T_1. Then steady reaction prevails and a slight pressure rise is observed. When conditions such as those at point 2 prevail, as discussed in Chapter 3, the rate of chain carrier generation exceeds the rate of chain termination; hence the reaction rate becomes progressively greater, and subsequently an explosion—or, in the context here, ignition—occurs. Generally,

325

pressure is used as a measure of the extent of reaction, although, of course, other measures can be used as well. The sensitivity of the measuring device determines the point at which a change in initial conditions is first detected. Essentially, this change in initial conditions (pressure) is not noted until after some time interval and, as discussed in Chapter 3, this interval can be related to the time required to reach the degenerate branching stage or some other stage in which chain branching begins to demonstrably affect the overall reaction. This time interval is considered to be an induction period and to correspond to the ignition concept. This induction period will vary considerably with temperature. Increasing the temperature increases the rates of the reactions leading to branching, thereby shortening the induction period. The isothermal events discussed in this paragraph essentially define chemical chain ignition.

Now if one begins at conditions similar to point 1 in Fig. 7 of Chapter 3—except that the experimental configuration is such that the heat of reaction is not dissipated at the walls of the vessel; i.e., the system is adiabatic—the reaction will self-heat until the temperature of the mixture moves the system into the explosive reaction regime. This type of event is called two-stage ignition and there are two induction periods, or ignition times, associated with it. The first is associated with the time (τ_c, chemical time) to build to the degenerate branching step or the critical concentration of radicals (or, for that matter, any other chain carriers), and the second (τ_t, thermal time) is associated with the subsequent steady reaction step and is the time it takes for the system to reach the thermal explosion (ignition) condition. Generally, $\tau_c > \tau_t$.

If the initial thermal condition begins in the chain explosive regime, such as point 2, the induction period τ_c still exists; however, there is no requirement for self-heating, so the mixture immediately explodes. In essence, $\tau_t \to 0$.

In many practical systems, one cannot distinguish the two stages in the ignition process since $\tau_c > \tau_t$; thus the time that one measures is predominantly the chemical induction period. Any errors in correlating experimental ignition data in this low-temperature regime are due to small changes in τ_t.

Sometimes point 2 will exist in the cool-flame regime. Again, the physical conditions of the nonadiabatic experiment can be such that the heat rise due to the passage of the cool flame can raise the temperature so that the flame condition moves from a position characterized by point 1 to one characterized by point 4. This phenomenon is also called two-stage ignition. The region of point 4 is not a chain branching explosion, but a self-heating explosion. Again, an induction period τ_c is associated with the initial cool-flame stage and a subsequent time τ_t is associated with the self-heating aspect.

If the reacting system is initiated under conditions similar to point 4, pure thermal explosions develop and these explosions have thermal induction or ignition times associated with them. As will be discussed in subsequent paragraphs, thermal explosion (ignition) is possible even at low temperatures, both under the nonadiabatic conditions utilized in obtaining hydrocarbon–air explosion limits and under adiabatic conditions.

The concepts just discussed concern premixed fuel–oxidizer situations. In reality these ignition types do not arise frequently in practical systems. However, one can use these concepts to gain a better understanding of many practical combustion systems, such as, for example, the ignition of liquid fuels.

Many ignition experiments have been performed by injecting liquid and gaseous fuels into heated stagnant and flowing air streams [1, 2]. It is possible from such experiments to relate an ignition delay (or time) to the temperature of the air. If this temperature is reduced below a certain value, no ignition occurs even after an extended period of time. This temperature is one of interest in fire safety and is referred to as the spontaneous or autoignition temperature (AIT). Figure 1 shows some typical data from which the spontaneous ignition temperature is obtained. The AIT is fundamentally the temperature at which elements of the fuel–air system enter the explosion regime. Thus, the AIT must be a function of pressure as well; however, most reported data, such as those given in Appendix F are for a pressure of 1 atm. As will be shown later, a plot of the data in Fig. 1 in the form of log(time) versus $(1/T)$ will give a straight line. In the experiments

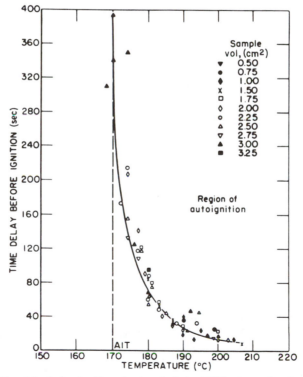

FIGURE 1   Time delay before ignition of *n*-propyl nitrate at 1000 psig as a function of temperature (from Zabetakis [2]).

mentioned, in the case of liquid fuels, the fuel droplet attains a vapor pressure corresponding to the temperature of the heated air. A combustible mixture close to stoichiometric forms irrespective of the fuel. It is this mixture that enters the explosive regime, which in actuality has an induction period associated with it. Approximate measurements of this induction period can be made in a flowing system by observing the distance from point of injection of the fuel to the point of first visible radiation, then relating this distance to the time through knowledge of the stream velocity.

In essence, droplet ignition is brought about by the heated flowing air stream. This type of ignition is called "forced ignition" in contrast to the "self-ignition" conditions of chain and thermal explosions. The terms *self-ignition*, *spontaneous ignition*, and *autoignition* are used synonymously. Obviously, forced ignition may also be the result of electrical discharges (sparks), heated surfaces, shock waves, flames, etc. Forced ignition is usually considered a local initiation of a flame that will propagate; however, in certain instances, a true explosion is initiated. After examination of an analytical analysis of chain spontaneous ignition and its associated induction time, this chapter will concentrate on the fundamental concepts of self- or spontaneous ignition, after which it will encompass aspects of forced ignition.

B. CHAIN SPONTANEOUS IGNITION

In Chapter 3 the conditions for a chain branching explosion were developed on the basis of a steady-state analysis. It was shown that when the chain branching factor α at a given temperature and pressure was greater than some critical value α_{crit}, the reacting system exploded. Obviously, in that development no induction period or critical chain ignition time τ_c evolved.

In this section consideration is given to an analytical development of this chain explosion induction period that has its roots in the early work on chain reactions carried out by Semenov [3] and Hinshelwood [4] and reviewed by Zeldovich *et al.* [5].

The approach considered as a starting point is a generalized form of Eq. (6) of Chapter 3, but not as a steady-state expression. Thus, the overall rate of change of the concentration of chain carriers (R) is expressed by the equation

$$d(\mathrm{R})/dt = \dot\omega_0 + k_{\mathrm{b}}(\mathrm{R}) - k_t(\mathrm{R}) = \dot\omega_0 + \phi(\mathrm{R}) \tag{1}$$

where $\dot\omega_0$ is the initiation rate of a very small concentration of carriers. Such initiation rates are usually very slow. Rate expressions k_{b} and k_t are for the overall chain branching and termination steps, respectively, and ϕ is simply the difference $k_{\mathrm{b}} - k_t$.

Constants k_{b}, k_t, and, obviously, ϕ are dependent on the physical conditions of the system; in particular, temperature and pressure are major factors in their explicit values. However, one must realize that k_b is much more temperature-dependent than k_t. The rates included in k_t are due to recombination (bond formation) reactions of very low activation energy that exhibit little temperature dependence, whereas most chain branching and propagating reactions can have significant values of activation energy. One can conclude, then, that ϕ can change sign as the temperature is raised. At low temperatures it is negative, and at high temperatures it is positive. Then at high temperatures $d(R)/dt$ is a continuously and rapidly increasing function. At low temperatures as $[d(R)/dt] \rightarrow 0$, (R) approaches a fixed limit $\dot{\omega}/\phi$; hence there is no runaway and no explosion. For a given pressure the temperature corresponding to $\phi = 0$ is the critical temperature below which no explosion can take place.

At time zero the carrier concentration is essentially zero, and (R) = 0 at $t = 0$ serves as the initial condition for Eq. (1). Integrating Eq. (1) results in the following expression for (R):

$$(R) = (\dot{\omega}_0/\phi)[\exp(\phi t) - 1] \tag{2}$$

If as a result of the chain system the formation of every new carrier is accompanied by the formation of j molecules of final product (P), the expression for the rate of formation of the final product becomes

$$\dot{\omega} = [d(P)/dt] = jk_b(R) = (jk_b\dot{\omega}_0/\phi)[\exp(\phi t) - 1] \tag{3}$$

An analogous result is obtained if the rate of formation of carriers is equal to zero ($\dot{\omega}_0 = 0$) and the chain system is initiated due to the presence of some initial concentration $(R)_0$. Then for the initial condition that at $t = 0$, $(R) = (R)_0$, Eq. (2) becomes

$$(R) = (R)_0 \exp(\phi t) \tag{4}$$

The derivations of Eqs. (1)–(4) are valid only at the initiation of the reaction system; k_b and k_t were considered constant when the equations were integrated. Even for constant temperature, k_b and k_t will change because the concentration of the original reactants would appear in some form in these expressions.

Equations (2) and (4) are referred to as Semenov's law, which states that in the initial period of a chain reaction the chain carrier concentration increases exponentially with time when $k_b > k_t$.

During the very early stages of the reaction the rate of formation of carriers begins to rise, but it can be below the limits of measurability. After a period of time, the rate becomes measurable and continues to rise until the system becomes explosive. The explosive reaction ceases only when the reactants are consumed. The time to the small measurable rate $\dot{\omega}_{ms}$ corresponds to the induction period τ_c.

For the time close to τ_c, $\dot{\omega}_{ms}$ will be much larger than $\dot{\omega}_0$ and $\exp(\phi t)$ much greater than 1, so that Eq. (3) becomes

$$\dot{\omega}_{ms} = (jk_b\dot{\omega}_0/\phi)\exp(\phi\tau) \tag{5}$$

The induction period then becomes

$$\tau_c \simeq (1/\phi)\ln(\dot{\omega}_{ms}\phi/jk_b\dot{\omega}_0) \tag{6}$$

If one considers either the argument of the logarithm in Eq. (6) as a nearly constant term, or k_b as much larger than k_t so that $\phi \cong k_b$, one has

$$\tau_c = \text{const}/\phi \tag{7}$$

so that the induction time depends on the relative rates of branching and termination. The time decreases as the branching rate increases.

C. THERMAL SPONTANEOUS IGNITION

The theory of thermal ignition is based upon a very simple concept. When the rate of thermal energy release is greater than the rate of thermal energy dissipation (loss), an explosive condition exists. When the contra condition exists, thermal explosion is impossible. When the two rates are equal, the critical conditions for ignition (explosion) are specified. Essentially, the same type of concept holds for chain explosions. As was detailed in Section B of Chapter 3, when the rate of chain branching becomes greater than the rate of chain termination ($\alpha > \alpha_{crit}$), an explosive condition arises, whereas $\alpha < \alpha_{crit}$ specifies steady reaction. Thus, when one considers the external effects of heat loss or chain termination, one finds a great deal of commonality between chain and thermal explosion.

In consideration of external effects, it is essential to emphasize that under some conditions the thermal induction period could persist for a very long period of time, even hours. This condition arises when the vessel walls are thermally insulated. In this case, even with a very low initial temperature, the heat of the corresponding slow reaction remains in the system and gradually self-heats the reactive components until ignition (explosion) takes place. If the vessel is not insulated and heat is transferred to the external atmosphere, equilibrium is rapidly reached between the heat release and heat loss, so thermal explosion is not likely. This point will be refined in Section 7.C.2.a.

It is possible to conclude from the preceding that the study of the laws governing thermal explosions will increase understanding of the phenomena controlling the spontaneous ignition of combustible mixtures and forced ignition in general.

The concepts discussed were first presented in analytical forms by Semenov [3] and later in more exact form by Frank-Kamenetskii [6]. Since the Semenov approach offers easier physical insight, it will be considered first, and then the details of the Frank-Kamenetskii approach will be presented.

1. Semenov Approach of Thermal Ignition

Semenov first considered the progress of the reaction of a combustible gaseous mixture at an initial temperature T_0 in a vessel whose walls were maintained at the same temperature. The amount of heat released due to chemical reaction per unit time ($\dot{q}_r$) then can be represented in simplified overall form as

$$\dot{q}_r = V\dot{\omega}Q = VQc^n \exp(-E/RT)$$
$$= VQA\rho^n\varepsilon^n \exp(-E/RT) \qquad (8)$$

where V is the volume of the vessel, $\dot{\omega}$ is the reaction rate, Q is the thermal energy release of the reactions, c is the overall concentration, n is the overall reaction order, A is the pre-exponential in the simple rate constant expression, and T is the temperature that exists in the gaseous mixture after reaction commences. As in Chapter 2, the concentration can be represented in terms of the total density ρ and the mass fraction ε of the reacting species. Since the interest in ignition is in the effect of the total pressure, all concentrations are treated as equal to $\rho\varepsilon$. The overall heat loss (q_l) to the walls of the vessel, and hence to the medium that maintains the walls of the vessel at T_0, can be represented by the expression

$$\dot{q}_l = hS(T - T_0) \qquad (9)$$

where h is the heat transfer coefficient and S is the surface area of the walls of the containing vessel.

The heat release $\dot{q}_r$ is a function of pressure through the density term and $\dot{q}_l$ is a less sensitive function of pressure through h, which, according to the heat transfer method by which the vessel walls are maintained at T_0, can be a function of the Reynolds number.

Shown in Fig. 2 is the relationship between $\dot{q}_r$ and $\dot{q}_l$ for various initial pressures, a value of the heat transfer coefficient h, and a constant wall temperature of T_0. In Eq. (8) $\dot{q}_r$ takes the usual exponential shape due to the Arrhenius kinetic rate term and $\dot{q}_l$ is obviously a linear function of the mixture temperature T. The $\dot{q}_l$ line intersects the $\dot{q}_r$ curve for an initial pressure p_3 at two points, a and b.

For a system where there is simultaneous heat generation and heat loss, the overall energy conservation equation takes the form

$$c_v\rho V(dT/dt) = \dot{q}_r - \dot{q}_l \qquad (10)$$

where the term on the left-hand side is the rate of energy accumulation in the containing vessel and c_v is the molar constant volume heat capacity of the gas mixture. Thus, a system whose initial temperature is T_0 will rise to point a spontaneously. Since $\dot{q}_r = \dot{q}_l$ and the mixture attains the steady, slow-reacting rate $\dot{\omega}(T_i)$ or $\dot{\omega}(T_a)$, this point is an equilibrium point. If the conditions of the mixture are somehow perturbed so that the temperature reaches a value greater than T_a, then $\dot{q}_l$ becomes greater than $\dot{q}_r$ and the system moves back to the equilibrium condition represented by point a. Only if there is a very great perturbation so that

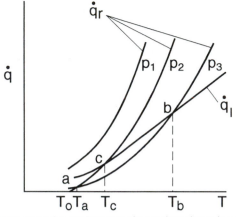

FIGURE 2   Rate of heat generation and heat loss of a reacting mixture in a vessel with pressure and thermal bath variations.

the mixture temperature becomes a value greater than that represented by point b will the system self-heat to explosion. Under this condition $\dot{q}_r > \dot{q}_l$.

If the initial pressure is increased to some value p_2, the heat release curve shifts to higher values, which are proportional to p^n (or ρ^n). The assumption is made that h is not affected by this pressure increase. The value of p_2 is selected so that the $\dot{q}_l$ becomes tangent to the $\dot{q}_r$ curve at some point c. If the value of h is lowered, $\dot{q}_r$ is everywhere greater than $\dot{q}_l$ and all initial temperatures give explosive conditions. It is therefore obvious that when the $\dot{q}_l$ line is tangent to the $\dot{q}_l$ curve, the critical condition for mixture self-ignition exists.

The point c represents an ignition temperature T_i (or T_c); and from the conditions there, Semenov showed that a relationship could be obtained between this ignition temperature and the initial temperature of the mixture—i.e., the temperature of the wall (T_0). Recall that the initial temperature of the mixture and the temperature at which the vessel's wall is maintained are the same (T_0). It is important to emphasize that T_0 is a wall temperature that may cause a fuel–oxidizer mixture to ignite. This temperature can be hundreds of degrees greater than ambient, and T_0 should not be confused with the reference temperature taken as the ambient (298 K) in Chapter 1.

The conditions at c corresponding to T_i (or T_c) are

$$\dot{q}_r = \dot{q}_l, \qquad (d\dot{q}_r/dT) = (d\dot{q}_l/dT) \tag{11}$$

or

$$VQ\rho^n \varepsilon^n A \, \exp(-E/RT_i) = hS(T_i - T_0) \tag{12}$$

$$(d\dot{q}_r/dT) = (E/RT_i^2)VQ\rho^n \varepsilon^n A \, \exp(-E/R_i) = (d\dot{q}_l/dT) = hS \tag{13}$$

Since the variation in T is small, the effect of this variation on the density is ignored for simplicity's sake. Dividing Eq. (12) by Eq. (13), one obtains

$$(RT_i^2/E) = (T_i - T_0) \tag{14}$$

Equation (14) is rewritten as

$$T_i^2 - (E/R)T_i + (E/R)T_0 = 0 \tag{15}$$

whose solutions are

$$T_i = (E/2R) \pm [(E/2R)^2 - (E/R)T_0]^{1/2} \tag{16}$$

The solution with the positive sign gives extremely high temperatures and does not correspond to any physically real situation. Rewriting Eq. (16)

$$T_i = (E/2R) - (E/2R)[1 - (4RT_0/E)]^{1/2} T_0 \tag{17}$$

and expanding, one obtains

$$T_i = (E/2R) - (E/2R)[1 - (2RT_0/E) - 2(RT_0/E)^2 - \cdots]$$

Since (RT_0/E) is a small number, one may neglect the higher-order terms to obtain

$$T_i = T_0 + (RT_0^2/E), \qquad (T_i - T_0) = (RT_0^2/E) \tag{18}$$

For a hydrocarbon–air mixture whose initial temperature is 700 K and whose overall activation energy is about 160 kJ/mol, the temperature rise given in Eq. (18) is approximately 25°K. Thus, for many cases it is possible to take T_i as equal to T_0 or $T_0 + (RT_0^2/E)$ with only small error in the final result. Thus, if

$$\dot{\omega}(T_i) = \rho^n \varepsilon^n A \exp(-E/RT_i) \tag{19}$$

and

$$\dot{\omega}(T_0) = \rho^n \varepsilon^n A \exp(-E/RT_0) \tag{20}$$

and the approximation given by Eq. (18) is used in Eq. (19), one obtains

$$\begin{aligned}
\dot{\omega}(T_i) &= \rho^n \varepsilon^n A \exp\left\{-E/R\left[T_0 + \left(RT_0^2/E\right)\right]\right\} \\
&= \rho^n \varepsilon^n A \exp\left\{-E/RT_0\left[1 + (RT_0/E)\right]\right\} \\
&= \rho^n \varepsilon^n A \exp\left\{-(E/RT_0)\left[1 - (RT_0/E)\right]\right\} \\
&= \rho^n \varepsilon^n A \exp\left[(-E/RT_0) + 1\right] \\
&= \rho^n \varepsilon^n A \left[\exp(-E/RT_0)\right] e \\
&= [\dot{\omega}(T_0)] e \tag{21}
\end{aligned}$$

That is, the rate of chemical reaction at the critical ignition condition is equal to the rate at the initial temperature times the number e. Substituting this result and the approximation given by Eq. (18) into Eq. (12), one obtains

$$eVQ\rho^n\varepsilon^n A \exp(-E/RT_0) = hSRT_0^2/E \tag{22}$$

Representing ρ in terms of the perfect gas law and using the logarithmic form, one obtains

$$\ln(P^n/T_0^{n+2}) = +(E/RT_0) + \ln(hSR^{n+1}/eVQ\varepsilon^n AE) \tag{23}$$

Since the overall order of most hydrocarbon oxidation reactions can be considered to be approximately 2, Eq. (23) takes the form of the so-called Semenov expression

$$\ln(P/T_0^2) = +(E/2RT_0) + \mathcal{B}, \qquad \mathcal{B} = \ln(hSR^3/eVQ\varepsilon^2 AE) \tag{24}$$

Equations (23) and (24) define the thermal explosion limits, and a plot of $\ln(P/T_0^2)$ versus $(1/T_0)$ gives a straight line as is found for many gaseous hydrocarbons. A plot of P versus T_0 takes the form given in Fig. 3 and shows the similarity of this result to the thermal explosion limit (point 3 to point 4 in Fig. 5 of Chapter 3) of hydrocarbons. The variation of the correlation with the chemical and physical terms in $\mathcal{B}$ should not be overlooked. Indeed, the explosion limits are a function of the surface area to volume ratio (S/V) of the containing vessel.

Under the inherent assumption that the mass fractions of the reactants are not changing, further interesting insights can be obtained by rearranging Eq. (22). If the reaction proceeds at a constant rate corresponding to T_0, a characteristic

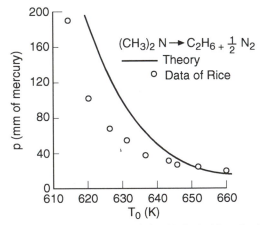

FIGURE 3   Critical pressure–temperature relationship for ignition of a chemical process.

reaction time τ_r can be defined as

$$\tau_r = \rho/[\rho^n \varepsilon^n A \exp(-E/RT_0)] \tag{25}$$

A characteristic heat loss time τ_l can be obtained from the cooling rate of the gas as if it were not reacting by the expression

$$V c_v \rho (dT/dt) = -hS(T - T_0) \tag{26}$$

The characteristic heat loss time is generally defined as the time it takes to cool the gas from the temperature $(T - T_0)$ to $[(T - T_0)/e]$ and is found to be

$$\tau_l = (V\rho c_v/hS) \tag{27}$$

By substituting Eqs. (18), (25), and (27) into Eq. (22) and realizing that (Q/c_v) can be approximated by $(T_f - T_0)$, the adiabatic explosion temperature rise, one obtains following expression:

$$(\tau_r/\tau_l) = e(T_f - T_0)/(T_i - T_0) = e(\Delta T_f/\Delta T_i) = e\Delta T_f/(RT_0^2/E)$$
$$= e(Q/c_v)(RT_0^2/E) \tag{28}$$

Thus, if (τ_r/τ_l) is greater than the value obtained from Eq. (28), thermal explosion is not possible and the reaction proceeds at a steady low rate given by point a in Fig. 2. If $(\tau_r/\tau_l) > (e\Delta T_f/\Delta T_i)$ and ignition still takes place, the explosion proceeds by a chain rather than a thermal mechanism.

With the physical insights developed from this qualitative approach to the thermal ignition problem, it is appropriate to consider the more quantitative approach of Frank-Kamenetskii [6].

2. Frank-Kamenetskii Theory of Thermal Ignition

Frank-Kamenetskii first considered the nonsteady heat conduction equation. However, since the gaseous mixture, liquid, or solid energetic fuel can undergo exothermic transformations, a chemical reaction rate term is included. This term specifies a continuously distributed source of heat throughout the containing vessel boundaries. The heat conduction equation for the vessel is then

$$c_v \rho \, dT/dt = \text{div}(\lambda \, \text{grad} \, T) + \dot{q}' \tag{29}$$

in which the nomenclature is apparent, except perhaps for $\dot{q}'$, which represents the heat release density.

The solution of this equation would give the temperature distribution as a function of the spatial distance and the time. At the ignition condition, the character of this temperature distribution changes sharply. There should be an abrupt transition from a small steady rise to a large and rapid rise. Although computational methods of solving this equation are available, much insight into overall practical ignition phenomena can be gained by considering the two approximate methods of

Frank-Kamenetskii. These two approximate methods are known as the stationary and nonstationary solutions. In the stationary theory, only the temperature distribution throughout the vessel is considered and the time variation is ignored. In the nonstationary theory, the spatial temperature variation is not taken into account, a mean temperature value throughout the vessel is used, and the variation of the mean temperature with time is examined. The nonstationary problem is the same as that posed by Semenov; the only difference is in the mathematical treatment.

a. The Stationary Solution—The Critical Mass and Spontaneous Ignition Problems

The stationary theory deals with time-independent equations of heat conduction with distributed sources of heat. Its solution gives the stationary temperature distribution in the reacting mixture. The initial conditions under which such a stationary distribution becomes impossible are the critical conditions for ignition.

Under this steady assumption, Eq. (29) becomes

$$\text{div}(\lambda \text{ grad } T) = -\dot{q}' \tag{30}$$

and, if the temperature dependence of the thermal conductivity is neglected,

$$\lambda \nabla^2 T = -\dot{q}' \tag{31}$$

It is important to consider the definition of q'. Defined as the amount of heat evolved by chemical reaction in a unit volume per unit time, $\dot{q}'$ is the product of the terms involving the energy content of the fuel and its rate of reaction. The rate of the reaction can be written as $Ze^{-E/RT}$. Recall that Z in this example is different from the normal Arrhenius pre-exponential term in that it contains the concentration terms and therefore can be dependent on the mixture composition and the pressure. Thus,

$$\dot{q}' = QZe^{-E/RT} \tag{32}$$

where Q is the volumetric energy release of the combustible mixture. It follows then that

$$\nabla^2 T = -(Q/\lambda)Ze^{-E/RT} \tag{33}$$

and the problem resolves itself to first reducing this equation under the boundary condition that $T = T_0$ at the wall of the vessel.

Since the stationary temperature distribution below the explosion limit is sought, in which case the temperature rise throughout the vessel must be small, it is best to introduce a new variable

$$v = T - T_0$$

where $v \ll T_0$. Under this condition, it is possible to describe the cumbersome exponential term as

$$\exp\left(\frac{-E}{RT}\right) = \exp\left[\frac{-E}{R(T_0 + v)}\right] = \exp\left\{\frac{E}{RT_0}\left[\frac{1}{1 + (v/T_0)}\right]\right\}$$

If the term in brackets is expanded and the higher-order terms are eliminated, this expression simplifies to

$$\exp\left(\frac{-E}{RT}\right) \cong \exp\left[-\frac{E}{RT_0}\left(1 - \frac{v}{T_0}\right)\right] = \exp\left[-\frac{E}{RT_0}\right]\exp\left[\frac{Ev}{RT_0^2}\right]$$

and (Eq. 33) becomes

$$\nabla^2 v = -\frac{Q}{\lambda}Z\,\exp\left[-\frac{E}{RT_0}\right]\exp\left[\frac{E}{RT_0^2}v\right] \tag{34}$$

In order to solve Eq. (34), new variables are defined

$$\theta = (E/RT_0^2)v, \qquad \eta_x = x/r$$

where r is the radius of the vessel and x is the distance from the center. Equation (34) then becomes

$$\nabla_\eta^2\theta = -\{(Q/\lambda)(E/RT_0^2)r^2 Z e^{-E/RT_0}\}e^\theta \tag{35}$$

and the boundary conditions are $\eta = 1, \theta = 0$, and $\eta = 0, d\theta/d\eta = 0$.

Both Eq. (35) and the boundary conditions contain only one nondimensional parameter δ:

$$\delta = (Q/\lambda)(E/RT_0^2)r^2 Z e^{-E/RT_0} \tag{36}$$

The solution of Eq. (35), which represents the stationary temperature distribution, should be of the form $\theta = f(\eta, \delta)$ with one parameter, i.e., δ. The condition under which such a stationary temperature distribution ceases to be possible, i.e., the critical condition of ignition, is of the form $\delta = \text{const} = \delta_{\text{crit}}$. The critical value depends upon T_0, the geometry (if the vessel is nonspherical), and the pressure through Z. Numerical integration of Eq. (35) for various δ's determines the critical δ. For a spherical vessel, $\delta_{\text{crit}} = 3.32$; for an infinite cylindrical vessel, $\delta_{\text{crit}} = 2.00$; and for infinite parallel plates, $\delta_{\text{crit}} = 0.88$, where r becomes the distance between the plates.

As in the discussion of flame propagation, the stoichiometry and pressure dependence are in Z and $Z \sim p^n$, where n is the order of the reaction. Equation (36) expressed in terms of δ_{crit} permits the relationship between the critical parameters to be determined. Taking logarithms,

$$\ln rp^n \sim (+E/RT_0)$$

If the reacting medium is a solid or liquid undergoing exothermic decomposition, the pressure term is omitted and

$$\ln r \sim (+E/RT_0)$$

These results define the conditions for the critical size of storage for compounds such as ammonium nitrate as a function of the ambient temperature T_0 [7]. Similarly, it represents the variation in mass of combustible material that will spontaneously ignite as a function of the ambient temperature T_0. The higher the ambient temperature, the smaller the critical mass has to be to prevent disaster. Conversely, the more reactive the material, the smaller the size that will undergo spontaneous combustion. A few linseed or tung oil rags piled together can spontaneously ignite, whereas the size of a pile of damp leaves or pulverized coal must be large for the same type of ignition.

b. The Nonstationary Solution

The nonstationary theory deals with the thermal balance of the whole reaction vessel and assumes the temperature to be the same at all points. This assumption is, of course, incorrect in the conduction range where the temperature gradient is by no means localized at the wall. It is, however, equivalent to a replacement of the mean values of all temperature-dependent magnitudes by their values at a mean temperature, and involves relatively minor error.

If the volume of the vessel is designated by V and its surface area by S, and if a heat transfer coefficient h is defined, the amount of heat evolved over the whole volume per unit time by the chemical reaction is

$$V Q Z e^{-E/RT} \tag{37}$$

and the amount of heat carried away from the wall is

$$h S(T - T_0) \tag{38}$$

Thus, the problem is now essentially nonadiabatic. The difference between the two heat terms is the heat that causes the temperature within the vessel to rise a certain amount per unit time,

$$c_v \rho V (dT/dt) \tag{39}$$

These terms can be expressed as an equality,

$$c_v \rho V (dT/dt) = V Q Z e^{-E/RT} - h S(T - T_0) \tag{40}$$

or

$$dT/dt = (Q/c_v \rho) Z e^{-E/RT} - (h S/c_v \rho V)(T - T_0) \tag{41}$$

Equations (40) and (41) are forms of Eq. (29) with a heat loss term. Nondimensionalizing the temperature and linearizing the exponent in the same manner as in the previous section, one obtains

$$d\theta/dt = (Q/c_v\rho)(E/RT_0^2)Ze^{-E/RT_0}e^{\theta} - (hS/c_v\rho V)\theta \tag{42}$$

with the initial condition $\theta = 0$ at $t = 0$.

The equation is not in dimensionless form. Each term has the dimension of reciprocal time. In order to make the equation completely dimensionless, it is necessary to introduce a time parameter. Equation (42) contains two such time parameters:

$$\tau_1 = [(Q/c_v\rho)(E/RT_0^2)Ze^{-E/RT_0}]^{-1}, \qquad \tau_2 = (hS/c_v\rho V)^{-1}$$

Consequently, the solution of Eq. (42) should be in the form

$$\theta = f(t/\tau_{1,2}, \tau_2/\tau_1)$$

where $\tau_{1,2}$ implies either τ_1 or τ_2.

Thus, the dependence of dimensionless temperature θ on dimensionless time $t/\tau_{1,2}$ contains one dimensionless parameter τ_2/τ_1. Consequently, a sharp rise in temperature can occur for a critical value τ_2/τ_1.

It is best to examine Eq. (42) written in terms of these parameters, i.e.,

$$d\theta/dt = (e^{\theta}/\tau_1) - (\theta/\tau_2) \tag{43}$$

In the ignition range the rate of energy release is much greater than the rate of heat loss; that is, the first term on the right-hand side of Eq. (43) is much greater than the second. Under these conditions, removal of heat is disregarded and the thermal explosion is viewed as essentially adiabatic.

Then for an adiabatic thermal explosion, the time dependence of the temperature should be in the form

$$\theta = f(t/\tau_1) \tag{44}$$

Under these conditions, the time within which a given value of θ is attained is proportional to the magnitude τ_1. Consequently, the induction period in the instance of adiabatic explosion is proportional to τ_1. The proportionality constant has been shown to be unity. Conceptually, this induction period can be related to the time period for the ignition of droplets for different air (or ambient) temperatures. Thus τ can be the adiabatic induction time and is simply

$$\tau = \frac{c_v\rho}{Q} \frac{RT_0^2}{E} \frac{1}{Z} e^{+E/RT_0} \tag{45}$$

Again, the expression can be related to the critical conditions of time, pressure, and ambient temperature T_0 by taking logarithms:

$$\ln(\tau P^{n-1}) \sim (+E/RT_0) \tag{46}$$

The pressure dependence, as before, is derived not only from the perfect gas law for ρ, but from the density–pressure relationship in Z as well. Also, the effect of the stoichiometry of a reacting gas mixture would be in Z. But the mole fraction terms would be in the logarithm, and therefore have only a mild effect on the induction time. For hydrocarbon–air mixtures, the overall order is approximately 2, so Eq. (46) becomes

$$\ln(\tau P) \sim (E/RT_0) \tag{47}$$

which is essentially the condition used in bluff-body stabilization considerations in Chapter 4, Section F. This result gives the intuitively expected answer that the higher the ambient temperature, the shorter is the ignition time. Hydrocarbon droplet and gas fuel injection ignition data correlate well with the dependences as shown in Eq. (47) [8, 9].

In a less elegant fashion, Todes [10] (see Jost [11]) obtained the same expression as Eq. (45). As Semenov [3] has shown by use of Eq. (25), Eq. (45) can be written as

$$\tau_i = \tau_r (c_v RT_0^2 / QE) \tag{48}$$

Since $(E/RT_0)^{-1}$ is a small quantity not exceeding 0.05 for most cases of interest and $(c_v T_0/Q)$ is also a small quantity of the order 0.1, the quantity $(c_v RT_0^2/QE)$ may be considered to have a range from 0.01 to 0.001. Thus, the thermal ignition time for a given initial temperature T_0 is from a hundredth to a thousandth of the reaction time evaluated at T_0. Since from Eq. (28) and its subsequent discussion,

$$\tau_r \leq (Q/c_v)(E/RT_0^2)e\tau_l \tag{49}$$

then

$$\tau_i \leq e\tau_l \tag{50}$$

which signifies that the induction period is of the same order of magnitude as the thermal relaxation time.

Since it takes only a very small fraction of the reaction time to reach the end of the induction period, at the moment of the sudden rapid rise in temperature (i.e., when explosion begins), not more than 1% of the initial mixture has reacted. This result justifies the inherent approximation developed that the reaction rate remains constant until explosion occurs. Also justified is the earlier assumption that the original mixture concentration remains the same from T_0 to T_i. This observation is important in that it reveals that no significantly different results would be obtained if the more complex approach using both variations in temperature and concentration were used.

D. FORCED IGNITION

Unlike the concept of spontaneous ignition which is associated with a large condensed-phase mass of reactive material, the concept of forced ignition is essentially associated with gaseous materials. The energy input into a condensed-phase reactive mass may be such that the material vaporizes and then ignites, but the phenomena that lead to ignition are those of the gas-phase reactions. There are many means to force ignition of a reactive material or mixture, but the most commonly studied concepts are those associated with various processes that take place in the spark-ignition, automotive engine.

The spark is the first and most common form of forced ignition. In the automotive cylinder it initiates a flame that travels across the cylinder. The spark is fired before the piston reaches top dead center and, as the flame travels, the combustible mixture ahead of this flame is being compressed. Under certain circumstances the mixture ahead of the flame explodes, in which case the phenomenon of knock is said to occur. The gases ahead of the flame are usually ignited as the temperature rises due to the compression or some hot spot on the metallic surfaces in the cylinder. As discussed in Chapter 2, knock is most likely an explosion, but not a detonation. The physical configuration would not permit the transformation from a deflagration to a detonation. Nevertheless, knock, or premature forced ignition, can occur when a fuel–air mixture is compressively heated or when a hot spot exists. Consequently, it is not surprising that the ignitability of a gaseous fuel–air mixture—or, for that matter, any exoergic system—has been studied experimentally by means approaching adiabatic compression to high temperature and pressure, by shock waves (which also raise the material to a high temperature and pressure), or by propelling hot metallic spheres or incandescent particles into a cold reactive mixture.

Forced ignition can also be brought about by pilot flames or by flowing hot gases, which act as a jet into the cold mixture to be ignited. Or it may be engendered by creating a boundary layer flow parallel to the cold mixture, which may also be flowing. Indeed, there are several other possibilities that one might evoke. For consideration of these systems, the reader is referred to Ref. [12].

It is apparent, then, that an ignition source can lead either to a pure explosion or to a flame (deflagration) that propagates. The geometric configuration in which the flame has been initiated can be conducive to the transformation of the flame into a detonation. There are many elements of concern with respect to fire and industrial safety in these considerations. Thus, a concept of a minimum ignition energy has been introduced as a test method for evaluating the ignitability of various fuel–air mixtures or any system that has exoergic characteristics.

Ignition by near adiabatic compression or shock wave techniques creates explosions that are most likely chain carrier, rather than thermal, initiated. This aspect of the subject will be treated at the end of this chapter. The main concentration in this section will be on ignition by sparks based on a thermal approach by Zeldovich [13]. This approach, which gives insights not only into the parameters that give spark ignition, but also into forced ignition systems that lead to flames, has applicability to the minimum ignition energy.

1. Spark Ignition and Minimum Ignition Energy

The most commonly used spark systems for mobile power plants are capacitance sparks, which are developed from discharged condensers. The duration of these discharges can be as short as 0.01 μs or as long as 100 μs for larger engines. Research techniques generally employ two circular electrodes with flanges at the tips. The flanges have a parallel orientation and a separation distance greater than the quenching distance for the mixture to be ignited. (Reference [12] reports extensive details about spark and all other types of forced ignition.) The energy in a capacitance spark is given by

$$E = \tfrac{1}{2}c_f(v_2^2 - v_1^2) \tag{51}$$

where E is the electrical energy obtained in joules, c_f is the capacitance of the condenser in farads, v_2 is the voltage on the condenser just before the spark occurs, and v_1 is the voltage at the instant the spark ceases.

In the Zeldovich method of spark ignition, the spark is replaced by a point heat source, which releases a quantity of heat. The time-dependent distribution of this heat is obtained from the energy equation

$$(\partial T / \partial t) = \alpha \nabla^2 T \tag{52}$$

When this equation is transformed into spherical coordinates, its boundary conditions become

$$r = \infty, T = T_0 \quad \text{and} \quad r = 0, (\partial T / \partial r) = 0$$

The distribution of the input energy at any time must obey the equality

$$Q'_v = 4\pi c_p \rho \int_0^\infty (T - T_0) r^2 \, dr \tag{53}$$

The solution of Eq. (52) then becomes

$$(T - T_0) = \{Q'_v / [c_p \rho (4\pi \alpha t)]^{3/2}\} \exp(-r^2 / 4\alpha t) \tag{54}$$

The maximum temperature (T_M) must occur at $r \to 0$, so that

$$(T_M - T_0) = \{Q'_v / [c_p \rho (4\pi \alpha t)^{3/2}] \tag{55}$$

Considering that the gaseous system to be ignited exists everywhere from $r = 0$ to $r = \infty$, the condition for ignition is specified when the cooling time (τ_c) associated with T_M is greater than the reaction duration time τ_r in the combustion zone of a laminar flame.

This characteristic cooling time is the period in which the temperature at $r = 0$ changes by the value θ. This small temperature difference θ is taken as (RT_M^2/E); i.e.,

$$\theta = RT_M^2/E \tag{56}$$

This expression results from the same type of analysis that led to Eq. (18). A plot of T_M versus t [Eq.(55)] is shown in Fig. 4. From this figure the characteristic cooling time can be taken to a close approximation as

$$\tau_c = \theta \mid dT_M/dt \mid_{T_M=T_r} \tag{57}$$

The slope is taken as a time when the temperature at $r = 0$ is close to the adiabatic flame temperature of the mixture to be ignited. By differentiating Eq. (55), the denominator of Eq. (57) can be evaluated to give

$$\tau_c = [\theta/6\alpha\pi(T_f - T_0)]\{Q_v'/[c_p\rho(T_f - T_0)]\}^{2/3} \tag{58}$$

where Q_v' is now given a specific definition as the amount of external input energy required to heat a spherical volume of radius r_f uniformly from T_0 to T_f; i.e.,

$$Q_v' = \left(\tfrac{4}{3}\right)\pi r_f^3 c_p\rho(T_f - T_0) \tag{59}$$

Thus, Eq. (58) becomes

$$\tau_c = 0.14[\theta/(T_f - T_0)](r_f^2/\alpha) \tag{60}$$

Considering that the temperature difference θ must be equivalent to $(T_f - T_i)$ in the Zeldovich–Frank-Kamenetskii–Semenov thermal flame theory, the reaction time corresponding to the reaction zone δ in the flame can also be approximated by

$$\tau_f \cong [2\theta/(T_f - T_0)](\alpha/S_L^2) \tag{61}$$

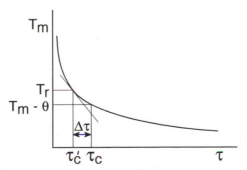

FIGURE 4   Variation of the maximum temperature with time for energy input into a spherical volume of a fuel–air mixture.

where (α/S_L^2) is the characteristic time associated with the flame and

$$a = (\alpha/S_L) \qquad (62)$$

specifies the thermal width of the flame.

Combining Eqs. (60), (61), and (62) for the condition $\tau_c \geq \tau_r$ yields the condition for ignition as

$$r_f \geq 3.7a \qquad (63)$$

Physically, Eq. (63) specifies that for a spark to lead to ignition of an exoergic system, the corresponding equivalent heat input radius must be several times the characteristic width of the laminar flame zone. Under this condition, the nearby layers of the initially ignited combustible material will further ignite before the volume heated by the spark cools.

The preceding developments are for an idealized spark ignition system. In actual systems, much of the electrical energy is expended in radiative losses, shock wave formation, and convective and conductive heat losses to the electrodes and flanges. Zeldovich [13] has reported for mixtures that the efficiency

$$\eta_s = (Q_v'/E) \qquad (64)$$

can vary from 2 to 16%. Furthermore, the development was idealized by assuming consistency of the thermophysical properties and the specific heat. Nevertheless, experimental results taking all these factors into account [13, 14] reveal relationships very close to

$$r_f \geq 3a \qquad (65)$$

The further importance of Eqs. (63) and (65) is in the determination of the important parameters that govern the minimum ignition energy. By substituting Eqs. (62) and (63) into Eq. (59), one obtains the proportionality

$$Q_{v,min}' \sim (\alpha/S_L)^3 c_p \rho (T_f - T_0) \qquad (66)$$

Considering $\alpha = (\lambda/\rho c_p)$ and applying the perfect gas law, the dependence of $Q_{v,min}'$ on P and T is found to be

$$Q_{v,min}' \sim [\lambda^3 T_0^2 (T_f - T_0)]/S_L^3 P^2 c_p^2 \qquad (67)$$

The minimum ignition energy is also a function of the electrode spacing. It becomes asymptotic to a very small spacing below which no ignition is possible. This spacing is the quenching distance discussed in Chapter 4. The minimum ignition energy decreases as the electrode spacing is increased, reaches its lowest value at some spacing, then begins to rise again. At small spacings the electrode removes large amounts of heat from the incipient flame, and thus a large minimum ignition energy is required. As the spacing increases, the surface area to volume ratio decreases, and, consequently, the ignition energy required decreases. Most experimental investigations [12, 15] report the minimum ignition energy for the electrode spacing that gives the lowest value.

An interesting experimental observation is that there appears to be an almost direct relation between the minimum ignition energy and the quenching distance [15, 16]. Calcote *et al.* [15] have reported significant data in this regard, and their results are shown in Fig. 5. These data are for stoichiometric mixtures with air at 1 atm.

The general variation of minimum ignition energy with pressure and temperature would be that given in Eq. (67), in which one must recall that S_L is also a function of the pressure and the T_f of the mixture. Figure 6 from Blanc *et al.* [14] shows the variation of Q'_v as a function of the equivalence ratio. The variation is very similar to the variation of quenching distance with the equivalence ratio

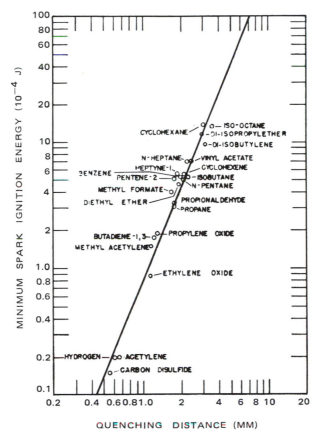

FIGURE 5   Correlation of the minimum spark ignition energy with quenching diameter (from Calcote *et al.* [15]).

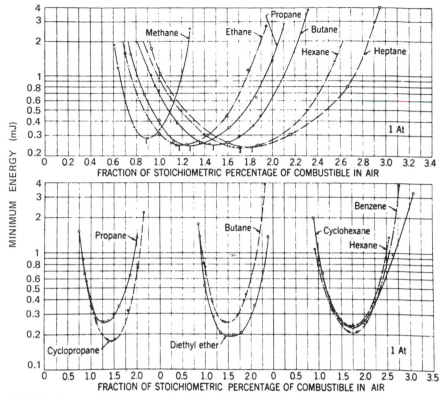

FIGURE 6   Minimum ignition energy of fuel–air mixtures as a function of stoichiometry (from Blanc *et al.* [14]).

ϕ [11] and is, to a degree, the inverse of S_L versus ϕ. However, the increase of Q'_v from its lowest value for a given ϕ is much steeper than the decay of S_L from its maximum value. The rapid increase in Q'_v must be due to the fact that S_L is a cubed term in the denominator of Eq. (67). Furthermore, the lowest Q'_v is always found on the fuel-rich side of stoichiometric, except for methane [13, 15]. This trend is apparently attributable to the difference in the mass diffusivities of the fuel and oxygen. Notice the position of the methane curve with respect to the other hydrocarbons in Fig. 6. Methane is the only hydrocarbon shown whose molecular weight is appreciably lower than that of the oxygen in air.

Many [12, 15] have tried to determine the effect of molecular structure on Q'_v. Generally the primary effect of molecular structure is seen in its effect on T_f (in S_L) and α.

Appendix G lists minimum ignition energies of many fuels for the stoichiometric condition at a pressure of 1 atm. The Blanc data in this appendix are taken from Fig. 6. It is remarkable that the minima of the energy curves for the various compounds occur at nearly identical values.

In many practical applications, sparks are used to ignite flowing combustible mixtures. Increasing the flow velocity past the electrodes increases the energy required for ignition. The flow blows the spark downstream, lengthens the spark path, and causes the energy input to be distributed over a much larger volume [12]. Thus the minimum energy in a flow system is greater than that under a stagnant condition.

From a safety point of view, one is also interested in grain elevator and coal dust explosions. Such explosions are not analyzed in this text, and the reader is referred to the literature [17]. However, many of the thermal concepts discussed for homogeneous gas-phase ignition will be fruitful in understanding the phenomena that control dust ignition and explosions.

2. Ignition by Adiabatic Compression and Shock Waves

Ignition by sparks occurs in a very local region and spreads by flame characteristics throughout the combustible system. If an exoergic system at standard conditions is adiabatically compressed to a higher pressure and hence to a higher temperature, the gas-phase system will explode. There is little likelihood that a flame will propagate in this situation. Similarly, a shock wave can propagate through the same type of mixture, rapidly compressing and heating the mixture to an explosive condition. As discussed in Chapter 5, a detonation will develop under such conditions only if the test section is sufficiently long.

Ignition by compression is similar to the conditions that generate knock in a spark-ignited automotive engine. Thus it would indeed appear that compression ignition and knock are chain-initiated explosions. Many have established the onset of ignition with a rapid temperature rise over and above that expected due to compression. Others have used the onset of some visible radiation or, in the case of shock tubes, a certain limit concentration of hydroxyl radical formation identified by spectroscopic absorption techniques. The observations and measurement techniques are interrelated. Ignition occurs in such systems in the 1000 K temperature range. However, it must be realized that in hydrocarbon–air systems the rise in temperature due to exothermic energy release of the reacting mixture occurs most sharply when the carbon monoxide which eventually forms is converted to carbon dioxide. This step is the most exothermic of all the conversion steps of the fuel–air mixture to products [18]. Indeed, the early steps of the process are overall isoergic owing to the simultaneous oxidative pyrolysis of the fuel, which is endothermic, and the conversion of some of the hydrogen formed to water, which is an exothermic process [18].

Shock waves are an ideal way of obtaining induction periods for high-temperature–high-pressure conditions. Since a shock system is nonisentropic, a system at some initial temperature and pressure condition brought to a final pressure by the shock wave will have a higher temperature than a system in which the same mixture at the same initial conditions is brought by adiabatic compression to the same pressure. Table 1 compares the final temperatures for the same ratios of shock and adiabatic compression for air.

TABLE 1   Compression versus Shock-Induced Temperature[a]

Shock and adiabatic compression ratio	Shock wave velocity (m/s)	T (K) Behind shock	T(K) After compression
2	452	336	330
5	698	482	426
10	978	705	515
50	2149	2261	794
100	3021	3861	950
1000	9205	19,089	1711
2000	12,893	25,977	2072

[a] Initial temperature, 273 K.

PROBLEMS

1. The reported decomposition of ammonium nitrate indicates that the reaction is unimolecular and that the rate constant has an A factor of $10^{13.8}$ and an activation energy of 170 kJ/mol. Using this information, determine the critical storage radius at 160°C. Report the calculation so that a plot of r_{crit} versus T_0 can be obtained. Take a temperature range from 80 to 320°C.
2. Concisely explain the difference between chain and thermal explosions.
3. Are liquid droplet ignition times appreciably affected by droplet size? Explain.

REFERENCES

1. Mullins, H. P., "Spontaneous Ignition of Liquid Fuels," Chap. 11, Agardograph No. 4. Butterworth, London, 1955.
2. Zabetakis, M. G., Bull.—U. S. Bur. Mines No. 627 (1965).
3. Semenov, N. N., "Chemical Kinetics and Chain Reactions." Oxford Univ. Press, London, 1935.
4. Hinshelwood, C. N., "The Kinetics of Chemical Change." Oxford Univ. Press, London, 1940.
5. Zeldovich, Y. B., Barenblatt, G. I., Librovich, V. B., and Makhviladze, G. M., "The Mathematical Theory of Combustion and Explosions," Chap. 1. Consultants Bureau, New York, 1985.
6. Frank-Kamenetskii, D. "Diffusion and Heat Exchange in Chemical Kinetics." Princeton Univ. Press, Princeton, New Jersey, 1955.
7. Hainer, R. M., Int. Symp. Combust., 5th p. 224. Van Nostrand-Reinhold, Princeton, New Jersey, 1955.
8. Mullins, B. P., Fuel 32, 343 (1953).
9. Mullins, B. P., Fuel 32, 363 (1953).
10. Todes, O. M., Acta Physicochem. URRS 5, 785 (1936).
11. Jost, W., "Explosions and Combustion Processes in Gases," Chap. 1, McGraw-Hill, New York, 1946.

12. Belles, F. E., and Swett, C C., *Natl. Advis. Comm. Aeronaut. Rep.* No. 1300 (1959).
13. Zeldovich, Y. B., *Zh. Eksp. Teor. Fiz.* **11** 159 (1941); see Shchelinkov, Y. S.,"The Physics of the Combustion of Gases," Chap. 5. Edited Transl. FTD-HT-23496-68, Transl. Revision, For. Technol. Div., Wright-Patterson AFB, Ohio, 1969.
14. Blanc, M. V., Guest, P. G., von Elbe, G., and Lewis, B., *Int. Symp. Combust., 3rd* p. 363. Williams & Wilkins, Baltimore, Maryland, 1949.
15. Calcote, H. F., Gregory, C. A., Jr., Barnett, C. M., and Gilmer, R. B., *Ind. Eng. Chem.* **44**, 2656 (1952).
16. Kanury, A. M., "Introduction to Combustion Phenomena," Chap. 4. Gordon & Breach, New York, 1975.
17. Hertzberg, M., Cashdollar, K. L., Contic, R. S., and Welsch, L. M., *Bur. Mines Rep.* (1984).
18. Dryer, F. L., and Glassman, I., *Prog. Astronaut Aeronaut.* **26**, 255 (1978).

8

Environmental Combustion Considerations

A. INTRODUCTION

In the mid 1940s, symptoms now attributable to photochemical air pollution were first encountered in the Los Angeles area. Several researchers recognized that the conditions there were producing a new kind of smog caused by the action of sunlight on the oxides of nitrogen and subsequent reactions with hydrocarbons. This smog was different from the "pea-soup" conditions prevailing in London in the early twentieth century and the polluted-air disaster that struck Donora, Pennsylvania, in the 1930s. It was also different from the conditions revealed by the opening of Eastern Europe in the last part of this century. In Los Angeles, the primary atmosphere source of nitrogen oxides, CO, and hydrocarbons was readily shown to be the result of automobile exhausts. The burgeoning population and industrial growth in U.S. urban and exurban areas were responsible for the problem of smog, which led to controls not only on automobiles, but also on other mobile and stationary sources.

Atmospheric pollution has become a worldwide concern. With the prospect of supersonic transports flying in the stratosphere came initial questions as to how the water vapor ejected by the power plants of these planes would affect the stratosphere. This concern led to the consideration of the effects of injecting large amounts of any species on the ozone balance in the atmosphere. It then became evident that the major species that would affect the ozone balance were the oxides of nitrogen. The principal nitrogen oxides found to be present in the atmosphere are nitric oxide (NO) and nitrogen dioxide (NO_2)—the combination of which is referred to as NO_x—and nitrous oxide (N_2O). As Bowman [1] has reported, the global emissions of NO_x and N_2O into the atmosphere have been increasing steadily since the middle of the nineteenth century. And, although important natural sources of the oxides of nitrogen exist, a significant amount of this increase is attributed to human activities, particularly those involving combustion of fossil and biomass fuels. For details as to the sources of combustion-generated nitrogen oxide emissions, one should refer to Bowman's review [1].

Improvement of the atmosphere continues to be of great concern. The continual search for fossil fuel resources can lead to the exploitation of coal, shale, and secondary and tertiary oil recovery schemes. For instance, the industrialization of China, with its substantial resource of sulfur coals, requires consideration of the effect of sulfur oxide emissions. Indeed, the sulfur problem may be the key in the more rapid development of coal usage worldwide. Furthermore, the fraction of aromatic compounds in liquid fuels derived from such natural sources or synthetically developed is found to be large, so that, in general, such fuels have serious sooting characteristics.

This chapter seeks not only to provide better understanding of the oxidation processes of nitrogen and sulfur and the processes leading to particulate (soot) formation, but also to consider appropriate combustion chemistry techniques for regulating the emissions related to these compounds. The combustion—or, more precisely, the oxidation—of CO and aromatic compounds has been discussed in earlier chapters. This information and that to be developed will be used to examine the emission of other combustion-generated compounds thought to have detrimental effects on the environment and on human health. How emissions affect the atmosphere is treated first.

B. THE NATURE OF PHOTOCHEMICAL SMOG

Photochemical air pollution consists of a complex mixture of gaseous pollutants and aerosols, some of which are photochemically produced. Among the gaseous compounds are the oxidizing species ozone, nitrogen dioxide, and peroxyacyl nitrate:

O_3 NO_2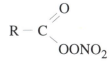

ozone   nitrogen dioxide   peroxyacyl nitrate

The member of this series most commonly found in the atmosphere is peroxyacetyl nitrate (PAN)

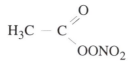

peroxyacetyl nitrate (PAN)

The three compounds O_3, NO_2, and PAN are often grouped together and called photochemical oxidant.

Photochemical smog comprises mixtures of particulate matter and noxious gases, similar to those that occurred in the typical London-type "pea soup" smog. The London smog was a mixture of particulates and oxides of sulfur, chiefly sulfur dioxide. But the overall system in the London smog was chemically reducing in nature. This difference in redox chemistry between photochemical oxidant and SO_x-particulate smog is quite important in several respects. Note in particular the problem of quantitatively detecting oxidant in the presence of sulfur dioxide. Being a reducing agent SO_x tends to reduce the oxidizing effects of ozone and thus produces low quantities of the oxidant.

In dealing with the heterogeneous gas–liquid–solid mixture characterized as photochemical smog, it is important to realize from a chemical, as well as a biological, point of view that synergistic effects may occur.

1. Primary and Secondary Pollutants

Primary pollutants are those emitted directly to the atmosphere while secondary pollutants are those formed by chemical or photochemical reactions of primary pollutants after they have been admitted to the atmosphere and exposed to sunlight. Unburned hydrocarbons, NO, particulates, and the oxides of sulfur are examples of primary pollutants. The particulates may be lead oxide from the oxidation of tetraethyllead in automobiles, fly ash, and various types of carbon formation. Peroxyacyl nitrate and ozone are examples of secondary pollutants.

Some pollutants fall in both categories. Nitrogen dioxide, which is emitted directly from auto exhaust, is also formed in the atmosphere photochemically from NO. Aldehydes, which are released in auto exhausts, are also formed in the photochemical oxidation of hydrocarbons. Carbon monoxide, which arises primarily from autos and stationary sources, is likewise a product of atmospheric hydrocarbon oxidation.

2. The Effect of NO_x

It has been well established that if a laboratory chamber containing NO, a trace of NO_2, and air is irradiated with ultraviolet light, the following reactions occur:

$$NO_2 + h\nu \left(3000 \text{ Å} \leq \lambda \leq 4200 \text{ Å}\right) \rightarrow NO + O \left(^3P\right) \tag{1}$$

$$O + O_2 + M \rightarrow O_3 + M \tag{2}$$

$$O_3 + NO \rightarrow O_2 + NO_2 \tag{3}$$

The net effect of irradiation on this inorganic system is to establish the dynamic equilibrium

$$NO_2 + O_2 \xleftrightarrow{h\nu} NO + O_3 \tag{4}$$

However, if a hydrocarbon, particularly an olefin or an alkylated benzene, is added to the chamber, the equilibrium represented by reaction (4) is unbalanced and the following events take place:

1. The hydrocarbons are oxidized and disappear.
2. Reaction products such as aldehydes, nitrates, PAN, etc., are formed.
3. NO is converted to NO_2.
4. When all the NO is consumed, O_3 begins to appear. On the other hand, PAN and other aldehydes are formed from the beginning.

Basic rate information permits one to examine these phenomena in detail. Leighton [2], in his excellent book *Photochemistry of Air Pollution*, gives numerous tables of rates and products of photochemical nitrogen oxide–hydrocarbon reactions in air; this early work is followed here to give fundamental insight into the photochemical smog problem. The data in these tables show low rates of photochemical consumption of the saturated hydrocarbons, as compared to the unsaturates, and the absence of aldehydes in the products of the saturated hydrocarbon reactions. These data conform to the relatively low rate of reaction of the saturated hydrocarbons with oxygen atoms and their inertness with respect to ozone.

Among the major products in the olefin reactions are aldehydes and ketones. Such results correspond to the splitting of the double bond and the addition of an oxygen atom to one end of the olefin.

Irradiation of mixtures of an olefin with nitric oxide and nitrogen dioxide in air shows that the nitrogen dioxide rises in concentration before it is eventually consumed by reaction. Since it is the photodissociation of the nitrogen dioxide that initiates the reaction, it would appear that a negative quantum yield results. More likely, the nitrogen dioxide is being formed by secondary reactions more rapidly than it is being photodissociated.

The important point is that this negative quantum yield is realized only when an olefin (hydrocarbon) is present. Thus, adding the overall step

$$\left.\begin{array}{c} O \\ O_3 \end{array}\right\} + \text{olefin} \rightarrow \text{products} \tag{5}$$

to reactions (1)–(3) would not be an adequate representation of the atmosphere photochemical reactions. However, if one assumes that O_3 attains a steady-state concentration in the atmosphere, then one can perform a steady-state analysis (see Chapter 2, Section B) with respect to O_3. Furthermore, if one assumes that O_3 is largely destroyed by reaction (3), one obtains a very useful approximate relationship:

$$(O_3) = -(j_1/k_3)(NO_2)/(NO)$$

where j is the rate constant for the photochemical reaction. Thus, the O_3 steady-state concentration in a polluted atmosphere is seen to increase with decreasing concentration of nitric oxide and vice versa. The ratio of j_1/k_3 approximately equals 1.2 ppm for the Los Angeles noonday condition [2]. Reactions such as

$$O + NO_2 \longrightarrow NO + O_2$$

$$O + NO_2 + M \longrightarrow NO_3 + M$$

$$NO_3 + NO \longrightarrow 2NO_2$$

$$O + NO + M \longrightarrow NO_2 + M$$

$$2NO + O_2 \longrightarrow 2NO_2$$

$$NO_3 + NO_2 \longrightarrow N_2O_5$$

$$N_2O_5 \longrightarrow NO_3 + NO_2$$

do not play a part. They are generally too slow to be important.

Furthermore, it has been noted that when the rate of the oxygen atom–olefin reaction and the rate of the ozone–olefin reaction are totaled, they do not give the complete hydrocarbon consumption. This anomaly is also an indication of an additional process.

An induction period with respect to olefin consumption is also observed in the photochemical laboratory experiments, thus indicating the buildup of an intermediate. When illumination is terminated in these experiments, the excess rate over the total of the O and O_3 reactions disappears. These and other results suggest that the intermediate formed is photolyzed and contributes to the concentration of the major species of concern.

Possible intermediates that fulfill the requirements of the laboratory experiments are alkyl and acyl nitrites and pernitrites. The second photolysis effect eliminates the possibility that aldehydes serve as the intermediate.

Various mechanisms have been proposed to explain the aforementioned laboratory results. The following low-temperature (atmospheric) sequence based on isobutene as the initial fuel was first proposed by Leighton [2] and appears to account for most of what has been observed:

$$O + C_4H_8 \longrightarrow CH_3 + C_3H_5O \tag{6}$$

$$CH_3 + O_2 \longrightarrow CH_3OO \tag{7}$$

$$CH_3OO + O_2 \longrightarrow CH_3O + O_3 \tag{8}$$

$$O_3 + NO \longrightarrow NO_2 + O_2 \tag{9}$$

$$CH_3O + NO \longrightarrow CH_3ONO \tag{10}$$

$$CH_3ONO + h\nu \longrightarrow CH_3O^* + NO \tag{11}$$

$$CH_3O^* + O_2 \longrightarrow H_2CO + HOO \tag{12}$$

$$HOO + C_4H_8 \longrightarrow H_2CO + (CH_3)_2CO + H \tag{13}$$

$$M + H + O_2 \longrightarrow HOO + M \tag{14}$$

$$HOO + NO \longrightarrow OH + NO_2 \tag{15}$$

$$OH + C_4H_8 \longrightarrow (CH_3)_2CO + CH_3 \tag{16}$$

$$CH_3 + O_2 \longrightarrow CH_3OO \quad \text{(as above)} \tag{17}$$

$$2HOO \longrightarrow H_2O_2 + O_2 \tag{18}$$

$$2OH \longrightarrow H_2 + O_2 \tag{19}$$

$$HOO + H_2 \longrightarrow H_2O + OH \tag{20}$$

$$HOO + H_2 \longrightarrow H_2O_2 + H \tag{21}$$

There are two chain-propagating sequences [reactions (13) and (14) and reactions (15)–(17)] and one chain-breaking sequence [reactions (18) and (19)]. The intermediate is the nitrite as shown in reaction (10). Reaction (11) is the required additional photochemical step. For every NO_2 used to create the O atom of reaction (6), one is formed by reaction (9). However, reactions (10), (11), and (15) reveal that for every two NO molecules consumed, one NO and one NO_2 form—hence the negative quantum yield of NO_2.

With other olefins, other appropriate reactions may be substituted. Ethylene would give

$$O + C_2H_4 \longrightarrow CH_3 + HCO \tag{22}$$

$$HOO + C_2H_4 \longrightarrow 2H_2CO + H \tag{23}$$

$$OH + C_2H_4 \longrightarrow H_2CO + CH_3 \tag{24}$$

Propylene would add

$$OH + C_3H_6 \longrightarrow CH_3CHO + CH_3 \tag{25}$$

Thus, PAN would form from

$$CH_3CHO + O_2 \xrightarrow{h\nu} CH_3CO + HOO \tag{26}$$

$$CH_3CO + O_2 \longrightarrow CH_3(CO)OO \tag{27}$$

$$CH_3(CO)OO + NO_2 \longrightarrow CH_3(CO)OONO_2 \tag{28}$$

And an acid could form from the overall reaction

$$CH_3(CO)OO + 2CH_3CHO \longrightarrow CH_3(CO)OH + 2CH_3CO + OH \tag{29}$$

Since pollutant concentrations are generally in the parts-per-million range, it is not difficult to postulate many types of reactions and possible products.

3. The Effect of SO_x

Historically, the sulfur oxides have long been known to have a deleterious effect on the atmosphere, and sulfuric acid mist and other sulfate particulate matter are well established as important sources of atmospheric contamination. However, the atmospheric chemistry is probably not as well understood as the gas-phase photoxidation reactions of the nitrogen oxides–hydrocarbon system. The pollutants form originally from the SO_2 emitted to the air. Just as mobile and stationary combustion sources emit some small quantities of NO_2 as well as NO, so do they emit some small quantities of SO_3 when they burn sulfur-containing fuels. Leighton [2] also discusses the oxidation of SO_2 in polluted atmospheres and an excellent review by Bulfalini [3] has appeared. This section draws heavily from these sources.

The chemical problem here involves the photochemical and catalytic oxidation of SO_2 and its mixtures with the hydrocarbons and NO; however the primary concern is the photochemical reactions, both gas-phase and aerosol-forming.

The photodissociation of SO_2 into SO and O atoms is markedly different from the photodissociation of NO_2. The bond to be broken in the sulfur compound requires about 560 kJ/mol. Thus, wavelengths greater than 2180 Å do not have sufficient energy to initiate dissociation. This fact is significant in that only solar radiation greater than 2900 Å reaches the lower atmosphere. If a photochemical effect is to occur in the SO_2–O_2 atmospheric system, it must be that the radiation electronically excites the SO_2 molecule but does not dissociate it.

There are two absorption bands of SO_2 within the range 3000–4000 Å. The first is a weak absorption band and corresponds to the transition to the first excited state (a triplet). This band originates at 3880 Å and has a maximum around 3840 Å. The second is a strong absorption band and corresponds to the excitation to the second excited state (a triplet). This band originates at 3376 Å and has a maximum around 2940 Å.

Blacet [4], who carried out experiments in high O_2 concentrations, reported that ozone and SO_3 appear to be the only products of the photochemically induced reaction. The following essential steps were postulated:

$$SO_2 + h\nu \longrightarrow SO_2^* \tag{30}$$

$$SO_2^* + O_2 \longrightarrow SO_4 \tag{31}$$

$$SO_4 + O_2 \longrightarrow SO_3 + O_3 \tag{32}$$

The radiation used was at 3130 Å, and it would appear that the excited SO_2^* in reaction (30) is a singlet. The precise roles of the excited singlet and triplet states in the photochemistry of SO_2 are still unclear [3]. Nevertheless, this point need not be one of great concern since it is possible to write the reaction sequence

$$SO_2 + h\nu \longrightarrow {}^1SO_2^* \tag{33}$$

$${}^1SO_2^* + SO_2 \longrightarrow {}^3SO_2^* + SO_2 \tag{34}$$

Thus, reaction (30) could specify either an excited singlet or triplet SO_2^*. The excited state may, of course, degrade by internal transfer to a vibrationally excited ground state which is later deactivated by collision, or it may be degraded directly by collisions. Fluorescence of SO_2 has not been observed above 2100 Å. The collisional deactivation steps known to exist in laboratory experiments are not listed here in order to minimize the writing of reaction steps.

Since they involve one species in large concentrations, reactions (30)–(32) are the primary ones for the photochemical oxidation of SO_2 to SO_3. A secondary reaction route to SO_3 could be

$$SO_4 + SO_2 \longrightarrow 2SO_3 \tag{35}$$

In the presence of water a sulfuric acid mist forms according to

$$H_2O + SO_3 \longrightarrow H_2SO_4 \tag{36}$$

The SO_4 molecule formed by reaction (31) would probably have a peroxy structure; and if SO_2^* were a triplet, it might be a biradical.

There is conflicting evidence with respect to the results of the photolysis of mixtures of SO_2, NO_x, and O_2. However, many believe that the following should be considered with the NO_x photolysis reactions:

$$SO_2 + NO \longrightarrow SO + NO_2 \tag{37}$$

$$SO_2 + NO_2 \longrightarrow SO_3 + NO \tag{38}$$

$$SO_2 + O + M \longrightarrow SO_3 + M \tag{39}$$

$$SO_2 + O_3 \longrightarrow SO_3 + O_2 \tag{40}$$

$$SO_3 + O \longrightarrow SO_2 + O_2 \tag{41}$$

$$SO_4 + NO \longrightarrow SO_3 + NO_2 \tag{42}$$

$$SO_4 + NO_2 \longrightarrow SO_3 + NO_3 \tag{43}$$

$$SO_4 + O \longrightarrow SO_3 + O_2 \tag{44}$$

$$SO + O + M \longrightarrow SO_2 + M \tag{45}$$

$$SO + O_3 \longrightarrow SO_2 + O_2 \tag{46}$$

$$SO + NO_2 \longrightarrow SO_2 + NO \tag{47}$$

The important reducing effect of the SO_2 with respect to different polluted atmospheres mentioned in the introduction of this section becomes evident from these reactions.

Some work [5] has been performed on the photochemical reaction between sulfur dioxide and hydrocarbons, both paraffins and olefins. In all cases, mists were found, and these mists settled out in the reaction vessels as oils with the characteristics of sulfuric acids. Because of the small amounts of materials formed, great problems arise in elucidating particular steps. When NO_x and O_2 are added to this system, the situation is most complex. Bufalini [3] sums up the status in this way: "The aerosol formed from mixtures of the lower hydrocarbons with NO_x and SO_2 is predominantly sulfuric acid, whereas the higher olefin hydrocarbons appear to produce carbonaceous aerosols also, possibly organic acids, sulfuric or sulfonic acids, nitrate-esters, etc."

C. FORMATION AND REDUCTION OF NITROGEN OXIDES

The previous sections help establish the great importance of the nitrogen oxides in the photochemical smog reaction cycles described. Strong evidence indicated that the major culprit in NO_x production was the automobile. But, as automobile emissions standards were enforced, attention was directed to power-generation plants that use fossil fuels. Given these concerns and those associated with supersonic flight in the stratosphere, great interest remains in predicting— and reducing—nitrogen oxide emissions; this interest has led to the formulation of various mechanisms and analytical models to predict specifically the formation and reduction of nitrogen oxides in combustion systems. This section offers some insight into these mechanisms and models, drawing heavily from the reviews by Bowman [1] and Miller and Bowman [6].

When discussing nitrogen oxide formation from nitrogen in atmospheric air, one refers specifically to the NO formed in combustion systems in which the original fuel contains no nitrogen atoms chemically bonded to other chemical elements such as carbon or hydrogen. Since this NO from atmospheric air forms most extensively at high temperatures, it is generally referred to as *thermal* NO.

One early controversy with regard to NO_x chemistry revolved around what was termed "prompt" NO. *Prompt* NO was postulated to form in the flame zone by mechanisms other than those thought to hold exclusively for NO formation from atmospheric nitrogen in the high-temperature zone of the flame or post-flame zone. Although the amount of prompt NO formed is quite small under most practical conditions, the fundamental studies into this problem have helped clarify much about NO_x formation and reduction both from atmospheric and fuel-bound nitrogen. The debate focused on the question of whether prompt NO formation resulted from reaction of hydrocarbon radicals and nitrogen in the flame or from nitrogen reactions with large quantities of O atoms generated early in the flame. Furthermore, it was suggested that superequilibrium concentrations of O atoms could, under certain conditions of pressure and stoichiometry, lead to the formation of nitrous oxide, N_2O, a subsequent source of NO. These questions are fully addressed later in this section.

The term "prompt" NO derives from the fact that the nitrogen in air can form small quantities of CN compounds in the flame zone. In contrast, thermal NO forms in the high-temperature post-flame zone. These CN compounds subsequently react to form NO. The stable compound HCN has been found in the flame zone and is a product in very fuel-rich flames. Chemical models of hydrocarbon reaction processes reveal that, early in the reaction, O atom concentrations can reach superequilibrium proportions; and, indeed, if temperatures are high enough, these high concentrations could lead to early formation of NO by the same mechanisms that describe thermal NO formation.

NO_x formation from fuel-bound nitrogen is meant to specify, as mentioned, the nitrogen oxides formed from fuel compounds that are chemically bonded to other elements. Fuel-bound nitrogen compounds are ammonia, pyridine, and many other amines. The amines can be designated as RNH_2, where R is an organic radical or H atom. The NO formed from HCN and the fuel fragments from the nitrogen compounds are sometimes referred to as *chemical* NO in terminology analogous to that of thermal NO.

Although most early analytical and experimental studies focused on NO formation, more information now exists on NO_2 and the conditions under which it is likely to form in combustion systems. Some measurements in practical combustion systems have shown large amounts of NO_2, which would be expected under the operating conditions. Controversy has surrounded the question of the extent of NO_2 formation in that the NO_2 measured in some experiments may actually have formed in the probes used to capture the gas sample. Indeed, some recent high-pressure experiments have revealed the presence of N_2O.

1. The Structure of the Nitrogen Oxides

Many investigators have attempted to investigate analytically the formation of NO in fuel–air combustion systems. Given of the availability of an enormous amount of computer capacity, they have written all the reactions of the nitrogen

TABLE 1 Structure of Gaseous Nitrogen Compounds

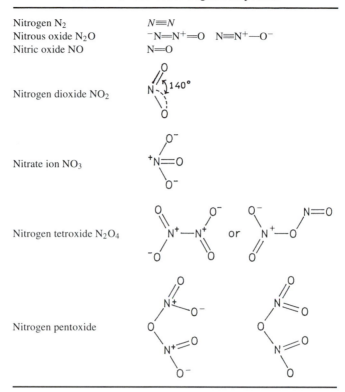

Nitrogen N_2	$N\equiv N$
Nitrous oxide N_2O	$^-N\!=\!N^+\!=\!O$ $N\equiv N^+\!-\!O^-$
Nitric oxide NO	$N\!=\!O$
Nitrogen dioxide NO_2	
Nitrate ion NO_3	
Nitrogen tetroxide N_2O_4	
Nitrogen pentoxide	

oxides they thought possible. Unfortunately, some of these investigators have ignored the fact that some of the reactions could have been eliminated because of steric considerations, as discussed with respect to sulfur oxidation. Since the structure of the various nitrogen oxides can be important, their formulas and structures are given in Table 1.

2. The Effect of Flame Structure

As the important effect of temperature on NO formation is discussed in the following sections, it is useful to remember that flame structure can play a most significant role in determining the overall NO_x emitted. For premixed systems like those obtained on Bunsen and flat flame burners and almost obtained in carbureted spark-ignition engines, the temperature, and hence the mixture ratio, is the prime parameter in determining the quantities of NO_x formed. Ideally, as in equilibrium systems, the NO formation should peak at the stoichiometric value and decline on both the fuel-rich and fuel-lean sides, just as the temperature does. Actually, because of kinetic (nonequilibrium) effects, the peak is found somewhat on the lean (oxygen-rich) side of stoichiometric.

However, in fuel-injection systems where the fuel is injected into a chamber containing air or an air stream, the fuel droplets or fuel jets burn as diffusion flames, even though the overall mixture ratio may be lean and the final temperature could correspond to this overall mixture ratio. The temperature of these diffusion flames is at the stoichiometric value during part of the burning time, even though the excess species will eventually dilute the products of the flame to reach the true equilibrium final temperature. Thus, in diffusion flames, more NO_x forms than would be expected from a calculation of an equilibrium temperature based on the overall mixture ratio. The reduction reactions of NO are so slow that in most practical systems the amount of NO formed in diffusion flames is unaffected by the subsequent drop in temperature caused by dilution of the excess species.

3. Reaction Mechanisms of Oxides of Nitrogen

Nitric oxide is the primary nitrogen oxide emitted from most combustion sources. The role of nitrogen dioxide in photochemical smog has already been discussed. Stringent emission regulations have made it necessary to examine all possible sources of NO. The presence of N_2O under certain circumstances could, as mentioned, lead to the formation of NO. In the following subsections the reaction mechanisms of the three nitrogen oxides of concern are examined.

a. Nitric Oxide Reaction Mechanisms

There are three major sources of the NO formed in combustion: (1) oxidation of atmospheric (molecular) nitrogen via the thermal NO mechanisms; (2) prompt NO mechanisms; and (3) oxidation of nitrogen-containing organic compounds in fossil fuels via the fuel-bound NO mechanisms [1]. The extent to which each contributes is an important consideration.

Thermal NO mechanisms   For premixed combustion systems a conservative estimate of the thermal contribution to NO formation can be made by consideration of the equilibrium system given by reaction (48):

$$N_2 + O_2 \rightleftharpoons 2NO \qquad\qquad (48)$$

As is undoubtedly apparent, the kinetic route of NO formation is not the attack of an oxygen molecule on a nitrogen molecule. Mechanistically, as described in Chapter 3, oxygen atoms form from the H_2–O_2 radical pool, or possibly from the dissociation of O_2, and these oxygen atoms attack nitrogen molecules to start the simple chain shown by reactions (49) and (50):

$$O + N_2 \rightleftharpoons NO + N \qquad k_f = 2 \times 10^{14} \exp(-315/RT) \qquad (49)$$

$$N + O_2 \rightleftharpoons NO + O \qquad k_f = 6.4 \times 10^9 \exp(-26/RT) \qquad (50)$$

where the activation energies are in kJ/mole. Since this chain was first postulated by Zeldovich [7], the thermal mechanism is often referred to as the Zeldovich mechanism. Common practice now is to include the step

$$N + OH \rightleftharpoons NO + H \qquad k_f = 3.8 \times 10^{13} \qquad (51)$$

in the thermal mechanism, even though the reacting species are both radicals and therefore the concentration terms in the rate expression for this step would be very small. The combination of reactions (49)–(51) is frequently referred to as the extended Zeldovich mechanism.

   If one invokes the steady-state approximation described in Chapter 2 for the N atom concentration and makes the partial equilibrium assumption also described in Chapter 2 for the reaction system

$$H + O_2 \rightleftharpoons OH + O$$

one obtains for the rate of formation of NO [8]

$$\frac{d(NO)}{dt} = 2k_{49f}(O)(N_2) \left\{ \frac{1 - [(NO)^2/K'(O_2)(N_2)]}{1 + [k_{49b}(NO)/k_{50f}(O_2)]} \right\};$$

$$K' = K_{49}/K_{50} = K_{c,f,NO}^2 \qquad (52)$$

where K is the concentration equilibrium constant for the specified reaction system and K' the square of the equilibrium constant of formation of NO.

   In order to calculate the thermal NO formation rate from the preceding expression, it is necessary to know the concentrations of O_2, N_2, O, and OH. But the characteristic time for the forward reaction (49) always exceeds the characteristic times for the reaction systems that make up the processes in fuel-oxidizer flame systems; thus, it would appear possible to decouple the thermal NO process from the flame process. Using such an assumption, the NO formation can be calculated from Eq. (52) using local equilibrium values of temperature and concentrations of O_2, N_2, O, and OH.

   From examination of Eq. (52), one sees that the maximum NO formation rate is given by

$$d(NO)/dt = 2k_{49f}(O)(N_2) \qquad (53)$$

which corresponds to the condition that $(NO) \ll (NO)_{eq}$. Due to the assumed equilibrium condition, the concentration of O atoms can be related to the concentration of O_2 molecules via

$$\tfrac{1}{2}O_2 \rightleftharpoons O$$

$$K_{c,f,O,T_{eq}} = (O)_{eq}/(O_2)_{eq}^{1/2}$$

and Eq. (53) becomes

$$d(NO)/dt = 2k_{49f}K_{c,f,O,T_{eq}}(O_2)_{eq}^{1/2}(N_2)_{eq} \tag{54}$$

The strong dependence of thermal NO formation on the combustion temperature and the lesser dependence on the oxygen concentration is evident from Eq. (54). Thus, considering the large activation energy of reaction (49), the best practical means of controlling NO is to reduce the combustion gas temperature and, to a lesser extent, the oxygen concentration. For a condition of constant temperature and varying pressure Eq. (53) suggests that the O atom concentration will decrease as the pressure is raised according to Le Chatelier's principle and the maximum rate will decrease. Indeed, this trend is found in fluidized bed reactors.

In order to determine the errors that may be introduced by the Zeldovich model, Miller and Bowman [6] calculated the maximum (initial) NO formation rates from the model and compared them with the maximum NO formation rates calculated from a detailed kinetics model for a fuel-rich ($\phi = 1.37$) methane–air system. To allow independent variation of temperature, an isothermal system was assumed and the type of prompt NO reactions to be discussed next were omitted. Thus, the observed differences in NO formation rates are due entirely to the nonequilibrium radical concentrations that exist during the combustion process. Their results are shown in Fig. 1, which indicates a noticeable acceleration of the maximum NO formation rate above that calculated using the Zeldovich model during the initial stages of the reaction due to nonequilibrium effects, with the departures from the Zeldovich model results decreasing with increasing temperature. As the lower curve in Fig. 1 indicates, while nonequilibrium effects are evident over a wide range of temperature, the accelerated rates are sufficiently low that very little NO is formed by the accelerated nonequilibrium component. Examining the lower curve, as discussed in Chapter 1, one sees that most hydrocarbon–air combustion systems operate in the range of 2100–2600 K.

Prompt NO mechanisms   In dealing with the presentation of prompt NO mechanisms, much can be learned by considering the historical development of the concept of prompt NO. With the development of the Zeldovich mechanism, many investigators followed the concept that in premixed flame systems, NO would form only in the post-flame or burned-gas zone. Thus, it was thought possible to experimentally determine thermal NO formation rates and, from these rates, to

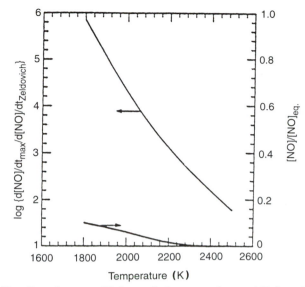

FIGURE 1   The effect of superequilibrium radical concentrations on NO formation rates in the isothermal reaction of 13% methane in air ($\phi = 1.37$). The upper curve is the ratio of the maximum NO formation rate calculated using the detailed reaction mechanism of Ref. [6] to the initial NO formation rate calculated using the Zeldovich model. The lower curve is the ratio of the NO concentration at the time of the maximum NO formation rate calculated using the detailed reaction mechanism to the equilibrium NO concentration (from Miller and Bowman [6]).

find the rate constant of Eq. (49) by measurement of the NO concentration profiles in the post-flame zone. Such measurements can be performed readily on flat flame burners. Of course, in order to make these determinations, it is necessary to know the O atom concentrations. Since hydrocarbon–air flames were always considered, the nitrogen concentration was always in large excess. As discussed in the preceding subsection, the O atom concentration was taken as the equilibrium concentration at the flame temperature and all other reactions were assumed very fast compared to the Zeldovich mechanism.

   These experimental measurements on flat flame burners revealed that when the NO concentration profiles are extrapolated to the flame-front position, the NO concentration goes not to zero, but to some finite value. Such results were most frequently observed with fuel-rich flames. Fenimore [9] argued that reactions other than the Zeldovich mechanism were playing a role in the flame and that some NO was being formed in the flame region. He called this NO, "prompt" NO. He noted that prompt NO was not found in nonhydrocarbon CO–air and H_2–air flames, which were analyzed experimentally in the same manner as the hydrocarbon flames. The reaction scheme he suggested to explain the NO found in the flame zone involved a hydrocarbon species and atmospheric nitrogen. The nitrogen compound was formed via the following mechanism:

$$CH + N_2 \rightleftarrows HCN + N \tag{55}$$

$$C_2 + N_2 \rightleftarrows 2CN \tag{56}$$

The N atoms could form NO, in part at least, by reactions (50) and (51), and the CN could yield NO by oxygen or oxygen atom attack. It is well known that CH exists in flames and indeed, as stated in Chapter 4, is the molecule that gives the deep violet color to a Bunsen flame.

In order to verify whether reactions other than the Zeldovich mechanism were effective in NO formation, various investigators undertook the study of NO formation kinetics by use of shock tubes. The primary work in this area was that of Bowman and Seery [10] who studied the $CH_4–O_2–N_2$ system. Complex kinetic calculations of the $CH_4–O_2–N_2$ reacting system based on early kinetic rate data at a fixed high temperature and pressure similar to those obtained in a shock tube [11] for $T = 2477$ K and $P = 10$ atm are shown in Fig. 2. Even though more recent kinetic rate data would modify the product–time distribution somewhat, it is the general trends of the product distribution which are important and they are relatively unaffected by some changes in rates. These results are worth considering in their own right, for they show explicitly much that has been implied. Examination of Fig. 2 shows that at about 5×10^{-5} s, all the energy-release reactions will have equilibrated before any significant amounts of NO have formed; and, indeed, even at 10^{-2} s the NO has not reached its equilibrium concentration for $T = 2477$ K. These results show that for such homogeneous or near-homogeneous reacting systems, it would be possible to quench the NO reactions, obtain the chemical heat release, and prevent NO formation. This procedure has been put in practice in certain combustion schemes.

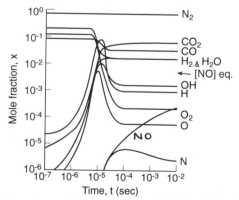

FIGURE 2   Concentration–time profiles in the kinetic calculation of the methane–air reaction at an inlet temperature of 1000 K. $P_2 = 10$ atm, $\phi = 1.0$, and $T_c = 2477$ (from Martenay [11]).

Equally important is the fact that Fig. 2 reveals large overshoots within the reaction zone. If these occur within the reaction zone, the O atom concentration could be orders of magnitude greater than its equilibrium value, in which case this condition could lead to the prompt NO found in flames. The mechanism analyzed to obtain the results depicted in Fig. 2 was essentially that given in Section 3.D.3.a with the Zeldovich reactions. Thus it was thought possible that the Zeldovich mechanism could account for the prompt NO.

The early experiments of Bowman and Seery appeared to confirm this conclusion. Some of their results are shown in Fig. 3. In this figure the experimental points compared very well with the analytical calculations based on the Zeldovich mechanisms alone. The same computational program as that of Marteney [11] was used. Figure 3 also depicts another result frequently observed: fuel-rich systems approach NO equilibrium much faster than do fuel-lean systems [12].

Although Bowman and Seery's results would, at first, seem to refute the suggestion by Fenimore that prompt NO forms by reactions other than the Zeldovich mechanism, one must remember that flames and shock-tube–initiated reacting systems are distinctively different processes. In a flame there is a temperature profile that begins at the ambient temperature and proceeds to the flame temperature. Thus, although flame temperatures may be simulated in shock tubes, the reactions in flames are initiated at much lower temperatures than those in shock tubes. As stressed many time before, the temperature history frequently determines the kinetic route and the products. Therefore shock tube results do not prove that the Zeldovich mechanism alone determines prompt NO formation. The prompt NO could arise from other reactions in flames, as suggested by Fenimore.

Bachmeier *et al.* [13] appear to confirm Fenimore's initial postulates and to shed greater light on the flame NO problem. These investigators measured the prompt NO formed as a function of equivalence ratio for many hydrocarbon

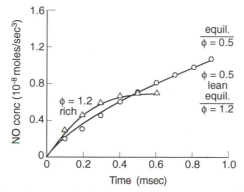

FIGURE 3 Comparison of measured and calculated NO concentration profiles for a CH_4–O_2–N_2 mixture behind reflected shocks. Initial post-shock conditions: $T = 2960$ K, $P = 3.2$ (from Bowman [12]).

compounds. Their results are shown in Fig. 4. What is significant about these results is that the maximum prompt NO is reached on the fuel-rich side of stoichiometric, remains at a high level through a fuel-rich region, and then drops off sharply at an equivalence ratio of about 1.4.

Bachmeier *et al.* also measured the HCN concentrations through propane–air flames. These results, which are shown in Fig. 5, show that HCN concentrations rise sharply somewhere in the flame, reach a maximum, and then decrease sharply. However, for an equivalence ratio of 1.5, a fuel-rich condition for which little prompt NO is found, the HCN continues to rise and is not depleted. The explanation offered for this trend is that HCN forms in all the rich hydrocarbon flames; however, below an equivalence ratio of 1.4, O radicals are present in sufficient abundance to deplete HCN and form the NO. Since the sampling and analysis techniques used by Bachmeier *et al.* [13] did not permit the identification of the cyanogen radical CN, the HCN concentrations found most likely represent the sum of CN and HCN as they exist in the flame. The CN and HCN in the flame are related through the

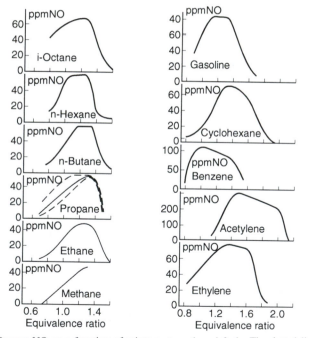

FIGURE 4   Prompt NO as a function of mixture strength and fuel. The dotted lines show the uncertainty of the extrapolation at the determination of prompt NO in propane flames; similar curves were obtained for the other hydrocarbons (from Bachmeier *et al.* [13]).

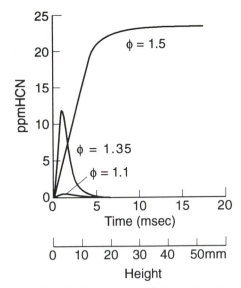

FIGURE 5 HCN profiles of fuel-rich propane–air flames (from Bachmeier *et al.* [13]).

rapid equilibrium reactions [14]

$$CN + H_2 \rightleftharpoons HCN + N \tag{57}$$

$$CN + H_2O \rightleftharpoons HCN + OH \tag{58}$$

The HCN concentration is probably reduced mainly by the oxidation of the CN radicals [14, 15].

From other more recent studies of NO formation in the combustion of lean and slightly rich methane–oxygen–nitrogen mixtures as well as lean and very rich hydrocarbon–oxygen–nitrogen mixtures, it must be concluded that some of the prompt NO is due to the overshoot of O and OH radicals above their equilibrium values, as the Bowman and Seery results suggested. But even though O radical overshoot is found on the fuel-rich side of stoichiometric, this overshoot cannot explain the prompt NO formation in fuel-rich systems. It would appear that both the Zeldovich and Fenimore mechanisms are feasible.

Some very interesting experiments by Eberius and Just [16] seem to clarify what is happening in the flame zone with regard to NO formation. Eberius and Just's experiments were performed on a flat flame burner with propane as the fuel. Measurements were made of the prompt NO at various fuel–oxygen equivalence ratios whose flame temperatures were controlled by dilution with nitrogen. Thus a range of temperatures could be obtained for a given propane–oxygen equivalence ratio. The results obtained are shown in Fig. 6. The highest temperature point for each equivalence ratio corresponds to zero dilution.

The shapes of the plots in Fig. 6 are revealing. At both the low- and high-temperature ends, all the plots seem nearly parallel. The slopes at the low-temperature end are very much less than the slopes at the high-temperature end, thereby indicating two mechanisms for the formation of prompt NO. The two mechanisms are not related solely to the fuel-rich and fuel-lean stoichiometry, as many investigators thought, but also to the flame temperature. These results suggest two routes—a high-temperature, high-activation route and a lower-temperature, low-activation route.

The systematic appearance of these data led Eberius and Just to estimate the activation energy for the two regions. Without correcting for diffusion, they obtained an activation energy of the order of 270 kJ/mol for the high-temperature zone. This value is remarkably close to the 315 kJ/mol activation energy required for the initiating step in the Zeldovich mechanism. Furthermore, diffusion corrections would raise the experimental value somewhat. The low-temperature region has an activation energy of the order of 50–60 kJ/mol. As will be shown later, radical attack on the cyano species is faster than oxygen radical attack on hydrogen. The activation energy of $O + H_2$ is about 33 kJ/mol; therefore, the HCN reaction should be less. Again, diffusion corrections for the oxygen atom concentration could lower the apparent activities of activation energies of 50–60 kJ/mol to below 34 kJ/mol. This crude estimate of the activation energy from Eberius and Just's low-temperature region, together with the formation of HCN found by the same

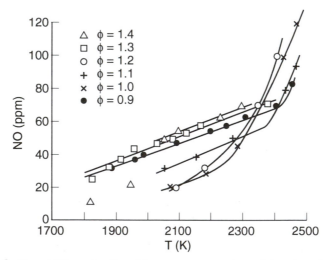

FIGURE 6   Prompt NO as a function of the temperature at various mixture strengths ϕ in adiabatic propane–synthetic air flames (from Eberius and Just [16]).

group [13] in their other flame studies with propane (Fig. 5), appears to indicate that the Fenimore mechanism [9] would hold in the lower-temperature region.

The kinetic details for prompt NO formation must begin with the possible reactions between N_2 and hydrocarbon fragments, as Fenimore [9] originally suggested. Hayhurst and Vance [17] suggest that two other likely candidate reactions may be added to those posited by Fenimore. The four candidate reactions would then be

$$C_2 + N_2 \longrightarrow CN + CN \tag{59}$$

$$C_2H + N_2 \longrightarrow HCN + CN \tag{60}$$

$$CH + N_2 \longrightarrow HCN + N \tag{61}$$

$$CH_2 + N_2 \longrightarrow HCN + NH \tag{62}$$

As discussed in the introduction to Chapter 4, the existence of C_2 and CH in hydrocarbon–air flames is well established. Methylene (CH_2) arises in most combustion systems by OH and H attack on the methyl radical (CH_3). Similar attack on CH_2 creates CH.

Hayhurst and Vance [17] established that the amount of prompt NO in moderately fuel-rich systems is proportional to the number of carbon atoms present per unit volume and is independent of the original parent hydrocarbon identity. This result indicates that reactions (59) and (60) are not primary contributors because it is unlikely that C_2 or C_2H could derive from CH_4 with an efficiency one-half that from C_2H_2 or one-third that from C_3H_6 or C_3H_8. In their complete model Miller and Bowman [6] introduced the reactions

$$CH_2N_2 \rightleftharpoons H_2CN + N \tag{63}$$

and

$$C + N_2 \rightleftharpoons CN + N \tag{64}$$

They conclude from estimated rates that reaction (63) is an insignificant contributor to prompt NO. However, they also point out that the reverse reaction (64) is very fast at room temperature and under shock tube conditions. Hence this step is a minor, but nonnegligible contributor to prompt NO and, because of the large endothermicity of reaction (64), its importance with respect to reaction (61) increases with increasing temperature.

The major products of reactions (61) and (62) are HCN and NH. It is to be noted that, experimentally, HCN and CN are indistinguishable in hydrocarbon flames because of the equilibrium reactions (57) and (58) mentioned with respect to the work of Bachmeier *et al.* [13]. In consideration of the work of Ref. [1], the major kinetic route to prompt NO would then appear to be

$$HCN + O \rightleftarrows NCO + H \tag{65}$$

$$NCO + H \rightleftarrows NH + CO \tag{66}$$

$$NH + (H, OH) \rightleftarrows N + (H_2, H_2O) \tag{67}$$

$$N + OH \rightleftarrows NO + H \tag{68}$$

$$N + O_2 \rightleftarrows NO + O \tag{69}$$

Following the conclusions of Bowman [1], then, from the definition of prompt NO, these sources of prompt NO in hydrocarbon fuel combustion can be identified: (1) nonequilibrium O and OH concentrations in the reaction zone and burned gas, which accelerate the rate of the thermal NO mechanism; (2) a reaction sequence, shown in Fig. 7, that is initiated by reactions of hydrocarbon radicals, present in and near the reaction zone, with molecular nitrogen (the Fenimore prompt-NO mechanism); and (3) reaction of O atoms with N_2 to form N_2O via the three-body recombination reaction,

$$O + N_2 + M \rightarrow N_2O + M$$

and the subsequent reaction of the N_2O to form NO via

$$N_2O + O \rightarrow NO + NO$$

The relative importance of these three mechanisms in NO formation and the total

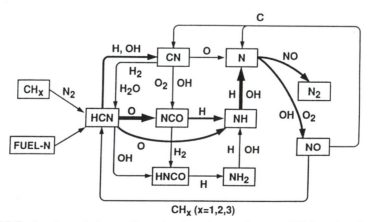

FIGURE 7   Reaction path diagram illustrating the major steps in prompt NO formation, the conversion of fuel nitrogen in flames, and reburning (from Bowman [12]).

amount of prompt NO formed depend on conditions in the combustor. Acceleration of NO formation by nonequilibrium radical concentrations appears to be more important in non-premixed flames, in stirred reactors for lean conditions, and in low- pressure premixed flames, accounting for up to 80% of the total NO formation. Prompt NO formation by the hydrocarbon radical–molecular nitrogen mechanism is dominant in fuel-rich premixed hydrocarbon combustion and in hydrocarbon diffusion flames, accounting for greater than 50% of the total NO formation. Nitric oxide formation by the N_2O mechanism increases in importance as the fuel–air ratio decreases, as the burned-gas temperature decreases, or as pressure increases. The N_2O mechanism is most important under conditions where the total NO formation rate is relatively low [1].

Fuel-bound nitrogen NO mechanisms   In several recent experiments, it has been shown that NO emissions from combustion devices that operate with nitrogen-containing compounds in the fuel are high; in other words, fuel-bound nitrogen is an important source of NO. The initial experiments of Martin and Berkau [18] commanded the greatest interest. These investigators added 0.5% pyridine to base oil and found almost an order of magnitude increase over the NO formed from base oil alone. Their results are shown in Fig. 8.

During the combustion of fuels containing bound nitrogen compounds, the nitrogen compounds most likely undergo some thermal decomposition prior to entering the combustion zone. Hence, the precursors to NO formation will, in general, be low-molecular-weight, nitrogen-containing compounds or radicals (NH_3, NH_2, NH, HCN, CN, etc.). All indications are that the oxidation of fuel-bound

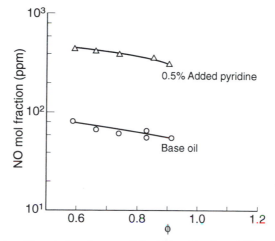

FIGURE 8   Nitric oxide emissions from an oil-fired laboratory furnace (from Martin and Berkau [18]).

nitrogen compounds to NO is rapid and occurs on a time scale comparable to the energy-release reactions in the combustion systems. This conclusion arises from the fact that the NH and CN oxidation reactions discussed in the previous section are faster [19, 20] than the important chain-branching reaction

$$O + H_2 \rightarrow OH + H \tag{70}$$

Thus, the reaction system cannot be quenched to prevent NO formation from fuel-bound nitrogen, as is the case with atmospheric nitrogen. In fact, in the vicinity of the combustion zone, observed NO concentrations significantly exceed calculated equilibrium values. In the post-combustion zone, the NO concentration decreases relatively slowly for fuel-lean mixtures and more rapidly for fuel-rich mixtures. Recall Bowman and Seery's results (Fig. 3) showing that fuel-rich systems approach equilibrium faster. When fuel-nitrogen compounds are present, high NO yields are obtained for lean and stoichiometric mixtures and relatively lower yields are found for fuel-rich mixtures. The NO yields appear to be only slightly dependent on temperature, thus indicating a low-activation-energy step. This result should be compared to the strong temperature dependence of NO formation from atmospheric nitrogen.

The high yields on the lean side of stoichiometric pose a dilemma. It is desirable to operate lean to reduce hydrocarbon and carbon monoxide emissions; but with fuel containing bound nitrogen, high NO yields would be obtained. The reason for the superequilibrium yields is that the reactions leading to the reduction of NO to its equilibrium concentration, namely,

$$O + NO \rightarrow N + O_2 \tag{71}$$

$$NO + NO \rightarrow N_2O + O \tag{72}$$

$$NO + RH \rightarrow products \tag{73}$$

are very slow. NO can be reduced under certain conditions by CH and NH radicals, which can be present in relatively large concentrations in fuel-rich systems. These reduction steps and their application will be discussed later.

The extent of conversion of fuel nitrogen to NO is nearly independent of the parent fuel molecule, but is strongly dependent on the local combustion environment and on the initial fuel nitrogen in the reactant. Unlike sulfur in the fuel molecule, nitrogen is much more tightly bound in the molecule and, for the most part, in an aromatic ring [21]. Regardless, all fuel-nitrogen compounds exhibit solely carbon–nitrogen or nitrogen–hydrogen bonding. Thus, it is not surprising that in the oxidation of fuel-nitrogen compounds, the major intermediates are HCN and CN and amine radicals stemming from an ammonia structure, i.e., NH_2, NH, and N.

In a large radical pool, there exists an equilibrium

$$NH_i + X \rightleftharpoons NH_{i-1} + XH \tag{74}$$

which essentially establishes an equilibrium between all NH compounds, i.e.,

$$NH_3 \rightleftarrows NH_2 \rightleftarrows NH \rightleftarrows N \qquad (75)\text{–}(77)$$

Consequently, reactions (61) and (62) can be written as a generalized reaction

$$CH_i + N_2 \rightarrow HCN + NH_{i-1} \qquad (78)$$

Thus, there is great similarity between the prompt-NO reactions discussed and the fuel-nitrogen reactions.

In the combustion of fuel-nitrogen compounds, the equilibrium

$$HCN + H \rightleftarrows CN + H_2 \qquad (57)$$

will certainly exist. Thus, the conversion of all relevant intermediates to NO has essentially been discussed, and only the NH_2 reactions remain to be considered. These reactions follow the sequence

$$NH_2 + H \rightarrow NH + H_2 \qquad (79)$$

$$NH_2 + OH \rightarrow NH + H_2O \qquad (80)$$

$$NH + \left\{ \begin{array}{c} H \\ OH \end{array} \right\} \rightarrow N + \left\{ \begin{array}{c} H_2 \\ H_2O \end{array} \right\} \qquad (81)$$

The general scheme of the fuel-nitrogen reactions is also represented in Fig. 7.

In fuel-rich systems, there is evidence [8, 22, 23] that the fuel-nitrogen intermediate reacts not only with oxidizing species in the manner represented, but also competitively with NO (or another nitrogen intermediate) to form N_2. This second step, of course, is the reason that NO yields are lower in fuel-rich systems. The fraction of fuel nitrogen converted to NO in fuel-rich systems can be as much as an order of magnitude less than that of lean or near-stoichiometric systems. One should realize, however, that even in fuel-rich systems, the exhaust NO concentration is substantially greater than its equilibrium value at the combustion temperature.

Haynes *et al.* [14] have shown that when small amounts of pyridine are added to a premixed, rich ($\phi = 1.68$; $T = 2030$ K) ethylene–air flame, the amount of NO increases with little decay of NO in the post-flame gases. However, when larger amounts of pyridine are added, significant decay of NO is observed after the reaction zone. When increasingly higher amounts of pyridine are added, high concentrations of NO leave the reaction zone, but this concentration drops appreciably in the post-flame gases to a value characteristic of the flame, but well above the calculated equilibrium value. Actual experimental results are shown in Fig. 9.

In fuel-rich systems, the conversion reactions of the fuel-nitrogen intermediates are subject to doubt, mainly because the normal oxidizing species O_2, O, and OH are present only in very small concentrations, particularly near the end of the

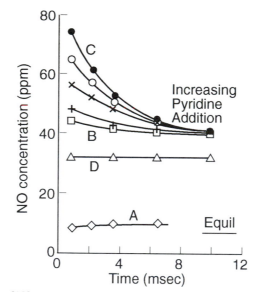

FIGURE 9   Effect of NO concentrations leaving the reaction zones of an ethylene–air flame (ϕ = 1.68, T = 2030 K) with various pyridine additions. Curve A, no pyridine addition; curves B and C, 0.1–0.5 N by weight of fuel; and curve D, NO addition to the fuel–air mixture (from Haynes *et al.* [14]).

reaction zone. Haynes *et al.* [14] offer the interesting suggestion that the CN can be oxidized by CO_2 since the reaction

$$CN + CO_2 \rightarrow OCN + CO \qquad (82)$$

is 85 kJ/mol exothermic and estimated to be reasonably fast.

b. Nitrogen Dioxide Reaction Mechanisms

Significant concentrations of NO_2 have been reported in the exhaust of gas turbines and in the products of range-top burners [21]. These results are surprising because chemical equilibrium considerations reveal that the NO_2/NO ratio should be negligibly small for typical flame temperatures. Furthermore, when kinetic models are modified to include NO_2 formation and reduction, they show that the conversion of NO to NO_2 can be neglected in practical devices.

However, in the case of sampling from gas turbines, NO_2 can vary from 15 to 50% of the total NO_x, depending on the NO level [21, 24, 25]. In the case of range-type burners, the NO_x has been reported as high as 15 to 20 times the NO levels in parts of the flame surrounding the burner top [26, 27].

Merryman and Levy [28] examined both NO and NO_2 formation in a flat flame burner operated near stoichiometric. In the low-temperature regime of visible flames, they found large concentrations of HO_2 that can react with the NO formed in the high-temperature regime and diffuse back to the lower-temperature zone. Their measurements showed that NO_2 is produced in the visible regime of all air flames (with and without fuel-bound nitrogen) and that NO is observed only in the visible region when fuel-bound nitrogen is present. Furthermore, these investigators found that NO_2 is consumed rapidly in the near–post-flame zone, whereupon the NO concentration rises correspondingly. They postulated the following scheme to represent their findings:

$$\left.\begin{array}{l} HN \\ CN \end{array}\right\} + O_2 \rightarrow NO + \cdots \tag{83}$$

$$NO + HO_2 \rightarrow NO_2 + OH \tag{84}$$

$$NO_2 + O \rightarrow NO + O_2 \tag{85}$$

In light of Fig. 7, whether reaction (83) should be the representative reaction for NO formation is irrelevant.

The significant step is represented by reaction (84). One should recall that there can be appreciable amounts of HO_2 in the early parts of a flame. The appearance of the NO_2 is supported further by the fact that reaction (84) is two orders of magnitude faster than reaction (85). The importance of the hydroperoxy radical attack on NO appeared to be verified by the addition of NO to the cold-fuel mixtures in some experiments. In these tests, the NO disappeared before the visible region was reached in oxygen-rich and stoichiometric flames, i.e., flames that would produce HO_2. The NO_2 persists because, as mentioned previously, its reduction to N_2 and O_2 is very slow. The role of HO_2 would not normally be observed in shock-tube experiments owing to the high temperatures at which they usually operate.

The Merryman–Levy sequence could explain the experimental results that show high NO_2/NO ratios. For the experiments in which these high ratios were found, it is quite possible that reaction (85) is quenched, in which case the NO_2 is not reduced. Cernansky and Sawyer [29], in experiments with turbulent diffusion flames, also concluded that the high levels of NO_2 found were due to the reactions of NO with HO_2 and O atoms.

The experimental efforts reporting high NO_2 levels have come into question because of the possibility that much of the NO_2 actually forms in sampling tubes [30, 31]. Optical techniques are now being applied; but, unfortunately, the low concentrations of NO_2 that exist make resolution of the controversy very difficult.

c. Nitrous Oxide Formation Mechanisms

Quoting directly from Bowman [1],

> [T]he principal gas-phase reactions forming N_2O in fossil fuel combustion are

$$NCO + NO \rightarrow N_2O + CO$$

$$NH + NO \rightarrow N_2O + H$$

In natural gas combustion an increasingly important contribution from

$$O + N_2 + M \rightarrow N_2O + M$$

occurs in fuel-lean mixtures and at low temperature and elevated pressures. The primary N_2O removal steps are

$$H + N_2O \rightarrow N_2 + OH$$

and

$$O + N_2O \rightarrow N_2 + O_2$$
$$\rightarrow NO + NO$$

Calculated lifetimes of N_2O in combustion products indicate that for temperatures above 1500 K, the lifetime of N_2O typically is less than 10 ms, suggesting that except for low temperature combustion, as found in fluidized bed combustors and in some post-combustion NO removal systems, N_2O emissions should not be significant, a conclusion that is in agreement with the most recent measurements of N_2O emissions from combustion devices.

4. The Reduction of NO_x

Because of the stringent emissions standards imposed on both mobile and stationary power sources, methods for reducing NO_x must be found; moreover, such methods should not impair the efficiency of the device. The simplest method of reducing NO_x, particularly from gas turbines, is by adding water to the combustor can. Water vapor can reduce the O radical concentration by the following scavenging reaction:

$$H_2O + O \rightarrow 2OH \tag{86}$$

Fortunately, OH radicals do not attack N_2 efficiently. However, it is more likely that the effect of water on NO_x emissions is through the attendant reduction in combustion temperature. NO_x formation from atmospheric nitrogen arises primarily from the very temperature-sensitive Zeldovich mechanism.

The problem of NO_x reduction is more difficult in heterogeneous systems such as those which arise from direct liquid fuel injection and which are known to burn as diffusion flames. One possible means is to decrease the average droplet size formed from injection. Kesten [32] and Bracco [33] have shown that the amount of NO formed from droplet diffusion flames can be related to the droplet size; viz., one large droplet will give more NO than can be obtained from a group of smaller droplets whose mass is equal to that of the larger droplet. Any means of decreasing the heterogeneity of a flame system will decrease the NO_x. Another possible practical scheme is to emulsify the fuel with a higher vapor-pressure, nonsoluble component such as water. It has been shown [34] that droplets from such emulsified fuels explode after combustion has been initiated. These microexplosions occur when the superheated water within the fuel droplet vaporizes, hence appreciably decreasing the heterogeneity of the system. A further benefit is obtained not only because the water is available for dilution, but also because the water is present in the immediate vicinity of the diffusion flame.

If it is impossible to reduce the amount of NO_x in the combustion section of a device, the NO_x must be removed somewhere in the exhaust. Myerson [35] has shown that it is possible to reduce NO_x by adding small concentrations of fuel and oxygen. The addition of about 0.1% hydrocarbon (isobutane) and 0.4% O_2 to a NO_x-containing system at 1260 K reduced the NO_x concentration by a factor of 2 in about 125 ms. Myerson [35] found that the ratio of O_2/HC was most important. When the concentrations of O_2 and the hydrocarbon are large, a HCN-formation problem could arise. This procedure is feasible only for slightly fuel-lean or fuel-rich systems. The oxygen is the creator and the destroyer of other species involved in the NO reduction. This fact, in turn, means that the initial addition of O_2 to the hydrocarbon–NO mixture promotes the production of the strongly reducing species CH and CH_2 and similar substituted free radicals that otherwise must be produced by slower pyrolysis reactions.

Continued addition of O_2 beyond one-half the stoichiometric value with the hydrocarbons present encourages a net destruction of the hydrocarbon radicals. For the temperature range 1200–1300 K, production of the hydrocarbon radicals via hydrogen abstraction by O_2 is rapid, even assuming an activation energy of 520 kJ/mol, and more than adequate to provide sufficient radicals for NO reduction in the stay time range of 125 ms.

Myerson postulated that the following reactions are involved:

$$CH + NO \rightarrow HCO + N + 217\ kJ \tag{87}$$

$$CH + NO \rightarrow HCN + O + 305\ kJ \tag{88}$$

The exothermicity of reaction (87) is sufficient to fragment the formyl radical and could be written as

$$CH + NO \rightarrow H + CO + N + 104\ kJ \tag{89}$$

In the absence of O_2, the N radicals in these fuel-rich systems can react rapidly with NO via

$$N + NO \rightarrow N_2 + O + 305\ kJ \tag{90}$$

Another technique currently in practice is known as the thermal $DeNO_x$ process [36], which uses ammonia as the NO_x reduction agent. The ammonia is injected into the exhaust gases of stationary power plants burning fossil fuels. The process is effective in a narrow temperature range, about $T \sim 1250$ K. Below about 1100 K, the reaction takes place too slowly to be of value, and about 1400 K more NO is formed. Miller *et al.* [37] discovered that if H_2 is added to the system, the center of the temperature window moves to a lower value without changing the width of the window. They also found that slightly lean combustion products appear to be required for the reduction reaction to be effective; that is, the process is implemented under excess oxygen conditions. Increasing the NH_3 concentration to a comparable O_2 concentration inhibits the process under certain conditions, and the presence of water slightly inhibits the NO reduction because the optimum temperature is increased slightly [6].

Explaining these effects has been one of the successes of kinetic modeling [6]. The ammonia in the process is considered to form the amine radical NH_2, which reacts with the NO to form an intermediate that decays into products:

$$NH_2 + NO \rightarrow \quad \overset{H}{\underset{H}{\diagdown}}N{-}N{=}O \rightarrow products \tag{91}$$

There are various possibilities as to the fate of the intermediate. It could form $HNNO + H$. However, this route is unlikely because no H atoms are found at low temperatures; nor is there any N_2O, which would inevitably have to form. In addition, this decay step is endothermic. Another possibility is the formation of N_2O and H_2, which is exothermic. But there is a large energy barrier involved in forming H_2. Also, of course, no N_2O is found. The formation of $H_2N{=}N$ and O is very endothermic and is not conceivable.

The possibility exists of a migration of a H atom. Because the migration step

$$\overset{H}{\underset{H}{\diagdown}}N{=}N{-}O \rightarrow \quad \overset{H}{\diagdown}N{-}N\overset{O}{\underset{H}{\diagup}} \rightarrow \quad \overset{H}{\diagdown}N{=}N + O\underset{H}{\diagdown} \tag{92}$$

is also very endothermic, it is ruled out. But what appears to be most feasible is the migration of H to the O atom in the following step:

$$\overset{H}{\underset{H}{\diagdown}}N{=}N{-}O \rightarrow \quad \overset{H}{\diagdown}N{=}N\underset{OH}{\diagdown} \rightarrow HNN + OH \tag{93}$$

The product HNN provides a plausible route for the overall NO reduction mechanism, which permits the determination of the temperature window. Miller and Bowman [6] proposed the competitive channels shown in Table 2 as the explanation.

The thermal $DeNO_x$ system removes NO in practical systems because the $NH_2 + NO$ initiates a significant chain-branching system, thereby allowing the overall reaction sequence to be self-sustaining. Following the general scheme in

TABLE 2   NO Reduction Scheme

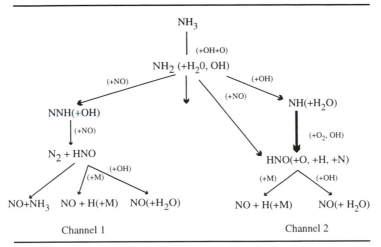

Table 2, the conversion of NH_3 to NH_2 occurs principally by reaction with OH:

$$NH_3 + OH \rightleftharpoons NH_2 + H_2O$$

But, in the absence of water vapor, it may occur by reaction with O atoms:

$$NH_3 + O \rightleftharpoons NH_2 + OH$$

The required chain branching to regenerate OH and O and hence to continue the conversion is accomplished [6] by the reaction sequence for Channel 1:

$$NH_2 + NO \rightleftharpoons NNH + OH$$

$$NNH + NO \rightleftharpoons N_2 + HNO$$

$$HNO + M \rightleftharpoons H + NO + M$$

The H atom produced in the last step reacts with O_2 in the familiar chain-branching step

$$H + O_2 \rightleftharpoons OH + O$$

giving an overall chain-branching radical pool. In the presence of the water inherent in the product composition of most combustion systems, the system is further augmented by the reaction of O atoms with water via

$$O + H_2O \rightleftharpoons OH + OH$$

and OH becomes the dominant species converting NH_3 to NH_2.

In their model, Miller and Bowman [6] consider the important chain-termination steps to be

$$NH_2 + NO \rightleftharpoons N_2 + H_2O$$

$$NH_2 + HNO \rightleftharpoons NH_3 + NO$$

and

$$OH + HNO \rightleftharpoons H_2O + NO$$

They conclude that, at the low-temperature end of the effective temperature window, the NO reduction effectiveness is limited principally by the rates of the chain-termination reactions that compete with the preceding branching sequence. In addition, below about 1100 K, hydrogen abstraction by OH is so slow that little NH_2 forms and $H + O_2 + M \rightarrow HO_2 + M$ becomes a competitive step for $H + O_2 \rightarrow OH + O$. In the temperature range 1100–1400 K, the mix of branching and termination is proper for the conversion of NH_3 to NH_2 at a sufficiently rapid and sustaining rate. However, above about 1400 K, Channel 2 becomes dominant. At such temperatures, there is more chain branching, which leads to a higher concentration of OH. Thus, the Channel 2 reaction

$$NH_2 + OH \rightleftharpoons NH + H_2O$$

is favored over the Channel 1 reaction

$$NH_2 + NO \rightarrow NNH + OH$$

These two reactions are competitive around 1250 K. Note that in Channel 1, two NO molecules react to form one N_2 and one NO so that an overall reduction in NO is obtained. The role of added H_2 manifests itself by increasing the amount of chain branching at the lower temperatures and increasing the overall concentration of OH. Thus the window shifts to lower temperatures.

Other post-combustion NO_x removal techniques include the injection of urea ($[NH_2]_2CO$) and cyanuric acid ($[HOCN]_3$). The latter is termed the RAPRENO process [37a]. When heated, cyanuric acid sublimes and decomposes to form isocyanic acid HNCO. Similarly, urea reacts to form NH_3 and HNCO. Thus its NO_x reduction path follows that of the thermal DeNO$_x$ route as well as that of the RAPRENO route to be discussed.

The major route in the RAPRENO process is the radical attack on the isocyanuric acid by H and OH via

$$H + HNCO \rightleftharpoons NH_2 + CO$$

$$OH + HNCO \rightleftharpoons NCO + H_2O$$

The first of these two steps forms the amine radical NH_2 and it acts as in the thermal DeNO$_x$ process. The importance of the CO is that its oxidation produces H atoms from the well known step

$$CO + OH \rightleftharpoons CO_2 + H$$

The importance of the second step is that it provides the primary NO removal step [6]

$$NCO + NO \rightleftharpoons N_2O + CO$$

for the process. The N$_2$O decays, as discussed earlier, to form N$_2$. The advantage of the RAPRENO process may be its ability to remove NO$_x$ at much lower temperatures than the thermal DeNO$_x$ process. There is some indication that this lower-temperature effect may be to decompose cyanuric acid on surfaces.

Considering that wet scrubbers are in place in many facilities and more are planned for the future, another efficient means for NO$_x$ removal could be considered. These scrubbing methods are limited by the relatively inert nature of NO. It has been proposed that this difficulty can be overcome by the conversion of NO to the much more active NO$_2$ through reaction (84), as discussed earlier:

$$NO + HO_2 \rightarrow NO_2 + OH \qquad (84)$$

The question arises as to how to produce HO$_2$ radicals. For post-combustion conversion, the obvious candidate would be an aqueous solution of hydrogen peroxide H$_2$O$_2$. Although hydrogen peroxide readily converts to water and oxygen through a heterogeneous decomposition, its homogeneous decomposition route follows the simple steps

$$H_2O_2 + M \rightarrow 2OH + M$$

$$OH + H_2O_2 \rightarrow H_2O + HO_2$$

to form the necessary hydroperoxy radical HO$_2$. However, one must realize that this simple reaction sequence is so slow that it is ineffective below 600 K. At high temperatures, particularly at atmospheric pressure, HO$_2$ dissociates. Moreover, it would appear that at temperatures above 1100 K, the general radical pool is large enough that recombination reactions become too competitive. Thus, this aqueous process has an effective temperature window between 600 and 1100 K [6].

D. SO$_x$ EMISSIONS

Sulfur compounds pose a dual problem. Not only do their combustion products contribute to atmospheric pollution, but these products are also so corrosive that they cause severe problems in the operation of gas turbines and industrial power plants. Sulfur pollution and corrosion were recognized as problems long before the nitrogen oxides were known to affect the atmosphere. For a time, the general availability of low-sulfur fuels somewhat diminished the general concern with respect to the sulfur. However, the possibility that China is developing its huge coal resources has again raised the specter of massive sulfur oxide emissions. Sulfur may be removed from residual oils by catalytic hydrodesulfurization techniques, but the costs of this process are high and the desulfurized residual oils have a tendency to become "waxy" at low temperatures. To remove sulfur from coal is an even more imposing problem. It is possible to remove pyrites from coal, but this approach is limited by the size of the pyrite particles.

Unfortunately, pyrite sulfur makes up only half the sulfur content of coal, while the other half is organically bound. Coal gasification is the only means by which this sulfur mode can be removed. Of course, it is always possible to eliminate the deleterious effects of sulfur by removing the major product oxide SO_2 by absorption processes. These processes impose large initial capital investments.

The presence of sulfur compounds in the combustion process can affect the nitrogen oxides, as well. Thus, it is important to study sulfur compound oxidation not only to find alternative or new means of controlling the emission of objectionable sulfur oxides, but also to understand their effect on the formation and concentration of other pollutants, especially NO_x.

There are some very basic differences between the sulfur problem and that posed by the formation of the nitrogen oxides. Nitrogen in any combustion process can be either atmospheric or organically bound. Sulfur can be present in elemental form or organically bound, or it may be present as a species in various inorganic compounds. Once it enters the combustion process, sulfur is very reactive with oxidizing species and, in analogy with fuel nitrogen, its conversion to the sulfurous oxides is fast compared to the other energy-releasing reactions.

Although sulfur oxides were recognized as a problem in combustion processes well before the concern for photochemical smog and the role of the nitrogen oxides in creating this smog, much less is understood about the mechanisms of sulfur oxidation. Indeed, the amount of recent work on sulfur oxidation has been minimal. The status of the field has been reviewed by Levy *et al.* [38] and Cullis and Mulcahy [39] and much of the material from the following subsections has been drawn from Cullis and Mulcahy's article.

1. The Product Composition and Structure of Sulfur Compounds

When elemental sulfur or a sulfur-bearing compound is present in any combustion system, the predominant product is sulfur dioxide. The concentration of sulfur trioxide found in combustion systems is most interesting. Even under very lean conditions, the amount of sulfur trioxide formed is only a few percent of that of sulfur dioxide. Generally, however, the sulfur trioxide concentration is higher than its equilibrium value, as would be expected from the relation

$$SO_2 + \tfrac{1}{2}O_2 \rightleftharpoons SO_3 \qquad (94)$$

These higher-than-equilibrium concentrations may be attributable to the fact that the homogeneous reactions which would reduce the SO$_3$ to SO$_2$ and O$_2$ are slow. This point will be discussed later in this section.

It is well known that SO$_3$ has a great affinity for water and that at low temperatures it appears as sulfuric acid H$_2$SO$_4$. Above 500°C, sulfuric acid dissociates almost completely into sulfur trioxide and water.

Under fuel-rich combustion conditions, in addition to sulfur dioxide, the stable sulfur products are found to be hydrogen sulfide, carbonyl sulfide, and elemental sulfur.

Owing to their reactivity, other oxides of sulfur may appear only as intermediates in various oxidation reactions. These are sulfur monoxide SO, its dimer (SO)$_2$, and disulfur monoxide S$_2$O. Some confusion has attended the identification of these oxides; for example, what is now known to be S$_2$O was once thought to be SO or (SO)$_2$. The most important of these oxides is sulfur monoxide, which is the crucial intermediate in all high-temperature systems. SO is a highly reactive radical whose ground state is a triplet which is electronically analogous to O$_2$. According to Cullis and Mulcahy [39], its lifetime is seldom longer than a few milliseconds. Spectroscopic studies have revealed other species in flames, such as CS, a singlet molecule analogous to CO and much more reactive; S$_2$, a triplet analogous to O$_2$ and the main constituent of sulfur vapor above 600°C; and the radical HS. Johnson *et al.* [40] calculated the equilibrium concentration of the various sulfur species for the equivalent of 1% SO, in propane–air flames. Their results, as a function of fuel–air ratio, are shown in Fig. 10. The dominance of SO$_2$ in the product composition for these equilibrium calculations, even under deficient air conditions, should be noted. As reported earlier, practical systems reveal SO$_3$ concentrations that are higher (1–2 %) than those depicted in Fig. 10.

Insight into much that has and will be discussed can be obtained by the study of the structures of the various sulfur compounds given in Table 3.

2. Oxidative Mechanisms of Sulfur Fuels

Sulfur fuels characteristically burn with flames that are pale blue, sometimes very intensely so. This color comes about from emissions as a result of the reaction

$$O + SO \rightarrow SO_2 + h\nu \qquad (95)$$

and, since it is found in all sulfur-fuel flames, this blue color serves to identify SO as an important reaction intermediate in all cases.

Most studies of sulfur-fuel oxidation have been performed using hydrogen sulfide, H$_2$S, as the fuel. Consequently, the following material will concentrate on understanding the H$_2$S oxidation mechanism. Much of what is learned from this mechanism can be applied to understanding the combustion of COS and CS$_2$ as well as elemental and organically bound sulfur.

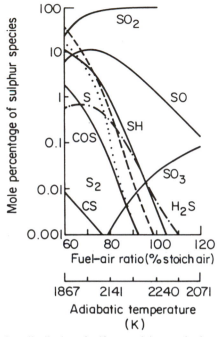

FIGURE 10   Equilibrium distribution of sulfur-containing species in propane–air flames with unburned gases initially containing 1% SO_2 (from Johnson *et al.* [40]).

TABLE 3   Structure of Gaseous Sulfur Compounds

Sulfur$_8$	*Rhombic*		
Sulfur monoxide SO	S=O		
Sulfur dioxide SO_2(OSO)	S⟨118°	$^+$S⟨	
Sulfur superoxide SO_2(SOO)	$^+$S=O—O$^-$		
Sulfur trioxide SO_3	O=S⟨120°	O=S^{++}	
Sulfur suboxide S_2O	S=S=O		
Carbonyl sulfide COS	S=C=O	$^-$S—C≡O$^+$	$^+$S≡C—O$^-$
Carbon disulfide CS_2	S=C=S	$^-$S—C≡S$^+$	
Organic thiols	R′—SH		
Organic sulfides	R′—S—R′		
Organic disulfides	R′—S—S—R′		

a. H$_2$S

Figure 11 is a general representation of the explosion limits of H$_2$S/O$_2$ mixtures. This three-limit curve is very similar to that shown for H$_2$/O$_2$ mixtures. However, there is an important difference in the character of the experimental data that determine the H$_2$S/O$_2$ limits. In the H$_2$S/O$_2$ peninsula and in the third limit region, explosion occurs after an induction period of several seconds.

The main reaction scheme for the low-temperature oxidation of H$_2$S, although not known explicitly, would appear to be

$$H_2S + O_2 \rightarrow SH + HO_2 \qquad 176 \text{ kJ/mol} \qquad (96)$$

$$SH + O_2 \rightarrow SO + OH \qquad -88 \text{ kJ/mol} \qquad (97)$$

$$H_2S + SO \rightarrow S_2O + H_2 \qquad -29 \text{ kJ/mol} \qquad (98)$$

$$OH + H_2S \rightarrow H_2O + HS \qquad -125 \text{ kJ/mol} \qquad (99)$$

The addition of reaction (98) to this scheme is necessary because of the identification of S$_2$O in explosion limit studies. More importantly, Merryman and Levy

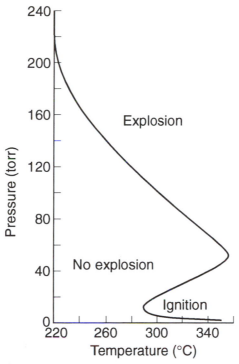

FIGURE 11 Approximate explosion limits for stoichiometric mixtures of hydrogen sulfide and oxygen.

[41] in burner studies showed that S_2O occurs upstream from the peak of the SO concentration and that elemental sulfur is present still further upstream in the preignition zone.

The most probable system for the introduction of elemental sulfur is

$$SH + SH \rightarrow H_2S + S \quad -13 \text{ kJ/mol} \tag{100}$$

$$S + SH \rightarrow S_2 + H \quad -63 \text{ kJ/mol} \tag{101}$$

Given the preignition zone temperatures and overall pressures at which the flame studies were carried out, it does not seem kinetically feasible that the reaction

$$S_2 + O + M \rightarrow S_2O + M \tag{102}$$

could account for the presence of S_2O. The disproportionation of SO would have to give SO_2 as well as S_2O. Since SO_2 is not found in certain experiments where S_2O can be identified, disproportionation would not be feasible, so reaction (98) appears to be the best candidate for explaining the presence of S_2O.

Reaction (97) is the branching step in the mechanism. It has been suggested that

$$SH + O_2 + M \rightarrow HSO_2 + M \tag{103}$$

competes with reaction (97), thus determining the second limit. Cullis and Mulcahy [39] suggested the reaction

$$S_2O + O_2 \rightarrow SO_2 + SO \tag{104}$$

as the degenerate branching step. The explicit mechanism for forming S_2O and its role in flame processes must be considered an uncertainty.

At higher temperatures, the reaction

$$O_2 + SO \rightarrow O + SO_2 \tag{105}$$

becomes competitive with reaction (98) and introduces O radicals into the system. The presence of O radicals gives another branching reaction, namely,

$$O + H_2S \rightarrow OH + SH \tag{106}$$

The branching is held in check by reaction (98), which removes SO, and the fast termolecular reaction

$$O + SO + M \rightarrow SO_2 + M \tag{107}$$

which removes both O radicals and SO.

In shock tube studies, SO_2 is formed before OH radicals appear. To explain this result, it has been postulated that the reaction

$$O + H_2S \rightarrow SO + H_2 \quad 222 \text{ kJ/mol} \tag{108}$$

is possible. This reaction and reaction (98) give the overall step

$$O_2 + H_2S \rightarrow SO_2 + H_2 \tag{109}$$

Detailed sampling in flames by Sachjan *et al.* [42] indicates that the H$_2$S is oxidized in a three-step process. During the first stage, most of the H$_2$S is consumed, and the products are mainly sulfur monoxide and water. In the second stage, the concentration of SO decreases, the concentration of OH reaches its maximum value, the SO$_2$ reaches its final concentration, and the concentration of the water begins to build as the hydrogen passes through a maximum.

The interpretation given to these results is that, during the first stage, the H$_2$S and O$_2$ are consumed mainly by reactions (108) and (105)

$$O + H_2S \rightarrow SO + H_2 \tag{108}$$

$$SO + O_2 \rightarrow SO_2 + O \tag{105}$$

with some degree of chain branching by reaction (106)

$$O + H_2S \rightarrow OH + SH \tag{106}$$

In the second stage, reaction (105) predominates over reaction (108) because of the depletion of the H$_2$S; then the OH concentration rises via reaction (106) and begins the oxidation of the hydrogen

$$O + H_2 \rightarrow OH + H \tag{110}$$

Of course, the complete flame mechanism must include

$$H + O_2 \rightarrow OH + O \tag{111}$$

$$OH + H_2 \rightarrow H_2O + H \tag{112}$$

Reactions (108) and (110) together with the fast reaction at the higher temperature

$$SO + OH \rightarrow SO_2 + H \tag{113}$$

explain the known fact that H$_2$S inhibits the oxidation of hydrogen.

Using laser fluorescence measurements on fuel-rich H$_2$/O$_2$/N$_2$ flames seeded with H$_2$S, Muller *et al.* [43] determined the concentrations of SH, S$_2$, SO, SO$_2$, and OH in the post-flame gases. From their results and an evaluation of rate constants, they postulated that the flame chemistry of sulfur under rich conditions could be described by the eight fast bimolecular reactions and the two three-body recombination reactions given in Table 4.

Reactions 1 and 6 in Table 4 were identified as the reactions that control the S$_2$ and SO$_2$ concentrations, respectively, while reactions 2 and 4 control the S concentration with some contribution from reaction 1. Reaction 6 was the major one for the SO flux rate. The three reactions involving H$_2$S were said to play an important role. SH was found to be controlled by reactions 1–5. Because

TABLE 4   Major Flame Chemistry Reactions of Sulfur under Rich Conditions

(1)	$H + S_2 \rightleftarrows SH + S$
(2)	$S + H_2 \rightleftarrows SH + H$
(3)	$SH + H_2 \rightleftarrows H_2S + H$
(4)	$S + H_2S \rightleftarrows SH + SH$
(5)	$OH + H_2S \rightleftarrows H_2O + SH$
(6)	$H + SO_2 \rightleftarrows SO + OH$
(7)	$S + OH \rightleftarrows SO + H$
(8)	$SH + O \rightleftarrows SO + H$
(9)	$H + SO_2 + M \rightleftarrows HSO_2 + M$
(10)	$O + SO_2 + M \rightleftarrows SO_3 + M$

the first eight reactions were found to be fast, it was concluded that they rapidly establish and maintain equilibrium so that the species S, S_2, H_2S, SH, SO, and SO_2 are efficiently interrelated. Thus relative concentrations, such as those of SO and SO_2, can be calculated from thermodynamic considerations at the local gas temperature from the system of reactions

$$SO_2 + H \rightleftarrows SO + OH \tag{114}$$

$$OH + H_2 \rightleftarrows H + H_2O \tag{112}$$

$$SO_2 + H_2 \rightleftarrows SO + H_2O \tag{115}$$

The three-body recombustion reactions listed in Table 4 are significant in sulfur-containing flames because one provides the homogeneous catalytic recombination of the important H_2–O_2 chain carriers H and OH via

$$H + SO_2 + M \rightleftarrows HSO_2 + M \tag{116}$$

$$H + HSO_2 \longrightarrow H_2 + SO_2 \tag{117}$$

$$OH + HSO_2 \longrightarrow H_2O + SO_2 \tag{118}$$

while the other

$$SO_2 + O + M \longrightarrow SO_3 + M \tag{119}$$

is the only major source of SO_3 in flames. Under rich conditions, the SO_3 concentration is controlled by other reactions with H and SO, i.e.,

$$SO + SO_3 \longrightarrow SO_2 + SO_2 \tag{120}$$

$$H + SO_3 \longrightarrow OH + SO_2 \tag{121}$$

and under lean conditions by

$$O + SO_3 \rightarrow SO_2 + O_2 \tag{122}$$

b. COS and CS$_2$

Even though there have been appreciably more studies of CS_2, COS is known to exist as an intermediate in CS_2 flames. Thus it appears logical to analyze the COS oxidation mechanism first. Both substances show explosion limit curves which indicate that branched-chain mechanisms exist. Most of the reaction studies used flash photolysis; hence very little information exists on what the chain-initiating mechanism for thermal conditions would be.

COS flames exhibit two zones. In the first zone, carbon monoxide and sulfur dioxide form; and in the second zone, the carbon monoxide is converted into carbon dioxide. Since these flames are hydrogen-free, it is not surprising that the CO conversion in the second zone is rapidly accelerated by adding a very small amount of water to the system.

Photolysis initiates the reaction by generating sulfur atoms

$$COS + h\nu \rightarrow CO + S \tag{123}$$

The S atom then engenders the chain-branching step

$$S + O_2 \rightarrow SO + O \quad -21 \text{ kJ/mol} \tag{124}$$

which is followed by

$$O + COS \rightarrow CO + SO \quad -213 \text{ kJ/mol} \tag{125}$$

$$SO + O_2 \rightarrow SO_2 + O \quad -13 \text{ kJ/mol} \tag{105}$$

At high temperatures, the slow reaction

$$O + COS \rightarrow CO_2 + S \quad -226 \text{ kJ/mol} \tag{126}$$

must also be considered.

Although the initiation step under purely thermally induced conditions such as those imposed by shocks has not been formulated, it is expected to be a reaction which produces O atoms. The high-temperature mechanism would then be reactions (105), and (124)–(126), with termination by the elimination of the O atoms.

For the explosive reaction of CS_2, Myerson *et al.* [44] suggested

$$CS_2 + O_2 \rightarrow CS + SOO \tag{127}$$

as the chain-initiating step. Although the existence of the superoxide, SOO, is not universally accepted, it is difficult to conceive a more logical thermal initiating step, particularly when the reaction can be induced in the 200–300°C range. The introduction of the CS by reaction (127) starts the following chain scheme:

$$CS + O_2 \rightarrow CO + SO \qquad -347\ kJ/mol \qquad (128)$$

$$SO + O_2 \rightarrow SO_2 + O \qquad -54\ kJ/mol \qquad (105)$$

$$O + CS_2 \rightarrow CS + SO \qquad -326\ kJ/mol \qquad (129)$$

$$CS + O \rightarrow CO + S \qquad -322\ kJ/mol \qquad (130)$$

$$S + O_2 \rightarrow SO + O \qquad -21\ kJ/mol \qquad (131)$$

$$S + CS_2 \rightarrow S_2 + CS \qquad -25\ kJ/mol \qquad (132)$$

$$O + CS_2 \rightarrow COS + S \qquad (133)$$

$$O + COS \rightarrow CO + SO \qquad (126)$$

$$O + S_2 \rightarrow SO + S$$

The high flammability of CS_2 in comparison to COS is probably due to the greater availability of S atoms. At low-temperatures, branching occurs in both systems via

$$S + O_2 \rightarrow SO + O \qquad (124)$$

Even greater branching occurs since in CS_2 one has

$$O + CS_2 \rightarrow CS + SO \qquad (129)$$

The comparable reaction for COS is reaction (125)

$$O + COS \rightarrow CO + SO \qquad (125)$$

which is not chain-branching.

c. Elemental Sulfur

Elemental sulfur is found in the flames of all the sulfur-bearing compounds discussed in the previous subsections. Generally, this sulfur appears as atoms or the dimer S_2. When pure sulfur is vaporized at low temperatures, the vapor molecules are polymeric and have the formula S_8. Vapor-phase studies of pure sulfur oxidation around 100°C have shown that the oxidation reaction has the characteristics of a chain reaction. It is interesting to note that in the explosive studies the reaction must be stimulated by the introduction of O atoms (spark, ozone) in order for the explosion to proceed.

Levy *et al.* [38] reported that Semenov suggested the following initiation and branching reactions:

$$S_8 \rightarrow S_7 + S \qquad (134)$$

$$S + O_2 \rightarrow SO + O \tag{124}$$

$$S_8 + O \rightarrow SO + S + S_6 \tag{135}$$

with the products produced by

$$SO + O \rightarrow SO_2^* \rightarrow SO_2 + h\nu \tag{95}$$

$$SO + O_2 \rightarrow SO_2 + O \tag{105}$$

$$SO_2 + O_2 \rightarrow SO_3 + O \tag{136}$$

$$SO_2 + O + M \rightarrow SO_3 + M \tag{137}$$

A unique feature of the oxidation of pure sulfur is that the percentage of SO_3 formed is a very much larger (about 20%) fraction of the SO_x than is generally found in the oxidation of sulfur compounds.

d. Organic Sulfur Compounds

It is more than likely that when sulfur occurs in a crude oil or in coal (other than the pyrites), it is organically bound in one of the three forms listed in Table 3—the thiols, sulfides, or disulfides. The combustion of these compounds is very much different from that of other sulfur compounds in that a large portion of the fuel element is a pure hydrocarbon fragment. Thus in explosion or flame studies, the branched-chain reactions that determine the overall consumption rate or flame speed would follow those chains characteristic of hydrocarbon combustion rather than the CS, SO, and S radical chains which dominate in H_2S, CS_2, COS, and S_8 combustion.

A major product in the combustion of all organic sulfur compounds is sulfur dioxide. Sulfur dioxide has a well-known inhibiting effect on hydrocarbon and hydrogen oxidation and, indeed, is responsible for a self-inhibition in the oxidation of organic sulfur compounds. This inhibition most likely arises from its role in the removal of H atoms by the termolecular reaction

$$H + SO_2 + M \rightarrow HSO_2 + M \tag{116}$$

HSO_2, a known radical which has been found in H_2–O_2–SO_2 systems, is sufficiently inert to be destroyed without reforming any active chain carrier. In the lean oxidation of the thiols, even at temperatures around 300°C, all the sulfur is converted to SO_2. At lower temperatures and under rich conditions, disulfides form and other products such as aldehydes and methanol are found. The presence of the disulfides suggests a chain-initiating step very similar to that of low-temperature hydrocarbon oxidation,

$$RSH + O_2 \rightarrow RS + HO_2 \tag{138}$$

Cullis and Mulcahy reported that this step is followed by

$$RS + O_2 \rightarrow R + SO_2 \tag{139}$$

to form the hydrocarbon radical and sulfur dioxide. One must question whether the SO_2 in reaction (139) is sulfur dioxide or not. If the O_2 strips the sulfur from the RS radical, it is more likely that the SO_2 is the sulfur superoxide, which would decompose or react to form SO. The SO is then oxidized to sulfur dioxide as described previously. The organic radical is oxidized, as discussed in Chapter 3. The radicals formed in the subsequent oxidation, of course, attack the original fuel to give the RS radical, and the initiating step is no longer needed.

The SH bond is sufficiently weaker than the CH bonds so that the RS radical would be the dominant species formed. At high temperatures, it is likely that the RS decay leads to the thioaldehyde

$$RS^{\cdot} + M \rightarrow R' - C{\overset{\displaystyle H}{\underset{\displaystyle S}{\big\langle}}} + H \tag{140}$$

The disappearance of the thioaldehyde at these temperatures would closely resemble that of the aldehydes; namely,

$$R - C{\overset{\displaystyle H}{\underset{\displaystyle S}{\big\langle}}} + X \rightarrow R\dot{C}S + XH \tag{141}$$

$$M + R\dot{C}S \rightarrow R + CS + M \tag{142}$$

Then the CS radical is oxidized, as indicated in the previous discussion on CS_2.

The disulfide forms in the thiol oxidation from the recombination of the two RS radicals

$$M + RS + RS \rightarrow RSSR + M \tag{143}$$

The principal products in the oxidation of the sulfides are sulfur dioxide and aldehydes. The low-temperature initiating step is similar to reaction (138), except that the hydrogen abstraction is from the carbon atom next to the sulfur atom; i.e.,

$$RCH_2SCH_2R + O_2 \rightarrow RCH_2S{-}CHR + HO_2 \tag{144}$$

The radical formed in reaction (144) then decomposes to form an alkyl radical and a thioaldehyde molecule; i.e.,

$$M + RCH_2S{-}CHR \rightarrow RCH_2 + RCHS + M \tag{145}$$

Both products in reaction (145) are then oxidized, as discussed.

The oxidation of the disulfides follows a similar route to the sulfide with an initiating step

$$RCH_2SSCH_2R + O_2 \rightarrow RCH_2S\!-\!S\!-\!CHR + HO_2 \qquad (146)$$

followed by radical decomposition

$$RCH_2S\!-\!SCH\!-\!R \rightarrow RCH_2S + RCHS \qquad (147)$$

The thiol is then formed by hydrogen abstraction

$$RCH_2S + RH \rightarrow RCH_2SH + R \qquad (148)$$

and the oxidation proceeds as described previously.

e. Sulfur Trioxide and Sulfates

As was pointed out earlier, the concentration of sulfur trioxide found in the combustion gases of flames, though small, is greater than would be expected from equilibrium calculations. Indeed, this same phenomenon exists in large combustors, such as furnaces, in which there is a sulfur component in the fuel used. The equilibrium represented by Eq. (94)

$$SO_2 + \tfrac{1}{2}O_2 \rightleftarrows SO_3 \qquad (94)$$

is shifted strongly to the left at high temperatures, so one would expect very little SO$_3$ in a real combustion environment. It is readily apparent, then, that the combustion chemistry involved in oxidizing sulfur dioxide to the trioxide is such that equilibrium cannot be obtained.

Truly, the most interesting finding is that the superequilibrium concentrations of SO$_3$ are very sensitive to the original oxygen concentration. Under fuel-rich conditions approaching even stoichiometric conditions, practically no SO$_3$ is found. In proceeding from stoichiometric to 1% excess air, a sharp increase in the conversion of SO$_2$ to SO$_3$ is found. Further addition of air causes only a slight increase; however, the effect of the excess nitrogen in reducing the temperature could be a moderating factor in the rate of increase. Figure 12, taken from the work of Barrett *et al.* [45] on hydrocarbon flames, characterizes the results generally found both in flame studies and in furnaces. Such results strongly indicate that the SO$_2$ is converted into SO$_3$ in a termolecular reaction with oxygen atoms:

$$O + SO_2 + M \rightarrow SO_3 + M \qquad (119)$$

It is important to note that the superequilibrium results are obtained with sulfur fuels, small concentrations of sulfur fuels added to hydrocarbons, SO$_2$ added to hydrocarbon, and so forth. Further confirmation supporting reaction (119) as the conversion route comes from the observation that in carbon monoxide flames the amount of SO$_3$ produced is substantially higher than in all other cases. It is well known that, since O atoms cannot attack CO directly, the SO$_3$ concentration is

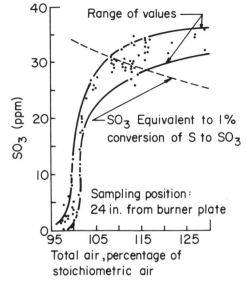

FIGURE 12   Effect of excess air on the formation of SO_3 in a hydrocarbon–air flame (after Barrett *et al.* [45]).

much higher in CO flames than in any other flames. The fact that in all cases the SO_3 concentration also increases with pressure supports a termolecular route such as reaction (119).

It is well known that the thermal dissociation of SO_3 is slow and that the concentration of SO_3 is therefore frozen within its stay time in flames and furnaces. The thermal dissociation rates are known, but one can also calculate the superequilibrium concentration of oxygen atoms in flames. If one does so, the SO_3 concentration should correspond to the equilibrium concentration given by reaction (119), in which the oxygen atom superequilibrium concentration is used. However, the SO_3 concentrations are never this high; thus, one must conclude that some SO_3 is being reduced by routes other than thermal decomposition. The two most likely routes are by oxygen and hydrogen atom attack on the sulfur trioxide via

$$O + SO_3 \rightarrow O_2 + SO_2 \qquad - 152 \text{ kJ/mol} \qquad (122)$$

$$H + SO_3 \rightarrow OH + SO_2 \qquad - 79 \text{ kJ/mol} \qquad (121)$$

Evidence supports this contention as well as the suggestion that reaction (122) would be more important than reaction (121) in controlling the SO_3 concentration with reaction (119). Furthermore, one must recognize that reactions (119) and (120) are effective means of reducing the O radical concentration. Since reaction (116) has been shown to be an effective means of reducing H radical concentrations, one can draw the important general conclusion that compounds reduce the extent of superequilibrium concentration of the characteristic chain-carrying radicals which exist in hydrocarbon flames.

In furnaces using residual oils, heterogeneous catalysis is a possible route for the conversion of SO$_2$ to SO$_3$. Sulfur dioxide and molecular oxygen will react catalytically on steel surfaces and vanadium pentoxide (deposited from vanadium compounds in the fuel). Catalytic reactions may also occur at lower temperatures where the equilibrium represented by reaction (94) favors the formation of SO$_3$.

If indeed SO$_2$ and SO$_3$ are effective in reducing the superequilibrium concentration of radicals in flames, sulfur compounds must play a role in NO formation from atmospheric nitrogen in flame systems. Since SO$_2$ and SO$_3$ form no matter what type of sulfur compound is added to combustion systems, these species should reduce the oxygen atom concentration and hence should inhibit NO formation. Wendt and Ekmann [46] have reported flame data which appear to substantiate this conclusion.

In examining reactions (121) and (122), one realizes that SO$_2$ plays a role in catalyzing the recombination of oxygen atoms. Indeed, this homogeneous catalytic recombination of oxygen atoms causes the decrease in the superequilibrium concentration of the oxygen atoms. SO$_3$ also plays a role in the recombination of hydrogen radicals through the route

$$H + SO_2 + M \rightarrow HSO_2 + M \qquad (116)$$

$$H + HSO_2 \rightarrow H_2 + SO_2 \qquad (117)$$

and in the recombination of hydrogen and hydroxyl radicals through the route of reaction (116) and

$$OH + HSO_2 \rightarrow H_2O + SO_2 \qquad (118)$$

Combining the preceding considerations of SO$_3$ formation with the realization that SO$_2$ and SO are the major sulfur oxide species in flames, one may use he term SO$_x$ to specify the sum of SO, SO$_2$, and SO$_3$. Also, considering the fact that in any combustion system sulfur can appear in the parent hydrocarbon fuel in various forms, it becomes evident that the details of the reaction mechanism for SO$_x$ formation are virtually impossible to specify. By virtue of the rapidity of the SO$_x$ formation process, Bowman [1] has argued that the need for a detailed fuel-sulfur oxidation model may be circumvented by postulating approximate models to estimate the gaseous SO$_x$ product distribution in the exhaust products. It has been suggested [46] that three principal assumptions would be involved in the proposed model: (1) the fuel-sulfur compounds should be considered minor species so that the major stable species concentrations would be those due to combustion of the hydrocarbon fuel; (2) the bimolecular H$_2$–O$_2$ reactions and reaction (115) would be partially equilibrated in the post-flame gases; (3) the SO$_3$ concentration would be calculated from reactions (119), (121), and (122). The SO$_x$ pool would then vary as the overall reaction approaches equilibrium by means of the H_2–O_2 radical recombination reactions, the reactions that determine the equilibrium between SO_2 and SO_3, and those that affect the catalytic recombination of H, O, and OH.

f. SO_x–NO_x Interactions

The reactions of fuel-sulfur and fuel-nitrogen are closely coupled to the fuel oxidation; moreover, sulfur-containing radicals and nitrogen-containing radicals compete for the available H, O, OH radicals with the hydrocarbons [21]. Because of this close coupling of the sulfur and nitrogen chemistry and the H, O, OH radical pool in flames, interactions between fuel-sulfur and fuel-nitrogen chemistry are to be expected.

The catalytic reduction of the radicals, particularly the O atom, by sulfur compounds will generally reduce the rates of reactions converting atmospheric nitrogen to NO by the thermal mechanism. However, experiments do not permit explicit conclusions [21]. For example, Wendt and Eckmann [46] showed that high concentrations of SO_2 and H_2S have an inhibiting effect on thermal NO in premixed methane–air flames, while deSoete [47] showed the opposite effect. To resolve this conflict, Wendt *et al.* [48] studied the influence of fuel-sulfur on fuel-NO in rich flames, whereupon they found both enhancement and inhibition.

A further interaction comes into play when the thermal DeNO$_x$ process is used to reduce NO$_x$. When stack gases cool and initial sulfur is present in the fuel, the SO_3 that forms reacts with water to form a mist of sulfuric acid, which is detrimental to the physical plant. Furthermore, the ammonia from the thermal DeNO$_x$ process reacts with water to form NH_4HSO_2—a glue-like, highly corrosive compound. These SO_3 conditions can be avoided by reducing the SO_3 back to SO_2. Under stack (post-combustion) temperatures, the principal elementary reactions for SO_3 to SO_2 conversion are

$$HO_2 + SO_3 \rightarrow HSO_3 + O_2$$

$$HSO_3 + M \rightarrow SO_2 + OH + M$$

Since the key to this sequence is the HO_2 radical, the aqueous hydrogen peroxide process discussed for NO to NO_2 conversion in the stack would be an appropriate approach. The SO_2 forms no corrosive liquid mist in the stack and could be removed by wet scrubbing of the exhaust.

E. PARTICULATE FORMATION

In earlier sections of this chapter, the role that particulates play in a given environmental scenario was identified. This section will be devoted exclusively to combustion-generated particulates whose main constituent is carbon. Those carbonaceous particulates that form from gas-phase processes are generally referred to as soot, and those that develop from pyrolysis of liquid hydrocarbon fuels are generally referred to as coke or cenospheres.

Although various restrictions have been placed on carbon particulate emissions from different types of power plants, these particles can play a beneficial, as well as a detrimental, role in the overall plant process. The detrimental effects are well known. The presence of particulates in gas turbines can severely affect the lifetime of the blades; soot particulates in diesel engines absorb carcinogenic materials, thereby posing a health hazard. It has even been postulated that, after a nuclear blast, the subsequent fires would create enormous amounts of soot whose dispersal into the atmosphere would absorb enough of the sun's radiation to create a "nuclear winter" on Earth. Nevertheless, particulates can be useful. In many industrial furnaces, for example, the presence of carbon particulates increases the radiative power of the flame, and thus can increase appreciably the heat transfer rates.

The last point is worth considering in more detail. Most hydrocarbon diffusion flames are luminous, and this luminosity is due to carbon particulates that radiate strongly at the high combustion gas temperatures. As discussed in Chapter 6, most flames appear yellow when there is particulate formation. The solid-phase particulate cloud has a very high emissivity compared to a pure gaseous system; thus, soot-laden flames appreciably increase the radiant heat transfer. In fact, some systems can approach black-body conditions. Thus, when the rate of heat transfer from the combustion gases to some surface, such as a melt, is important—as is the case in certain industrial furnaces—it is beneficial to operate the system in a particular diffusion flame mode to ensure formation of carbon particles. Such particles can later be burned off with additional air to meet emission standards. But some flames are not as luminous as others. Under certain conditions the very small particles that form are oxidized in the flame front and do not create a particulate cloud.

It is well known that the extent of soot formation is related to the type of flame existing in a given process. Diesel exhausts are known to be smokier than those of spark-ignition engines. Diffusion flame conditions prevail in fuel-injected diesel engines, but carburetted spark-ignition engines entail the combustion of nearly homogeneous premixed fuel–air systems.

The various phenomena involved in carbon particulate formation have been extensively studied. The literature is abundant and some extensive review articles [49–51] are available. Most of the subsequent material in this chapter will deal with soot formation while a brief commentary on the coke-like formation from liquid fuels will be given at the end.

1. Characteristics of Soot

The characteristics of soot are well described in the article by Palmer and Cullis [49], who provide detailed references on the topic. Aspects of their review are worth summarizing directly. They report the detailed physical characteristics of soot as follows:

> The carbon formed in flames generally contains at least 1% by weight of hydrogen. On an atomic basis this represents quite a considerable proportion of this element and corresponds approximately to an empirical formula of C_8H. When examined under the electron microscope, the deposited carbon appears to consist of a number of roughly spherical particles, strung together rather like pearls on a necklace. The diameters of these particles vary from 100 to 2000 Å and most commonly lie between 100 and 500 Å. The smallest particles are found in luminous but nonsooting flames, while the largest are obtained in heavily sooting flames. X-ray diffraction shows that each particle is made up of a large number (10^4) of crystallites. Each crystallite is shown by electron diffraction to consist of 5–10 sheets of carbon atoms (of the basic type existing in ideal graphite), each containing about 100 carbon atoms and thus having length and breadth of the order of 20–30 Å. But the layer planes, although parallel to one another and at the same distance apart, have a turbostratic structure, i.e., they are randomly stacked in relation to one another, with the result that the interlayer spacing (3.44 Å) is considerably greater than in ideal graphite (3.35 Å). It may readily be calculated on this picture of dispersed carbon deposits that an "average" spherical particle contains from 10^5 to 10^6 carbon atoms.

Investigators have used the words "carbon" and "soot" to describe a wide variety of carbonaceous solid materials, many of which contain appreciable amounts of hydrogen as well as other elements and compounds that may have been present in the original hydrocarbon fuel. The properties of the solids change markedly with the conditions of formation; and, indeed, several quite well-defined varieties of solid carbon may be distinguished. One of the most obvious and important differences depends on how the carbon is formed: carbon may be formed by a homogeneous vapor-phase reaction; it may be deposited on a solid surface that is present in or near the reaction zone; or it may be generated by a liquid-phase pyrolysis.

2. Soot Formation Processes

Determining the relative tendency of hydrocarbon fuels to soot, explaining why this relative tendency between fuels exists, and discovering how to control or limit soot production in a particular combustion process—these are the elements of greatest importance to the practicing engineer. Since the amount of soot formed from a particular fuel has a complex dependence on the overall combustion process, no single characteristic parameter can define the amount formed per unit weight of fuel consumed. Both the flame type and various physical parameters determine the extent of soot formation from a given fuel. Moreover, depending upon the combustion process, the final mass of particulates emitted from the system could be reduced by a particle afterburning process.

Examining the detailed chemical processes of soot formation and oxidation, one notes how very complex the overall system is. Regardless of the flame type, the fuel undergoes either pure or oxidative pyrolysis. If the fuel is nonaromatic, the precursors undergo cyclization to create an aromatic ring. The ring structure is thought to add alkyl groups, developing into a polynuclear aromatic (PAH) structure that grows owing to the presence of acetylene and other vapor-phase soot precursors. Under certain flame conditions, the elements forming large aromatic structures are simultaneously oxidized. The precursors and all subsequent structures must be conjugated so that they are resonance-stabilized at the high temperatures in which they grow. Eventually, the aromatic structures reach a large enough size to develop into particle nuclei. Such condensed-phase carbon particles contain large amounts of hydrogen. The particles dehydrogenate (age) in a high-temperature combustion field while physically and chemically absorbing gaseous hydrocarbon species, thereby effecting a large increase of soot mass. The growing particles also agglomerate and conglomerate. The absorbed species, to a large degree, undergo chemical reformation, which results in a carbonaceous soot structure. And while these events are occuring, oxidative attack on the particles continues to form gaseous products. Nevertheless, the chemical steps in the soot formation process appear to be similar, regardless of the initial fuel. As Palmer and Cullis (49) have noted, "With diffusion flames and premixed flames investigations have been made of the properties of the carbon formed and of the extent of carbon formation under various conditions. In general, however, the properties of the carbon formed in flames are remarkably little affected by the type of flame, the nature of the fuel being burnt and the other conditions under which they are produced."

Although many current investigations have shown that the physical and chemical processes noted above are both complex and various, the generalized model to be described below allows one to identify certain controlling steps, thus gaining some important insights into the relative tendency of fuels to soot under particular flame configurations. And, using this insight, one may develop means to control the amount of particulate emission.

3. Experimental Systems and Soot Formation

Estimates of fuel sooting tendency have been made using various types of flames and chemical systems. In the context to be used here, the term *sooting tendency* generally refers to a qualitative or quantitative measure of the incipient soot particle formation rate. In many cases, this tendency varies strongly with the type of flame or combustion process under investigation. This variation is important because the incipient soot particle formation rate determines the soot volume fraction formed in a combustion system.

For premixed fuel–air systems, results are reported in various terms that can be related to a critical equivalence ratio at which the onset of some yellow flame luminosity is observed. Premixed combustion studies have been performed primarily with Bunsen-type flames [52, 53], flat flames [54], and stirred reactors [55, 56]. The earliest work [57, 58] on diffusion flames dealt mainly with axisymmetric coflow (coannular) systems in which the smoke height or the volumetric or mass flow rate of the fuel at this height was used as the correlating parameter. The smoke height is considered to be a measure of the fuel's particulate formation rate, but is controlled by the soot particle burnup. The specific references to this early work and that mentioned in subsequent paragraphs can be found in Ref. 50.

Work on coflowing Wolfhard–Parker burners [59, 60], axisymmetric inverse coflowing configurations (oxidizer is the central jet)[61, 62], and counterflow (opposed-jet) diffusion flames [63–65] has employed chemical sampling and laser diagnostics for scattering, extinction, and fluorescence measurements to determine chemical precursors, soot number density, volume fraction, and average particle size. These experiments not only obtained a measurement of where and when incipient soot formation takes place, but also followed the variation of the number density, volume fraction, and size with time after incipient formation. An important feature of most of these flame configurations is that they separate any particle burnup from the formation rate. As described in Chapter 4, counterflow experiments are designed to be directly opposing gas jets; a flow stream, normally the oxidizer, opposing the fuel flow emanating from a horizontal porous cylinder; or a directly opposing jet emanating from a cylindrical tube. Thus the porous cylinder condition creates flames that emulate the wakes of burning liquid droplets in various types of streams. A small liquid droplet burning in perfect spherical symmetry is nonluminous; larger droplets—the result of a longer diffusion time from the liquid surface to the flame front, thus facilitating fuel pyrolysis and growth—have mildly luminous flames. In another approach, shock tube experiments in which fuel pyrolysis leads to soot also provide a very interesting means of measuring the relative sooting tendency. Confusion can arise, however, in interpreting soot formation results unless one understands how the structures of the processes taking place in premixed flames—normal coflowing, inverse coflowing, and counterflowing diffusion flames; or shock tubes—differ. This point cannot be overemphasized.

In premixed flames the formation of soot precursors through fuel pyrolysis is competitive with oxidative attack on these precursors, but no oxidative attack occurs on the precursors formed during the fuel pyrolysis in diffusion flames. However, in normal coflowing diffusion flames, the fuel stream, its pyrolysis products, and other constituents added to it are convected by fluid motion across the flame front; that is, convection dominates molecular diffusion. These constituents flow into an environment of increasing temperature. Recall that, in counterflow diffusion flames, all elements in the fuel stream reach the flame front only by molecular diffusion across the stagnation streamline. Soot particles form only on the fuel side of the flame front; then they are convected toward the stagnation streamline and are not oxidized. The flow pattern in inverse axisymmetric coflowing flames is such that the precursors and particulates formed are also convected away from the flame zone. In shock tubes, fuel pyrolysis (and soot formation) takes place at a relatively fixed temperature and pressure. Thus, it is possible to measure a relative sooting parameter at a fixed temperature. Although stoichiometric temperatures similar to the temperatures reached in shock tubes can be established with coflowing diffusion flames, soot formation in these flames occurs at much lower temperatures prior to reaching the stoichiometric flame surface. In diffusion flames the temperature–time history of a fuel element leading to pyrolysis, particulate formation, and growth varies; in shock tube experiments it generally does not.

Flame turbulence should not affect soot formation processes under premixed combustion conditions, and the near correspondence of the results from Bunsen flames [52] and stirred reactors [55] tends to support this contention. However, the effect of turbulence on sooting diffusion flames can be very complex and unclear in most experiments unless the effect of the intensity (and scale) of turbulence on the flame structure, the temperature–time history of the pyrolyzing fuel, the rate of incipient particle formation and particle growth, and, in the case of some fuels, the transport of oxygen to the fuel stream are known.

4. Sooting Tendencies

Most of the early work on soot formation under premixed conditions was conducted with air as the oxidizer. These early results reported data as a critical sooting equivalence ratio ϕ_c, in which the higher the ϕ_c, the lower the tendency to soot. As shown in Table 5, the trend observed followed the order

$$\text{aromatics} > \text{alkanes} > \text{alkenes} > \text{alkynes}$$

that is, ethane was reported to have a greater sooting tendency than ethene or acetylene.

The most extensive early data of sooting under laminar diffusion flame conditions, as measured by the smoke-height method, were obtained by Schalla and co-workers [57, 58]. The general trend observed followed the order

$$\text{aromatics} > \text{alkynes} > \text{alkenes} > \text{alkanes}$$

Upon comparison, the trends of the alkyl compounds in the premixed case revealed an inconsistency with the oxidation reaction mechanism of ethane under fuel-rich conditions: it was found that the alkane was rapidly converted first to the alkene and then to the alkyne [66, 67]. Thus one would expect the alkynes to exhibit the greatest sooting propensity under premixed conditions. The answer to this inconsistency was developed from the early work of Milliken [68], who studied the soot emission from ethene on a cooled flat flame burner. Milliken found that the cooler the flame, the greater the tendency to soot (i.e., the lower the critical sooting equivalence ratio). He explained this temperature trend by evaluating the effect of temperature on the two competing processes occurring in the sooting flame: the pyrolysis rate of the fuel intermediate (acetylene) leading to the precursors and the rate of oxidative (hydroxyl radical, OH) attack on these precursors. Both rates increase with temperature, but the oxidative rate increases faster. Thus the tendency of premixed flames to soot diminishes as the temperature

TABLE 5   Critical Sooting Equivalence Ratios (ϕ_c) of Various Fuels Pre-mixed with Air

	ϕ_c			
Fuel	60	61	67	70
Ethane	1.67			1.70
Propane	1.56	1.73		
n-Pentane	1.47			
Isopentane	1.49			
n-Hexane	1.45		1.61	
Isohexane	1.45	1.73		
n-Octane	1.39	1.73		
Isooctane	1.45		1.70	
n-Cetane	1.35			
Acetylene	2.08	2.00		2.00
Ethylene	1.82		2.00	1.75
Propylene	1.67			
Butylene	1.56			
Amylene	1.54			
Cyclohexane	1.52	1.74	1.70	
n-Heptene	1.45			
Ethyl alcohol	1.52			
Propyl alcohol	1.37			
Isopropyl alcohol	1.39			
Isoamyl alcohol	1.54			
Acetaldehyde	1.82			
Propionaldehyde	1.75			
Methyl ethyl ketone	1.69			
Diethyl ether	1.72			
Diisopropyl ether	1.56			
Benzene	1.43	1.54		
Toluene	1.33	1.46	1.39	
Xylene	1.27	1.45	1.30	
Cumene	1.28	1.56	1.40	
Methyl naphthalene	1.03		1.21	

rises. Other investigators [67, 69, 70] verified Milliken's calculations and extended the concept by showing that in diffusion flames, where there is no oxidative attack on the precursors, the soot volume fraction created increases with temperature. Therefore, for a proper comparison of the effect of fuel structure on the sooting tendency, the flame temperature must be controlled. In diffusion flames, radical diffusion processes are important. Thus radicals, particularly H, diffusing into fuel-rich zones accelerate normal pyrolysis reactions.

Early work on premixed Bunsen flames and coflowing diffusion flames [69–71] was repeated in experiments where the temperature was controlled by varying the N_2/O_2 ratio in the oxidizer used for the Bunsen experiments and by adding inerts, particularly nitrogen, to the fuel in the diffusion flame experiments. The critical sooting equivalence ratio in premixed flames was determined by the visual observation of the onset of soot luminosity. Varying the nitrogen concentration permitted the determination of a critical sooting equivalence ratio ψ_c (whereas ϕ_c is based on a stoichiometry referenced to carbon dioxide and water as products; ψ_c is based on a stoichiometry referenced to carbon monoxide and water as products since all conditions of operation are fuel rich) as a function of the calculated adiabatic flame temperature. The results are shown in Fig. 13. The values corresponding to air agree exactly with the early work of Street and Thomas [52]. However, one will note that at a fixed temperature, ψ_c (acetylene) $< \psi_c$ (ethene) $< \psi_c$ (ethane), as one would expect. Also, there is no specific trend with respect to homologous series.

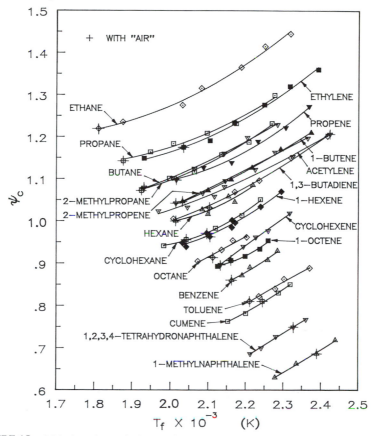

FIGURE 13 Critical sooting equivalence ratios (ψ_c) based on CO and H_2O stoichiometry of various hydrocarbon fuels as a function of calculated adiabatic flame temperature T_f.

Following the conceptual idea introduced by Milliken [68], Takahashi and Glassman [53] have shown, with appropriate assumptions, that, at a fixed temperature, ψ_c could correlate with the "number of C—C bonds" in the fuel and that a plot of the log ψ_c vs number of C—C bonds should give a straight line. This parameter, number of C—C bonds, serves as a measure of both the size of the fuel molecule and the C/H ratio. In pyrolysis, since the activation energies of hydrocarbon fuels vary only slightly, molecular size increases the radical pool size. This increase can be regarded as an increase in the Arrhenius pre-exponential factor for the overall rate coefficient and hence in the pyrolysis and precursor formation rates so that the C/H ratio determines the OH concentration [72]. The ψ_c vs C—C bond plot is shown in Fig. 14. When these data are plotted as log ψ_c vs C—C bonds,

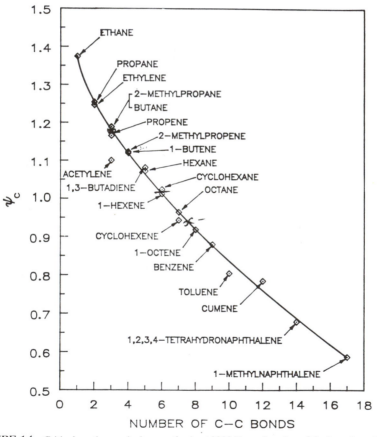

FIGURE 14   Critical sooting equivalence ratio ψ_c at 2200 K as a function of the "number of C—C bonds" in hydrocarbon fuels. +, 0, and − indicate ethene/1-octane mixtures in molar ratios of 5 to 1, 2 to 1, and 1 to 2, respectively; χ, acetylene/benzene at a molar ratio of 1 to 3. The O symbol for 2 to 1, falls on top of the butene symbol.

a straight line is obtained [53]. Also plotted on Fig. 14, as a test of the correlation, are the results of some fuel mixtures [73]. These data lead to the important conclusion that specific fuel structures do not play a role under premixed conditions; for example, decane has the same sooting tendency as benzene at the same premixed flame temperature. Thus, under premixed conditions, all fuels must break down to the same essential species—postulated to be acetylene [53]—which builds into a soot particle. This postulate is consistent with the results of Harris and Weiner [74, 75], who showed that increments of carbon introduced as ethene in premixed flames were just as effective in producing soot as increments of carbon introduced as toluene and that the greatest mass of soot came about by growth on particles in which the dominant species contributing to this particle growth was acetylene.

Similar temperature-control procedures were applied to a highly over-ventilated concentric coflowing diffusion flame [51, 76] in which 29 fuels were evaluated with respect to their sooting tendency. In these experiments the mass flow rate of the fuel jet is increased until soot breaks through the top of the "luminous visible flame." This "luminous visible flame" is not a flame; its edge defines the region in which the soot particles formed were completely burned. The stoichiometric (Burke–Schumann) flame front lies within this luminous envelope. The initial breakthrough of this luminous boundary is referred to as the sooting or smoke height, and most data are reported in terms of the fuel mass flow rate at this height.

If nitrogen, or any inert species, is added to the fuel jet when the smoke height is reached, the luminous zone closes and soot no longer emanates from the top of the flame. If fuel mass flow is then increased, another smoke height is reached. Additional inert again suppresses the soot emanation and additional fuel flow is necessary for the flame to smoke. In this manner the smoke height, or mass flow rate at the smoke height, of a single fuel may be obtained at different flame temperatures.

How this smoke effect varies with inert addition is best explained by considering the results of many investigators who reported that incipient soot formation occurred in a very narrow temperature range. The various results are shown in Table 6. Since, as stated earlier, the incipient particle formation mechanisms for various fuels follow quite similar routes, it seems appropriate to conclude that a high activation energy process or processes control the incipient particle formation. It is likely that the slight variation of temperatures shown in Table 6 is attributable to the different experimental procedures used. The experimental procedure of Ref. [77] sheds some light on the concept proposed here. In these experiments, a given sooting flame was diluted until all particle radiation just disappeared, then the temperature of the flame front was measured on the centerline of the coannuluar configuration. These measurements are shown in Fig. 15. Since different amounts of dilution were required to eliminate particle continuum radiation for the different fuels, it should not be surprising that there is some variation for different fuels in the temperature limit for incipient particle formation.

TABLE 6   Soot Inception Temperature (K) by Various Investigators

Fuel	A	B	C	D	E	F
Methane	1913					
Propane	1783					
Ethene	1753			1750	1460	
n-Butane	1723					
Cyclobutane	1643					
Butene	1623	1684	1348	1600		
Acetylene		1665	1390	1650		
Butadiene	1623	1650	1353			
Allene	1613	1586		1750	1430	
Benzene		1580	1332			1750
Toluene		1570				

A. Dilution of fuel jet until no luminosity [Smith, C. A., M.S. E. Thesis, Dep. Mech. Aerosp. Eng., Princeton Univ., Princeton, New Jersey, 1990].

B. Dilution of fuel jet until no luminosity (Glassman *et al.* [77]).

C. Centerline observation of first particulates (Gomez *et al.* [86]).

D. Optical measurements of inception and temperature [Boedecker, L. R., and Dobbs, G. M., *Int. Symp. Combust., 21st* p. 1097. Combust. Ins., Pittsburgh, Pennsylvania, 1986].

E. Critical sooting C/O ratio in premixed flames as a function of controlled flame temperature [Bohm, H., Hesse, D., Jander, H., Luers, B., Pietscher, J., Wagner, H. Gg., and Weiss, M., *Int. Symp. Combust., 22nd* p. 403. Combust. Inst., Pittsburgh, Pennsylvania, 1988].

F. Flat premixed flame observations and measurements [Olson, D. B., and Madronvich, S., *Combust. Flame* **60**, 203 (1985)].

If the temperature limit is taken as the average of those reported in Fig. 15 ($\sim$1600 K), it can be assumed that incipient particle formation occurs along a 1600 K isotherm in the coannular diffusion flame structure. Then, at the initial smoke height with no inert addition, the 1600 K isotherm within the fuel flow designates the point of incipient particle formation. Thus there is enough growth along flow streamlines that coincides with and closely parallels the 1600 K isotherm to cause particles to penetrate the flame and smoke. This flow coincidence will, of course, be near the sides of the flame. It is for this reason that the smoke height is observed as a flame breakthrough, lending a wing-like appearance to the flame luminous envelope [51]. Thus what determines the smoke height is the growth in mass of the incipient particles created at about the 1600 K isotherm. When the fuel is diluted with an inert, the flame temperature drops; and since the situation represents a moving-boundary (quasi-steady) heat-transfer problem, the 1600 K isotherm moves closer to the flame temperature isotherm. As a result, particle growth diminishes and the flame front closes. That is, because the mass of soot formed is smaller, it is consumed in the flame front, so no smoke height is observed. However, if the fuel mass is increased, more incipient particles form and, although growth is slower under dilute conditions, the soot mass again becomes

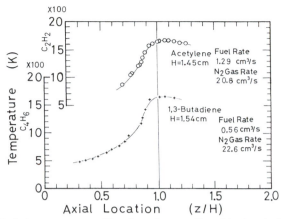

FIGURE 15 Typical temperature profiles along the centerline of laminar hydrocarbon fuel jets diluted with N_2 to the point of no luminosity when burning in overventilated air streams. H is the height of the flame; Z is the distance from jet exit along the centerline.

large enough to exhibit a smoke height. By repeating the dilution procedure to further decrease the flame temperature, one can eliminate the smoke penetration, which can be induced again by adding more fuel.

Consider, for example, the fact that a coannular diffusion flame with acetylene will have a much higher flame temperature than that of butene. When considered at the centerline, the incipient-particle temperature isotherm for acetylene is much further from the flame temperature isotherm at the same position than is that of butene; i.e, the acetylene particles formed have a larger growth path. Thus, although more incipient particles may form at about 1600 K for butene, the overall soot volume fraction from acetylene may be greater due to a longer growth period. The depth of the thermal wave (or isotherm) is of the order α/u where α is the thermal diffusivity and u is the linear velocity. There are self-compensating terms in each of these parameters even when there is dilution. Recall that the flame height for a diluted fuel varies to the first order solely with the volumetric flow rate of the fuel, even under buoyancy conditions. Figure 16 shows some results of this soot elimination procedure in which diluents other than nitrogen were chosen. Argon requires greater dilution to achieve the same temperature as nitrogen, and carbon dioxide requires less. Yet the variation in the incipient particle temperature for the four fuels shown in Fig. 16 is slight and well within experimental error.

Because different fuels produce different amounts of incipient soot particles at about 1600 K, the smoke height test becomes a measure of the fuel's relative propensity to soot under flame-like conditions. However, if the smoke height is to be used for fuel comparisons, the diffusion flame temperature must be adjusted to be the same for all fuels. The results of such experiments [51] are reported in Fig. 17. The data in this figure are plotted as the log of the reciprocal of the fuel

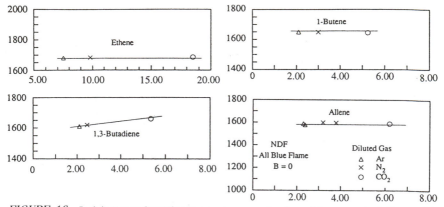

FIGURE 16   Incipient soot formation temperature as a function of the amount of diluent. The ordinate is the temperature in kelvins; the abscissa is $f = (Q_f/Q_f + Q_{dg}) \times 100$, where Q_f is the volumetric flow rate of the fuel and Q_{dg} is that of the diluent.

mass flow rate at the smoke height (1/FFM) versus the calculated adiabatic flame temperature. Plotting in this manner appears to be appropriate since the depth of the thermal wave, and hence the growth rate, is exponential. The calculated adiabatic flame temperature has been found to be a good surrogate temperature for the flame isotherm. The slopes of the data lines should be approximately the same; and, except for the heavy aromatic fuels, they are. The differences in the slopes are probably due to the variation of (α/u) for the different fuels and dilution conditions.

Inherent in the preceding discussion is the assumption that the growth rate of particles is similar for all fuels that produce particles. Opposed-jet systems permit good estimates of particle growth times along the jet stagnation streamline. The results of Vandsburger *et al.* [63, 64], who used such a system, are shown in Fig. 18. These results reveal, as expected, that increasing the temperature (increasing the oxygen mole fraction in the opposed stream, i.e., the oxygen index) increases the soot yield and that ethene produces more soot than propane. Figure 18 shows the same variational trend in soot volume fraction with time as that obtained with Wolfhard–Parker burners [60] and in premixed flames [74, 75]. Particle number density trends are also the same for various burner systems. For similar fuels, the general agreement between coflow and counterflow results for soot volume fractions, number densities, and particle size ranges is reasonably good. Since residence time is reduced near the flame in counterflow burners in comparison with coflow systems, the maximum soot formation rates in counterflow systems are found to be somewhat lower than those in coflow systems for comparable fuels and flame temperatures. When soot growth rates are normalized by the available soot surface area to give what Harris and Weiner [74, 75] call the "specific surface

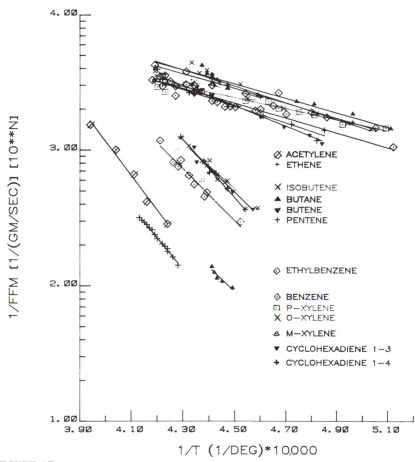

FIGURE 17 Sooting tendency of some hydrocarbon fuels plotted as the log of the reciprocal of the fuel mass flow rate at the smoke height versus the reciprocal of the calculated adiabatic flame temperature.

growth rate," the results of coflow burners agree not only with those of counter flow burners, but also with those of rich premixed flames [51].

Frenklach *et al.* [78], who evaluated the sooting tendency of fuels by shock tube pyrolysis at various temperatures, found that the sooting rate increased with temperature, reached a maximum, and then decreased. The maxima occur in the range 1900–2300 K. The pyrolysis in coflow diffusion flames is initiated at temperatures much lower than the stoichiometric temperature, so that the soot forms prior to reaching the maximum temperatures Frenklach *et al.* created in their shock tube experiments. In shock tubes the fuel is instantaneously exposed to very high temperatures; thus the precursors that form decompose and the soot

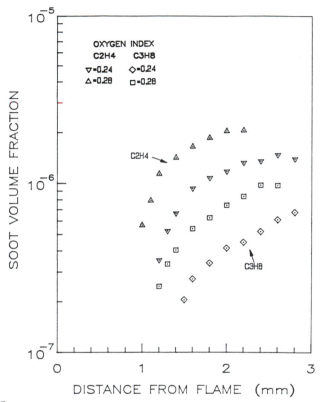

FIGURE 18 Soot volume fraction for propane and ethane opposed-jet diffusion flames versus distance from the flame front (from Vandsburger *et al.* [63]).

production rate is lowered. Shock tube pyrolysis data do, however, show the same trends as the diffusion flame results reported in Fig. 17. Both types of experiments show the following sooting tendency:

benzene > allene > butadiene > butenes > acetylene

Note that at a fixed flame temperature 1-butene has a greater flame tendency to soot than acetylene.

The results presented in Fig. 17 for diffusion flames and those from shock tubes clearly indicate that fuel structure does indeed play a role in a fuel's tendency to form particulates—in significant contrast to the results observed in premixed flames. One may conclude, then, that a fundamental knowledge of a fuel's pyrolysis chemistry [51, 76] will allow one to predict its relative tendency to soot with respect to the results presented in Fig. 17. For example, cyclohexadienes are known to dehydrogenate to benzene during pyrolysis; and, indeed, the data in

Fig. 17 show a nearly exact correspondence between benzene and the cyclohexadienes. Similarly, cyclopropane is known to pyrolyze initially to propene; and again, a correspondence with smoke height results is seen in Fig. 17. Surprisingly, cyclopentene, a C_5 compound, soots more extensively than cyclohexene, a C_6 compound. Kinetic data reveal that cyclopentene forms cyclopentadiene, which then forms benzene during pyrolysis, and that cyclohexenes pyrolyze primarily to ethene and butadiene. Thus, the cyclic-C_5 smoke-height results lie near the aromatics, and the C_6 results lie between ethene and butadiene. Numerous other examples are given in Ref. [51]. Thus, the important conclusion is reached that the fuel pyrolysis rate and mechanism determine the extent of incipient particle formation and hence a fuel's tendency to soot. The general fuel trends shown in the preceding paragraph will be explained in the next section.

5. Detailed Structure of Sooting Flames

The critical sooting equivalence ratios for premixed flames and the smoke heights for diffusion flames discussed so far can serve only as qualitative measurements that offer excellent insights into the controlling factors in the soot incipient formation process. In order to gain greater understanding of the soot processes in flames, it has been necessary to undertake extensive chemical and physical measurements of temperature and velocity as well as laser extinction, scattering, and fluorescence in premixed flames [79, 80], Wolfhard–Parker burners [81–83], coannular burners [84–86], and opposed-jet diffusion flames [63, 64, 87, 88] (see Ref. [51]).

Harris and Weiner [74, 75] have contributed much to the study of soot particle history in premixed flames. They used laser light scattering and extinction measurements to determine particle size, number density, and volume fraction in experiments with flat flame burners stabilized by a water-cooled porous plug. Their studies showed that richer ethene flames produce more soot owing to a higher nucleation rate, increased surface growth in these richer flames is induced by the increased surface area available for growth, and depletion of growth species does not occur in these flames. Therefore, these investigators concluded that the final particle size reached, when surface growth has virtually stopped, is determined not by growth species depletion, but rather by a decrease in the reactivity of the particles. They further concluded that the results are applicable to other fuels because the post-flame gases of aliphatic fuels are all quite similar, acetylene and methane being the prominent hydrocarbons present. Thus the similarity in surface growth in aliphatic fuels is related to the dominant role that acetylene plays in these processes. Although the post-flame gases of aromatic fuels have about 100 times more PAH than the post-flame gases of aliphatic fuels [89], acetylene remains the principal hydrocarbon in the burned gases of benzene flames [90]. This finding, combined with the fact that growth rates in benzene and ethylene are nearly identical [81], suggests that the mechanism of growth in benzene flames is also dominated by acetylene. These conclusions lend great validity to the qualitative

TABLE 7   Flow Rates^a for Flames in Fig. 19 [74]

	C_2H_4	Ar	O_2	C_7H_8	Total	C/O
Flame 1	1.68	8.89	2.05	0	12.62	0.82
Flame 2	1.68	8.89	2.05	0.044	12.66	0.89
Flame 3	1.83	8.89	2.05	0	12.77	0.89
Flame 4	2.17	9.12	2.44	0	12.73	0.89

^{a}In liters per minute (STP).

results, postulates, and correlation given previously for premixed flames [53, 91]. As stated, this surface growth pattern has been found to be the same in diffusion flames as well.

For the flames described in Table 7, Harris and Weiner [74] obtained the results shown in Fig. 19, where the increase of soot volume fraction is plotted as a function of time. Thus, temperature measurements revealed that Flame 3 has the lowest temperature, Flames 2 and 4 are of equal, and somewhat higher, temperature, and Flame 1 has the highest temperature. These results supply quantitative proof that in premixed flames the tendency to soot decreases with increasing flame temperature. Also of importance is the fact that Flame 2, which has toluene as a fuel constituent and the same equivalence ratio as that of Flame 4 for pure ethene, gives the same soot volume fraction as Flame 4 (Fig. 19). A larger initial soot volume fraction indicates a larger initial incipient particle formation. The particle number density decreases with time due to coagulation; however, fuels that have a larger initial number density (incipient particle formation) have the largest final number density. Similar results for particle histories were found in opposed-jet diffusion flames [63, 64]. These quantitative results support the use of the qualitative critical sooting equivalence ratio results obtained on Bunsen flames as an excellent means for evaluating the sooting tendency of fuels under premixed conditions.

One of the earliest detailed diagnostic efforts on sooting of diffusion flames was that of Wagner and co-workers [81–83], who made laser scattering and extinction measurements, profile determinations of velocity by LDV, and temperature measurements by thermocouples on a Wolfhard–Parker burner using ethene as the fuel. Their results show quite clearly that soot particles are generated near the reaction zone and are convected farther toward the center of the fuel stream as they travel up the flame. The particle number densities and generation rates decline with distance from the flame zone. The soot formation rate appeared to peak at values about 2–3 mm on the fuel side of the reaction zone. The particle number density decreases by coagulation and gradually levels off. This process leads to an increase in particle size and a simultaneous increase in soot volume fraction as the particles absorb hydrocarbon species. Thus the process is very similar to that found by Harris and Weiner in premixed flames. Smyth et al. [60] drew essentially

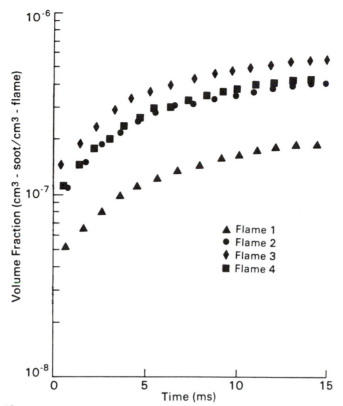

FIGURE 19   Soot volume fraction profiles for premixed flames. See Table 7. (From Harris and Weiner [74]).

the same conclusions in characterizing a sooting methane Wolfhard–Parker burner by detailed species profiles.

Santoro and co-workers [84, 85] performed some early extensive quantitative studies of coannular diffusion flames. They found that a striking feature of the coannular laminar diffusion flame is the existence of a roughly toroidal zone near the base of the flame where intense nucleation coexists with intense agglomeration to produce moderate-size particles. The soot formed in this region is convected upward along streamlines to form an annular region in which further gas-to-particle conversion occurs at moderate rates. Indeed, it is because of the intense particle formation in this toroidal region that the smoke height is identified as that point at which "wings" appear in the laminar diffusion flame [70]. The results of Santoro *et al.* [85], depicted in Fig. 20, also indicate that the major soot formation regime is 2–3 mm on the fuel side of the flame, whose position is specified by the maximum temperature. In support of the concept postulated previously, one notes that the

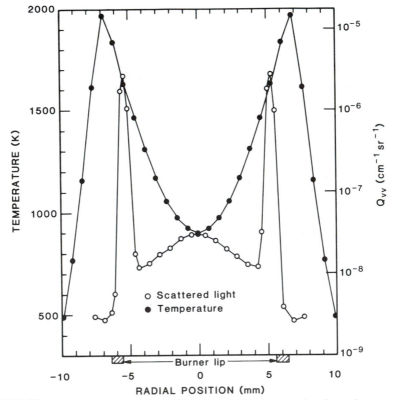

FIGURE 20   Comparison of the radial profiles for scattering cross-section Q_{vv} and temperature (uncorrected) as a function of radial position for a coannular ethene diffusion flame (from Santoro *et al.* [85]).

point of maximum particle density occurs at about 1600 K. The fact that particles exist at other temperatures along the analysis profile in Fig. 20 is attributable to thermophoresis.

Furthermore, fluorescence measurements aimed at the detection of PAHs showed that the maxima clearly precede the maxima of soot volume fraction. No apparent difference in fluorescence levels was observed in experiments with the butene and benzene fuels, in contrast with the findings of other investigators [81]; thus, PAHs apparently have the same role in diffusion flames for both aliphatic and aromatic fuels. PAHs may, in fact, be involved in the soot nucleation stage (as has often been hypothesized and will be discussed next); such a finding, together with the relative constancy of the temperature at the soot onset, would suggest that, even though the extent of fuel-to-soot conversion may vary significantly from fuel to fuel, a common soot formation mechanism exists for all fuels.

Kent and Wagner [92] made some very interesting quantitative measurements of overall axial soot concentrations as well as soot, flame zone, and centerline temperatures of coannular flames that have great significance. These results for ethene at different flow rates are shown in Fig. 21. The data indicate that flames emit smoke when the soot temperature in the oxidation zone falls below 1300 K. As the temperature of the soot decreases, the flow proceeds downstream because of cooling and radiation losses due to the soot mass formed interior to the flame zone. Thus the flow rates indicated in Fig. 21 for 3.4 and 5.7 ml/s would have flames below the smoke height, and the actual smoke height would exist for a volumetric flow rate between 5.7 and 6.1 ml/s. Establishing that smoke is emitted when burnout ceases as the soot temperature drops below 1300 K, regardless of other conditions, is a very significant and practical result. Thus, if incipient particle formation occurs in the 1600 K range under flame conditions and particle burnout ceases below about 1300 K range, a soot formation window exists for combustion processes. If one wishes to prevent soot from emanating from a combustion system operating in normal ranges, the temperature must be kept below about 1600 K. If this procedure is not possible, the exhaust must be kept lean and above 1300 K. With this understanding, one can visualize the "Catch-22" problem facing diesel engine operators. When diesel engines are operated lean to reduce the thermal NO_x, the manifold exhaust temperature becomes too low to burn the soot particles that will form; but when they are operated less lean to burn the soot, the NO_x increases.

6. Chemical Mechanisms of Soot Formation

The quantitative and qualitative results presented in the previous sections confirm some original arguments that the chemistry of fuel pyrolysis, and hence fuel structure, plays an important, and possibly dominant, role in sooting diffusion flames. The identical chemical character of the soot formed from diffusion or premixed controlled systems seems to suggest a commonality in the chemical mechanisms for soot formation. Apparently, there is an underlying fuel-independent general mechanism that is modified only with respect to alternative routes to intermediates. These routes are simply affected by the combustion system temperature and the general character of the initial fuel. Essentially, this concept implies that the relative propensity of one fuel to soot compared to another is primarily determined by differences in the initial rate of formation of the first and second ring structures; moreover, while the mechanisms controlling the growth of large condensed-ring aromatics, soot nucleation, soot growth, etc., remain essentially unchanged, the growth steps in large aromatic structures that lead to soot nucleation are significantly faster than the initial-ring formation [93]. Thus the formation of the initial aromatic rings controls the rate of incipient soot formation. As is well established now, the incipient soot formation particle concentration determines the soot volume fraction, i.e., the total amount of soot formed. In support of this general contention that initial-ring formation is the soot-controlling step are the results correlating the sooting tendencies of several fuels as measured by time-of-flight

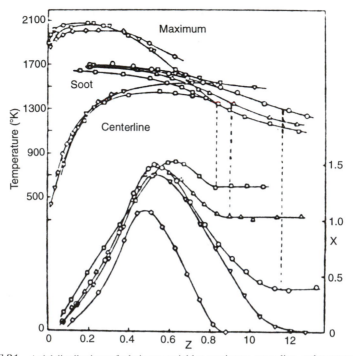

FIGURE 21   Axial distributions of relative soot yield x; maximum, centerline, and soot temperatures for ethene at flow rates around soot point in coaxial diffusion flames. Flow rates (ml/s): $\diamond$ 3.4, ∇ 5.2, $\bigcirc$ 6.1, $\triangle$ 7.2, $\square$ 10.9 (from Kent and Wagner [92]).

mass spectrometry coupled to a shock tube and chemical sampling in coannular normal and inverse diffusion flames [93]. These cases revealed the presence of allene, which was established kinetically to form benzene directly through a C_3 route [94]. Miller and Melius [94a] also have formulated a chemical kinetic model of soot formation in which they argue that the first aromatic ring is most likely formed by reaction of two propynyl (C_3H_3) radicals; correlations [95] of experimental results on the propensity of C_3 hydrocarbon fuels to soot appear to support this route.

To explain the underlying principles of molecular growth to soot formation, mechanisms related to ions [96], ring growth [97, 98], polyacetylene chains [99], Diels–Alder reactions, and neutral radicals [100] have been proposed [101]. However, the experimental results discussed on sooting tendencies of fuels under various flame configurations can be explained very well by a generalized mechanism based on soot precursor growth through a system dependent upon neutral radicals.

The detailed modeling of soot formation in the shock tube pyrolysis of acetylene [102] and other fuels [103] provides the central basis for the fuel-independent general mechanisms suggested here. It must be noted, as well, that a large body of work by Howard and co-workers [104, 105] on premixed flames with regard to formation of aromatic species provides direct tests of the proposed mechanisms and are key to understanding and modeling soot formation.

The dominant route, as proposed by Frenklach *et al.* [102], is the focus of this development. Indeed, this mechanism (Fig. 22) would occur mostly at high temperatures [102] and serves as the preferred route in premixed flames where the original fuel structure breaks down to acetylene and soot is formed in the high-temperature post-flame zone. Shock tube results by Colket [106] strongly suggest that formation of benzene from acetylene pyrolysis is largely attributable to acetylene addition to the *n*-butadienyl radical (C_4H_5), not C_4H_3. Thus, one could conclude that in diffusion flames where soot begins forming about 1600 K, the chemical route is alternate route A in Fig. 22. Colket's results [106] for vinylacetylene in the same temperature regime indicate that the main route to benzene is through vinyl radical addition to the vinylacetylene as specified in Fig. 22 by alternate route B.

These three paths offer excellent insight into the sooting tendencies measured in premixed, coflowing, and counterflowing diffusion flames and shock tubes. The fastest and most abundant route to the formation of the appropriate intermediates for the growth of the first aromatic ring determines the incipient particle formation rate and the relative propensity of a fuel to soot. By examining this fuel propensity to soot, it becomes evident that C_4 hydrocarbons have a greater propensity to soot because they most readily form the butadienyl radical [99]. This hypothesis certainly explains the very early results of Schalla *et al.* [57] for the *n*-olefin series in which butene had the smallest smoke height. Temperature control was not necessary in Schalla's tests with olefins because all olefins have the same C/H ratio and hence near identical flame temperature histories. Butene, then, soots more than acetylene (Fig. 17) because it creates the butadienyl radical more readily. And butadiene soots more than either butene or acetylene, again because it more readily forms the butadienyl radical.

The fact that most alkylated benzenes show the same tendency to soot is also consistent with a mechanism that requires the presence of phenyl radicals, concentrations of acetylene that arise from the pyrolysis of the ring, and the formation of a fused-ring structure. As mentioned, acetylene is a major pyrolysis product of benzene and all alkylated aromatics. The observation that 1-methylnaphthalene is one of the most prolific sooting compounds is likely explained by the immediate presence of the naphthalene radical during pyrolysis (see Fig. 22).

Sampling in inverse coannular diffusion flames [95] in which propene was the fuel has shown the presence of large quantities of allene. Schalla *et al.* [57] also have shown that propene is second to butene as the most prolific sooter of the *n*-olefins. Indeed, this result is consistent with the data for propene and allene in Fig. 17. Allene and its isomer methylacetylene exhibit what at first glance appears to be an unusually high tendency to soot. However, Wu and Kern [107] have shown that both pyrolyze relatively rapidly to form benzene. This pyrolysis step is represented as alternate route C in Fig. 22.

Detailed examination of the chemical reactions in the general mechanism in Fig. 22 reveals the importance of the pyrolysis fragments of all the fuels considered, especially the significance of H atoms, vinyl radicals, and acetylene. In particular, it becomes apparent that H atom and acetylene concentrations are important because they regulate, by abstraction and addition, respectively, the rapidity with which the large soot precursors can grow. The presence of large concentrations of H atoms in fuel pyrolysis is due to the thermal decay of the alkyl radicals present. Since the unimolecular cleavage of these radicals is not fast until temperatures well above 1100 K are reached [108], it is quite possible that this kinetic step and H atom diffusion set the range of the critical temperature for incipient soot formation in diffusion flames. Since there is no radical diffusion in shock tube studies of soot formation, it is not surprising, as mentioned, that the temperature at which soot formation is first identified is somewhat higher than that for diffusion flames.

7. The Influence of Physical and Chemical Parameters on Soot Formation

The evidence suggests that temperature is the major parameter governing the extent to which a given fuel will soot under a particular flame condition or combustion process. As emphasized earlier, increasing the temperature under premixed fuel–air conditions decreases soot production, whereas the opposite is true for diffusion flames. The main effect of varying pressure must be manifested through its effect on the system temperature. Fuel pyrolysis rates are a function of concentration, but the actual dependence on the fractional fuel conversion depends on the overall order of the pyrolysis process. Unfortunately, varying the pressure alters the structure of the diffusion flame and the diffusivity of the fuel. As discussed, a variation in the diffusivity can shorten the growth period of the incipient particles formed. The diffusivity varies inversely with the pressure.

Regrettably, the particular effect of an additive is difficult to assess. A chosen diluent cannot lower the flame temperature without altering the effective thermal diffusivity of the fuel–diluent jet. Thus, in examining the possible effect of an additive it is imperative to consider how the measurements of the extent of sooting were made. With respect to physical parameters, the effect of temperature is clear. However, the data reported on pressure variation must be viewed with caution since, as stated, pressure variations cause variations in temperature, thermal diffusivity, flow velocity, and flame structure.

It now seems evident that additives which alter the early concentrations of H atoms can have a recognizable effect on the soot output. Again, the extent of the effect depends on the type of flame, the specific fuel, and the addition. Perhaps the best example is the addition of modest amounts of oxygen to the fuel in a diffusion flame configuration. The addition of oxygen to tightly bound fuels such as ethene, ethyne, and benzene noticeably affects the sooting tendency of these fuels. Similarly, a modest amount of oxygen added to the paraffins tends to act

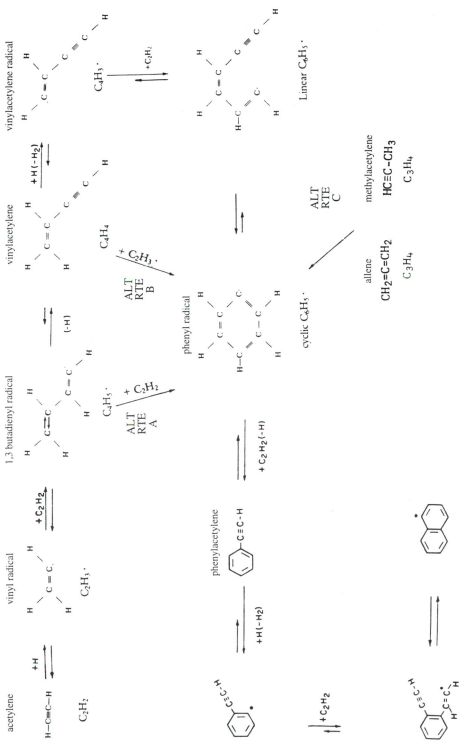

FIGURE 22 General mechanisms for soot formation.

as an inert diluent [65]. Chemical kinetic analyses of the pyrolysis rates of these fuels at modest temperatures confirm that the effect in the case of the tightly bound fuels is to rapidly increase the H atom concentration, whereas rapid creation of H atoms due to their thermal homolysis of paraffins is little affected by the presence of oxygen. The general effect of continuously increasing the oxygen concentration on the fuel side of ethene and propane in opposed-jet diffusion flame experiments supports these conclusions and gives an overall picture of the effect of oxygen addition [65]. In these experiments, as the fuel-oxygen ratio was increased, the oxygen concentration in the opposed jet was reduced to keep the flame temperature constant. The particular results are shown in Fig. 23. The higher the extinction coefficients, the greater are the soot volume fractions. Ethene's propensity to soot rises with decreasing ϕ (small oxygen addition). As ϕ decreases further, a chemical reaction releases energy, altering the temperature profile and increasing the sooting tendency still further. At a given ϕ, oxidizer species become present in sufficient abundance that the soot precursors are attacked. Eventually, oxidizer attack exceeds the thermal effect, a maximum is reached, and the conditions approach those of a premixed flame. The soot formed then decreases with decreasing ϕ, vanishing at a value near the critical sooting equivalence ratio determined by the conventional methods discussed. The same trends hold for propane except that the oxygen initially acts as a diluent, lowering soot production until the reaction is sufficiently initiated to raise the overall thermal profile level.

The results in Fig. 23 are significant in that, for values of ϕ *less* than the maximum shown, premixed flame conditions exist. Since the flammability limit lies between the critical sooting equivalence ratio and the maximum ϕ for sooting, it is possible to conclude that, under all premixed flame conditions, fuel structure plays no role in the sooting tendency. Also, when ϕ decreases near infinity, the sooting tendency rises for ethene and decreases not only for propane, but would also for *n*-butane and isobutane [65]. The fact that these opposing trends can be explained by a natural radical reaction mechanism tends to support the notion that such mechanisms control the soot formation rate. It is difficult to see how the trends could be explained on the basis of an ion mechanism. Furthermore, the wide variation of sooting tendency with ϕ indicates the difficulty of extracting fundamental information from studies of turbulent fuel jets.

The presence of halogen additives substantially increases the tendency of all fuels to soot under diffusion flame conditions [69]. The presence of H atoms increases the soot pyrolysis rate because the abstraction reaction of H+RH is much faster than R + RH, where R is a hydrocarbon radical. Halogenated compounds added to fuels generate halogen atoms (X) at modest temperatures. The important point is that X+RH abstraction is faster than H+RH, so that the halogen functions as a homogeneous catalyst through the system

$$X + RH \rightarrow R + XH$$
$$H + XH \rightarrow H_2 + X$$

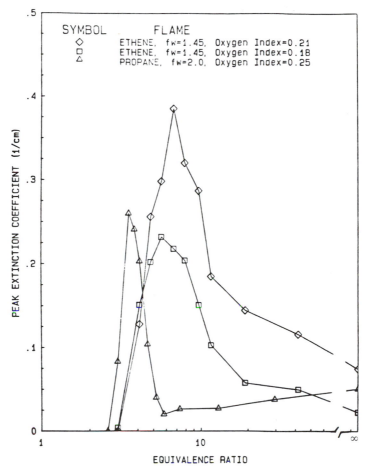

FIGURE 23   Peak extinction coefficient versus equivalence ratio of the fuel/oxygen stream mixture in propane and ethene opposed-jet diffusion flames. f_w is the fuel injection parameter.

Sulfur trioxide is known to suppress soot in diffusion flames and increase soot in premixed flames. These opposite effects can be explained in the context of the section on sulfur oxidation (Section 2e). In diffusion flames, the slight suppression can be attributed to the reaction of H atoms via the step

$$H + SO_3 \rightarrow OH + SO_2$$

If this step occurs late in the pyrolysis process, the hydroxyl radicals that form could attack the soot precursors. Thermal diffusivity may also have an effect. In premixed flames the SO_3 must dissociate into SO_2, which removes H atoms by

$$H + SO_2 + M \rightarrow HSO_2 + M$$

The reduction in H concentration leads to a decrease in OH and hence to an increase in soot production.

As additives to reduce soot output in flames, metal and organometallic compounds, such as ferrocene, have attracted the attention of many investigators (see Ref. [109]). The effect in premixed flames is best described by Bonczyk [110], who reported that the efficiency with which a given metal affects soot production characteristics depends almost exclusively on temperature and the metal atom's ionization potential. Thus the alkaline-earth metals Ba > Sr > Ca are the most effective. The metal additives reduce the soot particle size and volume fraction and increase the number density. Essentially, ion charge transfer reduces agglomeration, and, under the appropriate conditions, particle burnup is augmented. Indeed, the burnup effect is prevalent in diffusion flames since the smaller particles must pass through the flame front. The conditions describing the smoke height are pertinent. There is no clear evidence that metal additives affect the incipient particle formation in diffusion flames. The difficulty in understanding the effect of ferrocene in diffusion flames is compounded by the fact that it decomposes into cyclopendienyl, which readily forms soot.

F. STRATOSPHERIC OZONE

The ozone balance in the stratosphere is determined through complex interactions of solar radiation, meteorological movements within the stratosphere, transport to and from the troposphere, and the concentration of species based on elements other than oxygen that enter the stratosphere by natural or artificial means (such as flight of aircraft).

It is not difficult to see that ozone initially forms from the oxygen present in the air. Chapman [111] introduced the photochemical model of stratospheric ozone and suggested that the ozone mechanism depended on two photochemical and two chemical reactions:

$$O_2 + h\nu \rightarrow O + O \tag{149}$$

$$O + O_2 + M \rightarrow O_3 + M \tag{150}$$

$$O_3 + h\nu \rightarrow O_2 + O \tag{151}$$

$$O + O_3 \rightarrow O_2 + O_2 \tag{152}$$

Reactions (149) and (150) are the reactions by which the ozone is generated. Reactions (151) and (152) establish the balance, which is the ozone concentration in the troposphere. If one adds reactions (151) and (152), one obtains the overall rate of destruction of the ozone: namely,

$$O_3 + h\nu \rightarrow O_2 + O \tag{151}$$

$$O + O_3 \rightarrow O_2 + O_2 \tag{152}$$

$$\text{net} \quad 2O_3 + h\nu \rightarrow 3O_2 \tag{153}$$

The rates of reactions (149)–(152) vary with altitude. The rate constants of reactions (149) and (151) are determined by the solar flux at a given altitude, and the rate constants of the other reactions are determined by the temperature at that altitude. However, precise solar data obtained from rocket experiments and better kinetic data for reactions (150)–(152), coupled with recent meteorological analysis, have shown that the Chapman model was seriously flawed. The concentrations predicted by the model were essentially too high. Something else was affecting the ozone.

1. The HO_x Catalytic Cycle

Hunt [112, 113] suggested that excited electronic states of O atoms and O_2 might account for the discrepancy between the Chapman model and the measured meteorological ozone distributions. But he showed that reactions based on these excited species were too slow to account for the differences sought. Realizing that water could enter the stratosphere, Hunt considered the reactions of free radicals (H, HO, HOO) derived from water. Consistent with the shorthand developed for the oxides of nitrogen, these radicals are specified by the chemical formula HO_x. The mechanism that Hunt postulated was predicated on the formation of hydroxyl radicals. The photolysis of ozone by ultraviolet radiation below 310 nm produces excited singlet oxygen atoms which react rapidly with water to form hydroxyl radicals:

$$O_3 + h\nu \rightarrow O_2 + O(1D) \tag{154}$$

$$O(1D) + H_2O \rightarrow 2OH \tag{155}$$

Only an excited singlet oxygen atom could react with water at stratospheric temperatures to form hydroxyl radicals.

At these temperatures, singlet oxygen atoms could also react with hydrogen or methane to form OH. The OH reacts with O_3 to produce hydroperoxy radicals HO_2. Both HO and HO_2 destroy ozone by an indirect reaction which sometimes involves O atoms:

$$HO + O_3 \rightarrow HO_2 + O_2 \tag{156}$$

$$HO_2 + O_3 \rightarrow HO + O_2 + O_2 \tag{157}$$

$$HO + O \rightarrow H + O_2 \qquad (158)$$

$$HO_2 + O \rightarrow HO + O_2 \qquad (159)$$

$$H + O_3 \rightarrow HO + O_2 \qquad (160)$$

There are numerous reactions of HO_2 radicals possible in the stratosphere. The essential reactions for the discussion of the ozone balance are

$$HO + O_3 \rightarrow HO_2 + O_2 \qquad (156)$$

$$HO_2 + O_3 \rightarrow HO + O_2 + O_2 \qquad (157)$$

$$\text{net} \quad 2O_3 \rightarrow 3O_2$$

The reaction sequence (156)–(157) is a catalytic chain for ozone destruction and contributes to the net destruction. However, even given the uncertainty possible in the rates of these reactions and the uncertainty of the air motions, this system could not explain the imbalance in the ozone throughout the stratosphere.

2. The NO_x Catalytic Cycle

In the late 1960s, direct observations of substantial amounts (3 ppb) of nitric acid vapor in the stratosphere were reported. Crutzen [114] reasoned that if HNO_3 vapor is present in the stratosphere, it could be broken down to a degree to the active oxides of nitrogen NO_x (NO and NO_2) and that these oxides could form a catalytic cycle [or the destruction of the ozone]. Johnston and Whitten [115] first realized that if this were so, then supersonic aircraft flying in the stratosphere could wreak harm to the ozone balance in the stratosphere. Much of what appears in this section is drawn from an excellent review by Johnston and Whitten [115]. The most pertinent of the possible NO_x reactions in the atmosphere are

$$NO + O_3 \rightarrow NO_2 + O_2 \qquad (153)$$

$$NO_2 + O \rightarrow NO + O_2 \qquad (154)$$

$$NO_2 + h\nu \rightarrow NO + O \qquad (155)$$

whose rate constants are now quite well known. The reactions combine in two different competing cycles. The first is catalytic destructive:

$$NO + O_3 \rightarrow NO_2 + O_2 \qquad (3)$$

$$NO_2 + O \rightarrow NO + O_2 \qquad (158)$$

$$\text{net} \quad O_3 + O \rightarrow O_2 + O_2$$

and the second, parallel cycle is essentially a "do-nothing" one:

$$NO + O_3 \rightarrow NO_2 + O_2 \tag{3}$$

$$NO_2 + h\nu \rightarrow NO + O \tag{1}$$

$$O + O_2 + M \rightarrow O_3 + M \tag{150}$$

<hr/>

net   no chemical change

The rate of destruction of ozone with the oxides of nitrogen relative to the rate in "pure air" (Chapman model) is defined as the catalytic ratio, which may be expressed either in terms of the variables (NO_2) and (O_3) or (NO) and (O). These ratio expressions are

$$\beta = \frac{\text{rate of ozone destruction with } NO_x}{\text{rate of ozone destruction in pure air}} \tag{161}$$

$$\beta = 1 + \{k_{156}(NO_2)/k_{147}(O_3)\} \tag{162}$$

$$\beta = 1 + \{(k_{153}k_{154}/k_{147}j_{155})(NO)\} / \{1 + k_{154}(O)/j_{155}\} \tag{163}$$

Here, as throughout this book, the k's are the specific rate constants of the chemical reactions, while the j's are the specific rate constants of the photochemical reactions.

At low elevations, where the oxygen atom concentration is low and the NO_2 cycle is slow, another catalytic cycle derived from the oxides of nitrogen may be important:

$$NO_2 + O_3 \rightarrow NO_3 + O_2 \tag{164}$$

$$NO_3 + h\nu(\text{visible, day}) \rightarrow NO + O_2 \tag{165}$$

$$NO + O_3 \rightarrow NO_2 + O_2 \tag{3}$$

<hr/>

net $2O_3 + h\nu \rightarrow 3O_2$ (day)

The radiation involved here is red light, which is abundant at all elevations. Reaction (157) permits another route at night (including the polar night), which converts a few percent of NO_2 to N_2O_5:

$$NO_2 + O_3 \rightarrow NO_3 + O_2 \tag{166}$$

$$NO_2 + NO_3 + M \rightarrow N_2O_5 + M \quad \text{(night)} \tag{167}$$

The rate of reaction (152) is known accurately only at room temperature, and extrapolation to stratospheric temperature is uncertain; nevertheless, the extrapolated values indicate that the NO_3 catalytic cycle [reactions (159) and (160)] destroys ozone faster than the NO_2 cycle below 22 km and in the region where the temperature is at least 220 K.

The nitric acid cycle is established by the reactions

$$HO + NO_2 + M \rightarrow HNO_3 + M \qquad (168)$$

$$HNO_3 + h\nu \rightarrow OH + NO_2 \qquad (169)$$

$$HO + HNO_3 \rightarrow H_2O + NO_3 \qquad (170)$$

The steady-state ratio of nitrogen dioxide concentration to nitric acid can be readily found to be

$$[(NO_2)/(HNO_3)]_{ss} = [k_{164}/k_{162}] + [j_{163}/k_{162}(OH)] \qquad (171)$$

For the best data available for the hydroxyl radical concentration and the rate constants, the ratio has the values

$$0.1 \quad \text{at 15 km}, \qquad 1 \quad \text{at 25 km}, \qquad > 1 \quad \text{at 35 km}$$

Thus it can be seen that nitric acid is a significant reservoir or sink for the oxides of nitrogen. In the lowest stratosphere, the nitric acid predominates over the NO_2 and a major loss of NO_x from the stratosphere occurs by diffusion of the acid into the troposphere where it is rained out.

By using the HO_x and NO_x cycles just discussed and by assuming a NO_x concentration of 4.2×10^9 molecules/cm^3 distributed uniformly through the stratosphere, Johnston and Whitten [115] were able to make the most reasonable prediction of the ozone balance in the stratosphere. Measurements of the concentration of NO_x in the stratosphere show a range of 2–8 $\times 10^9$ molecules/cm^3.

It is possible to similarly estimate the effect of the various cycles upon ozone destruction. The results can be summarized as follows: between 15 and 20 km, the NO_3 catalytic cycle dominates; between 20 and 40 km, the NO_2 cycle dominates; between 40 and 45 km, the NO_2, HO_x, and O_x mechanisms are about equal; and above 45 km, the HO_x reactions are the controlling reactions.

It appears that between 15 and 35 km, the oxides of nitrogen are by far the most important agents for maintaining the natural ozone balance. Calculations show that the natural NO_x should be about 4×10^9 molecules/cm^3. The extent to which this concentration would be modified by anthropogenic sources such as supersonic aircraft determines the extent of the danger to the normal ozone balance. It must be stressed that this question is a complex one, since both concentration and distribution are involved (see Johnston and Whitten [115]).

3. The ClO_x Catalytic Cycle

Molina and Rowland [116] pointed out that fluorocarbons that diffuse into the stratosphere could also act as potential sinks for ozone. Cicerone *et al.* [117] show that the effect of these synthetic chemicals could last for decades. This possibly major source of atmospheric contamination arises by virtue of the widespread use of fluorocarbons as propellants and refrigerants. Approximately 80% of all fluorocarbons released to the atmosphere derive from these sources. There is no natural source.

Eighty-five percent of all fluorocarbons are Fl 1 (CCl_3F) or Fl 2 (CCl_2F_2). According to Molina and Rowland [116], these fluorocarbons are removed from the stratosphere by photolysis above altitudes of 25 km. The primary reactions are

$$CCl_3F + h\nu \rightarrow CCl_2F + Cl \tag{172}$$

$$CCl_2F + h\nu \rightarrow CClF + Cl \tag{173}$$

Subsequent chemistry leads to release of additional chlorine, and for purposes of discussion, it is here assumed that all of the available chlorine is eventually liberated in the form of compounds such as HCl, ClO, ClO_2, and Cl_2. The catalytic chain for ozone which develops is

$$Cl + O_3 \rightarrow ClO + O_2 \tag{174}$$

$$ClO + O \rightarrow Cl + O_2 \tag{175}$$

$$\text{net} \quad O_3 + O \rightarrow O_2 + O_2$$

Other reactions that are important in affecting the chain are

$$OH + HO_2 \rightarrow H_2O + O_2 \tag{176}$$

$$Cl + HO_2 \rightarrow HCl + O_2 \tag{177}$$

$$ClO + NO \rightarrow Cl + NO_2 \tag{178}$$

$$ClO + O_3 \rightarrow ClO_2 + O_2 \tag{179}$$

$$Cl + CH_4 \rightarrow HCl + CH_3 \tag{180}$$

$$Cl + NO_2 + M \rightarrow ClNO_2 + M \tag{181}$$

$$ClNO_2 + O \rightarrow ClNO + O_2 \tag{182}$$

$$ClNO_2 + h\nu \rightarrow ClNO + O \tag{183}$$

$$ClO + NO_2 + M \rightarrow ClONO_2 + M \tag{184}$$

The unique problem that arises here is that Fl 1 and Fl 2 are relatively inert chemically and have no natural sources or sinks, as CCl_4 does. The lifetimes of these fluorocarbons are controlled by diffusion to the stratosphere, where photodissociation takes place as designated by reactions (166) and (167). The lifetimes of halogen species in the atmosphere are given in Ref. [118]. These values are reproduced in Table 8. The incredibly long lifetimes of Fl 1 and Fl 2 and their

TABLE 8   Residence Time of Halocarbons in the Troposphere[a]

Halocarbon	Average residence time in years
Chloroform ($CHCl_3$)	0.19
Methylene chloride ($CHCl_3$)	0.30
Methyl chloride (CH_3Cl)	0.37
1,1,1-Trichloroethane (CH_3CCl_3)	1.1
F1 2	330 or more
Carbon tetrachloride (CCl_4)	330 or more
F1 1	1000 or more

[a] Based on reaction with OH radicals.

gradual diffusion into the stratosphere pose the problem. Even though use of these materials has virtually stopped today, their effects are likely to be felt for decades.

Some recent results indicate, however, that the rate of reaction (178) may be much greater than initially thought. If so, the depletion of ClO by this route could reduce the effectiveness of this catalytic cycle in reducing the O_3 concentration in the stratosphere.

PROBLEMS

1. Explain what is meant by *atmospheric*, *prompt*, and *fuel* NO.
2. Does prompt NO form in carbon monoxide–air flames? Why or why not?
3. Which is most sensitive to temperature—the formation of atmospheric, prompt, or fuel NO?
4. Order the tendency to soot of the following fuels under premixed combustion with air: hexadecane, ethyl benzene, cycloheptane, heptane, and heptene.

REFERENCES

1. Bowman, C. T., *Int. Symp. Combust., 24th* p. 859. Combust. Inst., Pittsburgh, Pennsylvania, 1992.
2. Leighton, P. A., "Photochemistry of Air Pollution," Chap. 10. Academic Press, New York, 1961; Heicklen, J., "Atmospheric Chemistry." Academic Press, New York, 1976.
3. Bulfalini, M., *Environ. Sci. Technol.* **8**, 685 (1971).
4. Blacet, F. R., *Ind. Eng. Chem.* **44**, 1339 (1952).
5. Dainton, F. S., and Irvin, K. J., *Trans. Faraday Soc.* **46**, 374, 382 (1950).
6. Miller, J. A., and Bowman, C. T., *Prog. Energy Combust. Sci.* **15**, 287 (1989).
7. Zeldovich, Y. B., *Acta Physicochem. USSR* **21**, 557 (1946).
8. Bowman, C. T., *Prog. Energy Combust. Sci.* **1**, 33 (1975).
9. Fenimore, C. P., *Int. Symp. Combust., 13th* p. 373. Combust. Inst., Pittsburgh, Pennsylvania, 1971.
10. Bowman, C. T., and Seery, D. V., "Emissions from Continuous Combustion Systems," p. 123. Plenum, New York, 1972.

11. Marteney, P. J., *Combust. Sci. Technol.* **1**, 461 (1970).

12. Bowman, C. T., *Int. Symp. Combust., 14th* p. 729. Combust. Inst., Pittsburgh, Pennsylvania, 1973.

13. Bachmeier, F., Eberius, K. H., and Just, T., *Combust. Sci. Technol.* **7**, 77 (1973).

14. Haynes, D. S., Iverach, D., and Kirov, N. Y., *Int. Symp. Combust., 15th* p. 1103. Combust. Inst., Pittsburgh, Pennsylvania, 1975.

15. Leonard, R. A., Plee, S. L., and Mellor, A. M., *Combust. Sci. Technol.* **14**, 183 (1976).

16. Eberius, K. H., and Just, T., Atmospheric Pollution by Jet Engines, *AGARD Conf. Proc.* **AGARD-CP**-125, p. 16-1 (1973).

17. Hayhurst, A. N., and Vance, I. M., *Combust. Flame* **50**, 41 (1983).

18. Martin, G. B., and Berkau, E. E., *AIChE Meet., San Francisco, 1971.*

19. Flagan, R. C., Galant, S., and Appleton, J. P., *Combust. Flame* **22**, 299 (1974).

20. Mulvihill, J. N., and Phillips, L. F., *Int. Symp. Combust., 15th* p. 1113. Combust. Inst., Pittsburgh, Pennsylvania, 1975.

21. Levy, A., *Int. Symp. Combust., 19th* p. 1223. Combust. Inst., Pittsburgh, Pennsylvania, 1982.

22. Fenimore, C. P., *Combust. Flame* **19**, 289 (1972).

23. deSoete, C. C., *Int. Symp. Combust., 15th* p. 1093. Combust. Inst., Pittsburgh, Pennsylvania, 1975.

24. Diehl, L. A., *NASA Tech. Memo.* **NASA TM-X-2726** (1973).

25. Hazard, H. R., *J. Eng. Power, Trans. ASME* **96**, 235 (1974).

26. Hargraves, K. J. A., Harvey, R., Roper, F. G., and Smith, D. B., *Int. Symp. Combust., 18th* p. 133. Combust. Inst., Pittsburgh, Pennsylvania, 1981.

27. Courant, R. W., Merryman, E. L., and Levy, A., *Int. Symp. Air Pollut., Health Energy Conserv.*, Univ. Mass., 1981.

28. Merryman, E. L., and Levy, A., *Int. Symp. Combust., 15th* p. 1073. Combust. Inst., Pittsburgh, Pennsylvania, 1975.

29. Cernansky, N. P., and Sawyer, R. F., *Int. Symp. Combust., 15th* p. 1039. Combust. Inst., Pittsburgh, Pennsylvania, 1975.

30. Allen, J. P., *Combust. Flame* **24**, 133 (1975).

31. Johnson, G. M., Smith, M. Y., and Mulcahy, M. F. R., *Int. Symp. Combust., 17th* p. 647. Combust. Inst., Pittsburgh, Pennsylvania, 1979.

32. Kesten, A. S., *Combust. Sci. Technol.* **6**, 115 (1972).

33. Bracco, F. V., *Int. Symp. Combust., 14th* p. 831. Combust. Inst., Pittsburgh, Pennsylvania, 1973.

34. Dryer, F. L., Mech. Aerosp. Sci. Rep. No. 1224. Princeton Univ., Princeton, New Jersey (1975).

35. Myerson, A. L., *Int. Symp. Combust., 15th* p. 1085. Combust. Inst., Pittsburgh, Pennsylvania, 1975.

36. Lyon, R. K., U.S. Pat. 3,900,554; *Int. J. Chem. Kinet.* **8**, 315 (1976).

37. Miller, J. A., Branch, M. C., and Kee, R. J., *Combust. Flame* **43**, 18 (1981).

37a. Perry, R. A., and Siebers, D. L., *Nature (London)* **324**, 657 (1986).

38. Levy, A., Merryman, E. L., and Ried, W. T., *Environ. Sci. Technol.* **4**, 653 (1970).

39. Cullis, C. F., and Mulcahy, M. F. R., *Combust. Flame* **18**, 222 (1972).

40. Johnson, G. M., Matthews, C. J., Smith, M. Y., and Williams, D. V., *Combust. Flame* **15**, 211 (1970).

41. Merryman, E. L., and Levy, A., *Int. Symp. Combust., 13th* p. 427. Combust. Inst., Pittsburgh, Pennsylvania, 1971.

42. Sachjan, G. A., Gershenzen, Y. M., and Naltandyan, A. B., *Dokl. Akad. Nauk SSSR* **175**, 647 (1971).

43. Muller, C. H., Schofield, K., Steinberg, M., and Broida, H. P., *Int. Symp. Combust., 17th* p. 687. Combust. Inst., Pittsburgh, Pennsylvania, 1979.

44. Myerson, A. L., Taylor, F. R., and Harst, P. L., *J. Chem. Phys.* **26**, 1309 (1957).

45. Barrett, R. E., Hummell, J. D., and Reid, W. T., *J. Eng. Power, Ser. A, Trans. ASME* **88**, 165 (1966).

46. Wendt, J. O. L., and Ekmann, J. M., *Combust. Flame* **25**, 355 (1975).

47. deSoete, G. G., Inst. Fr. Pet., Fuel Rep. No. 23306 (1975).

48. Wendt, J. O. L., Morcomb, J. T., and Corley, T. L., *Int. Symp. Combust., 17th* p. 671. Combust. Inst., Pittsburgh, Pennsylvania, 1979.
49. Palmer, H. B., and Cullis, H. F., "The Chemistry and Physics of Carbon," Vol. 1, p. 205. Dekker, New York, 1965.
50. Haynes, B. S., and Wagner, H. G., *Prog. Energy Combust. Sci.* **7**, 229 (1981).
51. Glassman, I., *Int. Symp. Combust., 22nd* p. 295. Combust. Inst., Pittsburgh, Pennsylvania, 1988.
52. Street, J. C., and Thomas, A., *Fuel* **34**, 4 (1955).
53. Takahashi, F., and Glassman, I., *Combust. Sci. Technol.* **37**, 1 (1984).
54. Miller, W. J., and Calcote, H. F., *East. States Sect. Combust. Inst. Meet., East Hartford, CT,* 1971; Calcote, H. F., Olson, D. B., and Kell, D. G., *Energy Fuels* **2**, 494 (1988).
55. Wright, W. J., *Int. Symp. Combust., 12th* p. 867. Combust. Inst., Pittsburgh, Pennsylvania, 1969.
56. Blazowski, W. S., *Combust. Sci. Technol.* **21**, 87 (1980).
57. Schalla, R. L., Clark, T. P., and McDonald, G. E., *Nat. Advis. Comm. Aeronaut. Rep.* No. 1186 (1954).
58. Schalla, R. L., and Hibbard, R. R., *Natl. Advis. Comm. Aeronaut. Rep.* No. 1300, Chap. 9 (1959).
59. Wolfhard, H. G., and Parker, W. G., *Proc. Phys. Soc.* **2**, A65 (1952).
60. Smyth, K. C., Miller, J. H., Dorfman, R. C., Mallard, W. G., and Santoro, R. J., *Combust. Flame* **62**, 157 (1985).
61. Wu, K. T., and Essenhigh, R. H., *Int. Symp. Combust., 20th* p. 1925. Combust. Inst., Pittsburgh, Pennsylvania, 1985.
62. Sidebotham, G., Ph.D. Thesis, Dep. Mech. Aerosp. Eng., Princeton Univ., Princeton, New Jersey, 1988.
63. Vandsburger, U., Kennedy, I., and Glassman, I., *Combust. Sci. Technol.* **39**, 263 (1984).
64. Vandsburger, U., Kennedy, I., and Glassman, I., *Int. Symp. Combust., 20th* p. 1105. Combust. Inst., Pittsburgh, Pennsylvania, 1985.
65. Hura, H. S., and Glassman, I., *Int. Symp. Combust., 22nd* p. 371. Combust. Inst., Pittsburgh, Pennsylvania, 1988.
66. Dryer, F. L., and Glassman, I., *Prog. Astronaut. Aeronaut.* **62**, 225 (1978).
67. Glassman, I., AFOSR TR-79-1147 (1979).
68. Milliken, R. C., *J. Phys. Chem.* **66**, 794 (1962).
69. Schug, K. P., Manheimer-Timnat, Y., Yaccarino, P., and Glassman, I., *Combust. Sci. Technol.* **22**, 235 (1980).
70. Glassman, I., and Yaccarino, P., *Int. Symp. Combust., 18th* p. 1175. Combust. Inst., Pittsburgh, Pennsylvania, 1981.
71. Gomez, A., Sidebotham, G., and Glassman, I., *Combust. Flame* **58**, 45 (1984).
72. Glassman, I., "Combustion," 2nd Ed. Academic Press, Orlando, Florida, 1987.
73. Takahashi, F., Bonini, J., and Glassman, I., *East. States Sect. Combust. Inst. Meet. Pap.* No. 98 (1984).
74. Harris, S. J., and Weiner, A. M., *Combust. Sci. Technol.* **31**, 155 (1983).
75. Harris, S. J., and Weiner, A. M., *Int. Symp. Combust., 20th* p. 969. Combust. Inst., Pittsburgh, Pennsylvania, 1985.
76. Glassman, I., Brezinsky, K., Gomez, A., and Takahashi, F., "Recent Advances in the Aerospace Sciences"(C. Casci, ed.), p. 345. Plenum, New York, 1985.
77. Glassman, I., Nishida, O., and Sidebotham, G., "Soot Formation in Combustion," p. 316. Springer-Verlag, Berlin, 1989.
78. Frenklach, M., Clary, D. W., and Ramachandran, K. M., *NASA [Contract. Rep.] CR* **NASA CR**-174880 (1985).
79. D'Alessio, A., DiLorenzo, A., Borghese, A., Beretta, F., and Masi, S., *Int. Symp. Combust., 16th* p. 695. Combust. Inst., Pittsburgh, Pennsylvania, 1977.
80. D'Alessio, A., "Particulate Carbon Formation During Combustion" (N.C. Siegla and G.W. Smith, eds.), p. 207. Plenum. New York, 1981.
81. Haynes, B. S., and Wagner, H. G., *Ber. Bunsenges. Phys. Chem.* **84**, 499 (1980).

82. Kent, J. H., Jandes, H., and Wagner, H. G., *Int. Symp. Combust., 18th* p. 1117. Combust. Inst., Pittsburgh, Pennsylvania, 1981.

83. Kent, J. H., and Wagner, H. G., *Combust. Flame* **47**, 52 (1982).

84. Santoro, R. J., Semerjian, H. G., and Dobbins, R. A., *Combust. Flame* **51**, 157 (1983).

85. Santoro, R. J., Yeh, T. T., and Semerjian, H. G., *Combust. Sci. Technol.* **53**, 89 (1987).

86. Gomez, A., Littman, M., and Glassman, I., *Combust. Flame* **70**, 225 (1987).

87. Hura, H. S., and Glassman, I., *Combust. Sci. Technol.* **53**, 1 (1987).

88. Hura, H. S., and Glassman, I., *Int. Symp. Combust., 22nd* p. 295. Combust. Inst., Pittsburgh, Pennsylvania, 1988.

89. Homann, K. H., and Wagner, H., G., *Int. Symp. Combust., 11th* p. 371. Combust. Inst., Pittsburgh, Pennsylvania, 1967.

90. Bittner, J. D., and Howard, J. B., *Int. Symp. Combust., 18th* p. 1105. Combust. Inst., Pittsburgh, Pennsylvania, 1981.

91. Takahashi, F., and Glassman, I., *East. States Sect. Combust. Inst. Meet.* Pap. No. 12 (1982).

92. Kent, J. H., and Wagner, H. G., *Int. Symp. Combust., 20th* p. 1007. Combust. Inst., Pittsburgh, Pennsylvania, 1985.

93. Stein, S. E., and Kafafi, S. A., *East. States Sect. Combust. Inst. Meet.* Invited Pap. A (1987).

94. Kern, R. D., Wu, C. H., Yong, J. N., Pamedemukkala, K. M., and Singh, H. J., *Energy Fuels* **2**, 454 (1988).

94a. Miller, J. A., and Melius, C. F., *Combust. Flame* **91**, 21 (1992).

95. Sidebotham, G., Ph. D. Thesis, Dep. Mech. Aerosp. Eng., Princeton Univ., Princeton, New Jersey, 1988.

96. Calcote, H. F., *Combust. Flame* **42**, 215 (1981).

97. Gordon, A. S., Smith, S. R., and Nesby, J. R., *Int. Symp. Combust., 7th* p. 317. Combust. Inst., Pittsburgh, Pennsylvania, 1959.

98. Crittenden, B. D., and Long, R., *Combust. Flame* **20**, 239 (1973).

99. Glassman, I., AFOSR TR-79-1147 (1979).

100. Benson, S. W., *Int. Symp. Combust., 19th* p. 1350. Combust. Inst., Pittsburgh, Pennsylvania, 1983.

101. Smyth, K. C., and Miller, J. H., *Science* **236**, 1540 (1987).

102. Frenklach, M., Clary, D. W., Gardiner, W. C., Jr., and Stein, S. E., *Int. Symp. Combust., 12th* p. 889. Combust. Inst., Pittsburgh, Pennsylvania, 1985.

103. Frenklach, M., Clary, D. W., Yuan, T., Gardiner, W. C., Jr., and Stein, S. E., *Combust. Sci. Technol.* **50**, 79 (1986).

104. Cole, J. A., Bittner, J. D., Longwell, J. P., and Howard, J. B., *Combust. Flame* **56**, 51 (1984).

105. Westmoreland, P. R., Howard, J. B., and Longwell, J. P., *Int. Symp. Combust., 21st* p. 773. Combust. Inst., Pittsburgh, Pennsylvania, 1988.

106. Colket, M. B., III, *Int. Symp. Combust., 21st* p. 851. Combust. Inst., Pittsburgh, Pennsylvania, 1988.

107. Wu, C. H., and Kern, R. D., *J. Phys. Chem.* **91**, 6291 (1987).

108. Benson, S.W., *Int. Symp. Combust., 21st* p. 703. Combust. Inst., Pittsburgh, Pennsylvania, 1988.

109. Zhang, J., and Megaridis, C.M., *Int. Symp. Combust., 25th* p. 593. Combust. Inst., Pittsburgh, Pennsylvania, 1994.

110. Bonczyk, P.A., *Combust. Flame* **87**, 233 (1991).

111. Chapman, S., *Philos. Mag.* **10**, 369 (1930).

112. Hunt, B. G., *J. Atmos. Sci.* **23**, 88 (1965).

113. Hunt, B. G., *J. Geophys. Res.* **71**, 1385 (1966).

114. Crutzen, P. J., *R. Meteorol. Soc. Q. J.* **96**, 320 (1970).

115. Johnston, H., and Whitten, G., *AGARD Conf. Proc.* **AGARD-CP**-125, p. 2-1 (1973).

116. Molina, M. I., and Rowland, F. S., *Nature (London)*, **249**, 810 (1974).

117. Cicerone, R. J., Stolarski, R. S., and Walters, S., *Science* **185**, 1165 (1974).

118. "Fluorocarbons and the Environment." Counc. Environ. Qual., Washington, D.C., 1975.

9

Combustion of Nonvolatile Fuels

A. CARBON CHAR, SOOT, AND METAL COMBUSTION

The final stages of coal combustion produce a nonvolatile char which must be consumed to obtain good combustion efficiencies. The combustion of this char—a factor that has not yet been considered—is essentially a surface burning process similar to that occurring when carbon graphite burns. Coal is a natural solid fuel that contains carbon, moisture, ash-generating minerals, and a large number of different hydrocarbons that volatilize when combustion is initiated. The volatiles in coal contribute a substantial amount to the overall energy release. But the volatiles burn rapidly compared to the solid carbonaceous char that remains after devolatilization. It is the surface burning rate of this remaining nonvolatile solid carbonaceous fuel that determines the efficiency of the coal combustion process.

Similarly, the emission of soot from many practical devices, as well as from flames, is determined by the rate of oxidation of these carbonaceous particles as they pass through a flame zone and into the post-combustion gases. As mentioned in the previous chapter, the soot that penetrates the reaction zone of a co-annular diffusion flame normally burns if the temperatures remain above 1300 K. This soot combustion process takes place by surface oxidation.

Heterogeneous surface burning also arises in the combustion of many metals. Since the energy release in combustion of metals is large, many metals are used as additives in solid propellants. Indeed, in the presence of high oxygen concentrations, metal containers, particularly aluminum, have been known to burn, thereby leading to serious conflagrations.

Not all metals burn heterogeneously. The determination of which metals will burn in a heterogeneous combustion mode can be made from a knowledge of the thermodynamic and physical properties of the metal and its oxide product [1].

In the field of high-temperature combustion synthesis, metals have been reacted with nitrogen, both in the gaseous and liquid phases, to form refractory nitrides [2]. In most cases, this nitriding process is heterogeneous.

B. METAL COMBUSTION THERMODYNAMICS

1. The Criterion for Vapor-Phase Combustion

What is unique about metal particle burning in oxygen is that the flame temperature developed is a specific known value—the vaporization–dissociation or volatilization temperature of the metal oxide product. This interesting observation is attributable to the physical fact that the heat of vaporization–dissociation or decomposition of the metal oxide formed is greater than the heat available to raise the condensed state of the oxide above its boiling point. That is, if Q_R is the heat of reaction of the metal at the reference temperature 298 K and $(H^\circ_{T,\text{vol}} - H^\circ_{298})$ is the enthalpy required to raise the product oxide to its volatilization temperature at the pressure of concern, then

$$\Delta H_{\text{vap-dissoc}} > Q_R - (H^\circ_{T,\text{vol}} - H^\circ_{298}) = \Delta H_{\text{avail}} \qquad (1)$$

where $\Delta H_{\text{vap-dissoc}}$ is the heat of vaporization–dissociation of the metal oxide [3]. Note that Q_R is the negative of the heat of formation of the metal oxide. Equation (1) assumes conservatively that no vaporization or decomposition takes place between the reference temperature and the volatilization temperature.

The developments in Chapter 6 show that in the steady state the temperature of the fuel particle undergoing combustion approaches its boiling point (saturation) temperature at the given pressure. Characteristically, it is a few degrees less. For a condensed-phase fuel to burn in the vapor phase, the flame temperature must be greater than the fuel saturation temperature so that the fuel will vaporize and diffuse toward the oxidizing atmosphere to react. For liquid hydrocarbon fuels, the droplet flame temperature is substantially higher than the fuel saturation temperature. However, many metals have very high saturation temperatures. Thus, for a metal to burn as a vapor, the oxide volatilization temperature must be greater than the temperature of the metal boiling point. This statement is known as Glassman's criterion for the vapor-phase combustion of metals. The metals that will burn in the vapor phase in oxygen can then be determined by comparing the second and last columns of Table 1 for the appropriate oxide.

TABLE 1   Various Properties of Metal Oxides and Nitrides

Metallic compound	$T_{vol}^{a,c}$ (K)	$\Delta H_{vol}^{a,c}$ (kJ/mol)	$(H_{T_{vol}}^{\circ} - H_{298}^{\circ}) + \Delta H_{vol}$ (kJ/mol)	$\Delta H_{f,298}^{\circ}{}^{a}$ (kJ/mol)	*Metal* T_b^b (K)
AlN	2710 (2710)	620 (620)	740 (730)	−318	2791
Al$_2$O$_3$	4000	1860	2550	−1676	2791
BN	4010 (2770)	840 (300)	950 (410)	−251	4139
B$_2$O$_3$	2340	360	640	−1272	4139
BeO	4200	740	1060	−608	2741
Cr$_2$O$_3$	3280	1160	1700	−1135	2952
FeO	3400	610	830	−272	3133
Li$_2$O	2710	400	680	−599	1620
MgO	3430	670	920	−601	1366
Si$_3$N$_4$	3350 (2150)	2120 (870)	2420 (1160)	−745	3505
TiN	3540 (3450)	700 (450)	950 (710)	−338	3631
Ti$_3$O$_5$	4000	1890	2970	−2459	3631
ZrO$_2$	4280	920	1320	−1097	4703

T_{vol} = volatilization temperature (or stoichiometric adiabatic combustion temperature creating compound under ambient conditions T = 298 K, P = 1 atm).
T_b = metal boiling point at 1 atm.
T_e = decomposition temperature (see text).
^{a}Values reported are rounded to the nearest integer.
^{b}Values from JANAF Tables.
^{c}Value in parentheses are (or are based on) decomposition temperatures, T_d, or enthalpies of decomposition, ΔH_d.

2. Thermodynamics of Metal–Oxygen Systems

Modern computational programs [4] and thermodynamic tables [5] now make it possible to explicitly calculate metal–oxygen flame temperatures, thereby opening up a unique aspect of combustion thermodynamics that could be important in the consideration of metal as fuels and as reactants in combustion synthesis.

As early as 1953, Brewer [6] elaborated on the fact that an equilibrium thermodynamic boiling point is difficult to define for many metal oxides. Indeed, the experimental evidence demonstrated that many metal oxides volatilize exclusively by decomposition to gaseous suboxides and elements such as O$_2$, or by dissociation to species such as O. Few actually formed gaseous molecules of the original metallic oxide. Subsequent investigators [1, 7] recognized the importance of dissociation in metal combustion systems. But despite the understanding of the thermodynamic definition of boiling point implied by this recognition, the term was used to describe the vaporization–decomposition process of metal oxides. The JANAF tables [5] recognize the difficulty of defining a true boiling point for metallic oxides. In a few cases, a "decomposition temperature" is defined for a particular reaction equilibrium. For MgO, for example, a "decomposition or boiling temperature (3533 K)" is defined at 1 atm, where the free energy change of the reaction MgO(l) → MgO(g) approaches zero. Similarly, a heat of vaporization is defined as the difference between $\Delta H_{f,3533}^\circ$ for MgO(g) and MgO(l), where ΔH_f° is the standard state heat of formation [5]. This definition, however, is not acceptable in the case of many metal oxides, including MgO. The term "boiling temperature" would be correct for the Mg–O$_2$ system if MgO(l) were vaporized congruently to MgO(g). But as the JANAF tables note, "vaporization near 2000 K is primarily to the elements Mg(g) and presumably O(g) and O$_2$(g) rather than to MgO(g)"[5]. The term "vaporization–decomposition" temperature would probably be more appropriate to describe the volatilization of metallic oxides. In fact, this temperature was later defined as a "volatilization temperature, exhibiting characteristics of a transition temperature or boiling point" [6]. Furthermore, in order to distinguish between the condition of complete decomposition of a nitride and the condition in which all the decomposed products are gases, the decomposition temperature and volatilization temperatures will be defined more explicitly.

Von Grosse and Conway [7] were the first to realize that the evaluation of the adiabatic combustion temperatures for metal–oxygen systems was largely dependent on the metal oxide volatilization characteristics. Recognizing that many metallic oxides do not exert any appreciable vapor pressure until they reach very high temperatures, these investigators concluded that the enthalpy required for complete volatilization was large. Citing aluminum oxide as an example, von Grosse and Conway noted that "only about 140 kcal/mol are required to heat aluminum oxide to its boiling point· · · · The heat amounting to the difference between 400.2 (the enthalpy of reaction) and 140 kcal must be interpreted as being used to vaporize and decompose the aluminum oxide at its boiling point (aluminum oxide decomposes on vaporization). *The combustion temperature is thus limited to the boiling point of the oxide* [emphasis added]. This is the case in many metal–oxygen systems, but each reaction must be evaluated independently."

The current ability to readily calculate combustion temperatures has confirmed the concept of a "limiting temperature" [1, 7] in metal processes. The term *limit temperature* is now used as a formal statement that the enthalpy of combustion of the reactants at the stoichiometric ambient (298 K) condition is insufficient to volatilize all the condensed-phase metallic oxide formed; i.e., the metallic oxide undergoes a phase change. This concept of a limit temperature can be applied to metal–nitrogen systems as well [8]. Shown in Fig. 1 are the adiabatic combustion temperatures at 1 atm, calculated for the oxidation and nitriding of many metals at 298 K as a function of the equivalence ratio ϕ. Of course, a value of ϕ greater or less than 1 signifies an excess of one or the other of the reactants. Note that the temperature variation over a wide range (0.5–1.5) of equivalence ratios varies by only a few percent for many metal–oxygen/nitrogen systems. Boron is an exception because, at the stoichiometric value with oxygen, no condensed phases exist in the combustion products. Off stoichiometric for other metal–oxygen/nitrogen systems, a temperature is reached that will vary minimally as stoichiometric is approached. The only reason there is a variation with equivalence ratio is that the equilibria of the decomposed metal oxide or nitride species are influenced by an

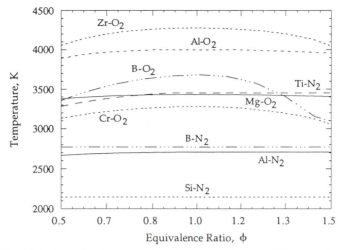

FIGURE 1   Adiabatic combustion temperature of various metal–oxygen/nitrogen systems as a function of equivalence ratio ϕ. Initial conditions 298 K and 1 atm.

excess of one of the reactants. Of course, it is possible to be so far off stoichiometric that the temperature drops appreciably to a value indicative of transition temperatures for other condensed-phase products formed. The point is that the very weak dependence of the combustion temperature over a wide range of equivalence ratios near stoichiometric further confirms the reasoning that the adiabatic combustion temperature in pure oxygen and nitrogen is limited due to the volatilization of the condensed-phase product formed. It would be possible, of course, to exceed this limiting temperature for a stoichiometric metal–oxygen/nitrogen system if the reactants were initially at very high temperatures.

The equilibrium combustion temperatures and compositions of many metal–oxygen/nitrogen systems were calculated by assigning various values of the total enthalpy, H_T°, of these reactants at a given, fixed total pressure [8]. One should recognize that this procedure is analogous to varying the total enthalpy or the enthalpy of formation of the product metal oxide or nitride. Performing these calculations for stoichiometric proportions of metal and oxidizer provides a means for the determination of the enthalpies needed to completely decompose and/or volatilize the metal oxide or nitride in question. Shown in Figs. 2–9 are the results for most of the systems designated by the product compounds listed in Table 1. An input enthalpy of zero is a condition equivalent to one in which the reactants enter at the ambient (298 K) state. Examining Fig. 2 for the Al–O_2 system, one notes that the product composition shows a 0.216 mole fraction of Al_2O_3 liquid at this initial reactant state. As the input enthalpy increases, the temperature remains the same, but the amount of Al_2O_3 liquid decreases. When the assigned enthalpy reaches 8.56 kJ/g reactants, condensed-phase Al_2O_3 no longer exists and any further increase in the total assigned enthalpy raises the temperature. It is also possible to

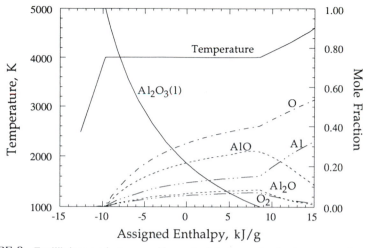

FIGURE 2   Equilibrium product composition and adiabatic combustion temperature for stoichio-metric Al and O_2 at 298 K and 1 atm as a function of assigned enthalpy. An assigned enthalpy of zero corresponds to the true ambient condition. (*l*) specifies a liquid product, (s) specifies a solid product, all other products are gases.

withdraw enthalpy from the system. Again, the temperature does not change, but the amount of Al_2O_3 liquid increases. When an enthalpy of 9.68 kJ/g reactants is withdrawn, the temperature begins to drop and the product composition contains only condensed-phase Al_2O_3. These results verify that a "pseudo–phase change" exists and that the limiting combustion equilibrium temperature has characteristics "similar to a boiling point." An enthalpy of volatilization can thus be calculated for Al_2O_3 (MW 102) to be [{8.56 − (−9.68)}102] = 1860 kJ/mol. Obviously, this enthalpy of volatilization/decomposition holds only for the stoichiometric condition, or more precisely, when the ratio of metal to oxygen (nitrogen) reactants is equal to that of the oxide (nitride) in question. This value of 1860 kJ/mol plus the enthalpy 688 kJ/mol needed to raise Al_2O_3 liquid to the limiting temperature is clearly greater than 1676 kJ/mol, the overall enthalpy of reaction or the enthalpy of formation of the metal oxide at the ambient condition (see Table 1). This example and the results shown in Figs. 2–9 appear to verify that limiting combustion temperatures are reached for many metal–oxygen/nitrogen systems.

It is necessary to distinguish between the limiting decomposition temperature and the limiting volatilization temperature as well as their corresponding enthalpies of decomposition and volatilization. For example, depending on the system and system pressure, a decomposing refractory product may produce a species in a condensed phase. There is a temperature associated with this condition desig-nated as the limiting decomposition temperature. As the assigned enthalpy is increased past the value that produced the limiting decomposition temperature, the condensed-phase product of decomposition volatilizes further. The tempera-

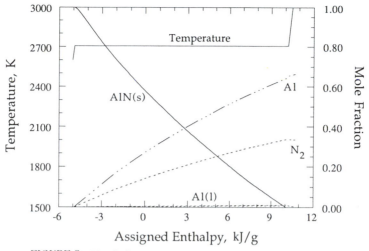

FIGURE 3   Plot similar to Fig. 2 for the stoichiometric Al–N$_2$ system.

ture increases gradually with respect to this assigned enthalpy increase until the condensed-phase product species is completely vaporized, at which point the temperature rises sharply. The point of this sharp rise is designated the volatilization temperature. At 1 atm, none of the oxide systems exhibits a condensed-phase decomposition product so that the decomposition and volatilization temperatures, T_d and T_{vol}, are the same. These two limiting temperatures and their corresponding

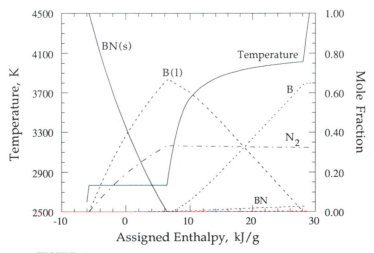

FIGURE 4   Plot similar to Fig. 2 for the stoichiometric B–N$_2$ system.

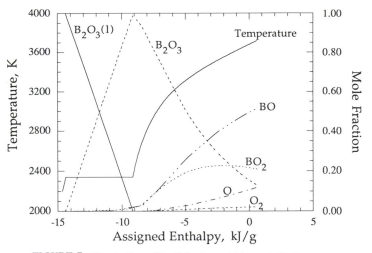

FIGURE 5   Plot similar to Fig. 2 for the stoichiometric B–O$_2$ system.

enthalpies are readily seen in Fig. 7 for the Ti–N$_2$ system at 1 atm. This consideration could be particularly important in various combustion synthesis approaches.

A volatilization temperature for the B–O$_2$ system can be defined only if sufficient enthalpy is withdrawn to allow for B$_2$O$_3$(l) [see Fig. 5]. This value of 2340 K corresponds exactly to the "boiling point" (2340 K) reported in the JANAF tables as the temperature at which the fugacity is 1 atm for B$_2$O$_3$(l) $\rightarrow$ B$_2$O$_3$(g). Since

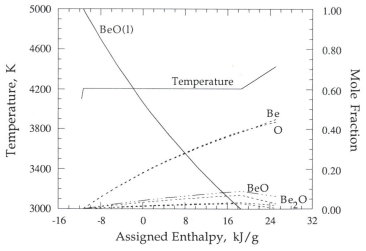

FIGURE 6   Plot similar to Fig. 2 for the stoichiometric Be–O$_2$ system.

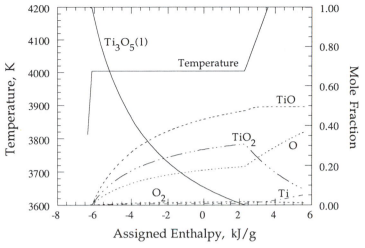

FIGURE 7 Plot similar to Fig. 2 for the stoichiometric Ti–N$_2$ system.

B$_2$O$_3$(l) vaporizes congruently to B$_2$O$_3$(g), the boiling point reported for this case should equal the calculated volatilization temperature.

The JANAF tables specify a volatilization temperature of a condensed-phase material to be where the standard-state free energy ΔG_f° approaches zero for a given equilibrium reaction, i.e., M$_x$O$_y$(l) $\rightleftharpoons$ M$_x$O$_y$(g). One can obtain a heat of vaporization of materials such as Li$_2$O(l), FeO(l), BeO(l), and MgO(l), which

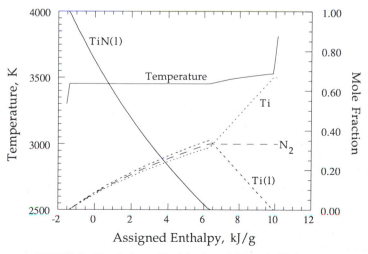

FIGURE 8 Plot similar to Fig. 2 for the stoichiometric Ti–O$_2$ system.

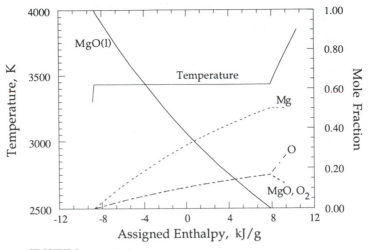

FIGURE 9   Plot similar to Fig. 2 for the stoichiometric Mg–O$_2$ system.

also exist in the gas phase, by the difference in the ΔH_f°'s of the condensed and gas phases at this volatilization temperature. This type of thermodynamic calculation attempts to specify a true equilibrium thermodynamic volatilization temperature and enthalpy of volatilization at 1 atm. Values determined in this manner would not correspond to those calculated by the approach described simply because the procedure discussed takes into account the fact that some of the condensed-phase species dissociate upon volatilization.

Examination of Fig. 4 for the B–N$_2$ system reveals that BN decomposes into gaseous nitrogen and liquid boron. Since these elements are in their standard states at 1 atm and the decomposition temperature, the ΔH_d° must equal the enthalpy of formation ΔH_f° of the BN at the decomposition temperature. Indeed, the ΔH_d°, (300 kJ/mol) calculated by the means described agrees with the value of $\Delta H_{f,T_d}^\circ$ (300 kJ/mol) given in the JANAF tables, as it should. The same condition holds for the Si–N$_2$ system.

As noted in Fig. 7, over the range of assigned enthalpies that define the limiting decomposition temperature, the major TiN decomposition products are Ti(l), Ti(g), and N$_2$. Since both Ti(g) and Ti(l) exist, the products of decomposition are not in their standard states. For TiN, then, unlike BN, the enthalpy of decomposition ΔH_d° will not equal the enthalpy of formation at the decomposition temperature. When the assigned enthalpy is increased to 9.9 kJ/(g) reactants for the Ti–N$_2$ system, the decomposition products are only gases [Ti(g) and N$_2$]. This condition specifies a volatilization temperature of 3540 K and a partial Ti(g) pressure of 0.666 atm. At 0.666 atm, the vaporization temperature for titanium has been found to be 3530 K. Indeed, neglecting dissociation, the values should be the same because the vaporization of Ti becomes the limiting condition. The enthalpy of volatilization for TiN, as determined by the procedure described here, is then found to be 700 kJ/mol. A value for the enthalpy of volatilization can be estimated as the heat of formation of TiN at 3630 K [7], the vaporization temperature of Ti at 1 atm. This value is of the order of 690 kJ/mol. Since the enthalpy of vaporization of Ti is not a function of pressure or temperature, the agreement between the value of enthalpy of volatilization of TiN at 3540 K and the estimated value of the enthalpy of formation of TiN from the JANAF tables at 3630 K, where the value has been determined at 1 atm, is not surprising.

The volatilization process of the metal oxides behaves like a "pseudo–boiling point." For the volatilization of liquid Al_2O_3 alone, Brewer and Searcy [9] clearly pointed out that $AlO(g)$ is the principal species, so that $Al_2O_3 \rightarrow 2AlO + O$ is the principal reaction of concern. From Fig. 2, one sees clearly that as enthalpy is increased in the stoichiometric $Al–O_2$ system, the equilibrium shifts to favor more $AlO(g)$ and $O(g)$ and less $Al_2O_3(l)$. As in the case of Al_2O_3, such dissociated species as O are important in the volatilization process of many metal oxides. Dissociation is also evident with the nitrides. Therefore, if the pressure is increased, as dictated by Le Chatelier's principle, less dissociation occurs; thus, a smaller enthalpy of volatilization or decomposition and a higher adiabatic combustion temperature are found for most of the metal–oxygen and nitrogen systems examined. Figure 10 depicts the temperature variation of the $Al–O_2$ system over a tenfold variation in pressure. Indeed, if the total pressure is increased from 1 to 10 atm in the stoichiometric $Al–O_2$ system, the heat of volatilization decreases by approximately 7% and the temperature increases by 15%. Similarly, there is an approximate 4% decrease in the volatilization enthalpy and a 17% increase in the temperature for the $Mg–O_2$ system. With respect to the compositions shown in Figs. 2–9, the variations in the species composition can be explained by the combination of temperature and pressure effects, not just by pressure alone.

The volatilization or pseudo–boiling point temperature and the corresponding enthalpies of decomposition and volatilization of all the systems examined as a function of total pressure are reported in Table 2. The variation in the enthalpy of volatilization as a function of pressure is generally less than 7%. The $Ti–O_2$ system (Fig. 8) shows an anomalous rise in ΔH_{vol}° with pressure. When the $Al–N_2$ system is raised from 0.1 to 10 atm, the $Al(g)$ product condenses and the species of decomposition are no longer in their standard states. When the assigned enthalpy is increased further so that the Al returns to the gaseous state, an enthalpy of volatilization ΔH_{vol}° based on this new assigned enthalpy can be defined. The values of ΔH_{vol}° specify these new values based on the condition that the elements are in their gaseous states at the pressure and volatilization temperature. When values of ΔH_{vol}° for the metal–nitrogen systems are calculated for the condition in which the decomposition products are completely in the gaseous state, the pressure variation of ΔH_{vol}°, unlike ΔH_d^o, is minimal.

The apparently anomalous result that the ΔH_{vol}° of the $Ti–O_2$ system rises with pressure is explained by examining the volatilization product compositions

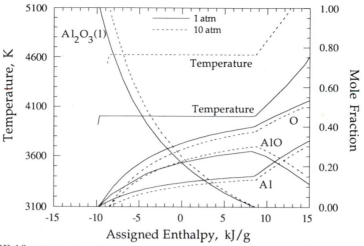

FIGURE 10   Comparison of the plot of adiabatic combustion temperature and product composition of the stoichiometric Al–O$_2$ system at total pressure of 1 atm (Fig. 2) and that at 10 atm.

as a function of pressure. The following equilibria exist for this system:

$$Ti_3O_5(l) \rightleftharpoons 2TiO_2 + TiO \tag{2}$$

$$TiO_2 \rightleftharpoons TiO + O \tag{3}$$

As the pressure is increased in the system, the equilibrium designated as reaction (2) shifts to Ti$_3$O$_5$(l) because the pressure effect overrides the increase in temperature with pressure that would make the shift to the gaseous products. Indeed, at a given assigned enthalpy, the Ti$_3$O$_5$(l) mol fraction increases as the total pressure increases. For the equilibrium designated as reaction (3), however, the increase of temperature with pressure overrides the shift that would occur due to pressure; so there is more dissociation, resulting in a greater quantity of O atoms. This trend explains the increase in ΔH_{vol}° with pressure for the Ti–O$_2$ system and for other systems that form the more complex stable oxides.

Considering that dissociation occurs upon volatilization, the temperatures can be correlated extremely well on a ln P vs ($1/T_{d,vol}$) plot, where P is the total system pressure and $T_{d,vol}$ is the volatilization or decomposition temperature, as the case dictates. Such a plot is shown in Fig. 11. Since the Clausius–Clapeyron relation for vapor pressure of pure substances shows an exponential dependence on temperature, T_{vol} was considered a pseudo–boiling point at the respective system pressure. For a substance that vaporizes congruently to its gaseous state, the slope of lines on a ln P vs ($1/T_{vol}$) plot represents the enthalpy of vaporization. Indeed, the enthalpy of vaporization calculated from the slope on a ln P vs ($1/T_{vol}$) plot for the B–O$_2$ system (360 kJ/mol) agrees exactly with the value calculated by using

TABLE 2   Temperatures and Heats of Volatilization of Various Oxide and Nitride Products at Various Pressures*

Metallic compound	$T_{vol}^{a,b}$ (K)			$\Delta H_{vol}^{a,b}$ (kJ/mol)		
	0.1 atm	1 atm	10 atm	0.1 atm	1 atm	10 atm
AlN	2410(2410)	2710(2710)	3290(3030)	630(630)	620(620)	620(370)
Al_2O_3	3540	4000	4620	1970	1860	1750
BN	3470(2540)	4010(2770)	4770(3040)	830(300)	840(300)	850(300)
B_2O_3	2080	2340	2670	370	360	360
BeO	3640	4200	4980	740	740	730
Cr_2O_3	2910	3280	3760	1180	1160	1140
FeO	2940	3400	4050	610	610	610
Li_2O	2330	2710	3270	410	400	400
MgO	(s)	3430	4020	(s)	670	640
Si_3N_4	2880(1960)	3350(2150)	4000(2370)	2120(870)	2120(870)	2140(860)
TiN	(s)	3540(3450)	4250(3960)	(s)	700(460)	700(370)
Ti_3O_5	3470	4000	4740	1800	1890	1930
ZrO_2	3790	4280	4920	930	920	910

*Temperatures are those obtained by the reacting systems creating the products under stoichiometric conditions for initial temperature of 298 K.

[a] Values reported are rounded to nearest integer.

[b] Values in parentheses are decomposition temperatures, T_d, or enthalpies of decomposition, ΔH_d. (s) = solid forms.

the procedure outlined. Since the other metal–oxides/nitrides examined do not vaporize congruently to their gaseous state, it is quite apparent that enthalpies of dissociation play a role in determining the slope of the ln P vs ($1/T_{vol}$) plots.

Table 1 and Fig. 11 also depict vaporization temperatures of the metals in each product composition and give a graphical representation of Glassman's criterion. When T_{vol} (or T_d, as the case dictates) of the refractory compound formed is greater than the vaporization temperature, T_b, of the metal reactant, small metal particles will vaporize during combustion and burn in the vapor phase. When the contra condition holds, much slower surface reactions will take place. This temperature condition could change with pressure; however, change is not likely to occur over a large pressure variation [3]. Thus in pure oxygen, Al, Be, Cr, Li, Mg, and Ti fit the criterion for vapor-phase combustion while B and Zr do not. The temperatures for the criterion for vapor-phase combustion at 1 atm for Fe and Ti are close (about 300 K for Fe and 400 K for Ti); consequently, thermal losses from the flame front would make the actual flame temperature less than the volatilization temperature of FeO and Ti_3O_5, just to complicate matters. With regard to combustion synthesis processes, it would appear that, at least for nitride formation, heterogeneous surface reactions would dominate the synthesis procedure.

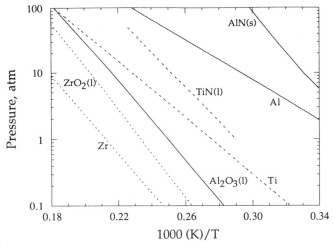

FIGURE 11   Vapor pressure–temperature relationships of various vaporization and combustion temperatures of various metal–oxygen/nitrogen systems as a function of total pressure, both plotted in the form ln P vs $1/T$.

3. Thermodynamics of Metal–Air Systems

In the metal reaction systems described in the preceding section, the gaseous atmosphere was either pure oxygen or pure nitrogen. Two questions now arise: Would a metal burning in air have a flame temperature equal to the vaporization–dissociation temperature of the metal oxide at the total pressure of the system? And would a temperature plateau exist over a range of equivalence ratios or over a range of assigned enthalpies?

Figure 12 details the same type of stoichiometric calculations as shown in earlier figures, except that given amounts of inert (argon) are added to an aluminum–oxygen mixture [10]. In one case, 8.46 mol of argon is added to the stoichiometric amount of oxygen, and in another 2.82 mol. Argon was considered instead of nitrogen because the formation of gaseous nitrogen oxides and aluminum nitride would obscure the major thermodynamic point to be made. As one can see from Fig. 12, the results with inert addition do not show a complete temperature plateau as the assigned enthalpy is varied. Next, a question arises as to whether the volatilization of the condensed-phase product oxide controls the combustion temperature. Analysis of points 1, 2, 3, and 4 specified on this figure verifies that it does. Table 3 lists the data for the explicit calculations at the four points.

It is evident from the earlier discussion of the titanium–nitrogen system that the final volatilization temperature (3540 K) was controlled by the complete vaporization of the titanium formed due to the dissociation of the product titanium nitride. The partial pressure of titanium vapor was equal to 0.666 atm at the final volatilization temperature of 3540 K (see Fig. 7). Indeed, the vaporization temperature of titanium at 0.666 atm is 3530 K. Consequently, to analyze the results

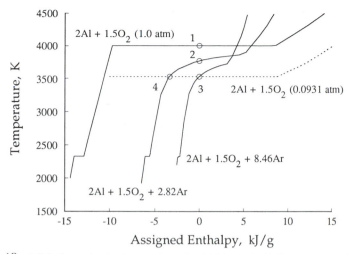

FIGURE 12 Adiabatic combustion temperature of a stoichiometric Al–O$_2$ system containing various amounts of the inert diluent argon as a function of assigned enthalpy, other conditions being the same as those for Fig. 2.

in Fig. 12, the conditions represented by points 3 and 4 were selected in order to consider what the gaseous partial pressures of the aluminum oxide vaporization components would be at the same temperature of 3500 K. Even though different amounts of assigned enthalpies were used in each case, the partial pressures of the oxide volatilization gases (the total pressure less the partial pressure of argon) are equal to 0.0931 atm in both cases.

Figure 13 shows a plot of the calculated adiabatic flame temperature for the stoichiometric Al–O$_2$ system as a function of pressure in the form used previously. The solid line labeled 2Al+1.5 O2 is based on the condition that no inert was added. What is significant is that points 3 and 4 on Fig. 13 were found to have a partial pressure of 0.0931 atm for the dissociated gases—the remaining gas contributing to the total pressure of 1 atm being argon. For a pressure of 0.0931 atm, Fig. 12 gives a temperature of 3500 K for a stoichiometric Al–O$_2$

TABLE 3 Summary Data for Figure 11

Point	H_0° [kJ/g]	T (K)	$P_{tot} - P_{Ar}$ (atm)	Al$_2$O$_3$(l) (mole fraction)	Ar (mole fraction)
1	0.000	4005	1.0000	0.2160	0.0000
2	0.000	3769	0.3350	0.1178	0.5868
3	−0.000	3527	0.0931	0.0736	0.8402
4	−3.351	3527	0.0931	0.2271	0.7009

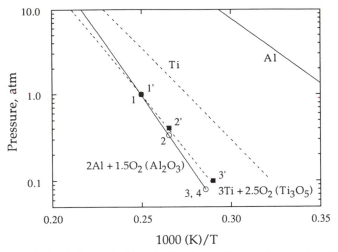

FIGURE 13   A plot similar to Fig. 11 to elucidate the significance of the points on Fig. 12.

mixture. Similarly, for the nonargon gases, the partial pressure for point 2 is 0.3350 atm and the corresponding temperature is 3769 K. Point 2 falls directly on the ln P vs ($1/T$) line in Fig. 13. The significance of these correspondences is that the volatilization—or, more explicitly, the enthalpy of vaporization–dissociation determined from Fig. 2 of the condensed-phase Al_2O_3 that forms—controls the flame temperature even when the gaseous mixture reacting with the metal is not pure oxygen.

Figure 14 reports the data for titanium reacting with air and another O_2–N_2 mixture at 1 atm under stoichiometric conditions. The temperature–pressure variation for the pure O_2–Ti system is also detailed on Fig. 13. Table 4 reports data similar to those in Table 3 for the aluminum system. Note that point $3'$ (Ti–air) has a combustion temperature of 3450 K and a partial pressure of the nonnitrogen decomposition gases of 0.10 atm. The Ti–O_2 system data in Fig. 13 reveal that at a pressure of 0.10 atm the adiabatic flame temperature should be 3471 K. For point $2'$, where the amount of nitrogen is one-third that in air, the combustion temperature is 3768 K and the partial pressure of nonnitrogen decomposition gases is 0.4069 atm. The temperature for this pressure on Fig. 13 is 3777 K. Thus, it is concluded that the formation of nitrogen oxides has a minimal effect on the controlling vaporization characteristics of the condensed-phase Ti_3O_5 that forms. When the N_2 was replaced with the same molar concentration of Ar for the Ti–air case ($3Ti + 2.5O_2 + 9.4N_2$), even though the specific heats of N_2 and Ar are greatly different, the corresponding noninert partial pressure was 0.1530 atm and the adiabatic flame temperature was 3558 K. The flame temperature calculated for the pure O_2–Ti system at 0.1530 atm is also 3558 K, and the condition falls exactly on the P–T relationship for the Ti–O_2 system in Fig. 13.

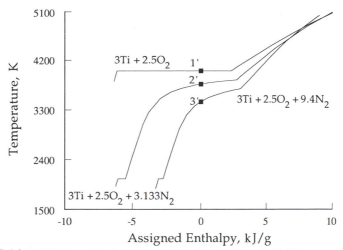

FIGURE 14 Adiabatic combustion temperatures of a stoichiometric Ti–N$_2$ system with various amounts of the diluent nitrogen as a function of assigned enthalpy. See Figs. 12 and 13.

These results for the metal reactions in air and oxygen–argon mixtures have great practical significance. In solid propellant rocket operation, the temperature of the flame around a burning aluminum particle will correspond to the volatilization temperature of the oxide at a pressure less than the total pressure; consequently, the radiative contribution to the propellant grain would not correspond to the known volatilization temperature at the total pressure, but to some lower temperature that would vary with the propellant composition. Furthermore, it is possible that some metals that would burn in the vapor phase in pure oxygen at 1 atm may burn heterogeneously in air.

4. Combustion Synthesis

The increasing importance of refractory materials has stimulated the search for new, economical techniques for synthesizing a number of substances that are

TABLE 4 Summary Data for Figure 14

Point	H_0° [kJ/g]	T (K)	$P_{tot} - P_{Ar}$ (atm)	Al$_2$O$_3$(l) (mole fraction)	Ar (mole fraction)
1′	0.0	4004	1.0000	0.0923	0.0000
2′	0.0	3768	0.4069	0.0771	0.5474
3′	0.0	3450	0.1000	0.0661	0.8453

not plentiful or pure enough to be useful. These materials are needed because new technologies require components capable of withstanding higher temperatures, more corrosive atmospheres, and increasingly abrasive environments. Material properties are important limiting factors in the design of cutting tools, turbine blades, engine nozzles, superconductors, semiconductors, electrodes, and certain components of nuclear power facilities. For those materials currently of practical importance, industrial production techniques require large amounts of external heat and involve complex, time-consuming processes. For example, the methods of forming the refractory compound titanium nitride (TiN) employ temperatures from 1400 to 1900 K and require anywhere from several hours to days. Additionally, most of these techniques require significant post-processing to achieve reasonable purity levels.

A new development in the area of refractory material production began to receive considerable attention in the former Soviet Union in the mid-1970s. Led by Merzhanov and Borovenskaya [2] this work has been recognized as a major advancement and has created a new and important field of study in combustion involving mostly solid-phase (heterogeneous) reactions of metals. This new technique, designated self-propagating high-temperature synthesis (SHS), makes use of the exothermic nature of reactions that form ceramics and similar materials from their constituent elements. Using this process, carbides, borides, selenides, silicides, and sulfides have been fabricated by igniting a compacted mixture of their constituent powders (see Table 5). Nitrides and hydrides have been produced by igniting powder compacts under nitrogen or hydrogen gas [11]. SHS offers many advantages in comparison with current commercial techniques. It requires much less external heat and far shorter reaction times; moreover, the purity of its products can be better than that of the initial reactants owing to the vaporization of impurities [12]. Initially, very poor yields were recovered using SHS when gaseous reactants N_2 and H_2 were involved. However, variations of the initial SHS process have improved yields. Such variations include replacing the gaseous nitrogen with a solid source of nitrogen (specifically, sodium azide) [12] and using high pressures in combination with dilution of the initial reactants with the product being produced [12]. Indeed, a slurry of liquid nitrogen and titanium powder will, when ignited, propagate a thermal wave and create titanium nitride [14]. Unfortunately, the total mass of titanium necessary to cause thermal propagation of the wave is such that it is in excess stoichiometrically with the nitrogen; thus Ti cannot be completely converted to TiN.

Many SHS processes include an oxide as one of the reactant materials. As a convenience, all the elements to be discussed in the various SHS processes will be referred to as metals. The selection of a metal–metal oxide reacting combination is readily made without detailed thermodynamic considerations. Nevertheless, it is fruitful to examine the overall thermodynamics that govern the choice of a particular SHS combination. Consider the classic thermite reaction

$$Fe_2O_3 + 2Al \rightleftharpoons Al_2O_3 + 2Fe + 850 \text{ kJ} \tag{4}$$

TABLE 5 Compounds Produced by Self-Propagating, High-Temperature Synthesis (SHS)[a]

Compounds	*Group of metals of the periodic table*						
	II	III	IV	V	VI	VII	VIII
Nitrides	Mg_3N_2	BN, AlN	TiN, ZrN, HfN	VN, NbN, $TaN_{(cub)}$ $TaN_{(hex)} Ta_2N$	–	–	–
Carbides	—	–	TiC_x, ZrC_x, HfC ($x = 0.6$–1.0)	VC, NbC, Nb_2C, TaC	WC	–	–
Borides	MgB_2 MgB_4 MgB_6	—	TiB, TiB_2, ZrB_2, ZrB_{12} HfB_2	VB, VB_2, NbB, NbB_2 TaB, TaB_2	CrB, CrB_2 MoB, Mo_2B Mo_2B_5, W_2B WB, WB_2	MnB	FeB, NiB
Silicides	–	–	TiSi, $TiSi_2$ ZrSi, $ZrSi_2$	–	$MoSi_2$	–	–
Chalco- genides	–	–	$TiSe_2$	$NbSe_2$, $TaSe_2$	$MoSe_2$, WSe_2	–	–
Solid solutions	TiC–WC, Ni–Al, MoS_2–NbS_2, BC–BN, NbC–NbN, TiC–TiN, TaC–TaN,$Nb_xZr_{1-x}C_yN_{1-y}$						

[a]Reported by Merzhanov and Borovenskaya [2].

Intuitively, one knows that this equilibrium reaction will proceed to the right to form Al_2O_3 and release heat. What largely determines the direction is the free energy change of the reacting system:

$$\Delta G^\circ = \Delta H^\circ - T \Delta S^\circ$$

where the symbols are defined in Chapter 1. In most SHS processes the difference in the total number of moles of product formed compared to the total number of moles of reactant is small, particularly in comparison to that which occurs in the oxidation of a hydrocarbon fuel. Consequently, in SHS processes the order–disorder change is small and indicates that the entropy change must be small. In fact, the entropy change is quite small in these processes so that the $T\Delta S$ term in the free energy equation above can be neglected even at the elevated temperatures that occur in SHS. Thus the free energy change is directly proportional to the enthalpy change and the enthalpy change becomes a measure of the direction a SHS reaction will proceed. Since an oxide product is always forming in a SHS system, it is evident that if a metal oxide with a smaller negative heat of formation is reacted with a metal whose oxide has a higher negative heat of formation, an exothermic reaction will occur, in which case the SHS process will proceed. If the opposite condition with respect to the heats of formation exists, the reaction will be endothermic and conversion will not occur. A good thermodynamic screening method for selecting a SHS process from among all the exothermic, and hence possible, metal–metal oxide systems is to consider the order of the heats of formation in terms of kilojoules per atom of oxygen as originally proposed by Venturini [15]. This procedure is feasible because $\Delta H°$ is determined from the heats of formation of the reactant and product oxides. Since the heats of formation of the elements are zero, the molar differences between the two oxides can be accounted for by dealing with the heats of formation per oxygen atom. An example of a molar difference would be

$$3TiO_2 + 4Al \rightarrow 2Al_2O_3 + 3Ti \qquad (5)$$

A list of oxides in these terms is presented as Table 6. Following the logic described, one obtains an exothermic system by choosing a metal whose oxide has a $-\Delta H_f°$ per O atom high on the list to react with a metal oxide that is lower on the list.

TABLE 6   Heats of Formation of Certain Oxides

Oxide	$\Delta H_f°$ at 298 K (kJ/mol)	Per O atom
CaO	−635	−635
ThO$_2$	−1222	−611
BeO	−608	−608
MgO	−601	−601
Li$_2$O	−599	−599
SrO	−592	−592
Al$_2$O$_3$	−1676	−559
ZrO$_2$	−1097	−549
BaO	−548	−548
UO$_2$	−1084	−542
CeO$_2$	−1084	−542
TiO$_2$	−945	−473
SiO$_2$	−903	−452
B$_2$O$_3$	−1272	−424
Cr$_2$O$_3$	−1135	−378
V$_2$O$_5$	−1490	−298
Fe$_2$O$_3$	−826	−275
WO$_3$	−789	−263
CuO	−156	−156

   The reaction sequence (4) must be ignited by an external heat source, even though the reaction is quite exothermic. The reason is that metals such as aluminum have a protective thin ($\sim$ 35 Å) oxide coat. Until this coat is destroyed by another thermal source, the thermite reaction cannot proceed. Once the protective oxide is broken, reaction is initiated and the energy release from the initial reaction phase is sufficient to ignite the next layer of reaction, and so forth. The thermal wave established in these solid-phase reactions propagates in much the same way that a premixed gaseous flame propagates. Although some metals that react are not protected by oxide coats, ignition energy is necessary to initiate the reaction for the intermetallics.

   For practical applications, one must calculate the temperature of the reaction in order to establish whether this temperature is so high that it will melt the metal that forms. Such melting causes metal flow which will prevent further reaction and hence propagation of the thermal wave. The actual temperature can be controlled by adding some of the product oxide to the reaction mixture. Consequently, for the system depicted as reaction (4), the initial reactants would be Fe_2O_3, Al, and Al_2O_3. Sometimes, the production of very high temperatures works to advantage. If the temperature of a SHS is high enough to vaporize the metal formed, the reaction proceeds well. This type of reaction permits the titanium–sodium azide reaction to proceed to completion. The SHS technique found application in the system used to seed the van Allen belts during the 1968 International Geophysical Year. In this instance, the SHS system chosen was the reaction of barium with copper oxide with an excess of barium. Thus the products were barium oxide, copper, and barium. The temperature of the reaction was sufficient to ionize the barium in the product. Of course, Table 6 reveals that the such a reaction will proceed.

   It is not necessary that one of the reactants be an oxide; it may, for example, be a chloride. Thus Table 7 represents a similar table to that for the oxides. The same logic for the choice of reacting systems prevails here [16]. Since the metal halides are readily vaporized, halogen exchange reactions can be used to liberate free metal atoms through gas-phase reactions, as opposed to the mostly heterogeneous "thermite-type" oxide reactions just discussed. In some early work by Olson *et al.* [17], a technique was developed for the production of solar-grade silicon based on this type of chemistry. These investigators examined the reaction between silicon tetrachloride and sodium vapor under conditions such that

$$SiCl_4(v) + 4Na(v) \rightarrow 4NaCl(v) + Si(l) \tag{6}$$

where (v) specifies vapor and (l) specifies liquid. The silicon liquid product comes from the nucleation of silicon atoms. Another SHS process is a modification of this halogen reaction to produce nitrides and other refractory and cermet powders. The simple addition of gaseous nitrogen to reactants similar to those above permits further reaction of the very reactive nascent metal atoms with the nitrogen to form the nitride. Table 8 gives the nitride information comparable to that in Table 7 for chlorides.

TABLE 7 Heats of Formation of Certain Chlorides

Chloride	ΔH_f° at 298 K (kJ/mol)	Per Cl atom
CsCl	−443	−443
KCl	−437	−437
BaCl$_2$	−859	−430
RbCl	−430	−430
SrCl$_2$	−829	−415
NaCl	−411	−411
LiCl	−408	−408
CaCl$_2$	−796	−398
CeCl$_3$	−1088	−362
AlCl$_3$	−706	−235
TiCl$_4$	−815	−204
SiCl$_4$	−663	−166

From this information one again observes that the metals which form useful metal nitrides have nitrides with large negative heats of formation and chlorides with relatively small negative heats of formation. Thermodynamically, because the heat of formation of nitrogen is zero, the titanium–nitrogen reaction must proceed, as has been found experimentally [18, 19]. Note that the opposite is true for the alkali metals, which have small nitride negative heats of formation and large chloride negative heats of formation. These comparisons suggest a unique method of forming nitrides via gas-phase reactions. For example, the reaction of an alkali-metal vapor, say sodium, with titanium tetrachloride, silicon tetrachloride, or aluminum tetrachloride vapor in the presence of nitrogen should produce titanium

TABLE 8 Heats of Formation of Certain Nitrides

Nitride	ΔH_f° at 298 K (kJ/mol)	Per N atom
HfN	−369	−369
ZrN	−365	−365
TiN	−338	−338
AlN	−318	−318
BN	−251	−251
Mg$_3$N$_2$	−461	−231
Si$_3$N$_4$	−745	−186
Li$_3$N	−165	−165
N$_2$	0	0
NaN$_3$	22	7

nitride, silicon nitride, or aluminum nitride according to the following overall reaction:

$$MCl_x + xNa + (y/2)N_2 \rightarrow xNaCl + MN_y \tag{7}$$

Mixed chlorides can also be used to produce intermetallics:

$$M'Cl_x + nM''Cl_y + (x + ny)Na \rightarrow (x + ny)NaCl + M'M''_n \tag{8}$$

where the M represents various metal species.

Thermodynamic calculations of the equilibrium product distribution from these alkali-vapor reactions reveal very poor yields with a large amount of sodium impurity owing to the very high flame temperatures involved and a correspondingly high degree of dissociation. Because of the very large latent heat of vaporization of sodium, the calculated results for liquid sodium were extremely promising, with conversion of metal to nitride in all three cases.

C. DIFFUSIONAL KINETICS

In the case of heterogeneous surface burning of a particle, consideration must be given to the question of whether diffusion rates or surface kinetic reaction rates are controlling the overall burning rate of the material. In many cases, it cannot be assumed that the surface oxidation kinetic rate is fast compared to the rate of diffusion of oxygen to the surface. The surface temperature determines the rate of oxidation and this temperature is not always known *a priori.* Thus, in surface combustion the assumption that chemical kinetic rates are much faster than diffusion rates cannot be made.

Consider, for example, a carbon surface burning in a concentration of oxygen in the free stream specified by $\rho m_{0\infty}$. The burning is at a steady mass rate. Then the concentration of oxygen at the surface is some value m_{os}. If the surface oxidation rate follows first-order kinetics, as Frank-Kamenetskii [20] assumed,

$$G_{ox} = G_f/i = k_s \rho m_{os} \tag{9}$$

where G is the flux in grams per second per square centimeter; k_s is the heterogeneous specific reaction rate constant for surface oxidation in units reflecting a volume to surface area ratio, i.e., centimeters per second; and i is the mass stoichiometric index. The problem is that m_{os} is unknown. But one knows that the consumption rate of oxygen must be equal to the rate of diffusion of oxygen to the surface. Thus, if $h_D\rho$ is designated as the overall convective mass transfer coefficient (conductance), one can write

$$G_{ox} = k_s \rho m_{os} = h_D\rho(m_{o\infty} - m_{os}) \tag{10}$$

What is sought is the mass burning rate in terms of $m_{o\infty}$. It follows that

$$h_D m_{os} = h_D m_{o\infty} - k_s m_{os} \tag{11}$$

$$k_s m_{os} + h_D m_{os} = h_D m_{o\infty} \tag{12}$$

$$m_{os} = \{h_D/(k_s + h_D)\} m_{o\infty} \tag{13}$$

or

$$G_{ox} = [\{\rho k_s h_D/(k_s + h_D)\} m_{o\infty}] = \rho K m_{o\infty} \tag{14}$$

$$K \equiv k_s h_D/(k_s + h_D) \tag{15}$$

$$1/K = (k_s + h_D)/k h_D = (1/h_D) + (1/k_s) \tag{16}$$

When the kinetic rates are large compared to the diffusion rates, $K = h_D$; when the diffusion rates are large compared to the kinetic rates, $K = k_s$. When $k \ll h_D$, $m_{os} \cong m_{o\infty}$ from Eq. (13); thus

$$G_{ox} = k_s \rho m_{o\infty} \tag{17}$$

When $k_s \gg h_D$, Eq. (13) gives

$$m_{os} = (h_D/k_s) m_{o\infty} \tag{18}$$

But since $k_s \gg h_D$, it follows from Eq. (18) that

$$m_{os} \ll m_{o\infty} \tag{19}$$

This result permits one to write Eq. (10) as

$$G_{ox} = h_D \rho (m_{o\infty} - m_{os}) \cong h_D \rho m_{o\infty} \tag{20}$$

Consider the case of rapid kinetics, $k_s \gg h_D$, further. In terms of Eq. (14), or examining Eq. (20) in light of K,

$$K = h_D \tag{21}$$

Of course, Eq. (20) also gives one the mass burning rate of the fuel

$$G_f/i = G_{ox} = h_D \rho m_{o\infty} \tag{22}$$

Here h_D is the convective mass transfer coefficient for an unspecified geometry. For a given geometry, h_D would contain the appropriate boundary layer thickness, or it would have to be determined by independent measurements giving correlations that permit h_D to be found from other parameters of the system. More interestingly, Eq. (22) should be compared to Eq. (179) in Chapter 6, which can be written as

$$G_{ox} \cong \frac{\lambda}{c_p \delta} m_{o\infty} \frac{H}{L_v} = \frac{\lambda}{c_p \rho} \frac{\rho}{\delta} m_{o\infty} \frac{H}{L_v} = \frac{\alpha}{\delta} \frac{H}{L_v} \rho m_{o\infty} = \frac{D}{\delta} \frac{H}{L_v} \rho m_{o\infty} \tag{23}$$

Thus one notes that the development of Eq. (22) is for a small B number, in which case

$$h_{\mathrm{D}} = \frac{D}{\delta}\frac{H}{L_{\mathrm{v}}} \tag{24}$$

where the symbols are as defined in Chapter 6. (H/L_{v}) is a simplified form of the B number. Nevertheless, the approach leading to Eq. (22) gives simple physical insight into surface oxidation phenomena where the kinetic and diffusion rates are competitive.

D. DIFFUSION-CONTROLLED BURNING RATE

This situation, as discussed in the last section, closely resembles that of the droplet diffusion flame, in which the oxygen concentration approaches zero at the flame front. Now, however, the flame front is at the particle surface and there is no fuel volatility. Of course, the droplet flame discussed earlier had a specified spherical geometry and was in a quiescent atmosphere. Thus, h_{D} must contain the transfer number term because the surface regresses and the carbon oxide formed will diffuse away from the surface. For the diffusion-controlled case, however, one need not proceed through the conductance h_{D}, as the system developed earlier is superior.

Recall for the spherical symmetric case of particle burning in a quiescent atmosphere that

$$G_{\mathrm{f}} = (D\rho/r_{\mathrm{s}})\,\ln(1+B) \tag{25}$$

The most convenient B in liquid droplet burning was

$$B_{\mathrm{oq}} = [im_{\mathrm{o}\infty}H + c_p(T_\infty - T_{\mathrm{s}})]/L_{\mathrm{v}} \tag{26}$$

since, even though T_{s} was not known directly, $c_p(T_\infty - T_{\mathrm{s}})$ could always be considered much less than $im_{\mathrm{o}\infty}H$ and hence could be ignored. Another form of B, however, is

$$B_{\mathrm{fo}} = \frac{(im_{\mathrm{o}\infty} + m_{\mathrm{fs}})}{(1 - m_{\mathrm{fs}})} \tag{27}$$

Indeed, this form of B was required in order to determine T_{s} and m_{fs} with the use of the Clausius–Clapeyron equation. This latter form is not frequently used because m_{fs} is essentially an unknown in the problem; thus it cannot be ignored as the $c_p(T_\infty - T_{\mathrm{s}})$ term was in Eq. (26). It is, of course, readily determined and necessary in determining G_{f}. But observe the convenience in the current problem. Since there is no volatility of the fuel, $m_{\mathrm{fs}} = 0$, so Eq. (27) becomes

$$B_{\mathrm{fo}} = im_{\mathrm{o}\infty} \tag{28}$$

Thus, a very simple expression is obtained for surface burning with fast kinetics:

$$G_f = (D\rho/r_s)\ln(1 + im_{o\infty}) \tag{29}$$

Whereas in liquid droplet burning B was not explicitly known because T_s is an unknown, in the problem of heterogeneous burning with fast surface reaction kinetics, B takes the simple form of $im_{o\infty}$ which is known provided the mass stoichiometric coefficient i is known. For small values of $im_{o\infty}$, Eq. (29) becomes very similar in form to Eq. (22) where for the quiescent case $h_D = D/r_s = \alpha/r_s$.

1. Burning of Metals in Nearly Pure Oxygen

The concept of the B number develops from the fact that in the quasi-steady approach used for droplet burning rates, the bulk flow was due not only to the fuel volatilization but also to the formation of product gases. This flow, represented by the velocity v in the conservation equations, is outward-directed; that is, the flow is in the direction of increasing r. In the case of metal combustion in pure oxygen and at relatively high pressures, the possibility arises that heterogeneous processes may occur with no product gas volatilization. Thus the B number effect disappears and the bulk velocity is inward-directed. Indeed, it has been noted experimentally [21] that under these conditions small amounts of impurities in oxygen can reduce the burning rates of metals appreciably.

The extent of this impurity effect is surprising and is worthy of examination. Consideration of Eq. (111) of Chapter 6 readily shows [22] for heterogeneous oxidation that there is no apparent gas-phase reaction. This equation is now written in the form

$$\frac{d}{dr}\left(4\pi r^2 \rho D \frac{dY_o}{dr} - 4\pi r^2 \rho v Y_o\right) = 0 \tag{30}$$

where the symbol Y_o is now used for the mass fraction of oxygen to distinguish this unique case of droplet burning from all the others. Integrating Eq. (30) yields

$$4\pi r^2 \rho D \frac{dY_o}{dr} - 4\pi r^2 \rho v Y_o = c = \dot{m}_o \tag{31}$$

where the constant of integration is by definition the net mass flow rate of oxygen $\dot{m}_o$. Oxygen is the only species that has a net mass flow in this case, so

$$\dot{m}_o = -4\pi r^2 \rho v = -4\pi (r^2 \rho v)_s \tag{32}$$

where the subscript s, as before, designates the particle surface. The negative sign in Eq. (32) indicates that $\dot{m}_o$ is inward-directed. Integrating Eq. (31) from $r = r_s$ to $r = \infty$ then yields

$$\tilde{m}_o = \ln\left[\frac{(1 - Y_{os})}{(1 - Y_{o\infty})}\right] \tag{33}$$

where $\tilde{m}_o = \dot{m}_o/(4\pi\rho D r_s)$ and $(1 - Y_{o\infty})$ represents the initial impurity mass fraction. The metal surface reaction rate $\dot{m}_f$ is now written as

$$\dot{m}_f = 4\pi r^2 \rho_s Y_{os} k_s \tag{34}$$

which is another representation of Eq. (17). Since $m_f = im_o$, it is possible to define from Eqs. (32) and (34) a nondimensional rate constant $\tilde{k}_s$ such that

$$\tilde{k}_s = \frac{4\pi^2 \rho_s k_s}{i(4\pi\rho D r_s)} \tag{35}$$

so that

$$\tilde{m}_o = \tilde{k}_s Y_{os} \tag{36}$$

One can see that $\tilde{k}_s$ is a form of a Damkohler number $[k_s/(D_s/r_s)]$ which indicates the ratio of the kinetic rate to the diffusion rate.

Substituting Y_{os} from Eq. (36) into Eq. (33) yields the solution

$$\tilde{m}_o = \ln\left[\frac{1 - (\tilde{m}_o/\tilde{k}_s)}{1 - Y_{o\infty}}\right] \tag{37}$$

Knowing $Y_{o\infty}$ and $\tilde{k}_s$, one can iteratively determine $\tilde{m}_o$ from Eq. (37); and knowing $\tilde{m}_o$, one can determine the metal burning rate from $\tilde{m}_f = i\tilde{m}_o$. The surface oxidizer concentration is given by

$$Y_{os} = \tilde{m}_o/\tilde{k}_s \tag{38}$$

The next consideration is how small amounts of inert affect the burning rate. Thus, Eq. (37) is considered in the limit $Y_{o\infty} \to 1$. In this limit, as can be noted from Eq. (37), $\tilde{m}_o < \tilde{k}_s$. Thus rewriting Eq. (37) as

$$e^{\tilde{m}_o}(1 - Y_{o\infty}) = 1 - (\tilde{m}_o/\tilde{k}_s) \cong e^{\tilde{k}_s}(1 - Y_{o\infty}) \tag{39}$$

and solving for $\tilde{m}_o$, one obtains

$$\tilde{m}_o \approx \tilde{k}_s[1 - (1 - Y_{o\infty})e^{\tilde{k}_s}] \tag{40}$$

Two observations can be made regarding Eq. (40). First, differentiating Eq. (40) with respect to the oxygen mass fraction, one obtains

$$\frac{\partial \tilde{m}_o}{\partial Y_{o\infty}} = \tilde{k}_s e^{\tilde{k}_s} \tag{41}$$

Thus one finds that $\tilde{m}_o$ varies in an exponentially sensitive manner with the ambient oxygen concentration, $Y_{o\infty}$, and consequently with the impurity level for a sufficiently fast surface reaction. Second, since $\tilde{k}_s$ is an exponential function of temperature through the Arrhenius factor, the sensitivity of the oxidation rate to the oxygen concentration, and hence the impurity concentration, depends on the

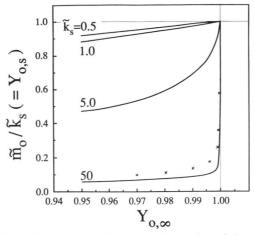

FIGURE 15 A plot of the oxygen mass fraction Y_{os} at the surface of a burning metal particle as a function of the ambient mass fraction $Y_{o\infty}$ for the condition of $Y_{o\infty \to 1}$.

metal surface condition temperature in an extremely sensitive, double exponentiation manner.

The variation of the oxygen concentration at the surface can be clearly seen by rearranging Eq. (41) after recalling Eq. (38) to give

$$\frac{\partial(\tilde{m}_o/\tilde{k}_s)}{\partial Y_{o\infty}} = \frac{\partial Y_{os}}{\partial Y_{o\infty}} = e^{\tilde{k}} \tag{42}$$

Figure 15 shows a plot of $[(\tilde{m}/\tilde{k})_s = Y_{os}]$ versus $Y_{o\infty}$ for different values of $\tilde{k}_s$. The points on this figure were extracted from experimental data obtained by Benning et al. [21] for the burning of aluminum alloy rods in oxygen with an argon impurity. These data correspond to a $\tilde{k}_s$ close to 36. Large values of $\tilde{k}_s$ specify very fast surface reaction rates. For a value of $\tilde{k}_s$ of 50, an impurity mass fraction of 0.5% reduces the oxygen mass fraction at the surface to 0.1, a tenfold decrease from the ambient.

2. The Burning of Carbon Char Particles

The value of i in Eq. (28) for carbon—or, for that matter, any heterogeneous combustion system—depends on the surface chemistry; consequently, its value is not readily apparent.

If the product of the carbon surface reaction is carbon dioxide

$$C + O_2 \to CO_2$$

then i is the stoichiometric coefficient defined as the number of grams of fuel that burns one gram of oxidizer. Thus for this reaction, $i = 12/32$ or 0.375. However, the structure of CO_2

$$O=C=O$$

suggests that CO_2 would be unlikely to form as a gaseous product on the surface. In order to consider the proper choice of i, it is appropriate to consider experimental studies [23, 24] on the reaction of carbonaceous particles with various concentrations of oxygen. These studies reveal an interesting trend with respect to the particle temperature. What was followed in these experiments was the fraction of carbon oxides such as CO_2 near the particle surface. The results depicted in Fig. 16 reveal that from approximately 630 to 760 K, the CO_2 fraction increases with increasing temperature, then decreases steadily to values less than 10% at temperatures below 1500 K. Normally, at about 1000 K, the CO ignites and the particle is in a true combustion mode [24]. To achieve this condition, a cloud of particles must be in a fluidized bed so that there is essentially no radiative heat loss from the surface burning particles. Any CO_2 found near or absorbed on the surface is most likely due to the reaction between the CO and O_2 in the absorbed layer, rather than a reaction between the carbon and oxygen to form CO_2 directly. The rise and subsequent decline of the CO_2 fraction as a function of temperature can be attributed to a competition between the rate of formation of CO_2 in the absorbed layer and the rate of desorption compared to CO [24]. This explanation is apparently confirmed by the results found at high temperatures which showed that the CO_2 fraction increases when some water vapor is present in the oxygen [23]. From a combustion point of view, it is important to consider that at about 1000 K the CO ignites to form CO_2 [24] and the penetration of O_2 to the carbon particle surface is negligible. This consideration is consistent with the early work of Coffin and Brokaw [25], who proposed that the carbon is oxidized by CO_2 which diffuses from the flame front created by the CO oxidation. The simple model of Coffin and Brokaw is shown in Fig. 17. The overall surface reaction is then

$$C + O_2 \rightarrow 2CO$$

Since the stoichiometric coefficient in the transfer number B, both for the overall case of CO ignition and for the establishment of a gaseous flame in which CO_2 is created (as depicted in the Coffin and Brokaw model), is related to the ambient mass fraction of oxygen, the reaction

$$C + O_2 \rightarrow CO_2$$

must be considered and the Law of Heat Summation (Chapter 1) applied. This overall consideration leads to

$$2C + O_2 \rightarrow 2CO$$

Consequently, the mass stoichiometric index $i = 24/32$ or 0.75—the value that would have been achieved if CO were assumed to be the sole product formed at the surface when the carbon surface reacts with oxygen.

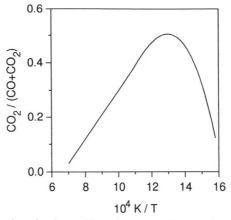

FIGURE 16   The fraction of carbon oxide products appearing as carbon dioxide as a function of reciprocal temperature. Data points represent the single particle experiments of Tognotti *et al.* [23] and the fluidized bed reactor results of Purzer *et al.* [24].

In order to verify that this value is the correct i for the sequence of reactions, one must proceed through the analytical development of graphite particle burning. In this problem, one is required to deal only with the oxygen diffusion equation because there is no combustion in the gas phase. Indeed, this is the reason for the simple result that $B = im_{o\infty}$. The procedure of Blackshear [26] is followed.

The mass diffusion equation developed earlier for droplet evaporation alone

$$4\pi r^2 \rho v \, dm_{ox}/dr = (d/dr)(4\pi r^2 \rho D \, dm_{ox}/dr) \tag{43}$$

now holds for the case where there is combustion on the surface; i.e., there is no reaction rate term in Eq. (43). In the gas phase, i' grams of CO react with 1 gram of O_2 to give $(1 + i')$ grams of CO_2; therefore,

$$m_{O_2} = [1/(1 + i')]m_{CO_2} \tag{44}$$

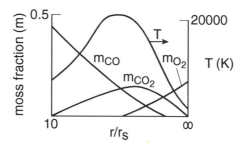

FIGURE 17   Distribution of gaseous species and temperature above a carbon sphere burning on the surface (from Coffin and Brokaw [25]).

A new variable, physically symbolizing and allowing for the diffusion of both the O_2 and CO_2,

$$m_H = m_{O_2} + [1/(1 + i')]m_{CO_2} \tag{45}$$

satisfies the fundamental differential equation for diffusion:

$$\rho v \, dm_H/dr = (d/dr)(\rho D \, dm_H/dr) \tag{46}$$

When this equation is integrated and evaluated at $r = r_s$, one obtains

$$\text{const} = r_s^2[\rho_s v_s(m_H)_s - D_s \rho_s(dm_H/dr)_s] \tag{47}$$

$4\pi r_s^2 \rho_s v_s$ is the mass consumption rate of the fuel and is a constant. At $r = r_s$, $m_H = [1/(1 + i')]m_{CO_2}$, since $m_{O_2} = 0$. Thus, the term in brackets is the flux of m_H into the surface, but it is, of course, $[1/(1 + i')]$ times the negative flux of CO_2 into the surface as well. But, by realizing that 1 gram of CO_2 can combine with i'' grams of C to form $(1 + i'')$ grams of CO, one can write the flux of CO_2 in terms of the flux of carbon. Thus, the flux of CO_2 must be $(1/i'')$ times the flux of C. The flux of carbon comes from the basic consumption rate of the carbon, and its form is very much like that of liquid droplet consumption, i.e.,

$$\rho_s v_s(m_H)_s = D\rho[d(m_H)/dr_s] \tag{48}$$

which is correct for a solid or gas as long as consistent values of density and velocity are chosen. Thus, the flux of CO_2 must be

$$\rho_s v_s(1/i'') \tag{49}$$

But in Eq. (47) the term in the brackets is $[1/(1 + i')]$ times the flux of CO_2. Then the term in the brackets is

$$-\rho_s v_s(1/i'')[1/(1 + i')] \tag{50}$$

and the equation is written

$$\text{const} = -r_s^2 \rho_s v_s(1/i'')[1/(1 + i')] \tag{51}$$

Integrating the second time, one has

$$\frac{r_s^2 \rho_s v_s}{r D\rho} = \ln\left(\frac{m_{H\infty} + [1/i''(1 + i')]}{m_H + [1/i''(1 + i')]}\right) \tag{52}$$

However, both CO_2 and O_2 must approach zero at $r = r_s$; therefore, $m_H = 0$ at $r = r_s$ and, of course, $m_H = m_{O_2\infty}$ at $r = \infty$. Conseqently,

$$r_s \rho_s v_s/\rho D = \ln[m_{O_2\infty}i''(1 + i') + 1] \tag{53}$$

Comparing this equation with the many forms that were obtained previously, one has

$$B = i''(1 + i')m_{O_2\infty} \tag{54}$$

The values of i'' and i' are $i'' = 12/44$ and $i' = 28/16$. Then

$$i''(1 + i') = (12/44)(1 + 28/16) \tag{55}$$

and, thus

$$i''(1 + i') = 12/16 = i$$

Thus, irrespective of the mechanism of removing carbon from the surface, the main considerations are as follows: (1) only CO must form on the surface, and (2) the flux of oxygen from the quiescent atmosphere must be stoichiometric with respect to CO formation, regardless of the intermediate reactions that take place. This result is the same as that reached earlier, where the two primary reactions were added according to the Law of Heat Summation.

Recall this discussion has dealt with $k > h_D$, which, in the context of combustion reactions, means high-temperature particles. Of course, such high temperatures are created at the surface when the mass burning rate is accelerated by increasing the convective rates. The convective expression for the burning of the carbon particle has the same form as that of the burning liquid droplet, except that the expression for B is simpler. Various investigators have shown for relatively large particles that the combustion of carbon above 1200 K exhibits rates that are strictly proportional to the diffusional characteristics. For the small particles found in pulverized coal or metallic fuel additives, the Reynolds numbers are so small that the particles may be considered to burn under quiescent conditions, in which case the above analyses apply. However, the burning rates of very small particles may be controlled by chemical oxidation kinetics, as will be discussed later. It is now known [27] that above 1100 K, CO oxidation to CO_2 is very rapid. Thus, it would appear that all the assumptions made above are self-consistent. If the CO rates were not rapid, one would have to be concerned with oxygen penetrating to the carbon surface. The overall analysis would therefore become more complex algebraically, but the same overall result would be obtained. Indeed, for the combustion of small pulverized coal particles, CO appears not to be converted to carbon dioxide owing to the low temperatures involved as well as penetration of oxygen to the char surface.

3. The Burning of Boron Particles

In certain respects, the combustion of boron is different from that of carbon because, under normal temperature and pressure conditions, the product oxide, B_2O_3, is not a gas. Thus, a boron particle normally has an oxide coat that thickens as the particle is heated in an oxidizing atmosphere. This condition is characteristic of most metals, even those that will burn in the vapor phase. For the efficient combustion of the boron particle, the oxide coat must be removed. The practical means for removing the coat is to undertake the oxidation at temperatures greater than the saturation temperature of the boron oxide B_2O_3. This temperature is about 2300 K at 1 atm.

The temperature at which sufficient oxide is removed so that oxidation can take place rapidly is referred to as the metal ignition temperature. The rate of oxidation when the oxide coat persists has been discussed extensively in Refs. [28 and 29]. Nevertheless, what does control the burning time of a boron particle is the heterogeneous oxidation of the clean particle after the oxide has been evaporated. Thus, for efficient burning the particle and oxidizing medium temperatures must be close to the saturation temperature of the B_2O_3. Then the burning rate of the particle is given by Eq. (29), the same as that used for carbon except the mass stoichiometric coefficient i is different. Even though the chemical reaction steps for boron are quite different from those of carbon, i is a thermodynamic quantity and the atomic weight of boron of 10 is comparable to 12 for carbon; consequently, it is not surprising that the i values for both materials are nearly the same.

Just as the surface oxidation chemistry makes it unlikely that carbon would yield CO_2, it is also unlikely that boron would yield B_2O_3. Gaseous boron monoxide BO forms at the surface and this product is oxidized further to gaseous B_2O_3 by vapor-phase reactions. The gaseous B_2O_3 diffuses back to the clean boron surface and reacts to form three molecules of BO. The actual reaction order is most likely given by the sequence of reactions [29] discussed in the following paragraphs.

In a high-temperature atmosphere created by the combustion of a host hydrocarbon fuel, there will be an abundance of hydroxyl radicals. Thus, boron monoxide reacts with hydroxyl radicals to form gaseous metaboric oxide HOBO:

$$M + BO + OH \rightarrow HOBO + M \tag{56}$$

It is postulated that HOBO then reacts with BO to form gaseous boron oxide hydride HBO and boron dioxide BO_2 (OBO):

$$OBOH + BO \rightarrow OBO + HBO \tag{57}$$

The boron dioxide then reacts with another BO to form boron oxide B_2O_3:

$$OBO + BO \rightarrow B_2O_3 \tag{58}$$

This route is consistent with the structure of the various boron oxide compounds in the system [29].

In a hydrogen-free oxidizing atmosphere, a slower step forms the boron dioxide,

$$BO + O_2 \rightarrow OBO + O \tag{59}$$

whereupon B_2O_3 again forms via reaction (58).

After the gaseous reaction system is established, the B_2O_3 diffuses back to the nascent boron surface to form BO, just as CO_2 diffuses back to the carbon surface to form CO. The reaction is

$$B + B_2O_3 \rightarrow 3BO \tag{60}$$

Thus, the overall thermodynamic steps required to calculate the mass stoichiometric index in Eq. (29) are

$$2BO + \tfrac{1}{2}O_2 \rightarrow B_2O_3$$

$$B + B_2O_3 \rightarrow 3BO$$

$$B + \tfrac{1}{2}O_2 \rightarrow BO$$

and

$$i = 10/16 = 0.625$$

compared to the value of 0.75 obtained for the carbon system.

Boron does not meet Glassman's criterion for vapor-phase combustion of the metal. Thus, the boron surface remains coated with a vitreous B_2O_3 layer and boron consumption becomes extremely slow; consequently, boron is not burned efficiently in propulsion devices.

4. Oxidation of Very Small Particles—Pulverized Coal and Soot

Equation (29) shows that the rate of fuel consumption in a diffusion-controlled system is inversely proportional to the radius of the fuel particle. Hence, below some critical particle size, other conditions being constant, the rate of mass (oxygen) transfer will become faster than the rate of the surface chemical reaction. When this condition prevails, the kinetic rate controls the mass consumption rate of the fuel and the concentration of the oxidant close to the surface does not differ appreciably from its bulk (ambient) concentration. If the fuel particle is porous, the essential assumption is always that the chemical rate is fast enough to render the particle impervious to the oxygen concentration.

The surface chemical reaction rate is different from that of ordinary gaseous reactions. It is considered to be a two-step physical process: (1) attachment of the oxygen chemically to the fuel surface (absorption) and (2) desorption of the oxygen with the attached fuel component from the surface as a product. In the discussion to follow, the fuel is assumed to be carbonaceous. In principle, then, either adsorption of the oxygen or desorption of the gaseous fuel oxide can be controlling. Since these physical processes actually govern the chemical conversion rate and since the particular fuel atoms to which the oxygen attaches are the ones that generate the products, it is clear that either adsorption or desorption can control the number of active sites that will produce products.

This concept has been applied by Mulcahy and Smith [30] to determine whether the burning process of pulverized coal particles is controlled by chemical kinetic or diffusion rates. It is their work which is followed in the subsequent paragraphs.

The maximum rate of the absorption-controlled step may be calculated by assuming that every oxygen molecule that has the necessary activation energy

E_{ads} reacts immediately, upon striking the fuel surface, to form two molecules of the product CO. It makes no significant difference whether the oxidant molecule is O_2 or CO_2 since, in either case, the product evolved is two molecules of CO.

This rate of chemical adsorption is given by

$$R_{ads} = (2Z/N)\exp(-E_{ads}/RT) \qquad \text{(of carbon/cm}^2\text{ s)} \qquad (61)$$

where N is Avogadro's number and Z, the number of collisions per square centimeter per second, can be calculated from the kinetic theory of gases. Thus, the adsorption rate expression becomes

$$R_{ads} = [2M_{O_2}P_{O_2}/(2RTM_{O_2})^{0.5}]\exp(-E_{ads}/RT) \qquad (62)$$

where M_{O_2} is the molecular weight of oxygen. Thus, the adsorption rate is seen to be proportional to the oxygen particle pressure (or concentration) and is first-order with respect to the oxygen.

The maximum rate of the desorption process can be estimated by assuming that, in the steady state, an oxygen atom is attached to every carbon atom on the surface. The rate at which carbon evolves from the surface as CO is found to be

$$R_{des} = (12RTC_A/N^2h)\exp(-E_{des}/RT) \qquad (63)$$

where C_A is assumed to be the number of carbon atoms per square centimeter of the carbon lattice and h is Planck's constant. Thus, the desorption rate is seen to be independent of the oxygen concentration, that is, zero-order with respect to the oxygen consideration.

For a fixed fuel particle size, whether the fuel consumption rate is controlled by diffusional or chemical rates depends on the temperature range. Since the diffusion rate is essentially independent of temperature and Eqs. (62) and (63) show that the possible controlling chemical rates follow an exponential dependence, the overall rate of fuel consumption as a function of temperature can be represented in the form of an Arrhenius plot as shown in Fig. 18. It should be realized that the diffusional and kinetic processes in heterogeneous burning are sequential; thus, it is apparent from Fig. 18 that even for a small particle, if the temperature is high enough, the kinetic rates will again become faster than the diffusion rates. Figure 18 also shows the oxygen concentration profiles for each type of control.

Mulcahy and Smith [30] have shown how the transition temperature is determined for 40-μm particle burning at a pressure of 1 atm under an oxygen partial pressure of 0.1 atm. These results are shown in Fig. 19 and the chemical rates are those represented by Eqs. (62) and (63). Figure 19 reveals that the particle burning rate is controlled chemically when $E_{des} < 10$ kcal/mol. It is also interesting that when $E_{des} < 5$ kcal/mol, mass transfer to a 40-μm particle is incapable of sustaining the maximum chemical rate. For E_{ads} of 40 and 60 kcal/mol, the transition from chemical to mass transfer control takes place at 1150 and 1700 K, respectively.

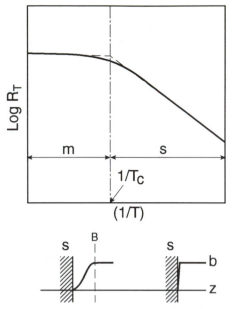

FIGURE 18 Rate-controlling regimes in gas–solid reactions for an impervious solid: *m* is the rate controlled by mass transfer of the oxidizer to the solid; *s* is the rate controlled by surface reaction, (S) solid; (B) boundary layer; (b) bulk concentration of oxidizer; (z) zero concentration of oxidizer (from Mulcahy and Smith [30]).

Thus, if the true E_{act} for the reaction is greater than a few kilocalories per mole, the observation (or assumption) of first-order kinetics is compatible with chemical control only in the range of particle sizes found in pulverized coal.

Since the mass diffusion rate varies inversely with the particle radius and the chemical rate is independent of diameter, the temperature at which the transition from one to the other takes place is a function of the particle size. A graphical representation of this approach [30] is given in Fig. 20. Considering that the pulverized coal particle range ends at 200-μm, one will note from Fig. 20 that their rate of burning is controlled by heterogeneous oxidation kinetics.

Inherent in the developments given is the assumption that all adsorption sites will yield a product oxide. In considering the heterogeneous oxidation of coal char or soot particles, however, it is most apparent not all sites are reactive, nor do they all have the same reactivity. In an effort to obtain a more detailed analysis of the burning rates of such materials as a function of temperature, Radcliffe and Appleton [31] proposed a mechanism that leads to the development of the semiempirical correlation of Nagle and Strickland-Constable [32]. Indeed, on the basis of structural similarities, these investigators have suggested that the rate of oxidation of soot and char-like materials should be equivalent. In their mechanism of surface oxidation, they posited two types of reaction sites in the exposed area

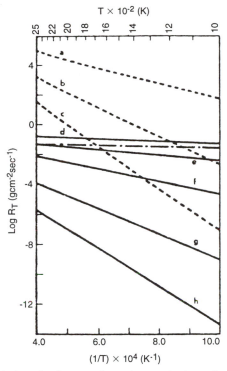

FIGURE 19 Theoretical combustion rates for various mechanisms of rate control. (- - - -) R_{MT}, (40-μ m particle); (——) $R_{chem,ads}$; (— — — —)$R_{chem,des}$. *E* (kcal/mol): (a) = 20; (b) = 40; (c) = 60; (d) = 5; (e) = 10; (f) = 20; (g) = 40; and (h) = 60 (from Mulcahy and Smith [30]).

of the particle: namely, an A site, which is reactive, and a B site, which is much less reactive. The fraction of the surface covered by A sites is assumed to be x and the remaining fraction $(1 - x)$ is assumed to be covered by B sites.

It is proposed that a steady-state fraction of the A sites are covered by a surface oxide and that this fraction, x_{Ao}, is given by a balance between the rate of activated adsorption of oxygen from the gas phase on the A sites to produce the surface oxide, i.e.,

$$A_{site} + O_2(g) \rightarrow \text{surface oxide}$$

$$\text{Rate} = k_A p_{O_2} x_A (1 - x_{Ao}) \tag{64}$$

and the rate of activated desorption of CO from the surface, i.e.,

$$\text{surface oxide} \rightarrow 2CO(g) + A_{site}$$

$$\text{Rate} = k' x_A x_{Ao} \tag{65}$$

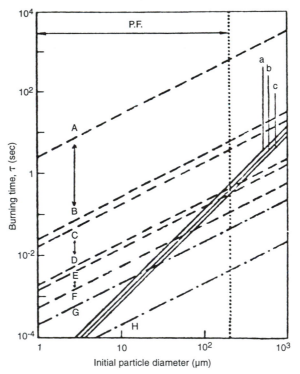

FIGURE 20   Theoretical particle burning times showing effects of particle diameter, temperature, and pressure (after Mulcahy and Smith [30]).

	$T(K)$	E (kcal/mol)	P (atm)
A	1000	20	1
B	2000	20	1
C	1000	10	1
D	2000	10	1
E	1000	5	1
F	2000	5	1
G	2000	10	10
H	2000	10	100

Solid lines indicate mass transfer control, all pressures: (a) 1000 K; (b) 1600 K; (c) 2000 K.

It follows, then, that

$$k_A p_{O_2} x_A (1 - x_{Ao}) = k' x_{Ao} x_A \tag{66}$$

which gives

$$x_{Ao} = (k_A p_{O_2})/(k' + k_A p_{O_2}) \tag{67}$$

Therefore, the rate at which carbon leaves the surface is given by the substitution of Eq. (67) into Eq. (65) to yield

$$\text{Rate} = x_A[k_A p_{O_2}/(1 + k_Z p_{O_2})] \tag{68}$$

where $k_Z = (k_A/k')$.

The model further assumes that oxygen reacts with B sites in a slow endothermic first-order reaction to yield an A site and CO, which is then desorbed from the surface; i.e.,

$$B_{site} + O_2(g) \rightarrow 2CO(g) + A_{site}$$

$$\text{Rate} = k_B p_{O_2}(1 - x_A) \tag{69}$$

Finally, it is assumed that A sites undergo a slow process of active thermal rearrangement to yield B sites; i.e.,

$$A_{site} \rightarrow B_{site} \tag{70}$$

$$\text{Rate} = k_T x_A$$

For a steady-state value of x_A, one obtains

$$x_A = [1 + (k_T/p_{O_2}k_B)]^{-1} \tag{71}$$

and the overall specific surface reaction rate is then given by

$$(\dot{\omega}/MW_C) = x_A[k_A p_{O_2}/(1 + k_Z p_{O_2})] + k_B p_{O_2}(1 - x_A) \tag{72}$$

where $\dot{\omega}$ is the specific reaction rate and MW_C is the atomic weight of carbon 12. This expression is the semiempirical correlation developed by Nagle and Strickland-Constable [32].

Nagle and Strickland-Constable [32] chose the following values for the rate constants:

$$k_A = 20 \exp(-30/RT) \quad (g\,cm^{-2}s^{-1}atm^{-1})$$

$$k_B = (4.46 \times 10^{-3}) \exp(-15.2/RT) \quad (g\,cm^{-2}s^{-1}atm^{-1})$$

$$k_T = (1.51 \times 10^5) \exp(-97/RT) \quad (g\,cm^{-2}s^{-1})$$

$$k_Z = 21.3 \exp(4.1/RT) \quad (atm^{-1})$$

where all activation energies are in kilocalories per mole. Their results, which are depicted in Fig. 21, have served as the standard for comparison of experimental burning rates of small carbonaceous particles including soot.

Using the values chosen by Nagle and Strickland-Constable, it is possible to explain the trends given in Fig. 21. One observes that at low temperatures $k_T \rightarrow 0$ and $x_A \rightarrow 1$, and for low O_2 partial pressure,

$$\dot{\omega} \cong (2.4 \times 10^2)p_{O_2} \exp(-30/RT)$$

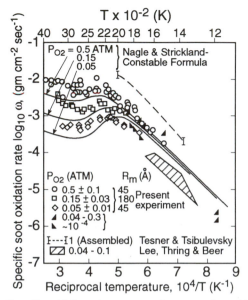

FIGURE 21   Log of specific oxidation rate measurements versus reciprocal temperature and oxygen partial pressure (from Radcliffe and Appleton [31]).

At high O_2 partial pressures, $x \to 1$, the second term of Eq. (72) approaches zero, and the rate becomes zero-order with respect to p_{O_2}:

$$\dot{\omega} \cong 11.3 \exp(-34/RT)$$

At even higher temperatures, where thermal rearrangements produce an increasing number of B sites,

$$x_A = (3 \times 10^{-8})p_{O_2} \exp(81.8/RT)$$

which means that the apparent activation energy of the A site oxidation term in the rate equation changes sign and value with the following net result: for a fixed p_{O_2}, the rate decreases with increasing temperature. At still higher temperatures, the A site oxidation term in the rate equation becomes negligibly small in comparison with the B site term, and again the rate begins to increase with temperature since the k_B term takes over to give

$$\dot{\omega} = (5.35 \times 10^{-2})p_{O_2} \exp(-15.2/RT)$$

Of particular interest in soot particle oxidation is the recent evidence [33–35] that OH radicals and O atoms are effective in gasifying carbon atoms. Roth and co-workers [34, 35], who studied the effectiveness of various oxidizing species in gasifying particles by following the rate of generation of CO, derived the effectiveness of the oxidizers from the expression

$$\frac{d(CO)}{dt} = \alpha Z a_p$$

where α is the reaction probability of the oxidizing species, Z is the collision number per time and per unit area, and a_p is the reacting particle area per unit volume. The collision number Z was calculated from the Hertz–Knudsen equation in which the number concentration or partial pressure of the oxidizing species is embodied. For the temperature range between 2000 and 3500 K, Roth found the following reaction probabilities [35]:

O atom	$\alpha_O = 0.23$
OH radicals	$\alpha_{OH} = 0.05 - 0.2$
NO molecule	$\alpha_{NO} = 1.82 \exp(-1500/T)$

In the range of temperatures found in hydrocarbon–air combustion systems, the reaction probabilities for O_2 and H_2O are at least an order of magnitude less than those for O and OH [32, 33]. The significance of the values for O and OH is that between 10 and 20% of all collisions with soot are effective in gasifying a carbon atom. Examination of the product composition of the stoichiometric propane–air flame in Table 3 of Chapter 1 reveals that OH concentrations are an order of magnitude greater than those of O; thus, though O may have a larger reaction probability than OH, it is most likely that OH radicals will dominate soot oxidation in flames even in the presence of large concentrations of O_2. Haynes [36] reports that this conclusion is confirmed when the experimental rates of soot oxidation by Neoh et al. [33] are compared with those predicted for O_2 attack according to the equation of Nagle and Strickland-Constable. Product composition calculations readily reveal that OH is the dominant species over all mixture ratios and that under lean conditions O_2 and H_2O contribute. Temperature, as well as mixture ratio, plays a significant role in determining the extent of O atom and OH mole fractions in a particular combustion system. As Haynes [36] emphasizes, soot oxidation rates can be expected to possess substantial apparent activation energies. The overall activation energy of the Nagle and Strickland-Constable equation is approximately that of k_A/k_Z or 140 kJ/mol. However, the equilibrium OH concentration defined by $H_2O + \frac{1}{2}O_2 \rightleftharpoons 2OH$ has a temperature dependence corresponding to 280 kJ/mol. Undoubtedly, the temperature of 1300 K reported in the previous chapter for the smoke height condition in diffusion flames is related to these considerations.

E. THE BURNING OF POROUS CHARS

Real coal particles have pores and thus are not like the ideal carbon particles discussed in the last sections. Indeed, one could analyze the pore situation by assuming that pores give increased surface area to the particle. Of course, if diffusion rates to the particle are controlling, the surface area of the particle does not play a significant role. What does play a role is the rate at which the oxygen reaches the surface, i.e., the molecular or convective diffusion rate.

Physically, it is better to consider a large surface area particle as one that has a great number of pores. However, one can use physical arguments to distinguish between ranges of applicability for particles having deep pores or a large, rough, external surface area. Following the pore concept, one must realize that oxygen can penetrate into these pores and that the reaction or depletion of carbon takes place within these pores as well.

Consider now, as Knorre *et al.* [37] have, the situation in which diffusion to the particle is so fast that it is not the controlling rate. For the porous medium, carbon is being consumed within the pores as well as on the surface. The surface consumption rates are therefore controlled by the kinetic rates; however, the consumption rates in the pores are controlled by the diffusion of oxygen into the pore. Thus the mass consumption rate of oxygen in terms of a flux of oxygen must be that consumed at the surface together with that diffused into the pores

$$G_{ox} = k(C_{O_2})_s + D_i(\partial C_{i,O_2}/\partial n)_s \tag{73}$$

where D_i is the internal diffusion coefficient and C_{i,O_2} is the oxygen concentration within the particle.

Following a convention established earlier, this equation is written in the form

$$G_{ox} = k(C_{O_2})_s + D_i(\partial C_{i,O_2}/\partial n)_s = k'(C_{O_2})_s \tag{74}$$

The easiest way to consider this problem is to assume that the pore is spherically symmetric, in which case the internal oxygen diffusion process is described by the equation

$$D_i\left(\frac{d^2C_i}{dr^2} + \frac{2}{r}\frac{dC_i}{dr}\right) - q_{O_2} = 0 \tag{75}$$

where q_{O_2} is the oxygen requirement (consumption) rate per unit particle volume. It is possible to express the quantity q_{O_2} as

$$q_{O_2} = kS_iC_i \tag{76}$$

where S_i is the internal surface area in a unit particle volume (m^2/m^3) such that a very porous particle would approximate the total surface area per unit volume.

The solution of Eq. (75) in terms of the expression given in Eq. (74) results in the following expression for k':

$$k' = k + \lambda D_i[\coth(\lambda R) - 1/\lambda R] \tag{77}$$

where R is the particle radius and

$$\lambda = (S_ik)^{1/2}/D_i \tag{78}$$

For the case of small values of $\lambda R(\lambda R < 0.55)$, which are physically representative of burning at low temperatures or the burning of small particles, the $\coth(\lambda R)$ can be expanded into a series in which only the first two terms are significant; i.e.,

$$\coth(\lambda R) \cong (1/\lambda R) + (\lambda R/3) \tag{79}$$

Substituting Eq. (64) in (62), one obtains

$$k' = k\{1 + (S_i R/3)\} \tag{80}$$

This expression, then, is the rate constant when the inner surface of the pores participate.

If in Eq. (80) the second term in parentheses is small compared to unity—i.e., if the internal surface area is small with respect to the external surface area of the particle—then

$$k' = k \tag{81}$$

Since S_i can be a very large number, tens of thousands of square microns per cubic micron, the condition $(S_i R/3) < 1$ may be satisfied only for very small particles which have radii of the order of tens of microns. Physically, one would not expect that very small particles would have a large internal surface area compared to the external surface.

For large values of λR, which correspond to high temperatures and large particles,

$$\coth(\lambda R) \approx 1 \tag{82}$$

and Eq. (67) becomes

$$k' = k + (S_i Dk)^{1/2} \tag{83}$$

As the temperature increases, the first term in Eq. (83) increases more rapidly than the second because the temperature dependence is only in k. Therefore, at high temperatures

$$k' = k \tag{84}$$

which simply means that the oxygen is completely consumed at the external surface, an important conclusion. For moderate temperatures, it is found that

$$k' \approx (S_i D_i k)^{1/2} \tag{85}$$

or

$$k'/k \approx (S_i D_i/k)^{1/2} \tag{86}$$

$S_i D_i/k$ is a dimensionless number that arises in diffusional kinetics problems. In reality, the best form for k' is Eq. (85), since Eq. (74) may now be written as

$$G_{ox} \approx (S_i D_i k)^{1/2}(C_{O_2})_s \approx (S_i D_i k)^{1/2} C_\infty \qquad (87)$$

because $(C_{O_2})_s \approx C_\infty$ when external diffusion is fast.

Heretofore, the limit conditions have been handled; however, one can make some experimental headway by recalling Eq. (14),

$$G_{ox} = (kh_D/(k + h_D))C_\infty = KC_\infty \qquad (88)$$

and Eq. (15)

$$K = kh_D/(k + h_D) \qquad (89)$$

More generally, for the possible porous problem, Eq. (89) can be written as

$$K = k'h_D/(k' + h_D)$$

where k' can take the values

$$k' = k[1 + (S_i R/3)] \qquad \left\{ \begin{array}{l} \text{low temperature or} \\ \text{small particles} \end{array} \right.$$

$$= k(\text{high temperature}) \qquad \text{all particles}$$

$$= k + (S_i D_i k)^{1/2} \qquad \left\{ \begin{array}{l} \text{high temperatures and} \\ \text{large porous particles} \end{array} \right.$$

$$= (S_i D_i k)^{1/2} \qquad \left\{ \begin{array}{l} \text{moderate temperatures and} \\ \text{large porous particles} \end{array} \right.$$

F. THE BURNING RATE OF ASH-FORMING COAL

Some coals contain an ash in addition to carbon, moisture, and volatiles. To obtain a conservative estimate, one should assume that a porous ash shell is retained during the burning of the combustible material. This ash may, of course, have a catalytic effect on the heterogeneous carbon combustion reactions; however, it is a cause for additional diffusion resistance.

It is apparent that this shell offers great resistance to oxygen diffusion from the free stream to the combustible material. It does not matter whether the oxygen actually diffuses to the carbonaceous surface. The actual mechanism by which the carbonaceous material is consumed is probably very much like that for the pure carbon particle except that the CO to CO_2 conversion is largely heterogeneous. In this problem, diffusion of oxygen to the particle is very much faster than diffusion through the ash; hence, one can assume that the oxygen concentration at the edge of the shell is the same as the atmospheric concentration. The oxygen (or CO_2) concentration at the fuel surface approaches zero.

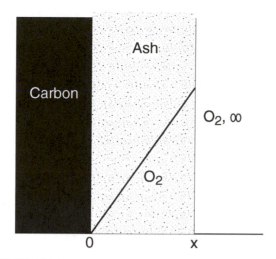

FIGURE 22   Thickness of ash shell in diffusion combustion of high-ash fuel.

Realizing that it does not matter whether O_2 or CO_2 reaches the surface, one can assume, as a first approximation, that a linear oxygen gradient determines the oxygen flux. Since it is not a convective problem, one can write a simplified expression as

$$G_{ox} = D_i(C_{O_{2,\infty}}/x) \tag{90}$$

where x is the thickness of the ash as shown in Fig. 22 and D_i is a diffusion coefficient through the ash.

The fuel, which recedes at a rate dx/dt, is related to the oxygen flux by the expression

$$C_{O_2} = (\rho_c/i)(dx/dt) \tag{91}$$

where ρ_c is the density of the carbonaceous material and i is the mass stoichiometric index. Thus, at any given instant

$$(\rho_c/i)(dx/dt) = D_i C_{O_{2,\infty}}/x \tag{92}$$

If $x = 0$ at $t = 0$, the solution of Eq. (92) is

$$x = \{(2D_i C_{O_{2,\infty}} i/\rho_c)t\}^{1/2} \tag{93}$$

Substituting in Eq. (90), one obtains

$$G_{O_2} = (D_i C_{O_{2,\infty}} \rho_c/i2)^{1/2}(1/t)^{1/2} \tag{94}$$

or

$$G_f \sim (C_{O_2,\infty})^{1/2} \qquad (95)$$

It is not surprising that the oxygen flux decreases or that the consumption of fuel decreases with time as the ash thickness increases. What one obtains, however, is an inverse square root dependence with time and a square root dependence with concentration. Thus for an ash-forming fuel in which ash remains firm throughout the combustion process, the burning rate is proportional to the square root of the oxygen concentration and is independent of the convective nature of the oxidizer stream.

For non–ash-forming coals, the burning rate according to Eq. (29) is

$$G_f \sim \ln(1 + im_{O_2}) \qquad (96)$$

However, the mass fraction of oxygen in air is 0.23 and i is 0.75, so the product is 0.172. Thus im_{O_2} is small compared to 1 and

$$G_f \sim m_{O_2,\infty} \sim C_{O_2,\infty} \qquad (97)$$

i.e., the burning rate is directly proportional to the oxygen concentration.

PROBLEMS

1. Consider a spherical metal particle that is undergoing a high-temperature surface oxidation process. The product of this reaction is a nonvolatile oxide that immediately dissolves in the metal itself. The surface reaction and oxide dissolving rates are very fast compared to the oxidizer diffusion rate. Calculate an expression for the burning rate of this metal.
2. Calculate the value of the transfer number for silicon combustion in air. Show all the stoichiometric relationships in the calculation.
3. A carbon particle is large enough so that the burning rate is diffusion-controlled. In one case the carbon monoxide leaving the surface burns to carbon dioxide in the gas phase; in another, no further carbon monoxide combustion takes place. Is the burning rate of the particle different in the two cases? If so, which is larger? Explain.

REFERENCES

1. Glassman, I., Am. Rocket Soc., Prepr. 938-59 (Nov. 1959).
2. Merzhanov, A. G., and Borovenskaya, I. P., *Combust. Sci. Technol.* **10**, 195 (1975).
3. Glassman,I., *in* "Solid Propellant Rocket Research," ARS Progress Series in Astronautics and Rocketry, Vol. 1, p. 253. Academic Press, New York, 1960.
4. Gordon, S., and McBride, B. J., *NASA [Spec. Publ.] SP* **NASA SP-272**, Interim Revision (Mar. 1976).

5. "JANAF Thermochemical Tables," 3rd Ed., *J. Phys. Chem. Ref. Data* **14** (1985).
6. Brewer, L., *Chem. Rev.* **52**, 1 (1953).
7. von Grosse, H., and Conway, J. B., *Ind. Eng. Chem.* **50**, 663 (1958).
8. Glassman, I., and Papas, P., *J. Mater. Sci. Process* **2**, 151 (1994).
9. Brewer, L., and Searcy, A. W., *J. Am. Chem. Soc.* **75**, 5308 (1951).
10. Glassman, I., and Papas, P., *East. States Meet./Combust. Inst., Clearwater, FL* Pap. No. 59 (1994).
11. Merzhanov, A. G., *Int. Symp. Combust. Flame Synth. High-Temp. Mater. San Francisco* Keynote Lect. (1988).
12. Holt, J. B., *Ind. Res. Dev.* p. 88 (Apr. 1983).
13. Gatica, J. E., and Hlavacek, V., *Ceram. Bull.* **69**, 1311 (1990).
14. Borovenskaya, I. P., *Arch. Procesow Spalania* **5**, 145 (1974).
15. Venturini, J., *Met. Corros.* **28**, 293 (1953).
16. Glassman, I., Davis, K. A., and Brezinsky, K., *Int. Symp. Combust., 24th* p. 1877. Combust. Inst., Pittsburgh, Pennsylvania, 1994.
17. Olson, D. B., Miller, W. J., and Gould, R. K., Aero Chem TP-395. AeroChem Corp., Princeton, New Jersey (1980).
18. Merzhanov, A. G., Borovenskaya, I. P., and Loryan, V. E., *Rep. USSR Acad. Sci.* **206**, 905 (1972).
19. Davis, K. A., Ph.D. Thesis, Dep. Mech. Aerospace Eng., Princeton Univ., Princeton, New Jersey, 1992.
20. Frank-Kamenetskii, D. A., "Diffusion and Heat Exchange in Chemical Kinetics," Chap. 2. Princeton Univ. Press, Princeton, New Jersey, 1955.
21. Benning, M. A., Zabrenski, J. S., and Le, N. B., "Flammability and Sensitivity of Materials in Oxygen Enriched Atmospheres" (D. W. Schroll, ed.), p. 54. ASTM STP 986, American Society of Testing Materials, Philadelphia, Pennsylvania, 1988.
22. Glassman, I., and Law, C. K., *Combust. Sci. Technol.* **80**, 151 (1991).
23. Tognotti, L., Longwell, J. P., and Sarofim, A. F., *Int. Symp. Combust., 23rd* p. 1207. Combust. Inst., Pittsburgh, Pennsylvania, 1990.
24. Purzer, N. R., Yetter, R. A., Dryer, F. L., and Lawson, R. J., *Combust. Sci. Technol.* **147**, 110 (1995).
25. Coffin, K. P., and Brokaw, R. S., *Natl. Advis. Comm. Aeronaut., Tech. Notes* **TN 3929** (1957).
26. Blackshear, P. L., Jr., "An Introduction to Combustion," Chap. 5. Dep. Mech. Eng., Univ. of Minnesota, Minneapolis, 1960.
27. Dryer, F. L., Naegeli, D. W., and Glassman, I., *Combust. Flame* **17**, 270 (1971).
28. Glassman, I., Williams, F. A., and Antaki, P., *Int. Symp. Combust., 20th* p. 2057. Combust. Inst., Pittsburgh, Pennsylvania, 1985.
29. King, M. K., *Spacecr. Rockets* **19**, 294 (1982).
30. Mulcahy, M. F. R., and Smith, I. W., *Rev. Pure Appl. Chem.* **19**, 81 (1969).
31. Radcliffe, S. W., and Appleton, J. P., *Combust. Sci. Technol.* **4**, 171 (1971).
32. Nagle, J., and Strickland-Constable, R. F., *Proc. Conf. Carbon, 5th* **1**, 154 (1962).
33. Neoh, K. G., Howard, J. B., and Sarofim, A. F., "Particulate Carbon: Formation during Combustion," (D. C. Siegla and G. W. Smith, eds.), p. 261. Plenum, New York, 1981.
34. Roth, P., Brandt, O., and Von Gersum, S., *Int. Symp. Combust., 23rd* p. 1484. Combust. Inst., Pittsburgh, Pennsylvaina, 1985.
35. Roth, P., "Turbulence and Molecular Processes in Combustion," p. 149. Elsevier, Amsterdam, 1993.
36. Haynes, B. S., "Fossil Fuel Combustion" (W. Bartok and A. F. Sarofim, eds.), p. 261. Wiley, New York, 1991.
37. Knorre, G. F., Aref'yev, K. M., and Blokh, A. G., "Theory of Combustion Processes," Chap. 24, WPAFB FTD-HT-23-495-68. Wright-Patterson AFB, Ohio, 1968.

Appendixes

Assembled in collaboration with and extended by

Richard A. Yetter

Department of Mechanical and Aerospace Engineering
Princeton University
Princeton, New Jersey

The data presented in these appendixes are provided as a convenience to assist in solving some of the problems in the text, to meet preliminary research needs, to make rapid estimates in evaluating physical concepts, and to serve as a reference volume for the combustion community. Data in certain areas are constantly changing and thus those presented are subject to change. Although judgement was used in selecting the data and the most reliable and recent values available were chosen, it is important to note further that those presented are not the result of an extensive, critical survey. The reader is cautioned to use the material with care and to seek out original sources when a given datum value is crucial to a research result.

Contents

Appendix A

Thermochemical Data
and Conversion Factors

The thermochemical data for the chemical compounds which follow in this appendix are extracted directly from the JANAF tables (JANAF Thermochemical Tables, Third Edition, M. W. Chase Jr., C. A. Davies, J. R. Davies, Jr., D. J. Fulrip, R. A. McDonald, and A. N. Syverud, *Journal of Physical and Chemical Reference Data*, 14, Supplement 1, 1985). The compounds chosen from the numerous ones given are those believed to be most frequently used and those required to solve some of the problem sets given in Chapter 1. Since SI units have been used in the JANAF tables, these units were chosen as the standard throughout. Conversion to cgs units is readily accomplished by use of the conversion factors in this appendix (Table 1). Table 2 contains the thermochemical data.

The ordered listing of the chemical compounds in Table 2 is the same as that in the JANAF tables and is alphabetical according to the chemical formula with the lowest-order letter in the formula determining the position. The thermochemical tables have the following order:

Al (cr,l)	B_2O_3 (g)	HO_2 (g)	SO (g)
Al (g)	Be (cr,l)	H_2 (g)	TiO (g)
AlN (cr)	Be (g)	H_2O (g)	O_2 (g)
AlN (g)	BeO (cr,l)	NH_3 (g)	SO_2 (g)
AlO (g)	BeO (g)	Mg (cr,l)	TiO_2 (cr,l)
Al_2O (g)	C graphite	Mg (g)	TiO_2 (g)
Al_2O_3 (cr,l)	C (g)	MgO (cr,l)	O_3 (g)
B (cr,l)	CH_4 (g)	MgO (g)	SO_3 (g)
B (g)	CO (g)	N (g)	Ti_3O_5 (cr,l)
BHO_2 (g)	CO_2 (g)	NO (g)	Ti_3O_5 (l)
BO (g)	C_2H_2 (g)	NO^+ (g)	S (cr,l)
BO_2 (g)	C_2H_4 (g)	NO_2 (g)	S (g)
B_2O_2 (g)	H (g)	N_2 (g)	Ti (cr,l)
B_2O_3 (cr,l)	OH (g)	O (g)	Ti (g)

The reader should refer to the original tables for the reference material on which the thermochemical data are based. The reference state used in Chapter 1 was chosen as 298 K; consequently, the thermochemical values at this temperature are identified from this listing. The logarithm of the equilibrium constant is to the base 10. Supplemental thermochemical data for species included in the reaction listing of Appendix B, and not given in Table 2, are listed in Table 3. These data, in combination with those of Table 2, may be used to calculate heats of reaction and reverse reaction rate constants as described in Chapter 2. References for the thermochemical data cited in Table 3 may be found in the respective references for the chemical mechanisms of Appendix B.

Detailed data on the higher-order hydrocarbons are not presented. Such data are obtained readily from NBS Circular C 461, "Selected Values of Properties of Hydrocarbons," 1947; from the work of Stull, D. R., Westrum, E. F., Jr., and Sinke, G.C., "The Chemical Thermodynamics of Organic Compounds," Wiley, New York, 1969; and from Reid, R. C., Prausnitz, J. M, and Poling, B.E., "The Properties of Gases and Liquids," McGraw Hill, New York, 1987.

Burcat ["Thermochemical Data for Combustion Calculations" in *Combustion Chemistry* (W. C. Gardiner, Jr., Ed.), Chapter 8, Wiley, New York, 1984] discusses in detail the various sources of thermochemical data and their adaptation for computer usage. Examples of thermochemical data fitted to polynomials for use in computer calculations are reported by McBride, B. J., Gordon, S., and Reno, M. A., "Coefficients for Calculating Thermodynamic and Transport Properties of Individual Species," NASA Technical Memorandum 4513, NASA, NASA Langley, VA, 1993; and Kee, R. J., Rupley, F. M., and Miller, J. A., "The Chemkin Thermodynamic Data Base," Sandia Technical Report SAND87-8215B, Sandia National Laboratories, Livermore, CA 9455, 1987.

TABLE 1   Conversion Factors and Physical Constants

$1 J = 1 W s = 1 N m = 10^7$ erg
1 cal (International Table) $= 4.1868$ J
1 cal (Thermochemical) $= 4.184$ J
1 cal/g K $= 1$ kcal/kg K $= 4.1868$ kJ/kg K
$1 N = 1$ kg m/s$^2 = 10^5$ dyne
1 Pa $= 1$ N/m^2
1 atm $= 1.0132 \times 10^5$ N/m$^2 = 1.0132 \times 10^6$ dyne/cm^2
1 bar $= 10^5$ N/m$^2 = 10^5$ Pa
g_s = gravitational acceleration conversion factor $= 1$ kg m/(N s^2) $= 9.80665$ m/s^2
R = universal gas constant $= 8.314$ J/(g mol K) $= 1.987$ cal/(g mol K) $= 82.05$ cm^3 atm/(g mol K)
σ = Stefan–Boltzmann constant $= 5.6697 \times 10^{-8}$ W/(m^2 K^4)
k_B = Boltzmann constant $= 1.38054 \times 10^{-16}$ erg/(molecule K)
N_A = Avogadro's number $= 6.02252 \times 10^{23}$ molecules/(g mol)
h = Planck's constant $= 6.256 \times 10^{-27}$ erg s
c = speed of light $= 2.997925 \times 10^8$ m/s

TABLE 2   Thermochemical Data of Selected Chemical Compounds
Aluminium (Al), crystal-liquid, mol. wt. $= 26.98154$

| | Enthalpy Reference Temperature = T_r = 298.15 K | | | | Standard State Pressure = p^o = 0.1 MPa | | |
| | J K^{-1}mol^{-1} | | | | kJ mol^{-1} | | |
T/K	C_p^o	S^o	-[G^o-H^o(T_r)]/T	H^o-H^o(T_r)	$\Delta_f H^o$	$\Delta_f G^o$	Log K_f
0	0.	0.	INFINITE	-4.539	0.	0.	0.
100	12.997	6.987	47.543	-4.056	0.	0.	0.
200	21.338	19.144	30.413	-2.254	0.	0.	0.
250	23.084	24.108	28.668	-1.140	0.	0.	0.
298.15	24.209	28.275	28.275	0.	0.	0.	0.
300	24.247	28.425	28.276	0.045	0.	0.	0.
350	25.113	32.231	28.574	1.280	0.	0.	0.
400	25.784	35.630	29.248	2.553	0.	0.	0.
450	26.335	38.699	30.130	3.856	0.	0.	0.
500	26.842	41.501	31.129	5.186	0.	0.	0.
600	27.886	46.485	33.283	7.921	0.	0.	0.
700	29.100	50.872	35.488	10.769	0.	0.	0.
800	30.562	54.850	37.663	13.749	0.	0.	0.
900	32.308	58.548	39.780	16.890	0.	0.	0.
933.450	32.959	59.738	40.474	17.982	___ CRYSTAL <--> LIQUID ___		
933.450	31.751	71.213	40.474	28.693	TRANSITION		
1000	31.751	73.400	42.594	30.806	0.	0.	0.
1100	31.751	76.426	45.534	33.981	0.	0.	0.
1200	31.751	79.189	48.225	37.156	0.	0.	0.
1300	31.751	81.730	50.706	40.331	0.	0.	0.
1400	31.751	84.083	53.007	43.506	0.	0.	0.
1500	31.751	86.273	55.153	46.681	0.	0.	0.
1600	31.751	88.323	57.162	49.856	0.	0.	0.
1700	31.751	90.247	59.052	53.031	0.	0.	0.
1800	31.751	92.062	60.836	56.207	0.	0.	0.
1900	31.751	93.779	62.525	59.382	0.	0.	0.
2000	31.751	95.408	64.129	62.557	0.	0.	0.
2100	31.751	96.957	65.656	65.732	0.	0.	0.
2200	31.751	98.434	67.112	68.907	0.	0.	0.
2300	31.751	99.845	68.505	72.082	0.	0.	0.
2400	31.751	101.196	69.839	75.257	0.	0.	0.
2500	31.751	102.493	71.120	78.432	0.	0.	0.
2600	31.751	103.738	72.350	81.607	0.	0.	0.
2700	31.751	104.936	73.535	84.782	0.	0.	0.
2790.812	31.751	105.986	74.574	87.665	------- FUGACITY = 1 bar ------		
2800	31.751	106.091	74.677	87.957	-293.901	0.968	-0.018
2900	31.751	107.205	75.780	91.132	-292.805	11.479	-0.207
3000	31.751	108.281	76.846	94.307	-291.710	21.953	-0.382

TABLE 2   Thermochemical Data of Selected Chemical Compounds
Aluminium (Al), ideal gas, mol. wt. = 26.98154

Enthalply Reference Temperature = T_r = 298.15 K				Standard State Pressure = p^0 = 0.1 MPa		
	J K^{-1} mol^{-1}			kJ mol^{-1}		
T/K	C_p^0	S^0	$-[G^0-H^0(T_r)]/T$	$H^0-H^0(T_r)$	$\Delta_f H^0$	$\Delta_f G^0$	Log K_f
0	0.	0.	INFINITE	-6.919	327.320	327.320	INFINITE
100	25.192	139.619	184.197	-4.458	329.297	316.034	-165.079
200	22.133	155.883	166.528	-2.129	329.824	302.476	-78.999
250	21.650	160.764	164.907	-1.036	329.804	295.639	-61.770
298.15	21.390	164.553	164.553	0.	329.699	289.068	-50.643
300	21.383	164.686	164.554	0.040	329.694	288.816	-50.287
350	21.221	167.969	164.813	1.104	329.524	282.016	-42.089
400	21.117	170.795	165.388	2.163	329.309	275.243	-35.943
450	21.046	173.278	166.130	3.217	329.060	268.499	-31.167
500	20.995	175.492	166.957	4.268	328.781	261.785	-27.348
600	20.930	179.314	168.708	6.363	328.141	248.444	-21.629
700	20.891	182.537	170.460	8.454	327.385	235.219	-17.552
800	20.866	185.325	172.147	10.542	326.492	222.112	-14.502
900	20.849	187.781	173.751	12.628	325.508	209.126	-12.137
1000	20.836	189.977	175.266	14.712	313.605	197.027	-10.292
1100	20.827	191.963	176.695	16.795	312.513	185.422	-8.805
1200	20.821	193.775	178.044	18.877	311.420	173.917	-7.570
1300	20.816	195.441	179.319	20.959	310.327	162.503	-6.529
1400	20.811	196.984	180.526	23.041	309.233	151.172	-5.640
1500	20.808	198.419	181.672	25.122	308.139	139.921	-4.872
1600	20.805	199.762	182.761	27.202	307.045	128.742	-4.203
1700	20.803	201.023	183.798	29.283	305.950	117.631	-3.614
1800	20.801	202.212	184.789	31.363	304.856	106.585	-3.093
1900	20.800	203.337	185.736	33.443	303.760	95.600	-2.628
2000	20.798	204.404	186.643	35.523	302.665	84.673	-2.211
2100	20.797	205.419	187.513	37.603	301.570	73.800	-1.836
2200	20.796	206.386	188.349	39.682	300.475	62.979	-1.495
2300	20.795	207.311	189.153	41.762	299.379	52.209	-1.186
2400	20.795	208.196	189.928	43.841	298.284	41.486	-0.903
2500	20.794	209.044	190.676	45.921	297.188	30.808	-0.644
2600	20.794	209.860	191.398	48.000	296.092	20.175	-0.405
2700	20.794	210.645	192.097	50.080	294.997	9.583	-0.185
2790.812	20.795	211.333	192.712	51.968	------- FUGACITY = 1 bar ------		
2800	20.795	211.401	192.773	52.159	0.	0.	0.
2900	20.796	212.131	193.428	54.239	0.	0.	0.
3000	20.798	212.836	194.063	56.318	0.	0.	0.
3100	20.800	213.518	194.680	58.398	0.	0.	0.
3200	20.804	214.178	195.279	60.478	0.	0.	0.
3300	20.808	214.818	195.861	62.559	0.	0.	0.
3400	20.815	215.440	196.428	64.640	0.	0.	0.
3500	20.823	216.043	196.980	66.722	0.	0.	0.
3600	20.833	216.630	197.518	68.805	0.	0.	0.
3700	20.846	217.201	198.042	70.889	0.	0.	0.
3800	20.862	217.757	198.553	72.974	0.	0.	0.
3900	20.881	218.299	199.053	75.061	0.	0.	0.
4000	20.904	218.828	199.541	77.150	0.	0.	0.
4100	20.932	219.345	200.017	79.242	0.	0.	0.
4200	20.964	219.849	200.484	81.337	0.	0.	0.
4300	21.002	220.343	200.940	83.435	0.	0.	0.
4400	21.046	220.826	201.386	85.537	0.	0.	0.
4500	21.088	221.299	201.823	87.642	0.	0.	0.
4600	21.143	221.763	202.252	89.753	0.	0.	0.
4700	21.206	222.219	202.672	91.870	0.	0.	0.
4800	21.276	222.666	203.084	93.995	0.	0.	0.
4900	21.352	223.105	203.488	96.125	0.	0.	0.
5000	21.439	223.537	203.885	98.264	0.	0.	0.
5100	21.535	223.963	204.274	100.413	0.	0.	0.
5200	21.641	224.382	204.657	102.572	0.	0.	0.
5300	21.757	224.795	205.033	104.742	0.	0.	0.
5400	21.884	225.203	205.403	106.924	0.	0.	0.
5500	22.021	225.606	205.766	109.119	0.	0.	0.
5600	22.170	226.004	206.124	111.328	0.	0.	0.
5700	22.330	226.398	206.476	113.553	0.	0.	0.
5800	22.496	226.787	206.823	115.792	0.	0.	0.
5900	22.680	227.173	207.165	118.050	0.	0.	0.
6000	22.836	227.552	207.501	120.306	0.	0.	0.

TABLE 2   Thermochemical Data of Selected Chemical Compounds
Aluminium Nitride (AlN), crystal, mol. wt. = 40.98824

| Enthalply Reference Temperature = T_r = 298.15 K | | | | Standard State Pressure = p^0 = 0.1 MPa | | | |
| | $J\ K^{-1}mol^{-1}$ | | | $kJ\ mol^{-1}$ | | | |
T/K	C_p^0	S^0	$-[G^0-H^0(T_r)]/T$	$H^0-H^0(T_r)$	$\Delta_f H^0$	$\Delta_f G^0$	Log K_f
0	0.	0.	INFINITE	-3.871	-312.980	-312.980	INFINITE
100	5.678	2.164	39.274	-3.711	-314.756	-306.283	159.986
200	19.332	10.267	22.581	-2.463	-316.764	-296.990	77.566
298.15	30.097	20.142	20.142	0.	-317.984	-286.995	50.280
300	30.254	20.329	20.143	0.056	-318.000	-286.803	49.937
400	36.692	29.987	21.416	3.428	-318.594	-276.301	36.081
500	40.799	38.647	24.013	7.317	-318.808	-265.697	27.757
600	43.538	46.341	27.106	11.541	-318.811	-255.072	22.206
700	45.434	53.201	30.353	15.994	-318.727	-244.455	18.241
800	46.791	59.361	33.601	20.608	-318.648	-233.850	15.269
900	47.792	64.932	38.777	25.339	-318.647	-223.252	12.957
1000	48.550	70.008	39.850	30.158	-329.363	-211.887	11.068
1100	49.136	74.664	42.807	35.043	-329.302	-200.142	9.504
1200	49.598	78.960	45.643	39.981	-329.214	-188.404	8.201
1300	49.970	82.945	48.361	44.960	-329.107	-176.674	7.099
1400	50.272	86.660	50.965	49.972	-328.986	-164.953	6.154
1500	50.521	90.137	53.462	55.012	-328.856	-153.240	5.336
1600	50.728	93.404	55.857	60.075	-328.717	-141.537	4.621
1700	50.903	96.485	58.157	65.157	-328.573	-129.843	3.990
1800	51.052	99.399	60.368	70.255	-328.425	-118.157	3.429
1900	51.180	102.162	62.496	75.368	-328.273	-106.480	2.927
2000	51.290	104.790	64.545	80.490	-328.119	-94.810	2.476
2100	51.385	107.295	66.522	85.624	-327.963	-83.149	2.068
2200	51.469	109.688	68.430	90.767	-327.805	-71.495	1.697
2300	51.543	111.977	70.274	95.917	-327.646	-59.848	1.359
2400	51.609	114.172	72.058	101.075	-327.486	-48.208	1.049
2500	51.666	116.280	73.785	106.239	-327.325	-36.574	0.764
2600	51.718	118.308	75.458	111.408	-327.165	-24.948	0.501
2700	51.765	120.260	77.082	116.582	-327.003	-13.327	0.258
2800	51.807	122.144	78.658	121.761	-620.743	-0.744	0.014
2900	51.845	123.962	80.189	126.943	-619.486	21.376	-0.385
3000	51.878	125.721	81.677	132.130	-618.229	43.454	-0.757

TABLE 2   Thermochemical Data of Selected Chemical Compounds
Aluminium Nitride (AlN), ideal gas, mol. wt. = 40.98824

| Enthalply Reference Temperature = T_r = 298.15 K | | | | Standard State Pressure = p^o = 0.1 MPa | | | |
| | J K^{-1}mol^{-1} | | | kJ mol^{-1} | | | |
T/K	C_p^o	S^o	-[G^o-H^o(T_r)]/T	H^o-H^o(T_r)	$\Delta_f H^o$	$\Delta_f G^o$	Log K_f
0	0.	0.	INFINITE	-8.942	522.933	522.933	INFINITE
100	29.134	195.617	255.951	-6.033	523.906	513.034	-267.981
200	30.316	216.062	231.448	-3.077	523.605	502.220	-131.166
250	31.378	222.940	229.080	-1.535	523.306	496.908	-103.823
298.15	32.367	228.553	228.553	0.	523.000	491.851	-86.170
300	32.403	228.753	228.553	0.060	522.988	491.658	-85.605
350	33.288	233.817	228.951	1.703	522.668	486.462	-72.600
400	34.019	238.311	229.846	3.386	522.348	481.311	-62.853
450	34.612	242.353	231.015	5.102	522.028	476.201	-55.276
500	35.093	246.026	232.335	6.845	521.704	471.126	-49.218
600	35.808	252.491	235.170	10.393	521.025	461.074	-40.140
700	36.304	258.051	238.051	14.000	520.263	451.141	-33.665
800	36.663	262.923	240.862	17.649	519.377	441.325	-28.816
900	36.934	267.258	243.558	21.330	518.328	431.630	-25.051
1000	37.147	271.160	246.127	25.034	506.497	422.821	-22.086
1100	37.320	274.709	248.566	28.758	505.396	414.507	-19.683
1200	37.465	277.963	250.882	32.497	504.286	406.293	-17.685
1300	37.589	280.967	253.082	36.250	503.167	398.172	-15.999
1400	37.699	283.756	255.175	40.014	502.040	390.138	-14.556
1500	37.797	286.361	257.168	43.789	500.905	382.184	-13.309
1600	37.886	288.803	259.070	47.573	499.765	374.307	-12.220
1700	37.969	291.102	260.887	51.366	498.620	366.501	-11.261
1800	38.046	293.275	262.627	55.167	497.471	358.762	-10.411
1900	38.119	295.334	264.294	58.975	496.320	351.087	-9.652
2000	38.189	297.291	265.896	62.791	495.166	343.473	-8.971
2100	38.257	299.156	267.435	66.613	494.011	335.917	-8.355
2200	38.322	300.937	268.918	70.442	492.855	328.416	-7.798
2300	38.387	302.642	270.347	74.278	491.698	320.967	-7.289
2400	38.450	304.277	271.727	78.119	490.542	313.569	-6.825
2500	38.514	305.848	273.061	81.968	489.387	306.219	-6.398
2600	38.577	307.360	274.351	85.822	488.234	298.915	-6.005
2700	38.641	308.817	275.601	89.683	487.082	291.655	-5.642
2800	38.706	310.223	276.812	93.550	192.031	285.406	-5.324
2900	38.772	311.583	277.988	97.424	191.979	288.742	-5.201
3000	38.840	312.898	279.130	101.305	191.930	292.079	-5.086
3100	38.910	314.173	280.240	105.192	191.884	295.418	-4.978
3200	38.981	315.409	281.320	109.087	191.842	298.759	-4.877
3300	39.056	316.610	282.371	112.989	191.804	302.101	-4.782
3400	39.133	317.777	283.395	116.898	191.770	305.444	-4.693
3500	39.212	318.913	284.394	120.815	191.741	308.787	-4.608
3600	39.294	320.018	285.368	124.741	191.716	312.132	-4.529
3700	39.379	321.096	286.319	128.674	191.696	315.477	-4.454
3800	39.467	322.148	287.248	132.616	191.681	318.823	-4.383
3900	39.558	323.174	288.157	136.568	191.671	322.169	-4.315
4000	39.652	324.177	289.045	140.528	191.665	325.515	-4.251
4100	39.748	325.157	289.913	144.498	191.665	328.861	-4.190
4200	39.847	326.116	290.764	148.478	191.670	332.207	-4.132
4300	39.949	327.055	291.597	152.468	191.679	335.553	-4.076
4400	40.053	327.974	292.413	156.468	191.693	338.899	-4.023
4500	40.159	328.876	293.214	160.478	191.713	342.244	-3.973
4600	40.268	329.759	293.999	164.500	191.735	345.589	-3.924
4700	40.378	330.627	294.769	168.532	191.760	348.934	-3.878
4800	40.490	331.478	295.525	172.575	191.788	352.277	-3.834
4900	40.604	332.314	296.267	176.630	191.819	355.621	-3.791
5000	40.720	333.135	296.996	180.696	191.851	358.963	-3.750
5100	40.836	333.943	297.713	184.774	191.884	362.305	-3.711
5200	40.954	334.737	298.417	188.864	191.917	365.646	-3.673
5300	41.073	335.518	299.110	192.965	191.948	368.987	-3.637
5400	41.192	336.287	299.791	197.078	191.978	372.327	-3.602
5500	41.313	337.044	300.462	201.203	192.005	375.667	-3.568
5600	41.433	337.790	301.122	205.341	192.028	379.006	-3.535
5700	41.554	338.524	301.771	209.490	192.046	382.344	-3.504
5800	41.675	339.248	302.411	213.651	192.060	385.682	-3.473
5900	41.795	339.961	303.042	217.825	192.064	389.021	-3.444
6000	41.916	340.665	303.663	222.010	192.082	392.357	-3.416

TABLE 2   Thermochemical Data of Selected Chemical Compounds
Aluminium Oxide (AlO), ideal gas, mol. wt. = 42.98094

| | Enthalpy Reference Temperature = T_r = 298.15 K | | | | Standard State Pressure = p^o = 0.1 MPa | | |
| | J K^{-1}mol^{-1} | | | | kJ mol^{-1} | | |
T/K	C_p^o	S^o	$-[G^o-H^o(T_r)]/T$	$H^o-H^o(T_r)$	$\Delta_f H^o$	$\Delta_f G^o$	Log K$_f$
0	0.	0.	INFINITE	-8.788	67.037	67.037	INFINITE
100	29.108	186.129	244.933	-5.880	68.009	58.780	-30.693
200	29.504	206.370	221.161	-2.958	67.673	49.577	-12.948
298.15	30.883	218.386	218.386	0.	66.944	40.845	-7.156
300	30.913	218.577	218.387	0.057	66.929	40.683	-7.084
400	32.489	227.691	219.619	3.229	66.107	32.057	-4.186
500	33.750	235.083	221.995	6.544	65.260	23.642	-2.470
600	34.685	241.323	224.710	9.968	64.369	15.401	-1.341
700	35.399	246.726	227.478	13.474	63.400	7.315	-0.546
800	36.001	251.492	230.187	17.044	62.321	-0.625	0.041
900	36.583	255.766	232.796	20.673	61.106	-8.422	0.489
1000	37.205	259.652	235.290	24.362	49.149	-15.315	0.800
1100	37.898	263.230	237.670	28.116	47.973	-21.704	1.031
1200	38.669	266.560	239.940	31.944	46.852	-27.988	1.218
1300	39.506	269.688	242.109	35.852	45.793	-34.182	1.373
1400	40.386	272.648	244.186	39.847	44.806	-40.296	1.503
1500	41.283	275.465	246.178	43.930	43.893	-46.343	1.614
1600	42.167	278.157	248.093	48.103	43.057	-52.331	1.708
1700	43.015	280.739	249.938	52.362	42.296	-58.269	1.790
1800	43.805	283.221	251.719	56.704	41.605	-64.164	1.862
1900	44.523	285.609	253.440	61.121	40.977	-70.023	1.925
2000	45.158	287.909	255.106	65.606	40.405	-75.850	1.981
2100	45.706	290.126	256.721	70.150	39.881	-81.649	2.031
2200	46.165	292.263	258.288	74.744	39.396	-87.425	2.076
2300	46.538	294.324	259.811	79.380	38.942	-93.179	2.116
2400	46.830	296.311	261.290	84.049	38.509	-98.914	2.153
2500	47.047	298.227	262.730	88.743	38.091	-104.632	2.186
2600	47.196	300.075	264.131	93.456	37.681	-110.333	2.217
2700	47.286	301.858	265.495	98.180	37.272	-116.018	2.244
2800	47.325	303.579	266.825	102.911	-257.042	-120.719	2.252
2900	47.321	305.240	268.121	107.644	-256.368	-115.863	2.087
3000	47.279	306.843	269.385	112.374	-255.706	-111.029	1.933
3100	47.209	308.393	270.619	117.099	-255.059	-106.217	1.790
3200	47.115	309.890	271.823	121.815	-254.430	-101.426	1.656
3300	47.003	311.338	272.998	126.521	-253.820	-96.654	1.530
3400	46.878	312.739	274.147	131.215	-253.231	-91.901	1.412
3500	46.744	314.096	275.269	135.896	-252.663	-87.164	1.301
3600	46.605	315.411	276.366	140.564	-252.118	-82.443	1.196
3700	46.464	316.686	277.438	145.217	-251.596	-77.737	1.097
3800	46.324	317.923	278.487	149.857	-251.096	-73.045	1.004
3900	46.187	319.125	279.514	154.482	-250.619	-68.365	0.916
4000	46.055	320.293	280.519	159.094	-250.163	-63.698	0.832
4100	45.930	321.428	281.503	163.694	-249.730	-59.042	0.752
4200	45.813	322.534	282.467	168.281	-249.318	-54.396	0.677
4300	45.704	323.610	283.411	172.857	-248.928	-49.760	0.604
4400	45.606	324.660	284.337	177.422	-248.557	-45.133	0.536
4500	45.518	325.684	285.244	181.978	-248.205	-40.514	0.470
4600	45.441	326.683	286.134	186.526	-247.874	-35.902	0.408
4700	45.376	327.660	287.008	191.067	-247.561	-31.297	0.348
4800	45.322	328.615	287.864	195.601	-247.268	-26.699	0.291
4900	45.280	329.549	288.706	200.131	-246.993	-22.106	0.236
5000	45.250	330.463	289.532	204.658	-246.736	-17.520	0.183
5100	45.232	331.359	290.343	209.182	-246.498	-12.937	0.133
5200	45.226	332.237	291.140	213.705	-246.278	-8.360	0.084
5300	45.231	333.099	291.924	218.227	-246.077	-3.787	0.037
5400	45.248	333.944	292.694	222.751	-245.894	0.783	-0.008
5500	45.276	334.775	293.452	227.277	-245.730	5.349	-0.051
5600	45.315	335.591	294.197	231.807	-245.585	9.913	-0.092
5700	45.365	336.394	294.930	236.341	-245.461	14.474	-0.133
5800	45.425	337.183	295.652	240.880	-245.354	19.034	-0.171
5900	45.494	337.960	296.363	245.426	-245.270	23.591	-0.209
6000	45.574	338.725	297.062	249.980	-245.186	28.145	-0.245

TABLE 2   Thermochemical Data of Selected Chemical Compounds
Aluminium Oxide (Al_2O), ideal gas, mol. wt. = 69.96248

Enthalpy Reference Temperature = T_r = 298.15 K					Standard State Pressure = p^o = 0.1 MPa		
		J K^{-1}mol^{-1}			kJ mol^{-1}		
T/K	C_p^o	S^o	-[G^o-$H^o(T_r)$]/T	H^o-$H^o(T_r)$	$\Delta_f H^o$	$\Delta_f G^o$	Log K$_f$
0	0.	0.	INFINITE	-12.710	-144.474	-144.474	INFINITE
100	40.424	201.813	295.494	-9.368	-143.552	-153.671	80.269
200	47.795	232.403	256.967	-4.913	-144.156	-163.630	42.736
250	50.162	243.332	253.179	-2.462	-144.662	-168.442	35.194
298.15	52.032	252.332	252.332	0.	-145.185	-172.975	30.304
300	52.098	252.654	252.333	0.096	-145.205	-173.147	30.148
350	53.696	260.809	252.973	2.742	-145.767	-177.760	26.529
400	55.010	268.068	254.415	5.461	-146.342	-182.291	23.805
450	56.087	274.611	256.301	8.239	-146.929	-186.750	21.677
500	56.971	280.568	258.435	11.067	-147.532	-191.142	19.968
600	58.297	291.080	263.023	16.834	-148.815	-199.746	17.389
700	59.213	300.139	267.693	22.713	-150.259	-208.123	15.530
800	59.863	308.091	272.256	28.668	-151.933	-216.278	14.121
900	60.337	315.171	276.638	34.679	-153.907	-224.206	13.013
1000	60.691	321.547	280.815	40.732	-177.417	-230.375	12.034
1100	60.961	327.345	284.786	46.815	-179.438	-235.573	11.186
1200	61.172	332.658	288.557	52.922	-181.456	-240.587	10.472
1300	61.339	337.562	292.140	59.048	-183.472	-245.433	9.862
1400	61.473	342.112	295.549	65.188	-185.488	-250.124	9.332
1500	61.583	346.357	298.796	71.341	-187.505	-254.670	8.868
1600	61.674	350.335	301.895	77.504	-189.526	-259.082	8.458
1700	61.750	354.076	304.855	83.676	-191.551	-263.368	8.092
1800	61.814	357.608	307.689	89.854	-193.581	-267.534	7.764
1900	61.869	360.951	310.405	96.038	-195.616	-271.587	7.466
2000	61.915	364.126	313.012	102.227	-197.658	-275.532	7.196
2100	61.956	367.148	315.519	108.421	-199.708	-279.376	6.949
2200	61.991	370.031	317.931	114.618	-201.764	-283.121	6.722
2300	62.022	372.787	320.257	120.819	-203.829	-286.773	6.513
2400	62.050	375.427	322.501	127.023	-205.902	-290.335	6.319
2500	62.075	377.961	324.669	133.229	-207.984	-293.810	6.139
2600	62.097	380.396	326.766	139.438	-210.073	-297.202	5.971
2700	62.118	382.740	328.796	145.649	-212.171	-300.513	5.814
2800	62.138	384.999	330.763	151.861	-802.079	-301.811	5.630
2900	62.156	387.180	332.671	158.076	-802.003	-283.946	5.114
3000	62.174	389.288	334.523	164.293	-801.934	-266.083	4.633
3100	62.192	391.327	336.323	170.511	-801.873	-248.222	4.183
3200	62.210	393.301	338.073	176.731	-801.820	-230.363	3.760
3300	62.229	395.216	339.776	182.953	-801.775	-212.506	3.364
3400	62.248	397.074	341.434	189.177	-801.737	-194.650	2.990
3500	62.269	398.879	343.049	195.403	-801.707	-176.795	2.639
3600	62.291	400.633	344.625	201.631	-801.684	-158.941	2.306
3700	62.315	402.340	346.162	207.861	-801.669	-141.087	1.992
3800	62.341	404.002	347.662	214.094	-801.661	-123.233	1.694
3900	62.369	405.622	349.127	220.329	-801.661	-105.380	1.411
4000	62.399	407.202	350.560	226.567	-801.669	-87.526	1.143
4100	62.433	408.743	351.960	232.809	-801.685	-69.673	0.888
4200	62.470	410.248	353.330	239.054	-801.710	-51.818	0.644
4300	62.509	411.718	354.671	245.303	-801.744	-33.963	0.413
4400	62.553	413.156	355.984	251.556	-801.788	-16.108	0.191
4500	62.600	414.562	357.270	257.814	-801.839	1.749	-0.020
4600	62.651	415.938	358.530	264.076	-801.904	19.607	-0.223
4700	62.705	417.286	359.766	270.344	-801.983	37.467	-0.416
4800	62.764	418.607	360.978	276.617	-802.075	55.328	-0.602
4900	62.827	419.902	362.168	282.897	-802.180	73.192	-0.780
5000	62.895	421.172	363.335	289.183	-802.303	91.058	-0.951
5100	62.967	422.418	364.481	295.476	-802.445	108.927	-1.116
5200	63.044	423.641	365.607	301.777	-802.606	126.798	-1.274
5300	63.125	424.843	366.714	308.085	-802.789	144.673	-1.426
5400	63.211	426.024	367.801	314.402	-802.995	162.551	-1.572
5500	63.301	427.184	368.870	320.727	-803.227	180.434	-1.714
5600	63.396	428.326	369.922	327.062	-803.487	198.321	-1.850
5700	63.496	429.449	370.956	333.407	-803.776	216.213	-1.981
5800	63.601	430.554	371.974	339.761	-804.092	234.110	-2.108
5900	63.710	431.642	372.977	346.127	-804.448	252.012	-2.231
6000	63.823	432.714	373.963	352.503	-804.796	269.918	-2.350

TABLE 2 Thermochemical Data of Selected Chemical Compounds
Aluminium Oxide (Al_2O_3), crystal-liquid, mol. wt. = 101.96128

| | Enthalply Reference Temperature = T_r = 298.15 K | | | | Standard State Pressure = p^0 = 0.1 MPa | | |
| | J K^{-1}mol^{-1} | | | | kJ mol^{-1} | | |
T/K	C_p^0	S^0	-[G^0-H^0(T_r)]/T	H^0-H^0(T_r)	$\Delta_f H^0$	$\Delta_f G^0$	Log K_f
0	0.	0.	INFINITE	-10.020	-1663.608	-1663.608	INFINITE
100	12.855	4.295	101.230	-9.693	-1668.606	-1641.642	857.506
200	51.120	24.880	57.381	-6.500	-1673.383	-1612.656	421.183
298.15	79.015	50.950	50.950	0.	-1675.692	-1582.275	277.208
300	79.416	51.440	50.951	0.147	-1675.717	-1581.696	275.398
400	96.086	76.779	54.293	8.995	-1676.342	-1550.226	202.439
500	106.131	99.388	61.098	19.145	-1676.045	-1518.718	158.659
600	112.545	119.345	69.177	30.101	-1675.300	-1487.319	129.483
700	116.926	137.041	77.632	41.586	-1674.391	-1456.059	108.652
800	120.135	152.873	86.065	53.447	-1673.498	-1424.931	93.038
900	122.662	167.174	94.296	65.591	-1672.744	-1393.908	80.900
1000	124.771	180.210	102.245	77.965	-1693.394	-1361.437	71.114
1100	126.608	192.189	109.884	90.535	-1692.437	-1328.286	63.075
1200	128.252	203.277	117.211	103.280	-1691.366	-1295.228	56.380
1300	129.737	213.602	124.233	116.180	-1690.190	-1262.264	50.718
1400	131.081	223.267	130.965	129.222	-1688.918	-1229.393	45.869
1500	132.290	232.353	137.425	142.392	-1687.561	-1196.617	41.670
1600	133.361	240.925	143.628	155.675	-1686.128	-1163.934	37.999
1700	134.306	249.039	149.592	169.060	-1684.632	-1131.342	34.762
1800	135.143	256.740	155.333	182.533	-1683.082	-1098.841	31.888
1900	135.896	264.067	160.864	196.085	-1681.489	-1066.426	29.318
2000	136.608	271.056	166.201	209.710	-1679.858	-1034.096	27.008
2100	137.319	277.738	171.354	223.407	-1678.190	-1001.849	24.920
2200	138.030	284.143	176.336	237.174	-1676.485	-969.681	23.023
2300	138.741	290.294	181.158	251.013	-1674.743	-937.593	21.293
2327.000	138.934	291.914	182.434	254.761	ALPHA <--> LIQUID		
2327.000	192.464	339.652	182.434	365.847	TRANSITION		
2400	192.464	345.597	187.307	379.896	-1557.989	-909.127	19.787
2500	192.464	353.454	193.796	399.143	-1550.905	-882.237	18.433
2600	192.464	361.002	200.083	418.389	-1543.853	-855.629	17.190
2700	192.464	368.266	206.179	437.636	-1536.832	-829.292	16.044
2800	192.464	375.265	212.093	456.882	-2117.645	-801.279	14.948
2900	192.464	382.019	217.837	476.128	-2108.494	-754.429	13.589
3000	192.464	388.544	223.419	495.375	-2099.372	-707.892	12.325
3100	192.464	394.855	228.848	514.621	-2090.279	-661.658	11.149
3200	192.464	400.965	234.132	533.868	-2081.214	-615.719	10.051
3300	192.464	406.888	239.277	553.114	-2072.175	-570.063	9.023
3400	192.464	412.633	244.292	572.360	-2063.162	-524.679	8.061
3500	192.464	418.212	249.182	591.607	-2054.176	-479.561	7.157
3600	192.464	423.634	253.953	610.853	-2045.213	-434.699	6.307
3700	192.464	428.908	258.610	630.100	-2036.276	-390.086	5.507
3800	192.464	434.040	263.160	649.346	-2027.363	-345.712	4.752
3900	192.464	439.040	267.606	668.592	-2018.474	-301.574	4.039
4000	192.464	443.912	271.953	687.839	-2009.609	-257.664	3.365

TABLE 2　Thermochemical Data of Selected Chemical Compounds

Boron (B), crystal-liquid, mol. wt. = 10.81

Enthalply Reference Temperature = T_r = 298.15 K					Standard State Pressure = p^0 = 0.1 MPa		
	— J K^{-1}mol^{-1} —				— kJ mol^{-1} —		
T/K	C_p^0	S^0	$-[G^0-H^0(T_r)]/T$	$H^0-H^0(T_r)$	$\Delta_f H^0$	$\Delta_f G^0$	Log K_f
0	0.	0.	INFINITE	-1.214	0.	0.	0.
100	1.076	0.308	12.207	-1.190	0.	0.	0.
200	5.998	2.419	6.704	-0.857	0.	0.	0.
250	8.821	4.063	6.007	-0.486	0.	0.	0.
298.15	11.315	5.834	5.834	0.	0.	0.	0.
300	11.405	5.904	5.834	0.021	0.	0.	0.
350	13.654	7.834	5.981	0.648	0.	0.	0.
400	15.693	9.794	6.335	1.384	0.	0.	0.
450	17.361	11.742	6.828	2.211	0.	0.	0.
500	18.722	13.644	7.415	3.115	0.	0.	0.
600	20.778	17.251	8.758	5.096	0.	0.	0.
700	22.249	20.570	10.212	7.251	0.	0.	0.
800	23.361	23.617	11.699	9.534	0.	0.	0.
900	24.245	26.421	13.181	11.915	0.	0.	0.
1000	24.978	29.014	14.637	14.378	0.	0.	0.
1100	25.606	31.425	16.055	16.908	0.	0.	0.
1200	26.161	33.677	17.430	19.496	0.	0.	0.
1300	26.663	35.792	18.762	22.138	0.	0.	0.
1400	27.125	37.785	20.051	24.828	0.	0.	0.
1500	27.557	39.671	21.296	27.562	0.	0.	0.
1600	27.966	41.463	22.501	30.338	0.	0.	0.
1700	28.356	43.170	23.667	33.155	0.	0.	0.
1800	28.732	44.801	24.796	36.009	0.	0.	0.
1900	29.097	46.365	25.891	38.901	0.	0.	0.
2000	29.452	47.866	26.952	41.828	0.	0.	0.
2100	29.799	49.312	27.983	44.791	0.	0.	0.
2200	30.140	50.706	28.984	47.788	0.	0.	0.
2300	30.475	52.053	29.958	50.819	0.	0.	0.
2350.000	30.641	52.710	30.435	52.346	——— BETA <--> LIQUID ———		
2350.000	31.750	74.075	30.435	102.554	TRANSITION		
2400	31.750	74.744	31.351	104.142	0.	0.	0.
2500	31.750	76.040	33.113	107.317	0.	0.	0.
2600	31.750	77.285	34.788	110.492	0.	0.	0.
2700	31.750	78.483	38.384	113.667	0.	0.	0.
2800	31.750	79.638	37.909	116.842	0.	0.	0.
2900	31.750	80.752	39.367	120.017	0.	0.	0.
3000	31.750	81.828	40.764	123.192	0.	0.	0.
3100	31.750	82.869	42.106	126.367	0	0.	0.
3200	31.750	83.877	43.398	129.542	0	0.	0.
3300	31.750	84.854	44.637	132.717	0.	0.	0.
3400	31.750	85.802	45.834	135.892	0.	0.	0.
3500	31.750	86.723	46.989	139.067	0.	0.	0.
3600	31.750	87.617	48.105	142.242	0.	0.	0.
3700	31.750	88.487	49.185	145.417	0.	0.	0.
3800	31.750	89.334	50.231	148.592	0.	0.	0.
3900	31.750	90.158	51.244	151.767	0.	0.	0.
4000	31.750	90.962	52.227	154.942	0.	0.	0.
4100	31.750	91.746	53.181	158.117	0.	0.	0.
4139.449	31.750	92.050	53.550	159.369	------ FUGACITY = 1 bar ------		
4200	31.750	92.511	54.109	161.292	-479.849	7.024	-0.087
4300	31.750	93.258	55.010	164.467	-478.782	18.604	-0.226
4400	31.750	93.988	55.888	167.642	-477.677	30.158	-0.358
4500	31.750	94.702	56.743	170.817	-476.593	41.687	-0.484

TABLE 2 Thermochemical Data of Selected Chemical Compounds
Boron (B), ideal gas, mol. wt. = 10.81

Enthalpy Reference Temperature = T_r = 298.15 K					Standard State Pressure = p^o = 0.1 MPa		
	——— J K^{-1}mol^{-1} ———				——— kJ mol^{-1} ———		
T/K	C_p^o	S^o	-[G^o-H^o(T_r)]/T	H^o-H^o(T_r)	$\Delta_f H^o$	$\Delta_f G^o$	Log K_f
0	0.	0.	INFINITE	-6.316	554.898	554.898	INFINITE
100	20.881	130.687	171.936	-4.125	557.065	544.027	-284.171
200	20.809	145.130	155.338	-2.042	558.815	530.273	-138.493
250	20.801	149.772	153.778	-1.001	559.485	523.057	-109.287
298.15	20.796	153.435	153.435	0.	560.000	515.993	-90.400
300	20.796	153.564	153.436	0.038	560.017	515.720	-89.795
350	20.793	156.769	153.689	1.078	560.430	508.302	-75.860
400	20.792	159.546	154.251	2.118	560.734	500.834	-65.402
450	20.790	161.995	154.978	3.157	560.946	493.332	-57.265
500	20.790	164.185	155.791	4.197	561.082	485.812	-50.752
600	20.789	167.975	157.516	6.276	561.180	470.745	-40.982
700	20.788	171.180	159.245	8.355	561.104	455.677	-34.003
800	20.787	173.956	160.914	10.433	560.900	440.628	-28.770
900	20.797	176.404	162.502	12.512	560.597	425.612	-24.702
1000	20.787	178.594	164.003	14.591	560.213	410.633	-21.449
1100	20.797	180.575	165.421	16.669	559.762	395.697	-18.790
1200	20.787	182.384	166.761	18.748	559.252	380.804	-16.576
1300	20.787	184.048	168.027	20.827	558.689	365.955	-14.704
1400	20.786	185.588	169.227	22.905	558.078	351.152	-13.102
1500	20.786	187.023	170.366	24.984	557.422	336.395	-11.714
1600	20.786	188.364	171.450	27.063	556.724	321.682	-10.502
1700	20.786	189.624	172.482	29.141	555.987	307.014	-9.433
1800	20.786	190.812	173.468	31.220	555.211	292.391	-8.485
1900	20.786	191.936	174.411	33.299	554.398	277.812	-7.638
2000	20.786	193.002	175.314	35.377	553.549	263.277	-6.876
2100	20.786	194.017	176.180	37.456	552.665	248.785	-6.188
2200	20.786	194.984	177.013	39.535	551.747	234.335	-5.564
2300	20.786	195.908	177.815	41.613	550.795	219.929	-4.995
2400	20.786	196.792	178.587	43.692	499.550	206.633	-4.497
2500	20.786	197.641	179.333	45.770	498.453	194.451	-4.063
2600	20.787	198.456	180.052	47.849	497.357	182.312	-3.663
2700	20.787	199.240	180.749	49.928	496.261	170.216	-3.293
2800	20.787	199.996	181.423	52.006	495.165	158.160	-2.951
2900	20.788	200.726	182.076	54.085	494.068	146.144	-2.632
3000	20.789	201.431	182.709	56.164	492.972	134.165	-2.336
3100	20.791	202.112	183.324	58.243	491.876	122.223	-2.059
3200	20.793	202.772	183.922	60.322	490.780	110.316	-1.801
3300	20.795	203.412	184.503	62.402	489.685	98.444	-1.558
3400	20.798	204.033	185.068	64.481	488.589	86.604	-1.331
3500	20.803	204.636	185.619	66.561	487.494	74.797	-1.116
3600	20.808	205.222	186.155	68.642	486.400	63.021	-0.914
3700	20.814	205.792	186.678	70.723	485.306	51.276	-0.724
3800	20.822	206.348	187.188	72.805	484.213	39.560	-0.544
3900	20.831	206.889	187.687	74.887	483.120	27.873	-0.373
4000	20.842	207.416	188.173	76.971	482.029	16.214	-0.212
4100	20.855	207.931	188.649	79.056	480.939	4.582	-0.058
4139.449	20.860	208.131	188.834	79.879	-------- FUGACITY = 1 bar ------		
4200	20.870	208.434	189.114	81.142	0.	0.	0.
4300	20.887	208.925	189.569	83.230	0.	0.	0.
4400	20.906	209.405	190.015	85.319	0.	0.	0.
4500	20.928	209.875	190.451	87.411	0.	0.	0.
4600	20.952	210.336	190.878	89.505	0.	0.	0.
4700	20.979	210.786	191.297	91.602	0.	0.	0.
4800	21.009	211.228	191.707	93.701	0.	0.	0.
4900	21.042	211.662	192.110	95.803	0.	0.	0.
5000	21.078	212.087	192.506	97.909	0.	0.	0.
5100	21.117	212.505	192.894	100.019	0.	0.	0.
5200	21.160	212.916	193.275	102.133	0.	0.	0.
5300	21.206	213.319	193.649	104.251	0.	0.	0.
5400	21.255	213.716	194.017	106.374	0.	0.	0.
5500	21.308	214.107	194.379	108.502	0.	0.	0.
5600	21.365	214.491	194.735	110.636	0.	0.	0.
5700	21.425	214.870	195.084	112.776	0.	0.	0.
5800	21.489	215.243	195.429	114.921	0.	0.	0.
5900	21.557	215.611	195.768	117.073	0.	0.	0.
6000	21.628	215.974	196.101	119.233	0.	0.	0.

TABLE 2   Thermochemical Data of Selected Chemical Compounds
Metaboric Acid (BHO$_2$), ideal gas, mol. wt. = 43.81674

Enthalpy Reference Temperature = T_r = 298.15 K					Standard State Pressure = p^o = 0.1 MPa		
	J K^{-1}mol^{-1}				kJ mol^{-1}		
T/K	C_p^o	S^o	$-[G^o-H^o(T_r)]/T$	$H^o-H^o(T_r)$	$\Delta_f H^o$	$\Delta_f G^o$	Log K_f
0	0.	0.	INFINITE	-10.687	-557.212	-557.212	INFINITE
100	33.404	200.044	273.638	-7.359	-558.313	-555.919	290.382
200	36.887	224.017	243.422	-3.881	-559.425	-553.107	144.457
250	39.583	232.535	240.414	-1.970	-560.041	-551.456	115.221
298.15	42.232	239.734	239.734	0.	-560.656	-549.748	96.313
300	42.332	239.995	239.735	0.078	-560.680	-549.680	95.708
350	45.007	246.723	240.260	2.262	-561.324	-547.796	81.754
400	47.553	252.901	241.459	4.577	-561.968	-545.819	71.277
450	49.941	258.641	243.053	7.015	-562.605	-543.762	63.118
500	52.157	264.019	244.883	9.568	-563.228	-541.635	56.584
600	56.073	273.885	248.910	14.985	-564.416	-537.203	46.768
700	59.364	282.783	253.124	20.762	-565.518	-532.579	39.742
800	62.136	290.896	257.345	26.841	-566.535	-527.803	34.462
900	64.489	298.354	261.493	33.175	-567.476	-522.904	30.349
1000	66.505	305.256	265.529	39.727	-568.349	-517.905	27.053
1100	68.244	311.678	269.436	46.467	-569.168	-512.820	24.352
1200	69.752	317.682	273.209	53.368	-569.944	-507.663	22.098
1300	71.066	323.318	276.849	60.411	-570.686	-502.443	20.188
1400	72.215	328.628	280.359	67.576	-571.406	-497.166	18.549
1500	73.223	333.645	283.746	74.849	-572.113	-491.839	17.127
1600	74.111	338.400	287.015	82.216	-572.814	-486.464	15.881
1700	74.894	342.917	290.171	89.668	-573.518	-481.046	14.781
1800	75.588	347.218	293.222	97.192	-574.231	-475.586	13.801
1900	76.203	351.321	296.173	104.782	-574.958	-470.086	12.924
2000	76.752	355.244	299.029	112.431	-575.705	-464.547	12.133
2100	77.242	359.001	301.796	120.131	-576.475	-458.970	11.416
2200	77.681	362.605	304.479	127.877	-577.273	-453.356	10.764
2300	78.075	366.067	307.082	135.666	-578.102	-447.705	10.168
2400	78.430	369.397	309.609	143.491	-629.224	-440.949	9.597
2500	78.751	372.605	312.065	151.350	-630.199	-433.084	9.049
2600	79.041	375.700	314.454	159.240	-631.179	-425.180	8.542
2700	79.305	378.688	316.778	167.158	-632.166	-417.238	8.072
2800	79.545	381.576	319.041	175.100	-633.161	-409.259	7.635
2900	79.764	384.372	321.245	183.066	-634.165	-401.246	7.227
3000	79.964	387.079	323.395	191.053	-635.179	-393.196	6.846
3100	80.148	389.704	325.492	199.058	-636.204	-385.113	6.489
3200	80.316	392.252	327.539	207.082	-637.240	-376.997	6.154
3300	80.472	394.725	329.537	215.121	-638.288	-368.849	5.838
3400	80.615	397.130	331.490	223.176	-639.349	-360.668	5.541
3500	80.747	399.469	333.399	231.244	-640.422	-352.456	5.260
3600	80.869	401.745	335.266	239.325	-641.508	-344.213	4.994
3700	80.982	403.962	337.093	247.417	-642.607	-335.940	4.743
3800	81.088	406.123	338.881	255.521	-643.719	-327.636	4.504
3900	81.185	408.231	340.632	263.635	-644.844	-319.303	4.277
4000	81.277	410.288	342.348	271.758	-645.982	-310.942	4.060
4100	81.362	412.296	344.030	279.890	-647.134	-302.551	3.855
4200	81.441	414.257	345.679	288.030	-1128.149	-287.109	3.571
4300	81.516	416.174	347.296	296.178	-1128.242	-267.082	3.244
4400	81.585	418.049	348.883	304.333	-1128.350	-247.055	2.933
4500	81.651	419.883	350.440	312.495	-1128.474	-227.025	2.635
4600	81.712	421.679	351.969	320.663	-1128.615	-206.991	2.350
4700	81.770	423.437	353.471	328.837	-1128.774	-186.954	2.078
4800	81.824	425.159	354.947	337.017	-1128.950	-166.913	1.816
4900	81.876	426.846	356.397	345.202	-1129.146	-146.868	1.566
5000	81.924	428.501	357.823	353.392	-1129.360	-126.819	1.325
5100	81.970	430.124	359.225	361.586	-1129.595	-106.766	1.094
5200	82.013	431.716	360.603	369.786	-1129.850	-86.709	0.871
5300	82.054	433.279	361.960	377.989	-1130.127	-66.645	0.657
5400	82.092	434.813	363.295	386.196	-1130.427	-46.577	0.451
5500	82.129	436.319	364.609	394.407	-1130.750	-26.504	0.252
5600	82.164	437.799	365.903	402.622	-1131.098	-6.422	0.060
5700	82.197	439.254	367.177	410.840	-1131.472	13.664	-0.125
5800	82.229	440.684	366.432	419.061	-1131.872	33.758	-0.304
5900	82.258	442.090	369.668	427.286	-1132.299	53.857	-0.477
6000	82.287	443.473	370.887	435.513	-1132.756	73.966	-0.644

TABLE 2 Thermochemical Data of Selected Chemical Compounds
Boron Monoxide (BO), ideal gas, mol. wt. = 26.8094

	Enthalpy Reference Temperature = T_r = 298.15 K					Standard State Pressure = p^o = 0.1 MPa		
	$\overline{\qquad J\ K^{-1}mol^{-1}\qquad}$				$\overline{\qquad kJ\ mol^{-1}\qquad}$			
T/K	C_p^o	S^o	$-[G^o-H^o(T_r)]/T$	$H^o-H^o(T_r)$	$\Delta_f H^o$	$\Delta_f G^o$	Log K_f
0	0.	0.	INFINITE	-8.674	-3.118	-3.118	INFINITE
100	29.105	171.662	229.370	-5.771	-1.692	-10.162	5.308
200	29.112	191.838	206.138	-2.860	-0.569	-19.105	4.990
250	29.133	198.336	203.952	-1.404	-0.213	-23.783	4.969
298.15	29.196	203.472	203.472	0.	0.	-28.343	4.966
300	29.200	203.652	203.472	0.054	0.006	-28.519	4.966
350	29.343	208.163	203.828	1.517	0.104	-33.283	4.967
400	29.574	212.096	204.621	2.990	0.094	-38.053	4.969
450	29.882	215.596	205.649	4.476	-0.007	-42.816	4.970
500	30.250	218.763	206.805	5.979	-0.178	-47.564	4.969
600	31.076	224.350	209.276	9.045	-0.673	-56.997	4.962
700	31.910	229.204	211.783	12.194	-1.306	-66.336	4.950
800	32.677	233.516	214.235	15.425	-2.027	-75.578	4.935
900	33.350	237.404	216.597	18.727	-2.809	-84.725	4.917
1000	33.926	240.949	218.858	22.091	-3.638	-93.783	4.899
1100	34.415	244.206	221.016	25.509	-4.505	-102.756	4.879
1200	34.829	247.219	223.075	28.972	-5.405	-111.648	4.860
1300	35.181	250.021	225.042	32.473	-6.337	-120.464	4.840
1400	35.481	252.639	226.920	36.006	-7.300	-129.208	4.821
1500	35.739	255.096	228.718	39.568	-8.294	-137.881	4.801
1600	35.961	257.410	230.439	43.153	-9.318	-146.487	4.782
1700	36.155	259.596	232.091	46.759	-10.375	-155.028	4.763
1800	36.324	261.667	233.677	50.383	-11.463	-163.505	4.745
1900	36.473	263.635	235.202	54.023	-12.584	-171.922	4.726
2000	36.606	265.510	236.671	57.677	-13.739	-180.278	4.708
2100	36.724	267.299	238.087	61.344	-14.928	-188.576	4.691
2200	36.831	269.010	239.454	65.022	-16.151	-196.816	4.673
2300	36.928	270.649	240.775	68.710	-17.409	-205.000	4.656
2400	37.016	272.222	242.053	72.407	-68.961	-212.060	4.615
2500	37.098	273.735	243.290	76.113	-70.368	-217.194	4.555
2600	37.173	275.192	244.489	79.826	-71.778	-223.871	4.498
2700	37.244	276.596	245.653	83.547	-73.190	-229.694	4.444
2800	37.310	277.952	246.782	87.275	-74.607	-235.464	4.393
2900	37.374	279.262	247.880	91.009	-76.026	-241.185	4.344
3000	37.435	280.530	248.947	94.749	-77.449	-246.855	4.298
3100	37.495	281.759	249.986	98.496	-78.876	-252.478	4.254
3200	37.554	282.950	250.997	102.248	-80.305	-258.056	4.212
3300	37.613	284.106	251.983	106.007	-81.737	-263.589	4.172
3400	37.672	285.230	252.945	109.771	-83.172	-269.078	4.134
3500	37.731	286.323	253.883	113.541	-84.609	-274.525	4.097
3600	37.792	287.387	254.799	117.317	-86.047	-279.930	4.062
3700	37.855	288.423	255.693	121.100	-87.487	-285.296	4.028
3800	37.920	289.433	256.568	124.888	-88.927	-290.623	3.995
3900	37.987	290.419	257.424	128.684	-90.368	-295.912	3.963
4000	38.057	291.382	258.261	132.486	-91.809	-301.164	3.933
4100	38.130	292.323	259.080	136.295	-93.249	-306.380	3.903
4200	38.206	293.242	259.882	140.112	-574.537	-304.537	3.787
4300	38.285	294.142	260.669	143.936	-574.887	-298.104	3.621
4400	38.368	295.023	261.439	147.769	-575.237	-291.663	3.462
4500	38.455	295.887	262.195	151.610	-575.586	-285.215	3.311
4600	38.546	296.733	262.937	155.460	-575.935	-278.758	3.165
4700	38.640	297.563	263.665	159.319	-576.284	-272.294	3.026
4800	38.738	298.377	264.380	163.188	-576.632	-265.823	2.893
4900	38.840	299.177	265.082	167.067	-576.980	-259.343	2.765
5000	38.946	299.963	265.771	170.956	-577.327	-252.858	2.642
5100	39.055	300.735	266.449	174.856	-577.674	-246.365	2.523
5200	39.169	301.495	267.116	178.768	-578.020	-239.865	2.409
5300	39.285	302.242	267.772	182.690	-578.368	-233.359	2.300
5400	39.406	302.977	268.417	186.625	-578.715	-226.846	2.194
5500	39.529	303.701	269.052	190.571	-579.064	-220.327	2.092
5600	39.656	304.415	269.677	194.531	-579.414	-213.801	1.994
5700	39.787	305.118	270.293	198.503	-579.766	-207.269	1.899
5800	39.920	305.811	270.899	202.488	-580.120	-200.731	1.808
5900	40.055	306.494	271.497	206.487	-580.477	-194.187	1.719
6000	40.194	307.169	272.086	210.499	-580.838	-187.637	1.634

TABLE 2   Thermochemical Data of Selected Chemical Compounds
Boron Dioxide (BO_2), ideal gas, mol. wt. $= 42.8088$

| | Enthalply Reference Temperature = T_r = 298.15 K | | | | Standard State Pressure = p^0 = 0.1 MPa | | |
| | J $K^{-1}mol^{-1}$ | | | | kJ mol^{-1} | | |
T/K	C_p^0	S^0	$-[G^0-H^0(T_r)]/T$	$H^0-H^0(T_r)$	$\Delta_f H^0$	$\Delta_f G^0$	Log K_f
0	0.	0.	INFINITE	-10.731	-285.346	-285.346	INFINITE
100	33.725	188.764	264.960	-7.620	-285.163	-286.678	149.745
200	38.528	213.526	233.623	-4.019	-284.807	-288.331	75.304
250	41.020	222.394	230.514	-2.030	-284.646	-289.231	60.432
298.15	43.276	229.814	229.814	0.	-284.512	-290.127	50.829
300	43.359	230.082	229.815	0.080	-284.507	-290.162	50.522
350	45.527	236.931	230.351	2.303	-284.388	-291.114	43.446
400	47.495	243.141	231.567	4.630	-284.292	-292.082	38.142
450	49.246	248.838	233.174	7.049	-284.217	-293.060	34.018
500	50.780	254.108	235.007	9.550	-284.160	-294.046	30.719
600	53.263	263.597	238.999	14.759	-284.093	-296.030	25.772
700	55.110	271.953	243.122	20.182	-284.080	-298.021	22.239
800	56.490	279.406	247.200	25.765	-284.116	-300.011	19.589
900	57.533	286.123	251.158	31.469	-284.200	-301.994	17.527
1000	58.333	292.228	254.964	37.264	-284.329	-303.964	15.877
1100	58.957	297.818	258.610	43.129	-284.502	-305.920	14.527
1200	59.451	302.970	262.094	49.051	-284.719	-307.857	13.401
1300	59.847	307.745	265.425	55.016	-284.978	-309.776	12.447
1400	60.169	312.192	268.608	61.018	-285.280	-311.672	11.629
1500	60.435	316.353	271.654	67.048	-285.624	-313.546	10.919
1600	60.655	320.260	274.571	73.103	-286.013	-315.395	10.297
1700	60.841	323.943	277.368	79.178	-286.446	-317.218	9.747
1800	60.998	327.425	280.053	85.270	-286.924	-319.015	9.258
1900	61.133	330.727	282.634	91.377	-287.449	-320.783	8.819
2000	61.250	333.866	285.118	97.496	-288.019	-322.523	8.423
2100	61.352	336.857	287.511	103.627	-288.637	-324.233	8.065
2200	61.443	339.713	289.819	109.766	-289.302	-325.912	7.738
2300	61.523	342.446	292.048	115.915	-290.016	-327.561	7.439
2400	61.597	345.066	294.203	122.071	-341.036	-328.109	7.141
2500	61.664	347.582	296.288	128.234	-341.923	-327.552	6.844
2600	61.728	350.002	298.308	134.404	-342.824	-326.959	6.569
2700	61.788	352.332	300.266	140.579	-343.741	-326.332	6.313
2800	61.847	354.581	302.166	146.761	-344.672	-325.670	6.075
2900	61.904	356.752	304.011	152.949	-345.617	-324.975	5.853
3000	61.961	358.852	305.804	159.142	-346.575	-324.247	5.646
3100	62.019	360.884	307.548	165.341	-347.547	-323.486	5.451
3200	62.078	362.854	309.246	171.546	-348.531	-322.695	5.267
3300	62.138	364.765	310.900	177.757	-349.526	-321.872	5.095
3400	62.201	366.621	312.511	183.974	-350.532	-321.019	4.932
3500	62.265	368.425	314.083	190.197	-351.548	-320.136	4.778
3600	62.333	370.180	315.617	196.427	-352.572	-319.224	4.632
3700	62.403	371.889	317.115	202.663	-353.604	-318.284	4.493
3800	62.475	373.554	318.579	208.907	-354.644	-317.315	4.362
3900	62.551	375.178	320.009	215.159	-355.690	-316.318	4.237
4000	62.629	376.763	321.408	221.418	-356.741	-315.296	4.117
4100	62.711	378.310	322.777	227.685	-357.798	-314.247	4.004
4200	62.795	379.822	324.117	233.960	-838.708	-306.148	3.808
4300	62.882	381.301	325.430	240.244	-838.686	-293.467	3.565
4400	62.971	382.747	326.716	246.536	-838.669	-280.789	3.333
4500	63.062	384.164	327.977	252.838	-838.657	-268.110	3.112
4600	63.156	385.551	329.214	259.149	-838.650	-255.432	2.901
4700	63.252	386.910	330.427	265.469	-838.649	-242.753	2.698
4800	63.350	388.243	331.618	271.799	-838.653	-230.074	2.504
4900	63.449	389.550	332.787	278.139	-838.664	-217.395	2.317
5000	63.549	390.833	333.935	284.489	-838.680	-204.715	2.139
5100	63.651	392.092	335.063	290.849	-838.705	-192.036	1.967
5200	63.754	393.329	336.172	297.219	-838.736	-179.357	1.802
5300	63.857	394.545	337.262	303.600	-838.777	-166.675	1.643
5400	63.961	395.739	338.333	309.991	-838.828	-153.994	1.490
5500	64.065	396.914	339.388	316.392	-838.889	-141.312	1.342
5600	64.169	398.069	340.425	322.804	-838.962	-128.627	1.200
5700	64.273	399.206	341.447	329.226	-839.048	-115.942	1.062
5800	64.377	400.324	342.452	335.658	-839.149	-103.254	0.930
5900	64.480	401.426	343.443	342.101	-839.266	-90.567	0.802
6000	64.582	402.510	344.418	348.554	-839.400	-77.875	0.678

TABLE 2 Thermochemical Data of Selected Chemical Compounds

Boron Monoxide, Dimeric $(BO)_2$, ideal gas, mol. wt. $= 53.6188$

| | Enthalpy Reference Temperature = T_r = 298.15 K | | | | Standard State Pressure = p^o = 0.1 MPa | | |
| | J K^{-1}mol^{-1} | | | | kJ mol^{-1} | | |
T/K	C_p^o	S^o	$-[G^o-H^o(T_r)]/T$	$H^o-H^o(T_r)$	$\Delta_f H^o$	$\Delta_f G^o$	Log K_f
0	0	0.	INFINITE	-12.398	-457.343	-457.343	INFINITE
100	34.367	193.295	286.969	-9.367	-457.265	-459.202	239.863
200	48.231	221.483	247.582	-5.220	-456.694	-461.326	120.486
250	53.477	232.836	243.520	-2.671	-456.345	-462.525	96.639
298.15	57.297	242.597	242.597	0.	-456.056	-463.743	81.246
300	57.424	242.952	242.598	0.106	-456.046	-463.791	80.753
350	60.445	252.040	243.309	3.056	-455.827	-465.100	69.412
400	62.871	260.275	244.923	6.141	-455.708	-466.434	60.910
450	64.921	267.802	247.053	9.337	-455.684	-467.777	54.298
500	66.722	274.737	249.479	12.629	-455.741	-469.118	49.008
600	69.820	287.184	254.750	19.461	-456.031	-471.770	41.071
700	72.409	298.147	260.182	26.576	-456.480	-474.359	35.397
800	74.575	307.962	265.551	33.928	-457.030	-476.877	31.137
900	76.378	316.853	270.765	41.479	-457.649	-479.321	27.819
1000	77.874	324.980	275.786	49.194	-458.320	-481.694	25.161
1100	79.116	332.462	280.603	57.045	-459.038	-483.996	22.983
1200	80.149	339.392	285.217	65.010	-459.800	-486.232	21.165
1300	81.013	345.842	289.635	73.069	-460.607	-488.402	19.624
1400	81.739	351.873	293.868	81.208	-461.461	-490.509	18.301
1500	82.353	357.534	297.925	89.413	-462.365	-492.552	17.152
1600	82.876	362.866	301.819	97.675	-463.323	-494.534	16.145
1700	83.323	367.904	305.560	105.986	-464.337	-496.454	15.254
1800	83.709	372.678	309.157	114.338	-465.410	-498.313	14.461
1900	84.042	377.213	312.621	122.726	-466.544	-500.110	13.749
2000	84.333	381.532	315.959	131.145	-467.743	-501.846	13.107
2100	84.588	385.653	319.180	139.591	-469.007	-503.520	12.524
2200	84.812	389.593	322.292	148.062	-470.339	-505.132	11.993
2300	85.010	393.367	325.301	156.553	-471.740	-506.683	11.507
2400	85.185	396.989	328.213	165.063	-573.730	-506.034	11.014
2500	85.342	400.470	331.034	173.589	-575.428	-503.178	10.513
2600	85.482	403.820	333.770	182.131	-577.133	-500.254	10.050
2700	85.608	407.048	336.424	190.685	-578.846	-497.265	9.620
2800	85.722	410.164	339.002	199.252	-580.567	-494.211	9.220
2900	85.824	413.174	341.508	207.829	-582.297	-491.098	8.846
3000	85.917	416.085	343.946	216.416	-584.037	-487.923	8.495
3100	86.002	418.903	346.319	225.012	-585.787	-484.690	8.167
3200	86.079	421.635	348.630	233.617	-587.546	-481.401	7.858
3300	86.150	424.285	350.883	242.228	-589.316	-478.057	7.567
3400	86.214	426.858	353.079	250.846	-591.095	-474.658	7.292
3500	86.274	429.358	355.223	259.471	-592.885	-471.208	7.032
3600	86.328	431.789	357.317	268.101	-594.684	-467.706	6.786
3700	86.379	434.155	359.361	276.736	-596.492	-464.154	6.553
3800	86.426	436.459	361.360	285.377	-598.311	-460.552	6.331
3900	86.469	438.705	363.315	294.021	-600.138	-456.903	6.120
4000	86.509	440.894	365.227	302.670	-601.975	-453.208	5.918
4100	86.547	443.031	367.099	311.323	-603.820	-449.466	5.726
4200	86.581	445.117	368.931	319.979	-1565.374	-431.630	5.368
4300	86.614	447.155	370.727	328.639	-1565.063	-404.640	4.915
4400	86.644	449.146	372.487	337.302	-1564.766	-377.657	4.483
4500	86.672	451.094	374.212	345.968	-1564.481	-350.681	4.071
4600	86.699	452.999	375.904	354.637	-1564.211	-323.711	3.676
4700	86.724	454.864	377.564	363.308	-1563.955	-296.746	3.298
4800	86.747	456.690	379.194	371.981	-1563.715	-269.787	2.936
4900	86.769	458.479	380.794	380.657	-1563.492	-242.831	2.589
5000	86.790	460.232	382.365	389.335	-1563.287	-215.881	2.255
5100	86.810	461.951	383.909	398.015	-1563.101	-188.935	1.935
5200	86.828	463.637	385.426	406.697	-1562.935	-161.993	1.627
5300	86.846	465.291	386.917	415.381	-1562.791	-135.052	1.331
5400	86.862	466.914	388.384	424.066	-1562.670	-108.116	1.046
5500	86.878	468.508	389.826	432.753	-1562.573	-81.181	0.771
5600	86.893	470.074	391.245	441.442	-1562.503	-54.246	0.506
5700	86.907	471.612	392.641	450.132	-1562.461	-27.314	0.250
5800	86.920	473.124	394.016	458.823	-1562.449	-0.381	0.003
5900	86.933	474.609	395.370	467.516	-1562.468	26.550	-0.235
6000	86.945	476.071	396.702	476.210	-1562.521	53.484	-0.466

TABLE 2   Thermochemical Data of Selected Chemical Compounds
Boron Oxide (B_2O_3), crystal-liquid, mol. wt. = 69.6182

Enthalpy Reference Temperature = T_r = 298.15 K					Standard State Pressure = p^o = 0.1 MPa		
	J $K^{-1}mol^{-1}$				kJ mol^{-1}		
T/K	C_p^o	S^o	$-[G^o-H^o(T_r)]/T$	$H^o-H^o(T_r)$	$\Delta_f H^o$	$\Delta_f G^o$	Log K_f
0	0.	0.	INFINITE	-9.293	-1265.776	-1265.776	INFINITE
100	20.740	10.983	96.353	-8.537	-1269.425	-1244.466	650.042
200	43.844	32.869	59.093	-5.245	-1271.165	-1218.726	318.299
298.15	62.588	53.953	53.953	0.	-1271.936	-1192.796	208.973
300	62.927	54.341	53.954	0.116	-1271.944	-1192.305	207.599
400	77.948	74.600	56.614	7.194	-1272.047	-1165.729	152.229
500	89.287	93.269	62.105	15.582	-1271.710	-1139.180	119.009
600	98.115	110.360	68.744	24.969	-1271.024	-1112.733	96.672
700	105.228	126.039	75.825	35.149	-1270.036	-1086.426	81.070
723.000	106.692	129.464	77.477	37.587	___ CRYSTAL <--> LIQUID ___		
723.000	129.704	162.757	77.477	61.657	TRANSITION		
800	129.704	175.883	86.328	71.644	-1243.112	-1062.927	69.402
900	129.704	191.160	97.144	84.615	-1240.014	-1040.593	60.394
1000	129.704	204.826	107.241	97.585	-1237.160	-1018.590	53.206
1100	129.704	217.188	116.683	110.556	-1234.514	-996.863	47.337
1200	129.704	228.474	125.535	123.526	-1232.044	-975.368	42.457
1300	129.704	238.855	133.858	136.496	-1229.732	-954.074	38.335
1400	129.704	248.468	141.706	149.467	-1227.561	-932.951	34.809
1500	129.704	257.416	149.125	162.437	-1225.521	-911.980	31.758
1600	129.704	265.787	156.157	175.408	-1223.604	-891.141	29.093
1700	129.704	273.650	162.840	188.378	-1221.804	-870.417	26.745
1800	129.704	281.064	169.204	201.348	-1220.116	-849.797	24.660
1900	129.704	288.077	175.277	214.319	-1218.538	-829.267	22.798
2000	129.704	294.730	181.085	227.289	-1217.066	-808.818	21.124
2100	129.704	301.058	186.649	240.260	-1215.699	-788.439	19.611
2200	129.704	307.092	191.987	253.230	-1214.435	-768.123	18.238
2300	129.704	312.857	197.118	266.200	-1213.272	-747.863	16.985
2400	129.704	318.378	202.056	279.171	-1312.728	-725.514	15.790
2500	129.704	323.672	206.816	292.141	-1311.920	-701.063	14.648
2600	129.704	328.759	211.409	305.112	-1311.144	-676.644	13.594
2700	129.704	333.654	215.846	318.082	-1310.400	-652.254	12.619
2800	129.704	338.372	220.139	331.052	-1309.686	-627.891	11.713
2900	129.704	342.923	224.294	344.023	-1309.002	-603.555	10.871
3000	129.704	347.320	228.322	356.993	-1308.347	-579.240	10.085

TABLE 2   Thermochemical Data of Selected Chemical Compounds
Boron Oxide (B_2O_3), ideal gas, mol. wt. = 69.6182

	Enthalply Reference Temperature = T_r = 298.15 K				Standard State Pressure = p^o = 0.1 MPa		
		J $K^{-1}mol^{-1}$			kJ mol^{-1}		
T/K	C_p^o	S^o	$-[G^o-H^o(T_r)]/T$	$H^o-H^o(T_r)$	$\Delta_f H^o$	$\Delta_f G^o$	Log K_f
0	0.	0.	INFINITE	-14.337	-834.847	-834.847	INFINITE
100	40.042	226.986	334.961	-10.797	-835.713	-832.354	434.777
200	55.330	259.361	289.548	-6.037	-835.985	-828.844	216.472
250	61.832	272.432	284.842	-3.102	-835.979	-827.059	172.805
298.15	66.855	283.768	283.768	0.	-835.963	-825.343	144.597
300	67.028	284.182	283.789	0.124	-835.963	-825.277	143.694
350	71.228	294.841	284.601	3.584	-835.972	-823.496	122.900
400	74.727	304.587	286.499	7.235	-836.033	-821.710	107.304
450	77.728	313.566	289.014	11.048	-836.152	-819.913	95.173
500	80.362	321.894	291.891	15.002	-836.317	-818.100	85.466
600	84.817	336.954	298.173	23.269	-836.752	-814.418	70.901
700	88.439	350.310	304.685	31.938	-837.275	-810.655	60.492
800	91.403	362.320	311.151	40.935	-837.849	-806.814	52.680
900	93.830	373.230	317.452	50.200	-838.455	-802.898	46.599
1000	95.822	383.223	323.537	59.686	-839.087	-798.914	41.731
1100	97.462	392.435	329.387	69.353	-839.743	-794.864	37.745
1200	98.819	400.975	335.001	79.169	-840.428	-790.755	34.421
1300	99.950	408.931	340.386	89.109	-841.146	-786.586	31.605
1400	100.897	416.374	345.550	99.153	-841.902	-782.361	29.190
1500	101.696	423.363	350.507	109.284	-842.701	-778.081	27.095
1600	102.375	429.949	355.269	119.488	-843.550	-773.746	25.260
1700	102.955	436.173	359.846	129.755	-844.453	-769.355	23.639
1800	103.454	442.072	364.252	140.076	-845.415	-764.911	22.197
1900	103.886	447.677	368.496	150.444	-846.440	-760.411	20.905
2000	104.262	453.016	372.590	160.852	-847.531	-755.855	19.741
2100	104.591	458.111	376.542	171.295	-848.691	-751.243	18.686
2200	104.880	462.983	380.361	181.769	-849.924	-746.573	17.726
2300	105.136	467.651	384.056	192.270	-851.230	-741.847	16.848
2400	105.363	472.131	387.633	202.795	-953.131	-734.924	15.995
2500	105.565	476.436	391.099	213.342	-954.747	-725.799	15.165
2600	105.746	480.580	394.462	223.907	-956.376	-716.609	14.397
2700	105.908	484.574	397.726	234.490	-958.019	-707.356	13.685
2800	106.054	488.428	400.897	245.088	-959.677	-698.041	13.022
2900	106.186	492.152	403.980	255.700	-961.351	-688.669	12.404
3000	106.306	495.754	406.979	266.325	-963.042	-679.237	11.827
3100	106.415	499.242	409.899	276.961	-964.749	-669.748	11.285
3200	106.515	502.622	412.744	287.608	-966.474	-660.205	10.777
3300	106.606	505.901	415.518	298.264	-968.214	-650.608	10.298
3400	106.689	509.085	418.223	308.929	-969.971	-640.956	9.847
3500	106.765	512.178	420.864	319.602	-971.744	-631.254	9.421
3600	106.836	515.187	423.442	330.282	-973.532	-621.500	9.018
3700	106.901	518.115	425.961	340.969	-975.337	-611.697	8.636
3800	106.961	520.967	428.424	351.662	-977.156	-601.843	8.273
3900	107.017	523.746	430.833	362.361	-978.991	-591.943	7.928
4000	107.069	526.456	433.190	373.065	-980.840	-581.996	7.600
4100	107.117	529.100	435.497	383.774	-982.703	-572.001	7.287
4200	107.161	531.682	437.756	394.488	-1944.280	-547.913	6.814
4300	107.203	534.204	439.970	405.206	-1943.998	-514.669	6.252
4400	107.242	536.669	442.140	415.929	-1943.734	-481.433	5.715
4500	107.279	539.080	444.267	426.655	-1943.488	-448.202	5.203
4600	107.313	541.438	446.354	437.384	-1943.261	-414.977	4.712
4700	107.345	543.746	448.402	448.117	-1943.055	-381.755	4.243
4800	107.375	546.006	450.412	458.853	-1942.871	-348.538	3.793
4900	107.403	548.221	452.386	469.592	-1942.709	-315.323	3.361
5000	107.430	550.391	454.324	480.334	-1942.570	-282.113	2.947
5100	107.455	552.518	456.229	491.078	-1942.457	-248.904	2.549
5200	107.479	554.605	458.101	501.825	-1942.370	-215.700	2.167
5300	107.502	556.653	459.941	512.574	-1942.312	-182.494	1.799
5400	107.523	558.662	461.750	523.325	-1942.285	-149.291	1.444
5500	107.543	560.636	463.530	534.078	-1942.289	-116.088	1.103
5600	107.562	562.574	465.282	544.834	-1942.328	-82.883	0.773
5700	107.580	564.477	467.005	555.591	-1942.403	-49.678	0.455
5800	107.597	566.349	468.702	566.350	-1942.517	-16.470	0.148
5900	107.613	568.188	470.373	577.110	-1942.672	16.736	-0.148
6000	107.629	569.997	472.018	587.872	-1942.870	49.949	-0.435

TABLE 2   Thermochemical Data of Selected Chemical Compounds
Beryllium (Be), crystal-liquid, mol. wt. = 9.01218

| | Enthalply Reference Temperature = T_r = 298.15 K | | | | Standard State Pressure = p^o = 0.1 MPa | | |
| | J K^{-1}mol^{-1} | | | | kJ mol^{-1} | | |
T/K	C_p^o	S^o	-[G^o-H^o(T$_r$)]/T	H^o-H^o(T$_r$)	$\Delta_f H^o$	$\Delta_f G^o$	Log K$_f$
0	0.	0.	INFINITE	-1.932	0.	0.	0.
100	1.819	0.503	19.434	-1.893	0.	0.	0.
200	9.984	4.174	10.751	-1.315	0.	0.	0.
250	13.574	6.801	9.695	-0.723	0.	0.	0.
298.15	16.380	9.440	9.440	0.	0.	0.	0.
300	16.472	9.542	9.440	0.030	0.	0.	0.
350	18.520	12.245	9.649	0.909	0.	0.	0.
400	19.965	14.817	10.136	1.872	0.	0.	0.
450	21.061	17.235	10.792	2.899	0.	0.	0.
500	21.943	19.501	11.551	3.975	0.	0.	0.
600	23.336	23.630	13.227	6.242	0.	0.	0.
700	24.463	27.314	14.981	8.634	0.	0.	0.
800	25.458	30.647	16.734	11.130	0.	0.	0.
900	26.384	33.699	18.452	13.723	0.	0.	0.
1000	27.274	36.525	20.119	16.406	0.	0.	0.
1100	28.147	39.166	21.732	19.177	0.	0.	0.
1200	29.016	41.652	23.290	22.035	0.	0.	0.
1300	29.886	44.009	24.793	24.980	0.	0.	0.
1400	30.763	46.256	26.247	28.013	0.	0.	0.
1500	31.650	48.409	27.653	31.133	0.	0.	0.
1527.000	31.892	48.975	28.025	31.991	ALPHA <--> BETA		
1527.000	30.000	53.461	28.025	38.840	TRANSITION		
1560.000	30.000	54.102	28.570	39.830	BETA <--> LIQUID		
1560.000	28.788	59.163	28.570	47.725	TRANSITION		
1600	28.874	59.893	29.344	48.879	0.	0.	0.
1700	29.089	61.650	31.193	51.777	0.	0.	0.
1800	29.304	63.319	32.932	54.696	0.	0.	0.
1900	29.519	64.909	34.573	57.638	0.	0.	0.
2000	29.734	66.429	36.128	60.600	0.	0.	0.
2100	29.949	67.884	37.606	63.584	0.	0.	0.
2200	30.164	69.283	39.014	66.590	0.	0.	0.
2300	30.379	70.628	40.360	69.617	0.	0.	0.
2400	30.594	71.926	41.648	72.666	0.	0.	0.
2500	30.809	73.179	42.885	75.736	0.	0.	0.
2600	31.024	74.391	44.073	78.828	0.	0.	0.
2700	31.239	75.566	45.218	81.941	0.	0.	0.
2741.437	31.328	76.043	45.680	83.237	FUGACITY = 1 bar		
2800	31.454	76.706	46.322	85.075	-290.957	6.223	-0.116
2900	31.669	77.814	47.389	88.232	-289.894	16.817	-0.303
3000	31.884	78.891	48.421	91.409	-288.814	27.375	-0.477
3100	32.099	79.940	49.421	94.608	-287.718	37.897	-0.639
3200	32.314	80.963	50.391	97.829	-286.609	48.383	-0.790
3300	32.529	81.960	51.333	101.071	-285.488	58.834	-0.931
3400	32.744	82.934	52.248	104.335	-284.356	69.251	-1.064
3500	32.959	83.887	53.138	107.620	-283.215	79.635	-1.188

TABLE 2   Thermochemical Data of Selected Chemical Compounds
Beryllium (Be), ideal gas, mol. wt. $= 9.01218$

	Enthalply Reference Temperature = T_r = 298.15 K				Standard State Pressure = p^0 = 0.1 MPa		
	$J \, K^{-1} mol^{-1}$				$kJ \, mol^{-1}$		
T/K	C_p^0	S^0	$-[G^0-H^0(T_r)]/T$	$H^0-H^0(T_r)$	$\Delta_f H^0$	$\Delta_f G^0$	Log K_f
0	0.	0.	INFINITE	-6.197	319.735	319.735	INFINITE
100	20.786	113.567	154.755	-4.119	321.774	310.468	-162.172
200	20.786	127.975	138.176	-2.040	323.275	298.515	-77.964
250	20.786	132.613	136.617	-1.001	323.723	292.269	-61.066
298.15	20.786	136.274	136.274	0.	324.000	286.184	-50.138
300	20.786	136.403	136.275	0.038	324.008	285.950	-49.788
350	20.786	139.607	136.528	1.078	324.169	279.592	-41.727
400	20.786	142.383	137.090	2.117	324.245	273.218	-35.679
450	20.786	144.831	137.817	3.156	324.257	266.839	-30.974
500	20.786	147.021	138.630	4.196	324.221	260.460	-27.210
600	20.786	150.811	140.354	6.274	324.032	247.724	-21.566
700	20.786	154.015	142.082	8.353	323.719	235.029	-17.538
800	20.786	156.791	143.751	10.431	323.301	222.386	-14.520
900	20.786	159.239	145.339	12.510	322.787	209.802	-12.177
1000	20.786	161.429	146.840	14.589	322.183	197.279	-10.305
1100	20.786	163.410	148.258	16.667	321.490	184.822	-8.776
1200	20.786	165.219	149.597	18.746	320.711	172.431	-7.506
1300	20.786	166.882	150.863	20.824	319.844	160.109	-6.433
1400	20.786	168.423	152.063	22.903	318.890	147.857	-5.517
1500	20.786	169.857	153.202	24.982	317.848	135.676	-4.725
1600	20.786	171.198	154.286	27.060	302.182	124.093	-4.051
1700	20.786	172.458	155.318	29.139	301.362	112.988	-3.472
1800	20.787	173.647	156.303	31.218	300.521	101.931	-2.958
1900	20.787	174.770	157.246	33.296	299.659	90.922	-2.500
2000	20.789	175.837	158.149	35.375	298.775	79.958	-2.088
2100	20.791	176.851	159.016	37.454	297.870	69.040	-1.717
2200	20.795	177.818	159.849	39.533	296.943	58.165	-1.381
2300	20.801	178.743	160.650	41.613	295.996	47.332	-1.075
2400	20.811	179.628	161.423	43.694	295.028	36.541	-0.795
2500	20.824	180.478	162.168	45.775	294.039	25.791	-0.539
2600	20.844	181.295	162.888	47.859	293.031	15.081	-0.303
2700	20.870	182.082	163.584	49.944	292.003	4.410	-0.085
2741.437	20.883	182.400	163.867	50.809	------ FUGACITY = 1 bar -----		
2800	20.905	182.842	164.259	52.033	0.	0.	0.
2900	20.949	183.576	164.912	54.126	0.	0.	0.
3000	21.006	184.287	165.546	56.223	0.	0.	0.
3100	21.075	184.977	166.162	58.327	0.	0.	0.
3200	21.159	185.648	166.761	60.439	0.	0.	0.
3300	21.259	186.300	167.343	62.559	0.	0.	0.
3400	21.376	186.937	167.910	64.691	0.	0.	0.
3500	21.512	187.558	168.462	66.835	0.	0.	0.
3600	21.668	188.166	169.001	68.994	0.	0.	0.
3700	21.844	188.762	169.527	71.170	0.	0.	0.
3800	22.041	189.347	170.041	73.364	0.	0.	0.
3900	22.259	189.923	170.544	75.578	0.	0.	0.
4000	22.500	190.489	171.035	77.816	0.	0.	0.
4100	22.761	191.048	171.516	80.079	0.	0.	0.
4200	23.045	191.600	171.988	82.369	0.	0.	0.
4300	23.349	192.146	172.451	84.689	0.	0.	0.
4400	23.674	192.686	172.904	87.040	0.	0.	0.
4500	24.019	193.222	173.350	89.424	0.	0.	0.
4600	24.383	193.754	173.788	91.844	0.	0.	0.
4700	24.764	194.282	174.218	94.301	0.	0.	0.
4800	25.162	194.808	174.642	96.798	0.	0.	0.
4900	25.576	195.331	175.058	99.334	0.	0.	0.
5000	26.003	195.852	175.469	101.913	0.	0.	0.
5100	26.443	196.371	175.874	104.535	0.	0.	0.
5200	26.894	196.889	176.273	107.202	0.	0.	0.
5300	27.354	197.405	176.667	109.914	0.	0.	0.
5400	27.822	197.921	177.056	112.673	0.	0.	0.
5500	28.296	198.436	177.440	115.479	0.	0.	0.
5600	28.776	198.950	177.819	118.333	0.	0.	0.
5700	29.258	199.464	178.195	121.234	0.	0.	0.
5800	29.743	199.977	178.566	124.184	0.	0.	0.
5900	30.227	200.489	178.933	127.183	0.	0.	0.
6000	30.711	201.001	179.296	130.230	0.	0.	0.

TABLE 2   Thermochemical Data of Selected Chemical Compounds
Beryllium Oxide (BeO), crystal-liquid, mol. wt. = 25.01158

| Enthalply Reference Temperature = T_r = 298.15 K | | | | | Standard State Pressure = p^0 = 0.1 MPa | | |
| J $K^{-1}mol^{-1}$ | | | | | kJ mol^{-1} | | |
T/K	C_p^0	S^0	$-[G^0-H^0(T_r)]/T$	$H^0-H^0(T_r)$	$\Delta_f H^0$	$\Delta_f G^0$	Log K_f
0	0.	0.	INFINITE	-2.835	-604.914	-604.914	INFINITE
100	2.636	0.824	28.539	-2.771	-606.343	-597.709	312.211
200	14.159	5.887	15.759	-1.974	-607.579	-588.573	153.719
298.15	25.560	13.770	13.770	0.	-608.354	-579.062	101.449
300	25.744	13.928	13.770	0.047	-608.364	-578.880	100.792
400	33.757	22.513	14.886	3.051	-608.688	-568.992	74.303
500	38.920	30.641	17.235	6.703	-608.668	-559.065	58.405
600	42.376	38.061	20.098	10.778	-608.440	-549.163	47.809
700	44.823	44.786	23.152	15.144	-608.092	-539.310	40.244
800	46.656	50.897	26.244	19.722	-607.680	-529.511	34.574
900	48.091	56.478	29.298	24.462	-607.235	-519.767	30.166
1000	49.262	61.607	32.276	29.331	-606.780	-510.073	26.643
1100	50.254	66.350	35.161	34.308	-606.329	-500.424	23.763
1200	51.116	70.760	37.946	39.377	-605.892	-490.815	21.365
1300	51.882	74.882	40.630	44.527	-605.478	-481.243	19.337
1400	52.580	78.753	43.216	49.751	-605.094	-471.701	17.599
1500	53.225	82.403	45.708	55.042	-604.744	-462.185	16.095
1600	53.827	85.857	48.111	60.394	-618.971	-452.166	14.762
1700	54.396	89.138	50.428	65.806	-618.303	-441.761	13.574
1800	54.944	92.262	52.666	71.273	-617.614	-431.396	12.519
1900	55.467	95.247	54.829	76.794	-616.904	-421.069	11.576
2000	55.974	98.105	56.922	82.366	-616.176	-410.781	10.729
2100	56.467	100.848	58.949	87.988	-615.430	-400.530	9.963
2200	56.948	103.486	60.914	93.659	-614.669	-390.314	9.267
2300	57.421	106.028	62.820	99.378	-613.893	-380.133	8.633
2373.001	57.756	107.827	64.177	103.582		ALPHA <--> BETA	
2373.001	57.758	110.649	64.177	110.276		TRANSITION	
2400	57.881	111.303	64.704	111.837	-606.409	-370.063	8.054
2500	58.338	113.675	66.616	117.648	-605.605	-360.232	7.527
2600	58.789	115.972	68.470	123.504	-604.789	-350.433	7.040
2700	59.233	118.199	70.271	129.406	-603.959	-340.665	6.591
2800	59.676	120.361	72.021	135.351	-894.074	-324.707	6.057
2821.220	59.770	120.812	72.386	136.618		BETA <--> LIQUID	
2821.220	79.496	148.834	72.386	215.676		TRANSITION	
2900	79.496	151.024	74.493	221.509	-811.558	-306.635	5.523
3000	79.496	153.719	77.089	229.888	-807.695	-289.290	5.037
3100	79.496	156.325	79.603	237.838	-803.847	-272.073	4.584
3200	79.496	158.849	82.041	245.787	-800.016	-254.980	4.162
3300	79.496	161.295	84.405	253.737	-796.202	-238.007	3.767
3400	79.496	163.669	86.702	261.687	-792.408	-221.149	3.398
3500	79.496	165.973	88.934	269.636	-788.635	-204.402	3.051
3600	79.496	168.213	91.105	277.586	-784.884	-187.763	2.724
3700	79.496	170.391	93.219	285.535	-781.157	-171.228	2.417
3800	79.496	172.511	95.278	293.485	-777.455	-154.792	2.128
3900	79.496	174.576	97.285	301.435	-773.782	-138.454	1.854
4000	79.496	176.588	99.242	309.384	-770.138	-122.211	1.596
4100	79.496	178.551	101.153	317.334	-766.525	-106.057	1.351
4200	79.496	180.467	103.018	325.283	-762.946	-89.992	1.119
4300	79.496	182.337	104.841	333.233	-759.403	-74.011	0.899
4400	79.496	184.165	106.624	341.183	-755.897	-58.112	0.690
4500	79.496	185.952	108.367	349.132	-752.431	-42.293	0.491
4600	79.496	187.699	110.072	357.082	-749.007	-26.550	0.301
4700	79.496	189.408	111.742	365.031	-745.626	-10.881	0.121
4800	79.496	191.082	113.378	372.981	-742.290	4.717	-0.051
4900	79.496	192.721	114.980	380.931	-739.001	20.246	-0.216
5000	79.496	194.327	116.551	388.880	-735.761	35.708	-0.373

TABLE 2 Thermochemical Data of Selected Chemical Compounds
Beryllium Oxide (BeO), ideal gas, mol. wt. = 25.01158

Enthalpy Reference Temperature = T_r = 298.15 K				Standard State Pressure = p^0 = 0.1 MPa			
	J K^{-1}mol^{-1}			kJ mol^{-1}			
T/K	C_p^o	S^o	-[G^o-H^o(T$_r$)]/T	H^o-H^o(T$_r$)	Δ_fH^o	Δ_fG^o	Log K$_f$
0	0.	0.	INFINITE	-8.688	133.984	133.984	INFINITE
100	29.108	165.763	223.605	-5.784	135.397	127.536	-66.618
200	29.139	185.943	200.307	-2.873	136.275	119.270	-31.150
298.15	29.481	197.625	197.625	0.	136.398	110.873	-19.425
300	29.493	197.807	197.626	0.055	136.395	110.715	-19.277
400	30.348	206.400	198.791	3.043	136.056	102.198	-13.346
500	31.421	213.286	201.023	6.131	135.513	93.793	-9.799
600	32.454	219.108	203.564	9.326	134.860	85.509	-7.444
700	33.345	224.179	206.155	12.617	134.133	77.340	-5.771
800	34.084	228.682	208.694	15.990	133.340	69.280	-4.524
900	34.704	232.733	211.144	19.430	132.485	61.324	-3.559
1000	35.252	236.418	213.490	22.928	131.569	53.465	-2.793
1100	35.784	239.803	215.730	26.480	130.595	45.702	-2.170
1200	36.357	242.941	217.869	30.086	129.569	38.029	-1.655
1300	37.023	245.876	219.911	33.755	128.501	30.444	-1.223
1400	37.825	248.648	221.866	37.496	127.403	22.942	-0.856
1500	38.788	251.290	223.740	41.325	126.291	15.519	-0.540
1600	39.927	253.829	225.542	45.259	110.646	8.696	-0.284
1700	41.236	256.288	227.278	49.316	109.959	2.346	-0.072
1800	42.698	258.685	228.957	53.511	109.377	-3.967	0.115
1900	44.285	261.036	230.584	57.860	108.914	-10.250	0.282
2000	45.956	263.350	232.164	62.371	108.582	-16.513	0.431
2100	47.670	265.633	233.704	67.052	108.386	-22.762	0.566
2200	49.379	267.890	235.206	71.905	108.329	-29.005	0.689
2300	51.042	270.122	236.676	76.927	108.408	-35.249	0.801
2400	52.616	272.328	238.116	82.110	108.617	-41.499	0.903
2500	54.070	274.506	239.528	87.446	108.944	-47.761	0.998
2600	55.375	276.653	240.915	92.919	109.378	-54.037	1.086
2700	56.514	278.765	242.277	98.515	109.902	-60.332	1.167
2800	57.474	280.838	243.618	104.216	-180.457	-60.425	1.127
2900	58.255	282.869	244.936	110.004	-178.741	-56.168	1.012
3000	58.856	284.854	246.234	115.861	-176.970	-51.972	0.905
3100	59.286	286.792	247.511	121.770	-175.163	-47.834	0.806
3200	59.556	288.679	248.768	127.713	-173.338	-43.757	0.714
3300	59.681	290.514	250.006	133.676	-171.511	-39.735	0.629
3400	59.676	292.295	251.223	139.645	-169.698	-35.769	0.550
3500	59.553	294.024	252.422	145.608	-167.911	-31.856	0.475
3600	59.345	295.699	253.601	151.554	-166.164	-27.994	0.406
3700	59.051	297.321	254.760	157.474	-164.466	-24.180	0.341
3800	58.691	298.891	255.901	163.362	-162.827	-20.410	0.281
3900	58.279	300.410	257.023	169.210	-161.254	-16.682	0.223
4000	57.827	301.880	258.126	175.016	-159.754	-12.995	0.170
4100	57.347	303.302	259.211	180.775	-158.332	-9.344	0.119
4200	56.848	304.678	260.277	186.485	-156.993	-5.726	0.071
4300	56.337	306.010	261.325	192.144	-155.740	-2.139	0.026
4400	55.822	307.299	262.356	197.752	-154.576	1.419	-0.017
4500	55.308	308.548	263.368	203.309	-153.503	4.952	-0.057
4600	54.801	309.758	264.364	208.814	-152.523	8.462	-0.096
4700	54.303	310.931	265.342	214.269	-151.636	11.952	-0.133
4800	53.818	312.069	266.304	219.675	-150.844	15.424	-0.168
4900	53.348	313.174	267.249	225.033	-150.146	18.881	-0.201
5000	52.895	314.247	268.178	230.345	-149.544	22.324	-0.233
5100	52.460	315.291	269.092	235.613	-149.035	25.756	-0.264
5200	52.044	316.305	269.990	240.838	-148.621	29.179	-0.293
5300	51.647	317.293	270.873	246.022	-148.301	32.595	-0.321
5400	51.271	318.255	271.742	251.168	-148.073	36.006	-0.348
5500	50.914	319.192	272.596	256.277	-147.937	39.413	-0.374
5600	50.577	320.106	273.436	261.351	-147.892	42.819	-0.399
5700	50.259	320.999	274.263	266.393	-147.936	46.225	-0.424
5800	49.961	321.870	275.076	271.404	-148.069	49.633	-0.447
5900	49.681	322.722	275.877	276.386	-148.289	53.042	-0.470
6000	49.418	323.555	276.665	281.340	-148.596	56.458	-0.492

TABLE 2   Thermochemical Data of Selected Chemical Compounds
Carbon (C), reference state-graphite, mol. wt. = 12.011

| | Enthalpy Reference Temperature = T_r = 298.15 K | | | | Standard State Pressure = p^0 = 0.1 MPa | | |
| | J K^{-1}mol^{-1} | | | | kJ mol^{-1} | | |
T/K	C_p^0	S^0	-[G^0-H^0(T_r)]/T	H^0-H^0(T_r)	$\Delta_f H^0$	$\Delta_f G^0$	Log K_f
0	0.	0.	INFINITE	-1.051	0.	0.	0.
100	1.674	0.952	10.867	-0.991	0.	0.	0.
200	5.006	3.082	6.407	-0.665	0.	0.	0.
250	6.816	4.394	5.871	-0.369	0.	0.	0.
298.15	8.517	5.740	5.740	0.	0.	0.	0.
300	8.581	5.793	5.741	0.016	0.	0.	0.
350	10.241	7.242	5.851	0.487	0.	0.	0.
400	11.817	8.713	6.117	1.039	0.	0.	0.
450	13.289	10.191	6.487	1.667	0.	0.	0.
500	14.623	11.662	6.932	2.365	0.	0.	0.
600	16.844	14.533	7.961	3.943	0.	0.	0.
700	18.537	17.263	9.097	5.716	0.	0.	0.
800	19.827	19.826	10.279	7.637	0.	0.	0.
900	20.824	22.221	11.475	9.672	0.	0.	0.
1000	21.610	24.457	12.662	11.795	0.	0.	0.
1100	22.244	26.548	13.831	13.989	0.	0.	0.
1200	22.766	28.506	14.973	16.240	0.	0.	0.
1300	23.204	30.346	16.085	18.539	0.	0.	0.
1400	23.578	32.080	17.167	20.879	0.	0.	0.
1500	23.904	33.718	18.216	23.253	0.	0.	0.
1600	24.191	35.270	19.234	25.658	0.	0.	0.
1700	24.448	36.744	20.221	28.090	0.	0.	0.
1800	24.681	38.149	21.178	30.547	0.	0.	0.
1900	24.895	39.489	22.107	33.026	0.	0.	0.
2000	25.094	40.771	23.008	35.525	0.	0.	0.
2100	25.278	42.000	23.883	38.044	0.	0.	0.
2200	25.453	43.180	24.734	40.581	0.	0.	0.
2300	25.618	44.315	25.561	43.134	0.	0.	0.
2400	25.775	45.408	26.365	45.704	0.	0.	0.
2500	25.926	46.464	27.148	48.289	0.	0.	0.
2600	26.071	47.483	27.911	50.889	0.	0.	0.
2700	26.212	48.470	28.654	53.503	0.	0.	0.
2800	26.348	49.426	29.379	56.131	0.	0.	0.
2900	26.481	50.353	30.086	58.773	0.	0.	0.
3000	26.611	51.253	30.777	61.427	0.	0.	0.
3100	26.738	52.127	31.451	64.095	0.	0.	0.
3200	26.863	52.978	32.111	66.775	0.	0.	0.
3300	26.986	53.807	32.756	69.467	0.	0.	0.
3400	27.106	54.614	33.387	72.172	0.	0.	0.
3500	27.225	55.401	34.005	74.889	0.	0.	0.
3600	27.342	56.170	34.610	77.617	0.	0.	0.
3700	27.459	56.921	35.203	80.357	0.	0.	0.
3800	27.574	57.655	35.784	83.109	0.	0.	0.
3900	27.688	58.372	36.354	85.872	0.	0.	0.
4000	27.801	59.075	38.913	88.646	0.	0.	0.
4100	27.913	59.763	37.462	91.432	0.	0.	0.
4200	28.024	60.437	38.001	94.229	0.	0.	0.
4300	28.134	61.097	38.531	97.037	0.	0.	0.
4400	28.245	61.745	39.051	99.856	0.	0.	0.
4500	28.354	62.381	39.562	102.685	0.	0.	0.
4600	28.462	63.006	40.065	105.526	0.	0.	0.
4700	28.570	63.619	40.560	108.378	0.	0.	0.
4800	28.678	64.222	41.047	111.240	0.	0.	0.
4900	28.785	64.814	41.526	114.114	0.	0.	0.
5000	28.893	65.397	41.997	116.997	0.	0.	0.
5100	28.999	65.970	42.462	119.892	0.	0.	0.
5200	29.106	66.534	42.919	122.797	0.	0.	0.
5300	29.211	67.089	43.370	125.713	0.	0.	0.
5400	29.317	67.636	43.814	128.640	0.	0.	0.
5500	29.422	68.175	44.252	131.577	0.	0.	0.
5600	29.528	68.706	44.684	134.524	0.	0.	0.
5700	29.632	69.230	45.110	137.482	0.	0.	0.
5800	29.737	69.746	45.531	140.451	0.	0.	0.
5900	29.842	70.255	45.945	143.429	0.	0.	0.
6000	29.946	70.758	46.355	146.419	0.	0.	0.

TABLE 2 Thermochemical Data of Selected Chemical Compounds
Carbon (C), ideal gas, mol. wt. = 12.011

| Enthalpy Reference Temperature = T_r = 298.15 K | | | | Standard State Pressure = p^0 = 0.1 MPa | | |
| J K^{-1}mol^{-1} | | | | kJ mol^{-1} | | |
T/K	C_p^0	S^0	$-[G^0-H^0(T_r)]/T$	$H^0-H^0(T_r)$	$\Delta_f H^0$	$\Delta_f G^0$	Log K_f
0	0.	0.	INFINITE	-6.536	711.185	711.185	INFINITE
100	21.271	135.180	176.684	-4.150	713.511	700.088	-365.689
200	20.904	149.768	160.007	-2.048	715.287	685.950	-179.152
250	20.861	154.427	158.443	-1.004	716.035	678.527	-141.770
298.15	20.838	158.100	158.100	0.	716.670	671.244	-117.599
300	20.838	158.228	158.100	0.039	716.693	670.962	-116.825
350	20.824	161.439	158.354	1.080	717.263	663.294	-98.991
400	20.815	164.219	158.917	2.121	717.752	655.550	-85.606
450	20.809	166.671	159.645	3.162	718.165	647.749	-75.189
500	20.804	168.863	160.459	4.202	718.507	639.906	-66.851
600	20.799	172.655	162.185	6.282	719.009	624.135	-54.336
700	20.795	175.861	163.916	8.362	719.315	608.296	-45.392
800	20.793	178.638	165.587	10.441	719.474	592.424	-38.681
900	20.792	181.087	167.175	12.520	719.519	576.539	-33.461
1000	20.791	183.278	168.678	14.600	719.475	560.654	-29.286
1100	20.791	185.259	170.097	16.679	719.360	544.777	-25.869
1200	20.793	187.068	171.437	18.758	719.188	528.913	-23.023
1300	20.796	188.733	172.704	20.837	718.968	513.066	-20.615
1400	20.803	190.274	173.905	22.917	718.709	497.237	-18.552
1500	20.814	191.710	175.044	24.998	718.415	481.427	-16.765
1600	20.829	193.053	176.128	27.080	718.092	465.639	-15.202
1700	20.850	194.317	177.162	29.164	717.744	449.871	-13.823
1800	20.878	195.509	178.148	31.250	717.373	434.124	-12.598
1900	20.912	196.639	179.092	33.340	716.984	418.399	-11.503
2000	20.952	197.713	179.996	35.433	716.577	402.694	-10.517
2100	20.999	198.736	180.864	37.530	716.156	387.010	-9.626
2200	21.052	199.714	181.699	39.633	715.722	371.347	-8.817
2300	21.110	200.651	182.503	41.741	715.277	355.703	-8.078
2400	21.174	201.551	183.278	43.855	714.821	340.079	-7.402
2500	21.241	202.417	184.026	45.976	714.357	324.474	-6.780
2600	21.313	203.251	184.750	48.103	713.884	308.888	-6.206
2700	21.387	204.057	185.450	50.238	713.405	293.321	-5.675
2800	21.464	204.836	186.129	52.381	712.920	277.771	-5.182
2900	21.542	205.591	186.787	54.531	712.429	262.239	-4.723
3000	21.621	206.322	187.426	56.689	711.932	246.723	-4.296
3100	21.701	207.032	188.047	58.856	711.431	231.224	-3.896
3200	21.780	207.723	188.651	61.030	710.925	215.742	-3.522
3300	21.859	208.394	189.239	63.212	710.414	200.275	-3.170
3400	21.936	209.048	189.812	65.401	709.899	184.824	-2.839
3500	22.012	209.685	190.371	67.599	709.380	169.389	-2.528
3600	22.087	210.306	190.916	69.804	708.857	153.968	-2.234
3700	22.159	210.912	191.448	72.016	708.329	138.561	-1.956
3800	22.230	211.504	191.968	74.235	707.797	123.169	-1.693
3900	22.298	212.082	192.477	76.462	707.260	107.791	-1.444
4000	22.363	212.648	192.974	78.695	706.719	92.427	-1.207
4100	22.426	213.201	193.461	80.934	706.173	77.077	-0.982
4200	22.487	213.742	193.937	83.180	705.621	61.740	-0.768
4300	22.544	214.272	194.404	85.432	705.065	46.416	-0.564
4400	22.600	214.791	194.861	87.689	704.503	31.105	-0.369
4500	22.652	215.299	195.310	89.951	703.936	15.807	-0.183
4600	22.702	215.797	195.750	92.219	703.363	0.521	-0.006
4700	22.750	216.286	196.182	94.492	702.784	-14.752	0.164
4800	22.795	216.766	196.605	96.769	702.199	-30.012	0.327
4900	22.838	217.236	197.022	99.051	701.607	-45.261	0.482
5000	22.878	217.698	197.431	101.337	701.009	-60.497	0.632
5100	22.917	218.151	197.832	103.626	700.404	-75.721	0.776
5200	22.953	218.597	198.227	105.920	699.793	-90.933	0.913
5300	22.987	219.034	198.616	108.217	699.174	-106.133	1.046
5400	23.020	219.464	196.998	110.517	698.548	-121.322	1.174
5500	23.051	219.887	199.374	112.821	697.914	-136.499	1.296
5600	23.080	220.302	199.744	115.127	697.273	-151.664	1.415
5700	23.107	220.711	200.108	117.437	696.625	-166.818	1.529
5800	23.133	221.113	200.467	119.749	695.968	-181.961	1.639
5900	23.157	221.509	200.820	122.063	695.304	-197.092	1.745
6000	23.181	221.898	201.168	124.380	694.631	-212.211	1.847

TABLE 2   Thermochemical Data of Selected Chemical Compounds
Methane (CH$_4$), ideal gas, mol. wt. = 16.04276

Enthalpy Reference Temperature = T_r = 298.15 K				Standard State Pressure = p^0 = 0.1 MPa		
	J K^{-1}mol^{-1}			kJ mol^{-1}		
T/K	C_p^0	S^0	-[G^0-H^0(T_r)]/T	H^0-H^0(T_r)	$\Delta_f H^0$	$\Delta_f G^0$	Log K$_f$
0	0.	0.	INFINITE	-10.024	-66.911	-66.911	INFINITE
100	33.258	149.500	216.485	-6.698	-69.644	-64.353	33.615
200	33.473	172.577	189.418	-3.368	-72.027	-58.161	15.190
250	34.216	180.113	186.829	-1.679	-73.426	-54.536	11.395
298.15	35.639	186.251	186.251	0.	- 74.873	-50.768	8.894
300	35.708	186.472	186.252	0.066	-74.929	-50.618	8.813
350	37.874	192.131	186.694	1.903	-76.461	-46.445	6.932
400	40.500	197.356	187.704	3.861	-77.969	-42.054	5.492
450	43.374	202.291	189.053	5.957	-79.422	-37.476	4.350
500	46.342	207.014	190.614	8.200	-80.802	-32.741	3.420
600	52.227	215.987	194.103	13.130	-83.308	-22.887	1.993
700	57.794	224.461	197.840	18.635	-85.452	-12.643	0.943
800	62.932	232.518	201.675	24.675	-87.238	-2.115	0.138
900	67.601	240.205	205.532	31.205	-88.692	8.616	-0.500
1000	71.795	247.549	209.370	38.179	-89.849	19.492	-1.018
1100	75.529	254.570	213.162	45.549	-90.750	30.472	-1.447
1200	78.833	261.287	216.895	53.270	-91.437	41.524	-1.807
1300	81.744	267.714	220.558	61.302	-91.945	52.626	-2.115
1400	84.305	273.868	224.148	69.608	-92.308	63.761	-2.379
1500	86.556	279.763	227.660	78.153	-92.553	74.918	-2.609
1600	88.537	285.413	231.095	86.910	-92.703	86.088	-2.810
1700	90.283	290.834	234.450	95.853	-92.780	97.265	-2.989
1800	91.824	296.039	237.728	104.960	-92.797	108.445	-3.147
1900	93.188	301.041	240.930	114.212	-92.770	119.624	-3.289
2000	94.399	305.853	244.057	123.592	-92.709	130.802	-3.416
2100	95.477	310.485	247.110	133.087	-92.624	141.975	-3.531
2200	96.439	314.949	250.093	142.684	-92.521	153.144	-3.636
2300	97.301	319.255	253.007	152.371	-92.409	164.308	-3.732
2400	98.075	323.413	255.854	162.141	-92.291	175.467	-3.819
2500	98.772	327.431	258.638	171.984	-92.174	186.622	-3.899
2600	99.401	331.317	261.359	181.893	-92.060	197.771	-3.973
2700	99.971	335.080	264.020	191.862	-91.954	208.916	-4.042
2800	100.489	338.725	266.623	201.885	-91.857	220.058	-4.105
2900	100.960	342.260	269.171	211.958	-91.773	231.196	-4.164
3000	101.389	345.690	271.664	222.076	-91.705	242.332	-4.219
3100	101.782	349.021	274.106	232.235	-91.653	253.465	-4.271
3200	102.143	352.258	276.498	242.431	-91.621	264.598	-4.319
3300	102.474	355.406	278.842	252.662	-91.609	275.730	-4.364
3400	102.778	358.470	281.139	262.925	-91.619	286.861	-4.407
3500	103.060	361.453	283.391	273.217	-91.654	297.993	-4.447
3600	103.319	364.360	285.600	283.536	-91.713	309.127	-4.485
3700	103.560	367.194	287.767	293.881	-91.798	320.262	-4.521
3800	103.783	369.959	289.894	304.248	-91.911	331.401	-4.555
3900	103.990	372.658	291.982	314.637	-92.051	342.542	-4.588
4000	104.183	375.293	294.032	325.045	-92.222	353.687	-4.619
4100	104.363	377.868	296.045	335.473	-92.422	364.838	-4.648
4200	104.531	380.385	298.023	345.918	-92.652	375.993	-4.676
4300	104.688	382.846	299.967	356.379	-92.914	387.155	-4.703
4400	104.834	385.255	301.879	366.855	-93.208	398.322	-4.729
4500	104.972	387.612	303.758	377.345	-93.533	409.497	-4.753
4600	105.101	389.921	305.606	387.849	-93.891	420.679	-4.777
4700	105.223	392.182	307.424	398.365	-94.281	431.869	-4.800
4800	105.337	394.399	309.213	408.893	-94.702	443.069	-4.822
4900	105.445	396.572	310.973	419.432	-95.156	454.277	-4.843
5000	105.546	398.703	312.707	429.982	-95.641	465.495	-4.863
5100	105.642	400.794	314.414	440.541	-96.157	476.722	-4.883
5200	105.733	402.847	316.095	451.110	-96.703	487.961	-4.902
5300	105.818	404.861	317.750	461.688	-97.278	499.210	-4.920
5400	105.899	406.840	319.382	472.274	-97.882	510.470	-4.938
5500	105.976	408.784	320.990	482.867	-98.513	521.741	-4.955
5600	106.049	410.694	322.575	493.469	-99.170	533.025	-4.972
5700	106.118	412.572	324.137	504.077	-99.852	544.320	-4.988
5800	106.184	414.418	325.678	514.692	-100.557	555.628	-5.004
5900	106.247	416.234	327.197	525.314	-101.284	566.946	-5.019
6000	106.306	418.020	328.696	535.942	-102.032	578.279	-5.034

TABLE 2   Thermochemical Data of Selected Chemical Compounds
Carbon Monoxide (CO), ideal gas, mol. wt. = 28.0104

Enthalply Reference Temperature = T_r = 298.15 K				Standard State Pressure = p^o = 0.1 MPa			
	J K^{-1}mol^{-1}			kJ mol^{-1}			
T/K	$C_p{}^o$	S^o	$-[G^o\text{-}H^o(T_r)]/T$	$H^o\text{-}H^o(T_r)$	$\Delta_f H^o$	$\Delta_f G^o$	Log K$_f$
0	0.	0.	INFINITE	-8.671	-113.805	-113.805	INFINITE
100	29.104	165.850	223.539	-5.769	-112.415	-120.239	62.807
200	29.108	186.025	200.317	-2.858	-111.286	-128.526	33.568
298.15	29.142	197.653	197.653	0.	-110.527	-137.163	24.030
300	29.142	197.833	197.653	0.054	-110.516	-137.328	23.911
400	29.342	206.238	198.798	2.976	-110.102	-146.338	19.110
500	29.794	212.831	200.968	5.931	-110.003	-155.414	16.236
600	30.443	218.319	203.415	8.942	-110.150	-164.486	14.320
700	31.171	223.066	205.890	12.023	-110.469	-173.518	12.948
800	31.899	227.277	208.305	15.177	-110.905	-182.497	11.916
900	32.577	231.074	210.628	18.401	-111.418	-191.416	11.109
1000	33.183	234.538	212.848	21.690	-111.983	-200.275	10.461
1100	33.710	237.726	214.967	25.035	-112.586	-209.075	9.928
1200	34.175	240.679	216.988	28.430	-113.217	-217.819	9.481
1300	34.572	243.431	218.917	31.868	-113.870	-226.509	9.101
1400	34.920	246.006	220.761	35.343	-114.541	-235.149	8.774
1500	35.217	248.426	222.526	38.850	-115.229	-243.740	8.488
1600	35.480	250.707	224.216	42.385	-115.933	-252.284	8.236
1700	35.710	252.865	225.839	45.945	-116.651	-260.784	8.013
1800	35.911	254.912	227.398	49.526	-117.384	-269.242	7.813
1900	36.091	256.859	228.897	53.126	-118.133	-277.658	7.633
2000	36.250	258.714	230.342	56.744	-118.896	-286.034	7.470
2100	36.392	260.486	231.736	60.376	-119.675	-294.372	7.322
2200	36.518	262.182	233.081	64.021	-120.470	-302.672	7.186
2300	36.635	263.809	234.382	67.683	-121.278	-310.936	7.062
2400	36.721	265.359	235.641	71.324	-122.133	-319.164	6.946
2500	36.836	266.854	236.860	74.985	-122.994	-327.356	6.840
2600	36.924	268.300	238.041	78.673	-123.854	-335.514	6.741
2700	37.003	269.695	239.188	82.369	-124.731	-343.638	6.648
2800	37.083	271.042	240.302	86.074	-125.623	-351.729	6.562
2900	37.150	272.345	241.384	89.786	-126.532	-359.789	6.480
3000	37.217	273.605	242.437	93.504	-127.457	-367.816	6.404
3100	37.279	274.827	243.463	97.229	-128.397	-375.812	6.332
3200	37.338	276.011	244.461	100.960	-129.353	-383.778	6.265
3300	37.392	277.161	245.435	104.696	-130.325	-391.714	6.200
3400	37.443	278.278	246.385	108.438	-131.312	-399.620	6.139
3500	37.493	279.364	247.311	112.185	-132.313	-407.497	6.082
3600	37.543	280.421	248.216	115.937	-133.329	-415.345	6.027
3700	37.589	281.450	249.101	119.693	-134.360	-423.165	5.974
3800	37.631	282.453	249.965	123.454	-135.405	-430.956	5.924
3900	37.673	283.431	250.811	127.219	-136.464	-438.720	5.876
4000	37.715	284.386	251.638	130.989	-137.537	-446.457	5.830
4100	37.756	285.317	252.449	134.762	-138.623	-454.166	5.786
4200	37.794	286.228	253.242	138.540	-139.723	-461.849	5.744
4300	37.832	287.117	254.020	142.321	-140.836	-469.506	5.703
4400	37.869	287.988	254.782	146.106	-141.963	-477.136	5.664
4500	37.903	288.839	255.529	149.895	-143.103	-484.741	5.627
4600	37.941	289.673	256.262	153.687	-144.257	-492.321	5.590
4700	37.974	290.489	256.982	157.483	-145.424	-499.875	5.555
4800	38.007	291.289	257.688	161.282	-146.605	-507.404	5.522
4900	38.041	292.073	258.382	165.084	-147.800	-514.908	5.489
5000	38.074	292.842	259.064	168.890	-149.009	-522.387	5.457
5100	38.104	293.596	259.733	172.699	-150.231	-529.843	5.427
5200	38.137	294.336	260.392	176.511	-151.469	-537.275	5.397
5300	38.171	295.063	261.039	180.326	-152.721	-544.681	5.368
5400	38.200	295.777	261.676	184.146	-153.987	-552.065	5.340
5500	38.074	296.476	262.302	187.957	-155.279	-559.426	5.313
5600	38.263	297.164	262.919	191.775	-156.585	-566.762	5.287
5700	38.296	297.842	263.525	195.603	-157.899	-574.075	5.261
5800	38.325	298.508	264.123	199.434	-159.230	-581.364	5.236
5900	38.355	299.163	264.711	203.268	-160.579	-588.631	5.211
6000	38.388	299.808	265.291	207.106	-161.945	-595.875	5.188

TABLE 2   Thermochemical Data of Selected Chemical Compounds
Carbon Dioxide (CO_2), ideal gas, mol. wt. = 44.0098

| Enthalply Reference Temperature = T_r = 298.15 K | | | | Standard State Pressure = p^o = 0.1 MPa | | | |
| | J $K^{-1}mol^{-1}$ | | | kJ mol^{-1} | | | |
T/K	C_p^o	S^o	-[G^o-$H^o(T_r)$]/T	H^o-$H^o(T_r)$	$\Delta_f H^o$	$\Delta_f G^o$	Log K_f
0	0.	0.	INFINITE	-9.364	-393.151	-393.151	INFINITE
100	29.208	179.009	243.568	-6.456	-393.208	-393.683	205.639
200	32.359	199.975	217.046	-3.414	-393.404	-394.085	102.924
298.15	37.129	213.795	213.795	0.	-393.522	-394.389	69.095
300	37.221	214.025	213.795	0.069	-393.523	-394.394	68.670
400	41.325	225.314	215.307	4.003	-393.583	-394.675	51.539
500	44.627	234.901	218.290	8.305	-393.666	-394.939	41.259
600	47.321	243.283	221.772	12.907	-393.803	-395.182	34.404
700	49.564	250.750	225.388	17.754	-393.983	-395.398	29.505
800	51.434	257.494	228.986	22.806	-394.168	-395.586	25.829
900	52.999	263.645	232.500	28.030	-394.405	-395.748	22.969
1000	54.308	269.299	235.901	33.397	-394.623	-395.886	20.679
1100	55.409	274.528	239.178	38.884	-394.838	-396.001	18.805
1200	56.342	279.390	242.329	44.473	-395.050	-396.098	17.242
1300	57.137	283.932	245.356	50.148	-395.257	-396.177	15.919
1400	57.802	288.191	248.265	55.896	-395.462	-396.240	14.784
1500	58.379	292.199	251.062	61.705	-395.668	-396.288	13.800
1600	58.886	295.983	253.753	67.569	-395.876	-396.323	12.939
1700	59.317	299.566	256.343	73.480	-396.090	-396.344	12.178
1800	59.701	302.968	258.840	79.431	-396.311	-396.353	11.502
1900	60.049	306.205	261.248	85.419	-396.542	-396.349	10.896
2000	60.350	309.293	263.574	91.439	-396.784	-396.333	10.351
2100	60.622	312.244	265.822	97.488	-397.039	-396.304	9.858
2200	60.865	315.070	267.996	103.562	-397.309	-396.262	9.408
2300	61.086	317.781	270.102	109.660	-397.596	-396.209	8.998
2400	61.287	320.385	272.144	115.779	-397.900	-396.142	8.622
2500	61.471	322.890	274.124	121.917	-398.222	-396.062	8.275
2600	61.647	325.305	276.046	128.073	-398.562	-395.969	7.955
2700	61.802	327.634	277.914	134.246	-398.921	-395.862	7.658
2800	61.952	329.885	279.730	140.433	-399.299	-395.742	7.383
2900	62.095	332.061	281.497	146.636	-399.695	-395.609	7.126
3000	62.229	334.169	283.218	152.852	-400.111	-395.461	6.886
3100	62.347	336.211	284.895	159.081	-400.545	-395.298	6.661
3200	62.462	338.192	286.529	165.321	-400.998	-395.122	6.450
3300	62.573	340.116	288.124	171.571	-401.470	-394.932	6.251
3400	62.681	341.986	289.681	177.836	-401.960	-394.726	6.064
3500	62.785	343.804	291.202	184.109	-402.467	-394.506	5.888
3600	62.884	345.574	292.687	190.393	-402.991	-394.271	5.721
3700	62.980	347.299	294.140	196.686	-403.532	-394.022	5.563
3800	63.074	348.979	295.561	202.989	-404.089	-393.756	5.413
3900	63.166	350.619	296.952	209.301	-404.662	-393.477	5.270
4000	63.254	352.219	298.314	215.622	-405.251	-393.183	5.134
4100	63.341	353.782	299.648	221.951	-405.856	-392.874	5.005
4200	63.426	355.310	300.955	228.290	-406.475	-392.550	4.882
4300	63.509	356.803	302.236	234.637	-407.110	-392.210	4.764
4400	63.588	358.264	303.493	240.991	-407.760	-391.857	4.652
4500	63.667	359.694	304.726	247.354	-408.426	-391.488	4.544
4600	63.745	361.094	305.937	253.725	-409.106	-391.105	4.441
4700	63.823	362.466	307.125	260.103	-409.802	-390.706	4.342
4800	63.893	363.810	308.292	266.489	-410.514	-390.292	4.247
4900	63.968	365.126	309.438	272.882	-411.242	-389.892	4.156
5000	64.046	366.422	310.565	279.283	-411.986	-389.419	4.068
5100	64.128	367.691	311.673	285.691	-412.746	-388.959	3.984
5200	64.220	368.937	312.762	292.109	-413.522	-388.486	3.902
5300	64.312	370.161	313.833	298.535	-414.314	-387.996	3.824
5400	64.404	371.364	314.888	304.971	-415.123	-387.493	3.748
5500	64.496	372.547	315.925	311.416	-415.949	-386.974	3.675
5600	64.588	373.709	316.947	317.870	-416.794	-386.439	3.605
5700	64.680	374.853	317.953	324.334	-417.658	-385.890	3.536
5800	64.772	375.979	318.944	330.806	-418.541	-385.324	3.470
5900	64.865	377.087	319.920	337.288	-419.445	-384.745	3.406
6000	64.957	378.178	320.882	343.779	-420.372	-384.148	3.344

TABLE 2 Thermochemical Data of Selected Chemical Compounds
Acetylene (C_2H_2), ideal gas, mol. wt. $= 26.03788$

	Enthalply Reference Temperature = T_r = 298.15 K				Standard State Pressure = p^0 = 0.1 MPa		
	J $K^{-1}mol^{-1}$				kJ mol^{-1}		
T/K	C_p^0	S^0	$-[G^0-H^0(T_r)]/T$	$H^0-H^0(T_r)$	Δ_fH^0	Δ_fG^0	Log K_f
0	0.	0.	INFINITE	-10.012	235.755	235.755	INFINITE
100	29.347	163.294	234.338	-7.104	232.546	236.552	-123.562
200	35.585	185.097	204.720	-3.925	229.685	241.663	-63.116
298.15	44.095	200.958	200.958	0.	226.731	248.163	-43.477
300	44.229	201.231	200.959	0.082	226.674	248.296	-43.232
400	50.480	214.856	202.774	4.833	223.568	255.969	-33.426
500	54.869	226.610	206.393	10.108	220.345	264.439	-27.626
600	58.287	236.924	210.640	15.771	216.993	273.571	-23.816
700	61.149	246.127	215.064	21.745	213.545	283.272	-21.138
800	63.760	254.466	219.476	27.992	210.046	293.471	-19.162
900	66.111	262.113	223.794	34.487	206.522	304.111	-17.650
1000	68.275	269.192	227.984	41.208	202.989	315.144	-16.461
1100	70.245	275.793	232.034	48.136	199.451	326.530	-15.506
1200	72.053	281.984	235.941	55.252	195.908	338.239	-14.723
1300	73.693	287.817	239.709	62.540	192.357	350.244	-14.073
1400	75.178	293.334	243.344	69.985	188.795	362.523	-13.526
1500	76.530	298.567	246.853	77.572	185.216	375.057	-13.061
1600	77.747	303.546	250.242	85.286	181.619	387.830	-12.661
1700	78.847	308.293	253.518	93.117	177.998	400.829	-12.316
1800	79.852	312.829	256.688	101.053	174.353	414.041	-12.015
1900	80.760	317.171	259.758	109.084	170.680	427.457	-11.752
2000	81.605	321.335	262.733	117.203	166.980	441.068	-11.520
2100	82.362	325.335	265.620	125.401	163.250	454.864	-11.314
2200	83.065	329.183	268.422	133.673	159.491	468.838	-11.132
2300	83.712	332.890	271.145	142.012	155.701	482.984	-10.969
2400	84.312	336.465	273.793	150.414	151.881	497.295	-10.823
2500	84.858	339.918	276.369	158.873	148.029	511.767	-10.693
2600	85.370	343.256	278.878	167.384	144.146	526.393	-10.575
2700	85.846	346.487	281.322	175.945	140.230	541.169	-10.470
2800	86.295	349.618	283.706	184.553	136.282	556.090	-10.374
2900	86.713	352.653	286.031	193.203	132.302	571.154	-10.288
3000	87.111	355.600	288.301	201.895	128.290	586.355	-10.209
3100	87.474	358.462	290.519	210.624	124.245	601.690	-10.138
3200	87.825	361.245	292.686	219.389	120.166	617.157	-10.074
3300	88.164	363.952	294.804	228.189	116.053	632.751	-10.016
3400	88.491	366.589	296.877	237.022	111.908	648.471	-9.963
3500	88.805	369.159	298.906	245.886	107.731	664.313	-9.914
3600	89.101	371.665	300.892	254.782	103.519	680.275	-9.871
3700	89.388	374.110	302.838	263.706	99.274	696.354	-9.831
3800	89.666	376.498	304.745	272.659	94.996	712.549	-9.795
3900	89.935	378.830	306.615	281.639	90.683	728.856	-9.762
4000	90.194	381.110	308.449	290.646	86.336	745.275	-9.732
4100	90.439	383.341	310.248	299.678	81.955	761.803	-9.705
4200	90.678	385.523	312.015	308.733	77.538	778.438	-9.681
4300	90.910	387.659	313.749	317.813	73.087	795.178	-9.660
4400	91.137	389.752	315.453	326.915	68.601	812.023	-9.640
4500	91.358	391.802	317.127	336.040	64.080	828.969	-9.622
4600	91.563	393.813	318.772	345.186	59.524	846.017	-9.607
4700	91.768	395.784	320.390	354.353	54.933	863.164	-9.593
4800	91.970	397.718	321.981	363.540	50.307	880.410	-9.581
4900	92.171	399.617	323.546	372.747	45.648	897.751	-9.570
5000	92.370	401.481	325.086	381.974	40.957	915.189	-9.561
5100	92.571	403.312	326.602	391.221	36.234	932.720	-9.553
5200	92.768	405.111	328.094	400.488	31.481	950.345	-9.546
5300	92.963	406.880	329.564	409.774	26.699	968.061	-9.541
5400	93.153	408.620	331.012	419.080	21.889	985.868	-9.536
5500	93.341	410.331	332.439	428.405	17.052	1003.763	-9.533
5600	93.525	412.014	333.845	437.748	12.189	1021.748	-9.530
5700	93.706	413.671	335.231	447.110	7.303	1039.819	-9.529
5800	93.883	415.302	336.597	456.489	2.393	1057.976	-9.528
5900	94.057	416.909	337.945	465.886	-2.537	1076.217	-9.528
6000	94.228	418.491	339.274	475.301	-7.488	1094.543	-9.529

TABLE 2   Thermochemical Data of Selected Chemical Compounds
Ethene (C_2H_4), ideal gas, mol. wt. = 28.05376

| Enthalply Reference Temperature = T_r = 298.15 K | | | | | Standard State Pressure = p^0 = 0.1 MPa | | |
| | J K^{-1}mol^{-1} | | | | kJ mol^{-1} | | |
T/K	$C_p{}^0$	S^0	$-[G^0-H^0(T_r)]/T$	$H^0-H^0(T_r)$	$\Delta_f H^0$	$\Delta_f G^0$	Log K_f
0	0.	0.	INFINITE	-10.518	60.986	60.986	INFINITE
100	33.270	180.542	252.466	-7.192	58.194	60.476	-31.589
200	35.359	203.955	222.975	-3.804	55.542	63.749	-16.649
250	38.645	212.172	220.011	-1.960	54.002	65.976	-13.785
298.15	42.886	219.330	219.330	0.	52.467	68.421	-11.987
300	43.063	219.596	219.331	0.079	52.408	68.521	-11.930
350	48.013	226.602	219.873	2.355	50.844	71.330	-10.645
400	53.048	233.343	221.138	4.882	49.354	74.360	-9.710
450	57.907	239.874	222.858	7.657	47.951	77.571	-9.004
500	62.477	246.215	224.879	10.668	46.641	80.933	-8.455
600	70.663	258.348	229.456	17.335	44.294	88.017	-7.663
700	77.714	269.783	234.408	24.763	42.300	95.467	-7.124
800	83.840	280.570	239.511	32.847	40.637	103.180	-6.737
900	89.200	290.761	244.644	41.505	39.277	111.082	-6.447
1000	93.899	300.408	249.742	50.665	38.183	119.122	-6.222
1100	98.018	309.555	254.768	60.266	37.318	127.259	-6.043
1200	101.626	318.242	259.698	70.252	36.645	135.467	-5.897
1300	104.784	326.504	264.522	80.576	36.129	143.724	-5.775
1400	107.550	334.372	269.233	91.196	35.742	152.016	-5.672
1500	109.974	341.877	273.827	102.074	35.456	160.331	-5.583
1600	112.103	349.044	278.306	113.181	35.249	168.663	-5.506
1700	113.976	355.898	282.670	124.486	35.104	177.007	-5.439
1800	115.628	362.460	286.922	135.968	35.005	185.357	-5.379
1900	117.089	368.752	291.064	147.606	34.938	193.712	-5.326
2000	118.386	374.791	295.101	159.381	34.894	202.070	-5.278
2100	119.540	380.596	299.035	171.278	34.864	210.429	-5.234
2200	120.569	386.181	302.870	183.284	34.839	218.790	-5.195
2300	121.491	391.561	306.610	195.388	34.814	227.152	-5.159
2400	122.319	396.750	310.258	207.580	34.783	235.515	-5.126
2500	123.064	401.758	313.818	219.849	34.743	243.880	-5.096
2600	123.738	406.596	317.294	232.190	34.688	252.246	-5.068
2700	124.347	411.280	320.689	244.595	34.616	260.615	-5.042
2800	124.901	415.812	324.006	257.058	34.524	268.987	-5.018
2900	125.404	420.204	327.248	269.573	34.409	277.363	-4.996
3000	125.864	424.463	330.418	282.137	34.269	285.743	-4.975
3100	126.284	428.597	333.518	294.745	34.102	294.128	-4.956
3200	126.670	432.613	336.553	307.393	33.906	302.518	-4.938
3300	127.024	436.516	339.523	320.078	33.679	310.916	-4.921
3400	127.350	440.313	342.432	332.797	33.420	319.321	-4.906
3500	127.650	444.009	345.281	345.547	33.127	327.734	-4.891
3600	127.928	447.609	348.074	358.326	32.800	336.156	-4.877
3700	128.186	451.118	350.812	371.132	32.436	344.588	-4.865
3800	128.424	454.539	353.497	383.962	32.035	353.030	-4.853
3900	128.646	457.878	356.130	396.816	31.596	361.482	-4.842
4000	128.852	461.138	358.715	409.691	31.118	369.947	-4.831
4100	129.045	464.322	361.252	422.586	30.600	378.424	-4.821
4200	129.224	467.434	363.743	435.500	30.041	388.915	-4.812
4300	129.392	470.476	366.190	448.430	29.441	395.418	-4.803
4400	129.549	473.453	368.594	461.378	28.799	403.937	-4.795
4500	129.696	476.366	370.957	474.340	28.116	412.470	-4.788
4600	129.835	479.218	373.280	487.317	27.390	421.019	-4.781
4700	129.965	482.012	375.563	500.307	26.623	429.584	-4.774
4800	130.087	484.749	377.810	513.309	25.813	438.167	-4.768
4900	130.202	487.433	380.020	526.324	24.962	446.766	-4.763
5000	130.311	490.064	382.194	539.349	24.069	455.384	-4.757
5100	130.413	492.646	384.335	552.386	23.136	464.019	-4.753
5200	130.510	495.179	386.442	565.432	22.162	472.673	-4.748
5300	130.602	497.666	388.517	578.488	21.149	481.346	-4.744
5400	130.689	500.108	390.561	591.552	20.097	490.040	-4.740
5500	130.771	502.507	392.575	604.625	19.008	498.751	-4.737
5600	130.849	504.864	394.559	617.706	17.884	507.485	-4.734
5700	130.923	507.180	396.515	630.795	16.724	516.238	-4.731
5800	130.993	509.458	398.442	643.891	15.531	525.012	-4.728
5900	131.060	511.698	400.343	656.993	14.306	533.806	-4.726
6000	131.124	513.901	402.217	670.103	13.051	542.621	-4.724

TABLE 2   Thermochemical Data of Selected Chemical Compounds
Hydrogen, Monatomic (H), ideal gas, mol. wt. $= 1.00794$

Enthalply Reference Temperature = T_r = 298.15 K					Standard State Pressure = p^o = 0.1 MPa		
	J K^{-1}mol^{-1}				kJ mol^{-1}		
T/K	C_p^o	S^o	$-[G^o-H^o(T_r)]/T$	$H^o-H^o(T_r)$	$\Delta_f H^o$	$\Delta_f G^o$	Log K_f
0	0.	0.	INFINITE	-6.197	216.035	216.035	INFINITE
100	20.786	92.009	133.197	-4.119	216.614	212.450	-110.972
200	20.786	106.417	116.618	-2.040	217.346	208.004	-54.325
250	20.786	111.055	115.059	-1.001	217.687	205.629	-42.964
298.15	20.786	114.716	114.716	0.	217.999	203.278	-35.613
300	20.786	114.845	114.717	0.038	218.011	203.186	-35.378
350	20.786	118.049	114.970	1.078	218.326	200.690	-29.951
400	20.786	120.825	115.532	2.117	218.637	198.150	-25.876
450	20.786	123.273	116.259	3.156	218.946	195.570	-22.701
500	20.786	125.463	117.072	4.196	219.254	192.957	-20.158
600	20.786	129.253	118.796	6.274	219.868	187.640	-16.335
700	20.786	132.457	120.524	8.353	220.478	182.220	-13.597
800	20.788	135.232	122.193	10.431	221.080	176.713	-11.538
900	20.786	137.681	123.781	12.510	221.671	171.132	-9.932
1000	20.786	139.871	125.282	14.589	222.248	165.485	-8.644
1100	20.786	141.852	126.700	16.667	222.807	159.782	-7.587
1200	20.786	143.660	128.039	18.746	223.346	154.028	-6.705
1300	20.786	145.324	129.305	20.824	223.865	148.230	-5.956
1400	20.786	146.865	130.505	22.903	224.361	142.394	-5.313
1500	20.786	148.299	131.644	24.982	224.836	136.522	-4.754
1600	20.786	149.640	132.728	27.060	225.289	130.620	-4.264
1700	20.786	150.900	133.760	29.139	225.721	124.689	-3.831
1800	20.786	152.088	134.745	31.217	226.132	118.734	-3.446
1900	20.786	153.212	135.688	33.296	226.525	112.757	-3.100
2000	20.786	154.278	136.591	35.375	226.898	106.760	-2.788
2100	20.786	155.293	137.458	37.453	227.254	100.744	-2.506
2200	20.786	156.260	138.291	39.532	227.593	94.712	-2.249
2300	20.786	157.184	139.092	41.610	227.916	88.664	-2.014
2400	20.786	158.068	139.864	43.689	228.224	82.603	-1.798
2500	20.786	158.917	140.610	45.768	228.518	76.530	-1.599
2600	20.786	159.732	141.330	47.846	228.798	70.444	-1.415
2700	20.786	160.516	142.026	49.925	229.064	64.349	-1.245
2800	20.786	161.272	142.700	52.004	229.318	58.243	-1.087
2900	20.786	162.002	143.353	54.082	229.560	52.129	-0.939
3000	20.786	162.706	143.986	56.161	229.790	46.007	-0.801
3100	20.786	163.388	144.601	58.239	230.008	39.877	-0.672
3200	20.786	164.048	145.199	60.318	230.216	33.741	-0.551
3300	20.786	164.688	145.780	62.397	230.413	27.598	-0.437
3400	20.786	165.308	146.345	64.475	230.599	21.449	-0.330
3500	20.786	165.911	146.895	66.554	230.776	15.295	-0.228
3600	20.786	166.496	147.432	68.632	230.942	9.136	-0.133
3700	20.786	167.066	147.955	70.711	231.098	2.973	-0.042
3800	20.786	167.620	148.465	72.790	231.244	-3.195	0.044
3900	20.786	168.160	148.963	74.868	231.381	-9.366	0.125
4000	20.786	168.686	149.450	76.947	231.509	-15.541	0.203
4100	20.786	169.200	149.925	79.025	231.627	-21.718	0.277
4200	20.786	169.700	150.390	81.104	231.736	-27.899	0.347
4300	20.786	170.190	150.845	83.183	231.836	-34.082	0.414
4400	20.786	170.667	151.290	85.261	231.927	-40.267	0.478
4500	20.786	171.135	151.726	87.340	232.009	-46.454	0.539
4600	20.786	171.591	152.153	89.418	232.082	-52.643	0.598
4700	20.786	172.038	152.571	91.497	232.147	-58.834	0.654
4800	20.786	172.476	152.981	93.576	232.204	-65.025	0.708
4900	20.786	172.905	153.383	95.654	232.253	-71.218	0.759
5000	20.786	173.325	153.778	97.733	232.294	-77.412	0.809
5100	20.786	173.736	154.165	99.811	232.327	-83.606	0.856
5200	20.786	174.140	154.546	101.890	232.353	-89.801	0.902
5300	20.786	174.536	154.919	103.969	232.373	-95.997	0.946
5400	20.786	174.924	155.286	106.047	232.386	-102.192	0.989
5500	20.786	175.306	155.646	108.126	232.392	-108.389	1.029
5600	20.786	175.680	156.001	110.204	232.393	-114.584	1.069
5700	20.786	176.048	156.349	112.283	232.389	-120.780	1.107
5800	20.786	176.410	156.692	114.362	232.379	-126.976	1.144
5900	20.786	176.765	157.029	116.440	232.365	-133.172	1.179
6000	20.786	177.114	157.361	118.519	232.348	-139.368	1.213

TABLE 2 Thermochemical Data of Selected Chemical Compounds
Hydroxyl (OH), ideal gas, mol. wt. = 17.0074

| | Enthalply Reference Temperature = T$_r$ = 298.15 K | | | | Standard State Pressure = p^0 = 0.1 MPa | | |
| | J K^{-1}mol^{-1} | | | | kJ mol^{-1} | | |
T/K	C$_p^0$	S^0	-[G^0-H^0(T$_r$)]/T	H^0-H^0(T$_r$)	Δ$_f$H^0	Δ$_f$G^0	Log K$_f$
0	0.	0.	INFINITE	-9.172	38.390	38.390	INFINITE
100	32.627	149.590	210.980	-6.139	38.471	37.214	-19.438
200	30.777	171.592	186.471	-2.976	38.832	35.803	-9.351
250	30.283	178.402	184.204	-1.450	38.930	35.033	-7.320
298.15	29.986	183.708	183.708	0.	38.987	34.277	-6.005
300	29.977	183.894	183.709	0.055	38.988	34.248	-5.963
350	29.780	188.499	184.073	1.549	39.019	33.455	-4.993
400	29.650	192.466	184.880	3.035	39.029	32.660	-4.265
450	29.567	195.954	185.921	4.515	39.020	31.864	-3.699
500	29.521	199.066	187.082	5.992	38.995	31.070	-3.246
600	29.527	204.447	189.542	8.943	38.902	29.493	-2.568
700	29.663	209.007	192.005	11.902	38.764	27.935	-2.085
800	29.917	212.983	194.384	14.880	38.598	26.399	-1.724
900	30.264	216.526	196.651	17.888	38.416	24.884	-1.444
1000	30.676	219.736	198.801	20.935	38.230	23.391	-1.222
1100	31.124	222.680	200.840	24.024	38.046	21.916	-1.041
1200	31.586	225.408	202.775	27.160	37.867	20.458	-0.891
1300	32.046	227.955	204.615	30.342	37.697	19.014	-0.764
1400	32.492	230.346	206.368	33.569	37.535	17.583	-0.656
1500	32.917	232.602	208.043	36.839	37.381	16.163	-0.563
1600	33.319	234.740	209.645	40.151	37.234	14.753	-0.482
1700	33.694	236.771	211.182	43.502	37.093	13.352	-0.410
1800	34.044	238.707	212.657	46.889	36.955	11.960	-0.347
1900	34.369	240.557	214.078	50.310	36.819	10.575	-0.291
2000	34.670	242.327	215.446	53.762	36.685	9.197	-0.240
2100	34.950	244.026	216.767	57.243	36.551	7.826	-0.195
2200	35.209	245.658	218.043	60.752	36.416	6.462	-0.153
2300	35.449	247.228	219.278	64.285	36.278	5.103	-0.116
2400	35.673	248.741	220.474	67.841	36.137	3.750	-0.082
2500	35.881	250.202	221.635	71.419	35.992	2.404	-0.050
2600	36.075	251.613	222.761	75.017	35.843	1.063	-0.021
2700	36.256	252.978	223.855	78.633	35.689	-0.271	0.005
2800	36.426	254.300	224.918	82.267	35.530	-1.600	0.030
2900	36.586	255.581	225.954	85.918	35.365	-2.924	0.053
3000	36.736	256.824	226.962	89.584	35.194	-4.241	0.074
3100	36.878	258.031	227.945	93.265	35.017	-5.552	0.094
3200	37.013	259.203	228.904	96.960	34.834	-6.858	0.112
3300	37.140	260.344	229.839	100.667	34.644	-8.158	0.129
3400	37.261	261.455	230.753	104.387	34.448	-9.452	0.145
3500	37.376	262.537	231.645	108.119	34.246	-10.741	0.160
3600	37.486	263.591	232.518	111.863	34.037	-12.023	0.174
3700	37.592	264.620	233.372	115.617	33.821	-13.300	0.188
3800	37.693	265.624	234.208	119.381	33.599	-14.570	0.200
3900	37.791	266.604	235.026	123.155	33.371	-15.834	0.212
4000	37.885	267.562	235.827	126.939	33.136	-17.093	0.223
4100	37.976	268.499	236.613	130.732	32.894	-18.346	0.234
4200	38.064	269.415	237.383	134.534	32.646	-19.593	0.244
4300	38.150	270.311	238.138	138.345	32.391	-20.833	0.253
4400	38.233	271.189	238.879	142.164	32.130	-22.068	0.262
4500	38.315	272.050	239.607	145.991	31.862	-23.297	0.270
4600	38.394	272.893	240.322	149.827	31.587	-24.520	0.278
4700	38.472	273.719	241.023	153.670	31.305	-25.737	0.286
4800	38.549	274.530	241.713	157.521	31.017	-26.947	0.293
4900	38.625	275.326	242.391	161.380	30.722	-28.152	0.300
5000	38.699	276.107	243.057	165.246	30.420	-29.350	0.307
5100	38.773	276.874	243.713	169.120	30.111	-30.542	0.313
5200	38.846	277.627	244.358	173.001	29.796	-31.729	0.319
5300	38.919	278.368	244.993	176.889	29.473	-32.909	0.324
5400	38.991	279.096	245.617	180.784	29.144	-34.083	0.330
5500	39.062	279.812	246.233	184.687	28.807	-35.251	0.335
5600	39.134	280.517	246.839	188.597	28.464	-36.412	0.340
5700	39.206	281.210	247.436	192.514	28.113	-37.568	0.344
5800	39.278	281.892	248.024	196.438	27.756	-38.716	0.349
5900	39.350	282.564	248.604	200.369	27.391	-39.860	0.353
6000	39.423	283.226	249.175	204.308	27.019	-40.997	0.357

TABLE 2   Thermochemical Data of Selected Chemical Compounds
Hydroperoxyl (HO_2), ideal gas, mol. wt. $= 33.00674$

| | Enthalpy Reference Temperature = T_r = 298.15 K | | | | Standard State Pressure = p^0 = 0.1 MPa | | |
| | J K^{-1} mol^{-1} | | | | kJ mol^{-1} | | |
T/K	C_p^0	S^0	$-[G^0-H^0(T_r)]/T$	$H^0-H^0(T_r)$	$\Delta_f H^0$	$\Delta_f G^0$	Log K_f
0	0.	0.	INFINITE	-10.003	5.006	5.006	INFINITE
100	33.258	192.430	259.201	-6.677	3.928	7.052	-3.683
200	33.491	215.515	232.243	-3.346	3.001	10.536	-2.752
250	34.044	223.040	229.676	-1.659	2.532	12.475	-2.606
298.15	34.905	229.106	229.106	0.	2.092	14.430	-2.528
300	34.943	229.322	229.107	0.065	2.076	14.506	-2.526
350	36.072	234.791	229.536	1.839	1.649	16.612	-2.479
400	37.296	239.688	230.504	3.673	1.260	18.777	-2.452
450	38.519	244.151	231.776	5.569	0.908	20.988	-2.436
500	39.687	248.271	233.222	7.524	0.591	23.236	-2.427
600	41.781	255.697	236.363	11.601	0.043	27.819	-2.422
700	43.558	262.275	239.603	15.870	-0.411	32.485	-2.424
800	45.084	268.193	242.813	20.304	-0.790	37.211	-2.430
900	46.418	273.582	245.937	24.880	-1.107	41.981	-2.436
1000	47.604	278.535	248.952	29.583	-1.368	46.783	-2.444
1100	48.672	283.123	251.853	34.397	-1.582	51.609	-2.451
1200	49.643	287.400	254.639	39.314	-1.754	56.452	-2.457
1300	50.535	291.410	257.315	44.323	-1.887	61.308	-2.463
1400	51.360	295.185	259.886	49.419	-1.988	66.173	-2.469
1500	52.128	298.755	262.360	54.593	-2.058	71.045	-2.474
1600	52.845	302.143	264.741	59.842	-2.102	75.920	-2.479
1700	53.518	305.367	267.037	65.161	-2.122	80.797	-2.483
1800	54.149	308.444	269.252	70.545	-2.121	85.674	-2.486
1900	54.742	311.388	271.393	75.989	-2.102	90.551	-2.489
2000	55.299	314.210	273.464	81.492	-2.067	95.427	-2.492
2100	55.820	316.921	275.469	87.048	-2.019	100.301	-2.495
2200	56.308	319.529	277.413	92.655	-1.960	105.172	-2.497
2300	56.763	322.042	279.299	98.309	-1.893	110.040	-2.499
2400	57.186	324.467	281.131	104.006	-1.818	114.905	-2.501
2500	57.578	326.809	282.911	109.745	-1.740	119.767	-2.502
2600	57.940	329.075	284.644	115.521	-1.659	124.625	-2.504
2700	58.274	331.268	286.330	121.332	-1.577	129.481	-2.505
2800	58.580	333.393	287.973	127.175	-1.497	134.334	-2.506
2900	58.860	335.453	289.575	133.047	-1.419	139.183	-2.507
3000	59.115	337.453	291.138	138.946	-1.346	144.030	-2.508
3100	59.347	339.395	292.663	144.869	-1.278	148.876	-2.509
3200	59.557	341.283	294.153	150.814	-1.218	153.718	-2.509
3300	59.745	343.118	295.609	156.780	-1.165	158.558	-2.510
3400	59.915	344.904	297.033	162.763	-1.122	163.398	-2.510
3500	60.066	346.643	298.426	168.762	-1.089	168.236	-2.511
3600	60.200	348.337	299.789	174.775	-1.067	173.074	-2.511
3700	60.318	349.989	301.123	180.801	-1.058	177.910	-2.512
3800	60.423	351.599	302.431	186.839	-1.061	182.748	-2.512
3900	60.513	353.169	303.711	192.886	-1.078	187.586	-2.512
4000	60.592	354.702	304.987	198.941	-1.109	192.423	-2.513
4100	60.659	356.199	306.199	205.004	-1.156	197.262	-2.513
4200	60.716	357.662	307.406	211.072	-1.217	202.102	-2.514
4300	60.764	359.091	308.592	217.146	-1.296	206.945	-2.514
4400	60.803	360.488	309.756	223.225	-1.391	211.788	-2.514
4500	60.834	361.855	310.898	229.307	-1.504	216.634	-2.515
4600	60.858	363.193	312.021	235.391	-1.634	221.482	-2.515
4700	60.876	364.502	313.123	241.478	-1.784	226.335	-2.515
4800	60.888	365.783	314.207	247.566	-1.952	231.190	-2.516
4900	60.895	367.039	315.272	253.656	-2.141	236.050	-2.516
5000	60.897	368.269	316.320	259.745	-2.350	240.913	-2.517
5100	60.894	369.475	317.351	265.835	-2.580	245.780	-2.517
5200	60.888	370.658	318.364	271.924	-2.831	250.652	-2.518
5300	60.879	371.817	319.362	278.012	-3.105	255.530	-2.518
5400	60.866	372.955	320.344	284.100	-3.402	260.412	-2.519
5500	60.851	374.072	321.311	290.185	-3.722	265.300	-2.520
5600	60.834	375.168	322.263	296.270	-4.067	270.195	-2.520
5700	60.814	376.245	323.200	302.352	-4.437	275.096	-2.521
5800	60.792	377.302	324.124	308.432	-4.832	280.004	-2.522
5900	60.769	378.341	325.034	314.511	-5.254	284.917	-2.522
6000	60.745	379.362	325.931	320.586	-5.702	289.839	-2.523

TABLE 2   Thermochemical Data of Selected Chemical Compounds
Hydrogen (H_2), ideal gas-reference state, mol. wt. $= 2.01588$

| Enthalply Reference Temperature = T_r = 298.15 K | | | | Standard State Pressure = p^o = 0.1 MPa | | | |
| | J K^{-1}mol^{-1} | | | kJ mol^{-1} | | | |
T/K	C_p^o	S^o	$-[G^o-H^o(T_r)]/T$	$H^o-H^o(T_r)$	$\Delta_f H^o$	$\Delta_f G^o$	Log K_f
0	0.	0.	INFINITE	-8.467	0.	0.	0.
100	28.154	100.727	155.408	-5.468	0.	0.	0.
200	27.447	119.412	133.284	-2.774	0.	0.	0.
250	28.344	125.640	131.152	-1.378	0.	0.	0.
298.15	28.836	130.680	130.680	0.	0.	0.	0.
300	28.849	130.858	130.680	0.053	0.	0.	0.
350	29.081	135.325	131.032	1.502	0.	0.	0.
400	29.181	139.216	131.817	2.959	0.	0.	0.
450	29.229	142.656	132.834	4.420	0.	0.	0.
500	29.260	145.737	133.973	5.882	0.	0.	0.
600	29.327	151.077	136.392	8.811	0.	0.	0.
700	29.441	155.606	138.822	11.749	0.	0.	0.
800	29.624	159.548	141.171	14.702	0.	0.	0.
900	29.881	163.051	143.411	17.676	0.	0.	0.
1000	30.205	166.216	145.536	20.680	0.	0.	0.
1100	30.581	169.112	147.549	23.719	0.	0.	0.
1200	30.992	171.790	149.459	26.797	0.	0.	0.
1300	31.423	174.288	151.274	29.918	0.	0.	0.
1400	31.861	176.633	153.003	33.082	0.	0.	0.
1500	32.298	178.846	154.652	36.290	0.	0.	0.
1600	32.725	180.944	156.231	39.541	0.	0.	0.
1700	33.139	182.940	157.743	42.835	0.	0.	0.
1800	33.537	184.846	159.197	46.169	0.	0.	0.
1900	33.917	186.669	160.595	49.541	0.	0.	0.
2000	34.280	188.418	161.943	52.951	0.	0.	0.
2100	34.624	190.099	163.244	56.397	0.	0.	0.
2200	34.952	191.718	164.501	59.876	0.	0.	0.
2300	35.263	193.278	165.719	63.387	0.	0.	0.
2400	35.559	194.785	166.899	66.928	0.	0.	0.
2500	35.842	196.243	168.044	70.498	0.	0.	0.
2600	36.111	197.654	169.155	74.096	0.	0.	0.
2700	36.370	199.021	170.236	77.720	0.	0.	0.
2800	36.618	200.349	171.288	81.369	0.	0.	0.
2900	36.856	201.638	172.313	85.043	0.	0.	0.
3000	37.087	202.891	173.311	88.740	0.	0.	0.
3100	37.311	204.111	174.285	92.460	0.	0.	0.
3200	37.528	205.299	175.236	96.202	0.	0.	0.
3300	37.740	206.457	176.164	99.966	0.	0.	0.
3400	37.946	207.587	177.072	103.750	0.	0.	0.
3500	38.149	208.690	177.960	107.555	0.	0.	0.
3600	38.348	209.767	178.828	111.380	0.	0.	0.
3700	38.544	210.821	179.679	115.224	0.	0.	0.
3800	38.738	211.851	180.512	119.089	0.	0.	0.
3900	38.928	212.860	181.328	122.972	0.	0.	0.
4000	39.116	213.848	182.129	126.874	0.	0.	0.
4100	39.301	214.816	182.915	130.795	0.	0.	0.
4200	39.484	215.765	183.686	134.734	0.	0.	0.
4300	39.665	216.696	184.442	138.692	0.	0.	0.
4400	39.842	217.610	185.186	142.667	0.	0.	0.
4500	40.017	218.508	185.916	146.660	0.	0.	0.
4600	40.188	219.389	186.635	150.670	0.	0.	0.
4700	40.355	220.255	187.341	154.698	0.	0.	0.
4800	40.518	221.106	188.035	158.741	0.	0.	0.
4900	40.676	221.943	188.719	162.801	0.	0.	0.
5000	40.829	222.767	189.392	166.876	0.	0.	0.
5100	40.976	223.577	190.054	170.967	0.	0.	0.
5200	41.117	224.374	190.706	175.071	0.	0.	0.
5300	41.252	225.158	191.349	179.190	0.	0.	0.
5400	41.379	225.931	191.982	183.322	0.	0.	0.
5500	41.498	226.691	192.606	187.465	0.	0.	0.
5600	41.609	227.440	193.222	191.621	0.	0.	0.
5700	41.712	228.177	193.829	195.787	0.	0.	0.
5800	41.806	228.903	194.427	199.963	0.	0.	0.
5900	41.890	229.619	195.017	204.148	0.	0.	0.
6000	41.965	230.323	195.600	208.341	0.	0.	0.

TABLE 2   Thermochemical Data of Selected Chemical Compounds
Water (H_2O), ideal gas, mol. wt. = 18.01528

T/K	C_p^o	S^o	$-[G^o-H^o(T_r)]/T$	$H^o-H^o(T_r)$	$\Delta_f H^o$	$\Delta_f G^o$	Log K_f
		J K^{-1}mol^{-1}			kJ mol^{-1}		
0	0.	0.	INFINITE	-9.904	-238.921	-238.921	INFINITE
100	33.299	152.388	218.534	-6.615	-240.083	-236.584	123.579
200	33.349	175.485	191.896	-3.282	-240.900	-232.766	60.792
298.15	33.590	188.834	188.834	0.	-241.826	-228.582	40.047
300	33.596	189.042	188.835	0.062	-241.844	-228.500	39.785
400	34.262	198.788	190.159	3.452	-242.846	-223.901	29.238
500	35.226	206.534	192.685	6.925	-243.826	-219.051	22.884
600	36.325	213.052	195.550	10.501	-244.758	-214.007	18.631
700	37.495	218.739	198.465	14.192	-245.632	-208.612	15.582
800	38.721	223.825	201.322	18.002	-246.443	-203.496	13.287
900	39.987	228.459	204.084	21.938	-247.185	-198.083	11.496
1000	41.268	232.738	206.738	26.000	-247.857	-192.590	10.060
1100	42.536	236.731	209.285	30.191	-248.460	-187.033	8.881
1200	43.768	240.485	211.730	34.506	-248.997	-181.425	7.897
1300	44.945	244.035	214.080	38.942	-249.473	-175.774	7.063
1400	46.054	247.407	216.341	43.493	-249.894	-170.089	6.346
1500	47.090	250.620	218.520	48.151	-250.265	-164.376	5.724
1600	48.050	253.690	220.623	52.908	-250.592	-158.639	5.179
1700	48.935	256.630	222.655	57.758	-250.881	-152.883	4.698
1800	49.749	259.451	224.621	62.693	-251.138	-147.111	4.269
1900	50.496	262.161	226.526	67.706	-251.368	-141.325	3.885
2000	51.180	264.769	228.374	72.790	-251.575	-135.528	3.540
2100	51.823	267.282	230.167	77.941	-251.762	-129.721	3.227
2200	52.408	269.706	231.909	83.153	-251.934	-123.905	2.942
2300	52.947	272.048	233.604	88.421	-252.092	-118.082	2.682
2400	53.444	274.312	235.253	93.741	-252.239	-112.252	2.443
2500	53.904	276.503	236.860	99.108	-252.379	-106.416	2.223
2600	54.329	278.625	238.425	104.520	-252.513	-100.575	2.021
2700	54.723	280.683	239.952	109.973	-252.643	-94.729	1.833
2800	55.089	282.680	241.443	115.464	-252.771	-88.878	1.658
2900	55.430	284.619	242.899	120.990	-252.897	-83.023	1.495
3000	55.748	286.504	244.321	126.549	-253.024	-77.163	1.344
3100	56.044	288.337	245.711	132.139	-253.152	-71.298	1.201
3200	56.323	290.120	247.071	137.757	-253.282	-65.430	1.068
3300	56.583	291.858	248.402	143.403	-253.416	-59.558	0.943
3400	56.828	293.550	249.705	149.073	-253.553	-53.681	0.825
3500	57.058	295.201	250.982	154.768	-253.696	-47.801	0.713
3600	57.276	296.812	252.233	160.485	-253.844	-41.916	0.608
3700	57.480	298.384	253.459	166.222	-253.997	-36.027	0.509
3800	57.675	299.919	254.661	171.980	-254.158	-30.133	0.414
3900	57.859	301.420	255.841	177.757	-254.326	-24.236	0.325
4000	58.033	302.887	256.999	183.552	-254.501	-18.334	0.239
4100	58.199	304.322	258.136	189.363	-254.684	-12.427	0.158
4200	58.357	305.726	259.252	195.191	-254.876	-6.516	0.081
4300	58.507	307.101	260.349	201.034	-255.078	-0.600	0.007
4400	58.650	308.448	261.427	206.892	-255.288	5.320	-0.063
4500	58.787	309.767	262.486	212.764	-255.508	11.245	-0.131
4600	58.918	311.061	263.528	218.650	-255.738	17.175	-0.195
4700	59.044	312.329	264.553	224.548	-255.978	23.111	-0.257
4800	59.164	313.574	265.562	230.458	-256.229	29.052	-0.316
4900	59.275	314.795	266.554	236.380	-256.491	34.998	-0.373
5000	59.390	315.993	267.531	242.313	-256.763	40.949	-0.428
5100	59.509	317.171	268.493	248.256	-257.046	46.906	-0.480
5200	59.628	318.327	269.440	254.215	-257.338	52.869	-0.531
5300	59.746	319.464	270.373	260.184	-257.639	58.838	-0.580
5400	59.864	320.582	271.293	266.164	-257.950	64.811	-0.627
5500	59.982	321.682	272.199	272.157	-258.268	70.791	-0.672
5600	60.100	322.764	273.092	278.161	-258.595	76.777	-0.716
5700	60.218	323.828	273.973	284.177	-258.930	82.769	-0.758
5800	60.335	324.877	274.841	290.204	-259.272	88.787	-0.799
5900	60.453	325.909	275.698	296.244	-259.621	94.770	-0.839
6000	60.571	326.926	276.544	302.295	-259.977	100.780	-0.877

Enthalpy Reference Temperature = T_r = 298.15 K   Standard State Pressure = p^o = 0.1 MPa

TABLE 2   Thermochemical Data of Selected Chemical Compounds
Ammonia (NH$_3$), ideal gas, mol. wt. = 17.0352

Enthalpy Reference Temperature = T$_r$ = 298.15 K					Standard State Pressure = p^o = 0.1 MPa		
	J K^{-1}mol^{-1}				kJ mol^{-1}		
T/K	C$_p$^{o}	S^o	-[G^o-H^o(T$_r$)]/T	H^o-H^o(T$_r$)	Δ$_f$H^o	Δ$_f$G^o	Log K$_f$
0	0.	0.	INFINITE	-10.045	-38.907	-38.907	INFINITE
100	33.284	155.840	223.211	-6.737	-41.550	-34.034	17.777
200	33.757	178.990	195.962	-3.394	-43.703	-25.679	6.707
298.15	35.652	192.774	192.774	0.	-45.898	-16.367	2.867
300	35.701	192.995	192.775	0.066	-45.939	-16.183	2.818
400	38.716	203.663	194.209	3.781	-48.041	-5.941	0.776
500	42.048	212.659	197.021	7.819	-49.857	4.800	-0.501
600	45.293	220.615	200.302	12.188	-51.374	15.879	-1.382
700	48.354	227.829	203.727	16.872	-52.618	27.190	-2.029
800	51.235	234.476	207.160	21.853	-53.621	38.662	-2.524
900	53.948	240.669	210.543	27.113	-54.411	50.247	-2.916
1000	56.491	246.486	213.849	32.637	-55.013	61.910	-3.234
1100	58.859	251.983	217.069	38.406	-55.451	73.625	-3.496
1200	61.048	257.199	220.197	44.402	-55.746	85.373	-3.716
1300	63.057	262.166	223.236	50.609	-55.917	97.141	-3.903
1400	64.893	266.907	226.187	57.008	-55.982	108.918	-4.064
1500	66.564	271.442	229.054	63.582	-55.954	120.696	-4.203
1600	68.079	275.788	231.840	70.315	-55.847	132.469	-4.325
1700	69.452	279.957	234.549	77.193	-55.672	144.234	-4.432
1800	70.695	283.962	237.184	84.201	-55.439	155.986	-4.527
1900	71.818	287.815	239.748	91.328	-55.157	167.725	-4.611
2000	72.833	291.525	242.244	98.561	-54.833	179.447	-4.687
2100	73.751	295.101	244.677	105.891	-54.473	191.152	-4.755
2200	74.581	298.552	247.048	113.309	-54.084	202.840	-4.816
2300	75.330	301.884	249.360	120.805	-53.671	214.509	-4.872
2400	76.009	305.104	251.616	128.372	-53.238	228.160	-4.922
2500	76.626	308.220	253.818	136.005	-52.789	237.792	-4.968
2600	77.174	311.236	255.969	143.695	-52.329	249.406	-5.011
2700	77.672	314.158	258.070	151.438	-51.860	261.003	-5.049
2800	78.132	316.991	260.124	159.228	-51.386	272.581	-5.085
2900	78.529	319.740	262.132	167.062	-50.909	284.143	-5.118
3000	78.902	322.409	264.097	174.933	-50.433	295.689	-5.148
3100	79.228	325.001	266.020	182.840	-49.959	307.218	-5.177
3200	79.521	327.521	267.903	190.778	-49.491	318.733	-5.203
3300	79.785	329.972	269.747	198.744	-49.030	330.233	-5.227
3400	80.011	332.358	271.554	206.734	-48.578	341.719	-5.250
3500	80.216	334.680	273.324	214.745	-48.139	353.191	-5.271
3600	80.400	336.942	275.060	222.776	-47.713	364.652	-5.291
3700	80.550	339.147	276.763	230.824	-47.302	376.101	-5.310
3800	80.684	341.297	278.433	238.886	-46.908	387.539	-5.327
3900	80.793	343.395	280.072	246.960	-46.534	398.967	-5.344
4000	80.881	345.441	281.680	255.043	-46.180	410.385	-5.359
4100	80.956	347.439	283.260	263.136	-45.847	421.795	-5.374
4200	81.006	349.391	284.811	271.234	-45.539	433.198	-5.388
4300	81.048	351.297	286.335	279.337	-45.254	444.593	-5.401
4400	81.065	353.161	287.833	287.442	-44.996	455.981	-5.413
4500	81.073	354.983	289.305	295.550	-44.764	467.364	-5.425
4600	81.057	356.765	290.752	303.656	-44.561	478.743	-5.436
4700	81.032	358.508	292.175	311.761	-44.387	490.117	-5.447
4800	80.990	360.213	293.575	319.862	-44.242	501.488	-5.457
4900	80.931	361.882	294.952	327.958	-44.129	512.856	-5.467
5000	80.856	363.517	296.307	336.048	-44.047	524.223	-5.477
5100	80.751	365.117	297.641	344.127	-43.999	535.587	-5.486
5200	80.751	366.685	298.954	352.202	-43.979	546.951	-5.494
5300	80.751	368.223	300.246	360.277	-43.982	558.315	-5.503
5400	80.751	369.732	301.519	368.352	-44.006	569.680	-5.511
5500	80.751	371.214	302.773	376.428	-44.049	581.044	-5.518
5600	80.751	372.669	304.008	384.503	-44.112	592.410	-5.526
5700	80.751	374.098	305.225	392.578	-44.193	603.778	-5.533
5800	80.751	375.503	306.425	400.653	-44.291	615.147	-5.540
5900	80.751	376.883	307.607	408.728	-44.404	626.516	-5.547
6000	80.751	378.240	308.773	416.803	-44.531	637.889	-5.553

TABLE 2　Thermochemical Data of Selected Chemical Compounds
Magnesium (Mg), crystal-liquid, mol. wt. = 24.305

| Enthalply Reference Temperature = T_r = 298.15 K | | | | | Standard State Pressure = p^0 = 0.1 MPa | | |
| | $J K^{-1}mol^{-1}$ | | | | $kJ mol^{-1}$ | | |
T/K	C_p^0	S^0	$-[G^0-H^0(T_r)]/T$	$H^0-H^0(T_r)$	$\Delta_f H^0$	$\Delta_f G^0$	Log K_f
0	0.	0.	INFINITE	-4.998	0.	0.	0.
100	15.762	9.505	53.066	-4.356	0.	0.	0.
200	22.724	23.143	34.888	-2.349	0.	0.	0.
250	24.018	28.364	33.076	-1.178	0.	0.	0.
298.15	24.869	32.671	32.671	0.	0.	0.	0.
300	24.897	32.825	32.671	0.046	0.	0.	0.
350	25.568	36.715	32.977	1.308	0.	0.	0.
400	26.144	40.167	33.664	2.601	0.	0.	0.
450	26.668	43.277	34.562	3.922	0.	0.	0.
500	27.171	46.113	35.578	5.268	0.	0.	0.
600	28.184	51.156	37.764	8.035	0.	0.	0.
700	29.279	55.581	39.999	10.907	0.	0.	0.
800	30.507	59.569	42.200	13.895	0.	0.	0.
900	31.895	63.241	44.336	17.014	0.	0.	0.
923.000	32.238	64.050	44.818	17.751	_____ CRYSTAL <--> LIQUID _____		
923.000	34.309	73.234	44.818	26.228	TRANSITION		
1000	34.309	75.983	47.113	28.870	0.	0.	0.
1100	34.309	79.253	49.888	32.301	0.	0.	0.
1200	34.309	82.238	52.462	35.732	0.	0.	0.
1300	34.309	84.984	54.859	39.163	0.	0.	0.
1366.104	34.309	86.686	56.358	41.431	------- FUGACITY = 1 bar ------		
1400	34.309	87.527	57.103	42.594	-127.409	3.167	-0.118
1500	34.309	89.894	59.211	46.024	-126.057	12.447	-0.433
1600	34.309	92.108	61.199	49.455	-124.705	21.636	-0.706
1700	34.309	94.188	63.079	52.886	-123.352	30.741	-0.945
1800	34.309	96.149	64.862	56.317	-122.000	39.766	-1.154
1900	34.309	98.004	66.558	59.748	-120.648	48.717	-1.339
2000	34.309	99.764	68.175	63.179	-119.296	57.596	-1.504

TABLE 2   Thermochemical Data of Selected Chemical Compounds
Magnesium (Mg), ideal gas, mol. wt. = 44.0098

Enthalply Reference Temperature = T_r = 298.15 K				Standard State Pressure = p^o = 0.1 MPa			
	J K^{-1}mol^{-1}			kJ mol^{-1}			
T/K	C_p^o	S^o	$-[G^o\text{-}H^o(T_r)]/T$	$H^o\text{-}H^o(T_r)$	$\Delta_f H^o$	$\Delta_f G^o$	Log K_f
0	0.	0.	INFINITE	-6.197	145.901	145.901	INFINITE
100	20.786	125.940	167.128	-4.119	147.337	135.694	-70.879
200	20.786	140.348	150.549	-2.040	147.409	123.968	-32.377
250	20.786	140.986	148.990	-1.001	147.277	118.122	-24.680
298.15	20.786	148.648	148.648	0.	147.100	112.522	-19.713
300	20.786	148.776	148.648	0.038	147.092	112.307	-19.554
350	20.786	151.980	148.901	1.078	146.870	106.527	-15.898
400	20.786	154.756	149.463	2.117	146.616	100.780	-13.161
450	20.786	157.204	150.190	3.156	146.335	95.067	-11.035
500	20.786	159.394	151.003	4.196	146.028	89.387	-9.338
600	20.786	163.184	152.727	6.274	145.339	78.122	-6.801
700	20.786	166.388	154.455	8.353	144.546	66.981	-4.998
800	20.786	169.164	156.124	10.431	143.636	55.961	-3.654
900	20.786	171.612	157.712	12.510	142.596	45.062	-2.615
1000	20.786	173.802	159.213	14.589	132.819	35.000	-1.828
1100	20.786	175.783	160.631	16.667	131.466	25.283	-1.201
1200	20.786	177.592	161.970	18.746	130.114	15.690	-0.683
1300	20.786	179.255	163.237	20.824	128.762	6.209	-0.249
1366.104	20.786	180.286	164.037	22.199	----- FUGACITY = 1 bar ----		
1400	20.786	180.796	164.437	22.903	0.	0.	0.
1500	20.786	182.230	165.575	24.982	0.	0.	0.
1600	20.786	183.571	166.659	27.060	0.	0.	0.
1700	20.786	184.832	167.691	29.139	0.	0.	0.
1800	20.787	186.020	168.677	31.218	0.	0.	0.
1900	20.787	187.144	169.619	33.296	0.	0.	0.
2000	20.789	188.210	170.522	35.375	0.	0.	0.
2100	20.791	189.224	171.389	37.454	0.	0.	0.
2200	20.795	190.192	172.222	39.533	0.	0.	0.
2300	20.802	191.116	173.023	41.613	0.	0.	0.
2400	20.812	192.002	173.796	43.694	0.	0.	0.
2500	20.826	192.851	174.541	45.776	0.	0.	0.
2600	20.846	193.669	175.261	47.859	0.	0.	0.
2700	20.874	194.456	175.958	49.945	0.	0.	0.
2800	20.909	195.216	176.632	52.034	0.	0.	0.
2900	20.956	195.950	177.285	54.127	0.	0.	0.
3000	21.014	196.661	177.920	56.226	0.	0.	0.
3100	21.085	197.352	178.535	58.331	0.	0.	0.
3200	21.172	198.022	179.134	60.443	0.	0.	0.
3300	21.275	198.675	179.716	62.566	0.	0.	0.
3400	21.396	199.312	180.283	64.699	0.	0.	0.
3500	21.537	199.934	180.836	66.845	0.	0.	0.
3600	21.697	200.543	181.375	69.007	0.	0.	0.
3700	21.879	201.140	181.901	71.186	0.	0.	0.
3800	22.083	201.726	182.415	73.384	0.	0.	0.
3900	22.310	202.303	182.918	75.603	0.	0.	0.
4000	22.559	202.871	183.409	77.846	0.	0.	0.
4100	22.832	203.431	183.891	80.116	0.	0.	0.
4200	23.128	203.985	184.363	82.413	0.	0.	0.
4300	23.447	204.533	184.825	84.742	0.	0.	0.
4400	23.789	205.076	185.279	87.103	0.	0.	0.
4500	24.152	205.614	185.725	89.500	0.	0.	0.
4600	24.537	206.149	186.164	91.934	0.	0.	0.
4700	24.944	206.681	186.594	94.408	0.	0.	0.
4800	25.372	207.211	187.018	96.923	0.	0.	0.
4900	25.820	207.739	187.436	99.483	0.	0.	0.
5000	26.287	208.265	187.847	102.088	0.	0.	0.
5100	26.773	208.790	188.253	104.741	0.	0.	0.
5200	27.276	209.315	188.653	107.443	0.	0.	0.
5300	27.794	209.839	189.048	110.196	0.	0.	0.
5400	28.329	210.364	189.437	113.002	0.	0.	0.
5500	28.878	210.888	189.823	115.862	0.	0.	0.
5600	29.442	211.414	190.204	118.778	0.	0.	0.
5700	30.020	211.940	190.580	121.751	0.	0.	0.
5800	30.610	212.467	190.953	124.782	0.	0.	0.
5900	31.213	212.996	191.322	127.873	0.	0.	0.
6000	31.827	213.525	191.688	131.025	0.	0.	0.

TABLE 2   Thermochemical Data of Selected Chemical Compounds
Magnesium Oxide (MgO), crystal-liquid, mol. wt. = 40.3044

Enthalply Reference Temperature = T_r = 298.15 K					Standard State Pressure = p^o = 0.1 MPa		
		J K^{-1}mol^{-1}			kJ mol^{-1}		
T/K	C_p^o	S^o	$-[G^o-H^o(T_r)]/T$	$H^o-H^o(T_r)$	$\Delta_f H^o$	$\Delta_f G^o$	Log K_f
0	0.	0.	INFINITE	-5.159	-597.060	-597.060	INFINITE
100	7.802	2.548	52.212	-4.966	-598.962	-589.601	307.976
200	26.681	14.096	30.037	-3.188	-600.646	-579.488	151.347
298.15	37.106	26.924	26.924	0.	-601.241	-568.945	99.677
300	37.244	27.154	26.925	0.069	-601.245	-568.745	99.027
400	42.561	38.678	28.460	4.087	-601.268	-557.898	72.854
500	45.544	48.523	31.513	8.505	-601.046	-547.078	57.153
600	47.430	57.006	35.072	13.160	-600.738	-536.312	46.690
700	48.748	64.420	38.746	17.972	-600.426	-525.600	39.221
800	49.740	70.996	42.374	22.898	-600.156	-514.930	33.621
900	50.539	76.902	45.888	27.913	-599.962	-504.289	29.268
1000	51.208	82.262	49.262	33.001	-608.462	-492.952	25.749
1100	51.794	87.171	52.488	38.151	-608.496	-481.399	22.880
1200	52.325	91.701	55.569	43.358	-608.495	-469.844	20.452
1300	52.810	95.908	58.512	48.615	-608.461	-458.291	18.414
1400	53.262	99.839	61.325	53.918	-735.804	-443.575	16.550
1500	53.693	103.528	64.017	59.266	-734.355	-422.752	14.722
1600	54.107	107.007	66.596	64.656	-732.877	-402.026	13.125
1700	54.509	110.299	69.071	70.087	-731.371	-381.394	11.719
1800	54.898	113.426	71.449	75.558	-729.837	-360.851	10.472
1900	55.278	116.404	73.737	81.067	-728.277	-340.395	9.358
2000	55.651	119.249	75.942	86.613	-726.690	-320.021	8.358
2100	56.019	121.973	78.070	92.197	-725.079	-299.727	7.455
2200	56.379	124.587	80.125	97.816	-723.442	-279.510	6.636
2300	56.738	127.101	82.114	103.472	-721.781	-259.368	5.890
2400	57.094	129.524	84.039	109.164	-720.097	-239.300	5.208
2500	57.445	131.862	85.905	114.891	-718.389	-219.301	4.582
2600	57.797	134.121	87.716	120.653	-716.659	-199.372	4.005
2700	58.145	136.309	89.476	126.450	-714.906	-179.509	3.473
2800	58.491	138.430	91.187	132.282	-713.132	-159.712	2.979
2900	58.836	140.489	92.851	138.148	-711.338	-139.979	2.521
3000	59.177	142.489	94.473	144.049	-709.524	-120.308	2.095
3100	59.516	144.436	96.053	149.987	-707.689	-100.697	1.697
3105.000	59.523	144.532	96.131	150.284	CRYSTAL < --> LIQUID		
3105.000	66.944	169.595	96.131	228.107	TRANSITION		
3200	66.944	171.613	98.342	234.466	-627.329	-83.538	1.364
3300	66.944	173.673	100.594	241.161	-624.772	-66.584	1.054
3400	66.944	175.671	102.773	247.855	-622.235	-49.708	0.764
3500	66.944	177.612	104.884	254.550	-619.719	-32.905	0.491
3600	66.944	179.498	106.930	261.244	-617.226	-16.174	0.235
3700	66.944	181.332	108.916	267.938	-614.757	0.487	-0.007
3800	66.944	183.117	110.846	274.633	-612.315	17.083	-0.235
3900	66.944	184.856	112.721	281.327	-609.901	33.615	-0.450
4000	66.944	186.551	114.546	288.022	-607.518	50.084	-0.654
4100	66.944	188.204	116.322	294.716	-605.167	66.495	-0.847
4200	66.944	189.817	118.053	301.410	-602.851	82.849	-1.030
4300	66.944	191.393	119.740	308.105	-600.572	99.149	-1.204
4400	66.944	192.932	121.386	314.799	-598.332	115.395	-1.370
4500	66.944	194.436	122.993	321.494	-596.133	131.591	-1.527
4600	66.944	195.907	124.562	328.188	-593.978	147.739	-1.678
4700	66.944	197.347	126.095	334.882	-591.869	163.840	-1.821
4800	66.944	198.756	127.595	341.577	-589.807	179.897	-1.958
4900	66.944	200.137	129.061	348.271	-587.796	195.912	-2.088
5000	66.944	201.489	130.496	354.966	-585.838	211.886	-2.214

TABLE 2   Thermochemical Data of Selected Chemical Compounds
Magnesium Oxide (MgO), ideal gas, mol. wt. = 40.3044

| Enthalpy Reference Temperature = T_r = 298.15 K | | | | | Standard State Pressure = p^0 = 0.1 MPa | | |
| | J K^{-1}mol^{-1} | | | | kJ mol^{-1} | | |
T/K	C_p^0	S^0	-[G^0-H^0(T_r)]/T	H^0-H^0(T_r)	$\Delta_f H^0$	$\Delta_f G^0$	Log K$_f$
0	0.	0.	INFINITE	-8.909	58.588	58.588	INFINITE
100	29.124	180.485	240.493	-6.001	59.402	50.970	-26.624
200	30.117	200.873	216.143	-3.054	58.886	42.689	-11.149
250	31.121	207.699	213.793	-1.524	58.517	38.682	-8.082
298.15	32.173	213.269	213.269	0.	58.158	34.895	-6.113
300	32.215	213.468	213.270	0.060	58.144	34.750	-6.051
350	33.401	218.522	213.666	1.699	57.783	30.880	-4.609
400	34.783	223.069	214.562	3.403	57.446	27.060	-3.534
450	36.452	227.259	215.743	5.182	57.147	23.280	-2.702
500	38.434	231.200	217.093	7.053	56.901	19.531	-2.040
600	43.078	238.608	220.068	11.124	56.624	12.088	-1.052
700	47.785	245.608	223.221	15.671	56.672	4.666	-0.348
800	51.560	252.250	226.439	20.649	56.994	-2.783	0.182
900	53.890	258.472	229.657	25.934	57.457	-10.282	0.597
1000	54.795	264.208	232.829	31.379	49.315	-17.121	0.894
1100	54.608	269.429	235.923	36.857	49.607	-23.779	1.129
1200	53.729	274.147	238.915	42.278	49.823	-30.461	1.326
1300	52.496	278.400	241.792	47.591	49.914	-37.156	1.493
1400	51.139	282.241	244.546	52.773	-77.551	-40.685	1.518
1500	49.802	285.723	247.177	57.820	-76.404	-38.093	1.327
1600	48.558	288.897	249.687	62.737	-75.399	-35.573	1.161
1700	47.439	291.807	252.080	67.535	-74.525	-33.111	1.017
1800	46.454	294.490	254.363	72.229	-73.768	-30.697	0.891
1900	45.599	296.978	256.541	76.831	-73.114	-28.323	0.779
2000	44.862	299.298	258.621	81.353	-72.552	-25.980	0.679
2100	44.230	301.471	260.611	85.806	-72.070	-23.664	0.589
2200	43.691	303.516	262.515	90.202	-71.658	-21.368	0.507
2300	43.231	305.447	264.340	94.547	-71.308	-19.091	0.434
2400	42.841	307.279	266.091	98.850	-71.012	-16.827	0.366
2500	42.509	309.021	267.774	103.117	-70.764	-14.574	0.305
2600	42.229	310.682	269.392	107.354	-70.559	-12.331	0.248
2700	41.993	312.272	270.951	111.565	-70.393	-10.095	0.195
2800	41.795	313.795	272.454	115.754	-70.262	-7.864	0.147
2900	41.629	315.259	273.905	119.925	-70.163	-5.637	0.102
3000	41.493	316.668	275.307	124.081	-70.094	-3.414	0.059
3100	41.381	318.026	276.664	128.224	-70.053	-1.191	0.020
3200	41.291	319.339	277.977	132.357	-70.040	1.029	-0.017
3300	41.220	320.608	279.250	136.483	-70.052	3.250	-0.051
3400	41.167	321.838	280.484	140.602	-70.090	5.472	-0.084
3500	41.128	323.030	281.683	144.717	-70.154	7.695	-0.115
3600	41.103	324.189	282.848	148.828	-70.243	9.921	-0.144
3700	41.091	325.315	283.980	152.938	-70.359	12.149	-0.172
3800	41.090	326.411	285.082	157.047	-70.503	14.381	-0.198
3900	41.099	327.478	286.156	161.156	-70.674	16.617	-0.223
4000	41.117	328.519	287.202	165.267	-70.874	18.858	-0.246
4100	41.144	329.534	288.222	169.380	-71.105	21.104	-0.269
4200	41.179	330.526	289.218	173.496	-71.367	23.356	-0.290
4300	41.222	331.496	290.190	177.616	-71.662	25.615	-0.311
4400	41.272	332.444	291.139	181.741	-71.992	27.881	-0.331
4500	41.329	333.372	292.067	185.871	-72.358	30.154	-0.350
4600	41.393	334.281	292.975	190.007	-72.761	32.437	-0.368
4700	41.463	335.172	293.864	194.149	-73.203	34.729	-0.386
4800	41.539	336.046	294.733	198.300	-73.686	37.030	-0.403
4900	41.621	336.903	295.585	202.458	-74.212	39.342	-0.419
5000	41.709	337.745	296.420	206.624	-74.781	41.665	-0.435
5100	41.802	338.572	297.238	210.800	-75.395	44.000	-0.451
5200	41.901	339.384	298.041	214.985	-76.057	46.347	-0.466
5300	42.004	340.183	298.829	219.180	-76.766	48.709	-0.480
5400	42.113	340.970	299.602	223.386	-77.525	51.083	-0.494
5500	42.227	341.743	300.361	227.603	-78.335	53.471	-0.508
5600	42.345	342.505	301.107	231.831	-79.198	55.876	-0.521
5700	42.467	343.256	301.840	236.072	-80.115	58.296	-0.534
5800	42.594	343.995	302.560	240.325	-81.087	60.733	-0.547
5900	42.725	344.725	303.269	244.591	-82.116	63.186	-0.559
6000	42.860	345.444	303.966	248.870	-83.202	65.658	-0.572

TABLE 2   Thermochemical Data of Selected Chemical Compounds
Nitrogen, Monatomic (N), ideal gas, mol. wt. = 14.0067

| | Enthalpy Reference Temperature = T_r = 298.15 K | | | | | Standard State Pressure = p^o = 0.1 MPa | | |
| | J K^{-1}mol^{-1} | | | | kJ mol^{-1} | | | |
T/K	C_p^o	S^o	-[G^o-H^o(T_r)]/T	H^o-H^o(T_r)	$\Delta_f H^o$	$\Delta_f G^o$	Log K_f
0	0.	0.	INFINITE	-6.197	470.820	470.820	INFINITE
100	20.786	130.593	171.780	-4.119	471.448	466.379	-243.611
200	20.786	145.001	155.201	-2.040	472.071	461.070	-120.419
250	20.786	149.639	153.642	-1.001	472.383	458.283	-95.753
298.15	20.786	153.300	153.300	0.	472.683	455.540	-79.809
300	20.788	153.429	153.300	0.038	472.694	455.434	-79.298
350	20.786	156.633	153.554	1.078	473.005	452.533	-67.537
400	20.786	159.408	154.116	2.117	473.314	449.587	-58.710
450	20.786	161.857	154.843	3.156	473.621	446.603	-51.840
500	20.786	164.047	155.655	4.196	473.923	443.584	-46.341
600	20.766	167.836	157.379	6.274	474.510	437.461	-38.084
700	20.786	171.041	159.108	8.353	475.067	431.242	-32.180
800	20.786	173.816	160.777	10.431	475.591	424.945	-27.746
900	20.786	176.264	162.364	12.510	476.081	418.584	-24.294
1000	20.786	178.454	163.866	14.589	476.540	412.171	-21.530
1100	20.786	180.436	165.284	16.667	476.970	405.713	-19.266
1200	20.786	182.244	166.623	18.746	477.374	399.217	-17.377
1300	20.788	183.908	167.889	20.824	477.756	392.688	-15.778
1400	20.786	185.448	169.089	22.903	478.118	386.131	-14.407
1500	20.786	186.882	170.228	24.982	478.462	379.548	-13.217
1600	20.766	188.224	171.311	27.060	478.791	372.943	-12.175
1700	20.786	189.484	172.344	29.139	479.107	366.318	-11.256
1800	20.787	190.672	173.329	31.218	479.411	359.674	-10.437
1900	20.788	191.796	174.272	33.296	479.705	353.014	-9.705
2000	20.790	192.863	175.175	35.375	479.990	346.339	-9.045
2100	20.793	193.877	176.042	37.454	480.266	339.650	-8.448
2200	20.797	194.844	176.874	39.534	480.536	332.947	-7.905
2300	20.804	195.769	177.676	41.614	480.799	326.233	-7.409
2400	20.813	196.655	178.448	43.695	481.057	319.507	-6.954
2500	20.826	197.504	179.194	45.777	481.311	312.770	-6.535
2600	20.843	198.322	179.914	47.860	481.561	306.024	-6.148
2700	20.864	199.109	180.610	49.945	481.809	299.268	-5.790
2800	20.891	199.868	181.285	52.033	482.054	292.502	-5.457
2900	20.924	200.601	181.938	54.124	482.299	285.728	-5.147
3000	20.963	201.311	182.572	56.218	482.543	278.946	-4.857
3100	21.010	202.000	183.188	58.317	482.789	272.155	-4.586
3200	21.064	202.667	183.786	60.420	483.036	265.357	-4.332
3300	21.126	203.317	184.368	62.530	483.286	258.550	-4.093
3400	21.197	203.948	184.935	64.646	483.540	251.736	-3.867
3500	21.277	204.564	185.487	66.769	483.799	244.915	-3.655
3600	21.365	205.164	186.025	68.902	484.064	238.086	-3.455
3700	21.463	205.751	186.550	71.043	484.335	231.249	-3.265
3800	21.569	206.325	187.063	73.194	484.614	224.405	-3.085
3900	21.685	206.887	187.564	75.357	484.903	217.554	-2.914
4000	21.809	207.437	188.054	77.532	485.201	210.695	-2.751
4100	21.941	207.977	188.534	79.719	485.510	203.829	-2.597
4200	22.082	208.508	189.003	81.920	485.830	196.955	-2.449
4300	22.231	209.029	189.463	84.136	486.164	190.073	-2.309
4400	22.388	209.542	189.913	86.367	486.510	183.183	-2.175
4500	22.551	210.047	190.355	88.614	486.871	176.285	-2.046
4600	22.722	210.544	190.788	90.877	487.247	169.379	-1.923
4700	22.899	211.035	191.214	93.158	487.638	162.465	-1.806
4800	23.081	211.519	191.632	95.457	488.046	155.542	-1.693
4900	23.269	211.997	192.043	97.775	488.471	148.610	-1.584
5000	23.461	212.469	192.447	100.111	488.912	141.670	-1.480
5100	23.658	212.935	192.844	102.467	489.372	134.721	-1.380
5200	23.858	213.397	193.235	104.843	489.849	127.762	-1.283
5300	24.061	213.853	193.619	107.238	490.345	120.794	-1.190
5400	24.266	214.305	193.998	109.655	490.860	113.817	-1.101
5500	24.474	214.752	194.371	112.092	491.394	106.829	-1.015
5600	24.682	215.195	194.739	114.550	491.947	99.832	-0.931
5700	24.892	215.633	195.102	117.028	492.519	92.825	-0.851
5800	25.102	216.068	195.460	119.528	493.110	85.808	-0.773
5900	25.312	216.499	195.813	122.049	493.720	78.780	-0.697
6000	25.521	216.926	196.161	124.590	494.349	71.742	-0.625

TABLE 2   Thermochemical Data of Selected Chemical Compounds
Nitrogen Oxide (NO), ideal gas, mol. wt. = 30.0061

| Enthalpy Reference Temperature = T_r = 298.15 K | | | | | Standard State Pressure = p^o = 0.1 MPa | | |
| | J K^{-1}mol^{-1} | | | | kJ mol^{-1} | | |
T/K	C_p^o	S^o	$-[G^o-H^o(T_r)]/T$	$H^o-H^o(T_r)$	$\Delta_f H^o$	$\Delta_f G^o$	Log K_f
0	0.	0.	INFINITE	-9.192	89.775	89.775	INFINITE
100	32.302	177.031	237.757	-6.073	89.991	88.944	-46.460
200	30.420	198.747	213.501	-2.951	90.202	87.800	-22.931
250	30.025	205.488	211.251	-1.441	90.256	87.193	-18.218
298.15	29.845	210.758	210.758	0.	90.291	86.600	-15.172
300	29.841	210.943	210.759	0.055	90.292	86.577	-15.074
350	29.823	215.540	211.122	1.546	90.316	85.955	-12.828
400	29.944	219.529	211.929	3.040	90.332	85.331	-11.143
450	30.175	223.068	212.974	4.542	90.343	84.705	-9.832
500	30.486	226.263	214.145	6.059	90.352	84.079	-8.784
600	31.238	231.886	216.646	9.144	90.366	82.822	-7.210
700	32.028	236.761	219.179	12.307	90.381	81.564	-6.086
800	32.767	241.087	221.652	15.548	90.398	80.303	-5.243
900	33.422	244.985	224.031	18.858	90.417	79.041	-4.587
1000	33.987	248.536	226.307	22.229	90.437	77.775	-4.063
1100	34.468	251.799	228.478	25.653	90.457	76.508	-3.633
1200	34.877	254.816	230.549	29.120	90.476	75.239	-3.275
1300	35.226	257.621	232.525	32.626	90.493	73.969	-2.972
1400	35.524	260.243	234.412	36.164	90.508	72.697	-2.712
1500	35.780	262.703	236.217	39.729	90.518	71.425	-2.487
1600	36.002	265.019	237.945	43.319	90.525	70.151	-2.290
1700	36.195	267.208	239.603	46.929	90.526	68.878	-2.116
1800	36.364	269.282	241.195	50.557	90.522	67.605	-1.962
1900	36.514	271.252	242.725	54.201	90.511	66.332	-1.824
2000	36.647	273.128	244.199	57.859	90.494	65.060	-1.699
2100	36.767	274.919	245.619	61.530	90.469	63.788	-1.587
2200	36.874	276.632	246.990	65.212	90.438	62.519	-1.484
2300	36.971	278.273	248.315	68.904	90.398	61.251	-1.391
2400	37.060	279.849	249.596	72.606	90.350	59.984	-1.306
2500	37.141	281.363	250.837	76.316	90.295	58.720	-1.227
2600	37.216	282.822	252.039	80.034	90.231	57.458	-1.154
2700	37.285	284.227	253.205	83.759	90.160	56.199	-1.087
2800	37.350	285.585	254.338	87.491	90.081	54.943	-1.025
2900	37.410	286.896	255.438.	91.229	89.994	53.689	-0.967
3000	37.466	288.165	256.508	94.973	89.899	52.439	-0.913
3100	37.519	289.395	257.549	98.722	89.798	51.192	-0.863
3200	37.570	290.587	258.563	102.477	89.689	49.948	-0.815
3300	37.617	291.744	259.551	106.236	89.574	48.708	-0.771
3400	37.663	292.867	260.514	110.000	89.451	47.472	-0.729
3500	37.706	293.960	261.454	113.768	89.323	46.239	-0.690
3600	37.747	295.022	262.372	117.541	89.189	45.010	-0.653
3700	37.787	296.057	263.269	121.318	89.049	43.784	-0.618
3800	37.825	297.065	264.145	125.098	88.903	42.563	-0.585
3900	37.862	298.048	265.902	128.883	88.752	41.346	-0.554
4000	37.898	299.008	265.840	132.671	88.596	40.132	-0.524
4100	37.933	299.944	266.660	136.462	88.434	38.922	-0.496
4200	37.966	300.858	267.464	140.257	88.268	37.717	-0.469
4300	37.999	301.752	268.251	144.056	88.097	36.515	-0.444
4400	38.031	302.626	269.022	147.857	87.922	35.318	-0.419
4500	38.062	303.481	269.778	151.662	87.741	34.124	-0.396
4600	38.092	304.318	270.520	155.469	87.556	32.934	-0.374
4700	38.122	305.137	271.248	159.280	87.366	31.749	-0.353
4800	38.151	305.940	271.962	163.094	87.171	30.568	-0.333
4900	38.180	306.727	272.664	166.910	86.970	29.391	-0.313
5000	38.208	307.499	273.353	170.730	86.765	28.218	-0.295
5100	38.235	308.256	274.030	174.552	86.553	27.049	-0.277
5200	38.262	308.998	274.695	178.377	86.336	25.884	-0.260
5300	38.289	309.728	275.349	182.204	86.112	24.724	-0.244
5400	38.316	310.443	275.993	186.034	85.881	23.568	-0.228
5500	38.342	311.147	276.625	189.867	85.644	22.416	-0.213
5600	38.367	311.838	277.248	193.703	85.399	21.269	-0.198
5700	38.393	312.517	277.861	197.541	85.146	20.125	-0.184
5800	38.418	313.185	278.464	201.381	84.884	18.987	-0.171
5900	38.443	313.842	279.058	205.224	84.613	17.853	-0.158
6000	38.468	314.488	279.643	209.070	84.331	16.724	-0.146

TABLE 2   Thermochemical Data of Selected Chemical Compounds
Nitrogen Oxide, Ion (NO$^+$), ideal gas, mol. wt. = 30.00555

| | Enthalply Reference Temperature = T$_r$ = 298.15 K | | | | Standard State Pressure = p^0 = 0.1 MPa | | |
| | J K^{-1}mol^{-1} | | | | kJ mol^{-1} | | |
T/K	C$_p^0$	S^0	-[G^0-H^0(T$_r$)]/T	H^0-H^0(T$_r$)	Δ$_f$H^0	Δ$_f$G^0	Log K$_f$
0	0.	0.	INFINITE	-8.670	983.995		
100	29.104	166.421	224.100	-5.768			
200	29.107	186.595	200.882	-2.857			
250	29.111	193.091	198.698	-1.402			
298.15	29.123	198.219	198.219	0.	990.185	983.978	-172.389
300	29.124	198.399	198.219	0.054	990.224	983.939	-171.319
350	29.163	202.891	198.574	1.511	991.253	982.811	-146.676
400	29.244	206.790	199.363	2.971	992.275	981.535	-128.175
450	29.378	210.242	200.383	4.436	993.288	980.131	-113.771
500	29.568	213.346	201.527	5.910	994.293	978.615	-102.235
600	30.089	218.781	203.962	8.891	996.282	975.293	-84.907
700	30.728	223.466	206.421	11.931	998.252	971.639	-72.505
800	31.403	227.613	208.816	15.038	1000.214	967.703	-63.185
900	32.059	231.350	211.115	18.211	1002.175	963.521	-55.921
1000	32.666	234.760	213.312	21.448	1004.139	959.121	-50.099
1100	33.213	237.899	215.406	24.742	1006.109	954.524	-45.327
1200	33.697	240.810	217.403	28.088	1008.085	949.747	-41.341
1300	34.124	243.525	219.309	31.480	1010.066	944.805	-37.963
1400	34.497	246.067	221.131	34.911	1012.053	939.711	-35.061
1500	34.825	248.459	222.874	38.378	1014.043	934.474	-32.541
1600	35.112	250.716	224.544	41.875	1016.036	929.104	-30.332
1700	35.365	252.852	226.147	45.399	1018.030	923.610	-28.379
1800	35.588	254.880	227.687	48.947	1020.024	917.998	-26.640
1900	35.786	256.810	229.170	52.516	1022.017	912.276	-25.080
2000	35.963	258.650	230.598	56.103	1024.008	906.449	-23.674
2100	36.121	260.408	231.976	59.708	1025.995	900.522	-22.399
2200	36.263	262.092	233.307	63.327	1027.979	894.501	-21.238
2300	36.391	263.707	234.594	66.960	1029.959	888.389	-20.176
2400	36.507	265.258	235.839	70.605	1031.933	882.191	-19.200
2500	36.612	266.750	237.046	74.261	1033.902	875.911	-18.301
2600	36.709	268.188	238.216	77.927	1035.865	869.553	-17.470
2700	36.798	269.575	239.352	81.602	1037.823	863.119	-16.698
2800	36.880	270.915	240.456	85.286	1039.774	856.613	-15.980
2900	36.955	272.211	241.529	88.978	1041.720	850.037	-15.311
3000	37.025	273.465	242.572	92.677	1043.659	843.395	-14.685
3100	37.091	274.680	243.589	96.383	1045.593	836.687	-14.098
3200	37.152	275.858	244.579	100.095	1047.520	829.917	-13.547
3300	37.209	277.003	245.544	103.813	1049.442	823.087	-13.028
3400	37.263	278.114	246.486	107.537	1051.358	816.200	-12.539
3500	37.313	279.195	247.405	111.266	1053.269	809.255	-12.077
3600	37.361	280.247	248.303	114.999	1055.174	802.257	-11.640
3700	37.407	281.271	249.180	118.738	1057.074	795.205	-11.226
3800	37.450	282.269	250.038	122.481	1058.969	788.102	-10.833
3900	37.491	283.243	250.877	126.228	1060.860	780.950	-10.460
4000	37.531	284.192	251.698	129.979	1062.745	773.748	-10.104
4100	37.569	285.120	252.502	133.734	1064.626	766.500	-9.765
4200	37.605	286.025	253.289	137.492	1066.502	759.206	-9.442
4300	37.640	286.911	254.061	141.255	1068.374	751.867	-9.133
4400	37.673	287.776	254.817	145.020	1070.241	744.485	-8.838
4500	37.705	288.623	255.559	148.789	1072.103	737.060	-8.556
4600	37.737	289.452	256.287	152.561	1073.961	729.594	-8.285
4700	37.767	290.264	257.001	156.337	1075.814	722.088	-8.025
4800	37.796	291.060	257.702	160.115	1077.662	714.542	-7.776
4900	37.825	291.839	258.391	163.896	1079.505	706.959	-7.536
5000	37.853	292.604	259.068	167.680	1081.342	699.337	-7.306
5100	37.880	293.354	259.733	171.466	1083.174	691.679	-7.084
5200	37.906	294.089	260.386	175.256	1084.999	683.984	-6.871
5300	37.932	294.812	261.029	179.048	1086.819	676.255	-6.665
5400	37.957	295.521	261.661	182.842	1088.631	668.492	-6.466
5500	37.982	296.218	262.283	186.639	1090.436	660.694	-6.275
5600	38.006	296.902	262.895	190.438	1092.234	652.865	-6.090
5700	38.030	297.575	263.498	194.240	1094.023	645.003	-5.911
5800	38.053	298.237	264.091	198.044	1095.803	637.110	-5.738
5900	38.076	298.887	264.675	201.851	1097.574	629.186	-5.570
6000	38.098	299.528	265.251	205.660	1099.334	621.233	-5.408

TABLE 2   Thermochemical Data of Selected Chemical Compounds
Nitrogen Dioxide (NO$_2$), ideal gas, mol. wt. = 46.0055

Enthalpy Reference Temperature = T$_r$ = 298.15 K				Standard State Pressure = p^0 = 0.1 MPa			
	J K^{-1}mol^{-1}			kJ mol^{-1}			
T/K	C$_p{}^0$	S^0	-[G^0-H^0(T$_r$)]/T	H^0-H^0(T$_r$)	Δ_fH^0	Δ_fG^0	Log K$_f$
0	0.	0.	INFINITE	-10.186	35.927	35.927	INFINITE
100	33.276	202.563	271.168	-6.861	34.898	39.963	-20.874
200	34.385	225.852	243.325	-3.495	33.897	45.422	-11.863
250	35.593	233.649	240.634	-1.746	33.460	48.355	-10.103
298.15	36.974	240.034	240.034	0.	33.095	51.258	-8.980
300	37.029	240.262	240.034	0.068	33.083	51.371	-8.944
350	38.583	246.086	240.491	1.958	32.768	54.445	-8.125
400	40.171	251.342	241.524	3.927	32.512	57.560	-7.517
450	41.728	256.164	242.886	5.975	32.310	60.703	-7.046
500	43.206	260.638	244.440	8.099	32.154	63.867	-6.672
600	45.834	268.755	247.830	12.555	31.959	70.230	-6.114
700	47.986	275.988	251.345	17.250	31.878	76.616	-5.717
800	49.708	282.512	254.840	22.138	31.874	83.008	-5.420
900	51.076	288.449	258.250	27.119	31.923	89.397	-5.188
1000	52.166	293.889	261.545	32.344	32.005	95.779	-5.003
1100	53.041	298.903	264.717	37.605	32.109	102.152	-4.851
1200	53.748	303.550	267.761	42.946	32.226	108.514	-4.724
1300	54.326	307.876	270.683	48.351	32.351	114.867	-4.615
1400	54.803	311.920	273.485	53.808	32.478	121.209	-4.522
1500	55.200	315.715	276.175	59.309	32.603	127.543	-4.441
1600	55.533	319.288	278.759	64.846	32.724	133.868	-4.370
1700	55.815	322.663	281.244	70.414	32.837	140.186	-4.307
1800	56.055	325.861	283.634	76.007	32.940	146.497	-4.251
1900	56.262	328.897	285.937	81.624	33.032	152.804	-4.201
2000	56.441	331.788	288.158	87.259	33.111	159.106	-4.155
2100	56.596	334.545	290.302	92.911	33.175	165.404	-4.114
2200	56.732	337.181	292.373	98.577	33.223	171.700	-4.077
2300	56.852	339.706	294.377	104.257	33.255	177.993	-4.042
2400	56.958	342.128	296.316	109.947	33.270	184.285	-4.011
2500	57.052	344.455	298.196	115.648	33.268	190.577	-3.982
2600	57.136	346.694	300.018	121.357	33.248	196.870	-3.955
2700	57.211	348.852	301.787	127.075	33.210	203.164	-3.930
2800	57.278	350.934	303.505	132.799	33.155	209.460	-3.908
2900	57.339	352.945	305.176	138.530	33.082	215.757	-3.886
3000	57.394	354.889	306.800	144.267	32.992	222.058	-3.866
3100	57.444	356.772	308.382	150.009	32.885	228.363	-3.848
3200	57.490	358.597	309.923	155.756	32.761	234.670	-3.831
3300	57.531	360.366	311.425	161.507	32.622	240.981	-3.814
3400	57.569	362.084	312.890	167.262	32.467	247.298	-3.799
3500	57.604	363.754	314.319	173.020	32.297	253.618	-3.785
3600	57.636	365.377	315.715	178.783	32.113	259.945	-3.772
3700	57.666	366.957	317.079	184.548	31.914	266.276	-3.759
3800	57.693	368.495	318.412	190.316	31.701	272.613	-3.747
3900	57.719	369.994	319.715	196.086	31.475	278.956	-3.736
4000	57.742	371.455	320.991	201.859	31.236	285.305	-3.726
4100	57.764	372.881	322.239	207.635	30.985	291.659	-3.716
4200	57.784	374.274	323.461	213.412	30.720	298.020	-3.706
4300	57.803	375.634	324.659	219.191	30.444	304.388	-3.698
4400	57.821	376.963	325.833	224.973	30.155	310.762	-3.689
4500	57.837	378.262	326.983	230.756	29.854	317.142	-3.681
4600	57.853	379.534	328.112	236.540	29.540	323.530	-3.674
4700	57.867	380.778	329.219	242.326	29.214	329.925	-3.667
4800	57.881	381.996	330.306	248.114	28.875	336.326	-3.660
4900	57.894	383.190	331.373	253.902	28.523	342.736	-3.654
5000	57.906	384.360	332.421	259.692	28.158	349.152	-3.648
5100	57.917	385.507	333.451	265.483	27.778	355.576	-3.642
5200	57.928	386.631	334.463	271.276	27.384	362.006	-3.636
5300	57.938	387.735	335.458	277.069	26.974	368.446	-3.631
5400	57.948	388.818	336.436	282.863	26.548	374.892	-3.626
5500	57.957	389.881	337.398	288.658	26.106	381.347	-3.622
5600	57.965	390.926	338.344	294.455	25.646	387.811	-3.617
5700	57.973	391.952	339.276	300.251	25.167	394.281	-3.613
5800	57.981	392.960	340.193	306.049	24.669	400.762	-3.609
5900	57.988	393.951	341.096	311.848	24.150	407.249	-3.606
6000	57.995	394.926	341.985	317.647	23.608	413.748	-3.602

TABLE 2 Thermochemical Data of Selected Chemical Compounds
Nitrogen (N_2), ideal gas-reference state, mol. wt. = 28.0134

| | Enthalply Reference Temperature = T_r = 298.15 K | | | | Standard State Pressure = p^0 = 0.1 MPa | | |
| | J $K^{-1}mol^{-1}$ | | | | kJ mol^{-1} | | |
T/K	C_p^0	S^0	$-[G^0-H^0(T_r)]/T$	$H^0-H^0(T_r)$	$\Delta_f H^0$	$\Delta_f G^0$	Log K_f
0	0.	0.	INFINITE	-8.670	0.	0.	0.
100	29.104	159.811	217.490	-5.768	0.	0.	0.
200	29.107	179.985	194.272	-2.857	0.	0.	0.
250	29.111	186.481	192.088	-1.402	0.	0.	0.
298.15	29.124	191.609	191.609	0.	0.	0.	0.
300	29.125	191.789	191.610	0.054	0.	0.	0.
350	29.165	196.281	191.964	1.511	0.	0.	0.
400	29.249	200.181	192.753	2.971	0.	0.	0.
450	29.387	203.633	193.774	4.437	0.	0.	0.
500	29.580	206.739	194.917	5.911	0.	0.	0.
600	30.110	212.176	197.353	8.894	0.	0.	0.
700	30.754	216.866	199.813	11.937	0.	0.	0.
800	31.433	221.017	202.209	15.046	0.	0.	0.
900	32.090	224.757	204.510	18.223	0.	0.	0.
1000	32.697	228.170	206.708	21.463	0.	0.	0.
1100	33.241	231.313	208.804	24.760	0.	0.	0.
1200	33.723	234.226	210.802	28.109	0.	0.	0.
1300	34.147	236.943	212.710	31.503	0.	0.	0.
1400	34.518	239.487	214.533	34.936	0.	0.	0.
1500	34.843	241.880	216.277	38.405	0.	0.	0.
1600	35.128	244.138	217.948	41.904	0.	0.	0.
1700	35.378	246.275	219.552	45.429	0.	0.	0.
1800	35.600	248.304	221.094	48.978	0.	0.	0.
1900	35.796	250.234	222.577	52.548	0.	0.	0.
2000	35.971	252.074	224.006	56.137	0.	0.	0.
2100	36.126	253.833	225.385	59.742	0.	0.	0.
2200	36.268	255.517	226.717	63.361	0.	0.	0.
2300	36.395	257.132	228.004	66.995	0.	0.	0.
2400	36.511	258.684	229.250	70.640	0.	0.	0.
2500	36.616	260.176	230.458	74.296	0.	0.	0.
2600	36.713	261.614	231.629	77.963	0.	0.	0.
2700	36.801	263.001	232.765	81.639	0.	0.	0.
2800	36.883	264.341	233.869	85.323	0.	0.	0.
2900	36.959	265.637	234.942	89.015	0.	0.	0.
3000	37.030	266.891	235.986	92.715	0.	0.	0.
3100	37.096	268.106	237.003	96.421	0.	0.	0.
3200	37.158	269.285	237.993	100.134	0.	0.	0.
3300	37.216	270.429	238.959	103.852	0.	0.	0.
3400	37.271	271.541	239.901	107.577	0.	0.	0.
3500	37.323	272.622	240.821	111.306	0.	0.	0.
3600	37.373	273.675	241.719	115.041	0.	0.	0.
3700	37.420	274.699	242.596	118.781	0.	0.	0.
3800	37.465	275.698	243.454	122.525	0.	0.	0.
3900	37.508	276.671	244.294	126.274	0.	0.	0.
4000	37.550	277.622	245.115	130.027	0.	0.	0.
4100	37.590	278.549	245.919	133.784	0.	0.	0.
4200	37.629	279.456	246.707	137.545	0.	0.	0.
4300	37.666	280.341	247.479	141.309	0.	0.	0.
4400	37.702	281.208	248.236	145.078	0.	0.	0.
4500	37.738	282.056	248.978	148.850	0.	0.	0.
4600	37.773	282.885	249.706	152.625	0.	0.	0.
4700	37.808	283.698	250.420	156.405	0.	0.	0.
4800	37.843	284.494	251.122	160.187	0.	0.	0.
4900	37.878	285.275	251.811	163.973	0.	0.	0.
5000	37.912	286.041	252.488	167.763	0.	0.	0.
5100	37.947	286.792	253.153	171.556	0.	0.	0.
5200	37.981	287.529	253.807	175.352	0.	0.	0.
5300	38.013	288.253	254.451	179.152	0.	0.	0.
5400	38.046	288.964	255.083	182.955	0.	0.	0.
5500	38.080	289.662	255.705	186.761	0.	0.	0.
5600	38.116	290.348	256.318	190.571	0.	0.	0.
5700	38.154	291.023	256.921	194.384	0.	0.	0.
5800	38.193	291.687	257.515	198.201	0.	0.	0.
5900	38.234	292.341	258.099	202.023	0.	0.	0.
6000	38.276	292.984	258.675	205.848	0.	0.	0.

TABLE 2　Thermochemical Data of Selected Chemical Compounds
Oxygen, Monatomic (O), ideal gas, mol. wt. = 15.9994

| Enthalpy Reference Temperature = T_r = 298.15 K | | | | | Standard State Pressure = p^o = 0.1 MPa | | |
| J K^{-1}mol^{-1} | | | | | kJ mol^{-1} | | |
T/K	C_p^o	S^o	$-[G^o-H^o(T_r)]/T$	$H^o-H^o(T_r)$	$\Delta_f H^o$	$\Delta_f G^o$	Log K_f
0	0.	0.	INFINITE	-6.725	246.790	246.790	INFINITE
100	23.703	135.947	181.131	-4.518	247.544	242.615	-126.729
200	22.734	152.153	163.085	-2.186	248.421	237.339	-61.986
250	22.246	157.170	161.421	-1.063	248.816	234.522	-49.001
298.15	21.911	161.058	161.058	0.	249.173	231.736	-40.599
300	21.901	161.194	161.059	0.041	249.187	231.628	-40.330
350	21.657	164.551	161.324	1.129	249.537	228.673	-34.128
400	21.482	167.430	161.912	2.207	249.868	225.670	-29.469
450	21.354	169.953	162.668	3.278	250.180	222.626	-25.842
500	21.257	172.197	163.511	4.343	250.474	219.549	-22.936
600	21.124	176.060	165.291	6.462	251.013	213.312	-18.570
700	21.040	179.310	167.067	8.570	251.494	206.990	-15.446
800	20.984	182.116	168.777	10.671	251.926	200.602	-13.098
900	20.944	184.585	170.399	12.767	252.320	194.163	-11.269
1000	20.915	186.790	171.930	14.860	252.682	187.681	-9.803
1100	20.893	188.782	173.373	16.950	253.018	181.165	-8.603
1200	20.877	190.599	174.734	19.039	253.332	174.619	-7.601
1300	20.864	192.270	176.019	21.126	253.627	168.047	-6.752
1400	20.853	193.816	177.236	23.212	253.906	161.453	-6.024
1500	20.845	195.254	178.390	25.296	254.171	154.840	-5.392
1600	20.838	196.599	179.486	27.381	254.421	148.210	-4.839
1700	20.833	197.862	180.530	29.464	254.659	141.564	-4.350
1800	20.830	199.053	181.527	31.547	254.884	134.905	-3.915
1900	20.827	200.179	182.479	33.630	255.097	128.234	-3.525
2000	20.826	201.247	183.391	35.713	255.299	121.552	-3.175
2100	20.827	202.263	184.266	37.796	255.488	114.860	-2.857
2200	20.830	203.232	185.106	39.878	255.667	108.159	-2.568
2300	20.835	204.158	185.914	41.962	255.835	101.450	-2.304
2400	20.841	205.045	186.693	44.045	255.992	94.734	-2.062
2500	20.851	205.896	187.444	46.130	256.139	88.012	-1.839
2600	20.862	206.714	188.170	48.216	256.277	81.284	-1.633
2700	20.877	207.502	188.871	50.303	256.405	74.551	-1.442
2800	20.894	208.261	189.550	52.391	256.525	67.814	-1.265
2900	20.914	208.995	190.208	54.481	256.637	61.072	-1.100
3000	20.937	209.704	190.846	56.574	256.741	54.327	-0.946
3100	20.963	210.391	191.466	58.669	256.838	47.578	-0.802
3200	20.991	211.057	192.068	60.767	256.929	40.826	-0.666
3300	21.022	211.704	192.653	62.867	257.014	34.071	-0.539
3400	21.056	212.332	193.223	64.971	257.094	27.315	-0.420
3500	21.092	212.943	193.777	67.079	257.169	20.555	-0.307
3600	21.130	213.537	194.318	69.190	257.241	13.794	-0.200
3700	21.170	214.117	194.845	71.305	257.309	7.030	-0.099
3800	21.213	214.682	195.360	73.424	257.373	0.265	-0.004
3900	21.257	215.234	195.862	75.547	257.436	-6.501	0.087
4000	21.302	215.772	196.353	77.675	257.496	-13.270	0.173
4100	21.349	216.299	196.834	79.808	257.554	-20.010	0.255
4200	21.397	216.814	197.303	81.945	257.611	-26.811	0.333
4300	21.445	217.318	197.763	84.087	257.666	-33.583	0.408
4400	21.495	217.812	198.213	86.234	257.720	-40.358	0.479
4500	21.545	218.295	198.654	88.386	257.773	-47.133	0.547
4600	21.596	218.769	199.086	90.543	257.825	-53.909	0.612
4700	21.647	219.234	199.510	92.705	257.876	-60.687	0.674
4800	21.697	219.690	199.925	94.872	257.926	-67.465	0.734
4900	21.748	220.138	200.333	97.045	257.974	-74.244	0.791
5000	21.799	220.578	200.734	99.222	258.021	-81.025	0.846
5100	21.849	221.010	201.127	101.405	258.066	-87.806	0.899
5200	21.899	221.435	201.514	103.592	258.110	-94.589	0.950
5300	21.949	221.853	201.893	105.784	258.150	-101.371	0.999
5400	21.997	222.264	202.267	107.982	258.189	-108.155	1.046
5500	22.045	222.668	202.634	110.184	258.224	-114.940	1.092
5600	22.093	223.065	202.995	112.391	258.255	-121.725	1.135
5700	22.139	223.457	203.351	114.602	258.282	-128.510	1.178
5800	22.184	223.842	203.701	116.818	258.304	-135.296	1.218
5900	22.229	224.222	204.046	119.039	258.321	-142.083	1.258
6000	22.273	224.596	204.385	121.264	258.332	-148.869	1.296

TABLE 2   Thermochemical Data of Selected Chemical Compounds
Sulfur Monoxide (SO), ideal gas, mol. wt. = 48.0594

| | Enthalply Reference Temperature = T_r = 298.15 K | | | | Standard State Pressure = p^0 = 0.1 MPa | | |
| | J K^{-1}mol^{-1} | | | | kJ mol^{-1} | | |
T/K	C_p^0	S^0	$-[G^0-H^0(T_r)]/T$	$H^0-H^0(T_r)$	$\Delta_f H^0$	$\Delta_f G^0$	Log K_f
0	0.	0.	INFINITE	-8.733	5.028	5.028	INFINITE
100	29.106	189.916	248.172	-5.826	5.793	-3.281	1.714
200	29.269	210.114	224.668	-2.911	5.610	-12.337	3.222
250	29.634	216.680	222.437	-1.439	5.333	-16.793	3.509
298.15	30.173	221.944	221.944	0.	5.007	-21.026	3.684
300	30.197	222.130	221.944	0.056	4.994	-21.187	3.689
350	30.867	226.835	222.314	1.582	4.616	-25.521	3.809
400	31.560	231.002	223.144	3.143	1.998	-29.711	3.880
450	32.221	234.758	224.230	4.738	0.909	-33.619	3.902
500	32.826	238.184	225.456	6.364	-0.238	-37.391	3.906
600	33.838	244.263	228.097	9.699	-2.067	-44.643	3.887
700	34.612	249.540	230.791	13.124	-3.617	-51.614	3.851
800	35.206	254.202	233.432	16.616	-5.005	-58.374	3.811
900	35.672	258.377	235.976	20.161	-59.419	-63.886	3.708
1000	36.053	262.155	238.408	23.748	-59.420	-64.382	3.363
1100	36.379	265.607	240.726	27.369	-59.424	-64.878	3.081
1200	36.672	268.785	242.933	31.022	-59.432	-65.374	2.846
1300	36.946	271.731	245.037	34.703	-59.446	-65.869	2.647
1400	37.210	274.479	247.043	38.411	-59.462	-66.362	2.476
1500	37.469	277.055	248.959	42.145	-59.482	-66.854	2.328
1600	37.725	279.482	250.791	45.905	-59.505	-67.345	2.199
1700	37.980	281.776	252.547	49.690	-59.528	-67.834	2.084
1800	38.232	283.954	254.232	53.501	-59.552	-68.323	1.983
1900	38.482	286.028	255.851	57.336	-59.575	-68.809	1.892
2000	38.727	288.008	257.410	61.197	-59.597	-69.294	1.810
2100	38.967	289.904	258.912	65.082	-59.618	-69.779	1.736
2200	39.200	291.722	260.363	68.990	-59.636	-70.262	1.668
2300	39.425	293.469	261.764	72.921	-59.652	-70.745	1.607
2400	39.641	295.152	263.121	76.875	-59.666	-71.227	1.550
2500	39.847	296.774	264.434	80.849	-59.678	-71.708	1.498
2600	40.043	298.341	265.709	84.844	-59.687	-72.189	1.450
2700	40.229	299.856	266.945	88.857	-59.694	-72.670	1.406
2800	40.404	301.322	268.147	92.889	-59.699	-73.150	1.365
2900	40.568	302.742	269.316	96.938	-59.702	-73.631	1.326
3000	40.721	304.120	270.453	101.002	-59.705	-74.111	1.290
3100	40.864	305.458	271.561	105.082	-59.707	-74.591	1.257
3200	40.996	306.757	272.640	109.175	-59.708	-75.072	1.225
3300	41.119	308.021	273.693	113.281	-59.710	-75.552	1.196
3400	41.232	309.250	274.721	117.398	-59.712	-76.032	1.168
3500	41.336	310.447	275.725	121.527	-59.716	-76.512	1.142
3600	41.432	311.613	276.706	125.665	-59.722	-76.991	1.117
3700	41.520	312.749	277.665	129.813	-59.730	-77.471	1.094
3800	41.601	313.857	278.603	133.969	-59.741	-77.950	1.071
3900	41.676	314.939	279.520	138.133	-59.755	-78.429	1.050
4000	41.745	315.995	280.419	142.304	-59.774	-78.908	1.030
4100	41.810	317.027	281.299	146.482	-59.796	-79.386	1.011
4200	41.871	318.035	282.162	150.666	-59.824	-79.863	0.993
4300	41.929	319.021	283.008	154.856	-59.857	-80.340	0.976
4400	41.986	319.985	283.837	159.051	-59.896	-80.816	0.959
4500	42.042	320.930	284.651	163.253	-59.941	-81.291	0.944
4600	42.098	321.854	285.450	167.460	-59.992	-81.765	0.928
4700	42.156	322.760	286.234	171.673	-60.049	-82.238	0.914
4800	42.217	323.648	287.004	175.891	-60.113	-82.709	0.900
4900	42.282	324.520	287.761	180.116	-60.183	-83.179	0.887
5000	42.352	325.374	288.505	184.348	-60.259	-83.648	0.874
5100	42.429	326.214	289.236	188.587	-60.342	-84.114	0.862
5200	42.514	327.039	289.955	192.834	-60.430	-84.580	0.850
5300	42.608	327.849	290.663	197.090	-60.524	-85.043	0.838
5400	42.712	328.647	291.359	201.356	-60.623	-85.505	0.827
5500	42.829	329.431	292.044	205.633	-60.726	-85.965	0.816
5600	42.959	330.204	292.718	209.922	-60.832	-86.423	0.806
5700	43.104	330.966	293.383	214.225	-60.941	-86.879	0.796
5800	43.265	331.717	294.037	218.543	-61.051	-87.333	0.787
5900	43.444	332.458	294.682	222.879	-61.162	-87.786	0.777
6000	43.642	333.190	295.318	227.233	-61.271	-88.235	0.768

TABLE 2   Thermochemical Data of Selected Chemical Compounds
Titanium Oxide (TiO), ideal gas, mol. wt. = 63.8794

Enthalply Reference Temperature = T_r = 298.15 K	J K^{-1}mol^{-1}				Standard State Pressure = p^o = 0.1 MPa	kJ mol^{-1}	
T/K	C_p^o	S^o	-[G^o-H^o(T_r)]/T	H^o-H^o(T_r)	$\Delta_f H^o$	$\Delta_f G^o$	Log K_f
0	0.	0.	INFINITE	-9.630	53.934	53.934	INFINITE
100	33.880	198.046	262.006	-6.396	55.155	44.842	-23.423
200	31.771	220.717	236.408	-3.138	55.040	34.491	-9.008
250	31.906	227.809	234.005	-1.549	54.737	29.387	-6.140
298.15	32.476	233.474	233.474	0.	54.392	24.535	-4.298
300	32.502	233.675	233.474	0.060	54.378	24.350	-4.240
350	33.263	238.741	233.873	1.704	53.997	19.375	-2.892
400	34.021	243.233	234.767	3.386	53.605	14.456	-1.888
450	34.702	247.280	235.937	5.105	53.208	9.586	-1.113
500	35.285	250.968	237.258	6.855	52.804	4.760	-0.497
600	36.176	257.485	240.100	10.431	51.975	-4.771	0.415
700	36.786	263.110	242.995	14.081	51.109	-14.161	1.057
800	37.212	268.051	245.824	17.782	50.217	-23.425	1.529
900	37.518	272.453	248.543	21.519	49.261	-32.573	1.891
1000	37.745	276.418	251.136	25.283	48.172	-41.609	2.173
1100	37.918	280.024	253.600	29.066	46.886	-50.527	2.399
1200	38.052	283.330	255.942	32.865	41.416	-59.198	2.577
1300	38.158	286.380	258.168	36.676	40.455	-67.545	2.714
1400	38.245	289.211	260.285	40.496	39.409	-75.814	2.829
1500	38.319	291.852	262.302	44.324	38.267	-84.005	2.925
1600	38.386	294.327	264.227	48.160	37.014	-92.116	3.007
1700	38.450	296.656	266.067	52.001	35.637	-100.145	3.077
1800	38.516	298.856	267.828	55.850	34.124	-108.089	3.137
1900	38.588	300.940	269.517	59.705	32.461	-115.945	3.188
2000	38.669	302.922	271.138	63.568	15.951	-123.256	3.219
2100	38.762	304.810	272.697	67.439	13.206	-130.149	3.237
2200	38.869	306.616	274.198	71.321	10.459	-136.911	3.251
2300	38.992	308.346	275.645	75.213	7.713	-143.548	3.260
2400	39.132	310.009	277.042	79.120	4.969	-150.066	3.266
2500	39.290	311.609	278.393	83.040	2.229	-156.469	3.269
2600	39.466	313.154	279.701	86.978	-0.505	-162.763	3.270
2700	39.661	314.647	280.967	90.934	-3.232	-168.952	3.269
2800	39.874	316.093	282.196	94.911	-5.948	-175.040	3.265
2900	40.105	317.496	283.389	98.910	-8.651	-181.032	3.261
3000	40.352	318.860	284.549	102.932	-11.341	-186.930	3.255
3100	40.614	320.187	285.677	106.981	-14.014	-192.738	3.248
3200	40.890	321.481	286.776	111.056	-16.670	-198.462	3.240
3300	41.179	322.744	287.847	115.159	-19.306	-204.102	3.231
3400	41.478	323.978	288.892	119.292	-21.921	-209.662	3.221
3500	41.787	325.184	289.911	123.455	-24.513	-215.146	3.211
3600	42.103	326.366	290.908	127.649	-27.082	-220.557	3.200
3700	42.425	327.524	291.882	131.876	-438.725	-218.108	3.079
3800	42.751	328.660	292.835	136.135	-440.010	-212.127	2.916
3900	43.080	329.774	293.768	140.426	-441.324	-206.113	2.761
4000	43.409	330.869	294.681	144.751	-442.666	-200.065	2.613
4100	43.738	331.945	295.577	149.108	-444.034	-193.983	2.471
4200	44.064	333.003	296.456	153.498	-445.427	-187.868	2.336
4300	44.386	334.044	297.318	157.921	-446.843	-181.719	2.207
4400	44.703	335.068	298.164	162.375	-448.281	-175.537	2.084
4500	45.015	336.076	298.996	166.861	-449.739	-169.322	1.965
4600	45.319	337.068	299.812	171.378	-451.216	-163.074	1.852
4700	45.614	338.046	300.616	175.925	-452.711	-156.794	1.743
4800	45.901	339.010	301.405	180.500	-454.222	-150.482	1.638
4900	46.178	339.959	302.183	185.104	-455.750	-144.138	1.537
5000	46.444	340.895	302.947	189.736	-457.293	-137.763	1.439
5100	46.699	341.817	303.701	194.393	-458.850	-131.357	1.345
5200	46.943	342.726	304.442	199.075	-460.421	-124.920	1.255
5300	47.174	343.622	305.173	203.781	-462.005	-118.453	1.167
5400	47.394	344.506	305.893	208.509	-463.602	-111.956	1.083
5500	47.602	345.378	306.603	213.259	-465.211	-105.430	1.001
5600	47.797	346.237	307.303	218.029	-466.833	-98.873	0.922
5700	47.980	347.085	307.994	222.818	-468.467	-92.288	0.846
5800	48.150	347.921	308.675	227.625	-470.113	-85.674	0.772
5900	48.309	348.745	309.347	232.448	-471.772	-79.032	0.700
6000	48.455	349.558	310.011	237.286	-473.442	-72.361	0.630

TABLE 2   Thermochemical Data of Selected Chemical Compounds
Oxygen (O_2), ideal gas-reference state, mol. wt. = 31.9988

T/K	C_p^o	S^o	$-[G^o-H^o(T_r)]/T$	$H^o-H^o(T_r)$	$\Delta_f H^o$	$\Delta_f G^o$	Log K_f
			Enthalpy Reference Temperature = T_r = 298.15 K		Standard State Pressure = p^o = 0.1 MPa		
		J K^{-1} mol^{-1}			kJ mol^{-1}		
0	0.	0.	INFINITE	-8.683	0.	0.	0.
100	29.106	173.307	231.094	-5.779	0.	0.	0.
200	29.126	193.485	207.823	-2.868	0.	0.	0.
250	29.201	199.990	205.630	-1.410	0.	0.	0.
298.15	29.376	205.147	205.147	0.	0.	0.	0.
300	29.385	205.329	205.148	0.054	0.	0.	0.
350	29.694	209.880	205.506	1.531	0.	0.	0.
400	30.106	213.871	206.308	3.025	0.	0.	0.
450	30.584	217.445	207.350	4.543	0.	0.	0.
500	31.091	220.693	208.524	6.084	0.	0.	0.
600	32.090	226.451	211.044	9.244	0.	0.	0.
700	32.981	231.466	213.611	12.499	0.	0.	0.
800	33.733	235.921	216.126	15.835	0.	0.	0.
900	34.355	239.931	218.552	19.241	0.	0.	0.
1000	34.870	243.578	220.875	22.703	0.	0.	0.
1100	35.300	246.922	223.093	26.212	0.	0.	0.
1200	35.667	250.010	225.209	29.761	0.	0.	0.
1300	35.988	252.878	227.229	33.344	0.	0.	0.
1400	36.277	255.556	229.158	36.957	0.	0.	0.
1500	36.544	258.068	231.002	40.599	0.	0.	0.
1600	36.796	260.434	232.768	44.266	0.	0.	0.
1700	37.040	262.672	234.462	47.958	0.	0.	0.
1800	37.277	264.796	236.089	51.673	0.	0.	0.
1900	37.510	266.818	237.653	55.413	0.	0.	0.
2000	37.741	268.748	239.160	59.175	0.	0.	0.
2100	37.969	270.595	240.613	62.961	0.	0.	0.
2200	38.195	272.366	242.017	66.769	0.	0.	0.
2300	38.419	274.069	243.374	70.600	0.	0.	0.
2400	38.639	275.709	244.687	74.453	0.	0.	0.
2500	38.856	277.290	245.959	78.328	0.	0.	0.
2600	39.068	278.819	247.194	82.224	0.	0.	0.
2700	39.276	280.297	248.393	86.141	0.	0.	0.
2800	39.478	281.729	249.558	90.079	0.	0.	0.
2900	39.674	283.118	250.691	94.036	0.	0.	0.
3000	39.864	284.466	251.795	98.013	0.	0.	0.
3100	40.048	285.776	252.870	102.009	0.	0.	0.
3200	40.225	287.050	253.918	106.023	0.	0.	0.
3300	40.395	288.291	254.941	110.054	0.	0.	0.
3400	40.559	289.499	255.940	114.102	0.	0.	0.
3500	40.716	290.677	256.916	118.165	0.	0.	0.
3600	40.868	291.826	257.870	122.245	0.	0.	0.
3700	41.013	292.948	258.802	126.339	0.	0.	0.
3800	41.154	294.044	259.716	130.447	0.	0.	0.
3900	41.289	295.115	260.610	134.569	0.	0.	0.
4000	41.421	296.162	261.485	138.705	0.	0.	0.
4100	41.549	297.186	262.344	142.854	0.	0.	0.
4200	41.674	298.189	263.185	147.015	0.	0.	0.
4300	41.798	299.171	264.011	151.188	0.	0.	0.
4400	41.920	300.133	264.821	155.374	0.	0.	0.
4500	42.042	301.076	265.616	159.572	0.	0.	0.
4600	42.164	302.002	266.397	163.783	0.	0.	0.
4700	42.287	302.910	267.164	168.005	0.	0.	0.
4800	42.413	303.801	267.918	172.240	0.	0.	0.
4900	42.542	304.677	268.660	176.488	0.	0.	0.
5000	42.675	305.538	269.389	180.749	0.	0.	0.
5100	42.813	306.385	270.106	185.023	0.	0.	0.
5200	42.956	307.217	270.811	189.311	0.	0.	0.
5300	43.105	308.037	271.506	193.614	0.	0.	0.
5400	43.262	308.844	272.190	197.933	0.	0.	0.
5500	43.426	309.639	272.864	202.267	0.	0.	0.
5600	43.599	310.424	273.527	206.618	0.	0.	0.
5700	43.781	311.197	274.181	210.987	0.	0.	0.
5800	43.973	311.960	274.826	215.375	0.	0.	0.
5900	44.175	312.713	275.462	219.782	0.	0.	0.
6000	44.387	313.457	276.089	224.210	0.	0.	0.

TABLE 2   Thermochemical Data of Selected Chemical Compounds
Sulfur Dioxide (SO_2), ideal gas, mol. wt. = 64.0588

| | Enthalply Reference Temperature = T_r = 298.15 K | | | | Standard State Pressure = p^o = 0.1 MPa | | |
| | J $K^{-1}mol^{-1}$ | | | | kJ mol^{-1} | | |
T/K	C_p^o	S^o	$-[G^o-H^o(T_r)]/T$	$H^o-H^o(T_r)$	$\Delta_f H^o$	$\Delta_f G^o$	Log K_f
0	0.	0.	INFINITE	-10.552	-294.299	-294.299	INFINITE
100	33.526	209.025	281.199	-7.217	-294.559	-296.878	155.073
200	36.372	233.033	251.714	-3.736	-295.631	-298.813	78.042
298.15	39.878	248.212	248.212	0.000	-296.842	-300.125	52.581
300	39.945	248.459	248.213	0.074	-296.865	-300.145	52.260
400	43.493	260.448	249.824	4.250	-300.257	-300.971	39.303
500	46.576	270.495	252.979	8.758	-302.736	-300.871	31.432
600	49.049	279.214	256.641	13.544	-304.694	-300.305	26.144
700	50.961	286.924	260.427	18.548	-306.291	-299.444	22.345
800	52.434	293.829	264.178	23.721	-307.667	-298.370	19.482
900	53.580	300.073	267.825	29.023	-362.026	-296.051	17.182
1000	54.484	305.767	271.339	34.428	-361.940	-288.725	15.081
1100	55.204	310.995	274.710	39.914	-361.835	-281.109	13.363
1200	55.794	315.824	277.937	45.464	-361.720	-274.102	11.931
1300	56.279	320.310	281.026	51.069	-361.601	-266.806	10.720
1400	56.689	324.496	283.983	56.718	-361.484	-259.518	9.683
1500	57.036	328.419	286.816	62.404	-361.372	-252.239	8.784
1600	57.338	332.110	289.533	68.123	-361.268	-244.967	7.997
1700	57.601	335.594	292.141	73.870	-361.176	-237.701	7.304
1800	57.831	338.893	294.647	79.642	-361.096	-230.440	6.687
1900	58.040	342.026	297.059	85.436	-361.031	-223.183	6.136
2000	58.229	345.007	299.383	91.250	-360.981	-215.929	5.639
2100	58.400	347.853	301.624	97.081	-360.948	-208.678	5.191
2200	58.555	350.573	303.787	102.929	-360.931	-201.427	4.782
2300	58.702	353.179	305.878	108.792	-360.930	-194.177	4.410
2400	58.840	355.680	307.902	114.669	-360.947	-186.927	4.068
2500	58.965	358.085	309.861	120.559	-360.980	-179.675	3.754
2600	59.086	360.400	311.761	126.462	-361.030	-172.422	3.464
2700	59.199	362.632	313.604	132.376	-361.095	-165.166	3.195
2800	59.308	364.787	315.394	138.302	-361.175	-157.908	2.946
2900	59.413	366.870	317.133	144.238	-361.270	-150.648	2.713
3000	59.513	368.886	318.825	150.184	-361.379	-143.383	2.497
3100	59.609	370.839	320.471	156.140	-361.502	-136.114	2.294
3200	59.706	372.733	322.075	162.106	-361.638	-128.842	2.103
3300	59.794	374.572	323.638	168.081	-361.786	-121.565	1.924
3400	59.881	376.358	325.162	174.065	-361.946	-114.283	1.756
3500	59.969	378.095	326.650	180.057	-362.118	-106.996	1.597
3600	60.053	379.786	328.103	186.058	-362.300	-99.704	1.447
3700	60.137	381.432	329.522	192.068	-362.493	-92.408	1.305
3800	60.216	383.037	330.909	198.086	-362.697	-85.105	1.170
3900	60.296	384.602	332.266	204.111	-362.911	-77.796	1.042
4000	60.375	386.130	333.593	210.145	-363.134	-70.484	0.920
4100	60.450	387.621	334.893	216.186	-363.368	-63.164	0.805
4200	60.530	389.079	336.166	222.235	-363.611	-55.840	0.694
4300	60.605	390.504	337.413	228.292	-363.864	-48.508	0.589
4400	60.676	391.898	338.636	234.356	-364.128	-41.172	0.489
4500	60.752	393.263	339.834	240.427	-364.401	-33.829	0.393
4600	60.823	394.599	341.010	246.506	-364.686	-26.480	0.301
4700	60.894	395.908	342.185	252.592	-364.981	-19.125	0.213
4800	60.969	397.190	343.298	258.685	-365.288	-11.763	0.128
4900	61.036	398.448	341.410	264.785	-365.607	- 4.394	0.047
5000	61.107	399.682	345.504	270.893	-365.938	2.981	-0.031
5100	61.178	400.893	346.578	277.007	-366.283	10.364	-0.106
5200	61.250	402.082	347.634	283.128	-366.641	17.751	-0.178
5300	61.317	403.249	348.672	289.257	-367.014	25.148	-0.248
5400	61.388	404.396	349.693	295.392	-367.403	32.550	-0.315
5500	61.455	405.523	350.698	301.534	-367.807	39.960	-0.380
5600	61.522	406.631	351.687	307.683	-368.230	47.379	-0.442
5700	61.588	407.720	352.661	313.838	-368.671	54.804	-0.502
5800	61.655	408.792	353.619	320.000	-369.131	62.238	-0.561
5900	61.727	409.846	354.563	326.169	-369.611	69.677	-0.617
6000	61.793	410.884	355.493	332.346	-370.113	77.128	-0.671

TABLE 2   Thermochemical Data of Selected Chemical Compounds
Titanium Dioxide (TiO_2), crystal-liquid, mol. wt. $= 79.8788$

Enthalply Reference Temperature = T_r = 298.15 K				Standard State Pressure = p^o = 0.1 MPa			
	J K^{-1}mol^{-1}			kJ mol^{-1}			
T/K	$C_p{}^o$	S^o	-[G^o-H^o(T_r)]/T	H^o-H^o(T_r)	$\Delta_f H^o$	$\Delta_f G^o$	Log K$_f$
0	0.000	0.	INFINITE	-8.636	-939.870	-939.870	INFINITE
100	18.502	10.142	89.638	-7.950	-942.649	-925.506	483.435
200	42.012	30.807	54.969	-4.833	-944.360	-907.579	237.035
298.15	55.103	50.292	50.292	0.000	-944.747	-889.406	155.820
500	67.203	82.201	57.077	12.562	-943.670	-852.157	89.024
600	69.931	94.712	62.331	19.429	-942.789	-833.936	72.601
700	71.764	105.638	67.754	26.519	-941.841	-815.868	60.881
800	73.078	115.311	73.106	33.764	-940.857	-797.939	52.100
900	74.057	123.977	78.285	41.122	-939.896	-780.132	45.278
1000	74.852	131.822	83.253	48.569	-939.032	-762.428	39.825
1100	75.479	138.986	87.999	56.086	-938.339	-744.803	35.368
1200	76.023	145.577	92.526	63.662	-941.807	-727.113	31.650
1300	76.525	151.683	96.844	71.290	-940.742	-709.265	28.499
1400	76.944	157.370	100.967	78.964	-939.741	-691.497	25.800
1500	77.320	162.691	104.906	86.677	-938.819	-673.798	23.464
1600	77.655	167.692	108.676	94.426	-937.992	-656.158	21.421
1700	77.990	172.410	112.287	102.209	-937.274	-638.566	19.621
1800	78.283	176.876	115.753	110.022	-936.679	-621.013	18.221
1900	78.576	181.117	119.082	117.865	-936.224	-603.489	16.591
2000	78.868	185.155	122.286	125.738	-950.606	-585.531	15.293
2100	79.161	189.010	125.372	133.640	-951.213	-567.263	14.110
2130.000	79.228	190.133	126.276	136.016	CRYSTAL <--> LIQUID		
2130.000	100.416	221.563	126.276	202.960	TRANSITION		
2200	100.416	224.809	129.360	209.989	-883.396	-551.188	13.087
2300	100.416	229.273	133.608	220.030	-881.909	-536.122	12.176
2400	100.416	233.547	137.683	230.072	-880.444	-521.120	11.342
2500	100.416	237.646	141.601	240.114	-879.001	-506.177	10.576
2600	100.416	241.584	145.371	250.155	-877.579	-491.293	9.870
2700	100.416	245.374	149.005	260.197	-876.179	-476.462	9.218
2800	100.416	249.026	152.512	270.238	-874.799	-461.682	8.613
2900	100.416	252.550	155.901	280.280	-873.438	-446.953	8.050
3000	100.416	255.954	159.180	290.322	-872.098	-432.269	7.526
3100	100.416	259.247	162.355	300.363	-870.775	-417.630	7.037
3200	100.416	262.435	165.433	310.405	-869.471	-403.034	6.579
3300	100.416	265.525	168.420	320.446	-868.184	-388.478	6.149
3400	100.416	268.522	171.320	330.488	-866.914	-373.960	5.745
3500	100.416	271.433	174.139	340.530	-865.660	-359.480	5.365
3600	100.416	274.262	176.881	350.571	-864.422	-345.035	5.006
3700	100.416	277.013	179.550	360.613	-1272.296	-322.836	4.558
3800	100.416	279.691	182.151	370.654	-1269.853	-297.207	4.085
3900	100.416	282.300	184.685	380.696	-1267.478	-271.642	3.638
4000	100.416	284.842	187.157	390.738	-1265.171	-246.138	3.214

TABLE 2   Thermochemical Data of Selected Chemical Compounds
Titanium Dioxide (TiO_2), ideal gas, mol. wt. = 79.8788

Enthalply Reference Temperature = T_r = 298.15 K					Standard State Pressure = p^o = 0.1 MPa		
	J $K^{-1}mol^{-1}$				kJ mol^{-1}		
T/K	C_p^o	S^o	$-[G^o-H^o(T_r)]/T$	$H^o-H^o(T_r)$	$\Delta_f H^o$	$\Delta_f G^o$	Log K_f
0	0.	0.	INFINITE	-11.357	-303.276	-303.276	INFINITE
100	35.950	217.025	296.659	-7.963	-303.347	-306.893	160.304
200	40.232	243.358	264.059	-4.140	-304.352	-310.081	80.985
250	42.214	252.547	260.864	-2.079	-304.913	-311.449	65.074
298.15	44.149	260.148	260.148	0.	-305.432	-312.660	54.777
300	44.222	260.422	260.149	0.082	-305.451	-312.705	54.447
350	46.106	267.382	260.695	2.341	-305.956	-313.873	46.843
400	47.777	273.651	261.929	4.689	-306.429	-314.971	41.131
450	49.210	279.363	263.553	7.114	-306.878	-316.012	36.682
500	50.419	284.612	265.400	9.606	-307.311	-317.004	33.117
600	52.280	293.979	269.402	14.746	-308.156	-318.863	27.759
700	53.591	302.143	273.509	20.044	-309.001	-320.581	23.922
800	54.532	309.363	277.548	25.452	-309.854	-322.177	21.036
900	55.222	315.828	281.448	30.942	-310.761	-323.664	16.785
1000	55.740	321.674	285.183	36.491	-311.795	-325.043	16.979
1100	56.136	327.006	288.747	42.086	-313.024	-326.310	15.495
1200	56.446	331.905	292.142	47.716	-318.438	-327.337	14.249
1300	56.693	336.433	295.377	53.373	-319.344	-328.042	13.181
1400	56.891	340.642	298.462	59.052	-320.337	-328.674	12.263
1500	57.053	344.573	301.406	64.750	-321.431	-329.233	11.465
1600	57.188	348.259	304.220	70.462	-322.641	-329.714	10.764
1700	57.300	351.730	306.914	76.187	-323.980	-330.116	10.143
1800	57.396	355.008	309.496	81.922	-325.465	-330.435	9.589
1900	57.478	356.113	311.973	87.666	-327.108	-330.667	9.091
2000	57.550	361.063	314.355	93.417	-343.612	-330.354	8.628
2100	57.614	363.873	316.646	99.175	-346.363	-329.624	8.199
2200	57.673	366.554	318.854	104.940	-349.130	-328.761	7.806
2300	57.727	369.119	320.985	110.710	-351.915	-327.773	7.444
2400	57.781	371.577	323.042	116.485	-354.716	-326.664	7.110
2500	57.833	373.937	325.031	122.266	-357.534	-325.437	6.800
2600	57.888	376.206	326.956	128.052	-360.368	-324.098	6.511
2700	57.945	378.392	328.820	133.843	-363.217	-322.648	6.242
2800	58.006	380.500	330.629	139.641	-366.081	-321.093	5.990
2900	58.072	382.537	332.384	145.445	-368.959	-319.436	5.754
3000	58.144	384.507	334.088	151.256	-371.849	-317.679	5.531
3100	58.223	386.415	335.746	157.074	-374.750	-315.825	5.322
3200	58.310	388.265	337.358	162.900	-377.661	-313.879	5.124
3300	58.405	390.060	338.928	168.736	-380.580	-311.841	4.936
3400	58.509	391.806	340.458	174.582	-383.505	-308.713	4.758
3500	58.622	393.503	341.949	180.438	-386.437	-307.501	4.589
3600	58.743	395.156	343.404	186.306	-389.371	-305.204	4.428
3700	58.874	396.768	344.825	192.187	-801.407	-295.037	4.165
3800	59.013	398.339	346.213	198.081	-803.110	-281.328	3.867
3900	59.161	399.874	347.569	203.990	-804.869	-267.574	3.584
4000	59.318	401.374	348.895	209.914	-806.679	-253.775	3.314
4100	59.483	402.841	350.193	215.854	-808.539	-239.929	3.057
4200	59.655	404.276	351.464	221.811	-810.446	-226.038	2.811
4300	59.834	405.682	352.709	227.785	-812.397	-212.100	2.577
4400	60.019	407.060	353.928	233.778	-814.390	-198.117	2.352
4500	60.210	408.411	355.124	239.789	-816.421	-184.088	2.137
4600	60.406	409.736	356.297	245.820	-818.489	-170.014	1.931
4700	60.607	411.037	357.448	251.871	-820.591	-155.894	1.733
4800	60.812	412.315	358.578	257.941	-822.725	-141.729	1.542
4900	61.019	413.571	359.687	264.033	-824.889	-127.518	1.359
5000	61.229	414.806	360.777	270.145	-827.081	-113.264	1.183
5100	61.440	416.021	361.849	276.279	-829.299	-98.966	1.014
5200	61.653	417.216	362.902	282.433	-831.542	-84.625	0.850
5300	61.865	418.392	363.938	288.609	-833.807	-70.238	0.692
5400	62.077	419.551	364.957	294.806	-836.095	-55.810	0.540
5500	62.289	420.692	365.960	301.025	-838.403	-41.340	0.393
5600	62.498	421.816	366.947	307.264	-840.731	-26.826	0.250
5700	62.706	422.924	367.920	313.524	-843.079	-12.272	0.112
5800	62.911	424.016	368.877	319.805	-845.444	2.325	-0.021
5900	63.113	425.093	369.821	326.106	-847.828	16.961	-0.150
6000	63.311	426.156	370.751	332.428	-850.230	31.640	-0.275

TABLE 2   Thermochemical Data of Selected Chemical Compounds
Ozone (O_3), ideal gas, mol. wt. = 47.9982

T/K	C_p^o	S^o	$-[G^o-H^o(T_r)]/T$	$H^o-H^o(T_r)$	$\Delta_f H^o$	$\Delta_f G^o$	Log K_f
		J K^{-1}mol^{-1}			kJ mol^{-1}		
0	0.	0.	INFINITE	-10.351	145.348	145.348	INFINITE
100	33.292	200.791	271.040	-7.025	144.318	150.235	-78.474
200	35.058	224.221	242.401	-3.636	143.340	156.541	-40.884
298.15	39.238	238.932	238.932	0.000	142.674	163.184	-28.589
300	39.330	239.175	238.933	0.073	142.665	163.311	-28.435
400	43.744	251.116	240.531	4.234	142.370	170.247	-22.232
500	47.262	261.272	243.688	8.792	142.340	177.224	-18.514
600	49.857	270.129	247.373	13.654	142.462	184.191	-16.035
700	51.752	277.963	251.194	18.738	142.665	191.130	-14.262
800	53.154	284.969	254.986	23.986	142.907	198.037	-12.931
900	54.208	291.292	258.674	29.356	143.169	204.913	-11.893
1000	55.024	297.048	262.228	34.819	143.439	211.759	-11.061
1100	55.660	302.323	265.637	40.355	143.711	218.578	-10.379
1200	56.174	307.189	268.899	45.947	143.980	225.372	-9.810
1300	56.593	311.702	272.020	51.586	144.245	232.144	-9.328
1400	56.948	315.909	275.007	57.264	144.502	238.896	-8.913
1500	57.245	319.849	277.866	62.974	144.750	245.629	-8.554
1600	57.501	323.551	280.607	68.711	144.987	252.347	-8.238
1700	57.722	327.044	283.237	74.473	145.211	259.050	-7.960
1800	57.919	330.349	285.763	80.255	145.419	265.740	-7.712
1900	58.095	333.485	288.193	86.056	145.611	272.419	-7.489
2000	58.250	336.469	290.533	91.873	145.784	279.089	-7.289
2100	58.396	339.315	292.789	97.705	145.938	285.750	-7.108
2200	58.526	342.035	294.966	103.552	146.072	292.406	-6.943
2300	58.647	344.639	297.069	109.410	146.185	299.054	-6.792
2400	58.764	347.137	299.104	115.281	146.276	305.698	-6.653
2500	58.869	349.538	301.073	121.163	146.346	312.339	-6.526
2600	58.969	351.849	302.982	127.055	146.393	318.978	-6.408
2700	59.066	354.077	304.833	132.956	146.419	325.616	-6.299
2800	59.158	356.226	306.631	138.868	146.424	332.253	-6.198
2900	59.245	358.304	308.377	144.788	146.408	338.889	-6.104
3000	59.329	360.314	310.075	150.716	146.371	345.527	-6.016
3100	59.409	362.260	311.727	156.653	146.314	352.167	-5.934
3200	59.488	364.148	313.336	162.598	146.238	358.808	-5.857
3300	59.563	365.980	314.904	168.551	146.144	365.451	-5.785
3400	59.639	367.759	316.432	174.511	146.033	372.100	-5.717
3500	59.714	369.489	317.923	180.479	145.905	378.750	-5.653
3600	59.781	371.172	319.379	186.453	145.761	385.405	-5.592
3700	59.852	372.811	320.801	192.435	145.601	392.063	-5.535
3800	59.919	374.408	322.191	198.424	145.427	398.728	-5.481
3900	59.986	375.965	323.550	204.419	145.239	405.396	-5.430
4000	60.053	377.485	324.879	210.421	145.038	412.069	-5.381
4100	60.120	378.968	326.181	216.429	144.823	418.747	-5.335
4200	60.183	380.418	327.455	222.445	144.597	425.431	-5.291
4300	60.245	381.835	328.703	228.466	144.358	432.121	-5.249
4400	60.308	383.220	329.926	234.494	144.106	438.815	-5.209
4500	60.371	384.576	331.126	240.528	143.844	445.516	-5.171
4600	60.434	385.904	332.302	246.568	143.568	452.222	-5.135
4700	60.492	387.204	333.457	252.614	143.281	458.935	-5.100
4800	60.555	388.479	334.590	258.667	142.981	465.655	-5.067
4900	60.614	389.728	335.702	264.725	142.667	472.381	-5.036
5000	60.672	390.953	336.795	270.789	142.340	479.113	-5.005
5100	60.731	392.155	337.869	276.859	141.999	485.853	-4.976
5200	60.789	393.335	338.924	282.935	141.643	492.597	-4.948
5300	60.848	394.493	339.962	289.017	141.270	499.351	-4.921
5400	60.906	395.631	340.982	295.105	140.880	506.110	-4.896
5500	60.965	396.749	341.986	301.199	140.472	512.876	-4.871
5600	61.024	397.848	342.974	307.298	140.045	519.653	-4.847
5700	61.078	398.929	343.946	313.403	139.596	526.434	-4.824
5800	61.137	399.992	344.903	319.514	139.126	533.226	-4.802
5900	61.191	401.037	345.846	325.630	138.631	540.023	-4.781
6000	61.250	402.066	346.774	331.752	138.111	546.832	-4.761

TABLE 2 Thermochemical Data of Selected Chemical Compounds
Sulfur Trioxide (SO₃), ideal gas, mol. wt. = 80.0582

| | Enthalply Reference Temperature = T_r = 298.15 K | | | | Standard State Pressure = p^o = 0.1 MPa | | |
| | J K⁻¹mol⁻¹ | | | | kJ mol⁻¹ | | |
T/K	C_p^o	S^o	-[G^o-$H^o(T_r)$]/T	H^o-$H^o(T_r)$	$\Delta_f H^o$	$\Delta_f G^o$	Log K_f
0	0.	0.	INFINITE	-11.697	-390.025	-390.025	INFINITE
100	34.076	212.371	295.976	-8.361	-391.735	-385.724	201.481
200	42.336	238.259	261.145	-4.577	-393.960	-378.839	98.943
250	46.784	248.192	257.582	-2.348	-394.937	-374.943	78.340
298.15	50.661	256.769	256.769	0.	-395.765	-371.016	65.000
300	50.802	257.083	256.770	0.094	-395.794	-370.862	64.573
350	54.423	265.191	257.402	2.726	-396.543	-366.646	54.719
400	57.672	272.674	258.849	5.530	-399.412	-362.242	47.304
450	60.559	279.637	260.777	8.487	-400.656	-357.529	41.501
500	63.100	286.152	262.992	11.580	-401.878	-352.668	36.843
600	67.255	298.041	267.862	18.107	-403.675	-342.647	29.830
700	70.390	308.655	272.945	24.997	-405.014	-332.365	24.801
800	72.761	318.217	278.017	32.160	-406.068	-321.912	21.019
900	74.570	326.896	282.973	39.531	-460.062	-310.258	18.007
1000	75.968	334.828	287.768	47.060	-459.581	-293.639	15.338
1100	77.065	342.122	292.382	54.714	-459.063	-277.069	13.157
1200	77.937	348.866	296.811	62.466	-458.521	-260.548	11.341
1300	78.639	355.133	301.060	70.296	-457.968	-244.073	9.807
1400	79.212	360.983	305.133	78.189	-457.413	-227.640	8.493
1500	79.685	366.465	309.041	86.135	-456.863	-211.247	7.356
1600	80.079	371.620	312.793	94.124	-456.323	-194.890	6.363
1700	80.410	376.485	316.398	102.149	-455.798	-178.567	5.487
1800	80.692	381.090	319.865	110.204	-455.293	-162.274	4.709
1900	80.932	385.459	323.203	118.286	-454.810	-146.009	4.014
2000	81.140	389.616	326.421	126.390	-454.351	-129.768	3.389
2100	81.319	393.579	329.525	134.513	-453.919	-113.549	2.824
2200	81.476	397.366	332.523	142.653	-453.514	-97.350	2.311
2300	81.614	400.990	335.422	150.807	-453.137	-81.170	1.843
2400	81.735	404.466	338.227	158.975	-452.790	-65.006	1.415
2500	81.843	407.805	340.944	167.154	-452.472	-48.855	1.021
2600	81.939	411.017	343.578	175.343	-452.183	-32.716	0.657
2700	82.025	414.111	346.133	183.541	-451.922	-16.587	0.321
2800	82.102	417.096	348.614	191.748	-451.690	-0.467	0.009
2900	82.171	419.978	351.026	199.961	-451.487	15.643	-0.282
3000	82.234	422.765	353.371	208.182	-451.311	31.748	-0.553
3100	82.290	425.462	355.653	216.408	-451.161	47.849	-0.806
3200	82.342	428.076	357.876	224.640	-451.038	63.943	-1.044
3300	82.389	430.610	360.042	232.876	-450.940	80.034	-1.267
3400	82.432	433.070	362.153	241.117	-450.866	96.124	-1.477
3500	82.472	435.460	364.214	249.363	-450.817	112.210	-1.675
3600	82.508	437.784	366.225	257.612	-450.792	128.297	-1.862
3700	82.542	440.045	368.190	265.864	-450.789	144.382	-2.038
3800	82.573	442.247	370.110	274.120	-450.809	160.469	-2.206
3900	82.601	444.392	371.987	282.379	-450.850	176.556	-2.365
4000	82.628	446.484	373.824	290.640	-450.914	192.643	-2.516
4100	82.653	448.524	375.621	298.904	-450.999	208.733	-2.659
4200	82.676	450.516	377.381	307.171	-451.105	224.825	-2.796
4300	82.697	452.462	379.104	315.439	-451.234	240.921	-2.927
4400	82.717	454.364	380.793	323.710	-451.383	257.019	-3.051
4500	82.735	456.223	382.449	331.983	-451.555	273.120	-3.170
4600	82.753	458.041	384.072	340.257	-451.749	289.225	-3.284
4700	82.769	459.821	385.665	348.533	-451.965	305.336	-3.393
4800	82.785	461.564	387.228	356.811	-452.205	321.452	-3.498
4900	82.799	463.271	388.763	365.090	-452.469	337.573	-3.599
5000	82.813	464.944	390.270	373.371	-452.757	353.700	-3.695
5100	82.825	466.584	391.750	381.653	-453.071	369.832	-3.788
5200	82.837	468.192	393.205	389.936	-453.412	385.969	-3.877
5300	82.849	469.770	394.634	398.220	-453.780	402.116	-3.963
5400	82.860	471.319	396.040	406.505	-454.178	418.268	-4.046
5500	82.870	472.840	397.423	414.792	-454.605	434.427	-4.126
5600	82.879	474.333	398.783	423.079	-455.065	450.597	-4.203
5700	82.889	475.800	400.121	431.368	-455.557	466.773	-4.277
5800	82.897	477.242	401.439	439.657	-456.084	482.960	-4.350
5900	82.906	478.659	402.735	447.947	-456.647	499.153	-4.419
6000	82.913	480.052	404.012	456.238	-457.248	515.359	-4.487

TABLE 2 Thermochemical Data of Selected Chemical Compounds
Titanium Oxide (Ti$_3$O$_5$), crystal-liquid, mol. wt. = 223.6370

| Enthalpy Reference Temperature = T_r = 298.15 K | | | | | Standard State Pressure = p^0 = 0.1 MPa | | |
| | J K^{-1}mol^{-1} | | | | kJ mol^{-1} | | |
T/K	C_p^0	S^0	$-[G^0-H^0(T_r)]/T$	$H^0-H^0(T_r)$	$\Delta_f H^0$	$\Delta_f G^0$	Log K_f
0	0.000	0.000	INFINITE	-23.108	-2446.057	-2446.057	INFINITE
100	46.116	21.062	237.166	-21.610	-2453.501	-2409.803	1258.752
200	114.445	75.659	142.352	-13.339	-2458.258	-2363.911	617.390
298.15	154.808	129.369	129.369	0.000	-2459.146	-2317.293	405.980
300	155.477	130.329	129.372	0.287	-2459.135	-2316.413	403.323
400	182.841	179.228	135.852	17.350	-2457.340	-2269.052	296.308
450	189.954	201.217	141.909	26.688	-2455.868	-2245.602	260.663
450.000	181.586	230.691	141.909	39.952	------ ALPHA <--> BETA -----		
500	184.096	249.952	151.765	49.094	-2441.465	-2223.774	232.316
600	189.117	283.961	171.037	67.754	-2439.180	-2180.450	189.825
700	194.138	313.490	189.323	86.917	-2436.818	-2137.514	159.503
800	199.158	339.742	206.514	106.582	-2434.270	-2094.928	136.785
900	204.179	363.489	222.657	126.749	-2431.589	-2052.670	119.134
1000	209.200	385.261	237.844	147.418	-2428.938	-2010.711	105.029
1100	214.221	405.436	252.173	168.589	-2426.485	-1969.009	93.500
1200	219.242	424.291	265.739	190.262	-2436.168	-1927.162	83.887
1300	224.262	442.038	278.624	212.437	-2431.892	-1884.917	75.737
1400	229.283	458.841	290.902	235.114	-2427.425	-1843.008	68.763
1500	234.304	474.831	302.635	258.294	-2422.799	-1801.425	62.731
1600	239.325	490.113	313.878	281.975	-2418.050	-1760.154	57.463
1700	243.346	504.773	324.679	306.159	-2413.214	-1719.183	52.824
1800	249.366	518.881	335.079	330.844	-2408.329	-1678.499	48.709
1900	254.387	532.496	345.113	356.032	-2403.433	-1638.087	45.034
2000	259.408	545.674	354.813	381.722	-2442.627	-1596.570	41.698
2050.000	261.918	552.111	359.547	394.755	BETA <--> LIQUID		
2050.000	267.776	635.791	359.547	566.299	TRANSITION		
2100	267.776	642.243	366.201	579.668	-2268.296	-1558.516	38.766
2200	267.776	654.700	379.034	606.465	-2265.210	-1524.788	36.203
2300	267.776	666.603	391.280	633.243	-2262.180	-1491.201	33.866
2400	267.776	678.000	402.991	660.020	-2259.206	-1457.745	31.727
2500	267.776	688.931	414.212	686.798	-2256.287	-1424.411	29.761
2600	267.776	699.433	424.981	713.576	-2253.421	-1391.193	27.949
2700	267.776	709.539	435.334	740.353	-2250.608	-1358.083	26.274
2800	267.776	719.278	445.302	767.131	-2247.846	-1325.077	24.720
2900	267.776	728.674	454.913	793.908	-2245.133	-1292.170	23.275
3000	267.776	737.752	464.190	820.686	-2242.470	-1259.354	21.927
3100	267.776	746.533	473.157	847.464	-2239.852	-1226.626	20.669
3200	267.776	755.034	481.834	874.241	-2237.280	-1193.984	19.490
3300	267.776	763.274	490.238	901.019	-2234.751	-1161.421	18.384
3400	267.776	771.268	498.387	927.796	-2232.264	-1128.932	17.344
3500	267.776	779.030	506.295	954.574	-2229.818	-1096.518	16.365
3600	267.776	786.574	513.976	981.352	-2227.409	-1064.171	15.441
3700	267.776	793.910	521.443	1008.129	-3452.333	-1008.528	14.238
3800	267.776	801.051	528.708	1034.907	-3446.295	-942.558	12.956
3900	267.776	808.007	535.780	1061.684	-3440.457	-876.746	11.743
4000	267.776	814.787	542.671	1088.462	-3434.815	-811.083	10.592

TABLE 2   Thermochemical Data of Selected Chemical Compounds
Titanium Oxide (Ti_3O_5), liquid, mol. wt. = 223.6370

T/K	C_p^o	S^o	$-[G^o-H^o(T_r)]/T$	$H^o-H^o(T_r)$	$\Delta_f H^o$	$\Delta_f G^o$	Log K_f
		Enthalpy Reference Temperature = T_r = 298.15 K				Standard State Pressure = p^o = 0.1 MPa	
		— J K^{-1}mol^{-1} —			— kJ mol^{-1} —		
0							
100							
200							
298.15	173.962	232.446	232.446	0.000	-2289.059	-2177.939	381.566
300	174.054	233.522	232.449	0.322	-2289.013	-2177.250	379.093
400	179.075	284.282	239.336	17.978	-2286.625	-2140.359	279.502
500	184.096	324.781	252.507	36.137	-2284.335	-2190.058	219.809
600	189.117	358.789	267.460	54.798	-2282.050	-2068.217	180.051
700	194.138	388.319	282.661	73.960	-2279.687	-2032.764	151.687
800	199.158	414.570	297.539	93.625	-2277.139	-1997.660	130.434
900	204.179	438.318	311.882	113.792	-2274.459	-1962.886	113.923
1000	209.200	460.090	325.629	134.461	-2271.808	-1928.410	100.730
1100	214.221	480.264	338.781	155.632	-2269.355	-1894.191	89.948
1200	219.242	499.119	351.365	177.305	-2279.038	-1859.826	80.956
1300	224.262	516.866	363.420	199.480	-2274.762	-1825.064	73.332
1400	229.283	533.669	371.985	222.158	-2270.295	-1790.638	66.810
1400 000	229.283	533.669	374.985	222.158	GLASS <--> LIQUID		
1400 000	267.776	533.669	374.985	222.158	TRANSITION		
1500	267.776	552.144	386.187	248.935	-2262.071	-1756.666	61.173
1600	267.776	569.426	397.105	275.713	-2254.226	-1723.230	56.258
1700	267.776	585.660	407.724	302.490	-2246.795	-1690.273	51.936
1800	267.776	600.965	418.039	329.268	-2239.818	-1657.741	48.106
1900	267.776	615.443	428.051	356.046	-2233.333	-1625.582	44.690
2000	267.776	629.178	437.767	382.823	-2271.438	-1592.390	41.589
2050.000	267.776	635.790	442.516	396.212	BETA <--> LIQUID		
2100	267.776	642.243	447.195	409.601	-2268.296	-1558.516	38.766
2200	267.776	654.700	456.346	436.378	-2265.210	-1524.788	36.203
2300	267.776	666.603	465.231	463.156	-2262.180	-1491.201	33.866
2400	267.776	678.000	473.861	489.934	-2259.206	-1457.745	31.727
2500	267.776	688.931	482.246	516.711	-2256.287	-1424.411	29.761
2600	267.776	699.433	490.399	543.489	-2253.421	-1391.193	27.949
2700	267.776	709.539	498.329	570.266	-2250.608	-1358.083	26.274
2800	267.776	719.278	506.048	597.044	-2247.846	-1325.077	24.720
2900	267.776	728.674	513.563	623.822	-2245.133	-1292.170	23.275
3000	267.776	737.752	520.886	650.599	-2242.470	-1259.354	21.927
3100	267.776	746.533	528.024	677.377	-2239.852	-1226.626	20.669
3200	267.776	755.034	534.986	704.154	-2237.280	-1193.984	19.490
3300	267.776	763.274	541.779	730.932	-2234.751	-1161.421	18.384
3400	267.776	771.268	548.412	757.710	-2232.264	-1128.932	17.344
3500	267.776	779.030	554.891	784.487	-2229.818	-1096.518	16.365
3600	267.776	786.574	561.222	811.265	-2227.409	-1064.171	15.441
3700	267.776	793.910	567.412	838.042	-3452.333	-1008.528	14.238
3800	267.776	801.051	573.467	864.820	-3446.295	-942.558	12.956
3900	267.776	808.007	579.392	891.598	-3440.457	-876.746	11.743
4000	267.776	814.787	585.193	918.375	-3434.815	-811.083	10.592

TABLE 2   Thermochemical Data of Selected Chemical Compounds
Sulfur (S), alpha-beta-liquid, mol. wt. $= 32.06$

Enthalply Reference Temperature = T_r = 298.15 K					Standard State Pressure = p^0 = 0.1 MPa		
	J K^{-1}mol^{-1}				kJ mol^{-1}		
T/K	C_p^0	S^0	$-[G^0-H^0(T_r)]/T$	$H^0-H^0(T_r)$	$\Delta_f H^0$	$\Delta_f G^0$	Log K$_f$
0	0.	0.	INFINITE	-4.412	0.	0.	0.
100	12.770	12.522	49.744	-3.722	0.	0.	0.
200	19.368	23.637	34.038	-2.080	0.	0.	0.
250	21.297	28.179	32.422	-1.061	0.	0.	0.
298.15	22.698	32.056	32.056	0.	0.	0.	0.
300	22.744	32.196	32.056	0.042	0.	0.	0.
350	23.870	35.789	32.337	1.208	0.	0.	0.
368.300	24.246	37.015	32.540	1.648	_____ ALPHA <--> BETA _____		
368.300	24.773	38.103	32.540	2.049	TRANSITION		
388.360	25.167	39.427	32.861	2.550	_____ BETA <--> LIQUID _____		
388.360	31.062	43.859	32.861	4.271	TRANSITION		
400	32.162	44.793	33.195	4.639	0.	0.	0.
432.020	53.808	47.431	34.151	5.737	_____ Cp LAMBDA MAXIMUM ___		
432.020	53.806	47.431	34.151	5.737	TRANSITION		
450	43.046	49.308	34.720	6.564	0.	0.	0.
500	37.986	53.532	36.398	8.567	0.	0.	0.
600	34.308	60.078	39.825	12.152	0.	0.	0.
700	32.681	65.241	43.099	15.499	0.	0.	0.
800	31.699	69.530	46.143	18.710	0.	0.	0.
882.117	31.665	72.624	48.467	21.310	---------- FUGACITY = 1 bar ----------		
900	31.665	73.260	48.952	21.877	-53.090	1.079	-0.063
1000	31.665	76.596	51.553	25.043	-51.780	7.028	-0.367
1100	31.665	79.614	53.969	28.209	-50.485	12.846	-0.610
1200	31.665	82.369	56.222	31.376	-49.205	18.546	-0.807
1300	31.665	84.904	58.332	34.542	-47.941	24.141	-0.970
1400	31.665	87.250	60.315	37.709	-46.693	29.639	-1.106
1500	31.665	89.435	62.185	40.875	-45.460	35.048	-1.220

TABLE 2   Thermochemical Data of Selected Chemical Compounds
Sulfur, Monatomic (S), ideal gas, mol. wt. = 32.06

	Enthalply Reference Temperature = T_r = 298.15 K $J K^{-1} mol^{-1}$				Standard State Pressure = p^0 = 0.1 MPa $kJ mol^{-1}$		
T/K	C_p^0	S^0	$-[G^0-H^0(T_r)]/T$	$H^0-H^0(T_r)$	$\Delta_f H^0$	$\Delta_f G^0$	Log K_f
0	0.	0.	INFINITE	-6.657	274.735	274.735	INFINITE
100	21.356	142.891	188.580	-4.569	276.133	263.096	-137.427
200	23.388	158.392	169.994	-2.320	276.740	249.789	-65.238
250	23.696	163.653	168.218	-1.141	276.899	243.031	-50.779
298.15	23.673	167.828	167.828	0.	276.980	236.500	-41.434
300	23.669	167.974	167.828	0.044	276.982	236.248	-41.135
350	23.480	171.610	168.116	1.223	276.995	229.458	-34.245
400	23.233	174.729	166.752	2.391	274.732	222.757	-29.089
450	22.979	177.451	169.571	3.546	273.962	216.297	-25.107
500	22.741	179.859	170.481	4.689	273.102	209.938	-21.932
600	22.338	183.968	172.398	6.942	271.770	197.436	-17.188
700	22.031	187.388	174.302	9.160	270.641	185.138	-13.815
800	21.800	190.314	176.125	11.351	269.621	172.994	-11.295
900	21.624	192.871	177.847	13.522	215.535	162.055	-9.405
1000	21.489	195.142	179.465	15.677	215.834	156.096	-8.154
1100	21.386	197.185	180.984	17.821	216.106	150.109	-7.128
1200	21.307	199.042	182.413	19.955	216.354	144.098	-6.272
1300	21.249	200.745	183.758	22.083	216.579	138.067	-5.548
1400	21.209	202.318	185.029	24.205	216.784	132.020	-4.926
1500	21.186	203.781	186.231	26.325	216.970	125.959	-4.386
1600	21.178	205.148	187.371	28.443	217.140	119.886	-3.914
1700	21.184	206.432	188.455	30.561	217.295	113.803	-3.497
1800	21.203	207.643	189.487	32.680	217.438	107.711	-3.126
1900	21.234	208.790	190.473	34.802	217.570	101.611	-2.793
2000	21.276	209.880	191.417	36.927	217.694	95.505	-2.494
2100	21.327	210.920	192.321	39.058	217.812	89.392	-2.224
2200	21.386	211.913	193.189	41.193	217.925	83.275	-1.977
2300	21.452	212.865	194.024	43.335	218.035	77.152	-1.752
2400	21.523	213.780	194.828	45.484	218.143	71.024	-1.546
2500	21.598	214.660	195.604	47.640	218.250	64.892	-1.356
2600	21.676	215.508	196.353	49.803	218.358	58.755	-1.180
2700	21.756	216.328	197.078	51.975	218.467	52.615	-1.018
2800	21.837	217.121	197.780	54.155	218.579	46.470	-0.867
2900	21.919	217.888	198.460	56.343	218.694	40.321	-0.726
3000	21.999	218.633	199.120	58.538	218.811	34.169	-0.595
3100	22.078	219.355	199.761	60.742	218.931	28.012	-0.472
3200	22.155	220.058	200.384	62.954	219.055	21.851	-0.357
3300	22.230	220.740	200.991	65.173	219.183	15.687	-0.248
3400	22.303	221.405	201.582	67.400	219.313	9.518	-0.146
3500	22.372	222.053	202.157	69.634	219.447	3.346	-0.050
3600	22.439	222.684	202.719	71.874	219.583	-2.830	0.041
3700	22.502	223.300	203.267	74.121	219.721	-9.011	0.127
3800	22.561	223.900	203.802	76.375	219.862	-15.194	0.209
3900	22.618	224.487	204.325	78.634	220.003	-21.382	0.286
4000	22.670	225.061	204.836	80.898	220.146	-27.573	0.360
4100	22.720	225.621	205.336	83.167	220.289	-33.768	0.430
4200	22.766	226.169	205.826	85.442	220.432	-39.966	0.497
4300	22.808	226.705	206.305	87.720	220.575	-46.168	0.561
4400	22.847	227.230	206.775	90.003	220.716	-52.373	0.622
4500	22.884	227.744	207.235	92.290	220.856	-58.581	0.680
4600	22.917	228.247	207.686	94.580	220.993	-64.792	0.736
4700	22.947	228.740	208.129	96.873	221.127	-71.006	0.789
4800	22.974	229.224	208.563	99.169	221.258	-77.223	0.840
4900	22.999	229.698	208.990	101.468	221.386	-83.443	0.890
5000	23.021	230.163	209.409	103.769	221.509	-89.665	0.937
5100	23.040	230.619	209.820	106.072	221.628	-95.890	0.982
5200	23.057	231.066	210.224	108.377	221.741	-102.117	1.026
5300	23.072	231.505	210.622	110.683	221.849	-108.346	1.068
5400	23.085	231.937	211.013	112.991	221.952	-114.577	1.108
5500	23.096	232.361	211.397	115.300	222.048	-120.810	1.147
5600	23.105	232.777	211.775	117.610	222.138	-127.044	1.185
5700	23.112	233.186	212.147	119.921	222.222	-133.281	1.221
5800	23.117	233.588	212.513	122.232	222.299	-139.518	1.256
5900	23.121	233.983	212.874	124.544	222.368	-145.757	1.290
6000	23.124	234.372	213.229	126.857	222.431	-151.997	1.323

TABLE 2 Thermochemical Data of Selected Chemical Compounds
Titanium (Ti), crystal-liquid, mol. wt. = 47.88

T/K	C_p^0	S^0	$-[G^0-H^0(T_r)]/T$	$H^0-H^0(T_r)$	$\Delta_f H^0$	$\Delta_f G^0$	Log K_f
			J K^{-1}mol^{-1}			kJ mol^{-1}	

Enthalply Reference Temperature = T_r = 298.15 K   Standard State Pressure = p^0 = 0.1 MPa

T/K	C_p^0	S^0	$-[G^0-H^0(T_r)]/T$	$H^0-H^0(T_r)$	$\Delta_f H^0$	$\Delta_f G^0$	Log K_f
0	0.	0.	INFINITE	-4.830	0.	0.	0.
100	14.334	8.261	50.955	-4.269	0.	0.	0.
200	22.367	21.227	32.989	-2.352	0.	0.	0.
250	24.074	26.414	31.169	-1.189	0.	0.	0.
298.15	25.238	30.759	30.759	0.	0.	0.	0.
300	25.276	30.915	30.760	0.047	0.	0.	0.
350	26.169	34.882	31.071	1.334	0.	0.	0.
400	26.862	38.423	31.772	2.660	0.	0.	0.
450	27.418	41.620	32.692	4.018	0.	0.	0.
500	27.877	44.534	33.733	5.401	0.	0.	0.
600	28.596	49.683	35.973	8.226	0.	0.	0.
700	29.135	54.134	38.257	11.114	0.	0.	0.
800	29.472	58.039	40.490	14.039	0.	0.	0.
900	30.454	61.561	42.639	17.030	0.	0.	0.
1000	32.074	64.848	44.697	20.151	0.	0.	0.
1100	34.334	68.006	46.673	23.466	0.	0.	0.
1166.000	36.175	70.058	47.938	25.791		ALPHA <--> BETA	
1166.000	29.245	73.636	47.936	29.963		TRANSITION	
1200	29.459	74.479	48.679	30.961	0.	0.	0.
1300	30.175	76.864	50.756	33.941	0.	0.	0.
1400	31.023	79.131	52.702	37.000	0.	0.	0.
1500	32.003	81.304	54.537	40.150	0.	0.	0.
1600	33.115	83.404	56.276	43.405	0.	0.	0.
1700	34.359	85.448	57.932	46.778	0.	0.	0.
1800	35.736	87.451	59.517	50.281	0.	0.	0.
1900	37.244	89.422	61.039	53.929	0.	0.	0.
1939.000	37.868	90.186	61.617	55.394		BETA <--> LIQUID	
1939.000	47.237	97.481	61.617	69.540		TRANSITION	
2000	47.237	98.944	62.734	72.421	0.	0.	0.
2100	47.237	101.249	64.513	77.145	0.	0	0.
2200	47.237	103.446	66.233	81.869	0.	0.	0.
2300	47.237	105.546	67.897	86.592	0.	0.	0.
2400	47.237	107.557	69.508	91.316	0.	0.	0.
2500	47.237	109.485	71.069	96.040	0.	0.	0.
2600	47.237	111.338	72.582	100.764	0.	0.	0.
2700	47.237	113.120	74.051	105.487	0.	0.	0.
2800	47.237	114.838	75.477	110.211	0.	0.	0.
2900	47.237	116.496	76.863	114.935	0.	0.	0.
3000	47.237	118.097	78.211	119.659	0.	0.	0.
3100	47.237	119.646	79.523	124.382	0.	0.	0.
3200	47.237	121.146	80.800	129.106	0.	0.	0.
3300	47.237	122.600	82.045	133.830	0.	0.	0.
3400	47.237	124.010	83.259	138.554	0.	0.	0.
3500	47.237	125.379	84.443	143.277	0.	0.	0.
3600	47.237	126.710	85.598	148.001	0.	0.	0.
3630.956	47.237	127.114	85.951	0.149		------ FUGACITY = 1 bar ------	
3700	47.237	128.004	86.727	152.725	-409.098	7.788	-0.110
3800	47.237	129.264	87.830	157.449	-407.864	19.039	-0.262
3900	47.237	130.491	88.908	162.172	-406.685	30.258	-0.405
4000	47.237	131.687	89.963	166.896	-405.560	41.447	-0.541
4100	47.237	132.853	90.995	171.620	-404.488	52.609	-0.670
4200	47.237	133.991	92.005	176.343	-403.467	63.745	-0.793
4300	47.237	135.103	92.994	181.067	-402.495	74.857	-0.909
4400	47.237	136.189	93.964	185.791	-401.570	85.948	-1.020
4500	47.237	137.250	94.914	190.515	-400.691	97.018	-1.126

TABLE 2   Thermochemical Data of Selected Chemical Compounds
Titanium (Ti), ideal gas, mol. wt. = 47.88

T/K	C_p^o	S^o	$-[G^o-H^o(T_r)]/T$	$H^o-H^o(T_r)$	$\Delta_f H^o$	$\Delta_f G^o$	Log K_f
0	0.	0.	INFINITE	-7.539	470.920	470.920	INFINITE
100	26.974	151.246	203.383	-5.214	472.684	458.386	-239.436
200	26.487	170.126	182.593	-2.493	473.488	443.708	-115.885
250	25.355	175.913	180.704	-1.198	473.620	436.245	-91.148
298.15	24.430	180.297	180.297	0.	473.629	429.044	-75.167
300	24.399	180.448	180.297	0.045	473.627	428.768	-74.655
350	23.661	184.151	180.591	1.246	473.541	421.297	-62.875
400	23.104	187.272	181.236	2.414	473.383	413.843	-54.042
450	22.683	189.968	182.060	3.558	473.169	406.413	-47.175
500	22.360	192.340	182.972	4.684	472.912	399.009	-41.684
600	21.913	196.374	184.881	6.896	472.299	384.284	-33.455
700	21.632	199.729	186.769	9.072	471.587	369.670	-27.585
800	21.454	202.605	188.573	11.226	470.816	355.163	-23.190
900	21.353	205.126	190.275	13.366	469.965	340.756	-19.777
1000	21.323	207.373	191.875	15.499	468.977	326.451	-17.052
1100	21.362	209.407	193.377	17.633	467.795	312.254	-14.828
1200	21.474	211.270	194.792	19.774	462.442	298.293	-12.984
1300	21.657	212.996	196.127	21.930	461.617	284.647	-11.437
1400	21.910	214.609	197.390	24.107	460.736	271.066	-10.114
1500	22.228	216.132	198.589	26.314	459.793	257.551	-8.969
1600	22.604	217.578	199.731	28.555	458.779	244.101	-7.969
1700	23.029	218.961	200.822	30.836	457.688	230.717	-7.089
1800	23.497	220.290	201.867	33.162	456.510	217.399	-6.309
1900	23.999	221.574	202.870	35.537	455.237	204.149	-5.612
2000	24.529	222.818	203.837	37.963	439.171	191.423	-4.999
2100	25.080	224.028	204.769	40.443	436.927	179.091	-4.455
2200	25.648	225.208	205.672	42.979	434.740	166.864	-3.962
2300	26.228	226.361	206.546	45.573	432.610	154.736	-3.514
2400	26.819	227.489	207.396	48.225	430.538	142.699	-3.106
2500	27.416	228.596	208.221	50.937	428.526	130.748	-2.732
2600	28.019	229.683	209.026	53.709	426.574	118.875	-2.388
2700	28.626	230.752	209.811	56.541	424.683	107.077	-2.072
2800	29.235	231.804	210.578	59.434	422.852	95.347	-1.779
2900	29.846	232.841	211.328	62.388	421.082	83.682	-1.507
3000	30.456	233.863	212.062	65.403	419.374	72.077	-1.255
3100	31.065	234.871	212.781	68.479	417.726	60.528	-1.020
3200	31.671	235.867	213.487	71.616	416.139	49.031	-0.800
3300	32.273	236.851	214.180	74.813	414.612	37.582	-0.595
3400	32.870	237.823	214.861	78.071	413.146	26.179	-0.402
3500	33.460	238.785	215.531	81.367	411.739	14.818	-0.221
3600	34.042	239.736	216.190	84.762	410.390	3.497	-0.051
3630.956	34.219	240.028	216.392	85.819	-------FUGACITY = 1 bar -----		
3700	34.613	240.676	216.840	88.195	0.	0.	0.
3800	35.173	241.607	217.479	91.685	0.	0.	0.
3900	35.721	242.527	218.110	95.229	0.	0.	0.
4000	36.254	243.439	218.731	98.828	0.	0.	0.
4100	36.772	244.340	219.345	102.480	0.	0.	0.
4200	37.273	245.232	219.951	106.182	0.	0.	0.
4300	37.757	246.115	220.549	109.934	0.	0.	0.
4400	38.223	246.988	221.140	113.733	0.	0.	0.
4500	38.670	247.852	221.724	117.578	0.	0.	0.
4600	39.097	248.707	222.301	121.466	0.	0.	0.
4700	39.505	249.552	222.872	125.397	0.	0.	0.
4800	39.892	250.388	223.437	129.367	0.	0.	0.
4900	40.259	251.214	223.995	133.374	0.	0.	0.
5000	40.606	252.031	224.548	137.418	0.	0.	0.
5100	40.933	252.839	225.095	141.495	0.	0.	0.
5200	41.240	253.637	225.636	145.604	0.	0.	0.
5300	41.528	254.425	226.172	149.742	0.	0.	0.
5400	41.796	255.204	226.702	153.909	0.	0.	0.
5500	42.046	255.973	227.227	158.101	0.	0.	0.
5600	42.277	256.733	227.747	162.317	0.	0.	0.
5700	42.490	257.483	228.263	166.556	0.	0.	0.
5800	42.686	258.224	228.773	170.815	0.	0.	0.
5900	42.866	258.955	229.278	175.092	0.	0.	0.
6000	43.029	259.677	229.779	179.387	0.	0.	0.

Enthalpy Reference Temperature = T_r = 298.15 K   Standard State Pressure = p^o = 0.1 MPa
J K^{-1}mol^{-1} kJ mol^{-1}

TABLE 3 Thermochemical Data of Species Included in Reaction List of Appendix B

Species		$\Delta H_{f\,298}$	S_{298}	$C_{p\,300}$	$C_{p\,400}$	$C_{p\,500}$	$C_{p\,600}$	$C_{p\,800}$	$C_{p\,1000}$
		kJ mol^{-1}				J K^{-1}mol^{-1}			
Ar	Argon	0.00	154.85	20.79	20.79	20.79	20.79	20.79	20.79
H	Hydrogen				Table 2				
O	Oxygen				Table 2				
OH	Hydroxyl				Table 2				
H$_2$	Hydrogen				Table 2				
O$_2$	Oxygen				Table 2				
H$_2$O	Water				Table 2				
HO$_2$	Hydroperoxyl	12.55			Table 2				
H$_2$O$_2$	Hydrogen peroxide	-136.11	232.99	43.22	48.45	52.55	55.69	59.83	62.84
CO	Carbon monoxide				Table 2				
CO$_2$	Carbon dioxide				Table 2				
HCO	Formyl	43.51	224.65	34.63	36.53	38.73	40.97	44.96	48.06
CH$_2$O	Formaldehyde	-115.90	218.95	35.46	39.26	43.77	48.22	55.98	62.00
HCOOH	Formic acid	-387.50	248.01	44.82	53.52	60.85	67.03	76.76	83.96
CH$_2$OH	hydroxy Methyl	-17.17	246.35	47.38	54.14	60.16	65.37	73.37	78.62
CH$_3$O	Methoxy	16.30	228.48	37.99	45.13	52.02	58.47	69.58	77.81
CH$_3$OH	Methanol	-201.10	239.65	43.98	51.90	59.64	66.98	79.80	89.53
C	Carbon				Table 2				
CH	Methylidyne	594.17	182.93	29.09	29.30	29.49	29.77	30.82	32.56
CH$_2$	Methylene	386.99	195.48	34.50	35.78	37.15	38.60	41.57	44.21
^{1}CH$_2$	singlet Methylene	424.72	188.71	33.79	34.72	36.00	37.58	41.20	44.38
CH$_3$	Methyl	145.69	194.17	38.75	42.04	45.25	48.29	53.93	58.95
CH$_4$	Methane				Table 2				
C$_2$H	Ethynyl	564.90	207.37	37.25	40.29	42.78	44.87	48.27	50.94
C$_2$H$_2$	Acetylene				Table 2				
C$_2$H$_3$	Vinyl	286.25	231.52	40.03	46.81	53.48	59.88	71.04	78.43
C$_2$H$_4$	Ethene				Table 2				
C$_2$H$_5$	Ethyl	117.23	242.64	47.37	56.90	66.73	76.54	94.48	106.71
C$_2$H$_6$	Ethane	-83.86	228.98	52.61	65.66	77.89	89.10	108.05	122.59
HCCO	Ketyl	177.60	254.16	52.93	56.34	59.53	62.44	67.25	70.43
HCCOH	Ethynol	85.49	245.64	55.30	61.84	67.62	72.59	80.10	84.94
CH$_2$CO	Ketene	-51.87	241.81	52.01	59.30	65.54	70.76	78.63	84.70
CH$_3$CO	Acetyl	-22.60	266.71	51.95	60.40	68.32	75.55	87.60	96.51
CH$_2$CHO	Formyl methylene	14.71	258.90	54.31	65.84	75.29	83.06	94.91	103.53
CH$_3$CHO	Acetaldehyde	-165.32	263.82	55.43	66.41	76.61	85.84	101.11	112.48
C$_2$H$_4$O	Oxirane	-52.64	243.01	48.16	62.40	75.41	86.31	102.97	114.96
C$_3$H$_2$		542.29	271.22	62.47	67.36	70.74	73.41	78.33	82.59
C$_3$H$_3$	Propargyl	347.47	257.28	66.27	74.24	81.48	87.89	98.02	104.62
C$_3$H$_4$	Allene	199.31	242.45	59.61	70.98	81.41	90.83	106.47	117.98
pC$_3$H$_4$	Propyne	191.51	246.45	60.73	71.36	81.19	90.14	105.27	116.72
cC$_3$H$_4$	cyclo-Propene	284.53	242.50	53.39	66.39	78.19	88.71	105.77	117.76
C$_3$H$_5$	Allyl	161.70	270.91	67.24	81.78	95.04	106.82	125.50	137.60
CH$_3$CCH$_2$	iso-Propenyl	255.62	289.77	64.66	76.89	88.75	99.96	119.25	132.23
CHCHCH$_3$	Propenyl	270.94	287.65	65.03	77.65	89.70	100.92	119.89	132.54
C$_3$H$_6$	Propene	20.46	257.40	64.70	80.61	95.10	107.96	128.80	144.44
iC$_3$H$_7$	iso-Propyl	76.15	251.50	75.35	93.09	108.88	122.61	144.30	160.74
nC$_3$H$_7$	n-Propyl	94.56	268.37	75.74	93.16	108.72	122.28	143.81	160.22
C$_3$H$_8$	Propane	-103.85	270.18	73.94	94.09	112.40	128.63	154.85	174.60
N	Nitrogen				Table 2				
N$_2$	Nitrogen				Table 2				

(continues)

TABLE 3   (*continued*)

Species		$\Delta H_{f\,298}$	S_{298}	$C_{p\,300}$	$C_{p\,400}$	$C_{p\,500}$	$C_{p\,600}$	$C_{p\,800}$	$C_{p\,1000}$	$C_{p\,1500}$
		kJ mol⁻¹					J K⁻¹mol⁻¹			
NO	Nitric oxide					Table 2				
NO₂	Nitrogen dioxide					Table 2				
N₂O	Nitrous oxide	82.05	219.96	38.70	42.68	45.83	48.39	52.24	54.87	58.41
HNO	Nitrosyl hydride	106.26	220.72	34.67	36.77	39.09	41.29	45.04	47.92	52.35
HONO	Nitrous acid	-76.74	249.42	45.47	51.36	56.12	59.90	65.42	69.24	74.96
NH	Imidogen	356.91	181.22	26.16	27.04	27.87	28.64	30.04	31.27	33.75
NH₂	Amidogen	129.05	194.99	30.90	33.06	35.10	37.01	40.49	43.54	49.61
NH₃	Ammonia					Table 2				
NNH	Diazenyl	249.51	224.50	37.78	39.63	41.34	42.91	45.69	48.01	52.19
N₂H₂	Diazene	212.97	218.60	36.62	40.94	45.51	49.71	56.68	62.05	70.66
N₂H₃	Hydrazyl	153.92	228.54	44.33	51.21	57.68	63.46	72.56	79.22	90.24
N₂H₄	Hydrazine	95.35	238.72	51.02	61.70	70.54	77.56	88.20	96.36	110.32
NO₃	Nitrogen trioxide	71.13	252.62	47.11	55.93	62.60	67.38	73.27	76.48	80.01
HNO₃	Nitric acid	-134.31	266.40	53.53	63.19	70.84	76.77	85.04	90.43	97.98
Cl	Chlorine	121.30	165.19	21.85	22.47	22.74	22.78	22.55	22.23	21.65
Cl₂	Chlorine	0.00	223.08	33.98	35.30	36.06	37.55	37.11	37.44	37.95
HCl	Hydrogen chloride	-92.31	186.90	29.14	29.18	29.30	29.58	30.49	31.63	34.06
ClO	Chlorine oxide	101.22	226.65	31.58	33.23	34.43	35.27	36.30	36.88	37.65
HOCl	Hydrochlorous acid	-74.48	236.50	37.34	40.08	42.31	44.05	46.60	48.54	51.98
COCl	Carbonyl chloride	-16.74	265.97	45.12	47.27	48.91	50.30	52.48	54.00	56.02
O₃	Ozone					Table 2				
S	Sulfur					Table 2				
SO	Sulfur oxide					Table 2				
SO₂	Sulfur dioxide					Table 2				
SO₃	Sulfur trixoxide					Table 2				
SH	Mercapto	139.33	195.63	32.43	31.71	31.28	31.23	31.84	32.81	34.92
H₂S	Hydrogen sulfide	-20.50	205.76	34.21	35.58	37.19	38.94	42.52	45.79	51.48
HSO	Mercapto-oxy	-22.59	241.84	37.74	41.55	44.89	47.53	51.13	53.26	55.81
HOS	hydroxy Sulfur	0.00	239.12	36.44	39.41	41.84	43.72	46.36	48.24	51.59
HSO₂		-141.42	266.68	49.96	57.24	62.72	66.82	72.30	75.52	79.41
HOSO		-241.42	270.62	49.66	56.19	60.92	64.27	68.62	71.38	75.69
HSOH		-119.24	245.43	45.31	51.13	56.36	60.50	66.19	69.54	73.64
H₂SO		-47.28	239.58	39.87	46.57	52.97	58.37	66.19	71.13	77.15
HOSHO		-269.87	269.78	56.82	66.27	73.85	79.54	87.11	91.84	98.49
HOSO₂		-391.20	295.89	70.04	78.49	84.22	88.16	92.97	95.98	100.50

Appendix *B*

Specific Reaction Rate Constants

The rate constant data for the reactions that follow in this appendix are presented as sets of chemical mechanisms for describing high-temperature oxidation of various fuels, increasing in complexity from hydrogen to ethane. The order of their presentation follows the hierarchical approach to combustion modeling described by Westbrook and Dryer (*Prog. Energy Combust. Sci.*, **10**, 1, 1084), and includes the following tables.

Table 1. H_2/O_2 Mechanism
Table 2. $CO/H_2/O_2$ Mechanism
Table 3. $CH_2O/CO/H_2/O_2$ Mechanism
Table 4. $CH_3OH/CH_2O/CO/H_2/O_2$ Mechanism
Table 5. $CH_4/CH_3OH/CH_2O/CO/H_2/O_2$ Mechanism
Table 6. $C_2H_6/CH_4/CH_3OH/CH_2O/CO/H_2/O_2$ Mechanism

References for the origin of each mechanism are listed at the end of each table. The reader should refer to these references for the original source of each rate constant quoted. The backward rate constant at a given temperature is determined through the equilibrium constant at the temperature. The units are in cm^3 mol s kJ for the expression $k = AT^n \exp(-E/RT)$.

Complete mechanisms for the high temperature oxidation of propane and larger hydrocarbons are available in the literature (e.g., Warnatz., J., *Twenty-fourth*

TABLE 1 H_2/O_2 Mechanism[a]

		A	n	E
H_2–O_2 Chain Reactions				
1.1	$H + O_2 \rightleftharpoons O + OH$	1.94×10^{14}	0.00	68.78
1.2	$O + H_2 \rightleftharpoons H + OH$	5.08×10^{4}	2.67	26.32
1.3	$OH + H_2 \rightleftharpoons H + H_2O$	2.16×10^{8}	1.51	14.35
1.4	$O + H_2O \rightleftharpoons OH + OH$	2.95×10^{6}	2.02	56.07
H_2–O_2 Dissociation/Recombination Reactions				
1.5	$H_2 + M \rightleftharpoons H + H + M$	4.57×10^{19}	−1.40	436.73
	$H_2 + Ar \rightleftharpoons H + H + Ar$	5.89×10^{18}	−1.10	436.73
1.6	$O + O + M \rightleftharpoons O_2 + M$	6.17×10^{15}	−0.50	0.00
	$O + O + Ar \rightleftharpoons O_2 + Ar$	1.91×10^{13}	0.00	−7.49
1.7	$O + H + M \rightleftharpoons OH + M$	4.72×10^{18}	−1.00	0.00
1.8	$H + OH + M \rightleftharpoons H_2O + M$	2.24×10^{22}	−2.00	0.00
	$H + OH + Ar \rightleftharpoons H_2O + Ar$	8.32×10^{21}	−2.00	0.00
HO_2 Reactions				
1.9	$H + O_2 + M \rightleftharpoons HO_2 + M$	6.70×10^{19}	−1.42	0.00
	$k_{H_2} = 2.5 \times k_M, k_{H_2O} = 12 \times k_M,$			
	$k_{CO} = 1.9 \times k_M, k_{CO_2} = 3.8 \times k_M$			
	$H + O_2 + Ar \rightleftharpoons HO_2 + Ar$	1.51×10^{15}	0.00	−4.18
	$H + O_2 \rightleftharpoons HO_2$	4.52×10^{13}	0.00	0.00
1.10	$HO_2 + H \rightleftharpoons H_2 + O_2$	6.62×10^{13}	0.00	8.91
1.11	$HO_2 + H \rightleftharpoons OH + OH$	1.69×10^{14}	0.00	3.64
1.12	$HO_2 + O \rightleftharpoons OH + O_2$	1.75×10^{13}	0.00	−1.67
1.13	$HO_2 + OH \rightleftharpoons H_2O + O_2$	1.90×10^{16}	−1.00	0.00
H_2O_2 Reactions				
1.14[b]	$HO_2 + HO_2 \rightleftharpoons H_2O_2 + O_2$	4.20×10^{14}	0.00	50.12
		1.30×10^{11}	0.00	−6.82
1.15[c]	$H_2O_2 + M \rightleftharpoons OH + OH + M$	1.20×10^{17}	0.00	190.37
	$H_2O_2 + Ar \rightleftharpoons OH + OH + Ar$	1.91×10^{16}	0.00	179.91
	$H_2O_2 \rightleftharpoons OH + OH$	3.00×10^{14}	0.00	202.51
	$F_c = 0.5$			
1.16	$H_2O_2 + H \rightleftharpoons H_2O + OH$	1.00×10^{13}	0.00	15.02
1.17	$H_2O_2 + H \rightleftharpoons H_2 + HO_2$	4.82×10^{13}	0.00	33.26
1.18	$H_2O_2 + O \rightleftharpoons OH + HO_2$	9.64×10^{6}	2.00	16.61
1.19[b]	$H_2O_2 + OH \rightleftharpoons H_2O + HO_2$	1.00×10^{12}	0.00	0.00
		5.80×10^{14}	0.00	40.00

[a] Reaction rates in cm^3 mol kJ units, $k = AT^n \exp(-E/RT)$

[b] Rate represented by sum of two Arrhenius expressions.

[c] The fall-off behavior of this reaction is expressed as $k = [k_0 k_\infty/(k_0 + k_\infty/M)] \times F$, and $\log(F) = \log(F_c)/[1 + \{\log(k_0 \times M/k_\infty)\}^2]$.

Source. Kim, T. J., Yetter, R. A., and Dryer, F. L., New Results on Moist CO Oxidation: High Pressure, High Temperature Experiments and Comprehensive Kinetic Modeling, *Twenty-Fifth Symposium (International) on Combustion*, The Combustion Institute, Pittsburgh, PA, 1994, pp. 759–766.

TABLE 2   CO/H$_2$/O$_2$ Mechanism[a]

		A	n	E
CO and CO$_2$ Reactions				
2.1	CO + O + M $\rightleftharpoons$ CO$_2$ + M	2.51×10^{13}	0.00	−19.00
	CO + O + Ar $\rightleftharpoons$ CO$_2$ + Ar	2.19×10^{13}	0.00	−19.00
2.2	CO + O$_2$ $\rightleftharpoons$ CO$_2$ + O	2.53×10^{13}	0.00	199.53
2.3	CO + OH $\rightleftharpoons$ CO$_2$ + H	1.50×10^{7}	1.30	−3.20
2.4	CO + HO$_2$ $\rightleftharpoons$ CO$_2$ + OH	6.02×10^{13}	0.00	96.02
HCO Reactions				
2.5	HCO + M $\rightleftharpoons$ H + CO + M	1.86×10^{17}	−1.00	71.13
	$k_{H_2} = k_{CO} = 1.9 \times k_M, k_{H_2O} = 8 \times k_M,$			
	$k_{CO_2} = 2.8 \times k_M (k_{CH_2O} = k_{CH_3OH} = k_{CH_4} = 8 \times k_M)$			
	HCO + Ar $\rightleftharpoons$ H + CO + Ar	1.86×10^{17}	−1.00	71.13
2.6	HCO + O$_2$ $\rightleftharpoons$ CO + HO$_2$	7.58×10^{12}	0.00	1.72
2.7	HCO + H $\rightleftharpoons$ CO + H$_2$	7.23×10^{13}	0.00	0.00
2.8	HCO + O $\rightleftharpoons$ CO + OH	3.00×10^{13}	0.00	0.00
2.9	HCO + O $\rightleftharpoons$ CO$_2$ + H	1.00×10^{13}	0.00	0.00
2.10	HCO + OH $\rightleftharpoons$ CO + H$_2$O	3.00×10^{13}	0.00	0.00
2.11	HCO + HO$_2$ $\rightleftharpoons$ CO$_2$ + OH + H	3.00×10^{13}	0.00	0.00
2.12	HCO + HCO $\rightleftharpoons$ H$_2$ + CO + CO	3.00×10^{12}	0.00	0.00

[a]Reaction rates in cm^3 mol kJ units, $k = AT^n \exp(-E/RT)$
Source.   Kim, T. J., Yetter, R. A., and Dryer, F. L., New Results on Moist CO Oxidation: High Pressure, High Temperature Experiments and Comprehensive Kinetic Modeling, *Twenty-Fifth Symposium (International) on Combustion*, The Combustion Institute, Pittsburgh, PA, 1994, pp. 759–766.

Symposium (International) on Combustion, The Combustion Institute, pp. 553–579, 1992; and Ranzi, E., Sogaro, A., Gaffuri, P., Pennati, G., Westbrook, C. K., and Pitz., W. J., *Combustion and Flame*, **99**, 201, 1994). Because of space limitations, only selected reactions for propane oxidation are presented here (Table 7).

Table 7. Selected Reactions of a C$_3$H$_8$ Oxidation Mechanism

Significant progress has also been made on the development of low and intermediate temperature hydrocarbon oxidation mechanisms, and the reader is again referred to the literature.

Rate constant data for reactions of post combustion gases including nitrogen oxides, hydrogen chloride, ozone, and sulfur oxides are presented in Tables 8–11.

Table 8. N$_x$O$_y$/CO/H$_2$/O$_2$ Mechanism
Table 9. HCl/N$_x$O$_y$/CO/H$_2$/O$_2$ Mechanism
Table 10. O$_3$/N$_x$O$_y$/CO/H$_2$/O$_2$ Mechanism
Table 11. SO$_x$/N$_x$O$_y$/CO/H$_2$/O$_2$ Mechanism

TABLE 3   CH$_2$O/CO/H$_2$/O$_2$ Mechanism[a]

	A	n	E
CH$_2$O Reactions			
3.1 CH$_2$O + M $\rightleftharpoons$ HCO + H + M	4.00×10^{23}	-1.66	382.71
3.2 CH$_2$O + M $\rightleftharpoons$ CO + H$_2$ + M	8.32×10^{15}	0.00	290.96
3.3 CH$_2$O + H $\rightleftharpoons$ HCO + H$_2$	5.18×10^{7}	1.66	7.66
3.4 CH$_2$O + O $\rightleftharpoons$ HCO + OH	1.82×10^{13}	0.00	12.89
3.5 CH$_2$O + OH $\rightleftharpoons$ HCO + H$_2$O	3.43×10^{9}	1.18	-1.88
3.6 CH$_2$O + O$_2$ $\rightleftharpoons$ HCO + HO$_2$	2.00×10^{13}	0.00	163.18
3.7 CH$_2$O + HO$_2$ $\rightleftharpoons$ HCO + H$_2$O$_2$	1.47×10^{13}	0.00	63.60
3.8 HCO + HCO $\rightleftharpoons$ CH$_2$O + CO	1.82×10^{13}	0.00	0.00

[a]Reaction rates in cm^3 mol kJ units, $k = AT^n \exp(-E/RT)$ *Source.* S. Hochgreb and F. L. Dryer (1992). A Comprehensive Study on CH$_2$O Oxidation Kinetics, *Combust. Flame*, **91**, 257–284.

While Table 8 includes reactions for the formation of thermal NO, it does not include those for prompt NO. Mechanisms and reaction rate data for prompt NO formation and various methods for the reduction of NO have been described by Miller and Bowman (*Prog. Energy Combust. Sci.*, **15**, 287, 1989). The GRIMECH reaction set (http://www.me.berkeley.edu/gri_mech) is an example of a high temperature methane oxidation mechanism that includes both thermal and prompt NO production.

Critical reviews of reaction rate data are constantly appearing in the literature and are an important source for mechanism construction. Some examples of recent reviews are given below.

1. "Chemical Kinetic Data for Combustion Chemistry. Part 1. Methane and Related Compounds," W. Tsang and R. F. Hampson, *J. Phys. Chem. Ref. Data*, **15**, 1087, 1986. "Part 2. Methanol," W. Tsang, *J. Phys. Chem. Ref. Data*, **16**, 471, 1987. "Part 3. Propane," W. Tsang, *J. Phys. Chem. Ref. Data*, **17**, 887, 1988. "Part 4. Isobutane," W. Tsang, *J. Phys. Chem. Ref. Data*, **19**, 1, 1990. "Part 5. Propene," *J. Phys. Chem. Ref. Data*, **20**, 221, 1991.
2. "Chemical Kinetic Data Base for Propellant Combustion I. Reactions Involving NO, NO$_2$, HNO, HNO$_2$, HCN, and N$_2$O," W. Tsang and J. T. Herron, *J. Phys. Chem. Ref. Data*, **20**, 609, 1991. "II. Reactions Involving CN, NCO, and HNCO," W. Tsang, *J. Phys. Chem. Ref. Data*, **21**, 750, 1992.
3. "Evaluated Kinetic Data for Combustion Modeling," D. L. Baulch, C. J. Cobos, R. A. Cox, C. Esser, P. Frank, Th. Just, J. A. Kerr, M. J. Pilling, J. Troe, R. W. Walker, and J. Warnatz, *J. Phys. Chem. Ref. Data*, **21**, 411, 1992. Supplement I, D. L. Baulch, C. Cobos, R. A. Cox, P. Frank, G. Hayman, Th. Just, J. A. Kerr, T. Murrells, M. J. Murrells, M. J. Pilling, J. Troe, R. W. Walker, and J. Warnatz, *J. Phys. Chem. Ref. Data*, **23**, 847, 1994.

TABLE 4 $CH_3OH/CH_2O/CO/H_2/O_2$ Mechanism[a]

		A	n	E
HCOOH Reactions				
4.1	$HCOOH + M \rightleftharpoons CO + H_2O + M$	2.09×10^{14}	0.00	169.00
4.2	$HCOOH + M \rightleftharpoons CO_2 + H_2 + M$	1.35×10^{15}	0.00	253.60
4.3	$HCOOH + OH \rightleftharpoons H_2O + CO_2 + H$	3.00×10^{11}	0.00	0.00
CH$_2$OH Reactions				
4.4	$CH_2OH + M \rightleftharpoons CH_2O + H + M$	5.22×10^{37}	-6.54	170.46
4.5	$CH_2OH + H \rightleftharpoons CH_2O + H_2$	6.00×10^{12}	0.00	0.00
4.6	$CH_2OH + O \rightleftharpoons CH_2O + OH$	4.20×10^{13}	0.00	0.00
4.7	$CH_2OH + OH \rightleftharpoons CH_2O + H_2O$	2.40×10^{13}	0.00	0.00
4.8	$CH_2OH + O_2 \rightleftharpoons CH_2O + HO_2$	2.41×10^{14}	0.00	21.00
4.9	$CH_2OH + HO_2 \rightleftharpoons CH_2O + H_2O_2$	1.20×10^{13}	0.00	0.00
4.10	$CH_2OH + HO_2 \rightleftharpoons HCOOH + OH + H$	3.00×10^{13}	0.00	0.00
4.11	$CH_2OH + HO_2 \rightleftharpoons HCOOH + H_2O$	3.00×10^{13}	0.00	0.00
4.12	$CH_2OH + HCO \rightleftharpoons CH_2O + CH_2O$	1.80×10^{14}	0.00	0.00
4.13	$CH_2OH + CH_2OH \rightleftharpoons HOC_2H_4OH$	6.00×10^{12}	0.00	0.00
CH$_3$O Reactions				
4.14	$CH_3O + M \rightleftharpoons CH_2O + H + M$	8.30×10^{17}	-1.20	64.85
4.15	$CH_3O + H \rightleftharpoons CH_2O + H_2$	2.00×10^{13}	0.00	0.00
4.16	$CH_3O + O \rightleftharpoons CH_2O + OH$	6.00×10^{12}	0.00	0.00
4.17	$CH_3O + OH \rightleftharpoons CH_2O + H_2O$	1.80×10^{13}	0.00	0.00
4.18[b]	$CH_3O + O_2 \rightleftharpoons CH_2O + HO_2$	9.03×10^{13}	0.00	50.12
		2.20×10^{10}	0.00	7.31
4.19	$CH_3O + HO_2 \rightleftharpoons CH_2O + H_2O_2$	3.00×10^{11}	0.00	0.00
CH$_3$OH Reactions				
4.20	$CH_3OH + M \rightleftharpoons CH_2OH + H + M$	4.23×10^{16}	0.00	318.40
	$CH_3OH \rightleftharpoons CH_2OH + H$	1.90×10^{17}	0.00	384.00
4.21	$CH_3OH + H \rightleftharpoons CH_2OH + H_2$	3.20×10^{13}	0.00	25.50
4.22	$CH_3OH + H \rightleftharpoons CH_3O + H_2$	8.00×10^{12}	0.00	25.50
4.23	$CH_3OH + O \rightleftharpoons CH_2OH + OH$	3.88×10^{5}	2.50	12.89
4.24	$CH_3OH + OH \rightleftharpoons CH_3O + H_2O$	1.77×10^{4}	2.65	-3.69
4.25	$CH_3OH + OH \rightleftharpoons CH_2OH + H_2O$	1.77×10^{4}	2.65	-3.69
4.26	$CH_3OH + O_2 \rightleftharpoons CH_2OH + HO_2$	2.05×10^{13}	0.00	187.86
4.27	$CH_3OH + HO_2 \rightleftharpoons CH_2OH + H_2O_2$	4.00×10^{13}	0.00	81.17
4.28	$CH_2OH + HCO \rightleftharpoons CH_3OH + CO$	1.20×10^{14}	0.00	0.00
4.29	$CH_3O + HCO \rightleftharpoons CH_3OH + CO$	9.00×10^{13}	0.00	0.00
4.30	$CH_3OH + HCO \rightleftharpoons CH_2OH + CH_2O$	9.63×10^{3}	2.90	54.85
4.31	$CH_2OH + CH_2OH \rightleftharpoons CH_3OH + CH_2O$	3.00×10^{12}	0.00	0.00
4.32	$CH_3O + CH_2OH \rightleftharpoons CH_3OH + CH_2O$	2.40×10^{13}	0.00	0.00
4.33	$CH_3O + CH_3O \rightleftharpoons CH_3OH + CH_2O$	6.00×10^{13}	0.00	0.00
4.34	$CH_3OH + CH_3O \rightleftharpoons CH_3OH + CH_2OH$	3.00×10^{11}	0.00	16.99

[a]Reaction rates in cm^3 mol kJ units, $k = AT^n \exp(-E/RT)$

[b]Rate represented by sum of two Arrhenius expressions.

Source. Held, T., The Oxidation of Methanol, Isobutene, and Methyl *tertiary*-Butyl Ether, No. 1978-T, Ph.D. Dissertation, Princeton University, Princeton, NJ, 1993. See also Held, T. J., and Dryer, F. L., A Comprehensive Mechanism for Methanol Oxidation, *Int. J. Chem. Kinet.*, 1996.

TABLE 5   CH$_4$/CH$_3$OH/CH$_2$O/CO/H$_2$/O$_2$ Mechanism^a

		A	n	E
C Reactions				
5.1	C + OH $\rightleftharpoons$ CO + H	5.00×10^{13}	0.00	0.00
5.2	C + O$_2$ $\rightleftharpoons$ CO + O	2.00×10^{13}	0.00	0.00
CH Reactions				
5.3	CH + H $\rightleftharpoons$ C + H$_2$	1.50×10^{14}	0.00	0.00
5.4	CH + O $\rightleftharpoons$ CO + H	5.70×10^{13}	0.00	0.00
5.5	CH + OH $\rightleftharpoons$ HCO + H	3.00×10^{13}	0.00	0.00
5.6	CH + O$_2$ $\rightleftharpoons$ HCO + O	3.30×10^{13}	0.00	0.00
5.7	CH + H$_2$O $\rightleftharpoons$ CH$_2$O + H	1.17×10^{15}	-0.75	0.00
5.8	CH + CO$_2$ $\rightleftharpoons$ HCO + CO	3.40×10^{12}	0.00	2.88
CH$_2$ Reactions				
5.9	CH$_2$ + H $\rightleftharpoons$ CH + H$_2$	1.00×10^{18}	-1.56	0.00
5.10	CH$_2$ + O $\rightleftharpoons$ CO + H + H	5.00×10^{13}	0.00	0.00
5.11	CH$_2$ + O $\rightleftharpoons$ CO + H$_2$	3.00×10^{13}	0.00	0.00
5.12	CH$_2$ + OH $\rightleftharpoons$ CH$_2$O + H	2.50×10^{13}	0.00	0.00
5.13	CH$_2$ + OH $\rightleftharpoons$ CH + H$_2$O	1.13×10^{7}	2.00	12.55
5.14	CH$_2$ + O$_2$ $\rightleftharpoons$ CO + H + OH	8.60×10^{10}	0.00	-2.10
5.15	CH$_2$ + O$_2$ $\rightleftharpoons$ CO + H$_2$O	1.90×10^{10}	0.00	-4.18
5.16	CH$_2$ + O$_2$ $\rightleftharpoons$ CO$_2$ + H + H	1.60×10^{13}	0.00	4.18
5.17	CH$_2$ + O$_2$ $\rightleftharpoons$ CO$_2$ + H$_2$	6.90×10^{11}	0.00	2.10
5.18	CH$_2$ + O$_2$ $\rightleftharpoons$ HCO + OH	4.30×10^{10}	0.00	-2.10
5.19	CH$_2$ + O$_2$ $\rightleftharpoons$ CH$_2$O + O	5.00×10^{13}	0.00	37.65
5.20	CH$_2$ + HO$_2$ $\rightleftharpoons$ CH$_2$O + OH	1.81×10^{13}	0.00	0.00
5.21	CH$_2$ + CO$_2$ $\rightleftharpoons$ CH$_2$O + CO	1.10×10^{11}	0.00	4.18
^{1}CH$_2$ Reactions				
5.22	^{1}CH$_2$ + M $\rightleftharpoons$ CH$_2$ + M	1.00×10^{13}	0.00	0.00
	$k_H = 0 \times k_M$			
5.23	^{1}CH$_2$ + H $\rightleftharpoons$ CH$_2$ + H	2.00×10^{14}	0.00	0.00
5.24	^{1}CH$_2$ + O $\rightleftharpoons$ CO + H$_2$	1.51×10^{13}	0.00	0.00
5.25	^{1}CH$_2$ + O $\rightleftharpoons$ HCO + H	1.51×10^{13}	0.00	0.00
5.26	^{1}CH$_2$ + OH $\rightleftharpoons$ CH$_2$O + H	3.00×10^{13}	0.00	0.00
5.27	^{1}CH$_2$ + O$_2$ $\rightleftharpoons$ CO + OH + H	3.00×10^{13}	0.00	0.00
CH$_3$ Reactions				
5.28	CH$_3$ + H $\rightleftharpoons$ CH$_2$ + H$_2$	9.00×10^{13}	0.00	63.18
5.29	^{1}CH$_2$ + H$_2$ $\rightleftharpoons$ CH$_3$ + H	7.00×10^{13}	0.00	0.00
5.30	CH$_3$ + O $\rightleftharpoons$ CH$_2$O + H	8.00×10^{13}	0.00	0.00
5.31	CH$_2$OH + H $\rightleftharpoons$ CH$_3$ + OH	9.64×10^{13}	0.00	0.00
5.32	CH$_3$ + OH $\rightleftharpoons$ CH$_3$O + H	5.74×10^{12}	-0.23	58.29
5.33	CH$_3$OH + M $\rightleftharpoons$ CH$_3$ + OH + M	4.23×10^{17}	0.00	318.40
	CH$_3$OH $\rightleftharpoons$ CH$_3$ + OH	1.90×10^{16}	0.00	384.00
5.34	CH$_3$ + OH $\rightleftharpoons$ CH$_2$ + H$_2$O	7.50×10^{6}	2.00	20.92
5.35	CH$_3$ + OH $\rightleftharpoons$ ^{1}CH$_2$ + H$_2$O	8.90×10^{19}	-1.80	33.75
5.36	CH$_3$ + O$_2$ $\rightleftharpoons$ CH$_3$O + O	1.99×10^{18}	-1.57	122.30

(continues)

TABLE 5   (*continued*)

		A	n	E
5.37	$CH_3 + HO_2 \rightleftharpoons CH_3O + OH$	2.00×10^{13}	0.00	0.00
5.38	$CH_3O + CO \rightleftharpoons CH_3 + CO_2$	1.60×10^{13}	0.00	49.37
CH₄ Reactions				
5.39	$CH_4 + M \rightleftharpoons CH_3 + H + M$	7.21×10^{30}	-3.49	443.09
	$CH_4 \rightleftharpoons CH_3 + H$	3.70×10^{15}	0.00	434.30
5.40	$CH_4 + H \rightleftharpoons CH_3 + H_2$	2.20×10^4	3.00	36.61
5.41	$CH_4 + O \rightleftharpoons CH_3 + OH$	1.02×10^9	1.50	36.00
5.42	$CH_4 + OH \rightleftharpoons CH_3 + H_2O$	1.60×10^6	2.10	10.29
5.43	$CH_4 + O_2 \rightleftharpoons CH_3 + HO_2$	7.90×10^{13}	0.00	234.30
5.44	$CH_4 + HO_2 \rightleftharpoons CH_3 + H_2O_2$	1.81×10^{11}	0.00	78.24
5.45	$CH_3 + HCO \rightleftharpoons CH_4 + CO$	1.20×10^{14}	0.00	0.00
5.46	$CH_3 + CH_2O \rightleftharpoons CH_4 + HCO$	5.54×10^3	2.81	24.53
5.47	$CH_3 + CH_3O \rightleftharpoons CH_4 + CH_2O$	2.41×10^{13}	0.00	0.00
5.48	$CH_3 + CH_3OH \rightleftharpoons CH_4 + CH_2OH$	3.19×10^1	3.17	30.01
5.49	$CH_4 + {}^1CH_2 \rightleftharpoons CH_3 + CH_3$	4.00×10^{14}	0.00	0.00

[a]Reaction rates in cm^3 mol kJ units, $k = AT^n \exp(-E/RT)$

Sources.   Miller, J. A. and Bowman, C. T. (1989) Mechanism and Modeling of Nitrogen Chemistry in Combustion, *Prog. Energy Combust. Sci.*, **15**, 287–338. Held, T., The Oxidation of Methanol, Isobutene, and Methyl *tertiary*-Butyl Ether, No. 1978-T, Ph.D. Dissertation, Princeton University, Princeton, NJ, 1993. Burgess, D. R. F., Jr., Zachariah, M. R., Tsang, W., and Westmoreland, P. R., Thermochemical and Chemical Kinetic Data for Fluorinated Hydrocarbons, NIST Technical Note 1412, NIST, Gaithersburg, MD, 1995. The CH_4 and C_2H_6 mechanisms are based on the high temperature Miller-Bowman mechanism updated and extended to intermediate temperatures with kinetic data from Burgess *et al.* and Held. GRIMECH (Bowman, C. T., Frenklach, M., Gardiner, W., Golden, D., Lissianski, V., Smith, G., and Wang, H., Gas Research Institute Report, 1995) is also a recent hydrocarbon mechanism for methane/air mixtures with nitric oxide chemistry (http://www.me.berkeley.edu/gri_mech).

4. "Rate Coefficients in the C/H/O System," J. Warnatz, Chapter 5, "Survey of Rate Constants in the N/H/O System," R. K. Hanson and S. Salimian, Chapter 6, in *Combustion Chemistry*, W. C. Gardiner, Jr., ed., Springer-Verlag, NY, 1985.

5. "Chemical Kinetic Data Sheets for High Temperature Reactions. Part I," N. Cohen and K. R. Westberg, *J. Phys. Chem. Ref. Data*, **12**, 531, 1983. "Part II." *J. Phys. Chem. Ref. Data*, **20**, 1211, 1991.

6. "Evaluated Kinetic and Photochemical Data for Atmospheric Chemistry. Supplement IV." IUPAC Subcommittee on Gas Kinetic Data Evaluation for Atmospheric Chemistry, R. Atkinson, D. L. Baulch, R. A. Cox, R. F. Hampson, Jr., J. A. Kerr, and J. Troe, *J. Phys. Chem. Ref. Data*, **21**, 1125, 1992. "Supplement III," *J. Phys. Chem. Ref. Data*, **18**, 881, 1989. "Supplement II," *J. Phys. Chem. Ref. Data*, **13**, 1259, 1984. "Supplement I," *J. Phys. Chem. Ref. Data*, **11**, 327, 1982. *J. Phys. Chem. Ref. Data*, **9**, 295, 1980.

TABLE 6 $C_2H_6/CH_4/CH_3OH/CH_2O/CO/H_2/O_2$ Mechanism[a]

	A	n	E
C_2H Reactions			
6.1 $CH_2 + C \rightleftharpoons C_2H + H$	5.00×10^{13}	0.00	0.00
6.2 $C_2H + O \rightleftharpoons CH + CO$	5.00×10^{13}	0.00	0.00
6.3 $C_2H + O_2 \rightleftharpoons HCO + CO$	2.41×10^{12}	0.00	0.00
HCCO Reactions			
6.4 $HCCO + H \rightleftharpoons {}^1CH_2 + CO$	1.00×10^{14}	0.00	0.00
6.5 $C_2H + OH \rightleftharpoons HCCO + H$	2.00×10^{13}	0.00	0.00
6.6 $HCCO + O \rightleftharpoons H + CO + CO$	1.00×10^{14}	0.00	0.00
6.7 $C_2H + O_2 \rightleftharpoons HCCO + O$	6.02×10^{11}	0.00	0.00
6.8 $HCCO + O_2 \rightleftharpoons CO + CO + OH$	1.60×10^{12}	0.00	3.58
C_2H_2 Reactions			
6.9 $C_2H_2 + M \rightleftharpoons C_2H + H + M$	7.46×10^{30}	−3.70	531.78
6.10 $CH_3 + C \rightleftharpoons C_2H_2 + H$	5.00×10^{13}	0.00	0.00
6.11 $CH_2 + CH \rightleftharpoons C_2H_2 + H$	4.00×10^{13}	0.00	0.00
6.12 $C_2H + H_2 \rightleftharpoons C_2H_2 + H$	4.09×10^5	2.39	3.61
6.13 $C_2H_2 + O \rightleftharpoons CH_2 + CO$	1.02×10^7	2.00	7.95
6.14 $C_2H_2 + O \rightleftharpoons C_2H + OH$	3.16×10^{15}	−0.60	62.76
6.15 $C_2H_2 + O \rightleftharpoons HCCO + H$	1.02×10^7	2.00	7.95
6.16 $C_2H_2 + OH \rightleftharpoons CH_3 + CO$	4.83×10^{-4}	4.00	−8.37
6.17 $C_2H_2 + OH \rightleftharpoons C_2H + H_2O$	3.38×10^7	2.00	58.58
6.18 $CH_2 + CH_2 \rightleftharpoons C_2H_2 + H_2$	4.00×10^{13}	0.00	0.00
6.19 $C_2H_2 + O_2 \rightleftharpoons C_2H + HO_2$	1.20×10^{13}	0.00	311.75
6.20 $C_2H_2 + O_2 \rightleftharpoons HCCO + OH$	2.00×10^8	1.50	125.94
6.21 $HCCO + CH \rightleftharpoons C_2H_2 + CO$	5.00×10^{13}	0.00	0.00
6.22 $HCCO + HCCO \rightleftharpoons C_2H_2 + 2CO$	1.00×10^{13}	0.00	0.00
C_2H_3 Reactions			
6.23 $C_2H_2 + H + M \rightleftharpoons C_2H_3 + M$	2.67×10^{27}	−3.50	10.08
$k_{H_2} = k_{CO} = 2 \times k_M, k_{H_2O} = 5 \times k_M, k_{CO_2} = 3.0 \times k_M$			
$C_2H_2 + H \rightleftharpoons C_2H_3$	5.54×10^{12}	0.00	10.08
6.24 $C_2H_3 + H \rightleftharpoons C_2H_2 + H_2$	1.20×10^{13}	0.00	0.00
6.25 $CH_3 + CH \rightleftharpoons C_2H_3 + H$	3.00×10^{13}	0.00	0.00
6.26 $C_2H_3 + OH \rightleftharpoons C_2H_2 + H_2O$	3.00×10^{13}	0.00	0.00
6.27 $C_2H_3 + O_2 \rightleftharpoons HCO + HCO + H$	3.27×10^{23}	−3.94	20.96
6.28[b] $C_2H_3 + O_2 \rightleftharpoons CH_2O + HCO$	4.48×10^{26}	−4.55	22.93
	1.05×10^{38}	−8.22	29.41
6.29 $C_2H_3 + O_2 \rightleftharpoons C_2H_2 + HO_2$	5.10×10^{21}	−3.24	23.68
6.30 $HCCO + CH_2 \rightleftharpoons C_2H_3 + CO$	3.00×10^{13}	0.00	0.00
6.31 $C_2H_3 + CH \rightleftharpoons C_2H_2 + CH_2$	5.00×10^{13}	0.00	0.00
6.32 $C_2H_3 + CH_3 \rightleftharpoons C_2H_2 + CH_4$	3.90×10^{11}	0.00	0.00
6.33 $C_2H_3 + C_2H_3 \rightleftharpoons C_2H_4 + C_2H_2$	9.60×10^{11}	0.00	0.00
HCCOH Reactions			
6.34 $C_2H_2 + OH \rightleftharpoons HCCOH + H$	5.04×10^5	2.30	56.49

(continues)

TABLE 6 *(continued)*

	A	n	E
CH$_2$CO Reactions			
6.35 CH$_2$CO + M $\rightleftarrows$ CH$_2$ + CO + M	3.60×10^{15}	0.00	247.99
CH$_2$CO + $\rightleftarrows$ CH$_2$ + CO	3.00×10^{14}	0.00	297.00
6.36 CH$_2$CO + H $\rightleftarrows$ CH$_3$ + CO	1.13×10^{13}	0.00	14.34
6.37 CH$_2$CO + H $\rightleftarrows$ HCCO + H$_2$	5.00×10^{13}	0.00	33.47
6.38 C$_2$H$_2$ + OH $\rightleftarrows$ CH$_2$CO + H	2.18×10^{-4}	4.50	-4.18
6.39 C$_2$H$_3$ + O $\rightleftarrows$ CH$_2$CO + H	3.00×10^{13}	0.00	0.00
6.40 CH + CH$_2$O $\rightleftarrows$ CH$_2$CO + H	9.46×10^{13}	0.00	-2.15
6.41 HCCOH + H $\rightleftarrows$ CH$_2$CO + H	1.00×10^{13}	0.00	0.00
6.42 CH$_2$CO + O $\rightleftarrows$ CH$_2$ + CO$_2$	1.75×10^{12}	0.00	5.65
6.43 CH$_2$CO + O $\rightleftarrows$ HCCO + OH	1.00×10^{13}	0.00	33.47
6.44 CH$_2$CO + OH $\rightleftarrows$ HCCO + H$_2$O	7.50×10^{12}	0.00	8.37
6.45 C$_2$H$_3$ + HO$_2$ $\rightleftarrows$ CH$_2$CO + OH + H	3.00×10^{13}	0.00	0.00
CH$_3$CO Reactions			
6.46 CH$_3$CO + M $\rightleftarrows$ CH$_3$ + CO + M	8.73×10^{42}	-8.62	93.72
CH$_3$CO $\rightleftarrows$ CH$_3$ + CO	1.20×10^{22}	-3.04	78.66
6.47 CH$_3$CO + H $\rightleftarrows$ CH$_3$ + HCO	9.60×10^{13}	0.00	0.00
6.48 CH$_3$CO + O $\rightleftarrows$ CH$_3$ + CO$_2$	9.60×10^{12}	0.00	0.00
6.49 CH$_3$CO + OH $\rightleftarrows$ CH$_3$ + CO + OH	3.00×10^{13}	0.00	0.00
6.50 CH$_3$CO + OH $\rightleftarrows$ CH$_2$CO + H$_2$O	1.20×10^{13}	0.00	0.00
6.51 CH$_3$CO + HO$_2$ $\rightleftarrows$ CH$_3$ + CO$_2$ + OH	3.00×10^{13}	0.00	0.00
C$_2$H$_4$ Reactions			
6.52 C$_2$H$_4$ + M $\rightleftarrows$ C$_2$H$_2$ + H$_2$ + M	1.50×10^{15}	0.44	233.48
6.53 C$_2$H$_4$ + M $\rightleftarrows$ C$_2$H$_3$ + H + M	1.40×10^{16}	0.00	344.61
6.54 C$_2$H$_4$ + H $\rightleftarrows$ C$_2$H$_3$ + H$_2$	1.10×10^{14}	0.00	35.57
6.55 CH$_4$ + CH $\rightleftarrows$ C$_2$H$_4$ + H	6.00×10^{13}	0.00	0.00
6.56 CH$_3$ + CH$_2$ $\rightleftarrows$ C$_2$H$_4$ + H	3.00×10^{13}	0.00	0.00
6.57 C$_2$H$_4$ + O $\rightleftarrows$ CH$_3$ + HCO	1.60×10^{9}	1.20	3.12
6.58 C$_2$H$_4$ + OH $\rightleftarrows$ C$_2$H$_3$ + H$_2$O	2.02×10^{13}	0.00	24.92
6.59 C$_2$H$_4$ + O$_2$ $\rightleftarrows$ C$_2$H$_3$ + HO$_2$	4.22×10^{13}	0.00	241.00
6.60 C$_2$H$_4$ + CH$_3$ $\rightleftarrows$ C$_2$H$_3$ + CH$_4$	6.62	3.70	39.75
6.61 CH$_3$ + CH$_3$ $\rightleftarrows$ C$_2$H$_4$ + H$_2$	1.00×10^{16}	0.00	139.91
C$_2$H$_4$O Reactions			
6.62 C$_2$H$_4$O $\rightleftarrows$ CH$_4$ + CO	3.16×10^{14}	0.00	238.49
6.63 C$_2$H$_4$ + HO$_2$ $\rightleftarrows$ C$_2$H$_4$O + OH	6.00×10^{9}	0.00	33.26
CH$_3$CHO Reactions			
6.64 CH$_3$CHO $\rightleftarrows$ CH$_3$ + HCO	7.08×10^{15}	0.00	342.10
6.65 C$_2$H$_3$ + OH $\rightleftarrows$ CH$_3$CHO	3.00×10^{13}	0.00	0.00
6.66 CH$_3$CHO + H $\rightleftarrows$ CH$_3$CHO + H$_2$	4.00×10^{13}	0.00	17.60
6.67 CH$_3$CHO + O $\rightleftarrows$ CH$_3$CO + OH	5.00×10^{12}	0.00	7.50
6.68 CH$_3$CHO + OH $\rightleftarrows$ CH$_3$CO + H$_2$O	1.00×10^{13}	0.00	0.00
6.69 CH$_3$CHO + O$_2$ $\rightleftarrows$ CH$_3$CO + HO$_2$	2.00×10^{13}	0.50	176.56
6.70 CH$_3$CHO + HO$_2$ $\rightleftarrows$ CH$_3$CO + H$_2$O$_2$	1.70×10^{12}	0.00	44.77
6.71 CH$_3$CHO + CH$_3$ $\rightleftarrows$ CH$_3$CO + CH$_4$	1.74×10^{12}	0.00	35.31

(continues)

TABLE 6   (*continued*)

	A	n	E
C_2H_5 Reactions			
6.72 $C_2H_5 + M \rightleftharpoons C_2H_4 + H + M$	5.10×10^{64}	-14.00	251.46
$C_2H_5 \rightleftharpoons C_2H_4 + H$	4.90×10^9	1.19	155.64
6.73 $CH_3 + CH_3 \rightleftharpoons C_2H_5 + H$	8.00×10^{15}	0.00	110.93
6.74 $C_2H_5 + H \rightleftharpoons C_2H_4 + H_2$	1.81×10^{12}	0.00	0.00
6.75 $C_2H_5 + O \rightleftharpoons CH_2O + CH_3$	1.60×10^{13}	0.00	0.00
6.76 $C_2H_5 + O \rightleftharpoons CH_3HCO + H$	9.60×10^{14}	0.00	0.00
6.77 $C_2H_5 + OH \rightleftharpoons CH_3 + CH_2O + H$	2.40×10^{13}	0.00	0.00
6.78 $C_2H_5 + OH \rightleftharpoons C_2H_4 + H_2O$	2.40×10^{13}	0.00	0.00
6.79 $C_2H_5 + O_2 \rightleftharpoons C_2H_4 + HO_2$	2.56×10^{19}	-2.77	8.27
6.80 $C_2H_5 + HO_2 \rightleftharpoons CH_3 + CH_2O + OH$	2.40×10^{13}	0.00	0.00
6.81 $C_2H_5 + HO_2 \rightleftharpoons C_2H_4 + H_2O_2$	3.00×10^{11}	0.00	0.00
C_2H_6 Reactions			
6.82^c $CH_3 + CH_3 + M \rightleftharpoons C_2H_6 + M$	3.18×10^{41}	-7.03	11.56
$k_{H_2} = k_{CO} = 2 \times k_M, k_{H_2O} = 5 \times k_M,$			
$k_{CO_2} = 3 \times k_M$			
$CH_3 + CH_3 \rightleftharpoons C_2H_6$	9.03×10^{16}	-1.18	2.74
$F_c = (-0.64)\exp(-T/6927) + 0.64$			
$\exp(-132/T)$			
6.83 $C_2H_6 \rightleftharpoons C_2H_5 + H$	2.08×10^{38}	-7.08	445.60
6.84 $C_2H_6 + H \rightleftharpoons C_2H_5 + H_2$	5.42×10^2	3.50	21.80
6.85 $C_2H_6 + O \rightleftharpoons C_2H_5 + OH$	3.00×10^7	2.00	21.40
6.86 $C_2H_6 + OH \rightleftharpoons C_2H_5 + H_2O$	5.13×10^6	2.06	3.57
6.87 $C_2H_6 + O_2 \rightleftharpoons C_2H_5 + HO_2$	4.00×10^{13}	0.00	213.00
6.88 $C_2H_6 + HO_2 \rightleftharpoons C_2H_5 + H_2O_2$	2.94×10^{11}	0.00	62.51
6.89 $C_2H_6 + {}^1CH_2 \rightleftharpoons C_2H_5 + CH_3$	1.20×10^{14}	0.00	0.00
6.90 $C_2H_6 + CH_3 \rightleftharpoons C_2H_5 + CH_4$	5.50×10^{-1}	4.00	34.73
6.91 $C_2H_6 + C_2H_3 \rightleftharpoons C_2H_5 + C_2H_4$	6.00×10^2	3.30	43.93
6.92 $C_2H_5 + C_2H_5 \rightleftharpoons C_2H_6 + C_2H_4$	1.40×10^{12}	0.00	0.00

^{a}Reaction rates in cm^3 mol kJ units, $k = AT^n \exp(-E/RT)$

^{b}Rate represented by the sum of two Arrhenius expressions.

^{c}The fall-off behavior of this reaction is expressed as $k = [k_0 k_\infty/(k_0 + k_\infty/M)] \times F$, and $\log(F) = \log(F_c)/[1 + \{\log(k_0 \times M/k_\infty)\}^2]$.

Sources.   Miller, J. A. and Bowman, C. T. (1989). Mechanism and Modeling of Nitrogen Chemistry in Combustion, *Prog. Energy Combust. Sci.*, **15**, 287–338. Held, T., The Oxidation of Methanol, Isobutene, and Methyl *tertiary*-Butyl Ether, No. 1978-T, Ph.D. Dissertation, Princeton University, Princeton, NJ, 1993. Burgess, D. R. F., Jr., Zachariah, M. R., Tsang, W., and Westmoreland, P. R., Thermochemical and Chemical Kinetic Data for Fluorinated Hydrocarbons, NIST Technical Note 1412, NIST, Gaithersburg, MD, 1995.

TABLE 7 Selected Reactions of a C_3H_8 Mechanisms[a]

		A	n	E
C_3H_8 Reactions				
7.1	$C_3H_8 \rightleftharpoons C_2H_5 + CH_3$	7.90×10^{22}	-1.80	371.10
7.2	$C_3H_8 + H \rightleftharpoons n\text{-}C_3H_7 + H_2$	1.33×10^6	2.54	28.27
7.3	$C_3H_8 + H \rightleftharpoons i\text{-}C_3H_7 + H_2$	1.30×10^6	2.40	18.71
7.4	$C_3H_8 + O \rightleftharpoons n\text{-}C_3H_7 + OH$	1.93×10^5	2.68	15.55
7.5	$C_3H_8 + O \rightleftharpoons i\text{-}C_3H_7 + OH$	4.76×10^4	2.71	8.81
7.6	$C_3H_8 + OH \rightleftharpoons n\text{-}C_3H_7 + H_2O$	1.05×10^{10}	0.97	13.19
7.7	$C_3H_8 + OH \rightleftharpoons i\text{-}C_3H_7 + H_2O$	4.67×10^7	1.61	-0.29
7.8	$C_3H_8 + O_2 \rightleftharpoons n\text{-}C_3H_7 + HO_2$	3.98×10^{13}	0.00	213.10
7.9	$C_3H_8 + O_2 \rightleftharpoons i\text{-}C_3H_7 + HO_2$	3.98×10^{13}	0.00	199.10
7.10	$C_3H_8 + HO_2 \rightleftharpoons n\text{-}C_3H_7 + H_2O_2$	4.75×10^4	2.55	69.00
7.11	$C_3H_8 + HO_2 \rightleftharpoons i\text{-}C_3H_7 + H_2O_2$	9.64×10^3	2.60	58.20
7.12	$C_3H_8 + CH_3 \rightleftharpoons n\text{-}C_3H_7 + CH_4$	9.03×10^{-1}	3.65	59.47
7.13	$C_3H_8 + CH_3 \rightleftharpoons i\text{-}C_3H_7 + CH_4$	1.51	3.46	45.56
$n\text{-}C_3H_7$ Reactions				
7.14	$n\text{-}C_3H_7 \rightleftharpoons C_2H_4 + CH_3$	1.26×10^{13}	0.00	252.75
7.15	$n\text{-}C_3H_7 \rightleftharpoons C_3H_6 + H$	1.12×10^{13}	0.00	293.23
7.16	$n\text{-}C_3H_7 + O_2 \rightleftharpoons C_3H_6 + HO_2$	1.17×10^{19}	-1.59	112.16
7.17	$n\text{-}C_3H_7 + O_2 \rightleftharpoons C_2H_4 + CH_2O + OH$	2.44×10^{16}	-0.95	178.58
$i\text{-}C_3H_7$ Reactions				
7.18	$i\text{-}C_3H_7 \rightleftharpoons C_2H_4 + CH_3$	2.00×10^{10}	0.00	245.26
7.19	$i\text{-}C_3H_7 \rightleftharpoons C_3H_6 + H$	4.00×10^{13}	0.00	317.59
7.20	$i\text{-}C_3H_7 + O_2 \rightleftharpoons C_3H_6 + HO_2$	2.69×10^{27}	-4.60	117.56
C_3H_6 Reactions				
7.21	$C_3H_6 \rightleftharpoons C_3H_5 + H$	2.50×10^{15}	0.00	362.68
7.22	$C_3H_6 \rightleftharpoons C_2H_3 + CH_3$	1.10×10^{21}	-1.20	408.80
7.23	$C_3H_6 + H \rightleftharpoons C_3H_5 + H_2$	1.73×10^5	2.50	20.70
7.24	$C_3H_6 + H \rightleftharpoons CHCHCH_3 + H_2$	8.07×10^5	2.50	102.10
7.25	$C_3H_6 + H \rightleftharpoons CH_3CCH_2 + H_2$	4.10×10^5	2.50	81.43
7.26	$C_3H_6 + O \rightleftharpoons C_3H_5 + OH$	1.75×10^{11}	0.70	48.91
7.27	$C_3H_6 + O \rightleftharpoons CHCHCH_3 + OH$	1.20×10^{11}	0.70	74.49
7.28	$C_3H_6 + O \rightleftharpoons CH_3CCH_2 + OH$	6.02×10^{10}	0.70	63.45
7.29	$C_3H_6 + O \rightleftharpoons CH_2CO + CH_3 + H$	7.80×10^7	1.83	3.12
7.30	$C_3H_6 + O \rightleftharpoons CH_3CHCO + 2H$	3.90×10^7	1.83	3.12
7.31[b]	$C_3H_6 + O \rightleftharpoons C_2H_5 + HCO$	3.49×10^7	1.83	-2.29
		-1.17×10^7	1.83	3.12
7.32	$C_3H_6 + OH \rightleftharpoons C_3H_5 + H_2O$	3.12×10^6	2.00	-2.47
7.33	$C_3H_6 + OH \rightleftharpoons CHCHCH_3 + H_2O$	2.14×10^6	2.00	11.62
7.34	$C_3H_6 + OH \rightleftharpoons CH_3CCH_2 + H_2O$	1.11×10^6	2.00	6.07
7.35	$C_3H_6 + O_2 \rightleftharpoons C_3H_5 + HO_2$	6.02×10^{13}	0.00	199.10
7.36	$C_3H_6 + HO_2 \rightleftharpoons C_3H_5 + H_2O_2$	9.64×10^3	2.60	58.20
7.37	$C_3H_6 + CH_3 \rightleftharpoons C_3H_5 + CH_4$	2.22	3.50	23.74
7.38	$C_3H_6 + CH_3 \rightleftharpoons CHCHCH_3 + CH_4$	1.35	3.50	53.76
7.39	$C_3H_6 + CH_3 \rightleftharpoons CH_3CCH_2 + CH_4$	8.42×10^{-1}	3.50	4.88

(*continues*)

TABLE 7   (*continued*)

	A	n	E
CHCHCH$_3$ Reactions			
7.40 CHCHCH$_3 \rightleftharpoons$ C$_2$H$_2$ + CH$_3$	1.59×10^{12}	0.00	156.90
7.41 CHCHCH$_3$ + H $\rightleftharpoons$ C$_3$H$_4$ + H$_2$	3.33×10^{12}	0.00	0.00
7.42 CHCHCH$_3$ + O$_2 \rightleftharpoons$ CH$_3$HCO + HCO	4.34×10^{12}	0.00	0.00
7.43 CHCHCH$_3$ + CH$_3 \rightleftharpoons$ C$_3$H$_4$ + CH$_4$	1.00×10^{11}	0.00	0.00
CH$_2$CCH$_3$ Reactions			
7.44 CH$_2$CCH$_3$ + H $\rightleftharpoons$ pC$_3$H$_4$ + H$_2$	3.33×10^{12}	0.00	0.00
7.45 CH$_2$CCH$_3$ + O$_2 \rightleftharpoons$ CH$_2$O + CH$_3$CO	4.34×10^{12}	0.00	0.00
7.46 CH$_2$CCH$_3$ + CH$_3 \rightleftharpoons$ pC$_3$H$_4$ + CH$_4$	1.00×10^{11}	0.00	0.00
C$_3$H$_5$ Reactions			
7.47 C$_3$H$_5 \rightleftharpoons$ C$_3$H$_4$ + H	1.40×10^{13}	0.00	251.00
7.48 C$_3$H$_5$ + H $\rightleftharpoons$ C$_3$H$_4$ + H$_2$	1.80×10^{13}	0.00	0.00
7.49 C$_3$H$_5$ + O $\rightleftharpoons$ C$_2$H$_3$HCO + H	6.02×10^{13}	0.00	0.00
7.50 C$_3$H$_5$ + OH $\rightleftharpoons$ C$_3$H$_4$ + H$_2$O	6.02×10^{12}	0.00	0.00
7.51 C$_3$H$_5$ + O$_2 \rightleftharpoons$ C$_3$H$_4$ + HO$_2$	1.33×10^{7}	0.00	0.00
7.52 C$_3$H$_5$ + HO$_2 \rightleftharpoons$ C$_2$H$_3$HCO + OH + H	1.92×10^{11}	0.00	0.00
7.53 C$_3$H$_5$ + CH$_3 \rightleftharpoons$ C$_3$H$_4$ + CH$_4$	3.00×10^{12}	-0.32	-1.10
C$_3$H$_4$ Reactions			
7.54 C$_3$H$_4 \rightleftharpoons$ pC$_3$H$_4$	1.20×10^{15}	0.00	386.60
7.55 C$_3$H$_4$ + M $\rightleftharpoons$ C$_3$H$_3$ + H + M	1.14×10^{17}	0.00	292.90
7.56 C$_3$H$_4$ + H $\rightleftharpoons$ C$_3$H$_3$ + H$_2$	3.36×10^{-7}	6.00	7.08
7.57 C$_3$H$_4$ + H $\rightleftharpoons$ CH$_2$CCH$_3$	8.50×10^{12}	0.00	8.37
7.58 C$_3$H$_4$ + O $\rightleftharpoons$ CH$_2$O + C$_2$H$_2$	3.00×10^{-3}	4.61	-17.75
7.59 C$_3$H$_4$ + O $\rightleftharpoons$ CO + C$_2$H$_4$	9.00×10^{-3}	4.61	-17.75
7.60 C$_3$H$_4$ + OH $\rightleftharpoons$ C$_3$H$_3$ + H$_2$O	1.45×10^{13}	0.00	17.45
7.61 C$_3$H$_4$ + OH $\rightleftharpoons$ CH$_2$CO + CH$_3$	3.12×10^{12}	0.00	-1.66
7.62 C$_3$H$_4$ + O$_2 \rightleftharpoons$ C$_3$H$_3$ + HO$_2$	4.00×10^{13}	0.00	257.32
7.63 C$_3$H$_4$ + HO$_2 \rightleftharpoons$ CH$_2$CO + CH$_2$ + OH	4.00×10^{12}	0.00	79.50
7.64 C$_3$H$_4$ + CH$_3 \rightleftharpoons$ C$_3$H$_3$ + CH$_4$	2.00×10^{12}	0.00	32.22
p-C$_3$H$_4$ Reactions			
7.65 pC$_3$H$_4$ + M $\rightleftharpoons$ C$_3$H$_3$ + H + M	1.00×10^{17}	0.00	292.88
7.66 pC$_3$H$_4 \rightleftharpoons$ C$_2$H + CH$_3$	4.20×10^{16}	0.00	418.40
7.67 pC$_3$H$_4$ + H $\rightleftharpoons$ C$_3$H$_3$ + H$_2$	1.00×10^{12}	0.00	6.28
7.68 pC$_3$H$_4$ + H $\rightleftharpoons$ CH$_2$CCH$_3$	6.50×10^{12}	0.00	8.37
7.69 pC$_3$H$_4$ + H $\rightleftharpoons$ CHCHCH$_3$	5.80×10^{12}	0.00	12.97
7.70 pC$_3$H$_4$ + O $\rightleftharpoons$ C$_2$H$_2$ + CO + H$_2$	1.50×10^{13}	0.00	8.83
7.71 pC$_3$H$_4$ + O $\rightleftharpoons$ CH$_2$CO + H$_2$	3.20×10^{12}	0.00	8.41
7.72 pC$_3$H$_4$ + O $\rightleftharpoons$ C$_2$H$_3$ + HCO	3.20×10^{12}	0.00	8.41
7.73 pC$_3$H$_4$ + O $\rightleftharpoons$ C$_2$H$_4$ + CO	3.20×10^{12}	0.00	8.41
7.74 pC$_3$H$_4$ + O $\rightleftharpoons$ HCCO + CH$_3$	6.30×10^{12}	0.00	8.41
7.75 pC$_3$H$_4$ + O $\rightleftharpoons$ HCCO + CH$_2$ + H	3.20×10^{11}	0.00	8.41
7.76 pC$_3$H$_4$ + OH $\rightleftharpoons$ C$_3$H$_3$ + H$_2$O	1.50×10^{3}	3.00	0.84
7.77 pC$_3$H$_4$ + OH $\rightleftharpoons$ CH$_2$CO + CH$_3$	5.00×10^{-4}	4.50	-4.18
7.78 pC$_3$H$_4$ + O$_2 \rightleftharpoons$ C$_3$H$_3$ + HO$_2$	2.50×10^{12}	0.00	213.38
7.79 pC$_3$H$_4$ + O$_2 \rightleftharpoons$ HCCO + OH + CH$_2$	1.00×10^{7}	1.50	125.94

(*continues*)

TABLE 7 (*continued*)

		A	n	E
7.80	$pC_3H_4 + HO_2 \rightleftarrows C_2H_4 + CO + OH$	3.00×10^{12}	0.00	79.50
7.81	$pC_3H_4 + CH_3 \rightleftarrows C_3H_3 + CH_4$	2.00×10^{12}	0.00	32.22
c-C_3H_4 Reactions				
7.82	$cC_3H_4 \rightleftarrows C_3H_4$	1.51×10^{14}	0.00	210.90
7.83	$cC_3H_4 \rightleftarrows pC_3H_4$	1.20×10^{15}	0.00	182.80
C_3H_3 Reactions				
7.84	$CH_2 + C_2H_2 \rightleftarrows C_3H_3 + H$	1.20×10^{13}	0.00	27.61
7.85	$^1CH_2 + C_2H_2 \rightleftarrows C_3H_3 + H$	3.00×10^{13}	0.00	0.00
7.86	$C_3H_3 + O \rightleftarrows CH_2O + C_2H$	2.00×10^{13}	0.00	0.00
7.87	$C_3H_3 + OH \rightleftarrows C_3H_2 + H_2O$	2.00×10^{13}	0.00	0.00
7.88	$C_3H_3 + O_2 \rightleftarrows CH_2CO + HCO$	3.00×10^{10}	0.00	12.00
7.89	$C_3H_3 + CH_3 \rightleftarrows C_2H_5 + C_2H$	1.00×10^{13}	0.00	156.90
7.90	$C_3H_3 + C_3H_3 \rightleftarrows C_2H_2 + C_2H_2 + C_2H_2$	5.00×10^{11}	0.00	0.00
C_3H_2 Reactions				
7.91	$CH + C_2H_2 \rightleftarrows C_3H_2 + H$	1.00×10^{14}	0.00	0.00
7.92	$C_3H_2 + O_2 \rightleftarrows HCCO + HCO$	1.00×10^{13}	0.00	0.00

[a] Reaction rates in cm^3 mol kJ units, $k = AT^n \exp(-E/RT)$.
[b] Rate represented by sum of two Arrhenius expressions.
Sources. W. Tsang (1988). *J. Phys. Chem. Ref. Data*, **17**, 887, and W. Tsang (1991). *J. Phys. Chem. Ref. Data*, **20**, 221. W. J. Pitz and C. K. Westbrook, CDAT data base for HCT, LNL, Livermore CA, 1995. P. Dagaut, M. Cathonnet, and J. C. Boettner, *Combust. Sci. and Tech.*, **71**, 111, 1990. Miller, J. A. and Bowman, C. T. (1989). Mechanism and Modeling of Nitrogen Chemistry in Combustion, *Prog. Energy Combust. Sci.*, **15**, 287–338.

TABLE 8 $N_xO_y/CO/H_2/O_2$ Mechanism

		A	n	E
N and NO Reactions				
8.1	$NO + M \rightleftarrows N + O + M$	9.64×10^{14}	0.00	620.91
8.2	$NO + H \rightleftarrows N + OH$	1.69×10^{14}	0.00	204.18
8.3	$NO + O \rightleftarrows N + O_2$	1.81×10^{9}	1.00	162.13
8.4	$N + HO_2 \rightleftarrows NO + OH$	1.00×10^{13}	0.00	8.39
8.5	$NO + N \rightleftarrows N_2 + O$	3.27×10^{12}	0.30	0.00
NO_2 Reactions				
8.6[c]	$NO + O + M \rightleftarrows NO_2 + M$	4.72×10^{24}	-2.87	6.49
	$NO + O + Ar \rightleftarrows NO_2 + Ar$	7.56×10^{19}	-1.41	0.00
	$NO + O \rightleftarrows NO_2$	1.30×10^{15}	-0.75	0.00
	$F_c = 0.95 - 1.0 \times 10^4 T$			
8.7	$NO_2 + H \rightleftarrows NO + OH$	1.32×10^{14}	0.00	1.51
8.8	$NO_2 + O \rightleftarrows NO + O_2$	3.91×10^{12}	0.00	-1.00
8.9	$NO_2 + OH \rightleftarrows NO + HO_2$	1.81×10^{13}	0.00	27.93
8.10	$NO_2 + CO \rightleftarrows NO + CO_2$	9.03×10^{13}	0.00	141.34

(*continues*)

TABLE 8   (continued)

		A	n	E
8.11	$NO_2 + HCO \rightleftharpoons NO + H + CO_2$	8.39×10^{15}	-0.75	8.06
8.12	$NO_2 + N \rightleftharpoons N_2 + O_2$	1.00×10^{12}	0.00	0.00
8.13	$NO_2 + N \rightleftharpoons NO + NO$	4.00×10^{12}	0.00	0.00
8.14	$NO_2 + NO_2 + \rightleftharpoons 2NO + O_2$	1.63×10^{12}	0.00	109.29

N_2O Reactions

		A	n	E
8.15[c]	$N_2O + M \rightleftharpoons N_2 + O + M$	9.13×10^{14}	0.00	241.42
	$k_{Ar} = 0.63 \times k_M, k_{H_2O} = 7.5 \times k_M$			
	$N_2O \rightleftharpoons N_2 + O$	7.91×10^{10}	0.00	234.39
8.16[b]	$N_2O + H \rightleftharpoons N_2 + OH$	2.53×10^{10}	0.00	19.04
		2.23×10^{14}	0.00	70.08
8.17	$N_2O + O \rightleftharpoons N_2 + O_2$	1.00×10^{14}	0.00	117.15
8.18	$N_2O + O \rightleftharpoons NO + NO$	1.00×10^{14}	0.00	117.15
8.19	$NO_2 + N \rightleftharpoons N_2O + O$	5.00×10^{12}	0.00	0.00
8.20	$N_2O + OH \rightleftharpoons N_2 + HO_2$	2.00×10^{12}	0.00	167.36
8.21	$N_2O + CO \rightleftharpoons N_2 + CO_2$	5.01×10^{13}	0.00	184.10
8.22	$N_2O + N \rightleftharpoons N_2 + NO$	1.00×10^{13}	0.00	83.14
8.23	$N_2O + NO \rightleftharpoons N_2 + NO_2$	1.00×10^{14}	0.00	209.20

HNO Reactions

		A	n	E
8.24[c]	$NO + H + M \rightleftharpoons HNO + M$	8.96×10^{19}	-1.32	3.08
	$k_{Ar} = 0.63 \times k_M$			
	$NO + H \rightleftharpoons HNO$	1.52×10^{15}	-0.41	0.00
	$F_c = 0.82$			
8.25	$HNO + H \rightleftharpoons NO + H_2$	4.46×10^{11}	0.72	2.74
8.26	$HNO + O \rightleftharpoons NO + OH$	1.81×10^{13}	0.00	0.00
8.27	$HNO + OH \rightleftharpoons NO + H_2O$	1.30×10^{7}	1.88	-4.00
8.28	$NO + HCO \rightleftharpoons HNO + CO$	7.23×10^{12}	0.00	0.00
8.29	$HNO + HCO \rightleftharpoons NO + CH_2O$	6.02×10^{11}	0.00	8.31
8.30	$HNO + N \rightleftharpoons N_2O + H$	5.00×10^{10}	0.50	12.55
8.31	$HNO + NO \rightleftharpoons N_2O + OH$	2.00×10^{12}	0.00	108.78
8.32	$HNO + HNO \rightleftharpoons N_2O + H_2O$	8.51×10^{8}	0.00	12.89

HONO Reactions

		A	n	E
8.33[c]	$NO + OH + M \rightleftharpoons HONO + M$	5.08×10^{23}	-2.51	-0.28
	$k_{Ar} = 0.63 \times k_M$			
	$NO + OH \rightleftharpoons HONO$	1.99×10^{12}	-0.05	-3.02
	$F_c = 0.62$			
8.34	$NO_2 + H_2 \rightleftharpoons HONO + H$	3.21×10^{12}	0.00	120.54
8.35	$HONO + O \rightleftharpoons OH + NO_2$	1.20×10^{13}	0.00	24.94
8.36	$HONO + OH \rightleftharpoons H_2O + NO_2$	1.26×10^{10}	1.00	0.57
8.37	$NO_2 + HCO \rightleftharpoons HONO + CO$	1.24×10^{23}	-7.29	9.85
8.38	$NO_2 + CH_2O \rightleftharpoons HONO + HCO$	7.83×10^{2}	2.77	57.45
8.39	$HNO + NO_2 \rightleftharpoons HONO + NO$	6.02×10^{11}	0.00	8.31

NH Reactions

		A	n	E
8.40	$NH + M \rightleftharpoons N + H + M$	2.65×10^{14}	0.00	315.93
8.41	$N + HO_2 \rightleftharpoons NH + O_2$	1.00×10^{13}	0.00	8.37
8.42	$NH + O_2 \rightleftharpoons NO + OH$	7.60×10^{10}	0.00	6.40

(continues)

TABLE 8   (*continued*)

		A	n	E
8.43	$NH + O_2 \rightleftharpoons HNO + O$	3.89×10^{13}	0.00	74.85
8.44	$N + H_2 \rightleftharpoons NH + H$	1.60×10^{14}	0.00	105.19
8.45	$NH + O \rightleftharpoons N + OH$	3.72×10^{13}	0.00	0.00
8.46	$NH + O \rightleftharpoons NO + H$	5.50×10^{13}	0.00	0.00
8.47	$NH + OH \rightleftharpoons N + H_2O$	5.00×10^{11}	0.50	8.37
8.48	$NH + OH \rightleftharpoons HNO + H$	2.00×10^{13}	0.00	0.00
8.49	$NH + N \rightleftharpoons N_2 + H$	3.00×10^{13}	0.00	0.00
8.50	$NH + NO \rightleftharpoons N_2 + OH$	2.16×10^{13}	-0.23	0.00
8.51[b]	$NH + NO \rightleftharpoons N_2O + H$	2.94×10^{14}	-0.40	0.00
		-2.16×10^{13}	-0.23	0.00
8.52	$HNO + N \rightleftharpoons NH + NO$	1.00×10^{13}	0.00	8.37
8.53	$NH + NO_2 \rightleftharpoons HNO + NO$	1.00×10^{11}	0.50	16.74
8.54	$NH + NH \rightleftharpoons N_2 + H + H$	5.10×10^{13}	0.00	0.00

NH$_2$ Reactions

		A	n	E
8.55	$NH_2 + M \rightleftharpoons NH + H + M$	3.98×10^{23}	-2.00	382.44
8.56	$NH_2 + H \rightleftharpoons NH + H_2$	7.20×10^5	2.32	6.65
8.57	$NH_2 + O \rightleftharpoons HNO + H$	6.63×10^{14}	-0.50	0.00
8.58	$NH_2 + O \rightleftharpoons NH + OH$	6.75×10^{12}	0.00	0.00
8.59	$NH_2 + OH \rightleftharpoons NH + H_2O$	4.00×10^{12}	2.00	4.18
8.60	$NH_2 + O_2 \rightleftharpoons HNO + OH$	1.78×10^{12}	0.00	62.34
8.61[b]	$NH_2 + NO \rightleftharpoons N_2 + H_2O$	1.30×10^{16}	-1.25	0.00
		-2.80×10^{13}	-0.55	0.00
8.62	$NH_2 + NO \rightleftharpoons N_2O + H_2$	5.00×10^{13}	0.00	103.09
8.63	$NH_2 + NO \rightleftharpoons HNO + NH$	1.00×10^{13}	0.00	167.36
8.64	$NH_2 + NO_2 \rightleftharpoons N_2O + H_2O$	2.84×10^{18}	-2.20	0.00

NH$_3$ Reactions

		A	n	E
8.65	$NH_3 + M \rightleftharpoons NH_2 + H + M$	2.20×10^{16}	0.00	391.08
8.66	$NH_3 + H \rightleftharpoons NH_2 + H_2$	6.38×10^5	2.39	42.68
8.67	$NH_3 + O \rightleftharpoons NH_2 + OH$	9.40×10^6	1.94	27.03
8.68	$NH_3 + OH \rightleftharpoons NH_2 + H_2O$	2.04×10^6	2.04	2.37
8.69	$NH_2 + HO_2 \rightleftharpoons NH_3 + O_2$	3.00×10^{11}	0.00	92.05
8.70	$NH_2 + NH_2 \rightleftharpoons NH_3 + NH$	5.00×10^{13}	0.00	41.84

NNH Reactions

		A	n	E
8.71	$NNH + M \rightleftharpoons N_2 + H + M$	1.00×10^{14}	0.00	12.47
8.72	$NNH + H \rightleftharpoons N_2 + H_2$	1.00×10^{14}	0.00	0.00
8.73	$NNH + OH \rightleftharpoons N_2 + H_2O$	5.00×10^{13}	0.00	0.00
8.74	$NH_2 + NO \rightleftharpoons NNH + OH$	2.80×10^{13}	-0.55	0.00
8.75	$NNH + NO \rightleftharpoons HNO + N_2$	5.00×10^{13}	0.00	0.00
8.76	$NNH + NH \rightleftharpoons N_2 + NH_2$	5.00×10^{13}	0.00	0.00
8.77	$NNH + NH_2 \rightleftharpoons N_2 + NH_3$	5.00×10^{13}	0.00	0.00

N$_2$H$_2$ Reactions

		A	n	E
8.78	$N_2H_2 + M \rightleftharpoons NNH + H + M$	1.00×10^{16}	0.00	207.94
8.79	$N_2H_2 + M \rightleftharpoons NH + NH + M$	3.16×10^{16}	0.00	415.89
8.80	$N_2H_2 + H \rightleftharpoons NNH + H_2$	1.00×10^{13}	0.00	4.16
8.81	$NH + NH_2 \rightleftharpoons N_2H_2 + H$	3.16×10^{13}	0.00	4.16
8.82	$N_2H_2 + O \rightleftharpoons NNH + OH$	1.00×10^{11}	0.50	0.00

(*continues*)

TABLE 8   (*continued*)

		A	n	E
8.83	$N_2H_2 + OH \rightleftarrows NNH + H_2O$	1.00×10^{13}	0.00	8.33
8.84	$NH_2 + NH_2 \rightleftarrows N_2H_2 + H_2$	3.98×10^{13}	0.00	49.79
8.85	$N_2H_2 + HO_2 \rightleftarrows NNH + H_2O_2$	1.00×10^{13}	0.00	8.33
8.86	$NNH + NNH \rightleftarrows N_2H_2 + N_2$	1.00×10^{13}	0.00	41.59
8.87	$N_2H_2 + NH \rightleftarrows NNH + NH_2$	1.00×10^{13}	0.00	4.16
8.88	$N_2H_2 + NH_2 \rightleftarrows NNH + NH_3$	1.00×10^{13}	0.00	16.61

N_2H_3 Reactions

8.89	$N_2H_3 + M \rightleftarrows N_2H_2 + H + M$	1.00×10^{16}	0.00	207.94
8.90	$N_2H_3 + M \rightleftarrows NH_2 + NH + M$	1.00×10^{16}	0.00	174.47
8.91	$N_2H_3 + H \rightleftarrows NH_2 + NH_2$	1.58×10^{12}	0.00	0.00
8.92	$N_2H_3 + H \rightleftarrows NH + NH_3$	1.00×10^{11}	0.00	0.00
8.93	$N_2H_3 + H \rightleftarrows N_2H_2 + H_2$	1.00×10^{12}	0.00	8.33
8.94	$N_2H_3 + O \rightleftarrows N_2H_2 + OH$	3.16×10^{11}	0.50	0.00
8.95	$N_2H_3 + O \rightleftarrows NNH + H_2O$	3.16×10^{11}	0.50	0.00
8.96	$N_2H_3 + OH \rightleftarrows N_2H_2 + H_2O$	1.00×10^{13}	0.00	8.33
8.97	$N_2H_3 + HO_2 \rightleftarrows N_2H_2 + H_2O_2$	1.00×10^{13}	0.00	8.33
8.98	$NH_3 + NH_2 \rightleftarrows N_2H_3 + H_2$	7.94×10^{11}	0.50	90.37
8.99	$N_2H_2 + NH_2 \rightleftarrows NH + N_2H_3$	1.00×10^{11}	0.50	141.42
8.100	$N_2H_3 + NH_2 \rightleftarrows N_2H_2 + NH_3$	1.00×10^{11}	0.50	0.00
8.101	$N_2H_2 + N_2H_2 \rightleftarrows NNH + N_2H_3$	1.00×10^{13}	0.00	41.59

N_2H_4 Reactions

8.102	$N_2H_4 + M \rightleftarrows NH_2 + NH_2 + M$	4.00×10^{15}	0.00	171.13
8.103	$N_2H_4 + M \rightleftarrows N_2H_3 + H + M$	1.00×10^{15}	0.00	266.10
8.104	$N_2H_4 + H \rightleftarrows N_2H_3 + H_2$	1.29×10^{13}	0.00	10.46
8.105	$N_2H_4 + H \rightleftarrows NH_2 + NH_3$	4.46×10^{9}	0.00	12.97
8.106	$N_2H_4 + O \rightleftarrows N_2H_2 + H_2O$	6.31×10^{13}	0.00	4.98
8.107	$N_2H_4 + O \rightleftarrows N_2H_3 + OH$	2.51×10^{12}	0.00	4.98
8.108	$N_2H_4 + OH \rightleftarrows N_2H_3 + H_2O$	3.98×10^{13}	0.00	0.00
8.109	$N_2H_4 + HO_2 \rightleftarrows N_2H_3 + H_2O_2$	3.98×10^{13}	0.00	8.33
8.110	$N_2H_4 + NH \rightleftarrows NH_2 + N_2H_3$	1.00×10^{12}	0.00	8.33
8.111	$N_2H_4 + NH_2 \rightleftarrows N_2H_3 + NH_3$	3.98×10^{11}	0.50	8.33
8.112	$N_2H_3 + N_2H_2 \rightleftarrows N_2H_4 + NNH$	1.00×10^{13}	0.00	41.59
8.113	$N_2H_4 + N_2H_2 \rightleftarrows N_2H_3 + N_2H_3$	2.50×10^{10}	0.50	124.68

NO_3 Reactions

8.114^c	$NO_2 + O + M \rightleftarrows NO_3 + M$	1.49×10^{28}	-4.08	10.32
	$k_{Ar} = 0.63 \times k_M$			
	$NO_2 + O \rightleftarrows NO_3$	1.33×10^{13}	0.00	0.00
	$F_c = 0.79 - 1.8 \times 10^{-4}T$			
8.115	$NO_2 + NO_2 \rightleftarrows NO_3 + NO$	9.64×10^{9}	0.73	87.53
8.116	$NO_3 + H \rightleftarrows NO_2 + OH$	6.00×10^{13}	0.00	0.00
8.117	$NO_3 + O \rightleftarrows NO_2 + O_2$	1.00×10^{13}	0.00	0.00
8.118	$NO_3 + OH \rightleftarrows NO_2 + HO_2$	1.40×10^{13}	0.00	0.00
8.119	$NO_3 + HO_2 \rightleftarrows NO_2 + O_2 + OH$	1.50×10^{12}	0.00	0.00
8.120	$NO_3 + NO_2 \rightleftarrows NO + NO_2 + O_2$	5.00×10^{10}	0.00	12.30

(*continues*)

TABLE 8 (*continued*)

		A	n	E
HNO$_3$ Reactions				
8.121[b]	NO$_2$ + OH + M $\rightleftarrows$ HNO$_3$ + M	6.42×10^{32}	-5.49	9.83
	$k_{Ar} = 0.63 \times k_M$			
	NO$_2$ + OH $\rightleftarrows$ HNO$_3$	2.41×10^{13}	0.00	0.00
	$F_c = 0.725 - 2.5 \times 10^{-4}T$			
8.122	NO + HO$_2$ + M $\rightleftarrows$ HNO$_3$ + M	2.23×10^{12}	-3.50	9.20
8.123	HNO$_3$ + OH $\rightleftarrows$ NO$_2$ + HO$_2$	1.03×10^{10}	0.00	-5.19
8.124	NO$_3$ + HO$_2$ $\rightleftarrows$ HNO$_3$ + O$_2$	5.60×10^{11}	0.00	0.00

[a]Reaction rates in cm^3 mol kJ units, $k = AT^n \exp(-E/RT)$.
[b]Rate represented by sum of two Arrhenius expressions.
[c]The fall-off behavior of this reaction is expressed as $k = [k_0 k_\infty/(k_0 + k_\infty/M)] \times F$, and log $(F) =$ log $(F_c)/[1 + \{\log(k_0 \times M/k_\infty)\}^2]$.
Sources. Allen, M. T., Yetter, R. A., Dryer, F. L. (1995). The Decomposition of Nitrous Oxide at $1.5 \leq P \leq 10.5$ atm and $1103 \leq T \leq 1173$ K, *Int. J. Chem. Kinet.* Vol. 27, 883–909. Allen, M. T., Yetter, R. A. and Dryer, F. L. (in press, 1996). High Pressure Studies of Moist Carbon Monoxide/ Nitrous Oxide Kinetics, *Combust. Flame.*

TABLE 9 HCl/N$_x$O$_y$/CO/H$_2$/O$_2$ Mechanism[a]

		A	n	E
HCl and Cl Reactions				
9.1	Cl + H + M $\rightleftarrows$ HCl + M	7.20×10^{21}	-2.00	0.00
9.2	Cl + HO$_2$ $\rightleftarrows$ HCl + O$_2$	1.08×10^{13}	0.00	-1.38
9.3	HCl + H $\rightleftarrows$ Cl + H$_2$	1.69×10^{13}	0.30	17.32
9.4	HCl + O $\rightleftarrows$ Cl + OH	3.37×10^{3}	2.87	14.67
9.5	HCl + OH $\rightleftarrows$ Cl + H$_2$O	2.71×10^{7}	1.65	-0.93
9.6	Cl + H$_2$O$_2$ $\rightleftarrows$ HCl + HO$_2$	6.62×10^{12}	0.00	8.16
9.7	Cl + HCO $\rightleftarrows$ HCl + CO	1.00×10^{14}	0.00	0.00
9.8	Cl + HNO $\rightleftarrows$ HCl + NO	9.00×10^{13}	0.00	4.16
9.9	Cl + HONO $\rightleftarrows$ HCl + NO$_2$	5.00×10^{13}	0.00	0.00
Cl$_2$ Reactions				
9.10	Cl + Cl + M $\rightleftarrows$ Cl$_2$ + M	4.68×10^{14}	0.00	-7.53
9.11	Cl$_2$ + H $\rightleftarrows$ Cl + HCl	8.59×10^{13}	0.00	4.90
ClO Reactions				
9.12	ClO + O $\rightleftarrows$ Cl + O$_2$	5.70×10^{13}	0.00	1.52
9.13	Cl + HO$_2$ $\rightleftarrows$ ClO + OH	2.42×10^{13}	0.00	9.62
9.14	ClO + CO $\rightleftarrows$ Cl + CO$_2$	6.03×10^{11}	0.00	30.96
9.15	Cl$_2$ + O $\rightleftarrows$ ClO + Cl	2.52×10^{12}	0.00	11.38
9.16	ClO + NO $\rightleftarrows$ Cl + NO$_2$	3.85×10^{12}	0.00	0.59
HOCl Reactions				
9.17	HOCl $\rightleftarrows$ Cl + OH	1.76×10^{20}	-3.01	237.32
9.18	HOCl $\rightleftarrows$ ClO + H	8.12×10^{14}	-2.09	392.00
9.19	HOCl $\rightleftarrows$ HCl + OH	9.55×10^{13}	0.00	31.88

(*continues*)

TABLE 9   (*continued*)

		A	n	E
9.20	$ClO + H_2 \rightleftharpoons HOCl + H$	6.03×10^{11}	0.10	59.00
9.21	$HOCl + O \rightleftharpoons ClO + OH$	6.03×10^{12}	0.00	18.28
9.22	$HOCl + OH \rightleftharpoons ClO + H_2O$	1.81×10^{12}	0.00	4.14
9.23	$HCO + ClO \rightleftharpoons HOCl + CO$	3.16×10^{13}	0.00	0.00
9.24	$HOCl + Cl \rightleftharpoons Cl_2 + OH$	1.81×10^{12}	0.00	1.09
9.25	$HOCl + Cl \rightleftharpoons ClO + HCl$	7.62×10^{12}	0.00	0.75
CClO Reactions				
9.26	$COCl + M \rightleftharpoons Cl + CO + M$	1.30×10^{14}	0.00	33.47
9.27	$COCl + O_2 \rightleftharpoons ClO + CO_2$	7.94×10^{10}	0.00	13.81
9.28	$COCl + H \rightleftharpoons HCl + CO$	1.00×10^{14}	0.00	0.00
9.29	$COCl + O \rightleftharpoons ClO + CO$	1.00×10^{14}	0.00	0.00
9.30	$COCl + O \rightleftharpoons Cl + CO_2$	1.00×10^{13}	0.00	0.00
9.31	$COCl + OH \rightleftharpoons HOCl + CO$	3.30×10^{12}	0.00	0.00
9.32	$COCl + Cl \rightleftharpoons Cl_2 + CO$	4.00×10^{14}	0.00	3.35
NOCl Reactions				
9.33	$NOCl + M \rightleftharpoons Cl + NO + M$	2.51×10^{15}	0.00	133.47
	$k_{H_2} = 1.6 \times k_M, k_{CO_2} = 3.5 \times k_M,$			
	$k_{NO} = 1.38 \times k_M$			
9.34	$NOCl + H \rightleftharpoons HCl + NO$	4.60×10^{13}	0.00	3.73
9.35	$NOCl + O \rightleftharpoons ClO + NO$	5.00×10^{12}	0.00	12.55
9.36	$NOCl + Cl \rightleftharpoons Cl_2 + NO$	2.41×10^{13}	0.00	0.00

[a]Reaction rates in cm^3 mol kJ units, $k = AT^n \exp(-E/RT)$

Source.   Roesler, J. F., Yetter, R. A., Dryer, F. L. (1995). Kinetic Interactions of CO, NO_x, and HCl Emissions in Postcombustion Gases, *Combust. Flame*, **100**, 495–504.

TABLE 10 $O_3/N_xO_y/CO/H_2/O_2$ Mechanism[a]

		A	n	E
O_3 Reactions				
10.1	$O + O_2 + M \rightleftharpoons O_3 + M$	1.78×10^{21}	−2.80	0.00
	$O + O_2 \rightleftharpoons O_3$	1.69×10^{12}	0.00	0.00
10.2	$O_3 + H \rightleftharpoons O_2 + OH$	8.43×10^{13}	0.00	3.91
10.3	$O_3 + O \rightleftharpoons O_2 + O_2$	4.81×10^{12}	0.00	17.13
10.4	$O_3 + OH \rightleftharpoons O_2 + HO_2$	1.15×10^{12}	0.00	8.31
10.5	$O_3 + H_2O \rightleftharpoons O_2 + H_2O_2$	6.20×10^{1}	0.00	0.00
10.6	$O_3 + HO_2 \rightleftharpoons 2O_2 + OH$	8.43×10^{9}	0.00	4.99
10.7	$O_3 + CO \rightleftharpoons O_2 + CO_2$	6.02×10^{2}	0.00	0.00
10.8	$O_3 + HCO \rightleftharpoons O_2 + H + CO_2$	5.00×10^{11} (@ 298 K)		
10.9	$O_3 + N \rightleftharpoons O_2 + NO$	6.00×10^{7} (@ 298 K)		
10.10	$O_3 + NO \rightleftharpoons O_2 + NO_2$	1.08×10^{12}	0.00	11.39
10.11	$O_3 + NO_2 \rightleftharpoons O_2 + NO_3$	7.22×10^{10}	0.00	20.37

[a]Reaction rates in cm^3 mol kJ units, $k = AT^n \exp(-E/RT)$.

Source.   R. Atkinson, D. L. Baulch, R. A. Cox, R. F. Hampson, Jr., J. A. Kerr, and J. Troe, Evaluated Kinetic and Photochemical Data for Atmospheric Chemistry. Supplement IV. IUPAC Subcommittee on Gas Kinetic Data Evaluation for Atmospheric Chemistry, *J. Phys. Chem. Ref. Data*, Vol. 21, No. 6, 1992, pp. 1125–1568.

TABLE 11 $SO_x/N_xO_y/CO/H_2/O_2$ Mechanism[a]

	A	n	E
S and SO Reactions			
11.1 $SO + M \rightleftharpoons S + O + M$	4.00×10^{14}	0.00	447.69
11.2 $S + OH \rightleftharpoons SO + H$	4.00×10^{13}	0.00	0.00
11.3 $S + O_2 \rightleftharpoons SO + O$	2.00×10^{6}	1.93	-5.86
SO_2 Reactions			
11.4 $SO + O + M \rightleftharpoons SO_2 + M$	1.10×10^{22}	-1.84	0.00
11.5 $SO + OH \rightleftharpoons SO_2 + H$	5.20×10^{13}	0.00	0.00
11.6 $SO + O_2 \rightleftharpoons SO_2 + O$	6.20×10^{3}	2.42	12.76
11.7 $SO_2 + CO \rightleftharpoons SO + CO_2$	2.70×10^{12}	0.00	202.09
11.8 $SO + NO_2 \rightleftharpoons SO_2 + NO$	8.40×10^{12}	0.00	0.00
11.9 $SO + SO \rightleftharpoons SO_2 + S$	2.00×10^{12}	0.00	16.74
SO_3 Reactions			
11.10 $SO_2 + O + M \rightleftharpoons SO_3 + M$	4.00×10^{28}	-4.00	21.97
11.11 $SO_2 + OH \rightleftharpoons SO_3 + H$	4.90×10^{2}	2.69	99.58
11.12 $SO_3 + O \rightleftharpoons SO_2 + O_2$	1.30×10^{12}	0.00	25.52
11.13 $SO_2 + NO_2 \rightleftharpoons SO_3 + NO$	6.30×10^{12}	0.00	112.97
11.14 $SO_3 + SO \rightleftharpoons SO_2 + SO_2$	1.00×10^{12}	0.00	16.74
SH Reactions			
11.15 $SH + O_2 \rightleftharpoons SO + OH$	1.00×10^{12}	0.00	41.84
11.16 $S + H_2 \rightleftharpoons SH + H$	6.00×10^{14}	0.00	100.42
11.17 $SH + O \rightleftharpoons SO + H$	1.00×10^{14}	0.00	0.00
11.18 $SH + OH \rightleftharpoons S + H_2O$	1.00×10^{13}	0.00	0.00
H_2S Reactions			
11.19 $H_2S + M \rightleftharpoons S + H_2 + M$	2.00×10^{14}	0.00	276.14
11.20 $H_2S + H \rightleftharpoons SH + H_2$	1.20×10^{7}	2.10	2.93
11.21 $H_2S + O \rightleftharpoons SH + OH$	6.40×10^{7}	1.78	11.88
11.22 $H_2S + OH \rightleftharpoons SH + H_2O$	2.70×10^{12}	0.00	0.00
11.23 $H_2S + S \rightleftharpoons SH + SH$	4.00×10^{14}	0.00	63.18
HSO Reactions			
11.24 $SO + H + M \rightleftharpoons HSO + M$	5.00×10^{15}	0.00	0.00
11.25 $HSO + O_2 \rightleftharpoons SO_2 + OH$	1.00×10^{12}	0.00	41.84
11.26 $HSO + H \rightleftharpoons SH + OH$	4.90×10^{19}	-1.86	6.53
11.27 $HSO + H \rightleftharpoons S + H_2O$	1.60×10^{9}	1.37	-1.42
11.28 $HSO + H \rightleftharpoons H_2S + O$	1.10×10^{6}	1.03	43.51
11.29 $HSO + H \rightleftharpoons SO + H_2$	1.00×10^{12}	0.00	0.00
11.30 $HSO + O \rightleftharpoons SO + OH$	1.40×10^{13}	0.15	1.26
11.31 $HSO + O \rightleftharpoons SO_2 + H$	4.50×10^{14}	-0.40	0.00
11.32 $HSO + OH \rightleftharpoons SO + H_2O$	1.70×10^{9}	1.03	1.67
11.33 $SH + HO_2 \rightleftharpoons HSO + OH$	1.00×10^{12}	0.00	0.00
HOS Reactions			
11.34 $HSO + O \rightleftharpoons HOS + O$	4.80×10^{8}	1.02	22.34
HSO_2 Reactions			
11.35 $HSO_2 + M \rightleftharpoons SO_2 + H + M$	1.20×10^{28}	-4.41	79.08
11.36 $HSO + O + M \rightleftharpoons HSO_2 + M$	1.10×10^{19}	-1.73	-0.21

(continues)

TABLE 11 *(continued)*

	A	n	E
HOSO Reactions			
11.37 $SO + OH + M \rightleftharpoons HOSO + M$	8.00×10^{21}	-2.16	3.47
11.38 $HSO + O + M \rightleftharpoons HOSO + M$	6.90×10^{19}	-1.61	6.65
11.39 $HOSO + M \rightleftharpoons HOS + O + M$	2.50×10^{30}	-4.80	498.00
11.40 $HOSO + M \rightleftharpoons SO_2 + H + M$	5.90×10^{34}	-5.67	212.97
11.41 $HSO_2 + M \rightleftharpoons HOSO + M$	1.10×10^{21}	-1.99	125.10
11.42 $HOSO + O_2 \rightleftharpoons SO_2 + HO_2$	1.00×10^{12}	0.00	4.18
11.43 $HOSO + H \rightleftharpoons SO + H_2O$	6.30×10^{-10}	6.29	-7.95
11.44 $HSO + OH \rightleftharpoons HOSO + H$	5.30×10^{7}	1.57	15.69
11.45 $SO_2 + OH \rightleftharpoons HOSO + O$	3.90×10^{8}	1.89	317.98
11.46 $SO_3 + H \rightleftharpoons HOSO + O$	2.50×10^{5}	2.92	210.46
11.47 $HOSO + OH \rightleftharpoons SO_2 + H_2O$	1.00×10^{12}	0.00	0.00
11.48 $HSO + NO_2 \rightleftharpoons HOSO + NO$	5.80×10^{12}	0.00	0.00
H$_2$SO Reactions			
11.49 $HSO + H \rightleftharpoons H_2SO$	1.80×10^{17}	-2.47	0.21
11.50 $H_2SO \rightleftharpoons H_2S + O$	4.90×10^{28}	-6.66	300.00
HSOH Reactions			
11.51 $HSO + H \rightleftharpoons HSOH$	2.50×10^{20}	-3.14	3.85
11.52 $HSOH \rightleftharpoons SH + OH$	2.80×10^{39}	-8.75	314.64
11.53 $HSOH \rightleftharpoons S + H_2O$	5.80×10^{29}	-5.60	228.03
11.54 $HSOH \rightleftharpoons H_2S + O$	9.80×10^{16}	-3.40	361.92
HOSHO Reactions			
11.55 $HSO + OH \rightleftharpoons HOSHO$	5.20×10^{28}	-5.44	13.26
11.56 $HOSHO \rightleftharpoons HOSO + H$	6.40×10^{30}	-5.89	308.78
11.57 $HOSHO + H \rightleftharpoons HOSO + H_2$	1.00×10^{12}	0.00	0.00
11.58 $HOSHO + O \rightleftharpoons HOSO + OH$	5.00×10^{12}	0.00	0.00
11.59 $HOSHO + OH \rightleftharpoons HOSO + H_2O$	1.00×10^{12}	0.00	0.00
HOSO$_2$ Reactions			
11.60 $SO_2 + OH \rightleftharpoons HOSO_2$	1.00×10^{25}	-4.34	12.76
11.61 $HOSO_2 \rightleftharpoons HOSO + O$	5.40×10^{18}	-2.34	444.76
11.62 $HOSO_2 \rightleftharpoons SO_3 + H$	1.40×10^{18}	-2.91	229.70
11.63 $HOSO_2 + O_2 \rightleftharpoons SO_3 + HO_2$	7.80×10^{11}	0.00	2.74
11.64 $HOSO_2 + H \rightleftharpoons SO_2 + H_2O$	1.00×10^{12}	0.00	0.00
11.65 $HOSO_2 + O \rightleftharpoons SO_3 + OH$	5.00×10^{12}	0.00	0.00
11.66 $HOSO_2 + OH \rightleftharpoons SO_3 + H_2O$	1.00×10^{12}	0.00	0.00

[a] Reaction rates in cm^3 mol kJ units, $k = AT^n \exp(-E/RT)$.

Source. P. Glarborg, D. Kubel, Kim Dam-Johnansen, H.-M. Chiang, and J. Bozzelli (in press). Impact of SO_x and NO on CO Oxidation Under Post-Flame Conditions, *Int. J. Chem. Kinet.*

Note. All pressure dependent rate constants (except 11.19) have been estimated for 300–1500 K, 1 atm.

Appendix C

Bond Dissociation Energies of Hydrocarbons

The bond dissociation energies which follow are taken from the review of McMillan and Golden [*Ann. Rev. Phys. Chem.* **33**, 493 (1982)]. The reader should refer to this publication for the methods of determining the values presented, their uncertainty, and the original sources. In the tables presented, all bond energies and heats of formation are expressed in kJ/mol. The values listed in the first column are the heats of formation at 298 K for the reference radical and those above the column heading for the associated radical. Thus, the tables presented are not only a source of bond energies, but also heats of formation of radicals.

McMillan and Golden employ the commonly invoked uncertainty of 5 kJ/mol (1 kcal/mol) for dissociation energies and most of the uncertainties for the heats of formation fall in the same range. The reader is urged to refer to McMillan and Golden (1982) for the specific uncertainty values.

TABLE 1   Bond Dissociation Energies of Alkanes[a]

ΔH_f^0 (R)	R	(218) H	CH_3	C_2H_5	i-C_3H_7	t-C_4H_9	(329) C_6H_5	(200) $PhCH_2$
147	CH_3	440	378	359	359	352	426	317
108	C_2H_5	411	359	344	339	331	408	300
88	n-C_3H_7	410	362	344	336	330	409	302
76	i-C_3H_7	398	359	339	331	316	401	298
54	s-C_4H_9	400	356	335	328	—	—	—
36	t-C_4H_9	390	352	331	316	298	—	291
36	$CH_2C(CH_3)_3$	418	346	346	—	—	—	—
280	cyclopropyl	445	—	—	—	—	—	—
214	cyclopropyl-methyl	408	—	—	—	—	—	—
214	cyclobutyl	404	—	—	—	—	—	—
102	cyclopentyl	395	—	—	—	—	—	—
58	cyclohexyl	400	—	—	—	—	—	—
51	cycloheptyl	387	—	—	—	—	—	—

[a] Note that in alkanes, values of 410, 397, and 389 kJ/mol characterize primary, secondary, and tertiary C—H bonds, respectively.

TABLE 2   Bond Dissociation Energies of Alkenes, Alkynes, and Aromatics

ΔH_f^0 (R)	R	H	CH_3
295	•C=C	460	421
329	•⬡	464	426
565	•C≡C	552	526
−548	•C_6F_5	477	—
164	•C–C=C̄	361	311
126	•C–C=C (with C substituent)	358	305
127	•C–C=C (with C substituent)	345	—
77	•C–C=C (with C substituents)	323	285
38	•C–C=C (with C substituents)	319	—

(*continues*)

TABLE 2   (*continued*)

ΔH_f^0 (R)	R	H	CH₃
40	•C–C=C (with C, C substituents and C below)	326	—
—	•C–C=C (with Cl)	371	—
161	• (cyclopentenyl)	344	—
205	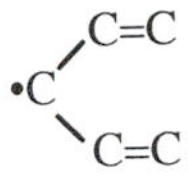 •C(C=C)(C=C)	318	—
205	•C–C=C–C=C	347	—
242	 • (cyclopentadienyl)	297	—
1977	• (phenyl)	305	—
271	• (cycloheptatrienyl)	305	—
440	• (cyclopropenyl)	379	—
200	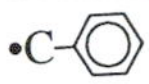 •C–(phenyl)	368	317
253	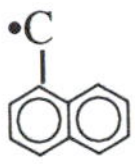 •C (naphthyl)	356	305
338	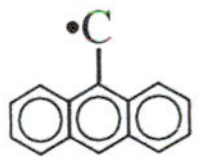 •C (anthracenyl)	342	283
311	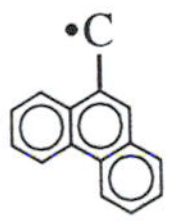 •C (phenanthrenyl)	356	305
169	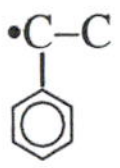 •C–C (phenyl)	357	312
139	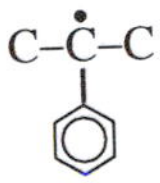 C–•C–C (phenyl)	353	308
—	•C–(furyl, O)	362	314
—	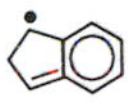 • (indenyl)	351	—
341	•C–C≡C	374	318

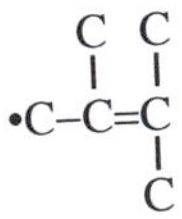

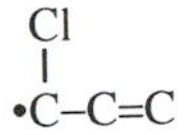

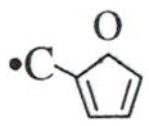

(*continues*)

TABLE 2   (*continued*)

ΔH_f^0 (R)	R	H	CH_3
294	•C–C≡C–C	365	308
273	$\overset{\textstyle C}{\overset{\textstyle \|}{\bullet C-C\equiv C-C}}$	365	321
222	$\overset{\textstyle C}{\overset{\textstyle \|}{\underset{\textstyle \|}{\underset{\textstyle C}{\bullet C-C\equiv C-C}}}}$	344	303
257	$\overset{\textstyle C}{\overset{\textstyle \|}{\underset{\textstyle \|}{\underset{\textstyle C}{\bullet C-C\equiv C}}}}$	339	296
295	$\underset{\textstyle \|}{\underset{\textstyle C}{\bullet C-C\equiv C}}$	348	305

TABLE 3   Bond Dissociation Energies of C/H/O Compounds

	Oxygen-centered radicals					
		(218)		(39)	(147)	(108)
ΔH_f^0(R)	R_1	H	R_1	OH	CH_3	C_2H_5
39	OH	498	213	213	386	383
18	OCH_3	437	157	—	349	342
−17	OC_2H_5	436	159	—	346	344
−41	$O\text{-}n\text{-}C_3H_7$	433	155	—	343	—
−63	$O\text{-}n\text{-}C_4H_9$	431	—	—	—	—
−52	$O\text{-}i\text{-}C_3H_7$	438	158	—	346	—
−69	$O\text{-}s\text{-}C_4H_9$	441	152	—	—	—
−91	$O\text{-}t\text{-}C_4H_9$	440	159	—	348	—
—	$O\text{-}t\text{-}C_5H_{11}$	—	164	—	—	—
—	$OCH_2C(CH_3)_3$	428	152	194	—	—
48	OC_6H_5	362	—	—	267	264
—	OCF_3	—	193	—	—	—
—	$OC(CF_3)_3$	—	149	—	—	—
10	O_2H	365	—	—	—	—
−208	O_2CCH_3	443	127	—	—	346
−228	$O_2CC_2H_5$	445	127	—	—	—
−249	$O_2C\text{-}nC_3H_7$	443	127	—	—	—

(*continues*)

TABLE 3   (*continued*)

ΔH_f^0 (R)	R_1	(218) H	(147) CH_3	(329) C_6H_5	R_1
		Carbon-centered radicals			
37	CHO	364	345	403	286
−24	$COCH_3$	360	340	391	282
72	$COCH=CH_2$	364	—	—	—
−43	COC_2H_5	366	337	395	—
109	COC_6H_5	364	343	377	278
—	$COCF_3$	381	—	—	—
−24	CH_2COCH_3	411	361	—	—
−70	$CH(CH_3)COCH_3$	386	—	—	—
−26	CH_2OH	393	—	403	336
−64	$CH(OH)CH_3$	389	—	—	355
−111	$C(OH)(CH_3)_2$	381	—	—	—
−12	CH_2OCH_3	389	361	—	—
−18	Tetrahydrofuran-2-yl	385	—	—	—
0.0	$CH(OH)CH=CH_2$	341	—	—	—
−169	$COOCH_3$	388	—	—	—
−70	$CH_2OCOC_6H_5$	419	—	—	—
−223	$COOH-CH_2C_6H_5$	280	—	—	—
248	$(C_6H_5)_2CH-COOH$	249	—	—	—
—	$C_6H_5CH_2CO-CH_2C_6H_5$	274	—	—	—
200	$C_6H_5CH_2-OH$	340	—	—	—
48	$C_6H_5CO-CF_3$	309	—	—	—

TABLE 4   Bond Dissociation Energies of Sulfur-Containing Compounds

R_1-R_2	D_{298}^0	$\Delta H_f^0(R_1)$
HS—H	381	141
CH_3S—H	379	139
RS—H	381	—
CH_3—SH	310	—
C_2H_5—SH	295	—
t-Bu—SH	286	—
C_6H_5—SH	362	—
CH_3S—CH_3	323	—
CH_3S—C_2H_5	307	—
CH_3S—n-C_3H_7	309	—
PhS—H	349	—
PhS—CH_3	290	230
$PhCH_2$—SCH_3	257	—
CS—S	431	272
OS—O	544	—
CH_3SO_2—CH_3	279	—
CH_3SO_2—$CH_2CH=CH_3$	208	—
CH_3SO_2—$CH_2C_6H_5$	221	—
RS_2—H	293	—
RS_2—CH_3	238	—
HS—SH	276	—
RS—SR	301	—

TABLE 5   Bond Dissociation Energies of Nitrogen-Containing Compounds

		Amines and nitriles					
		(218)	(147)	(108)	(200)	(329)	(189)
ΔH_f^0 (R)	R	H	CH_3	C_2H_5	$PhCH_2$	C_6H_5	NH_2
185	NH_2	449	355	341	297	427	275
177	$NHCH_3$	418	344	334	287	421	268
145	$N(CH_3)_2$	383	316	303	260	390	247
237	NHC_6H_5	368	299	289	—	339	219
233	$N(CH_3)C_6H_5$	366	296	—	—	—	—
33	NF_2	317	—	—	—	—	—
469	N_3	385	—	—	—	—	—
149	CH_2NH_2	390	344	332	285	390	—
126	$CH_2NH(CH_3)$	364	320	—	—	367	—
109	$CH_2N(CH_3)_2$	351	308	—	—	352	—
435	CN	518	510	495	—	548	—
245	CH_2CN	389	340	322	—	—	—
209	$CH(CH_3)CN$	376	330	—	—	—	—
167	$C(CH_3)_2CN$	362	313	—	—	—	—
249	$C(CH_3)(CN)C_6H_5$	—	251	—	—	—	—

		Nitro and nitroso compounds and nitrates		
		(90)	(33)	(71)
ΔH_f^0 (R)	R	NO	NO_2	ONO_2
218	H	—	328	423
39	OH	206	207	163
147	CH_3	167	254	—
108	i-C_3H_5	—	245	—
76	i-C_3H_7	153	247	—
36	t-C_4H_9	165	245	—
−467	CF_3	179	—	—
79	CCl_3	134	—	—
331	C_6H_5	213	298	—
−548	C_6F_5	208	—	—
—	$C(NO_2)R_2$	—	204	—
—	$C(NO_2)_2R$	—	183	—
—	$C(NO_2)_3$	—	169	—
—	RO	171	170	—
33	NO_2	41	57	—

TABLE 6 Bond Dissociation Energies of Halocarbons

ΔH_f^0 (R)	R	(218) H	(147) CH_3	(79) F	(121) Cl	(112) Br	(107) I	(−465) CF_3
147	CH_3	440	378	460	354	297	239	425
108	C_2H_5	411	—	451	334	284	223	—
76	i-C_3H_7	398	359	446	338	286	224	—
200	$CH_2C_6H_5$	368	317	—	302	241	202	—
−33	CH_2F	418	498	357	357	—	—	396
−248	CHF_2	423	400	527	—	289	—	—
−467	CF_3	446	425	546	361	295	230	413
118	CH_2Cl	422	—	—	335	—	—	—
101	$CHCl_2$	414	—	—	325	—	—	—
79	CCl_3	401	—	426	306	231	—	—
−269	CF_2Cl	425	—	515	318	270	—	—
−96	$CFCl_2$	—	—	460	305	—	—	—
174	CH_2Br	427	—	—	—	—	—	—
227	$CHBr_2$	434	—	—	—	—	—	—
—	CBr_3	402	—	—	—	235	—	—
−893	C_2F_5	430	—	531	346	287	214	—
—	n-C_3F_7	435	—	—	—	278	208	—
—	i-C_3F_7	431	—	—	—	274	—	—
−303	CF_2CH_3	416	—	522	—	—	218	—
−517	CH_2CF_3	446	—	458	—	—	236	—
—	$CHClCF_3$	426	—	—	—	275	—	—
—	$CClBrCF_3$	404	—	—	—	251	—	—
435	CN	518	510	470	422	367	303	561
39	OH	498	387	—	251	234	234	—
331	C_6H_5	466	428	526	400	337	274	—
−548	C_6F_5	487	—	—	383	—	277	—

Appendix D

Laminar Flame Speeds

The compilation of laminar flame speed data given in Tables 1 and 2 is due to Gibbs and Calcote [*J. Chem. Eng. Data* **4**, 226 (1959)]. The reader is referred to the quoted paper for details on the chosen values. The data are for premixed fuel–air mixtures at 25°C and 100°C and 1 atm pressure. Examples of more recent data (obtained from a counterflow, double-flame, burner configuration vs the Bunsen cone burner configuration) have also been included from Law [in *Reduced Kinetic Mechanisms for Applications in Combustion Systems,* N. Peters and B. Rogg, eds., Springer-Verlag, NY, 1993] and Vagelopoulos, Egolfopoulous, and Law [*Twenty-fifth Symposium (International) on Combustion,* The Combustion Institute, 1994, pp. 1341–1347]. The values of Law and co-workers are denoted by the letters (L) and (L2) following the flame speed. Table 3 from Law reports flame speed data as a function of pressure.

TABLE 1 Burning Velocities of Various Fuels at 25°C Air–Fuel Temperature (0.31 mol% H_2O in Air). Burning Velocity S as a Function of Equivalence Ratio ϕ in cm s^{-1}

Fuel	$\phi = 0.7$	0.8	0.9	1.0	1.1	1.2	1.3	1.4	S_{max}	ϕ at S_{max}
Saturated hydrocarbons										
Ethane	30.6, 22.0L	36.0, 29.0L	40.6, 36.5L	44.5, 42.5L	47.3, 43.0L	47.3, 42.5L	44.4, 40.0L	37.4, 27.5L	47.6	1.14
Propane	24.0L, 23.0L2	32.0L, 30.0L2	42.3, 39.5L, 37.0L2	45.6, 44.0L, 39.0L2	46.2, 45.0L, 41.0L2	42.4, 43.5L, 40.5L2	34.3, 37.0L, 33.5L2	28.0L, 25.0L2	46.4	1.06
n-Butane		38.0	42.6	44.8	44.2	41.2	34.4	25.0	44.9	1.03
Methane	20.5L, 17L2	30.0, 28.0L, 25.0L2	38.3, 36.0L, 33.0L2	43.4, 40.5L, 38L2	44.7, 42.0L, 38.5L2	39.8, 37.0L, 34.0L2	31.2, 27.0L, 24.0L2	17.5L, 13.5L2	44.8	1.08
n-Pentane		35.0	40.5	42.7	42.7	39.3	33.9	—	43.0	1.05
n-Heptane		37.0	39.8	42.2	42.0	35.5	29.4	—	42.8	1.05
2,2,4-Trimethylpentane		37.5	40.2	41.0	37.2	31.0	23.5	—	41.0	0.98
2,2,3-Trimethylpentane		37.8	39.5	40.1	39.5	36.2	—	—	40.1	1.00
2,2-Dimethylbutane		33.5	38.3	39.9	37.0	33.5	—	—	40.0	0.98
Isopentane		33.0	37.6	39.8	38.4	33.4	24.8	—	39.9	1.01
2,2-Dimethylpropane			31.0	34.8	36.0	35.2	33.5	31.2	36.0	1.10
Unsaturated hydrocarbons										
Acetylene		107, 107L	130	144, 136L	151	154, 151L	154	152, 155L	155	1.25
Ethylene	37.0, 37.0L	50.0, 48.0L	60.0, 60.0L	68.0, 66.0L	73.0, 70.0L	72.0, 72.0L	66.5, 71.0L	60.0, 65.0L	73.5	1.13
Propylene		62.0	66.6	70.2	72.2	71.2	61.0	—	72.5	1.14
1,3-Butadiene			42.6	49.6	55.0	57.0	56.9	55.4	57.2	1.23
n-1-Heptene		46.8	50.7	52.3	50.9	47.4	41.6	—	52.3	1.00
Propylene			48.4	51.2	49.9	46.4	40.8	—	51.2	1.00

(continues)

n-2-Pentene	—	35.1	42.6	47.8	46.9	42.6	34.9	—	48.0	1.03
2,2,4-Trimethyl-3-pentene	—	34.6	41.3	42.2	37.4	33.0	—	—	42.5	0.98
Substituted alkyls										
Methanol	—	34.5	42.0	48.0	50.2	47.5	44.4	42.2	50.4	1.08
	21.5L	31.0L	37.5L	48.0L	54.0L	53.5L	48.0L	42.0L	@45°C	
Isopropyl alcohol	—	34.4	39.2	41.3	40.6	38.2	36.0	34.2	41.4	1.04
Triethylamine	—	32.5	36.7	38.5	38.7	36.2	28.6	—	38.8	1.06
n-Butyl chloride	24.0	30.7	33.8	34.5	32.5	26.9	20.0	—	34.5	1.00
Allyl chloride	30.6	33.0	33.7	32.4	29.6	—	—	—	33.8	0.89
Isopropyl mercaptan	—	30.0	33.5	33.0	26.6	29.4	25.3	—	33.8	0.94
Ethylamine	—	28.7	31.4	32.4	31.8	27.7	—	—	32.4	1.00
Isopropylamine	—	27.0	29.5	30.6	29.8	—	—	—	30.6	1.01
n-Propyl chloride	—	24.7	28.3	27.5	24.1	—	—	—	28.5	0.93
Isopropyl chloride	—	24.8	27.0	27.4	25.3	—	—	—	27.6	0.97
n-Propyl bromide	No ignition									
Silanes										
Tetramethylsilane	—	39.5	57.3	58.2	57.7	54.5	47.5	—	58.2	1.01
Trimethylethoxysilane	—	34.7	47.4	50.3	46.5	41.0	35.0	—	50.3	1.00
Aldehydes										
Acrolein	47.0	58.0	66.6	65.9	56.5	—	—	—	67.2	0.95
Propionaldehyde	—	37.5	44.3	49.0	49.5	46.0	41.6	37.2	50.0	1.06
Acetaldehyde	—	26.6	35.0	41.4	41.4	36.0	30.0	—	42.2	1.05
Ketones										
Acetone	—	40.4	44.2	42.6	38.2	—	—	—	44.4	0.93
Methyl ethyl ketone	—	36.0	42.0	43.3	41.5	37.7	33.2	—	43.4	0.99
Esters										
Vinyl acetate	29.0	36.6	39.8	41.4	42.1	41.6	35.2	—	42.2	1.13
Ethyl acetate	—	30.7	35.2	37.0	35.6	30.0	—	—	37.0	1.00
Ethers										
Dimethyl ether	—	44.8	47.6	48.4	47.5	45.4	42.6	—	48.6	0.99
Diethyl ether	30.6	37.0	43.4	48.0	47.6	40.4	32.0	—	48.2	1.05

TABLE 1 (*continued*)

Fuel	φ = 0.7	0.8	0.9	1.0	1.1	1.2	1.3	1.4	S_{max}	φ at S_{max}
Dimethoxymethane	32.5	38.2	43.2	46.6	48.0	46.6	43.3	—	48.0	1.10
Diisopropyl ether	—	30.7	35.5	38.3	38.6	36.0	31.2	—	38.9	1.06
Thio ethers										
Dimethyl sulfide	—	29.9	31.9	33.0	30.1	24.8	—	—	33.0	1.00
Peroxides										
Di-*tert*-butyl peroxide	—	41.0	46.8	50.0	49.6	46.5	42.0	35.5	50.4	1.04
Aromatic Compounds										
Furan	48.0	55.0	60.0	62.5	62.4	60.0	—	—	62.9	1.05
Benzene	—	39.4	45.6	47.6	44.8	40.2	35.6	—	47.6	1.00
Thiophene	33.8	37.4	40.6	43.0	42.2	37.2	24.6	—	43.2	1.03
Cyclic compounds										
Ethylene oxide	57.2	70.7	83.0	88.8	89.5	87.2	81.0	73.0	89.5	1.07
Butadiene monoxide	—	36.6	47.4	57.8	64.0	66.9	66.8	64.5	67.1	1.24
Propylene oxide	41.6	53.3	62.6	66.5	66.4	62.5	53.8	—	67.0	1.05
Dihydropyran	39.0	45.7	51.0	54.5	55.6	52.6	44.3	32.0	55.7	1.08
Cyclopropane	—	40.6	49.0	54.2	55.6	53.5	44.0	—	55.6	1.10
Tetrahydropyran	44.8	51.0	53.6	51.5	42.3	—	—	—	53.7	0.93
Tetrahydrofuran	—	—	43.2	48.0	50.8	51.6	49.2	44.0	51.6	1.19
Cyclopentadiene	36.0	41.8	45.7	47.2	45.5	40.6	32.0	—	47.2	1.00
Ethylenimine	—	37.6	43.4	46.0	45.8	43.4	38.9	—	46.4	1.04
Cyclopentane	31.0	38.4	43.2	45.3	44.6	41.0	34.0	—	45.4	1.03
Cyclohexane	—	—	41.3	43.5	43.9	38.0	—	—	44.0	1.08
Inorganic compounds										
Hydrogen	102, 124L	120, 150L	145, 187L	170, 210L	204, 230L	245, 245L	213	290	325	1.80
Carbon disulfide	50.6	58.0	59.4	58.8	57.0	55.0	52.8	51.6	59.4	0.91
Carbon monoxide	—	—	—	—	28.5	32.0	34.8	38.0	52.0	2.05
Hydrogen sulfide	34.8	39.2	40.9	39.1	32.3	—	—	—	40.9	0.90

TABLE 2 Burning Velocities of Various Fuels at 100°C Air–Fuel Temperature (0.31 mol% H_2O in Air). Burning Velocity S as a Function of Equivalence Ratio ϕ in cm s^{-1}

Fuel	$\phi = 0.7$	0.8	0.9	1.0	1.1	1.2	1.3	1.4	S_{max}	ϕ at S_{max}
Propargyl alcohol	—	76.8	100.0	110.0	110.5	108.8	105.0	85.0	110.5	1.08
Propylene oxide	74.0	86.2	93.0	96.6	97.8	94.0	84.0	71.5	97.9	1.09
Hydrazine[a]	87.3	90.5	93.2	94.3	93.0	90.7	87.4	83.7	94.4	0.98
Furfural	62.0	73.0	83.3	87.0	87.0	84.0	77.0	65.5	87.3	1.05
Ethyl nitrate	70.2	77.3	84.0	86.4	83.0	72.3	—	—	86.4	1.00
Butadiene monoxide	51.4	57.0	64.5	73.0	79.3	81.0	80.4	76.7	81.1	1.23
Carbon disulfide	64.0	72.5	76.8	78.4	75.5	71.0	66.0	62.2	78.4	1.00
n-Butyl ether	—	67.0	72.6	70.3	65.0	—	—	—	72.7	0.91
Methanol	50.0,	58.5,	66.9,	71.2,	72.0,	66.4,	58.0,	48.8,	72.2	1.08
	31.5L	43.0L	59.5L	63.5L	66.0L	65.0L	61.5L	51.0L	@95°C	
Diethyl cellosolve	49.5	56.0	63.0	69.0	69.7	65.2	—	—	70.4	1.05
Cyclohexane monoxide	54.5	59.0	63.5	67.7	70.0	64.0	—	—	70.0	1.10
Epichlorohydrin	53.0	59.5	65.0	68.6	70.0	66.0	58.2	—	70.0	1.10
n-Pentane	—	50.0	55.0	61.0	62.0	57.0	49.3	42.4	62.9	1.05
n-Propyl alcohol	49.0	56.6	62.0	64.6	63.0	50.0	37.4	—	64.8	1.03
n-Heptane	41.5	50.0	58.5	63.8	59.5	53.8	46.2	38.8	63.8	1.00
Ethyl nitrite	54.0	58.8	62.6	63.5	59.0	49.5	42.0	36.7	63.5	1.00
Pinene	48.5	58.3	62.5	62.1	56.6	50.0	—	—	63.0	0.95
Nitroethane	51.5	57.8	61.4	57.2	46.0	28.0	—	—	61.4	0.92
Isooctane	—	50.2	56.8	57.8	53.3	50.5	—	—	58.2	0.98
Pyrrole	—	52.0	55.6	56.6	56.1	52.8	48.0	43.1	56.7	1.00
Aniline	—	41.5	45.4	46.6	42.9	37.7	32.0	—	46.8	0.98
Dimethylformamide	—	40.0	43.6	45.8	45.5	40.7	36.7	—	46.1	1.04

[a]Results are questionable because of an indication of decomposition in the stainless-steel feed system.

579

TABLE 3 Burning Velocities of Various Fuels in Air as a Function of Pressure for an Equivalence Ratio of 1 in cm s^{-1}

Fuel	$P = 0.25$ atm	0.5	1.0	2.0	3.0
Methane	59L	50L	40.5L	29L	22L
Ethane	54L	48L	42.5L	36.5L	30L
Propane	59L	51L	44.0L	35.5L	31L

Flammability Limits in Air

The data presented in Table 1 are for fuel gases and vapors and are taken almost exclusively from Zabetakis [*U.S. Bur. Mines Bulletin* 627 (1965)]. The conditions are for the fuel–air mixture at 25°C and 1 atm unless otherwise specified. As noted in the text, most fuels have a rich limit at approximately $\phi = 3.3$ and a lean limit at approximately $\phi = 0.5$. The fuels which vary most from the rich limit are those that are either very tightly bound as ammonia is or which can decompose as hydrazine or any monopropellant does.

There can also be a flammability limit associated with dust clouds. The flammability limits of combustible dusts are reported as the minimum explosion concentrations. The upper explosion limits for dust clouds have not been determined due to experimental difficulties. In the fourteenth edition of the Fire Protection Handbook [National Fire Protection Association (NFPA), Boston, MA, 1975], numerous results from the U.S. Bureau of Mines reports are listed. These results were obtained with dusts 74 μm or smaller. It should be noted that variations in minimum explosive concentrations will occur with change in particle diameter, i.e., the minimum explosive concentration is lowered as the diameter of the particle decreases. Other conditions which affect this limit are sample purity, oxygen concentration, strength of ignition source, turbulence, and uniformity of the dispersion. The NFPA tabulation is most extensive and includes data for dusts from agricultural materials, carbonaceous matter, chemicals, drugs, dyes, metals, pesticides, and various plastic resins and molding compounds. Except for metal dusts, it is rather remarkable that most materials have a minimum explosive concentration in the range 0.03–0.05 kg/m^3. It should be noted, however, that the variation according to the specific compound can range from 0.01 to 0.50 kg/m^3. For a specific value the reader should refer to the NFPA handbook.

TABLE 1   Flammability Limits of Fuel Gases and Vapors in Air at 25°C and 1 atm

| *Fuel* | *Lean limit* | | *Rich limit* | |
	Vol %	*(Vol %)/ (Vol %)$_{ST}$, [φ]*	*Vol %*	*(Vol %)/ (Vol %)$_{ST}$, [φ]*
Acetal	1.6		10	
Acetaldehyde	4.0	0.52	60	
Acetic acid	5.4 (100°C)			
Acetic anhydride	2.7 (47°C)		10 (75°C)	
Acetanilide	1.0 (calc)			
Acetone	2.6	0.52	13	2.6
Acetophenone	1.1 (calc)			
Acetylacetone	1.7 (calc)			
Acetyl chloride	5.0 (calc)			
Acetylene	2.5		100	
Acrolein	2.8		31	
Acrylonitrile	3.0			
Acetone cyanohydrin	2.2		12	
Adipic acid	1.6 (calc)			
Aldol	2.0 (calc)			
Allyl alcohol	2.5		18	
Allyl amine	2.2		22	
Allyl bromide	2.7 (calc)			
Allyl chloride	2.9			
o-Aminodiphenyl	0.66		4.1	
Ammonia	15	0.69 [0.63]	28	1.3 [1.4]
n-Amyl acetate	1.0 (100°C)	0.51	7.1 (100°C)	3.3
n-Amyl alcohol	1.4 (100°C)	0.51	10 (100°C)	
t-Amyl alcohol	1.4 (calc)			
n-Amyl chloride	1.6 (50°C)		8.6 (100°C)	
t-Amyl chloride	1.5 (85°C)			
n-Amyl ether	0.7 (calc)			
Amyl nitrite	1.0 (calc)			
n-Amyl propionate	1.0 (calc)			
Amylene	1.4		8.7	
Aniline	1.2 (140°C)		8.3 (140°C)	
Anthracene	0.65 (calc)			
n-Amyl nitrate	1.1			

(continues)

TABLE 1   (*continued*)

Fuel	Lean limit		Rich limit	
	Vol %	(Vol %)/ (Vol %)$_{ST}$, [ϕ]	Vol %	(Vol %)/ (Vol %)$_{ST}$, [ϕ]
Benzene	1.3 (100°C)	0.48	7.9 (100°C)	2.9
Benzyl benzoate	0.7 (calc)			
Benzyl chloride	1.2 (calc)			
Bicyclohexyl	0.65 (100°C)		5.1 (150°C)	
Biphenyl	0.70 (110°C)			
2-Biphenyl amine	0.8 (calc)			
Bromobenzene	1.6 (calc)			
Butadiene (1,3)	2.0	0.54	12	3.3
n-Butane	1.8	0.58 [0.57]	8.4	2.7 [2.8]
1,3-Butandiol	1.9 (calc)			
Butene-1	1.6	0.50	10	2.9
Butene-2	1.7	0.53	9.7	2.9
n-Butyl acetate	1.4 (50°C)	0.55	8.0 (100°C)	3.1
n-Butyl alcohol	1.7 (100°C)	0.5	12 (100°C)	
s-Butyl alcohol	1.7 (100°C)		9.8 (100°C)	
t-Butyl alcohol	1.9 (100°C)		9.0 (100°C)	
t-Butyl amine	1.7 (100°C)		8.9 (100°C)	
n-Butyl benzene	0.82 (100°C)		5.8 (100°C)	
s-Butyl benzene	0.77 (100°C)		5.8 (100°C)	
t-Butyl benzene	0.77 (100°C)		5.8 (100°C)	
n-Butyl bromide	2.5 (100°C)			
Butyl cellosolve	1.1 (150°C)		11 (175°C)	
n-Butyl chloride	1.8		10 (100°C)	
n-Butyl formate	1.7	0.54	8.2	2.6
n-Butyl stearate	0.3 (calc)			
Butyric acid	2.1 (calc)			
γ-Butyrolactone	2.0 (150°C)			
Carbon disulfide	1.3	0.2	50	7.7
Carbon monoxide	12.5		74	
Chlorobenzene	1.4			
m-Cresol	1.1 (150°C)			
Crotonaldehyde	2.1		16 (60°C)	
Cumene	0.88 (100°C)	0.51	6.5 (100°C)	3.8
Cyanogen	6.6			
Cyclobutane	1.8 (calc)	0.56		
Cycloheptane	1.1 (calc)	0.56	6.7 (calc)	3.4
Cyclohexane	1.3	0.57	7.8	3.4
Cyclohexanol	1.2 (calc)			
Cyclohexenel	1.2 (100°C)			
Cyclohexyl acetate	1.0 (calc)			
Cyclopentane	1.5 (calc)	0.55		
Cyclopropane	2.4	0.54	10.4	2.3
Cymene	0.85 (100°C)	0.56	6.5 (100°C)	3.6

(*continues*)

TABLE 1   (*continued*)

Fuel	Lean limit		Rich limit	
	Vol %	(Vol %)/ (Vol %)$_{ST}$, [ϕ]	Vol %	(Vol %)/ (Vol %)$_{ST}$, [ϕ]
Decaborane	0.2 (calc)	0.11		
Decalin	0.74 (100°C)		4.9 (100°C)	
n-Decane	0.75 (53°C)	0.56	5.6 (86°C)	4.2
Deuterium	4.9		75	
Diborane	0.8	0.12	88	13.5
Diethylamine	1.8		10	
Diethylaniline	0.8 (calc)			
1,4 Diethylbenzene	0.8 (100°C)			
Diethyl cyclohexane	0.75			
Diethyl ether	1.9	0.56	36	11
3,3-Diethylpentane	0.7 (100°C)			
Diethyl ketone	1.6 (calc)	0.55		
Diisobutyl carbinol	0.82 (100°C)		6.1 (175°C)	
Diisobutyl ketone	0.79 (100°C)		6.2 (100°C)	
Diisopropyl ether	1.4	0.57	7.9	3.5
Dimethylamine	2.8			
2,2-Dimethylbutane	1.2		7.0	
2,3-Dimethylbutane	1.2		7.0	
Dimethyl decalin	0.69 (100°C)		5.3 (110°C)	
Dimethyl dichlorosilane	3.4			
Dimethyl ether	3.4	0.52	27	4.1
N,*N*-Dimethylformamide	1.8 (100°C)		14 (100°C)	
2,3-Dimethylpentane	1.1		6.8	
2,2-Dimethylpropane	1.4		7.5	
Dimethyl sulfide	2.2	0.50	20	4.5
Dioxane	2.0		22	
Dipentene	0.75 (150°C)		6.1 (150°C)	
Diphenylamine	0.7 (calc)			
Diphenyl ether	0.8 (calc)			
Diphenylmethane	0.7 (calc)			
Divinyl ether	1.7	0.42	27	6.7
n-Dodecane	0.60 (calc)	0.54		
Ethane	3.0	0.53 [0.52]	12.4	2.2 [2.4]
Ethyl acetate	2.2	0.55	11	2.7
Ethyl alcohol	3.3	0.5	19 (60°C)	2.9
Ethylamine	3.5			
Ethyl benzene	1.0 (100°C)	0.51	6.7 (100°C)	3.4
Ethyl chloride	3.8			
Ethyl cyclobutane	1.2	0.53	7.7	3.4
Ethyl cyclohexane	0.95 (130°C)	0.56	6.6 (130°C)	3.9
Ethyl cyclopentane	1.1	0.56	6.7 (0.5 atm)	3.4
Ethyl formate	2.8	0.50	16	2.8

(*continues*)

TABLE 1   (*continued*)

Fuel	Lean limit		Rich limit	
	Vol %	(Vol %)/(Vol %)$_{ST}$, [ϕ]	Vol %	(Vol %)/(Vol %)$_{ST}$, [ϕ]
Ethyl lactate	1.5			
Ethyl mercaptan	2.8	0.63	18	4.4
Ethyl nitrate	4.0			
Ethyl nitrite	3.0		50	
Ethyl propionate	1.8	0.58	11	3.5
Ethyl propyl ether	1.7	0.62	9	3.3
Ethylene	2.7	0.41 [0.40]	36	5.5 [8.0]
Ethylenimine	3.6		46	
Ethylene glycol	3.5 (calc)			
Ethylene oxide	3.6		100	
Furfural alcohol	1.8 (72°C)		16 (117°C)	
Gasoline (100/130)	1.3		7.1	
Gasoline (115/145)	1.2		7.1	
n-Heptane	1.05	0.56	6.7	3.6
n-Hexadecane	0.43 (calc)	0.51		
n-Hexane	1.2	0.56	7.4	3.4
n-Hexyl alcohol	1.2 (100°C)	0.53		
n-Hexyl ether	0.6 (calc)			
Hydrazine	4.7	0.27	100	5.8
Hydrogen	4.0	0.14 [0.10]	75	2.54 [7.14]
Hydrogen cyanide	5.6		40	
Hydrogen sulfide	4.0	0.33	44	3.6
Isoamyl acetate	1.1 (100°C)		7.0 (100°C)	
Isoamyl alcohol	1.4 (100°C)		9.0 (100°C)	
Isobutane	1.8		8.4	
Isobutyl alcohol	1.7 (100°C)		11 (100°C)	
Isobutyl benzene	0.82 (100°C)		6.0 (175°C)	
Isobutyl formate	2.0		8.9	
Isobutylene	1.8	0.53	9.6	2.8
Isopentane	1.4			
Isophorone	0.84			
Isopropylacetate	1.7 (calc)			
Isopropyl alcohol	2.2			
Isopropyl biphenyl	0.6 (calc)			
Jet fuel (JP-4)	1.3		8	
Methane	5.0	0.53 [0.50]	15.0	1.6[1.7]
Methyl acetate	3.2	0.57	16	2.8
Methyl acetylene	1.7			
Methyl alcohol	6.7	0.55 [0.51]	36 (60°C)	2.9 [4.0]
Methylamine	4.2 (calc)			
Methyl bromide	10		15	
3-Methyl-1-butene	1.5	0.55	9.1	3.3
Methyl butyl ketone	1.2 (50°C)	0.58	8.0 (100°C)	3.3

(*continues*)

TABLE 1   (*continued*)

Fuel	Lean limit		Rich limit	
	Vol %	(Vol %)/ (Vol %)$_{ST}$, [ϕ]	Vol %	(Vol %)/ (Vol %)$_{ST}$, [ϕ]
Methyl cellosolve	2.5 (125°C)		20 (140°C)	
Methyl cellosolve acetate	1.7 (150°C)			
Methyl ethyl ether	2.2 (calc)			
Methyl chloride	7 (calc)			
Methyl cyclohexane	1.1	0.56	6.7 (calc)	3.4
Methyl cyclopentadiene	1.3 (100°C)		7.6 (100°C)	
Methyl ethyl ketone	1.9	0.52	10	2.7
Methyl formate	5.0	0.53	23	2.4
Methyl cyclohexanol	1.0 (calc)			
Methyl isobutyl carbinol	1.2 (calc)			
Methyl isopropenyl ketone	1.8 (50°C)		9.0 (50°C)	
Methyl lactate	2.2 (100°C)			
Methyl mercaptan	3.9	0.60	22	3.4
1-Methyl naphthalene	0.8 (calc)			
2-Methyl pentane	1.2 (calc)			
Methyl propionate	2.4	0.60	13	3.2
Methyl propyl ketone	1.6	0.55	8.2	2.8
Methyl styrene	1.0 (calc)			
Methyl vinyl ether	2.6		39	
Monoisopropyl bicyclohexyl	0.52		4.1 (200°C)	
2-Monoisopropyl biphenyl	0.53 (175°C)		3.2 (200°C)	
Monomethylhydrazine	4	0.52		
Naphthalene	0.88 (78°C)		5.9 (122°C)	
Nicotine	0.75 (calc)			
Nitroethane	3.4			
Nitromethane	7.3			
1-Nitropropane	2.2			
2-Nitropropane	2.5			
n-Nonane	0.85 (43°C)	0.58		
n-Octane	0.95	0.58		
Paraldehyde	1.3	0.48		
Pentaborane	0.42	0.12		
n-Pentadecane	0.50 (calc)	0.52		
n-Pentane	1.4	0.55	7.8	3.1
Phthalic anhydride	1.2 (140°C)		9.2 (195°C)	
3-Picoline	1.4 (calc)			
Pinane	0.74 (160°C)		7.2 (160°C)	
Propadiene	2.16			
Propane	2.1	0.57 [0.56]	9.5	2.5 [2.7]
1,2-Propanediol	2.5 (calc)			
β-Propiolactone	2.9 (75°C)			
Propionaldehyde	2.9	0.59	17	
n-Propyl acetate	1.8	0.58	8 (90°C)	2.6

(*continues*)

TABLE 1   (*continued*)

Fuel	Lean limit		Rich limit	
	Vol %	(Vol %)/ (Vol %)$_{ST}$, [ϕ]	Vol %	(Vol %)/ (Vol %)$_{ST}$, [ϕ]
n-Propyl alcohol	2.2 (53°C)	0.49	14 (100°C)	3.2
Propylamine	2.0			
Propyl chloride	2.4 (calc)			
n-Propyl nitrate	1.8 (125°C)		100 (125°C)	
Propylene	2.4	0.54 [0.53]	11	2.5 [2.7]
Propylene dichloride	3.1 (calc)			
Propylene glycol	2.6 (96°C)			
Propylene oxide	2.8		37	
Pyridine	1.8 (60°C)		12 (70°C)	
Propargyl alcohol	2.4 (50°C)			
Quinoline	1.0 (calc)			
Styrene	1.1 (29°C)			
Sulfur	2.0 (247°C)			
p-Terphenyl	0.96 (calc)			
Tetraborane	0.4 (calc)	0.11		
n-Tetradecane	0.5 (calc)	0.52		
Tetrahydrofuran	2.0			
Tetralin	0.84 (100°C)		5.0 (150°C)	
2,2,3,3-Tetramethyl pentane	0.8			
Toluene	1.2 (100°C)	0.53	7.1 (100°C)	3.1
Trichloroethylene	12 (30°C)		40 (70°C)	
n-Tridecane	0.55 (calc)	0.53		
Triethylamine	1.2		8.0	
Triethylene glycol	0.9 (150°C)		9.2 (203°C)	
2,2,3-Trimethyl butane	1.0			
Trimethyl amine	2.0		12	
2,2,4-Trimethyl pentane	0.95			
Trimethylene glycol	1.7 (calc)			
Trioxane	3.2 (calc)			
Turpentine	0.7 (100°C)			
n-Undecane	0.68 (calc)	0.56		
UDMH	2.0	0.40	95	19.1
Vinyl acetate	2.6			
Vinyl chloride	3.6		33	
m-Xylene	1.1 (100°C)	0.56	6.4 (100°C)	3.3
o-Xylene	1.1 (100°C)	0.56	6.4 (100°C)	3.3
p-Xylene	1.1 (100°C)	0.56	6.6 (100°C)	3.4

Spontaneous Ignition
Temperature Data

The greatest compilations of spontaneous ignition (autoignition) temperature are those of Mullins [*AGARDOgraph* No. 4 (1955)] and Zabetakis [*U.S. Bur. Mines Bulletin* 627 (1965)]. These data have been collated and are given with permission in this appendix. The largest compilation is that of Mullins, and consequently his general format of presenting the data is followed. There have been many methods of measuring ignition temperatures and the results from different measurements have not necessarily been self-consistent. Mullins lists the various results, a reference to the technique used, and the reporting investigators. Since the various techniques have not been discussed in the text, these references have been omitted from the data which are reproduced here, and only the reported ignition temperatures are presented.

All temperatures are reported in degrees Celsius. The delay period, where known, is in milliseconds and follows the temperature in parentheses. If no delay time appears, then the spontaneous ignition time was either not specifiable, not specified, or determined by the manner shown in Fig. 1 of Chapter 7, and probably always in excess of 1000 ms. All data are for atmospheric pressure with the exception of the vitiated air data for acetylene and hydrogen, which are for 0.9 atm. When there was a significant difference between the values reported by Mullins and Zabetakis, the values given by Zabetakis were added to the reorganized Mullins compilation. This value is designated by the letter z following the temperature. When the original Mullins listing reported two or more values within 2 degrees of each other, then only one value has been presented. If there are large differences between the reported values, the reader is urged to refer to Mullins and Zabetakis for the method and original source.

TABLE 1   Spontaneous Ignition Temperature Data

Fuel	Oxygen	Air	Vitiated air
Acetal	174	230	768 (20), 957 (1)
Acetaldehyde	140, 159	185, 275, 175z	869 (20), 1088 (1)
Acetanilide		546	
Acetic acid	570, 490	599, 550, 566, 465z	
Acetic anhydride	361	392, 401	
Acetone	568, 485	700, 727, 561, 538, 569, 465z	871 (20), 1046 (1)
Acetone cyanohydrin		688	
Acetonitrile			1000 (20), 1059 (10)
Acetonylacetone		493, 340z	816 (20), 996 (1)
Acetophenone		570z	
Acetylchloride		390z	
Acetylene	296	305, 335	623 (20), 826 (1)
Acetyl oxide	(see Acetic anhydride)		
Acrolein		278, 235z	712 (10), 859(1)
Acrylaldehyde	(see Acrolein)		
Acrylonitrile	460	481	
Adipic acid		420z	
Aldol		277, 248	
Allyl alcohol	348	389	767 (20), 979 (1)
Allylamine		374	
Allyl bromide		295	
Allyl chloride	404	487, 392	
Allyl ether	200		749 (10), 927 (1)
Aminobenzene	(see Aniline)		
o-Aminodiphenyl		450z	
2-Aminoethanol	(see Monoethanolamine)		
Aminoethylethanolamine		369	
Ammonia		651	
n-Amyl acetate		399, 360z, 378	
i-Amyl acetate		379, 360z	
n-Amyl alcohol	390, 332	409, 427, 327, 300z	806 (20), 990 (1)
i-Amyl alcohol		518, 343, 353, 350z	818 (20), 1013 (1)
s-Amyl alcohol		343–385	
t-Amyl alcohol		437	814 (20), 995 (1)
Amylbenzene	255		
n-Amyl chloride		259	
t-Amyl chloride		343	
n-Amylene		273	
n-Amyl ether		170z	

(continues)

TABLE 1   (*continued*)

Fuel	Oxygen	Air	Vitiated air
i-Amyl ether		428	
Amyl methyl ketone		311	
Amyl nitrate		195z	524 (20), 798 (1)
Amyl nitrite		210z	496 (20), 910 (1)
i-Amyl nitrite			437 (10), 918 (1)
n-Amyl propionate		380z	
Aniline	530	770, 628, 530, 617, 593 (6000)	907 (20), 1065 (2)
o-Anisidine			787 (20), 1039 (1)
Anisole	560		744 (20), 1025 (1)
Anthracene	580	472, 540z	
Antifebrin	(see Acetanilide)		
Banana oil	(see *i*-Amyl acetate)		
Benzaldehyde	168	180, 192	744 (20), 936 (1)
Benzene	662, 690, 566	740, 656, 580, 645, 592 (42000), 560z	814 (20), 1000 (1)
Benzene carbonal	(see Benzaldehyde)		
Benzoic acid	475, 556	573	
Benzyl acetate		588, 461	767 (20), 1019 (1)
Benzyl alcohol	373	502, 436	807 (20), 1007 (1)
Benzyl benzoate		480z	
Benzyl cellosolve	(see Ethyleneglycolmonobenzyl ether)		
Benzyl chloride		627, 585z	
Benzyl ethanoate	(see Benzyl acetate)		
Benzyl ethyl ether		496	
Bicyclohexyl		245z	
Biphenyl		577 (36000), 540z	
2-Biphenylamine		450z	
Bromobenzene		688, 565z	858 (20), 1046 (1)
1-Bromobutane	(see *n*-Butyl bromide)		
Bromoethane	(see Ethyl bromide)		
1,3-Butadiene	335	418	
n-Butaldehyde	(see *n*-Butyraldehyde)		
n-Butane	283	408, 430 (6000), 405z	
i-Butane	319	462, 543, 477 (18000)	
1,3-Butandiol		395z	
2,3-Butanedione	(see Diacetyl)		
1-Butanol	(see *n*-Butyl alcohol)		
2-Butanol	(see *s*-Butyl alcohol)		
2-Butanone	(see Methylethyl ketone)		
2-Butenal	(see Crotonaldehyde)		
1-Butene	310z	384	
2-Butene		435, 325z	
2-Butanol	(see *s*-Butyl alcohol)		
2-Butoxyethanol	(see Ethyleneglycolmonobutyl ether)		
n-Butyl acetate		423	793 (20), 1040 (1)
n-Butyl alcohol	385, 328	450, 503, 367, 359 (18000)	809 (20), 993 (1)

(*continues*)

TABLE 1   (*continued*)

Fuel	Oxygen	Air	Vitiated air
i-Butyl alcohol	364	542, 441, 414	794 (20), 1010 (1)
s-Butyl alcohol	377	414, 405z	833 (20), 990 (1)
t-Butyl alcohol	460	478, 483, 480z	
n-Butylamine		312	
i-Butylamine		374	
t-Butylamine		380z	
n-Butylbenzene		412, 444 (6000)	
i-Butylbenzene		428, 456 (12000)	
s-Butylbenzene		443, 420z, 447 (18000)	
t-Butylbenzene		448, 477 (72000)	779 (20), 1000 (1)
2-Butylbiphenyl		433 (12000)	
n-Butyl bromide		483, 316, 265z	
Butyl carbinol	(see *i*-Amyl alcohol)		
Butyl carbitol		228	
Butyl carbitol acetate		299	
Butyl cellosolve	(see Ethyleneglycolmonobutyl ether)		
n-Butyl chloride		460	
α-Butylene	(see 1-Butene)		
β-Butylene	(see 2-Butene)		
γ-Butylene	(see 2-Methylpropene)		
i-Butylene	(see 2-Methylpropene)		
β-Butylene glycol		377	
n-Butyl ether		194	
n-Butyl formate	308	322	
Butyl lactate		382	
i-Butyl methyl ketone		459	
n-Butyl nitrite			400 (4), 490 (1)
Butylphthalate		403	813 (20), 1021 (1)
n-Butyl propionate		426	
n-Butylstearate		355z	
n-Butyraldehyde	206	408, 230	
i-Butyraldehyde		254	
n-Butyric acid		552, 450z	
Camphor		466	
Carbon disulfide	107	149, 120, 125, 90z	610 (20), 842 (1)
Carbon monoxide	588	609, 651	758 (20), 848 (1)
Castor oil		449	
Cellosolve	(see Ethyleneglycolmonoethyl ether)		
Cetane		235	
Cetene			748 (20), 1036 (1)
o-Chloroaniline			885 (20), 1084 (2)
m-Chloroaniline			846 (20), 1080 (2)
Chlorobenzene		674, 640z	
Chloroethane	(see Ethyl chloride)		
2-Chloro-2-methyl chloride	318	343	
3-Chloro(trifluoromethyl) benzene		654	
Creosote oil		336	

(*continues*)

TABLE 1   (*continued*)

Fuel	Oxygen	Air	Vitiated air
o-Cresol		599	
m-Cresol		626	836 (20), 1100 (2)
Crotonaldehyde		232	703 (20), 924 (1)
Cumene		467 (6000), 425z	802 (20), 985 (1)
pseudo-Cumene			770 (20), 1025 (2)
Cyanogen		850	
Cyclohexadiene	360		
Cyclohexane	325, 296	259, 270 (102000), 245z	798 (20), 980 (1)
Cyclohexanol	350	300z	814 (20), 1030 (1)
Cyclohexanone	550	557, 453	816 (20), 1046 (1)
Cyclohexylamine		293	
Cyclohexyl acetate		335z	
Cyclohexene	325		781 (20), 972 (1)
Cyclopentadiene	510		
Cyclopentane		385 (6000)	
Cyclopentanone	540		
Cyclopropane	454	498	
p-Cymene		466, 494, 445, 435z	807 (20), 1050 (1)
Decahydronaphthalene	280	262, 272 (18000), 250z	
trans-Decahydronaphthalene			814 (20), 1002 (1)
Decalin	(see Decahydronaphthalene)		
n-Decane	202	463, 425, 250, 210z, 232 (54000), 236	
1-Decanol		291	
1-Decene		244 (78000)	
n-Decyl alcohol			793 (20), 960 (1)
Diacetone alcohol		603	805 (20), 1065 (1)
Diacetyl			748 (20), 930 (1)
1,2-Diacetylethane	(see Acetonylacetone)		
Diallyl	330		
Diallyl ether	(see Allyl ether)		
Dibutyl ether	(see *n*-Butyl ether)		
Dibutyl phthalate	(see Butyl phthalate)		
Di-*n*-butyl tartrate		284	
o-Dichlorobenzene		648	
1,2-Dichloro-*n*-butane	250	276	
Dichloro-1-(chlorotetrafluoroethyl)-4-(trifluoromethyl) benzene		591	
1,2-Dichloroethane	(see Ethylene dichloride)		
Dichloroethylene		441, 458	738 (20), 1079 (1)
2,2′-Dichloroethyl ether		369	766 (20), 953 (1)
Dichloromethane	(see Methylene chloride)		
1,2-Dichloropropane	(see Propylene dichloride)		
Dicyclopentadiene	510		

(*continues*)

TABLE 1   (*continued*)

Fuel	Oxygen	Air	Vitiated air
Di-*n*-decyl ether		217	
Diesel fuel (41 cetane)		233z	
Diesel fuel (55 cetane)		230z	
Diesel fuel (60 cetane)		225z	
Diesel fuel (68 cetane)		226z	
Diethanolamine		662	823 (20), 1000 (2)
1,1-Diethoxyethane	(see Acetal)		
Diethylamine		312	754 (20), 977 (1)
Diethylaniline		630z	762 (20), 965 (1)
1,2-Diethylbenzene		404 (6000)	
1,3-Diethylbenzene		455 (12000)	
1,4-Diethylbenzene		430, 451 (12000)	
Diethylcellosolve	(see Ethyleneglycoldiethyl ether)		
Dimethyl cyclohexane		240z	
1,4-Diethylene dioxide	(see Dioxane)		
Diethyl ether	(see Ethyl ether)		
Diethylene glycol		229	
Diethyleneglycolbenzoate-2-ethyl- hexoate		340	
Diethylene oxide	(see Dioxane)		
Diethylenetriamine		399	
Diethyl ketone		608, 450z	
3,3-Diethylpentane		322, 290z	
Diethyl peroxide		189	
Diethyl sulfate		436	
Dihexyl	(see Dodecane)		
Di-*n*-hexyl ether		200	
2,2'-Dihydroxyethylamine	(see Diethanolamine)		
Diisobutylenes		470	799 (20), 1064 (1)
Diisooctyladipate		366	
Diisopropyl	(see 2,3-Dimethylbutane)		
Diisopropylbenzene		449	
Diisopropyl ether		443, 416, 500	820 (20), 1037 (1)
Dimethylamine	346	402	
Dimethylaniline		371	780 (20), 960 (1)
2,2-Dimethylbutane		425, 440 (12000)	
2,3-Dimethylbutane	298	420, 421 (12000)	
2,3-Dimethyl-1-butene		369 (6000)	
2,3-Dimethyl-2-butene		407 (6000)	
Dimethylchloracetal		232	
Dimethyl decalin		235z	
2,4-Dimethyl-3-ethylpentane		390 (12000), 510	
Dimethyl ether	252	350	
trans-Dimethylethylene	(see 2-Methyl propene)		
Dimethylformamide		445, 435z	
Dimethylglyoxal	(see Diacetyl)		
3,3-Dimethylheptane		441 (3600)	

(*continues*)

TABLE 1 (*continued*)

Fuel	Oxygen	Air	Vitiated air
2,3-Dimethylhexane		438	
Dimethyl ketone	(see Acetone)		
2,3-Dimethyloctane		231 (72000)	
4,5-Dimethyloctane		388	
2,3-Dimethylpentane		337, 388 (6000)	
o-Dimethylphthalate		556	
2,2-Dimethylpropane		450,456 (3000)	
Dimethyl sulfide		206	
1,1-Dineopentylethane		500	
1,1-Dineopentylethylene		455	
Dioctylbenzenephosphonate		314	
Di-n-octyl ether		210	
Dioctylisooctenephosphonate		320	
1,4-Dioxane		266, 179	755 (20), 933 (1)
Dipentene		237z	
Diphenylamine		452, 635z	
1,1-Diphenylbutane		462 (6000)	
1,1-Diphenylethane		487 (6000)	
Diphenylether		620z	
Diphenylmethane		517 (18000), 485z	
Diphenyloxide		646 (12000)	
1,1-Diphenylpropane		466 (6000)	
Di-n-propyl ether		189	
Divinyl ether		360	
n-Dodecane		232, 205z	
i-Dodecane		534, 500	827 (20), 1010 (1)
1-Dodecanol		283	
n-Eicosane		240	
Ethanal	(see Acetaldehyde)		
Ethane	506z	472, 515	809 (20), 991 (1)
Ethanol	(see Ethyl alcohol)		
Ethene	(see Ethylene)		
Ether	(see Ethyl ether)		
p-Ethoxyaniline	(see p-Phenetidine)		
2-Ethoxyethanol	(see Ethyleneglycolmonoethyl ether)		
2-Ethoxylethanol acetate	(see Ethyleneglycolmonoethyl ether monoacetate)		
Ethyl acetate		610, 486	804 (20), 1063 (1)
Ethyl alcohol	425, 375	558, 426, 365z	814 (20), 1030 (1)
Ethylamine (70% aqueous solution)		384	
Ethylaniline		479	
Ethylbenzene	468	460 (18000), 553, 477, 430z	785 (20), 966 (1)
Ethylbenzoate		644	
2-Ethylbiphenyl		449 (18000)	
Ethyl bromide		588, 511	883 (20), 1055 (2)
2-Ethylbutane	273		
2-Ethyl-1-butanol	(see i-Hexyl alcohol)		

(*continues*)

TABLE 1   (*continued*)

Fuel	Oxygen	Air	Vitiated air
2-Ethyl-1-butene		324 (6000)	
Ethyl-*n*-butyrate	351	612, 463	
Ethyl caprate		493	
Ethyl-*n*-caproate		582	
Ethyl-*n*-caprylate		571	
Ethyl carbonate			782 (20), 1013 (1)
Ethyl chloride	468	516, 494	
Ethylcyclobutane		211	
Ethylcyclohexane		262, 264 (114000)	
Ethylcyclopentane		262	
Ethylene	485	490, 543	
Ethylene chlorhydrin	400	425	
Ethylene dichloride		413	
Ethyleneglycol	500	522, 413, 400z	
Ethyleneglycol diacetate		635	
Ethyleneglycoldiethyl ether		208	
Ethyleneglycolmonobenzyl ether		352	
Ethyleneglycolmonobutyl ether		244	792 (20), 964 (1)
Ethyleneglycolmonoethyl ether		238, 350z	790 (20), 954 (1)
Ethyleneglycolmonoethyl ether acetate		379	
Ethyleneglycolmonoethyl ether monoacetate			774 (20), 960 (1)
Ethyleneglycolmonomethyl ether		288, 382	780 (20), 933 (1)
Ethylene imine		322	
Ethylene oxide		429	
Ethyl ether	178, 182	343, 491, 186, 193, 160z	794 (20), 947 (1)
Ethyl formate		577, 455z	768 (20), 956 (1)
Bis(2-ethylhexyl)adipate		262	
Ethyl lactate		400	
Ethyl malonate		541	
Ethyl mercaptan	261	299	
Ethylmethyl ether	(see Methyl ethyl ether)		
Ethyl methyl ketone	(see Methyl ethyl ketone)		
1-Ethylnaphthalene		481 (6000)	
Ethyl nitrate			426 (20), 562 (1)
Ethyl nitrite		90	580 (20), 833 (1)
3-Ethyloctane		235	
4-Ethyloctane		237 (54000)	
Ethyloleate		353	
Ethyl oxalate			742 (10), 880 (1)
Ethyl palmitate		388	
Ethyl pelargonate		524	
Ethyl propionate	440	602, 476, 440z	
Ethyl propyl ketone		575	
Ethyl-*n*-valerianate		590	
Formaldehyde (37% solution)		430	

(*continues*)

TABLE 1 (*continued*)

Fuel	Oxygen	Air	Vitiated air
Formamide			969 (20), 1032 (10)
Formic acid		504	
Furan			783 (20), 982 (1)
2-Furancarbonal	(see Furfuraldehyde)		
Furfuraldehyde		391	696 (20), 880 (1)
Furfuran	(see Furan)		
Furfuryl alcohol	364	391, 491	775 (20), 944 (1)
Fusel oil	(see *n*-Amyl alcohol)		
Gas oil	270	336	
Gasoline (100/130)		440z	
Gasoline (115/145)		470z	
Glycerine		370z	
Glycerol	414, 320	500, 523, 393	
Glyceryl triacetate		433	
Glycol	(see Ethylene glycol)		
n-Heptane	300, 214, 209z	451, 230, 233, 250, 247 (30000), 215z, 223z	806 (20), 950 (1)
n-Heptanoic acid		523	
1-Heptene		332, 263 (66000)	
α-*n*-Heptylene	(see 1-Heptene)		
Hexachlorobutadiene		618 (6000)	
Hexachlorodiphenyl oxide		628, 600	
n-Hexadecane		230 (66000), 232, 205z	
i-Hexadecane		484	
1-Hexadecene		240 (78000), 253	
Hexahydrobenzene	(see Cyclohexane)		
Hexahydrophenol	(see Cyclohexanol)		
Hexamethylbenzene	375		
Hexamethylene	(see Cyclohexane)		
n-Hexane	296, 225z	487, 520, 248, 261, 225z, 261 (30000)	828 (20), 1015 (1)
i-Hexane	268, 284		
2,5-Hexanedione	(see Acetonylacetone)		
1-Hexene		272 (72000)	
Hexone	(see *i*-Butylmethyl ketone)		
n-Hexyl alcohol	300		801 (20), 970 (1)
i-Hexyl alcohol			800 (20), 963 (1)
Hexylene	325		
n-Hexylether		185z	
Hydrazine		270	
Hydrocyanic acid		538	
Hydrogen	560	572, 400z	610 (20), 700 (1)
Hydrogen sulfide	220	292	
Hydroquinone	630		
2-Hydroxyethylamine	(see Monoethanolamine)		

(continues)

TABLE 1   (*continued*)

Fuel	Oxygen	Air	Vitiated air
4-Hydroxy-4-methyl-2-pentanone	(see Diacetone alcohol)		
α-Hydroxytoluene	(see Benzyl alcohol)		
m-Hydroxytoluene	(see m-Cresol)		
Isophorone	322	462	
Isoprene	440		
JP-1		228z	
JP-3		238z	
JP-4		240z	
JP-6		230z	
Kerosine	270	295, 254, 210z, 249 (66000)	806 (20), 998 (1)
Ketohexahydrobenzene	(see Cyclohexanone)		
dl-Limonene		263 (30000)	
Linseed oil		438	
Mesitylene		621	
Mesityl oxide		344	823 (20), 1037 (1)
Methanamide	(see Formamide)		
Methane	556	632, 537, 540z	961 (20), 1050 (10)
Methanol	(see Methyl alcohol)		
Methone		506	
o-Methoxyaniline	(see o-Anisidine)		
2-Methoxyethanol	(see Ethyleneglycolmonomethyl ether)		
Methyl acetate		654, 502	816 (20), 1028 (2)
Methyl alcohol	555, 500, 461	574, 470, 464, 385z	820 (20), 1040 (1)
Methylamine	400	430	
Methylaniline	(see Toluidine)		
2-Methylbiphenyl		502 (12000)	
Methyl bromide		537	
2-Methylbutane	294	420, 427 (6000)	
2-Methyl-2-butanol	(see t-Amyl alcohol)		
3-Methyl-1-butanol	(see i-Amyl alcohol)		
3-Methyl-1-butene		374 (6000)	
1-Methyl-2-t-butylcyclohexane		314 (12000)	
1-Methyl-3-t-butylcyclohexane (high boiling isomer)		304 (12000)	
1-Methyl-3-t-butylcyclohexane (low boiling isomer)		291 (24000)	
Methyl butyl ketone		533	
Methyl cellosolve	(see Ethyleneglycolmonomethyl ether)		
Methyl chloride		632	
Methyl cyanide	(see Acetonitrile)		
Methylcyclohexane	285	265 (108000), 250z	
Methylcyclohexanone		598	
Methylcyclopentane	329	323 (6000)	812 (20), 1013 (1)
Methyl cyclopentadiene		445z	
2-Methyldecane		231	
Methylenedichloride	(see Methylene chloride)		

(*continues*)

TABLE 1   (*continued*)

Fuel	Oxygen	Air	Vitiated air
1-Methyl-3,5-diethylbenzene		461 (12000)	
Methylene chloride	606	642, 662, 615z	902 (20), 1085 (2)
Methyl ether	(see Dimethyl ether)		
1-Methyl-2-ethylbenzene		447 (18000)	
1-Methyl-3-ethylbenzene		485 (18000)	
1-Methyl-4-ethylbenzene		483 (12000)	
Methylethyl ether		190	
Methyl ethyl ketone		514, 505	804 (20), 975 (1)
Methylethylketone peroxide		390z	
2-Methyl-3-ethylpentane		461	
Methylformate		236, 449, 465z	797 (20), 1026 (2)
Methylheptenone		534	
Methylcyclohexanol		295z	
Methylhexyl ketone		572	
Methylisopropylcarbinol	(see *s*-Amyl alcohol)		
Methyl lactate		385	
1-Methylnaphthalene		566, 547 (24000), 553, 530z	
2-Methylnonane		214 (102000)	
2-Methylpropanal	(see *i*-Butyraldehyde)		
2-Methyloctane		227 (66000)	
3-Methyloctane		228 (60000)	
4-Methyloctane		232 (6000)	
2-Methylpentane	275	306, 307 (6000)	
3-Methylpentane		304 (12000)	
2-Methyl-1-pentene		306 (6000)	
4-Methyl-1-pentene		304 (12000)	
4-Methyl-3-penten-2-one	(see Mesityl oxide)		
2-Methyl-1-propanol	(see *i*-Butyl alcohol)		
α-Methylpropyl alcohol	(see *s*-Butyl alcohol)		
2-Methyl propane	(see *i*-Butane)		
2-Methylpropene		465	
Methylpropionate		469	
Methyl-*n*-propyl ketone		505	832 (20), 1020 (1)
2-Methylpyridine	(see α-Picoline)		
Methylsalicylate		454	
2-Methyltetrahydrofuran	(see Tetrahydrosylvan)		
Methylstyrene		495z	
Monoethanolamine			780 (20), 1006 (1)
Monoisopropyl bicyclohexyl		230z	
2-Monoisopropyl biphenyl		435z	
Monoisopropylxylenes			798 (20), 1040 (1)
Naphthalene	630, 560	587, 568, 526z	
Neatsfoot oil		442	
Neohexane	(see 2,2-Dimethylbutane)		
Nicotine	235	244	
Nitrobenzene		556, 482	706 (20), 884 (1)

(*continues*)

TABLE 1   (*continued*)

Fuel	Oxygen	Air	Vitiated air
Nitroethane		415	623 (20), 762 (1)
Nitroglycerol		270	
Nitromethane		419	684 (20), 784 (1)
1-Nitropropane		417	
2-Nitropropane		428	
o-Nitrotoluene			672 (20), 837 (1)
m-Nitrotoluene			711 (20), 911 (1)
n-Nonadecane		237	
n-Nonane		285, 234 (66000), 205z	
n-Octadecane		235	
1-Octadecene		251	
Octadecyl alcohol	270		
Octahydroanthracene	315		
n-Octane	208	458, 218, 240 (54000)	
i-Octane	(see 2,2,4-Trimethylpentane)		
1-Octene		256 (72000)	
Octyleneglycol		335	
Olive oil		441	
Oxalic acid	640		
Oxybutyric aldehyde	(see Aldol)		
Palm oil		343	
Paraldehyde		541, 242	846 (20), 1064 (1)
Pentamethylene glycol		335z	
n-Pentane	300, 258	579, 290, 284 (24000), 260z	
i-Pentane	(see 2-Methylbutane)		
1-Pentene		298 (18000)	
1-Pentanol	(see *n*-Amyl alcohol)		
2-Pentanone	(see Methylpropyl ketone)		
γ-Pentylene oxide	(see Tetrahydrosylvan)		
Perfluorodimethylcyclohexane		651 (6000)	
p-Phenetidine			822 (20), 1004 (1)
Phenol	574, 500	715	
Phenylamine	(see Aniline)		
Phenylaniline	(see Diphenylamine)		
Phenylbromide	(see Bromobenzene)		
Phenyl carbinol	(see Benzyl alcohol)		
Phenyl chloride	(see Chlorobenzene)		
n-Phenyldiethylamine	(see Diethylaniline)		
Phenylethane	(see Ethylbenzene)		
Phenylethylene	(see Styrene)		
Phenylmethyl ether	(see Anisole)		
Phenylmethyl ketone	(see Acetophenone)		
Phosphorus (yellow)		30	
Phosphorus (red)		260	

(*continues*)

TABLE 1   (*continued*)

Fuel	Oxygen	Air	Vitiated air
Phosphorus sesquisulfide		100	
Phthalic anhydride		584, 570z	
α-Picoline		538, 500z	846 (20), 1037 (2)
Picric acid		300	
Pinene	275	263 (60000)	797 (20), 1039 (1)
1,2-Propanediol		410z	
Propane	468	493, 450z, 466, 504 (6000)	
1-Propanol	(see *n*-Propyl alcohol)		
2-Propanol	(see *i*-Propyl alcohol)		
2-Propanone	(see Acetone)		
Propenal	(see Acrolein)		
Propene	423z	458	
Propen-1-ol	(see Allyl alcohol)		
Propene oxide	(see Propylene oxide)		
Propionaldehyde		419	
n-Propionic acid		596	
n-Propyl acetate	388	662, 450	828 (20), 1060 (2)
i-Propyl acetate	448	572, 460, 476	
n-Propyl alcohol	445, 370, 328	505, 540, 439, 433	811 (20), 1007 (1)
i-Propyl alcohol	512	590, 620, 456	811 (20), 1050 (1)
n-Propylamine		318	
i-Propylamine		402	
n-Propylbenzene		456 (12000)	
i-Propylbenzene	(see Cumene)		
2-Propylbiphenyl		452 (18000), 440z	
Propyl bromide	255	490	
Propyl chloride		520	
i-Propyl chloride		593	
Propylcyclopentane		285	
Propylene	(see Propene)		
Propylenealdehyde	(see Crotonaldehyde)		
Propylene dichloride		557	790 (20), 1012 (2)
Propylene glycol	392	421	
Propylene oxide			748 (10), 870 (1)
i-Propyl ether	(see Diisopropyl ether)		
n-Propyl formate		455	
n-Propyl nitrate		175z	
i-Propyl formate		485	
4-*i*-Propylheptane		288	
p-i-Propyltoluene	(see *p*-Cymene)		
Pseudocumene	(see *pseudo*-Cumene)		
Pulegon		426	
Pyridine		482, 574	745 (20), 1014 (1)
Pyrogallol	510		
Quinoline		480	
Quinone	575		

(*continues*)

TABLE 1   (*continued*)

Fuel	Oxygen	Air	Vitiated air
Rape seed oil		446	
Rosin oil		342	
Salicylicaldehyde			772 (20), 1015 (1)
Soya bean oil		445	
Stearic acid	250	395	
Styrene	450	490	777 (20), 1065 (1)
Sugar	378	385	
Sulfur		232	
Tannic acid		527	
Tartaric acid		428	
p-Terphenyl		535z	
Tetraaryl salicylate		577 (6000)	
n-Tetradecane		232, 200z	
1-Tetradecene		239 (66000), 255	
1,2,3,4-Tetrahydrobenzene	(see Cyclohexene)		
Tetrahydrofurfuryl alcohol	273	282	793 (20), 980 (1)
Tetrahydronaphthalene	420	423 (6000), 385z	819 (20), 1030 (1)
Tetrahydrosylvan			794 (20), 1025 (1)
Tetraisobutylene		415	
n-Tetradecane		200z	
Tetralin	(see Tetrahydronaphthalene)		
Tetramethylbenzene			791 (20), 1030 (1)
Tetramethylene glycol		390z	
2,2,3,3-Tetramethylpentane		430, 452 (42000), 516	
2,3,3,4-Tetramethylpentane		437 (24000), 514	
Toluene	552, 640, 516	810, 633, 552, 540, 480z, 568 (48000), 635	830 (20), 1057 (1)
o-Toluidine		537, 482	832 (20), 1040 (3)
m-Toluidine		580	846 (10), 1062 (2)
p-Toluidine		482	
Tributyl citrate		368	
Trichloroethane		500z	
Trichloroethylene	419	463, 420z	771 (10), 950 (1)
Trichloro-1-(pentafluoroethyl)-4-(trifluoromethyl) benzene		568	
Tricresyl phosphate		600	
Triethylene glycol	244	371	
Triethylenetramine		338	
Triisobutylenes		413	798 (20), 1060 (1)
1,2,3-Trimethylbenzene		479 (24000), 510	
1,2,4-Trimethylbenzene		521 (24000), 528	
1,3,5-Trimethylbenzene		559 (48000), 577	
2,2,3-Trimethylbutane		454 (18000), 420z	
2,3,3-Trimethyl-1-butene		383 (12000)	
Trimethylene glycol		400z	

(*continues*)

TABLE 1 (*continued*)

Fuel	Oxygen	Air	Vitiated air
2,5,5-Trimethylheptane		485	
2,2,3-Trimethylpentane		436 (24000)	
2,2,4-Trimethylpentane	283	561, 434, 447 (12000), 515, 415z	786 (20), 996 (2)
2,3,3-Trimethylpentane		430 (12000)	
2,3,4-Trimethyl-1-pentene		257 (12000)	
2,4,4-Trimethyl-1-pentene (cf. Diisobutylenes)		420 (12000)	
2,4,4-Trimethyl-2-pentene (cf. Diisobutylenes)		308 (30000)	
3,4,4-Trimethyl-2-pentene		330 (24000)	
2,4,6-Trimethyl-1,3,5-trioxane	(see Paraldehyde)		
Trinitrophenol	(see Picric acid)		
Trioxane		424, 414	
Tritolyl phosphate	(see Tricresyl phosphate)		
Tung oil		457	
Turkey red oil		445	
Turpentine		252, 255	780 (20), 996 (1)
Vinyl acetate		427	
Vinylcyclohexene		269	
Vinyl ether	(see Divinyl ether)		
Vinylethyl ether		201	
Vinyl-2-ethylhexyl ether		201	
Vinylisopropyl ether		272	
o-Xylene		496, 465z, 501 (30000), 551	
m-Xylene		563 (54000), 652, 530z	
p-Xylene		618, 564 (42000), 657, 530z	
Zinc stearate		421	

Appendix *G*

Minimum Spark Ignition Energies and Quenching Distances

Most of the data presented in Table 1 are taken from Calcote *et al.* [*Ind. Eng. Chem.* **44**, 2656 (1952)]. Additional data by Blanc *et al.* [*Third Symposium (International) on Combustion*, p. 363 (1949)] and Meltzer [NACA RM E53H31 (1953) and NACA RM E52 F27 (1952)] given in the table are designated by the letters B and M, respectively. Since the values for the least minimum ignition energy by Meltzer were extrapolated from low pressures, these values were not given if values by Calcote *et al.* or Blanc *et al.* were available.

The column labeled Plain contains data from 1/8-inch rod electrodes and that labeled Flange contains data for the negative electrode flanged and the other electrode an 1/8 inch rod. Values in parentheses are taken from a Calcote *et al.* correlation between the two different types of electrode sets.

Quenching distances can be obtained from the data presented in Table 1 and the correlation given as Fig. 5 of Chapter 7. It is interesting to note that most stable fuels have a least minimum ignition energy in the region of 0.2 mJ.

TABLE 1   Minimum Spark Ignition Energy Data for Fuels in Air at 1 atm Pressure

Fuel	Minimum ignition energy ($\phi = 1$), 10^{-4} J		Least minimum ignition energy, 10^{-4} J	ϕ Value for least minimum ignition energy
	Plain	Flange		
Acetaldehyde	3.76	(5.7)		
Acetone	11.5	(21.5)		
Acetylene	0.2	0.3	0.2	1.0
Acrolein	(1.37)	1.75	1.6 M	
Allyl chloride	7.75	13.5		
Benzene	5.5	(9.1), 7.8 B	2.25 B	1.8
1,3-Butadiene	(1.75)	2.35	1.25	1.4
n-Butane	7.6 M	7.0 B	2.6 B	1.5
Butane, 2,2-dimethyl-	16.4	(33)	2.5 M	1.4
Butane, 2-methyl- (Isopentane)	7.0	9.6	2.1 M	1.3
n-Butyl chloride	(12.4)	23.5		
Carbon disulfide	0.15	0.39		1.2
Cyclohexane	13.8	(26.5), 10 B	2.23 M	1.6
Cyclohexane, methyl-			2.7 M	1.8
Cyclohexene	(5.25)	8.6		
Cyclopentadiene	6.7	(11.4)		
Cyclopentane	(5.4)	8.3		
Cyclopropane	2.4	(3.4), 4.0 B	1.8 B	1.1
Diethylether	4.9	(7.9), 5.3 B	1.9 B	1.5
Dihydropyran	(3.65)	5.6		
Diisopropyl ether	11.4	(21.4)		
Dimethoxymethane	4.2	(6.6)		
Dimethyl ether	(3.3), 2.9	4.5		
Dimethyl sulfide	(4.8)	7.6		
Di-*t*-butyl peroxide	(4.1)	6.5		
Ethane	(2.85)	4.2, 3.2 B	2.4 B	1.2
Ethene	0.96	(1.14)	1.24 M	1.1
Ethylacetate	14.2	(28)	4.8	1.2
Ethylamine	24	52		
Ethylene oxide	0.87	1.05	0.62	1.3
Ethylenimine	4.8	(7.8)		
Furan	2.25	(3.28)		
Furan, tetrahydro-	5.4 M			
Furan, thio- (Thiophene)	(3.9)	6		
n-Heptane	7.0	11.5, 11 B	2.4 B	1.8
1-Heptyne	(5.6)	9.31		
Hexane	9.5 M	9.7 B	2.48 B	1.7
Hydrogen	0.2	0.3	0.18	0.8
Hydrogen sulfide	(0.68)	0.77		
Isopropyl alcohol	6.5	(11.1)		
Isopropylamine	(20)	41		
Isopropyl chloride	(15.5)	31		
Isopropyl mercaptan	(5.3)	8.7		
Methane	4.7, 3.3 M	(7.1), 3.3 B	2.8 B	0.9
Methanol	2.15	(3.0)	1.4 M	1.3

(continues)

TABLE 1 (*continued*)

Fuel	Minimum ignition energy ($\phi = 1$), 10^{-4} J		Least minimum ignition energy, 10^{-4} J	ϕ Value for least minimum ignition energy
	Plain	Flange		
Methyl acetylene	1.52	(2)	1.2	1.4
Methyl ethyl ketone	(6.8), 5.3	11	2.8	1.4
Methyl formate	(4.0)	6.2		
n-Pentane	(5.1), 4.9	8.2	2.2 M	1.3
n-Pentane, 2,4,4-trimethyl- (Isooctane)	13.5	29	2.8 M	
1-Pentene, 2,4,4-trimethyl- (Diisobutylene)	(9.6)	17.5		
2-Pentene	(5.1), 4.7	8.2	1.8 M	1.6
Propane	3.05	5, 4.0 (B)	2.5	1.3
Propane, 2,2-dimethyl- (Neopentane)	15.7	(31)		
Propane, 2-methyl- (Isobutane)	(5.2)	8.5		
Propene	2.82	(4.18), 4.1		
Propionaldehyde	(3.25)	4.9		
n-Propyl chloride	(10.8)	20		
Propylene oxide	1.9	2.1	1.4	1.4
Tetrahydropyran	12.1	(23)	2.2 M	1.7
Triethylamine	(7.5), 11.5	13		
Triptane	10	(18.2)		
Vinyl acetate	(7.0)	12.0		
Vinyl acetylene	0.822	(0.95)		

Programs for Combustion Kinetics

The increase in thermochemical and kinetic data bases and the development of fast and affordable personal computers and workstations has enabled the use of many programs for studying combustion kinetics problems. In this appendix, a listing of some of the available programs for studying combustion phenomena is provided.

Some of these programs, as well as many others, may be found and retrieved from the Internet by performing a search using key words such as "chemical kinetics computer codes" or by visiting the home page of various laboratories and universities, e.g., the home page of the National Institute of Standards and Technology (NIST).

I. THERMOCHEMICAL PARAMETERS

THERM: "Thermodynamic Property Estimation for Radicals and Molecules." Edward R. Ritter and Joseph Bozzelli, *Int. J. Chem. Kinet.*, **23**, 767–778, 1991. A computer program for IBM PC and compatibles for estimating, editing, and entering thermodynamic property data for gas-phase radicals and molecules using Benson's group additivity method.

RADICALC: J. W. Bozzelli and E. R. Ritter, *Chemical and Physical Processes in Combustion*. The Combustion Institute, Pittsburgh, PA, 1993, p. 453.

A computer code to calculate entropy and heat capacity contributions to transition states and radical species from changes in vibrational frequencies, barriers, moments of inertia, and internal rotations.

FITDAT: R. J. Kee, F. Rupley, and J. A. Miller, Sandia National Laboratories, Livermore, CA. A FORTRAN computer code (fitdat.f) that is part of the CHEMKIN package for fitting of species thermodynamic data (c_p, h, s) to polynomials in NASA format for usage in computer programs.

II. KINETIC PARAMETERS

UNIMOL: "Calculation of Rate Coefficients for Unimolecular and Recombination Reactions." R. G. Gilbert, T. Jordan, and S. C. Smith, Department of Theoretical Chemistry, Sydney, NSW 2006, Australia, 1990. FORTRAN computer code for calculating the pressure and temperature dependence of unimolecular and recombination (association) rate coefficients. Theory based on RRKM and numerical solution of the master equation. See "Theory of Unimolecular and Recombination Reactions," by R. G. Gilbert and S. C. Smith, Blackwell Scientific Publications, Oxford, 1990.

CHEMACT: "A Computer Code to Estimate Rate Constants for Chemically-Activated Reactions." A. M. Dean, J. W. Bozzelli, and E. R. Ritter, *Combust. Sci. Tech.*, **80**, 63–85, 1991. A computer code based on the QRRK treatment of chemical activation reactions to estimate apparent rate constants for the various channels that can result in addition, recombination, and insertion reactions.

NIST: Chemical Kinetics Database, N. G. Mallard, F. Westley, J. T. Herron, R. F. Hampson, and D. H. Frizzell, NIST, NIST Standard Reference Data, Gaithersburg, MD, 1993. A computer program for IBM PC and compatibles for reviewing kinetic data by reactant, product, author, and citation searches and for comparing existing data with newly evaluated data.

III. TRANSPORT PARAMETERS

TRANFIT: "A FORTRAN Computer Code Package for the Evolution of Gas-Phase Multicomponent Transport Properties." R. J. Kee, G. Dixon-Lewis, J. Warnatz, M. E. Coltrin, J. A. Miller. Sandia National Laboratories, Livermore, CA, Sandia Report SAND86–8246, 1986. TRANFIT is a FORTRAN computer code (tranlib.f, tranfit.f, and trandat.f) that allows for the evaluation and polynomial fitting of gas-phase multicomponent viscosities, thermal conductivities, and thermal diffusion coefficients.

IV. REACTION MECHANISMS

CHEMKIN-II: "A FORTRAN Chemical Kinetics Package for the Analysis of Gas-Phase Chemical Kinetics." R. J. Kee, F. M. Rupley, and J. A. Miller. Sandia National Laboratories, Livermore, CA, Sandia Report SAND89–8009, 1989. A FORTRAN computer program (cklib.f and ckinterp.f) designed to facilitate the formulation, solution, and interpretation of problems involving elementary gas-phase chemical kinetics. The software consists of two programs: an Interpreter and a Gas-Phase Subroutine Library. The Interpreter program converts a user-supplied text file of a reaction mechanism into vectorized binary output which forms a link with the gas-phase library. The two files can then be used in conjunction with user-supplied driver and solution codes for different kinetic related problems. See for example the PREMIX, SENKIN, PSR, and EQUIL codes to be described.

SURFACE CHEMKIN: "A FORTRAN Package for Analyzing Heterogeneous Chemical Kinetics at a Solid-Surface–Gas-Phase Interface." M. E. Coltrin, R. J. Kee, and F. M. Rupley. Sandia National Laboratories, Livermore, CA, Sandia Report SAND90–8003C, 1990. A FORTRAN computer program (sklib.f and skinterp.f) used with CHEMKIN-II for the formulation, solution, and interpretation of problems involving elementary heterogeneous and gas-phase chemical kinetics in the presence of a solid surface. The user format is similar to CHEMKIN-II.

SURFTHERM: M. E. Coltrin and H. K. Moffat, Sandia National Laboratories. SURFTHERM is a FORTRAN program (surftherm.f) that is used in combination with CHEMKIN (and SURFACE CHEMKIN) to aid in the development and analysis of chemical mechanisms by presenting in tabular form detailed information about the temperature and pressure dependence of chemical reaction rate constants and their reverse rate constants, reaction equilibrium constants, reaction thermochemistry, chemical species thermochemistry, and transport properties.

CHEMKIN REAL-GAS: "A FORTRAN Package for Analysis of Thermodynamic Properties and Chemical Kinetics in Nonideal Systems." R. G. Schmitt, P. B. Butler, and N. B. French. The University of Iowa, Iowa City, Iowa, Report UIME PBB 93–006, 1993. A FORTRAN program (rglib.f and rginterp.f) used in connection with CHEMKIN-II that incorporates several real-gas equations of state into kinetic and thermodynamic calculations. The real-gas equations of state provided include the van der Waals, Redlich–Kwong, Soave, Peng–Robinson, Becker–Kistiakowsky–Wilson, and Nobel–Abel.

V. THERMODYNAMIC EQUILIBRIUM

CEC: "Computer Program for Calculation of Complex Chemical Equilibrium Compositions, Rocket Performance, Incident and Reflected Shocks, and Chapman–Jouguet Detonations." S. Gordon and B. J. McBride, NASA Lewis Research Center, NASA Report NASA SP-273. NASA, Washington, D.C. A FORTRAN computer program for calculating (1) chemical equilibrium for assigned thermodynamic states (T, P), (H, P), (S, P), (T, V), (U, V), or (S, V); (2) theoretical rocket performance for both equilibrium and frozen compositions during expansion; (3) incident and reflected shock properties; and (4) Chapman–Jouguet detonation properties. The approach is based on minimization of free energy and considers condensed-phase as well as gaseous species. A useful program for obtaining thermodynamic input is given in the report "Computer Program for Calculating and Fitting Thermodynamic Functions," B. J. McBride and S. Gordon, NASA Lewis Research Center, NASA RP-1271(1992), NASA, Washington, D.C.

STANJAN: "The Element Potential Method for Chemical Equilibrium Analysis: Implementation in the Interactive Program STANJAN." W. C. Reynolds, Thermosciences Division, Department of Mechanical Engineering, Stanford University, Stanford, CA, 1986. A computer program for IBM PC and compatibles for making chemical equilibrium calculations in an interactive environment. The equilibrium calculations use a version of the method of element potentials in which exact equations for the gas-phase mole fractions are derived in terms of Lagrange multipliers associated with the atomic constraints. The Lagrange multipliers (the "element potentials") and the total number of moles are adjusted to meet the constraints and to render the sum of mole fractions unity. If condensed phases are present, their populations also are adjusted to achieve phase equilibrium. However, the condensed-phase species need not be present in the gas phase, and this enables the method to deal with problems in which the gas-phase mole fraction of a condensed-phase species is extremely low, as with the formation of carbon particulates.

EQUIL: A. E. Lutz and F. Rupley, Sandia National Laboratories, Livermore, CA. A FORTRAN computer program (eqlib.f) for calculating chemical equilibrium using a modified solution procedure of STANJAN (stanlib.f, W. C. Reynolds, Standord U.) and CHEMKIN data files for input.

VI. TEMPORAL KINETICS (STATIC AND FLOW REACTORS)

SENKIN: "A FORTRAN Program for Predicting Homogeneous Gas Phase Chemical Kinetics With Sensitivity Analysis." A. E. Lutz, R. J. Kee, and J. A. Miller, Sandia National Laboratories, Livermore, CA, Sandia Report SAND87–8248, 1987. A FORTRAN program (senkin.f) that solves the time evolution of a homogeneous reacting mixture. The program can solve adiabatic problems at constant pressure, at constant volume, or with a volume specified as a function of time. It can also solve constant-pressure problems at constant temperature or with the temperature specified as a function of time. The code uses CHEMKIN-II for mechanism construction and DASAC software (dasac.f, M. Caracotsios and W. E. Stewart—U. Wisconsin) for solving the nonlinear ordinary differential equations. The program also performs a kinetic sensitivity analysis with respect to reaction rate constants.

CONP: R. J. Kee, F. Rupley, and J. A. Miller, Sandia National Laboratories, Livermore, CA 94550. A FORTRAN program (conp.f) that solves the time-dependent kinetics of a homogeneous, constant pressure, adiabatic system. The program runs in conjunction with CHEMKIN and a stiff ordinary differential equation solver such as LSODE (lsode.f, A. C. Hindmarsh, "LSODE and LSODI, Two Initial Value Differential Equation Solvers," *ACM SIGNUM Newsletter*, **15**, 4, 1980). The simplicity of the code is particularly valuable for those not familiar with CHEMKIN.

HCT (Hydrodynamics, Chemistry, and Transport): "A General Computer Program for Calculating Time-Dependent Phenomena Including One-Dimensional Hydrodynamics, Transport, and Detailed Chemical Kinetics." C. M. Lund, Lawrence Livermore National Laboratory, Report No. UCRL-52504, Livermore, CA, 1978. Revised by L. Chase, 1989. A general purpose model capable of modeling in detail one-dimensional time-dependent combustion of gases. The physical processes that are modeled are chemical reactions, thermal conduction, species diffusion, and hydrodynamics. Difference equations are solved by a generalized Newton's iteration scheme. Examples of the types of problems which can be solved include homogeneous temporal kinetics, premixed freely propagating flames, and stirred reactors. This computer code accepts both elementary, detailed chemical kinetic models, and empirical global kinetic models as input.

LSENS: "A General Chemical Kinetics and Sensitivity Analysis Code for Homogeneous Gas-Phase Reactions." K. Radhakrishnan. NYMA, Inc., Lewis Research Center Group, Brook Park, Ohio, NASA Reference Publication 1328, 1994. A FORTRAN computer code for homogeneous, gas-phase chemical kinetics computations with sensitivity analysis. A variety of chemical models can be considered: static system; steady, one-dimensional, inviscid flow; reaction behind an incident shock wave, including boundary layer correction; and perfectly stirred (highly backmixed) reactor. In addition, the chemical equilibrium state can be computed for the assigned states of temperature and pressure, enthalpy and pressure, temperature and volume, and internal energy and volume. Any reaction problem can be adiabatic, have an assigned heat transfer profile, or for static and flow problems, have an assigned temperature profile. For static problems, either the density is constant or the pressure-versus-time profile is assigned. For flow problems, either the pressure or area can be assigned as a function of time or distance.

VII. STIRRED REACTORS

PSR: "A FORTRAN Program for Modeling Well-Stirred Reactors." P. Glarborg, R. J. Kee, J. F. Grcar, and J. A. Miller. Sandia National Laboratories, Livermore, CA, Sandia Report SAND86-8209, 1986. PSR is a FORTRAN computer program (psr.f) that predicts the steady-state temperature and species composition in a perfectly stirred reactor. Input parameters include the reactor volume, residence time or mass flow rate, pressure, heat loss or temperature, and the incoming mixture composition and temperature. The equations are a system of nonlinear algebraic equations that are solved by using a hybrid Newton/time-integration method. The corresponding transient equations are solved because they do not require as good an initial estimate as the algebraic equations. The program runs in conjunction with the CHEMKIN-II package. In addition, first-order sensitivity coefficients of the mass fractions and temperature with respect to rate constants are calculated. SURPSR (surpsr.f, M. E. Moffat, P. Glarborg, R. J. Kee, J. F. Grcar, and J. A. Miller, Sandia National Laboratories) is a version of PSR that incorporates surface reactions via SURFACE CHEMKIN.

LENS: Listed under Temporal Kinetics Calculations.

VIII. SHOCK TUBES

SHOCK: "A General-Purpose Computer Program for Predicting Kinetic Behavior Behind Incident and Reflected Shocks." R. E. Mitchell and R. J. Kee, Sandia National Laboratories, Livermore, CA, Sandia Report SAND82–8205, 1982, reprinted 1990. A FORTRAN computer code (shock.f) for predicting the chemical changes which occur after the shock heating of reactive gas mixtures. Both incident and reflected shock waves can be treated with allowances for real gas behavior, boundary layer effects, and finite rate chemistry. The program runs in conjunction with CHEMKIN as a preprocessor of the gas-phase mechanism and LSODE (lsode.f, A. C. Hindmarsh, "LSODE and LSODI, Two Initial Value Differential Equation Solvers," *ACM SIGNUM Newsletter*, **15**, 4, 1980) to solve the stiff ordinary differential equations.

LENS: Listed under Temporal Kinetics Calculations.

IX. PREMIXED FLAMES

PREMIX: "A FORTRAN Program for Modeling Steady Laminar One-Dimensional Premixed Flames." R. J. Kee, J. F. Grcar, M. D. Smooke, and J. A. Miller. Sandia National Laboratories, Livermore, CA, Sandia Report SAND85–8240, 1985. A FORTRAN computer program that computes species and temperature profiles in steady-state burner-stabilized and freely propagating premixed laminar flames with detailed elementary chemistry and molecular transport. For the burner-stabilized problem, the temperature profile can either be specified or calculated from the energy conservation equation. For freely propagating flames, flame speeds are calculated as an eigenvalue. Solution of the two-point boundary value problem is obtained by finite difference discretation and the Newton method (twopnt.f, J. F. Grcar, Sandia National Laboratories). Sensitivity analysis can also be performed.

RUN-1DL: The Cambridge Universal Laminar Flamelet Computer Code: B. Rogg, in *Reduced Kinetic Mechanisms for Applications in Combustion Systems*. Appendix C, N. Peters and B. Rogg, eds., Springer (Lecture Notes in Physics m15), Berlin/Heidelberg/New York, 1993. SA FORTRAN computer code for the simulation of steady, laminar, one-dimensional and quasi one-dimensional, chemically reacting flows such as (a) unstrained, premixed freely propagating and burner-stabilized flames, (b) strained, premixed flames, diffusion flames, and partially premixed diffusion flames, (c) linearly, cylindrically, and spherically symmetrical flames, (d) tubular flames, and (e) two-phase flames involving single droplets and sprays. The code employs detailed multicomponent models of chemistry, thermodynamics, and molecular transport, but simpler models can also be implemented. For example, the code accepts various chemistry models including detailed mechanisms of elementary reactions, systematically reduced kinetic mechanisms, one-step global finite-rate reactions, and the flame-sheet model (diffusion flames). Also implemented are radiation models, viz., a simple model based on the optically thin limit and a detailed model valid for the entire optical range. User-specified transport equations can also be provided to solve the equations for soot volume fraction and number density or the equations for a particle phase in particle laden flames. See also the RUN-1DL user manual, B. Rogg and W. Wang, Lehrstuhl fur Stromungsmechanik, Institut fur Thermo- und Fluiddynamik, Ruhr-Universität Bochum, D-44780 Bochum, Germany, 1995.

HCT: Listed under Temporal Kinetics Calculations.

X. DIFFUSION FLAMES

OPPDIF: "A FORTRAN Program for Computing Opposed-Flow Diffusion Flames." A. E. Lutz, R. J. Kee, and J. F. Grcar. Sandia National Laboratories, Livermore, CA. A FORTRAN computer program for solving the one-dimensional axisymmetric diffusion flame between two opposing nozzles. OPPDIF solves for the temperature, species, normal and radial velocities, and the radial pressure gradient. The program uses CHEMKIN-II and the TWOPNT software package (J. F. Grcar, Sandia National Laboratories) to solve the two-point boundary value problem.

RUN-1DL: Listed under "Premixed Flames."

XI. BOUNDARY LAYER FLOW

CRESLAF (Chemically Reacting Shear Layer Flow): "A FORTRAN Program for Modeling Laminar, Chemically Reacting, Boundary Layer Flow in Cylindrical or Planar Channels." M. E. Coltrin, H. K. Moffat, R. J. Kee, and F. M. Rupley, Sandia National Laboratories, Livermore, CA, 1991 (see also M. E. Coltrin, R. J. Kee, and J. A. Miller, *J. Electrochem. Soc.*, **133**, 1206, 1986.). CRESLAF is a

FORTRAN computer program that predicts the species, temperature, and velocity profiles in two-dimensional (planar or axisymmetric) channels. The model uses the boundary layer approximations for the fluid flow equations, coupled to gas-phase and surface species continuity equations. The program runs in conjunction with CHEMKIN preprocessors (CHEMKIN, SURFACE CHEMKIN, and TRAN-FIT) for the gas-phase and surface chemical reaction mechanisms and transport properties. The finite difference representation of the defining equations forms a set of differential algebraic equations which are solved using the computer program DASSL (dassal.f, L. R. Petzold, Sandia National Laboratories Report, SAND 82–8637, 1982).

XII. DETONATIONS

CEC: Listed under Thermodynamic Equilibrium Calculations.

XIII. MODEL ANALYSIS AND MECHANISM REDUCTION

CSP: Lam, S. H., "Using CSP to Understand Complex Chemical Kinetics." *Combust. Sci. Technol.* **89**, 375–404, 1993. Lam, S. H., and Goussis, D. A., "The CSP Method for Simplifying Kinetics," *Int. J. Chem. Kinet.*, **26**, 461–486, 1994 (www.Princeton.EDU/~lam/CSPCST.html). Computational Singular Perturbation is a systematic mathematical procedure to do boundary-layer type singular perturbation analysis on massively complex chemical kinetic problems. It is a programmable algorithm using CHEMKIN that not only generates the numerical solution, but can be used to obtain physical insights of the underlying kinetics from inspection of the numerical CSP data. The types of question that may be addressed include: (1) How to reduce the size of a chemical mechanism, (2) what are the rate-controlling reactions, and (3) which species can be approximated by algebraic relationships because of steady-state or partial equilibrium conditions?

AIM: Kramer, M. A., Calo, J. M., and Rabitz, H. "An Improved Computational Method for Sensitivity Analysis: Green's Function Method with AIM." *Appl. Math. Modeling*, **5**, 432, 1981. Program for performing sensitivity analysis of temporal chemical kinetic problems. The program allows for the calculation of first- and second-order sensitivity coefficients as well as for Green's function coefficients. A detailed description of the usage and interpretation of these gradients is given by Yetter, R. A., Dryer, F. L., and Rabitz, H., "Some Interpretive Aspects of Elementary Sensitivity Gradients in Combustion Kinetics Modeling," *Combust. Flame*, **59**, 1985, 107–133. The program is used in combination with other computer codes such as CONP.

KINALC: T. Turányi, *Comput. Chem.*, **14**, 253, 1990. Central Research Institute for Chemistry H-1525, Budapest, PO Box 17, Hungary (1995). KINALC is a postprocessor FORTRAN computer program (kinalc.f) for CHEMKIN-based simulation programs including SENKIN, PSR, SHOCK, PREMIX, and EQUIL. The program performs three types of analysis: processing sensitivity analysis results, extracting information from reaction rates and stoichiometry, and providing kinetic information about species. Processing of sensitivity information includes calculating the sensitivity of objective functions formed from concentrations of several species. In addition, a principal component analysis of the sensitivity matrix can be performed. This eigenvector–eigenvalue analysis groups the parameters on the basis of their effect on the model output; e.g., it will show which parameters have to be changed simultaneously for a maximum change of the concentration of one or several species. KINALC carries out rate-of-production analysis and gives a summary of important reactions. The matrix of normed reaction rate contributions can be considered as the sensitivity of reaction rates, and the principal component analysis of this matrix can be used for mechanism reduction. KINALC can estimate the instantaneous error of assumed steady-state species, and thus guide the selection of such species (Turányi, T., Tomlin, A., and Pilling, M. J., *J. Phys. Chem.*, **97**, 163, 1993). KINALC accepts mechanisms with irreversible reactions only. MECHMOD is a FORTRAN computer program (mech-mod.f) that transforms a mechanism with reversible reactions to one with pairs of irreversible reactions.

XSENKPLOT: An Interactive, Graphics Postprocessor for Numerical Simulations of Chemical Kinetics: D. Burgess, Reacting Flows Group, Process Measurements Division, NIST, Gaithersburg, MD (http://fluid.nist.gov/836.03/xsenkplot/index.html and http://fluid.nist.gov/ckmech.html). An interactive, graphics postprocessor for numerical simulations of chemical kinetics calculations. This graphics postprocessor allows one to rapidly sort through and display species and reaction information generated in numerical simulations. Such interactive computations facilitate the development of a fundamental understanding of coupled chemically reacting systems by providing the ability to quickly probe the impact of process parameters and proposed mechanisms. The FORTRAN code is configured to be used in conjunction with the SENKIN and CHEMKIN computer codes.

Author Index

Subject Index